MOUSE MODELS OF HUMAN CANCER

MOUSE MODELS OF HUMAN CANCER

Edited by

ERIC C. HOLLAND, M.D., PH.D.
Departments of Surgery (Neurosurgery), Neurology, and Cancer Biology and Genetics
Memorial Sloan Kettering Cancer Center
New York, New York

Mouse Models of Human Cancer is available Online
in full color at www.mrw.interscience.wiley.com/mmc

WILEY-LISS
A JOHN WILEY & SONS, INC., PUBLICATION

Cover illustrations:
Ventral view of a mouse to orient feet (see Chapter 1).
Summary of CGH and SKY analyses of human breast carcinomas and mouse mammary gland adenocarcinomas (see Chapter 7).

Published by John Wiley & Sons, Inc., Hoboken, New Jersey.
Published simultaneously in Canada.

Library of Congress Cataloging-in-Publication Data

Mouse models of cancer / edited by Eric C. Holland.
p. ; cm.
Includes bibliographical references and index.
ISBN 0-471-44460-X (cloth : alk. paper)
1. Cancer–Animal models. 2. Mice–Diseases–Pathophysiology.
[DNLM: 1. Disease Models, Animal. 2. Neoplasms–physiopathology. 3. Animals, Genetically Modified. 4. Antineoplastic Agents. 5. Drug Evaluation, Preclinical. 6. Mice, Inbred Strains. QZ 200 M932 2004]
I. Holland, Eric C.
RC267 .M68 2004
616.99'4027–dc22

2003026920

Printed in the United States of America

10 9 8 7 6 5 4 3 2 1

CONTENTS

PREFACE

In 1909, C. C. Little, founder of The Jackson Laboratory, became the first scientist to derive inbred mouse strains to investigate the genetic origins of cancer. When asked to edit a book on mouse models of cancer, I faced the challenge of showing how far the field has advanced since that time by highlighting approaches used by the world's top cancer researchers who work on genetically engineered mice. Finding the appropriate contents that would succinctly describe such a broad field—and provide a valuable, longstanding contribution—appeared to be a very daunting task. Furthermore, the field's leading thinkers and innovators are without question seriously overcommitted, raising the possibility that it would be impossible to convince all of the appropriate authors to make a contribution. Thankfully, nearly every individual who accepted my invitation came through in the end. The fact that my peers have validated the merits of putting together such a book has been particularly rewarding. As I read through their chapters now, I am pleased, grateful, and proud to have had the opportunity of taking part in this project.

There are two main reasons to model cancer in animals, both of which we address in this book. The first is to determine the causal relationships between specific genetic alterations found in cancer. The second purpose is to engineer mice with defined tumor models for use in studies involving imaging technologies and testing novel therapies. Although cell culture experiments certainly provide some information as to the mechanisms being studied, experiments in living animals are required to determine the complex interactions between tumor cells and the host environment.

Legitimate concerns about the use of animals in biomedical research exist. However, most people can reconcile the use of animals if the goal is to combat lethal human disease. Mice were originally used for such studies because of the ease with which they are genetically manipulated and because mammals with smaller brains pose fewer ethical problems than those with more intellectual complexity. Over time, the number of available genetic alterations in mice and the expanding number of technologies developed to produce them further reinforced the advantages of using this species. Even today, cancer is still one of the greatest threats to human health. Moreover, new and evolving technologies in modeling and imaging are helping to significantly reduce the number of mice needed for cancer research. For all of these reasons, mice remain the primary mammalian species used in genetic studies of cancer and are the focus of this book.

Mouse Models of Human Cancer is divided into four parts. The first deals with mouse handling and genetic engineering. The second part highlights modeling efforts in eight organ sites. These chapters review the mouse models and compare them to the human disease that these models represent. The third part is a mixture of topics that are critical to cancer biology in general but not organ site specific. Finally the last section of the book relates to imaging technologies for mouse models and the use of mouse models in preclinical trials.

Mouse modeling of human cancer is a highly competitive and rapidly evolving field. It is not possible for a textbook on this subject to contain everything or be completely up to date, even at the moment of its publication. Instead, we have aimed to (1) present the basic concepts and fundamental methods, (2) show good examples of what can be done with this technology, and (3) provide a perspective on what is to come in this field. I believe we have achieved these goals and, in doing so, hope that this text proves useful to you, the reader, as a springboard for future innovative exploration.

A Note About Internet Resources

Onc of thc main forccs that propcllcd thc ficld of mouse modeling of cancer into the limelight was the establishment of the National Cancer Institute's Mouse Models of Human Cancer Consortium (MMHCC) headed by Cheryl Marks, Dinah Singer, and Betty Tarnowski. This entity is both a source of funding for mouse modeling research and a venue for interaction between some of the most prominent investigators in the field. One of the many positive outcomes of the MMHCC is the establishment of collaborations between investigators and the generation of several websites that provide updated and extensive information on modeling research. These can be accessed via links from the main MMHCC website (http://emice.nci.nih.gov). In addition, other useful websites are referenced throughout the text. Students of this field are strongly encouraged to visit these websites for a more complete view of modeling activities and resources developed by investigators both inside and outside the MMHCC.

ACKNOWLEDGMENTS

I would like to thank Christina Glaster for her help and expertise in editing and organizing this book. In addition, Harold Varmus, Robert Cardiff, Terry Van Dyke, and Luis Parada provided helpful discussions early on in the book's development. I especially thank my wife, Dimitra, and our family, for their patience with me over the last year (and in general).

CONTRIBUTORS

Melvin E. Andersen, Department of Biomathematics and Physical Sciences, CIIT Centers for Health Research, Research Triangle Park, North Carolina

Dalit Barkan, National Cancer Institute, National Institutes of Health, Bethesda, Maryland

Stephen W. Barthold, Center for Comparative Medicine, School of Medicine and Veterinary Medicine, University of California—Davis, Davis, California

Robert Benezra, Department of Cell Biology, Memorial Sloan-Kettering Cancer Center, New York, New York

Dawnalyn Boggess, The Jackson Laboratory, Bar Harbor, Maine

Marcus Bosenberg, Department of Pathology, University of Vermont, Burlington, Vermont

Robert D. Cardiff, Mutant Mouse Pathology Laboratory, Center for Comparative Medicine, University of California—Davis, Davis, California

Dawn S. Chandler, Department of Molecular Genetics, University of Texas M.D. Anderson Cancer Center, Houston, Texas

Thomas L. Chenevert, Center for Molecular Imaging, University of Michigan, School of Medicine, Ann Arbor, Michigan

Christopher H. Contag, Departments of Pediatrics, Radiology, and Microbiology and Immunology, Stanford School of Medicine, Stanford, California

Franco J. DeMayo, Department of Molecular and Cellular Biology, Baylor College of Medicine, Houston, Texas

James E. Dennison, Quantitative and Computational Toxicology Group, Center for Environmental Toxicology and Technology, Department of Environmental and Radiological Health Sciences, Colorado State University, Fort Collins, Colorado

Katharine Ellwood-Yen, Department of Hematology and Oncology, University of California—Los Angeles, Los Angeles, California

Michael F. W. Festing, Medical Research Council Toxicology Unit, University of Leicester, Leicester, United Kingdom

Jeffrey E. Green, National Cancer Institute, National Institutes of Health, Bethesda, Maryland

Silvia Grisendi, Molecular Biology Program and Department of Pathology, Memorial Sloan-Kettering Cancer Center, New York, New York

David H. Gutmann, Department of Neurology, Washington University School of Medicine, St. Louis, Missouri

Stephen S. Hecht, Department of Laboratory Medicine and Pathology, University of Minnesota Cancer Center, Minneapolis, Minnesota

Eric C. Holland, Departments of Surgery, Neurology, and Cancer Biology and Genetics, Memorial Sloan-Kettering Cancer Center, New York, New York

Alan N. Houghton, Immunology Program, Memorial Sloan-Kettering Cancer Center, New York, New York

Kent W. Hunter, Laboratory of Population Genetics, National Cancer Institute, National Institutes of Health, Bethesda, Maryland

Scott C. Kogan, Comprehensive Cancer Center and Department of Laboratory Medicine, University of California—San Francisco, San Francisco, California

Jason A. Koutcher, Departments of Medical Physics, Radiology, and Medicine, Memorial Sloan-Kettering Cancer Center, New York, New York

Jonathan M. Kurie, Department of Thoracic and Head and Neck Medical Oncology, University of Texas M.D. Anderson Cancer Center, Houston, Texas

Carlisle P. Landel, The Jackson Laboratory, Bar Harbor, Maine

Andrew B. Lassman, Department of Neurology, Memorial Sloan-Kettering Cancer Center, New York, New York

Kai H. Liao, Quantitative and Computational Toxicology Group, Center for Environmental Toxicology and Technology, Department of Chemical Engineering, Colorado State University, Fort Collins, Colorado

Guillermina Lozano, Department of Molecular Genetics, University of Texas M.D. Anderson Cancer Center, Houston, Texas

David Lyden, Department of Pediatrics and the Children's Blood Foundation Laboratories, Cornell University Medical College, New York, New York

Kenneth F. Manly, Department of Cellular and Molecular Biology, Roswell Park Cancer Institute, Buffalo, New York

Alex Matter, Novartis Institute for Tropical Diseases, Novartis International, Basel, Switzerland

Patrick McConville, Molecular Imaging Research Inc., Ann Arbor, Michigan

Taha Merghoub, Immunology Program, Memorial Sloan-Kettering Cancer Center, New York, New York

Parviz Minoo, Department of Pediatrics, Women and Children's Hospital, University of Southern California, Los Angeles, California

Larry Mobraaten, The Jackson Laboratory, Bar Harbor, Maine

Bradford A. Moffat, Center for Molecular Imaging, University of Michigan, School of Medicine, Ann Arbor, Michigan

Cristina Montagna, National Cancer Institute, National Institutes of Health, Bethesda, Maryland

Jonathan Moody, Molecular Imaging Research Inc., Ann Arbor, Michigan

Manickam Muruganandham, Departments of Medical Physics, Radiology, and Medicine, Memorial Sloan-Kettering Cancer Center, New York, New York

Alexander Nikitin, Department of Biomedical Sciences, Cornell University, Ithaca, New York

Sandra Orsulic, Molecular Pathology Unit, Massachusetts General Hospital Cancer Center, Charlestown, Massachusetts

Ying C. Ou, Human Genome Sciences, Inc., Rockville, Maryland

Pier Paolo Pandolfi, Molecular Biology Program and Department of Pathology, Memorial Sloan-Kettering Cancer Center, New York, New York

Arie Perry, Division of Neuropathology, Washington University School of Medicine, St. Louis, Missouri

Shahin Rafii, Department of Hematology-Oncology and Genetics Medicine, Cornell University Medical College, New York, New York

Reshma Rangwala, Division of Neuroscience, Oregon National Primate Research Center, Oregon Health and Science University, Beaverton, Oregon

Alnawaz Rehemtulla, Center for Molecular Imaging, University of Michigan, School of Medicine, Ann Arbor, Michigan

Brad Reisfeld, Quantitative and Computational Toxicology Group, Center for Environmental Toxicology and Technology, Department of Chemical Engineering, Colorado State University, Fort Collins, Colorado

Thomas Ried, National Cancer Institute, National Institutes of Health, Bethesda, Maryland

Brian D. Ross, Center for Molecular Imaging, University of Michigan, School of Medicine, Ann Arbor, Michigan

Charles Sawyers, Department of Hematology and Oncology, University of California—Los Angeles, Los Angeles, California

William R. Sellers, Department of Medical Oncology, Dana-Farber Cancer Institute, and Department of Medicine, Brigham and Women's Hospital, Harvard Medical School, Boston, Massachusetts

Kevin Shannon, Department of Pediatrics, University of California—San Francisco, San Francisco, California

Larry S. Sherman, Division of Neuroscience, Oregon National Primate Research Center, Oregon Health and Science University, Beaverton, Oregon

Kathleen A. Silva, The Jackson Laboratory, Bar Harbor, Maine

John P. Sundberg, The Jackson Laboratory, Bar Harbor, Maine

Michael Teitell, Department of Pathology, David Geffen School of Medicine at University of California—Los Angeles, Los Angeles, California

David W. Threadgill, Department of Genetics and Lineberger Comprehensive Cancer Center, University of North Carolina, Chapel Hill, North Carolina

Lene Uhrbom, Rudbeck Laboratory, Uppsala University, Uppsala, Sweden

Jamey P. Weichert, Department of Radiology, University of Wisconsin Medical School, Madison, Wisconsin

Brian Weiss, Department of Pediatrics, University of California—San Francisco, San Francisco, California

Raymond S. H. Yang, Quantitative and Computational Toxicology Group, Center for Environmental Toxicology and Technology, Department of Environmental and Radiological Health Sciences, Colorado State University, Fort Collins, Colorado

Fei Zou, Department of Biostatistics, University of North Carolina, Chapel Hill, North Carolina

PART I

MOUSE HANDLING AND ENGINEERING

1

APPROACHES TO HANDLING, BREEDING, STRAIN PRESERVATION, GENOTYPING, AND DRUG ADMINISTRATION FOR MOUSE MODELS OF CANCER

DAWNALYN BOGGESS, KATHLEEN A. SILVA, CARLISLE P. LANDEL, LARRY MOBRAATEN, AND JOHN P. SUNDBERG

The Jackson Laboratory, Bar Harbor, Maine

Mouse models are extremely valuable biomedical tools for cancer research. Their value to individual investigators depends upon the investigator's ability to manage a colony and manipulate the mice. Basic genotyping to maintain a colony of genetically engineered mice is a critical technique. Understanding the differences between inbred and outbred (nonsegregating vs. segregating backgrounds) determines the reproducibility of models. Preserving valuable colonies for future use through cryopreservation of embryos, sperm, or just the constructs used to create the models enables future work to be done. This chapter provides a broad overview of many of the methods commonly used to perform these tasks as well as how and where to find more specific information.

INTRODUCTION

Inbred laboratory mice have long been the species of choice for biomedical research because (1) of the relatively low cost for maintaining a colony; (2) with inbreeding, mice become genetically identical to each other; (3) numerous reagents are available for very specialized testing; (4) there is a high degree of genetic homology between mice and humans; and (5) there are many sophisticated genetic tools available for working with mice. Mutations in laboratory mice do occur spontaneously. This may be due to inbreeding, where two alleles of a deleterious recessive allele become manifest. Alternatively, insertional mutagenesis may occur when a retrovirus integrates within the mouse genome. Genetic engineering has become technically accessible to many laboratories where overexpression (transgenesis), under- or nonexpression (targeted mutagenesis or the so-called knockout approaches), or selective temporal expression (so-called gene switch methods) of specific, known genes can now be done. Random point mutagenesis has come back into vogue using ethylnitrosourea (ENU) or ethylmethanesulfonate (EMS). All of these approaches provide an array of potential or real models for the study of many types of diseases, including cancer (Nakamura et al., 2002a, b).

Maintaining and working with all of these types of mice are challenging, even for those scientists familiar with conventional inbred strains. Detailed texts are available on systematic workup and characterization methods for mutant mice (Sundberg and Boggess, 2000), general background diseases, including cancer, in commonly used strains (Maronpot et al., 1999; Mikaelian et al., 2004; Mohr et al., 1996), or details on specific organ systems (Smith et al., 2002; Sundberg, 1994b; Ward et al., 2000) which are beyond the scope of this book. Much of this information, including color images, is coming on-line through sources such as the Mouse Tumor Biology Database (http://tumor.informatics.jax.org/FMPro?-db=TumorInstance&-format=mtdp.html&-view), Mouse-PhenomeDatabase (http://aretha.jax.org/pubcgi/phenome/mpdcgi?rtn=docs/home), and European Mouse Pathology

Mouse Models of Human Cancer, edited by Eric C. Holland
ISBN 0-471-44460-X

Table 1.1. Information on distributors of mice, databases on genetics, genomics, and pathology

Distributors of mutant mice and mouse production services	
B&K Universal, LTD.	http://www.bku.com/
Charles River, Inc.	http://www.criver.com/
MRC ENU-Mouse Mutagenesis Programme	http://www.mgu.har.mrc.ac.uk/mutabase/
The ENU-Mouse Mutagenesis Screen Project	http://www.gsf.de/isg/groups/enu-mouse.html
The European Mutant Mouse Archive	http://www.emmanet.org
Harlan Sprague Dawley, Inc.	http://www.harlan.com/
Harwell Mouse Stock List	http://www.mgu.har.mrc.ac.uk/stocklist/stocklist.html
Incyte Genomics, Mouse Knockout Services	http://www.incyte.com/reagents/knockouts/index.shtml
International Mouse Strains Resource	http://www.jax.org/pub-cgi/imsrlist
JAX Mice	http://jaxmice.jax.org/index.shtml
Lexicon Genetics, Inc.	http://www.lexgen.com/
National Institute of Genetics (Japan)	http://www.shigen.nig.ac.jp/mouse/mouse.default.html
Taconic, Inc.	http://www.taconic.com/
University of Michigan Transgenic Animal Model Core	http://www.med.umich.edu/tamc/
Database of Gene Knockouts	http://www.bioscience.org/knockout/knochome.htm
Induced Mutant Resource (IMR),	http://www.jax.org/resources/documents/imr/
Mouse Knockout & Mutation Database	http://research.bmn.com/mkmd
Mutant Mouse Regional Resource Centers (MMRRC)	http://www.mmrrc.org/
Transgenic and Targeted Mutation Database (TBASE)	http://tbase.jax.org/
Transgenic List	http://www.med.ic.ac.uk/db/dbbm/tglist.htm
U.C.D. Genetically Engineered Mouse Research	http://ccm.ucdavis.edu/tgmouse/
Cancer Genetics and Genomics	
Androgen Receptor Mutations Database	http://ww2.mcgill.ca/androgendb/
Atlas of Genetics and Cytogenetics in Oncology and Haematology	http://www.infobiogen.fr/services/chromcancer/
Ataxia-Telangiectasia Mutation Database	http://www.vmresearch.org/atm.htm
BRCA1	http://bioinfo.weizmann.ac.il/hotmolecbase/entries/brca1.htm
Breast Cancer Gene Database	http://condor.bcm.tmc.edu/ermb/bcgd/bcgd.html
Breast Cancer Information Core (BIC)	http://www.nhgri.nih.gov/Intramural_research/Lab_transfer/Bic/
Disease-specific Cancer Chromosome Aberration Project (CCAP)	http://www.ncbi.nlm.nih.gov/CCAP/
Human Cancer Gene Anatomy Project (CGAP)	http://www.ncbi.nlm.nih.gov/CGAP/
Human Cancer GeneticsWeb	http://spade39.ncl.ac.uk/geneweb/
Human Chromosomal Abnormalities in Cancer	http://www.waisman.wisc.edu/cytogenetics/Bmproject/CancerCyto.htmlx
Human Gene Cards	http://bioinformatics.weizmann.ac.il/cards/
Human Med. Gen. HotMolecBase	http://bioinformatics.weizmann.ac.il/hotmolecbase/
Human Med. Gen. cHmutDB,	http://www.ebi.ac.uk/mutations/central/
Human Med. Gen. Human Mutation Databases	http://www.hgmp.mrc.ac.uk/GenomeWeb/human-gen-db-mutation.html
Human Med. Gen. HUGO Mutation Database Initiative	http://www2.ebi.ac.uk/mutations/cotton/
Human Med. Gen. Human Gene Mutation Database (HGMD)	http://archive.uwcm.ac.uk/uwcm/mg/hgmd0.html
Human Med. Gen. The Mammary Transgene Database	http://mbcr.bcm.tmc.edu:80/BEP/ERMB/mtdb.html
Disease-specific Mouse Genome Database (MGD)	http://www.informatics.jax.org/
Mouse Tumor Biology Database (MTB)	http://tumor.informatics.jax.org/
MDM2 Database	http://www.infosci.coh.org/mdm2asp/default.asp
Gene-specific *NF1* GeneticMutation Analysis Consortium Web Site	http://www.nf.org/nf1gene/nf1gene.home.html
Gene-specific Oncogenes and Tumor Supressor Genes in dbEST	http://www.ncbi.nlm.nih.gov/dbEST/CancerGene.html
On-line Mendelian Inheritance in Man (OMIM)	http://www3.ncbi.nlm.nih.gov:80/Omim/
p53 Database	http://perso.curie.fr/Thierry.Soussi/p53_databaseWh.htm
p53 Database (IARC)	http://www.iarc.fr/p53/Index.html
p53 database (Human)	hhttp://www.ibiblio.org/dnam/des_p53.htm
p53 Mutation Database Analysis & Search (Human)	http://p53.genome.ad.jp/
Prostate Expression Database (PEDB)	http://www.pedb.org/
SAGEmap	http://www.ncbi.nlm.nih.gov/SAGE/index.cgi
The Tumor Gene Database	http://condor.bcm.tmc.edu/oncogene.html/
Universal Mutation Database Initiative	http://www.umd.necker.fr/UMD%20Home.html
International Collaborative Group on Hereditary Non-Polyposis Colorectal Cancer Mutation Database	http://www.nfdht.nl/
VHL Mutation Database	http://www.umd.necker.fr:2005/
WT1 Mutation Database	http://www.umd.necker.fr:2003/
Pathology	
Comparative Pathology of Human and Mouse Breast Cancer	http://www-p.ucdavis.edu/tgmice/cmpath/abstract.htm
NCI Veterinary Pathology	http://www.ncifcrf.gov/vetpath/
Transgenic Histopathology	http://www-mp.ucdavis.edu/tgmouse.html
A Tumor Atlas: Information, Resources, Images and Forms	http://bioscience.igh.cnrs.fr//atlases/tumpath/index.htm
Tumor Board	http://www.tumorboard.com/

Database (http://pathbase.net). Details on mouse genetics and gene mapping (Silver, 1995; Sundberg et al., 1998) and access to mouse strains and mutations are readily available through the Mouse Genome Informatics Database (http://informatics.jax.org). These and many other resources are listed in Table 1.1.

This chapter provides an overview of methodologies commonly used to maintain, manipulate, and preserve genetic-based mouse models of human diseases, especially cancer models.

METHODS FOR HANDLING LABORATORY MICE

Each institution has its own regulations dealing with maintaining a rodent colony, which includes various types of housing, bedding, feed, and water. Regardless, it is critical to optimize these to maintain a pathogen-free colony. This includes autoclaving feed, acidifying or treating water to account for specific issues in each locality, and avoiding bedding that releases aromatic hydrocarbons that might complicate cancer studies. It is beyond the scope of this chapter to describe all the variations, but these are described elsewhere (Clark et al., 1996). Maintaining a pathogen-free colony, or at least a specific pathogen-free colony whereby the pathogen status is known for each colony, is critical as well. Controversy over complications of phenotypes due to pathogen status of colonies has dictated that careful and continual microbiological monitoring be incorporated into colony maintenance protocols. Again, it is beyond the scope of this chapter to describe disease-monitoring selection and detection methods, but these are available (Lindsey et al., 1991a, b). This section will summarize some commonly used handling methods for mice to identify individuals as well as to collect bodily fluids and excretions.

Methods for individual identification of mice vary between institutions and laboratories. Some of the commonly used methods are summarized below (Seymour et al., 2004). All manipulations of mice require prior approval of the Institutional Animal Care and Use Committee (IACUC). Developing a good working relationship with the institutional attending veterinarian will be a great help in planning experiments, preparing documentation for IACUC protocol approval, and obtaining training in many of these procedures to optimize results.

Identifying Individual Mice

There are a number of methods that can be used for identification of individual mice in a colony or study. These range from ear notching using a small punch (National Band and TAg Company, Newport, KY) to making a hole or notch in the mouse's ear (Figs. 1.1 and 1.2), ear tags (National Band and TAg Company) attached to the ear, digit amputation using scissors (Fig. 1.3), foot (Fig. 1.4) or tail tattooing using various tattooing machines (Animal Identification and Marking Systems, AIMS Inc., Ottawa, Canada), and implanting electronic chips under the skin of the mice (Bio Medic Data Systems, Inc., Seaford, DE, or American Veterinary Identification Devices, Norco, CA). Wounding mice can result in abnormal healing such as keloids or development of specific types of cancer in some models. Therefore, attention to methods that might complicate procedures is important.

Collecting Blood and Body Fluids

Blood is collected by retro-orbital bleeding, tail tip amputation, cardiac puncture, or decapitation. The reasons to use each method varies with age, purpose of the study, volume needed, and methods approved at each

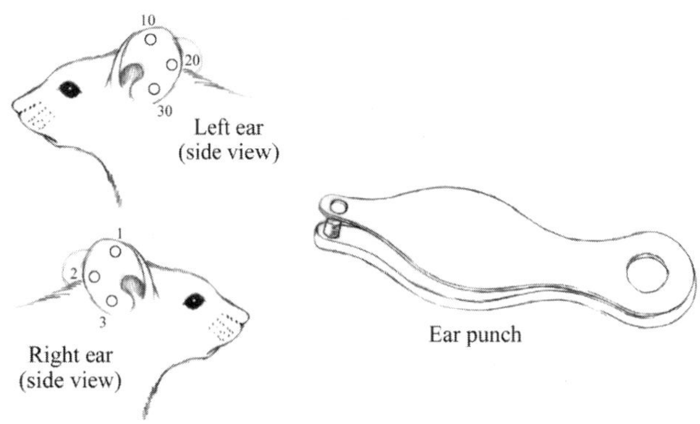

Figure 1.1. A small metal ear punch can be used to mark the ears of mice at specific locations to create a numerical code for individual mouse identification.

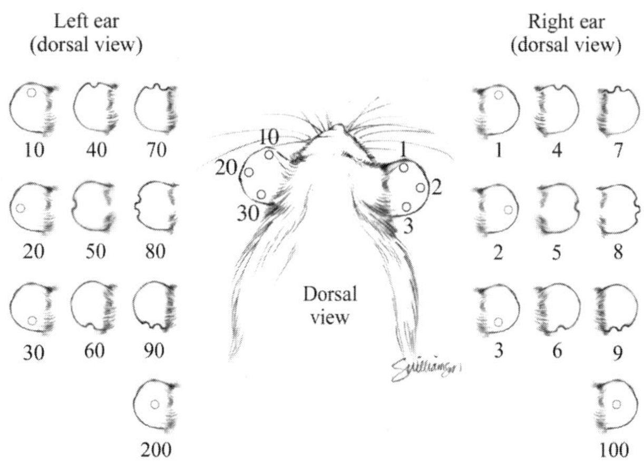

Figure 1.2. Dorsal view of a mouse showing commonly used numerical code approach for ear notching.

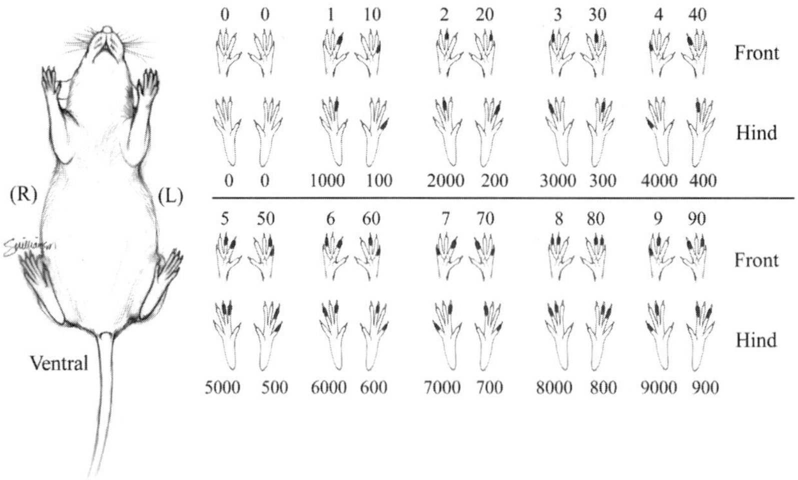

Figure 1.3. Ventral view of a mouse to orient feet. Digits can be amputated on neonatal mice as a form of numbering individuals.

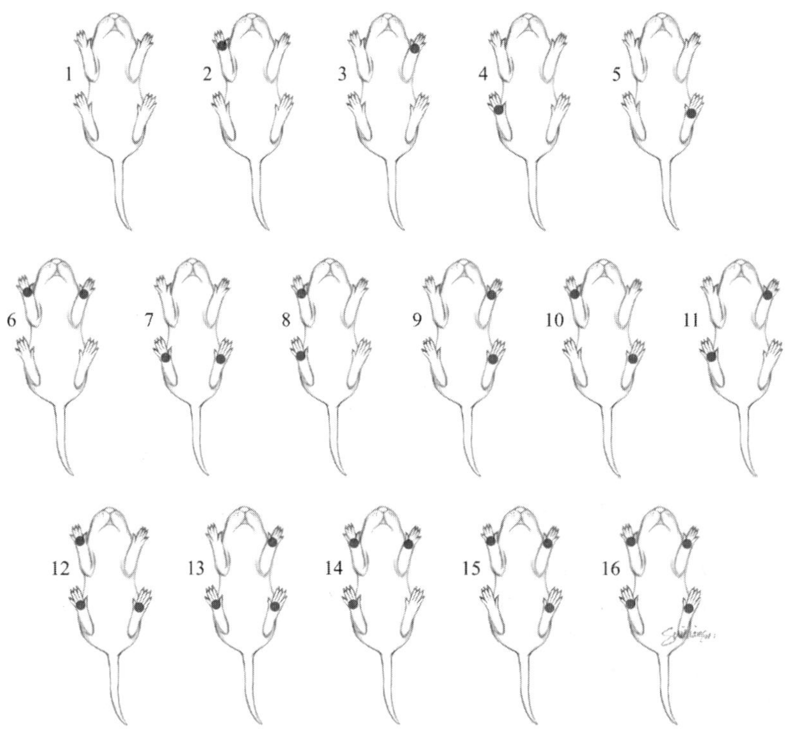

Figure 1.4. Tattoo ink can be injected into the skin at specific locations, as indicated, in neonatal mouse feet to create a numerical code to identify individuals.

institution. Methods are summarized below as modified from Seymour et al. (2004).

Retro-Orbital Bleeding This method is a valuable, nonterminal method for blood collection to obtain small volumes of blood. Retro-orbital bleeding is performed using manual restraint. Mice are grasped with thumb and index finger behind the ears placing the tail secured between the handlers' last finger and palm to immobilize them. A variety of anesthetics may be required by the IACUC at some institutions to perform this procedure.

The tip of an hematocrit tube is inserted into the medial canthus of the eye at a slight angle. Using a twisting motion with slight pressure on the tube, the retro-orbital venous plexus is entered. Once the blood begins flowing, the hematocrit tube should be withdrawn slightly to allow collection by capillary action. The hematocrit tube may be heparinized, contain ethylenediaminetetraacetic acid (EDTA), or other anticoagulants for plasma collection or contain no anticoagulant for serum collection.

Retro-orbital bleeding can also be used to collect larger volumes of blood if the mouse is going to be euthanized for necropsy. The mouse is anesthetized; then blood is collected in the same manner as for smaller volumes described above using heparinized hematocrit tubes. It is possible to fill 10–12 tubes using this method depending upon the size, age, and health status of the mouse. Each tube holds approximately 75 μL of blood. These tubes can be expressed into the appropriate collection tubes for the types of testing being done (see section on blood handling). This method decreases the variability in test results that may be caused by stress the mouse suffers during other methods of blood collection. This method also allows for more uniformity in sample volume when working with many mice.

Tail Bleeding Some technicians hold the mouse as described for the eye bleed for a manual tail bleed procedure. However, tail bleeding is often done when the mouse is placed in a restraint device with the tail protruding with an Eppendorf or glass tube for blood collection positioned below it (Fig. 1.5). Care must be taken not to contaminate blood with fecal matter or urine when using the tail-bleeding method. A tail bleed is accomplished by severing the ventral coccygeal vein on the ventral side of the base of the tail with a razor blade. Warming the mouse or its tail prior to bleeding will increase the blood flow.

Figure 1.5. A plastic restraint device can be used to collect blood from the tail. Note the tube holder and collecting tube below the tail.

Cardiac Puncture Cardiac puncture is a clean and simple way to get as much blood as possible from a mouse at the time of necropsy. It must be performed immediately after the mouse is euthanized. Many people perform cardiac puncture without opening up the thorax by inserting a 25- to 27-gauge needle through the skin and intercostal muscles and into the heart, then withdrawing blood into the syringe (a 1-mL syringe is large enough for most mice). This approach requires some technical skill but often results in sufficient amounts of blood being obtained. An alternative approach is to open the thoracic cavity immediately after euthanizing the mouse. This will expose the heart, making it easy to correctly insert a needle, obtaining a cleaner sample of larger volume. This is achieved by following normal necropsy procedures detailed elsewhere (Relyea et al., 2000; Seymour et al., 2004), being careful not to sever major blood vessels. Once the abdominal muscles have been incised, a pair of sharp/sharp scissors is used to cut the ribs in a triangular shape, starting equidistant from the xiphoid process on both sides and angling toward the clavicle. The sternum should not be severed since this may rupture major blood vessels near the thoracic inlet, causing rapid and excessive blood loss. The ribs should be retracted to reveal the heart and lungs. Using this approach, the right ventricle can be visualized and cleanly punctured. Blood is withdrawn slowly into the syringe. Gentle pressure applied to the visceral organs, working toward the heart, pushes abdominal blood into the right ventricle to optimize collection. Blood should be promptly and gently expressed into an appropriate collection tube before clotting occurs. The syringe plunger should be depressed slowly since high-velocity ejection may damage red blood cells, resulting in hemolysis.

Decapitation This method is seldom approved by animal-care-and-use committees unless there is sound scientific justification. This is the method of choice for some endocrine assays when hormones may be released during handling, altering results, or for mice less than 10 days old. A number of commercial guillotine-type instruments are available (Harvard Apparatus Inc., Holliston, MA). Alternatively, mice can have their heads disarticulated with a large pair of scissors (6.5–7 in., blunt/sharp). The mouse is firmly grasped by the skin at the back of the neck and, holding it over a blood collection tube, the head is swiftly and cleanly severed.

Blood Handling

For serum collection, blood should be collected into small-volume serum separator tubes (Microtainer Brand, stock number 365956, Becton Dickinson, Franklin, NJ). These can then be held at room temperature for 15 min to

1 h and then centrifuged. The serum should be decanted and stored in plastic tubes (Eppendorf tubes, Brinkmann Instruments, Inc., Westbury, NY; Nunc Tubes, Nalge Nunc International, Fisher Scientific, Pittsburgh, PA) for storage and frozen at −80°C until used. Plasma is obtained by collecting blood in tubes containing EDTA or heparin to prevent clotting. The blood is then centrifuged, the plasma is decanted into tubes, and it is frozen for future use. Whole blood should be handled according to the instructions provided by the diagnostic laboratory that will be doing the blood chemistry workup.

Feces

Feces are collected at the time of necropsy and can be frozen in plastic tubes for a variety of assays. Most mice defecate upon handling so a few fresh samples can be obtained from a defined individual. This is a simple resource for *Helicobacter* spp. surveillance using polymerase chain reaction (PCR) methods (Mahler et al., 1998) or for fecal IgA quantification (Bristol et al., 2000).

Urine

Urine is usually expelled when a mouse is handled. A person can pick up the mouse as described for the retro-orbital bleeding procedure, holding the tail back with the little finger. In this position most mice micturate a few drops of urine. If needed, the mouse can be manually restrained and the caudal abdomen gently massaged to induce urination. Urine can be collected in a clean plastic tube or tests done directly with a variety of chemically impregnated strips. Chemstrip (Boehringer Mannheim Diagnostics, Indianapolis, IN) and Ames Multistix (Miles Inc. Diagnostic Division, Tarrytown, NY) are two urine analysis reagent strips that test for numerous urine components, including glucose, ketones, protein content, and pH. For more specific tests that require larger volumes of urine or urine collected over defined intervals, metabolic cages are commercially available (Columbus Instruments, Columbus, OH). Urine specific gravity is measured using a hand-held refractometer. Some companies offer refractometers especially made for urine testing, such as the Fisherbrand UriSystem Refractometer (Fisher, Pittsburgh, PA).

GENERAL BREEDING STRATEGIES FOR MAINTAINING COLONIES OF GENETICALLY ENGINEERED MICE

When establishing a colony of potentially mutant mice, it is important to examine all mice as soon after birth as possible and at regular intervals (at least every other day during the first 2 weeks) thereafter until the phenotype (clinical features) are well defined. Defects may be observable at birth, as is seen in flaky skin (*fsn*) mice, which have a mild anemia at birth, that make them easy to identify (Sundberg et al., 1997b) or juvenile alopecia (*jal*) mice, which have easily identifiable abnormalities of the vibrissae as early as 2–3 days of age (McElwee et al., 1999). Other defects may develop as the mouse ages, as in harlequin ichthyosis (*ichq*), in which mice are normal until 5 days of age, when they develop thick scaling skin, then die by 10–12 days of age (Sundberg et al., 1997a); matted (*ma*), in which the hairs form matted clumps and mutants can be identified at 2–4 weeks of age (Sundberg, 1994c); and chronic proliferative dermatitis (*cpdm*), in which the skin lesions become apparent at 4–6 weeks of age (HogenEsch et al., 1993). Some defects may even disappear as the animal ages, as with flaky tail (*ft*), in which the skin defect is recognizable by 2–4 days of age and has usually disappeared and appears to be normal by 14–15 days of age (Presland et al., 2000; Sundberg, 1994a).

Establishing a colony of the suspected mutation is best done by obtaining the parents of the deviant pup or litter and allowing them to continue to produce litters. If the parents are unavailable or are too old to produce more litters, the siblings of the affected mouse can be mated together to produce more litters. If any of the offspring of the parental or sibling matings produce more of the phenotype in question, then it is possibly a genetic mutation.

When you have determined that the phenotype is reproducible, you can expand the colony. At this time you will also want to do a systematic diagnostic necropsy of at least one mutant and one normal littermate. This is to further rule out the possibility of infectious disease as well as to attempt to identify the disease processes that are causing the phenotype of the deviants. This will allow you to better direct the characterization and maintenance of the colony.

The colony can be expanded by simple brother-by-sister crosses. Matings of this type would be considered unknown test matings. If any of these matings produced at least one affected offspring in a litter, then the animals in that mating would be considered tested breeders. Matings are usually maintained until at least 18–20 pups are born, with no deviants found, before being deemed a negative test mating. Nonproductive breeders and breeders that produce no affected offspring should be remated to tested breeders or removed from the colony.

If recessive mutant mice can breed, it is a good idea to breed homozygous (m/m; $m =$ mutant) mutants to heterozygous ($+/m$) or wild-type ($+/+$) litter mates. With these types of matings your are assured of getting litters that will either have some homozygous, affected mice and some heterozygous, normal mice or, if none are affected, all heterozygous mice. There are some mutations, which, due to the phenotype, make the females poor mothers. Lethal milk mutation is an example of this type of mutation (Piletz and Ganschow, 1978a, b). For this reason it is usually best to mate mutant males to normal (heterozygous or wild-type) females.

In cases where the mutant mice do not breed, die in utero, or die soon after birth, it may be possible to expand and maintain the colony by use of ovarian transplantation. Ovaries from mutant mice are transplanted into oophorectomized, histocompatible recipients or severe combined immunodeficiency mutation ($Prkdc^{scid}/Prkdc^{scid}$) recipients that will accept grafts from nonhistocompatible strains (Boggess et al., 2000). These recipient females are then mated to homozygous mutant (if they can breed) or tested heterozygous males. This type of mating will produce litters the same as those mentioned above. When choosing the option of ovarian transplantation, one must remember to choose a recipient mouse that will enable you to verify that all the offspring born are indeed from the donated ovary and not from any chance remnant of ovary from the recipient mouse. This is usually done using a coat color marker. For example, when transplanting ovaries from a black mouse into a white (albino) recipient, any white offspring born would indicate the presence of ovarian remnants from the albino mouse. If a $Prkdc^{scid}/Prkdc^{scid}$ mutation is used as the recipient, you must be able to house this strain in a relatively pathogen-free environment and provide antibiotic treated water for part of each week. SulfaTrim (sulfamethoxazole and trimethoprin oral suspension, USP, Schein Pharmaceutical Inc., Florham, NJ), 10–15 mL/400 mL acid water, is commonly used (Shultz et al., 1989). The antibiotic water is put on the cages on alternate weeks with untreated acidified water. This is usually sufficient to keep the mice healthy. The immunodeficient recipient mice should not be kept on the antibiotic water continually since this will cause the microflora to develop resistance to the antibiotic.

Another method of expanding and maintaining the colony if the mutant mice have reduced viability or if the mutation was originally found on a segregating (outbred) background is to create a congenic strain by outcrossing the strain carrying the mutation (donor strain) with another inbred strain (recipient strain). By selecting mice homozygous (if fertile) or heterozygous for the mutation and outcrossing these animals to the recipient strain and then selecting affected offspring to backcross to the recipient strain, it is possible to "transfer" the mutation onto the selected inbred strain, thereby creating a new congenic strain (Silver, 1995). Creation of the congenic strain and the subsequent inbreeding done after the congenic is created can make profound changes to the phenotype as observed in the original mutation. Therefore, it is important to collect tissues for histology from animals with the original mutation for comparison with tissues from the congenic mice created to verify that the original phenotype was not lost or significantly changed.

Another important aspect of mutant mouse evaluation and colony management is the setting up of matings for genetic mapping purposes. If you are working with a mutant mouse colony with a phenotype that resembles another mutation which has already been characterized and mapped, you can save time by doing simple allelism tests (complimentation assays) between your mutant and any other mutant that it resembles. This type of testing is done by mating a homozygous or known heterozygous mouse with a known, characterized, and mapped mutation to a homozygous or known heterozygous mouse carrying the new mutation. If a mating between these animals produces offspring in the first generation with the same phenotype as both parents, it can then be interpreted that the new mutation is a remutation or allele (noncomplimentation) of the known mutation. This type of test is simple if the mutation is recessive but becomes much more complicated for a dominant mutation.

Setting up matings for interspecific and intersubspecific crosses, needed for gene mapping, comes with its own set of problems in colony management. *Mus musculus castaneous* (CAST/EiJ) mice are much more aggressive than most common inbred strains of mice and are known to kill mice smaller than themselves if the mice are introduced to each other much after weaning age (approximately 3 weeks of age). This is true of both males and females. CAST/EiJ mice are also slow to reach sexual maturity. A mating between two CAST/EiJ mice or a CAST/EiJ male and a female of another strain may not produce their first litter until the mating has been together for as long as 3 months. It is important not to deem a mating "nonproductive" until at least this amount of time has passed. Female CAST/EiJ mice mated to older males of another strain may have their first litter sooner than three months, but again it is important not to discontinue the mating too soon. An increased success rate in starting a colony of mutant × CAST/EiJ crosses can be achieved by using CAST/EiJ females mated to mutant or known heterozygous males rather than vice versa.

GENOTYPING GENETICALLY ENGINEERED MICE

Genotyping mutant mice, whether they be genetically engineered or spontaneous mutations, has become progressively easier as spontaneous mutations have been defined and protocols standardized for molecular assays. Protocols can be found on various websites (http://jaxmice.jax.org/info/index.html) or in the original papers that describe the constructs and genotyping used to create the model. Most of these methods utilize PCR methods with primers designed to detect the *Neo* cassette insert for targeted mutations, specific sequence changes within the mutated genes, restriction length polymorphisms with or without the use of Southern blots, or quantitative real-time PCR methods. Some mutations are easier to phenotype that to genotype. For example, immunodeficient mutant mice may be easier in some studies to phenotype by blood smear, fluorescence-activated cell sorter analysis, or other methods. The easiest way to proceed is to find a website that provides detailed protocols and information on primers and restriction sites used to genotype mice and follow those protocols. Some genetically engineered mice, such as B6.129P2-$Ncam^{tm1Cgn}$, can be quite difficult to genotype. Contacting the laboratory directly is sometimes warranted to learn specific protocols needed. Details on running the various assays are provided in standard methodology textbooks (Jackson and Abbott, 2000).

CRYOPRESERVATION OF EMBRYOS, SPERM, AND MOLECULAR CONSTRUCTS

Investigators who have produced a genetically engineered mouse strain useful as a model for human cancer and completed and published their experiments are faced with the question of what to do with these mice. On the one hand, these mice, which were produced with a significant investment of time and resources, might once again become a critical research tool as the field progresses. On the other hand, mice are expensive to maintain and animal facilities are often space limited, creating pressure to discard the animals. The solution is to cryopreserve the strain. This also provides security against loss of the strain through accident, disaster, malfeasance, genetic contamination, or disease. Cryopreservation also archives the genome of the mice, providing a kind of "genetic vintage" of a genome that will prevent change with time due to genetic drift.

Several repositories and distribution centers have been created worldwide that will accept, maintain, and distribute strains of mice at no cost to the donating investigator if oversight committees determine that these strains have potential future scientific value (Table 1.2). These repositories all cryopreserve the strains they accept as a matter of course, not only for the security this provides, but also because they too have only limited carrying capacity for mouse strains. In fact, a significant percentage of their inventories are held as cryopreserved material and not as live mice. Unfortunately, these repositories do not have the capacity to accept all the strains that are submitted to them.

The purpose of this section is to provide an overview of the requirements and considerations for cryopreserving one's own mouse strains. Explicit protocols are beyond the scope of this chapter, but they can be obtained from laboratory manuals (Nagy, 2002) or on the Web (The Jackson Laboratory Cryopreservation Resource, http://www.jax.org/cryo/index.html).

A major prerequisite for mouse strain cryopreservation is the ability to manipulate mouse embryos. These techniques include collecting embryos from superovulated donors, maintaining them in culture, and surgically transferring them to pseudopregnant foster mothers. The cryopreservation of mouse germplasm involves the following steps:

1. Collection of the material
2. Suspension in a concentrated nonphysiologic solution of cryoprotectant

Table 1.2. Mutant Mouse Repositories

Repository	Maintained By	Web Site
Mouse Models of Human Cancers Consortium (MMHCC)	National Cancer Institute	http://emice.nci.nih.gov/
Induced Mutant Resource (IMR)	The Jackson Laboratory, Bar Harbor, Maine, USA	http://www.jax.org/imr/index.html
Mutant Mouse Regional Resource Centers (MMRRC)	National Center for Research Resources University of California, Davis; University of North Carolina, Chapel Hill; Taconic Farms-University at Albany SUNY; Harlan Sprague Dawley-Missouri Consortium	http://www.mmrrc.org/
European Mouse Mutant Archive (EMMA)	European Commission Framework Programmes and various institutions	http://www.emmanet.org/
MRC Frozen Embryo & Sperm Archive (FESA)	Mammalian Genetics Unit, Medical Research Council, Harwell UK.	http://www.mgu.har.mrc.ac.uk/fesa/fesa.html

3. Cooling to a low subzero temperature in a way that the cells lose all intracellular water, thus avoiding the formation of intracellular ice that would physically disrupt the cells
4. Storage in liquid nitrogen
5. Warming to physiologic temperatures
6. Removing the cryoprotectant without damage due to osmotic shock or toxicity of the cryoprotectant
7. Production of live animals

The first consideration in cryopreserving a strain of mice is the choice of material to freeze. Embryos, sperm, and ovaries have all been frozen successfully. Less commonly, because they are technically difficult to deal with, oocytes and spermatogonia have also been successfully cryopreserved. Of these forms of germplasm, embryos have the advantage that they preserve the diploid genome of the strain of mice. Thus, the strain is regenerated by the direct recovery of live pups from the frozen embryos. The other forms of preserved haploid germplasm can only produce heterozygous or hybrid offspring unless sperm or oocyte donors of the same strain are available for fertilization with the frozen gametes. Since this is not usually practical, embryo cryopreservation is the best choice for inbred or congenic strains. In practical terms, this means that embryo cryopreservation is the only choice for inbred and congenic lines of mice. The other choices are only appropriate if one is not interested in preserving the genetic background but only the mutation or if one is reasonably confident that mice of the appropriate genetic background will always be available to act as sperm or oocyte donors for recovery.

Mouse embryos were first successfully cryopreserved in 1972 (Whittingham et al., 1972). Since then, many different protocols using a variety of cryoprotective agents, cooling rates, and storage containers have been published that permit the cryopreservation of mouse embryos at all preimplantation stages. Most of these work quite well, and the method chosen is simply a matter of personal preference. One method is outlined in Table 1.3. Recovery of pups is accomplished by the surgical transfer of embryos into pseudopregnant foster mothers, who carry them to term.

Sperm cryopreservation is a more recent development, and the most commonly used protocol was developed by Nakagata (Nakagata and Takeshima, 1993). In cases where a suitable oocyte donor is expected to be available, it is an attractive method for preserving a strain because it is extremely easy (sperm are collected from the vas deferentia and epididymides into a solution of 18% raffinose and 3% nonfat dried milk, cooled for 10 min at approximately $-110°C$ in the vapor phase of a liquid nitrogen storage tank, and plunged into liquid nitrogen). In addition, a single male produces a large amount of material (approximately 3×10^7 sperm). Recovery requires in vitro fertilization to produce embryos, which again are transferred to foster mothers to produce pups. The major drawback of this method, however, is that sperm from many inbred strains of mice, in particular C57BL/6 and 129, lose their ability to fertilize after freezing. Recovering pups at a reasonable efficiency requires manipulating the oocyte to disrupt the zona pellucida or using the technically difficult technique of intracytoplasmic sperm injection to fertilize the oocyte.

Ovary cryopreservation can be accomplished using a modification of an embryo-freezing protocol. Ovaries are collected from young females, cut in half, and frozen. One recovers live mice from this material by doing an ovarian transplant. This requires a histocompatible, immunosuppressed, or immunodeficient recipient host for the thawed ovary. Furthermore, a suitable sperm donor must be available to mate with the transplant recipient.

The amount of material one cryopreserves is driven in part by how often one expects to recover material. The large repositories typically store enough material to provide for 10 recoveries and restock a strain long before they consume all the frozen samples. Typically, this means 250–500 carrier embryos, at least a dozen vials of frozen sperm, or ovaries from 20 donors. An individual laboratory that is simply archiving a strain for personal use can reduce these amounts accordingly.

It is important to do a number of controls when cryopreserving a strain of mice. The first is to assure that one is able to recover live pups from the frozen material before discarding the colony of live mice. Second, when freezing embryos, it is wise to freeze a vial of control embryos on each day that archived embryos are frozen. The control embryos can then be thawed and cultured to the blastocyst stage to ensure that each day's freeze was successful.

Frozen material will last indefinitely at liquid nitrogen temperatures with no loss of viability over time. Thus, samples will recover with the same survival rate whether they are stored for minutes or for years. However, samples *must* be kept cold. Liquid nitrogen storage is relatively inexpensive but requires that the storage containers are always supplied with liquid nitrogen. Mechanical freezers avoid the necessity of maintaining a liquid nitrogen supply but are subject to mechanical failure or power loss. Some method of monitoring and logging temperature and, for liquid nitrogen, liquid levels is a good idea. Another good idea is to split samples between multiple storage containers. It is even better if one of these storage containers is in a physically separate space, ideally offsite.

Because of the advantages that cryopreservation provides in conserving mouse colony resources and protecting valuable strains against loss, cryopreservation is becoming more common as a service provided by institutional transgenic core facilities. It is also

Table 1.3. Mouse embryo cryopreservation protocol. This is the protocol used by the Jackson Laboratory Cryopreservation Resource to freeze 2-cell through 8-cell mouse embryos. It is a modification of the method published by Renard and Babinet (1984)

A. Freezing method.

1. Embryos are held in M2 medium at room temperature until ready for addition of cryoprotectant.
2. 0.25 ml semen straws are prepared as follows:
   a. Push the cotton/PVA plug into the straw until it is 75 mm from the end.
   b. Mark the straw 20 mm, 27 mm and 32 mm from the plug with an alcohol-proof marking pen

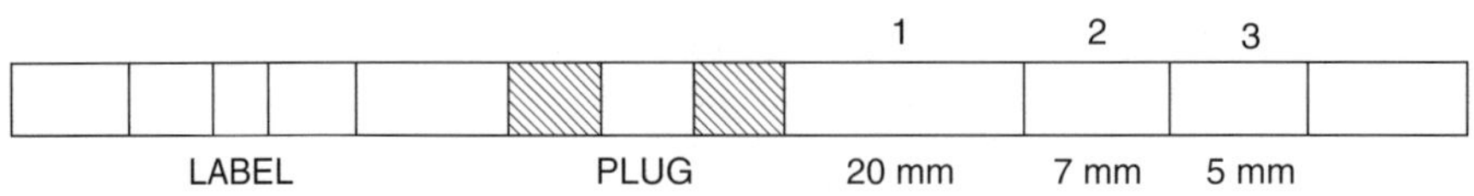

   c. Aspirate 1.0 M sucrose in M2 into the straw until it reaches mark 3.
   d. Aspirate air so that the sucrose meniscus reaches mark 2.
   e. Aspirate 1.5 M propanediol (PROH) in M2 until the sucrose meniscus reaches mark 1.
   f. Aspirate air until the sucrose column forms a seal with the plug.
3. Transfer the embryos into a drop of 1.5 M PROH in M2 and equilibrate for 15 minutes
4. Transfer the embryos into the PROH drop in the straw. This can be done under a dissecting microscope. Take care to avoid transferring a large amount of medium and large air bubbles along with the embryos. 25 embryos is a reasonable number to store in each straw.
5. Seal the straws using Critoseal.

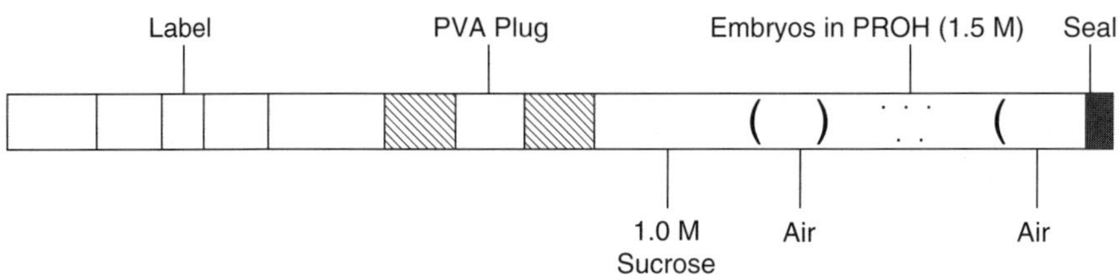

6. Place loaded straws into a controlled-rate alcohol bath freezer at −7°C so that the embryo-containing ends are submersed in the alcohol and equilibrate for 5 minutes.
7. Seed ice formation by touching the straws outside the sucrose column near the cotton plug with a cotton swab dipped in liquid nitrogen.
8. Wait 5 minutes and check that ice has migrated down into the embryo fraction. Re-seed straws if necessary.
9. Cool the straws at 0.5°C/minute until the temperature has reached −30°C.
10. Transfer the straws to liquid nitrogen for storage.

B. Thawing embryos

1. Transfer the straws from the liquid nitrogen storage freezer to a smaller container of liquid nitrogen
2. Grasp the straw near the label with forceps and hold in air for 40 seconds, then in room temperature water until the ice disappears.
3. Wipe the straw dry
4. Cut off the seal and cut through the PVA plug, leaving about half the cotton plug to act as a plunger. Use a metal rod to expel the contents into a Petri dish.
5. Wait 5 minutes. The embryos will shrink considerably as water and the PROH diffuse out.
6. Transfer the embryos to a drop of M2. The will rapidly assume a normal appearance as water diffuses back in.
7. Wash the embryos in fresh M2 and transfer to a pseudopregnant recipient or culture in KSOM medium.

commercially available from most of the large mouse vendors. Finally, comprehensive training in cryopreservation techniques is offered by The Jackson Laboratory (http://www.jax.org/courses/index.html) and by the European Mutant Mouse Archive (http://www.emmanet.org).

DRUG ADMINISTRATION IN LABORATORY MICE

Creation and characterization of mouse models for specific types of human cancer are the focus of most of the initial research in model development. The real value of many of these models is when they can be used to validate current treatments in human patients and then to test novel therapeutic approaches. Administration of drugs requires IACUC approval prior to testing. A number of methods are commonly used and are summarized below.

Intravenous Injections

The ventral coccygeal vein of the mouse tail is a common site for intravenous injections. The mouse is generally placed in a restraint device with the tail protruding. Warming the mouse or its tail will increase the flow of blood. Small-gauge (30 g) needles are used on syringes through the skin and into the vein.

Subcutaneous Injections

The mouse has very loosely attached skin that can be lifted and a small-gauge needle is inserted into the flap of skin over the back and neck once it is raised digitally.

Intradermal Injections

This method is similar to subcutaneous injections except the needle is positioned within the layers of the skin instead of through the skin. Mouse skin is extremely thin so this method is almost impossible to perform.

Intramuscular Injections

Generally, intramuscular injections, as in larger types of laboratory animals, are made with small-gauge needles into the epaxial muscles on either side of the spinal column or the quadricepts muscles of either rear leg. Due to the small size of mice, very small volumes should be injected.

Intralesional Injections

Small-gauge needles can be used on syringes to inject drugs into a neoplastic mass. The volume depends upon the size and number of the lesions.

Intraperitoneal Injections

The mouse is manually restrained with the head and body tilted downward. A small-gauge needle is inserted into the caudal left abdominal quarter, thereby avoiding injection into the cecum on the right side.

Osmotic Pumps

Osmotic pumps (ALZET Osmotic Pumps, Durect Corp., Cupertino, CA) can be surgically implanted subcutaneously or intraperitoneally to transport drugs over prolonged periods of time.

Slow-Release Subcutaneous Implants

Silastic capsules, consisting of 10-mm silastic tubing (Dow-Corning Corp., Midland, MI) packed with the drug or hormone and having the ends sealed with 3-mm glass beads, can be implanted through a small incision over the dorsal thoracic midline in the subcutaneous tissues.

CONCLUSIONS

In summary, identifying genetically engineered mice as models for human cancers, establishing and preserving colonies of these mice, and experimental manipulation of the mice with various therapeutic regiments are routine in many rodent facilities today. Understanding the basic approaches described in this chapter provides investigators with a working knowledge of the options available to address their specific research questions. Working with their institutional attending veterinarian assures that these methods will be used in a humane manner and meet IACUC requirements for the institution.

ACKNOWLEDGMENTS

This work was supported by grants from the National Institutes of Health (RR179, CA89713, and CA34196). The authors thank S. Williamson for the graphics and N. D. Buckley for assistance with the tables.

REFERENCES

Boggess D, Cunliffe-Beamer TL, Sundberg JP (2000): Colony establishment. In Boggess D (Ed), *Systematic Approach to Evaluation of Mouse Mutations*. CRC, Boca Raton, FL, pp 1–13.

Bristol IJ, Farmer MA, Cong Y, Zheng XX, Strom TB, Elson CO, Sundberg JP, Leiter EH (2000): Heritable susceptibility for colitis in mice induced by IL-10 deficiency. *Inflamm Bowel Dis* 6: 290–302.

Clark JD, Baldwin RL, Bayne KA, Brown MJ, Gebhart GF, Gonder JC, Gwathmey JK, Keeling ME, Kohn DF, Robb JW, Smith OA, Steggerda JAD, Vandenbergh JG, White WJ, Williams-Blangero S, VandeBerg JL (1996): *Guide for the Care and Use of Laboratory Animals*. National Academy Press, Washington, DC.

HogenEsch H, Gijbels M, Offerman E, Hooft Jv, Bekkum Dv, Zurcher C (1993): A spontaneous mutation characterized by chronic proliferative dermatitis in C57BL mice. *Am J Pathol* 143: 972–982.

Jackson IJ, Abbott CM (2000): In Hames BD (Ed), *Mouse Genetics and Transgenics*. Oxford University Press, New York.

Lindsey JR, Boorman GA, Collins MJ, Hsu C-K, VanHoosier GL, Wagner JE (1991a): *Companion Guide to Infectious Diseases of Mice and Rats*. National Academy Press, Washington, DC.

Lindsey JR, Boorman GA, Collins MJ, Hsu C-K, VanHoosier GL, Wagner JE (1991b): *Infectious Diseases of Mice and Rats*. National Academy Press, Washington, DC.

Mahler M, Bedigian HG, Burgett BL, Bates RJ, Hogan ME, Sundberg JP (1998): Comparison of four diagnostic methods for detection of Helicobacter species in laboratory mice. *Lab Anim Sci* 48: 85–91.

Maronpot RR, Boorman GA, Gaul BW (1999): *Pathology of the Mouse. Reference and Atlas*. Cache River, Vienna, IL.

McElwee KJ, Boggess D, King LE, Sundberg JP (1999): Alopecia areata versus juvenile alopecia in C3H/HeJ mice: Tools to dissect the role of inflammation in focal alopecia. *Exp Dermatol* 8: 354–355.

Mikaelian I, Nanney LB, Parman KS, Kusewitt DF, Ward JM, Naf D, Krupke DM, Eppig JT, Bult CJ, Seymour R, Ichiki T, Sundberg JP (2004): Antibodies that label paraffin-embedded mouse tissues: a collaborative endeavor. *Tox Pathol* 32: 181–191.

Mikaelian I, Ichiki T, Ward JM, Sundberg JP (2004): Diversity of spontaneous neoplasms in commonly used inbred strains and stocks of laboratory mice. In Hedrich HJ (Ed), *Handbook of Experimental Animals: The Laboratory Mouse*. Academic Press, London, pp. 345–354.

Mohr U, Dungworth DL, Capen CC, Carlton WW, Sundberg JP, Ward JM (1996): *Pathobiology of the Aging Mouse*. ILSI, Washington, DC.

Nagy A (2002): *Manipulating the Mouse Embryo*, 3rd ed. Cold Spring Harbor Laboratory, Cold Spring Harbor, NY.

Nakagata N, Takeshima T (1993): Cryopreservation of mouse spermatozoa from inbred and F1 hybrid strains. *Jikken Dobutsu* 42: 317–320.

Nakamura M, Sundberg JP, Paus R (2002a): Mutant laboratory mice with abnormalities in hair follicle morphogenesis, cycling, and/or structure: annotated tables. *Exp Dermatol* 10: 369–390.

Nakamura M, Tobin DJ, Richards-Smith B, Sundberg JP, Paus R (2002b): Mutant laboratory mice with abnormalities in pigmentation: annotated tables. *J Dermatol Sci* 28: 1–33.

Piletz JA, Ganschow RE (1978a): Lethal milk mutation results in dietary zinc deficiency in nursing mice. *Am J Clin Nutr* 31: 560–562.

Piletz JE, Ganschow RE (1978b): Zinc deficiency in murine milk underlies expression of the lethal milk (lm) mutation. *Science* 199: 181–183.

Presland RB, Boggess D, Lewis SP, Hull C, Fleckman P, Sundberg JP (2000): Loss of normal profilaggrin and filaggrin in flaky tail (*ft/ft*) mice: An animal model for the filaggrin-deficient skin disease ichthyosis vulgaris. *J Invest Dermatol* 115: 1072–1081.

Relyea MJ, Miller J, Boggess D, Sundberg JP (2000): Necropsy methods for laboratory mice: Biological characterization of a new mutation. In Boggess D (Ed), *Systematic approach to evaluation of mouse mutations*. CRC, Boca Raton, FL, pp 57–90.

Renard JP, Babinet C (1984): High survival of mouse embryos after rapid freezing and thawing inside plastic straws with 1-2 propanediol as cryoprotectant. *J Exp Zool* 230: 443–448.

Seymour R, Ichiki T, Mikaelian I, Boggess D, Silva KA, Sundberg JP (2004): Necropsy methods. In Hedrich HJ (Ed), *Handbook of Experimental Animals: The Laboratory Mouse*. Academic Press, London, pp. 495–516.

Shultz LD, Schweitzer PA, Hall EJ, Sundberg JP, Taylor S, Walzer PD (1989): *Pneumocystis carinii* pneumonia in scid/scid mice. *Curr Topics Microbiol Immunol* 152: 243–249.

Silver LM (1995): *Mouse Genetics: Concepts and Applications*. Oxford University Press, New York.

Smith RS, John SWM, Nashina PM, Sundberg JP (2002): In Sundberg JP (Ed), *Systematic Evaluation of the Mouse Eye. Anatomy, Pathology, and Biomethods*. CRC, Boca Raton, FL.

Sundberg JP (1994a): The flaky tail (*ft*) mutation, chromosome 3. In Sundberg JP (Ed), *Handbook of Mouse Mutations with Skin and Hair Abnormalities Animal Models and Biomedical Tools*. CRC, Boca Raton, FL, pp 269–273.

Sundberg JP (Ed) (1994b): *Handbook of Mouse Mutations with Skin and Hair Abnormalities. Animal Models and Biomedical Tools*. CRC, Boca Raton, FL.

Sundberg JP (1994c): The matted (*ma*) mutation, chromosome 3. In Sundberg JP (Ed), *Handbook of Mouse Mutations with Skin and Hair Abnormalities Animal Models and Biomedical Tools*. CRC, Boca Raton, FL, pp 345–349.

Sundberg JP, Boggess D (2000): *Systematic Characterization of Mouse Mutations*. CRC, Boca Raton, FL.

Sundberg JP, Boggess D, Hogan ME, Sundberg BA, Rourk MH, Harris B, Johnson K, Dunstan RW, Davisson MT (1997a): Harlequin ichthyosis. A juvenile lethal mouse mutation with ichthyosiform dermatitis. *Am J Pathol* 151: 293–310.

Sundberg JP, France M, Boggess D, Sundberg BA, Jenson AB, Beamer WG, Shultz LD (1997b): Development and progression of psoriasiform dermatitis and systemic lesions in the flaky skin (*fsn*) mouse mutant. *Pathobiology* 65: 271–286.

Sundberg JP, Montagutelli X, Boggess D (1998): Systematic approach to evaluation of mouse mutations with cutaneous appendage defects. In Chuong M (Ed), *Cutaneous Appendages*. Molecular Biology Intelligence Unit I, Landes Company, Austin, TX, pp 421–435.

Ward J, Mahler J, Maronpot R, Sundberg JP (2000): *Pathology of Genetically Engineered Mice*. Iowa State University Press, Ames, IA.

Whittingham DG, Leibo SP, Mazur P (1972): Survival of mouse embryos frozen to −196 degrees and −269 degrees C. *Science* 178: 411–414.

2

PATHOLOGY OF MOUSE MODELS OF CANCER

ROBERT D. CARDIFF

Mutant Mouse Pathology Laboratory, Center for Comparative Medicine, University of California—Davis, Davis, California

Analysis of the changes in structure (gross and histopathology) is essential for the validation of mouse models of human cancer. However, many investigators do not have immediate access to qualified pathologists who understand the unique features of genetically engineered mouse pathology. Further, the gross examination of the mouse and the preparation of tissues are usually done by the laboratory personnel and, most frequently, by relatively untrained graduate students or fellows. Some laboratories have the good fortune of having experienced laboratory technologists who perform the services and teach the incoming students and fellows the correct techniques. However, many laboratories do not have this advantage and frequently relegate the necropsy to the least experienced animal technician or the newest graduate student. Unfortunately, this often leads to disastrous results that cannot be overcome. Therefore, this chapter is not about how to be a pathologist but rather is about how to prepare a mouse for pathologic analysis and is meant to be a primer for the neophyte who needs basic help.

GENERAL TECHNICAL PRINCIPLES

What a Pathologist Is and Does

Pathology is the study of disease. A pathologist is a person who studies diseases. A person trained in pathology usually has special training in the interpretation of laboratory results (clinical pathology) and morphologic changes associated with disease (anatomic pathology). The pathologist is generally stereotyped as being the person who can look down a microscope and make a diagnosis. However, a pathologist is more than a high-throughput diagnostician. The well-trained, astute pathologist is a specialist in the integration of structural and functional changes associated with disease. The pathologist integrates the clinical, biological, demographic, structural, and functional information to provide an accurate interpretation of the disease state. At best pathology provides integrative biology. However, histopathologic interpretation without all the relevant animal information and without proper tissue handling is often misleading and mostly useless. The subsequent sections of this chapter will describe the materials, methods, information, and protocols that are relevant to the pathologist.

Planning Ahead

Genetically engineered mice (GEM) require months to develop and years to analyze. Anyone developing such a mouse should consult with a pathologist well in advance so that they can make the appropriate observations, keep useful records and samples, and prepare the appropriate tissues. Under any circumstance, this requires the integration of the pathologist into the research team. If the research organization does not have a qualified pathologist who is interested in helping, the research leadership should identify a pathologist who is willing to consult prior to initiating the experiments. Too much time, money, and energy are invested to risk mediocre pathology and equally mediocre results. The following sections discuss some of the variables that must be considered in planning the research project.

Knowing the Normal Anatomy and Histology of the Mouse. Before beginning the study of a new GEM line,

Mouse Models of Human Cancer, edited by Eric C. Holland
ISBN 0-471-44460-X

the normal anatomy, histology, and background pathology of the mouse strain(s) being used to create the GEM must be studied. It is a common experience among pathologists to have to tell an investigator that the samples submitted were from the wrong organ, such as salivary gland labeled thymus or pancreas thought to be lymph node or uterus. This type of mistake can be prevented by learning, in advance, where the organs are located and the normal changes that the mouse strain will undergo with aging. Although there is no single text that can be recommended for a complete description of the anatomy and histology of the mouse, this information can be found scattered in a variety of monographs. Moreover, most tumor-bearing mice are older animals. Fortunately, an excellent monograph on the pathology of aging mice is also available (Mohr et al., 1996).

Record Keeping. It may appear to be self-evident that good record keeping is essential (Brayton et al., 2001). No detail is too small to ignore (Blumberg, 1980). Details such as the specific brand and type of mouse chow that is being used by the laboratory animal care staff in the vivarium might cause subtle changes in phenotype. The types of microorganisms in the colony can influence the outcome of experiments (U.S., IoLAR et al., 1991). In some experiments the details of parity and suckling are significant (Nieto et al., 2003).

INVESTIGATOR'S RESPONSIBILITIES

Documentation

Records. The investigator is responsible for keeping extensive and accurate records. The records should include the molecular construct, background strain, vendor, size of the litter, proportion of males to females, birth date, date of sacrifice, age at tumor onset and sacrifice, mouse parity and breeding history, health status of the housing room, genotype, and any gross behavioral or physical differences between the GEM and the control (littermate, if possible). Ideally, each mouse is assigned an identification number that can be used throughout the experimental protocol to identify the individual mouse (Brayton et al., 2001).

Background Strain. The mice created by the investigator may come from hybrid backgrounds or may be from an inbred strain. Each inbred mouse strain has a background pathology that may influence interpretations. "Historical" controls (i.e., previous examples of wild-type mice from the same colony) are frequently used, but this is particularly fraught with error because of the mixture of genes in the hybrid mice. All primary surveys, also known as tier I (see the section on necropsy tiers, p. 17), of experimental animals must include age-matched and gender-matched controls from the same colony (Sundberg and Boggess, 2000).

Recognition of the background pathology of the aging mouse is important. Some stains commonly develop ovarian or testicular teratocarcinomas (Stevens, 1970, 1973) and yet others develop pituitary adenomas (Mahler et al., 1996; Wakefield et al., 2003). For example, C57Bl/6 mice are tumor resistant while the more susceptible FVB/N mice readily develop pulmonary tumors and are also blind (Mahler et al., 1996; Suwa et al., 2002). Hybrid mice are frequently even more difficult to interpret because of the mixture of genes. Some colonies of FVB/N have persistent mammary hyperplasia that is related to pituitary adenomas and hyperplasias (Nieto et al., 2003; Wakefield et al., 2003).

The origin of the mice and details of colony management must be recorded. Although the strains are closely managed, genetic drift occurs in isolated colonies that lead to differences in background pathology. This will be particularly true if the colony is not periodically replenished by breeding with mice from outside the colony.

Molecular Construct. The construct provides important information for the pathologist. Many pathologists do not have intimate knowledge of molecular biology. The investigator is the expert and must take the time and the responsibility for educating the pathologist about the details of the molecular or cellular biology created by the experimental conditions. Perhaps, the promoter is a slightly different version of a previous promoter. The pathologist, if properly instructed, might be able to recognize subtle shifts in phenotype that might go unobserved by the poorly informed pathologist. The detection of these subtleties may help the investigator with the understanding of their mice or help with the design of the next experiment.

Observation. In many institutions, animal technicians or investigators are the only people who observe the living animal. In such cases, the behavior and general appearance of the animal must be noted. The mouse should be thoroughly inspected for unusual gross findings. For example, rectal prolapse may indicate infections such as *Helicobacter* sp. (Ward et al., 1996). Runting might be because of misalignment of the teeth or oral abscesses rather than the result of genetic engineering (Brayton et al., 2001). The precise anatomic location of the tumor must be recorded. The use of shorthand notes might be useful to the primary investigator but become embarrassingly uninterpretable when reviewed by another person. The gross appearance of the tumors must be recorded (Brayton et al., 2001). The number, location, size, shape, color, consistency, and even odor are all valuable data about the gross appearance.

Palpation. Palpation of tumors is particularly important. The laboratory protocol should include a set schedule for palpation to accurately record the tumor latency. It is necessary to record the onset of any tumor and note the growth rate with accurate measurements. Palpation of the tumor is performed by grasping the tumor lightly between two fingers and gently compressing the mass. The relative consistency and shape of the tumor must be assessed. With experience, the investigator will be able to distinguish between solid and cystic masses. Palpation and measurement for tumor kinetics should be performed by a single individual. Different observers inevitably provide different measurements, leading to inconsistency.

Imaging. Many institutions have outstanding in vivo systems available for small animals. Data from in vivo imaging can provide numerous valuable clues to the biology of the lesion and must be shared with the pathologist. Since the in vivo systems record data about the entire tumor while the tissue section is a 10-μm, two-dimensional sample, the in vivo image may assist the pathologist in determining the size and location for the microscopic sample. It is often useful to validate the in vivo image with careful dissection and ex vivo imaging of the suspected tumor (Fig. 2.1). If the pathology does not correspond to the in vivo image, additional samples may need to be taken.

Microbiology. The microbiology is important and should be shared with the pathologist. (This will be discussed in a separate chapter.)

Preparations in the Research Laboratory

With the shortage of pathologists, many laboratories perform their own necropsies and prepare their own tissues. These tasks are frequently performed by relatively untrained personnel, sometimes the newest student or fellow in the laboratory. This section is designed to introduce the neophyte to the general principles of tissue selection and preparation. Even if investigators are not preparing their own tissues, they are inevitably responsible for some aspect of the tissue handling and need to know the principles involved.

Necropsy. A Necropsy involves detailed dissection and examination of the euthanized animal. It also involves the selection of tissues for fixation and histologic examination. Three levels of necropsies use different levels of organ examination, referred to as "tiers" (Brayton et al., 2001; Sundberg 2003). Books (Feldman and Seely, 1988) and several websites that demonstrate the mouse necropsy are available (http://ccm.ucdavis.edu/tvmouse, http://www.geocities.com/virtualbiology/necropsy.html, http://ccm.ucdavis.edu/tvmouse/strip.html, http://www.eulep.org/Necropsy_of_the_Mouse/index.php).

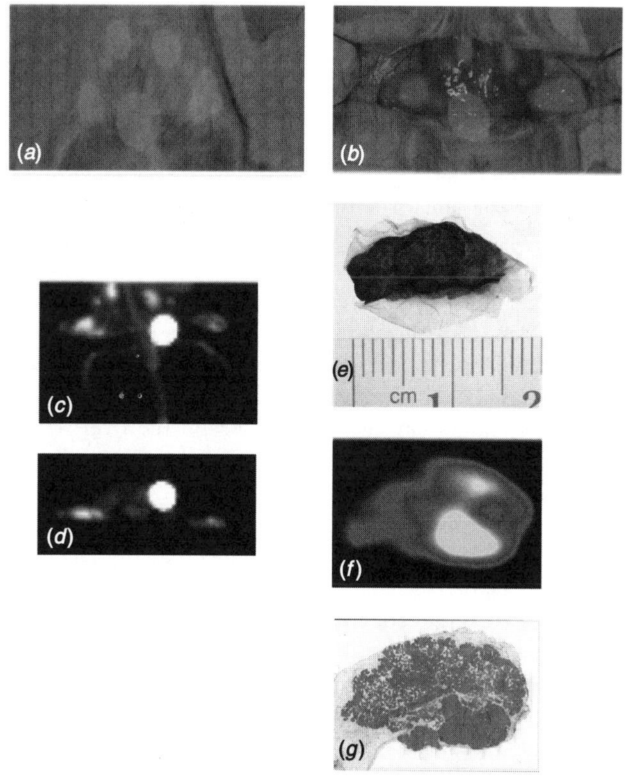

Figure 2.1. Ex vivo imaging. Validation of in vivo images may require detailed dissections to verify that the image detected in the animal represents a specific pathologic lesion. This figure illustrates how the in vivo image is removed and reimaged to prove that it represents a tumor mass. (*a*) Mammary tumors transplanted into the mammary fat pad are detected in vivo using micro–positron emission tomography (DET). (*b*) To determine whether the images correspond to the mammary mass, an incision was made and the skin that contains the mammary fat pad was spread to expose the tumors. The photographic and PET images are superimposed to demonstrate that the microPET images stay with the skin and the tumors. (*c*) Another animal went through the same procedure. The skin was spread. The microPET images of the tumor-bearing mouse show (*c*) the supine and (*d*) the coronal views. One of the tumors was (*e*) removed and placed on a slide and (*f*) reimaged to show that the in vivo image was the same as the ex vivo image. (*g*) The microscopic slide is used by the pathologist to verify the diagnosis. (Images courtesy of Simon Cherry, Craig Abbey, and Alexander Borowsky, University of California—Davis.) (See color insert).

Necropsy Tiers. The tier 1 necropsy is the most detailed survey of all tissues and is used when new GEM with an unknown or incompletely understood phenotype are being assessed. The tier 1 necropsies should include at least five GEM of each gender and five wild-type age- and gender-matched controls. The specific age of the mice necropsied will vary according to the circumstances and the time schedule should be developed with the pathologist. Samples of all organs should be taken and submitted for histologic examination. It is generally understood that at least 54 tissues will be examined for each animal.

Tier 2 necropsies are more limited, customized for the individual mouse, and generally focus on a known phenotype of interest. However, the necropsy should include the examination of all the organs. The dissector needs to understand the disease patterns in the animal and be alert for unusual occurrences. For example, most mouse tumors metastasize to the lungs. The gross tumors on the liver might be hepatomas and not metastases. The enlarged spleen might be a lymphoma or leukemia. These are better assessed by blood smears and samples of the liver. If the studies include mammary growth and development, the pituitary and ovaries must be examined.

Tier 3 necropsies are limited to the tumor and lungs and are used in the final phases of data collection or when involving a large-scale study. While only a limited number of organs are sampled, each organ is scrutinized in more detail and, generally, with more techniques. These types of samples are used to "validate" the model system.

Sampling of Tumors. All samples of tumors must include an equal and contiguous sample of the adjacent non-neoplastic host tissue (Fig. 2.2). Most of the important information about the tumor biology is at the host–tumor interface and not in the center of the tumor. The inexperienced dissector will frequently carefully dissect the tumor away from the host and proudly submit the sample and is crushed to learn that he or she has discarded the most important information. The adjacent tissue sample reveals the host response in terms of angiogenesis, inflammation, fibrosis, and other factors. The adjacent tissue also provides a measure of degree of differentiation for many organs. The early lesions that reveal the morphologic evolution of the neoplasm can frequently be found in an adequate sample of the attached adjacent host tissues. The center of the tumor is the most likely to be composed of necrotic debris that is not useful for either molecular or histologic examination. This can be seen in Figure 2.2*a*. The red-stained area is composed of necrotic debris. When given a choice, the center should be avoided.

Special Organ Preparation. In most cases, the animal experiment is designed to study tumors in a specific organ. Specialized techniques have been developed for the preparation of most organs. The research team will need to familiarize itself with the preparation of the target organ system. The sampling and fixation of the target organ needs to be optimized during the tier 2 and 3 necropsies. The faster the tissue is processed, the better the preparation. Examples of special techniques that warrant special attention include inflation for the lungs, the "Swiss roll" for the intestines (Boivin et al., 2003), perfusion for the brain (Weiss et al., 2002), and whole mounts for the mammary gland (Cardiff et al., 2000) (Fig. 2.3).

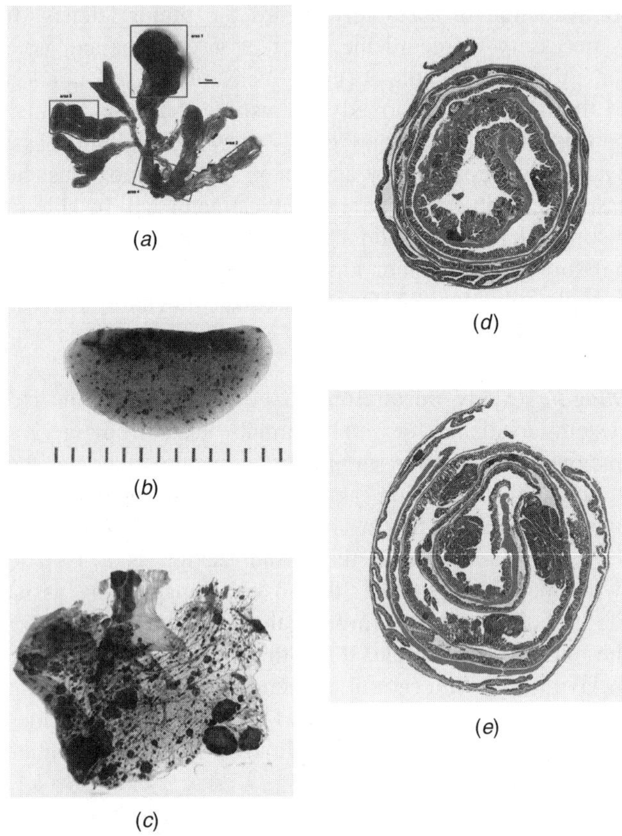

Figure 2.2. Special types of preparations. Special preparation techniques: whole-mount preparations of (*a*) prostate, (*b*) lung, and (*c*) mammary gland that allow the investigator to visualize number and distribution of tumors in the entire organ sampled. (*a*) Fresh, unstained prostate viewed via transmitted backlight. This image of a transgenic adenocarcinoma mouse prostate (TRAMP) was provided by A. A. Donjacour, University of California—San Francisco. (*b*) Fixed, hematoxylin-stained and cleared preparation of a mouse lung from a PyV-mT mouse. (Preparation and image by Erik T. McGoldrick, University of California—Davis.) The lung has numerous small dark spots that represent multiple metastatic masses. (*c*) Mammary gland whole mount from an *erbB* transgenic mouse. (Preparation and image by Erik T. McGoldrick, University of California—Davis). The normal architecture is disrupted by multiple small masses. (*d*, *e*) Histology obtained with Swiss roll preparations of the mouse gastrointestinal tract. Part (*d*) illustrates a normal mouse bowel and (*e*) illustrates multiple polyps in mice provided by Robert Coffee and Kay Washington, Vanderbilt Medical School. (See color insert).

Whole-mount techniques that allow "subgross" examination of the specimen are underutilized in mouse tumor biology and deserve greater emphasis. Mouse organs are

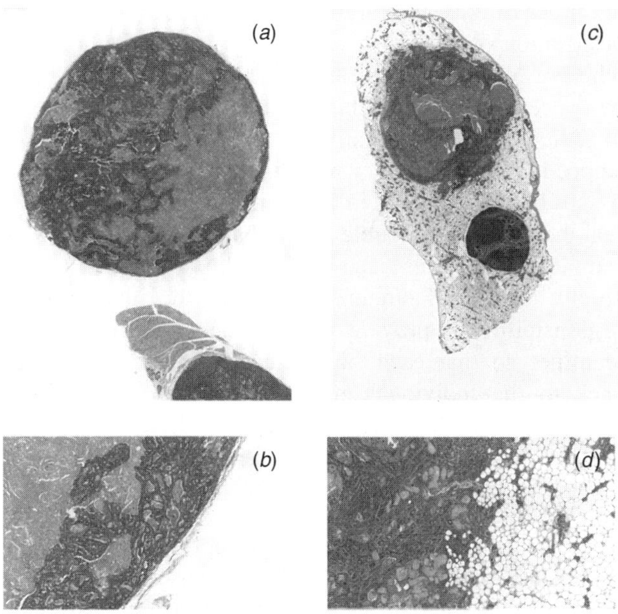

Figure 2.3. How to sample a tumor. Two sets of illustrations from different mouse mammary tumors: (*a*) Tumor was carefully dissected free from the surrounding host material. As a result, there is little host–tumor interface for study. Further, most of the tumor is necrotic. (*b*) Higher magnification of the rim of (*a*) illustrating the lack of host tissue. Note also that the center of the tumor has pink (eosinophilic), acellular debris (necrosis). The sample provides only minimal information about the biology of this tumor. (*c*, *d*) A properly sampled mouse mammary tumor. Note the abundant host tissue on all sides in (*c*). A higher magnification in (*d*) shows that the tumor is infiltrating into the surrounding mammary fat. This sample provides the pathologist with the context and with abundant biologic information.

relatively small and can be easily viewed under a dissecting microscope (Fig. 2.3). This can provide an overview of the distribution of the malignant or premalignant neoplasms in the entire organ in three dimensions while 5–10-μm tissue samples can be misleading. The techniques developed for the mammary gland can be readily adopted for examination of other organs such as lungs or prostate (Bhatia-Gaur et al., 1999; Park et al., 2002) (Fig. 2.3). In some organs, the early lesions can even be visualized in the live tissue without fixation (Fig. 2.3).

Fixatives, Fixation, and Processing

General Considerations. Fixation is critical to adequate histology and pathology. GEM research involves lengthy programs of development and observation. It is tragic to receive samples from a wonderfully designed and executed experiment only to find that the samples have not been adequately fixed and much of the morphologic detail lost. Simply dropping the sample into the fixative is not enough. The investigator needs to develop detailed protocols for fixation that are customized for the organ and the application. Many protocols can be found on the Internet.

The investigators, their colleagues, and their staff need to remember that the fixatives and organic solvents used to prepare tissues for histology are potentially dangerous chemicals. This author knows more than one pathologist who has lost the sense of smell and taste because of prolonged and careless use of formalin. Exposure to these chemicals should be limited and carefully controlled. The chemicals should only be handled in a fume hood. Gloves, safety glasses, and gowns should be used when handling the samples. All laboratory personnel should be required to take and pass safety classes.

In general, the volume of the fixative should be 10 times the volume of the tissue sample. In many laboratories, the samples are fixed in some type of conical or cylindrical test tube. If the tissue is bloody, such as spleen, liver, or many tumors, the fixative becomes cloudy brown. Cloudy fixative needs to be replaced by fresh fixative.

Most solutions fix by crosslinking or by coagulation. In most cases, the sample is just dropped into a solution of fixative. The fixative naturally fixes the surface of the tissue and creates a barrier to further diffusion of the fixative into the sample (Fig. 2.4). Therefore, the tissue sample should be relatively small, thin, and exposed to an adequate volume of fresh fixative. Large tumors or large organs are frequently soft and difficult to subdivide. In such cases, the best fixation is attained by letting the sample fix for an hour and then slicing the specimen into 2-mm-thick sections and placing the sections into fresh fixative.

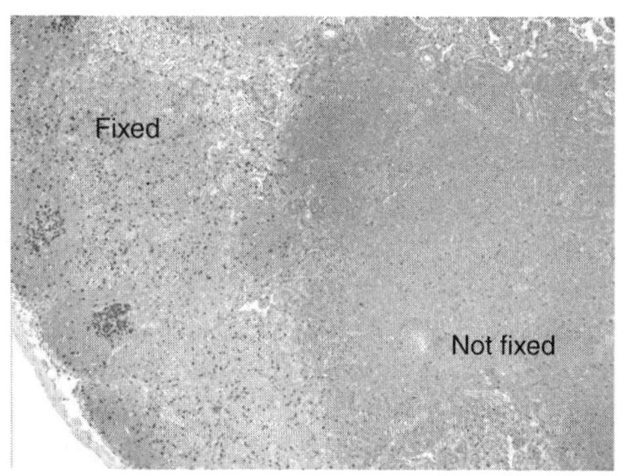

Figure 2.4. Fixation artifacts. Mouse lymph node that illustrates the effect of improper fixation. The outer rim of tissue is well fixed but the fixative did not penetrate adequately into the center of the lymph node. The disastrous immunohistochemistry results show that the Ki-67 stain is, as it should be, nicely located in the nuclei of cells in the germinal center of the fixed area. However, Ki-67 is a diffuse nuclear and cytoplasmic stain in the unfixed areas. If the lymph node were properly fixed, the Ki-67 stain would be limited to the nucleus.

The fixation time also needs attention. Too much fixation destroys antigens and nucleic acids. Too little fixation leads to autolysis and artifacts (Fig. 2.4). If the sample is not needed for immunohistochemistry, in situ hybridization, or nucleic acid extraction, it can be stored indefinitely in formalin or 70% ethanol. However, most investigators are interested in using adjunct technology to phenotype and genotype their samples. This requires careful attention to fixation time. Assuming optimal fixation of the tissue as outlined above, overnight exposure to the fixative is generally recommended. The tissue may be placed in 70% ethanol for short periods (one to four weeks) but should be processed as soon as possible.

Long-term storage of wet tissues can be a problem. Samples stored in some fixatives such as Bouin's solution or 100% ethanol for prolonged periods will become too brittle to section. In some cases, prolonged storage in Bouin's solution results in extensive exposure to picric acid, which is almost impossible to remove from the processed tissue. The standard formalin storage is probably the best, but one should be aware of the formation of small crystals in the tissues when stored in old formalin. All chemicals should be dated when the container is opened and expiration dates noted and chemicals discarded by the expiration date.

Special Techniques (What Fixative and Which Schedule). Formalin is the fixative most commonly used. It is a very good fixative for general purposes. However, formalin and paraformaldehyde are poor nuclear fixatives. Heavy-metal fixatives should be used to preserve cytologic detail. Other fixatives have specific advantages. One will need to tailor the fixative to the experiment. For example, paraformaldehyde results in less crosslinking than formalin and is preferred for in situ hybridization. Alcohol-based or organic solvent–based fixatives are preferred for tissues with fat such as mammary gland. Bouin's solution readily decalcifies the mouse bones and can be used when studying most mineralized tissues.

Immunohistochemistry has become routine in many laboratories. It is important to realize that different antigens and different antibodies will require different fixatives and fixation schedules. Many antigens will be better preserved by Bouin's solution than by formalin or paraformaldehyde. It is advisable to optimize the fixation and fixation schedule for each antigen and each antibody.

Fixatives have a limited shelf life. Care should be used to use fresh fixative and not a bottle that has been left on the shelf for several years. This is particularly true of paraformaldehyde and gluteraldehyde. Tissues fixed in outdated "paraformaldehyde" can become autolyzed because of inadequate fixation.

Shipping. Most pathologists prefer to process the tissues in their laboratory facilities because they can control the processing conditions. This generally results in better preparations or, at least, a standard artifact. Pathologists who are local will have their own system and generally will accept unembedded but fixed tissue. If the pathologist is remote, the investigator and the samples need to be shipped. The investigator will, first, need to dehydrate the tissues by placing them in 70% ethanol and then place and seal the samples in double plastic bags. It is illegal to ship samples in fixatives such as formalin without a special permit. Under all circumstances it is the investigator's responsibility to provide each container with a unique identifier so that each sample can be accurately traced back to the individual mouse. Paper and a number 2 lead pencil should be used for all record tags so that the number and record will not be dissolved by exposure to organic solvents.

Processing. Tissues should be promptly processed and embedded in paraffin. Most laboratories now use some type of automated processor. While these systems save time and money, they are only as good as the person taking care of them. Poor processing can be as damaging to the experiment as poor fixation. If the sample is not properly infiltrated with paraffin, it is impossible to section. The pathologist frequently has to reembed tissue blocks shipped from the outside. While this may give adequate sections, the histologic detail may be lost. The most common problems are outdated reagent solutions and improper timing. The outdated solutions are generally the most easily detected and corrected. The details are usually provided by the manufacturer of the processing equipment. However, the investigator needs to be aware of sloppy quality control within the context of the laboratory.

Sectioning. Many laboratories use commercial or institutional production facilities to cut and stain the slides. These facilities emphasize high-throughput, cost-effective slide production. While adequate for most purposes, the emphasis on turnaround time frequently produces sections that are far too thick with various chatter and wave artifacts. It is clear that these laboratories do not use adequate quality control or do not have anyone who can recognize adequate sections of mouse tissues. These problems are reflected in the embarrassingly poor histology images published in many molecular biology journals. The real tragedy is that the investigators have missed critical evidence. Slides for research document the morphology and are an important part of the research record. The investigator needs to have them sectioned correctly or learn to do it correctly.

Staining

1. Routine. Hematoxylin and eosin (H&E) are the standard, time-honored stains for histology and pathology (Luna, 1968). The hematoxylin stains nucleic acids and minerals blue while the eosin stains proteins pink or red. Any manual will provide the recipes for good staining. The most common problem observed is variable staining because different slide batches are processed at different times. Again, continuous monitoring, filtering, and refreshing of the stain solutions can prevent these problems. The use of different fixatives needs to be noted and transmitted to the technician in advance because they may result in different staining characteristics that require special processing. For example, mercury-based fixatives result in deposits that must be removed with iodine. Paraformaldehyde-fixed tissues stain much lighter than formalin-fixed tissues when processed at the same time.

2. *Special Stain Techniques*. The investigator has a large battery of special stain techniques available. These include immunohistochemistry and in situ hybridization. Examples using these special procedures appear throughout this book. Each special procedure stain requires detailed protocols that are best established in advance with a knowledgeable pathologist. Trial runs of new special stain protocols before performing the stain on experimental samples are very important. All reagents used in both routine and special staining procedures should be dated when opened and discarded when past the expiration date. One cannot find a better stain techniques manual than the Armed Forces Institute of Pathology (AFIP) manual (Luna, 1968).

Microscopy and Image Capture

Standard. Most laboratories now have a camera-equipped microscope and the student is encouraged to learn to use it. The proper use of this equipment requires an understanding of the physics of optics and cameras that are beyond the scope of this chapter. However, before beginning, the student is encouraged to learn to balance the optics by knowing about the numerical aperture of the objective, Kohler illumination and Nomarski optics, focusing the condenser, the difference between the field stop and the condenser diaphragm, and the light temperature versus intensity, and using neutral density filters.

Advanced Microscopy. Tremendous advances have been made in microscopy. The investigator has a number of marvelous tools for the analysis of cytologic-level morphology. Many laboratories are very well equipped for cell culture studies. These types of microscopes, however, are frequently not suitable for the lower resolution, lower magnification tissue-level patterns that are the foundation of pathology. The condensers used for phase contrast, confocal, and fluorescent images are not suitable for tissue-level pathology and need to be replaced by a standard optical condenser.

Image Capture and Postprocessing. The final phase of record keeping is image capture. Currently this is done using a digital camera and postprocessing the image with a software program such as PhotoShop. The quality of the image is a product of the quality of the slide and of the optics. Postprocessing cannot overcome defects in slides or optics. The most frequent problem is the shadows around the margins of the image (vignette). The problem is most frequently located in a poorly focused substage condenser or in poor objective quality. However, some camera systems have a built-in vignette. Manufacturers have been overcoming this with background whiteout software.

PATHOLOGIST'S RESPONSIBILITIES

Interpretation

Once the investigator has a mouse model with what appears to be a "tumor," the questions become: Is it a tumor? What kind of tumor? Does it "model" human cancer? These questions can, ideally, be answered by a qualified pathologist who can "validate" the model by studying, describing, classifying, and documenting the pathologic changes.

The pathologist should first examine the gross tissue or organ and provide the investigator with a gross impression (Brayton et al., 2001). Unfortunately, in many cases, the scientist will not have a pathologist at hand for the gross examination. The pathologist's role is reduced to the microscopic examination. While not ideal as the sole source of structural data analysis, the microscopic examination is essential for verification or validation. When properly performed, described, and documented, the microscopic examination becomes the most informative and critical part of the scientific record. The investigator, therefore, should expect the pathologist to provide an interpretation that includes a description of the microscopic slide, the classification of the pathologic changes (diagnosis), and documentation in the form of images.

Verification Versus Validation. All pathologists with training in either veterinary or human anatomic pathology should be able to look at a microscopic slide to verify whether the "tumor" seen grossly is or is not a neoplasm. This type of confirmation, provided by routine pathology screening, is considered "*verification*." However, the details needed by most cancer biologists requires a much more detailed examination, and the interpretation requires an understanding of rodent pathology and the unique aspects of pathology of genetically engineered mice. The more detailed, tier III, analysis is considered "model validation."

Since the term validation is frequently used in conjunction with GEM pathology, it is important to understand the roots of the term. Model validation refers to the process by which the attributes or characteristics of an animal model are compared to the attributes or characteristics of the human disease. In tumor biology, the scientist studies the function by inserting or deleting genes that alter the normal structural relationships to develop or interfere with neoplasms. The scientists wish to know the extent to which the murine neoplasms share the attributes of human cancers. The pathologist is expected not only to verify that the animal has a neoplasm but also to identify those characteristics that match characteristics of human neoplasms.

The ability to recognize matching attributes requires training and experience in both human and mouse pathology and an understanding of the pathobiology of the neoplastic process. In the case of the pathology of genetically engineered mice, the experience requires an awareness of the background pathology of the strain and the molecular biology and unique pathology of these mice. Pathologists with this type of experience or training are invaluable but relatively scarce.

Microscopic Interpretation. As stated earlier in this chapter, the pathologist generally has special training in microscopic anatomy that provides the final validation of the model. This skill requires visual recall and experience. The pathology of specific organ systems will be discussed in many of the other chapters in this book and in a number of reviews (Boivin et al., 2003; Borowsky et al., 2003; Cardiff, 2001, 2003; Cardiff et al., 2000; Kogan et al., 2002; Morse et al., 2002; Ward, 2000; Weiss et al., 2002). The background for these chapters should be appreciated by the reader and is provided in the following sections.

Microscopic Pathology. The short course in microscopic pathology for the beginner is based on blue (hematoxylin) and red (eosin). The hematoxylin stains the nuclei, nucleic acids, and minerals. Eosin stains proteins pink or red. A glance at H&E-stained slides without a microscope will reveal red- or blue-tinged tissues (Fig. 2.5). Blue generally means that the tissue has closely packed nuclei with relatively little cytoplasm. The lymphoid organs, such as spleen, lymph nodes, and thymus, are normally "blue" organs. In contrast, the solid organs, such as liver, heart, and nervous system, are pink or red organs because of the relative abundant cytoplasm. The beginner can also learn to recognize which organs are solid (kidney, liver, brain, lymphoid, and lung) and which are tubular (gastrointestinal, vessels, and genitourinary). Subdivisions of these normal organ patterns can be readily recognized without a microscope so that lungs are solid and lacy, brain is light pink, and kidney is two toned because of the cortex and hilus. With a little effort the neophyte can learn to identify all organs without a microscope.

The H&E pathology is a little more challenging but not mysterious. One looks for disruptions in the expected

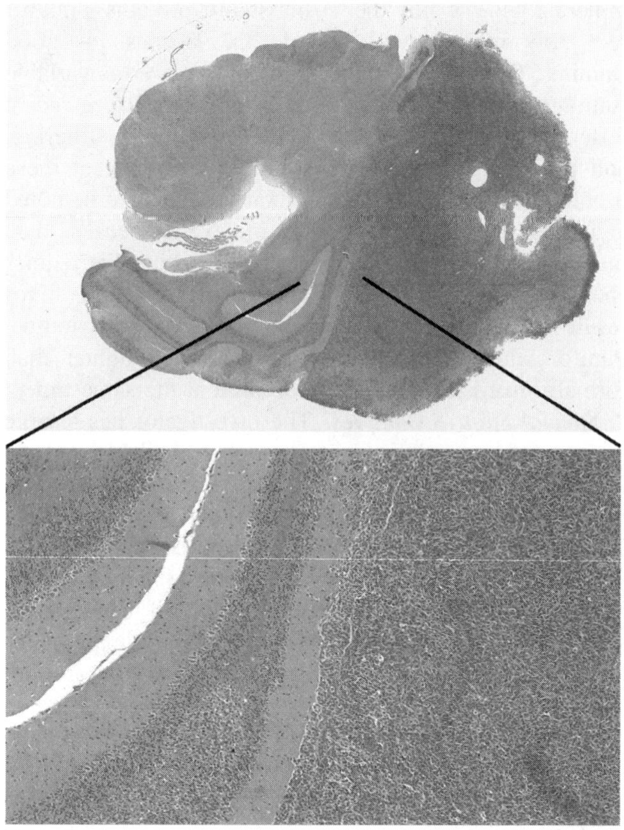

Figure 2.5. Red and blue pathology. "Red-and-blue" principle of H&E pathology. The image shows a mouse brain that has a large tumor mass. The normal brain is a light pink and has the topographic markings that are characteristic of the cerebellum. However, the entire right side of the sample is occupied by a bluish mass. The blue mass suggests that the cells in this area are primarily composed of blue nuclei with scanty pink cytoplasm. The image in the lower field is a higher magnification image that shows a poorly differentiated tumor with densely crowded nuclei. This slide is from a GFAP-Cre:RbLoxP/LoxP:p53LoxP/LoxP double knockout mouse kindly provided by Dr. Glen Marino, NCI.

normal configurations. However, the microscopist must distinguish between artifact and reality (Thompson and Luna, 1978). Neoplasms generally have an increased nuclear-to-cytoplasmic ratio and will appear on slides as blue foci (Fig. 2.5). Red foci might be tumors of the muscle or well-differentiated tumors such as hepatomas. However, space-occupying masses (tumors) that are red or pink could also be inflammatory, such as infarcts, abscesses, or granulomas.

The short course is not designed to teach pathology but rather to orient the beginner to the basic rules of pathologic examination. There are numerous exceptions and modifications to these very basic principles. For example, the host response may involve fibrotic stroma with so much protein that the tumor will be reddish.

By keeping the general rules in mind, the exceptions will make sense and "prove the rule." The more detailed pathology requires considerable training, but the interested investigator can begin appreciating the clues that the experienced pathologist uses for tissue and process recognition.

Integrative Pathology. The interpretation of the pathology requires the integration of the more detailed microscopic findings, the gross findings, the clinical or experimental context, and an understanding of the evolution of the disease process. Pathology is integrative biology. The integration begins with a knowledge of the background pathology of the mouse strain.

Many models develop tumors with long latency periods. The pathology related to the gene construct may be obscured by the spontaneous background pathology of the mouse strain. The pathology of the aging "wild-type" mouse has been extensively studied and described (Brayton et al., 2001; Frith and Ward, 1988; Green, 1966; Mahler et al., 1996; Mohr et al., 1996; Percy and Barthold, 2001; Suwa et al., 2002; U.S., IoLAR, et al., 1991; Wakefield et al., 2003). Since its development shortly after the turn of the last century, the mouse has been used as a model of mammalian tumorigenesis. The pathology of virus-infected, radiation-exposed, and carcinogen-induced mouse tumors has been recorded in extensive publications (Foster et al., 1982; Frith and Ward, 1988; Green, 1966; Maronpot, 1999; Mohr et al., 1996; Percy and Barthold, 2001; Sundberg, 2003; Turusov and Mohr, 1990; U.S., IoLAR, et al., 1991; Ward, 2000).

While these experimental and spontaneous tumor systems develop neoplasia in the same organs as humans, they have rarely resembled the morphology of human tumors. However, with the advent of the genetically altered mouse, the pathologist is seeing a unique tumor pathology not heretofore observed (Cardiff, 2001; Cardiff and Wellings, 1999). As the genes that are discovered in the context of human tumors are inserted into the mouse genome under targeted and controlled conditions, the morphology of the tumors increasingly resembles human tumors (Fig. 2.6). Since the GEM are going beyond the common experience, the pathologist is, literally, left wordless. The phenomenon exceeds the current terminology and challenges the pathologists to create new and better ways to classify, describe, and document the new pathology.

Diagnostic Terminology. The new pathology seen in GEM has required new diagnostic approaches. The GEM pathologist needs not only a background in both human and mouse pathology but also a new vocabulary (Cardiff, 2001; Cardiff et al., 2000). The standard vocabularies used in mouse and human pathology can be used as a framework, but most diagnostic vocabularies do not provide the level of detail required for research purposes. Thus, new vocabularies are being developed that are intended to meet the immediate needs but can be expanded to meet the research needs. The National Cancer Institute's Mouse Models of Human Cancers Consortium is developing and maintaining new vocabularies (http://emice.nci.nih.gov) (Boivin et al., 2003; Cardiff et al., 2000; Kogan et al., 2002; Morse et al., 2002; Weiss et al., 2002).

Hopefully, the newer vocabulary modifications will adhere to a strict hierarchical taxonomy that will be adaptable to the electronic media. One proposed hierarchical outline or "ontology" has been adapted from Rosse and associates (Rosse and Mejino, 2003) (Fig. 2.7). Although the Rosse ontology was developed for anatomy, it also provides a higher level of logic for a mouse diagnostic terminology. The ontology is adapted so that the relationship between the anatomic terminology (columns B in Fig. 2.7) relates to the diagnostic terminology (column C) and to the level of data collection and organization (column D).

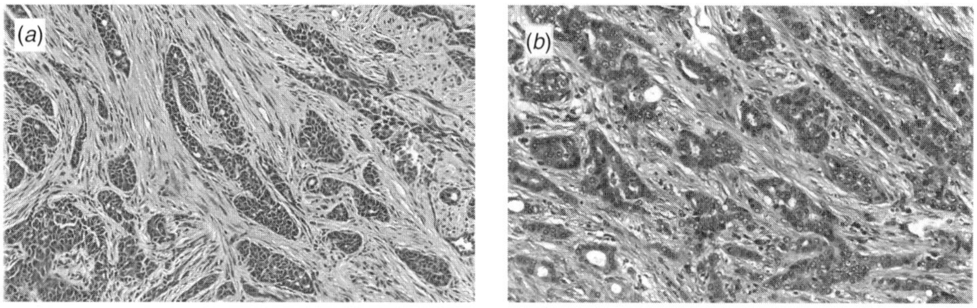

Figure 2.6. Comparative pathology. Morphologic similarity of mouse and human mammary tumors. Both are tumors composed of cords of cells growing in a densely fibrous connective tissue stroma. (*a*) Image from a transgenic mouse expressing an activated *src* transgene (mouse provided by William J. Muller, McGill University). (*b*) Image from a human with a scirrhous carcinoma. The pattern of tumor growth is similar. However, the discriminating reader will notice that the human cells are about one-third larger.

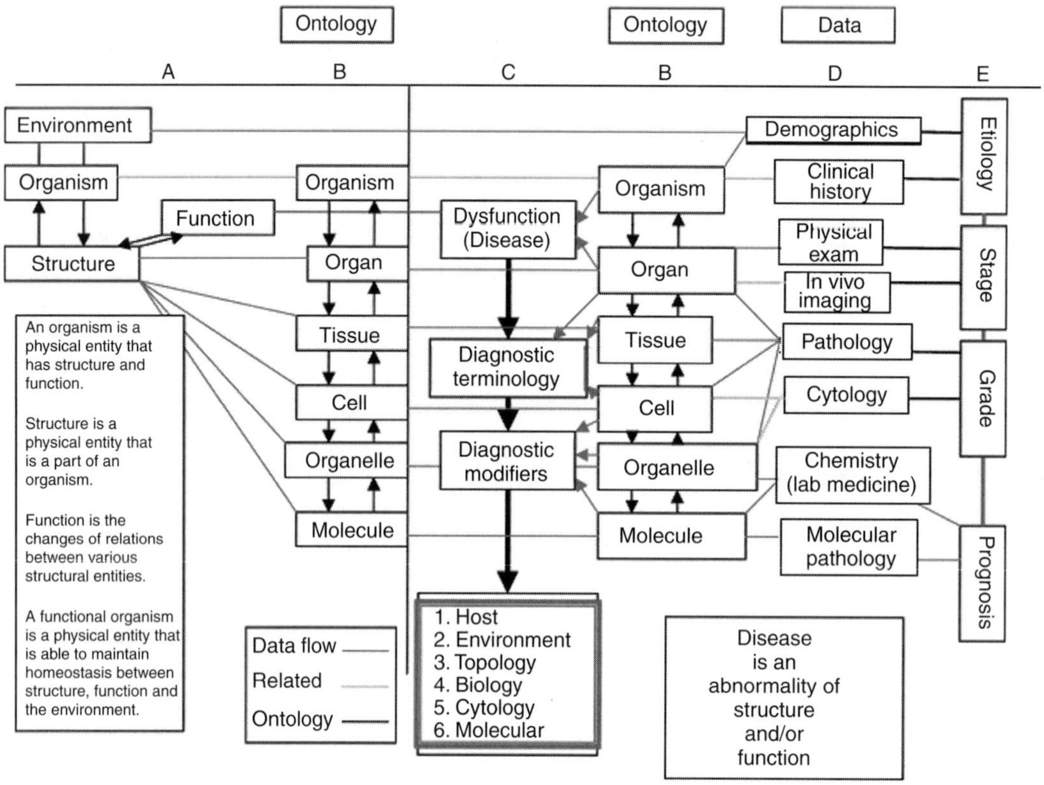

Figure 2.7. Ontology. Graph demonstrating the proposed ontology based on the Rosse model. The graph illustrates the hierarchical structure of anatomy (column B) and the proposed relationship between structure and function (column A). A disease is a dysfunction of structure and function. The diagnostic taxonomy (column C) provides the basis for the diagnostic taxonomy. The diagnosis is primarily based on the information from the organ, tissue, and cell (column C) that is collected by various medical disciplines (column D). Genetically engineered mice illustrate the striking effect of molecular dysfunction on the structure of tumors but have yet to be incorporated into the current diagnostic taxonomies (column C). The present recommendation is to use these elements as consistent modifiers of the diagnostic terminology. For example, the overexpression of erbB2 in the mouse mammary gland produces a characteristic tumor that may now be termed the erbB-type tumor, mammary gland. Column E indicates where the staging, grading, and prognostic information is collected.

When properly applied, these taxonomic systems permit the pathologist to place the tumor into the correct general category. The Mouse Models of Human Cancers Consortium Pathology workshops have recommended that GEM terminologies use descriptive vocabularies and avoid using clinical eponyms that are popular in human cancer pathology (Boivin et al., 2003; Cardiff et al., 2000; Kogan et al., 2002; Morse et al., 2002; Weiss et al., 2002).

For example, "lobular" carcinoma of the breast is a specific type of human breast cancer. The term originated from the characteristic microscopic appearance of the cells. The histology and biology of lobular carcinoma is well understood and it is now associated with silencing of e-cadherin (Cleton-Jansen, 2002). Neu-related tumors in transgenic mouse mammary glands have some of the characteristics of human lobular carcinomas but a different biology and are not associated with loss of e-cadherin (Andrechek et al., 2003). It may be premature and misleading to apply the appellation "lobular" to the mouse *erbB2*-related tumors. Therefore, the Annapolis Pathology Workshop recommended the use of a structurally related descriptive diagnostic terminology using terms that comply with the anatomic ontology of Rosse (Cardiff, 2001; Cardiff et al., 2000).

Descriptions. The diagnosis is an interpretation of the morphologic evidence but does not provide a description of the characteristics of the tumor and the host response. At best, the pathologist's descriptions will help integrate structure and function. Pathologists will also be able to infer a number of possibilities from their understanding of the construct of the GEM and the morphology leading to new avenues of research. These clues are embedded in the description. The description should contain the evidence for validation of the model. The pathologist should work with the investigator to provide the appropriate language for publication.

A good microscopic description is a linear algorithm that systematically provides the evidence that supports the diagnosis (Fig. 2.8). As a type of algorithm it will not only provide pertinent description but will also include pertinent negatives that allow the reader to understand the basis for the diagnostic decisions. In the case of GEM cancer models with a particular emphasis on validation of the model, details other than those that lead to the diagnosis become important to the investigator. If the investigator does not work closely with the pathologist, the pathologist may overlook the characteristics that *validate* the model. In turn, the investigator may not know enough to ask the right question or to ask that specific characteristics be described. Further, pathologists may have differing interests that lead to an overemphasis

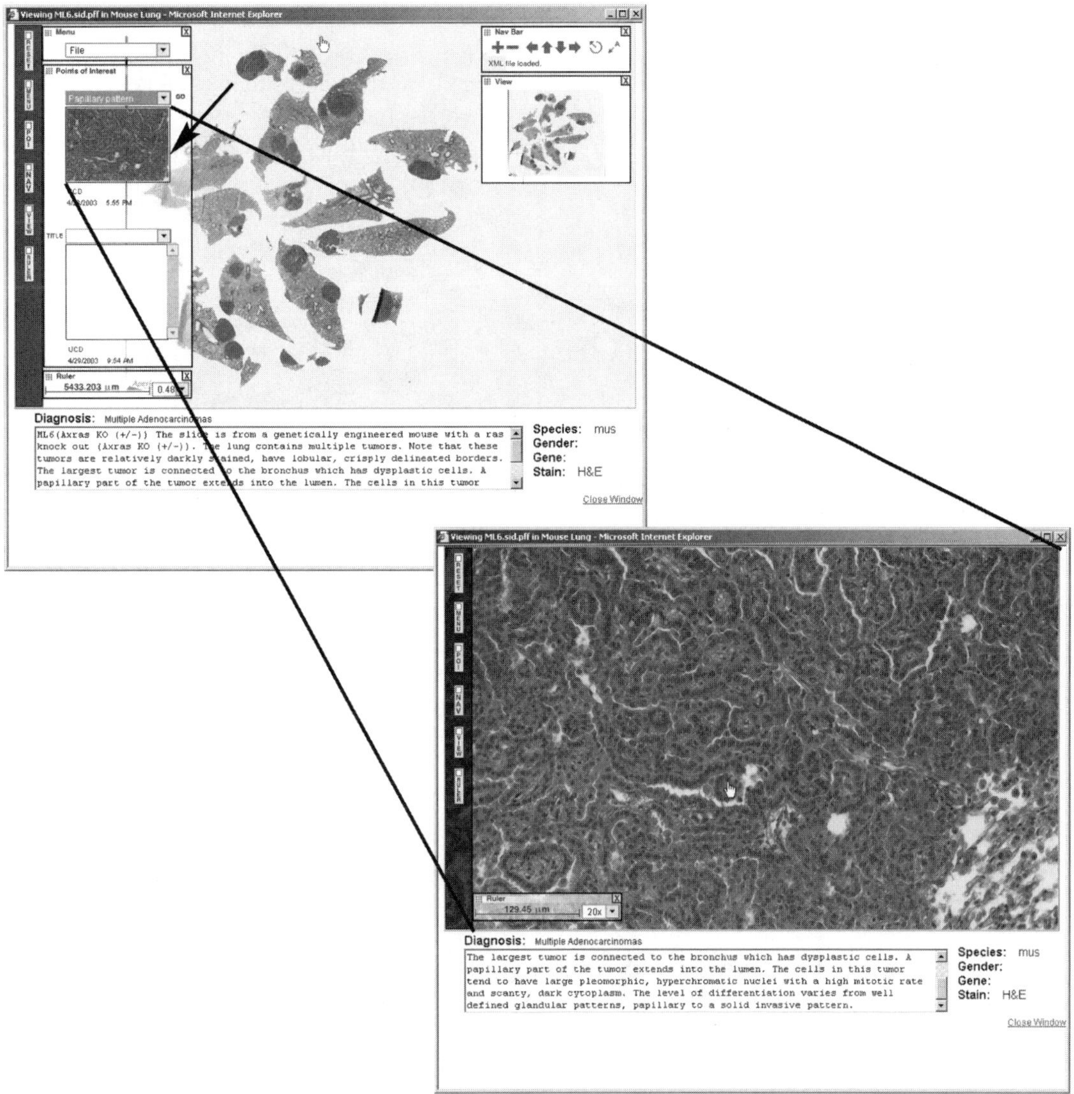

Figure 2.8. Virtual slide. Screen capture that illustrates one type of Web-based *virtual slide* image. The image illustrates a whole slide image from a transgenic *ras* mouse with multiple pulmonary papillomas (from a slide provided by Dr. Tyler Jacks, MIT). The entire slide is viewed in the upper frame with the annotations in the box in the lower portion of the screen. The user can use the Points of Interest (POI) menu to select a specific POI. For illustration purposes, the area selected by the viewer is circled. By clicking on the POI image, the virtual slide goes directly to the specific coordinates to view the POI now seen in the second, lower screen capture.

of one set of characteristics while disregarding others. This can result in variable quality and style reports.

The organized clinical trial systems have overcome some of the irregularities in reporting standards by insisting on cued or synoptic report forms. The pathologists filling out synoptic report forms are "cued" to record each critical observation. Generic synoptic report forms have been developed for some mouse organ systems (Fig. 2.9) (Cardiff et al., 2000; Kogan et al., 2002; Morse et al., 2002). All investigators should use these forms and work with their pathologist to make modifications that are appropriate for their specific model system.

Documentation

The histologic slide is the scientific evidence supporting the interpretation. For communication and publication purposes the documentation is provided in the form of images. Since images are used to document the characteristics of the pathology, they have long been emphasized in the scientific community. However, the reliance on still images introduces selection bias, and the advent of digital image capture permits image processing that can potentially alter the evidence. Therefore, images are discussed in detail in the next section.

IMAGES

Images are essential in the highly visual, image-based field of pathology. Scientists study phenomena by observing their characteristics using techniques described in standard protocols. In cancer biology, the phenomena are classified by a formal (diagnostic) terminology and documented

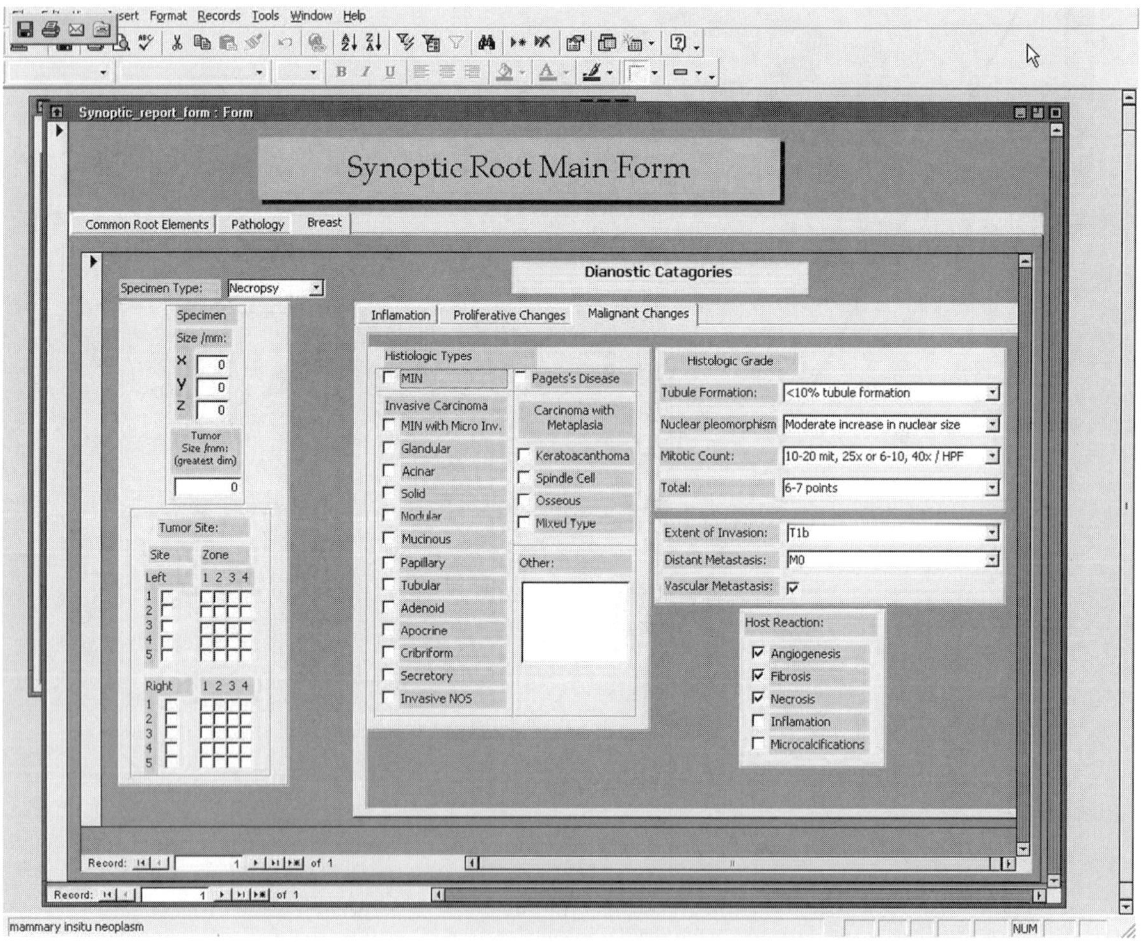

Figure 2.9. Synoptic report form. Screen capture of a computer monitor screen that illustrates one version of an electronic synoptic report. The synoptic report is a "cued" report format used in clinical trials and by some pathologists. It is designed to remind the pathologist of each important characteristic that needs to be recorded. This electronic format permits the user to make relatively simple yes/no decisions when reviewing a slide. The cued format assures the investigator that the pathologist records the appropriate observations.

by images. The *characteristics* are *documented* by images obtained using technical protocols and described using *terminology*. The problem is that the images and vocabulary of pathology are frequently difficult for the uninitiated to understand and interpret.

While words are wonderful, images are essential for documentation of the pathology. Images are, therefore, an essential part of the scientific record. In the modern era with groups of geographically dispersed scientists working together, the images of pathology have become even more important. In most current papers, the pathology is documented by high-magnification images that show a restricted area in the tumor that is out of context. These images consist of static images of individual fields at various magnifications that were limited by the selection bias of the photographer, the quality of the preparation, the quality of the camera, and the ability of the photographer.

The limitations to the current cameras and imaging systems are rapidly being overcome by recent advances in digital imaging. Equipment is now available that captures digital images of the whole slide in minutes at extremely high resolution (50,000 dpi). Compression algorithms reduce the image size. The compressed images can be displayed as a type of multitiered, geospatial image creating a "virtual slide." With appropriate viewing software that is largely available as free shareware, the investigator can now view images prepared at a remote site over the Internet. The software allows the viewers, themselves, to move the slide to any field at any magnification. These digital imaging systems are so powerful that for most purposes they are replacing the microscope for viewing of slides. Like many electronic applications, the technology is changing so rapidly that it is almost useless to give a detailed description of the current technology.

These digital tools allow the pathologist to annotate the image and provide exact coordinates that illustrate the specific example of the characteristic being described (Fig. 2.8). Without a doubt, teaching aids such as these will enhance understanding of what many consider obscure pathologic terminology and speed the dissemination of information. Many professional schools are no longer using microscopes to teach histology and pathology but are using laptops to display the virtual slides.

Examples of this technology are available at a number of sites, including commercial sites http://www.scanscope.com/scanscope/ and http://www.microbrightfield.com/products/virtualslice/summary.html); the National Cancer Institute (http://cancerimages.nci.nih.gov/caIMAGE/index.jsp); the University of Iowa, Department of Pathology (http://www.path.uiowa.edu/virtualslidebox/); and the UCD Center for Comparative Medicine (http://imagearchive.compmed.ucdavis.edu).

COMMUNICATION

The pathologist and investigator must communicate to share their respective data sources. This is another self-evident principle but frequently not effectively implemented. This chapter has emphasized the need to involve the pathologist in the initial planning phases of any scientific project. Effective communication will result in frequent interactions and sharing of data. It does little good to have the investigator spending months collecting the wrong data. In a like manner, the pathologist who does not immediately discuss important validating or novel histologic observations with the investigator is not effectively participating in the scientific process. Effective communication means frequent exchanges of information about the project.

Fortunately, the modern electronic era brings not only new technologies for sharing images across space and time but also a number of interactive tools that facilitate communication between the investigator and the pathologist. Several tools, including shareware applications, are now available that allow simultaneous sharing of images, annotations, and documents. These applications primarily use whiteboard formats and include audio voiceover. Thus, an investigator can now receive a virtual image of his or her whole slide and discuss it with a pathologist who is across the hall, across the city, across the country, or across the world. In part, the investigator is now equipped with all the tools to find and work with a qualified pathologist without regard to location. The pathologists can share their expertise with numerous and distant investigators. Separation by space or time is no longer a limiting factor. Truly effective scientists have learned to maximize their communication skills using these applications.

COMPARATIVE PATHOLOGY

Since the mouse has become the surrogate for human disease, the scientific community has gained a new appreciation for the role of the pathologist. The pathologist is the person who can integrate the structure and function to validate models of human disease and, in this context, models of human cancer.

The question is how closely do these models match human disease? In hierarchical ontologies, human and GEM tumors match to a remarkable degree. The same genes result in tumors of the same organs and the same cells thereby produce tumors which have very similar tissue organization and, in some cases, very similar cytology. Are they identical? No, mouse cells are smaller than human cells, the details of the host response are different, and the patterns of metastases are different. The anatomy of the mouse is obviously different from the human. The mouse has organs and organ configurations that the human does not. For example, the mouse has organs such as the Harderian gland and an extraocular lacrimal gland. The mouse prostate has six lobes and the mouse mammae have 10 glands that almost meet in the midline of the back. The mouse uterus is bicornuate and the ovaries are enclosed in a bursa. The mouse can be informative, but it is a *model*.

On the other hand, the remarkable aspect of GEM tumor biology is that the tumors in almost all organs are so unique that they more closely resemble human tumors than the classical "spontaneous" tumors induced by retroviruses, chemical carcinogenesis, or radiation. The introduction of activated genes first identified in human neoplasia has resulted in the induction of tumors with unique histology that have rarely, if ever, been seen in mice. In fact, when taken out of context, selected fields from mouse tumors are indistinguishable from selected fields from human cancers (Fig. 2.7).

REFERENCES

Andrechek ER, Laing MA, Girgis-Gabardo AA, Siegel PM, Cardiff RD, Muller WJ (2003): Gene expression profiling of neu-induced mammary tumors from transgenic mice reveals genetic and morphological similarities to ErbB2-expressing human breast cancers. *Cancer Res* 63: 4920–4926.

Bhatia-Gaur R, Donjacour AA, Sciavolino PJ, Kim M, Desai N, Young P, Norton CR, Gridley T, Cardiff RD, Cunha GR, Abate-Shen C, Shen MM (1999): Roles for Nkx3.1 in prostate development and cancer. *Genes Dev* 13: 966–977.

Blumberg BS (1980): Comments on scientific process in clinical research. *Perspect Biol Med* 24: 15–30.

Boivin GP, Washington K, Yang K, Ward JM, Pretlow TP, Russell R, Besselsen DG, Godfrey VL, Doetschman T, Dove WF, Pitot HC, Halberg RB, Itzkowitz SH, Groden J, Coffey RJ (2003): Pathology of mouse models of intestinal cancer: Consensus report and recommendations. *Gastroenterology* 124: 762–777.

Borowsky A, Galvez J, Munn R, Cardiff R (2003): Comparative pathology of mouse models of human cancers. *Compar Med* 53: 248–249.

Brayton C, Justice M, Montgomery CA (2001): Evaluating mutant mice: Anatomic pathology. *Vet Pathol* 38: 1–19.

Cardiff RD (2001): Validity of mouse mammary tumour models for human breast cancer: Comparative pathology. *Microsc Res Technol* 52: 224–230.

Cardiff RD (2003): Mouse models of human breast cancer. *Comp Med* 53: 250–253.

Cardiff RD, Anver MR, Gusterson BA, Hennighausen L, Jensen RA, Merino MJ, Rehm S, Russo J, Tavassoli FA, Wakefield LM, Ward JM, Green JE (2000): The mammary pathology of genetically engineered mice: The consensus report and recommendations from the Annapolis meeting. *Oncogene* 19: 968–988.

Cardiff RD, Wellings SR (1999): The comparative pathology of human and mouse mammary glands. *J Mammary Gland Biol Neoplasia* 4: 105–122.

Cleton-Jansen AM (2002): E-cadherin and loss of heterozygosity at chromosome 16 in breast carcinogenesis: Different genetic pathways in ductal and lobular breast cancer? *Breast Cancer Res* 4: 5–8.

Feldman DB, Seely JC (1988): *Necropsy Guide: Rodents and the Rabbit*. CRC, Boca Raton, FL.

Foster HL, Fox JG, Small JD (Eds) (1982): *Experimental Biology and Oncology*. Academic, New York.

Frith C, Ward J (1988): *Color Atlas of Neoplastic and Non-neoplastic Lesions in Aging Mice*. Elsevier, Amsterdam.

Green EL (Ed) (1966): *Biology of the Laboratory mouse, by the Staff of the Jackson Laboratory*. McGraw-Hill, New York.

Institute of Laboratory Animal Resources (U.S.) Committee on Infectious Diseases of Mice and Rats, National Cancer Institute (U.S.) Division of Cancer Treatment (1991): *Infectious Diseases of Mice and Rats*. National Academy Press, Washington, DC.

Kogan SC, Ward JM, Anver MR, Berman JJ, Brayton C, Cardiff RD, Carter JS, de Coronado S, Downing JR, Fredrickson TN, Haines DC, Harris AW, Harris NL, Hiai H, Jaffe ES, MacLennan IC, Pandolfi PP, Pattengale PK, Perkins AS, Simpson RM, Tuttle MS, Wong JF, Morse HC, III (2002): Bethesda proposals for classification of nonlymphoid hematopoietic neoplasms in mice. *Blood* 100: 238–245.

Luna LG (Ed) (1968): *Manual of Histologic Staining Methods of the Armed Forces Institute of Pathology*. Blakiston Division, New York.

Mahler JF, Stokes W, Mann PC, Takaoka M, Maronpot RR (1996): Spontaneous lesions in aging FVB/N mice. *Toxicol Pathol* 24: 710–716.

Maronpot R (1999): *Pathology of the Mouse*. Cache River, Vienna IL.

Mohr U, Dungworth D, Capen C, Carlton W, Sundberg J, Ward J (1996): *Pathobiology of the Aging Mouse*. ILSI, Washington DC.

Morse HC, Fredrickson TN, Jaffe ES, Kogan SC, Anver MR, Brayton C, Cardiff RD, Downing JR, Haines DC, Harris AW, Harris NL, Hiai H, Maclennan ICM, Paolo Pandolfi P, Pattengale PK, Perkins AS, Ward JM (2002): Bethesda proposals for classification of lymphoid neoplasms in mice. *Blood* 100: 246–258.

Nieto AI, Shyamala G, Galvez JJ, Cardiff RD (2003): Persistent hyperplasia in FVB/N. *Comp Med* 53: 361–366.

Park JH, Walls JE, Galvez JJ, Kim M, Abate-Shen C, Shen MM, Cardiff RD (2002): Prostatic intraepithelial neoplasia in genetically engineered mice. *Am J Pathol* 161: 727–735.

Percy D, Barthold S (2001): *Pathology of Laboratory Rodents & Rabbits*. Iowa State University Press, Ames, IA.

Rosse C, Mejino JLV (2003): A reference ontology for bioinformatics: The Foundational Model of Anatomy. *J Biomed Inform*.

Stevens LC (1970): Experimental production of testicular teratomas in mice of strains 129, A/He, and their F1 hybrids. *J Natl Cancer Inst* 44: 923–929.

Stevens LC (1973): A new inbred subline of mice (129-terSv) with a high incidence of spontaneous congenital testicular teratomas. *J Natl Cancer Inst* 50: 235–242.

Sundberg JP (2003): Necropsy procedures with histopathologic evaluation of tissues are fundamentally the same regardless of species. *Vet Pathol* 40: 222.

Sundberg JP, Boggess D (Eds) (2000): *Systematic Approach to Evaluation of Mouse Mutations*. CRC, Boca Raton, FL.

Suwa T, Nyska A, Haseman JK, Mahler JF, Maronpot RR (2002): Spontaneous lesions in control B6C3F1 mice and recommended sectioning of male accessory sex organs. *Toxicol Pathol* 30: 228–234.

Thompson SW, Luna LG (1978): *An Atlas of Artifacts Encountered in the Preparation of Microscopic Tissue Sections*. Thomas, Springfield, IL.

Turusov VS, Mohr U (1990): *Pathology of Tumours in Laboratory Animals*. International Agency for Research on Cancer, New York. Distributed in the United States Oxford University Press, Lyon, France.

Wakefield LM, Thordarson G, Anver M, Cardiff RD (2003): Spontaneous pituitary abnormalities and mammary hyperplasias in FVB/NCr mice: Implications for mouse modeling. *Comp Med* 53: 348–356.

Ward J (2000): *Pathology of Genetically Engineered Mice*. Iowa State University Press, Ames, IA.

Ward JM, Anver MR, Haines DC, Melhorn JM, Gorelick P, Yan L, Fox JG (1996): Inflammatory large bowel disease in immunodeficient mice naturally infected with Helicobacter hepaticus. *Lab Anim Sci* 46: 15–20.

Weiss WA, Israel M, Cobbs C, Holland E, James CD, Louis DN, Marks C, McClatchey AI, Roberts T, Van Dyke T, Wetmore C, Chiu IM, Giovannini M, Guha A, Higgins RJ, Marino S, Radovanovic I, Reilly K, Aldape K (2002): Neuropathology of genetically engineered mice: Consensus report and recommendations from an international forum. *Oncogene* 21: 7453–7463.

3

INTERCURRENT INFECTIONS IN GENETICALLY ENGINEERED MICE

STEPHEN W. BARTHOLD

Center for Comparative Medicine, School of Medicine and Veterinary Medicine, University of California—Davis, Davis, California

The laboratory mouse was domesticated from fancy mice that arose from geographically diverse genetic origins. The diverse origins of the mouse also contributed to a lengthy list of infectious agents that are carried by mice. Despite years of microbial quality control, the burgeoning use of mice for research is causing a resurgence of infectious disease among laboratory mice. Infections can be introduced through mice, but also through biologic products such as serum, hybridomas, and transplantable tumors and through iatrogenic introduction by scientists. Retroviruses represent an important group of agents with significant effects upon mouse research, but the transmissible exogenous viruses are no longer an issue among laboratory mice. The diversity of other infectious agents induce a wide variety of effects upon cancer research, including immunomodulation, tumor graft rejection, misinterpretation of phenotype, modified carcinogenesis, and zoonotic risk. Prudent research practices require that microbial status of research mice does not necessarily have to be standardized among laboratories, but it must be defined so that research results are reproducible and valid.

INTRODUCTION

Mice, cancer, genetics, and infectious agents have had a long and intimate history which is now compounded by the genomics revolution. A brief explanation is required to put these factors into perspective. The laboratory mouse began from a prehistoric association of its progenitors with mankind, starting as multiple species and subspecies of the *Mus musculus* genospecies group that coevolved and migrated with human civilization on literally a global scale. This commensal relationship, along with the fecundity and remarkable genetic plasticity of the mouse, gave rise to purposeful breeding of mice for coat color, with the first historical documentation of this activity being recorded over 3000 years ago in China. Selective breeding for behavioral characteristics also occurred, such as the Japanese Waltzing Mouse over 2000 years ago. "Fancy mice," the name applied to this hobby in the 1800s, became quite popular in Europe and Asia, with genetic contributions arising primarily from *Mus domesticus* in Europe and *Mus castaneus* and *Mus mollosinus* (a natural hybrid of *M. castaneus and M. musculus*) in Asia. Trade between Europe and Asia stimulated genetic mixing of fancy mice, with further contributions of genetic material from noncommensal species of the *M. musculus* genospecies complex.

The mouse as a laboratory animal began its popularity in the first decade of the last century with the rediscovery of Mendelian principles of inheritance and a desire to apply those principles to a mammalian system at a time when Darwin's theories on evolution were gaining interest. The rich variety of coat color among fancy mice, their ease of maintenance, and their propensity to readily reproduce favored this effort. Inheritance was soon found to be complex, with Clarence Cook Little recognizing the need for inbreeding to attain "homogeneity of genetic constitution." This resulted in his creation of the first inbred strain of mouse, DBA (dilute brown non-agouti) in 1908. This was followed by the development of other

Mouse Models of Human Cancer, edited by Eric C. Holland
ISBN 0-471-44460-X

inbred strains and outbred stocks of laboratory mice in the first third of the last century by Jacob Furth, Leonell C. Strong, Halsey Bagg, E. C. MacDowell, Clara Lynch, N. Dobrovolskaia-Zavadskaia, among others. Miss Abbey Lathrop, a mouse fancier and owner of the Granby Mouse Farm in Massachusetts, contributed much of the breeding stock for this early effort. These mouse strains, which were developed for reasons other than molecular genomics, nevertheless represent the mice that are used today.

Remarkably, although laboratory mice originated from an eclectic gene pool, they went through a genetic bottleneck during their domestication as laboratory animals. The early mouse genetics research community in the United States was quite small, largely centered in the Boston and New York regions, so that most of today's inbred strains of mice are closely related to one another. Indeed, many share the Y chromosome of *M. musculus* and are thus derived from a common paternal ancestor (Bishop, 1985; Tucker et al., 1992), and most carry identical mitochondrial genomes of *M. domesticus* origin and are thus derived from a single female ancestor (Ferris et al., 1982). Swiss mice are an exception, in that they are derived from *M. domesticus* subsp. *poschiavinus*. Excellent reviews on the history and origins of laboratory mice are available (Silver, 1995; Morse, 1981; Gross, 1970). A recent update on the phylogenetic tree that illustrates the genetic relatedness among laboratory mice has been published (Beck et al., 2000).

DIVERSE ORIGINS OF INFECTIOUS AGENTS

Given their diverse geographic and genetic origins and the fact that mice have been purposefully traded among fanciers and scientists, it is not surprising that laboratory mice carry an impressive list of overt and opportunistic pathogens that have been acquired from virtually all over the world for thousands of years (Table 3.1). Remarkably, most of these agents still exist in mouse colonies throughout the world. Cancer research promulgated the discovery of many of these agents. Early investigations into the genetic basis of cancer soon led to the realization that susceptibility of different lines of mice to neoplasia could not be completely explained by accepted laws of inheritance. This led to the discovery of murine retroviruses. Infectious disease also stimulated the development of inbred strains of mice that became

Table 3.1. Pathogens of Laboratory Mouse

Viruses

- Adenovirus 1 and 2
- Thymic virus
- Cytomegalovirus
- K virus
- Polyoma virus
- Minute virus of mice
- Mouse parvoviruses
- Ectromelia virus
- LCM virus
- Bunyavirus?
- Mouse hepatitis viruses
- Pneumonia virus of mice
- Sendai virus
- Cardiovirus (MEV)
- Reoviruses
- Rotavirus
- Retroviruses (MLV, MMTV)

Bacteria

- Cilia-associated respiratory bacillus
- *Chlamydia* species
- *Citrobacter rodentium*
- *Clostridium difficile*
- *Clostridium perfringens*
- *Clostridium piliforme*
- *Corynebacterium bovis*
- *Corynebacterium kutscheri*
- *Eperythrozoon coccoides*
- *Escherichia coli*
- *Helicobacter bilis*
- *Helicobacter ganmani*
- *Helicobacter hepaticus*
- *Helicobacter muris*
- *Helicobacter rodentium*
- *Helicobacter typhlonius*
- *Klebsiella oxytoca*
- *Lawsonia intracellularis*
- *Leptospira ballum*
- *Mycobacterium avium*
- *Mycobacterium chelonae*
- *Mycoplasma arthritidis*
- *Mycoplasma collis*
- *Mycoplasma neurolyticum*
- *Mycoplasma pulmonis*
- *Pasteurella pneumotropica*
- *Proteus mirabilis*
- *Pseudomonas aeruginosa*
- *Salmonella enterica*
- *Staphylococcus aureus*
- *Streptococcus* species

Fungi

- *Aspergillus terreus*
- *Candida albicans*
- *Paeciliomyces variotti*
- *Trichophyton mentagrophytes*

Protozoa

- *Cryptosporidium muris*
- *Cryptosporidium parvum*
- *Eimeria* species
- *Giardia muris*
- *Klossiella muris*
- *Sarcocystis muris*
- *Spironucleus muris*

Nematodes

- *Aspicularis tetraptera*
- *Cysticercus fasciolaris*
- *Rodentolepis dimunuta*
- *Rodentolepis nana*
- *Syphacia obvelata, muris*

Arthropods

- *Aspicularis tetraptera*
- *Demodex musculi*
- *Myobia musculi*
- *Myocoptes musculinus*
- *Ornithonyssus bacoti*
- *Polyplax serrata*

Note: LCM, lymphocytic choriomeningitis; MEV, mouse encephalomyelitis; MLV, murine leukemia virus; MMTV, mouse mammary tumor virus.

pivotal models for cancer research. An epizootic disease attributed to "paratyphoid" nearly destroyed early colonies of inbred dilute brown non agouti (DBA) mice, which stimulated the development of other inbred strains for cancer research. Cancer research during the last century also resulted in the discovery of many of the known murine viruses and other pathogens through their effects upon transplantation, cancer biology, and contamination of cell lines and tumors.

Despite years of cesarean rederivation, embryo transfer, husbandry practices, quarantine, microbiologic monitoring, Public Health Service (PHS) guidelines, Association for Assessment and Accreditation of Laboratory Animal Care (AAALAC) accreditation, rules, regulations, and other well-intentioned effort, nearly all of the agents listed in Table 3.1 continue to exist in laboratory mice. Some agents, such as polyoma virus, require the intimately filthy conditions of wild mice to sustain their infectious cycles and have therefore disappeared from laboratory mouse colonies, but still reappear in laboratory mouse populations periodically due to iatrogenic reintroductions (McCance et al., 1983; Sebesteny et al., 1980). Lactate dehydrogenase–elevating virus requires intimate contact for transmission among mice but has tenaciously maintained its status as a common contaminant of transplantable tumor lines (Collins and Parker, 1972; Riley, 1968).

A decade ago, the majority of laboratory mice that populated research institutions were purchased from commercial breeders, and the scientific community demanded high standards of microbial and genetic quality, resulting in a significant decline in infectious disease among laboratory mouse populations. The genomics revolution has caused a progressive reemergence of infectious diseases among laboratory mice. A case in point is *D. musculi*, which was discovered in a "specific pathogen-free" transgenic mouse colony in 1997. Remarkably, this agent had not been reported in laboratory mice since 1917 (Hill et al., 1999). Scientists in literally every major research institution are creating genetically engineered mice with considerable scientific value but minimal commercial value. Thus, mouse colonies have returned to the status of "cottage industries" much like the early 1900s, with unrestricted traffic among laboratories. These factors are further complicated by overcrowding of mouse facilities due to burgeoning populations of genetically engineered mice (which favors transmission) and by the fact that many of these animals have altered immune competence (which favors amplification), an intentional or unintentional consequence of genomic manipulation. The rising costs of mouse care, due to increasingly rigorous regulations, increasingly sophisticated and expensive housing systems to prevent infectious disease, and the increasingly difficult task of testing and controlling mouse traffic, have created pressure to cut expenses at a time when the need for infectious disease surveillance and control has never been greater. Feral and wild mice, always present in mouse facilities, contribute significantly to infectious diseases in caged animals.

Sources of infection go beyond mice "on the hoof," as evidenced by recent outbreaks of mouse pox (ectromelia virus) in multiple U.S. institutions due to importation of liter-quantity lots of untested mouse serum from China (Lipman et al., 2000). This virus has also been shown to contaminate lymphoma and hybridoma cell lines (Buller et al., 1987). Embryonic stem cells, transplantable tumors, and various biologic products of mouse origin, not to mention freezers full of archived material (which have initiated multiple mouse pox outbreaks), all contribute to reintroductions of infectious agents into mouse populations. A diverse array of agents, including lymphocytic choriomeningitis virus, lactate dehydrogenase–elevating virus, reovirus, mouse hepatitis virus, parvoviruses, and *M. pulmonis*, among others, have all been found to cryptically infect cell lines, hybridoma lines, and transplantable tumors, which can and do serve as sources of infection of mouse colonies. In one survey, nearly half of 135 mouse tumor lines tested were contaminated with one or more viruses, and this rate increased to over 70% when the tumors were propagated or passed through mice, including human xenografts passaged in athymic nude mice (Nicklas et al., 1993). Recently, human xenografts were shown to carry *H. hepaticus*, an enterohepatic bacterium, in severe combined immunodeficiency (SCID) mice (Goto et al., 2001). Embryonic stem cell lines have also been found to be infected with murine infectious agents (Nicklas and Weiss, 2000). In addition to cell substrates, leukemia virus stocks are frequently contaminated with murine viruses (Collins and Parker, 1972).

The litany of agents indigenous to the mouse has allowed use of many of these agents as model systems for research. Examples include the oncogenic effects of polyoma virus, the demyelinating effects of neurotropic mouse hepatitis and mouse encephalomyelitis viruses, the mechanisms of cell-mediated immune disease caused by lymphocytic choriomeningitis virus, the pathogenesis of *Salmonella* infection (typhoid model), and the mucosal proliferative effects of *Citrobacter* and *Helicobacter*, among many others. Even ectromelia virus, the fear of every mouse user, continues to be used as a model for pox viral infection, and its use as a model is likely to increase in this dawning age of biodefense research. Iatrogenic introduction of these agents into mouse populations is a constant threat, and that threat is amplified when immunodeficient populations of mice are at risk. Exposure of nude mice following iatrogenic introduction of polyoma virus into a mouse colony serves as a good example of experimental infectious agents getting access into mouse populations, resulting in natural disease that is fostered by the immunodeficient status of modern mouse strains (McCance et al., 1983; Sebesteny et al., 1980).

RETROVIRUSES: INTERFACE OF GENETICS AND INFECTIOUS DISEASE

Discussion of infectious agents of the laboratory mouse that impact upon cancer research cannot begin without mention of retroviruses, with their obvious and not so obvious effects on cancer. Mouse retroviruses include horizontally transmissible exogenous as well as the far more numerous vertically transmitted, replication-competent and defective endogenous proviruses and retrovirus (RV)–like elements that are maintained as Mendelian determinants within the genome of the laboratory mouse. Two major groups of replication-competent retroviruses include the MMTVs and the murine leukemia viruses (MuLVs) (reviewed in Coffin et al., 1997). Although these two major groups of viruses are genetically distinct, their pathogenesis overlaps and their names are misleading. The great majority of endogenous retroviruses do not induce neoplasia. Indeed, endogenous retroviruses play a role in mouse strain–related patterns of disease and other strain characteristics, including coat color, glomerulonephritis, neurologic disease, and diabetes (Gaskins et al., 1992; Izui et al., 1979; Jenkins et al., 1981; Morse et al., 1985; Stoye et al., 1988). Furthermore, despite their mammary or leukemia appellation, MMTVs can be associated with leukemia and lymphomas in some strains of mice, and MuLVs can be transmitted through the milk (Law and Moloney, 1961; Michalides et al., 1982; Zhang et al., 1996). The inefficiently transmitted exogenous leukemia and mammary tumor viruses have been eliminated by cesarean rederivation and husbandry practices, but not before they gave rise to the Friend, Moloney, and Rauscher leukemia viruses and the transmissible mouse mammary tumor virus, MMTV-S (milk factor, or the Bittner agent).

Exogenous MMTV-S is intentionally maintained in designated colonies of mice for the specific purpose of cancer research. This virus lent the first clues toward the existence of retroviruses in the mouse, which led to the revelation of the even more complex and still unraveling pathogenesis of the endogenous MMTVs and MuLVs. It took Ludwig Gross nearly a decade to successfully transmit oncogenic MuLV from Ak mice (Gross leukemia virus) and two more decades to discover that successful transmission required infection of specific strains of recipient mice. From this dogged determination, concepts of endogenous proviruses, temporal and organ-specific expression of these elements, ecotropism, xenotropism, polytropism, amphotropism, recombination, and susceptibility determinants (e.g., N and B tropism) were systematically understood, including the discovery that some proviral elements, such as *Fv-4*, prevent infection of other endogenous retroviruses. It is now known that various strains of laboratory mice have literally hundreds of copies of replication-competent endogenous retroviruses, including MMTVs and MuLVs, replication-incompetent MMTVs, MuLVs, and more ancient RV-like elements. The patterns of retrovirus integration within the mouse genome are so extensive that more than 85% of the murine genome can be mapped with only three nonecotropic proviral oligonucleotide probes (Frankel et al., 1990). Xenotropic proviruses, polytropic proviruses, and RV-like elements appear to be the most ancient acquisitions and are therefore widely dispersed among progenitor wild mice all over the world, whereas ecotropic viruses appear among wild mice of Asian origin and thus were acquired within the genome more recently (Kozak and O'Neill, 1987; Steffen et al., 1980). These Asian ecotropic determinants have been maintained, and often selected for, by inbreeding of laboratory mice, emphasizing the diverse geographic origins of the laboratory mouse.

Endogenous proviral elements are now well stabilized within the genome of all inbred mouse strains, with predicable and unique patterns of integration for each strain and predictable phenotypes (such as but not restricted to neoplasia) that are strain characteristics. Endogenous MMTVs, MuLVs, and RV-like elements, because they are integrated throughout the genome and inherited as mouse strain–specific determinants, are given gene designations (such as *Akv-1, Mtv-2*, and *Fv-4*). However, the mixing of mouse genomes through backcrossing or intercrossing that is now common in genomics research can lead to unpredictable consequences as the endogenous retrovirus cards are reshuffled within the genome and susceptibility and resistance alleles are mixed. Furthermore, RV-like elements are not simply effete or junk DNA but are rather actively recombining and transposing themselves within the genome, contributing to substrain variation, spontaneous mutations, and "genetic drift" (Coffin et al., 1997).

Retroviruses continue to serve as experimental oncogenic agents for cancer research, but they have different mechanisms of pathogenesis. The acute transforming viruses carry v-*onc* genes that induce proliferative lesions after short incubation periods. These agents are now largely laboratory stocks of viruses that have been maintained from early isolations of exogenous retroviruses from mice. The larger family of endogenous retroviruses and RV-like elements lack v-*onc* genes and exert their oncogenic effects, after a much longer incubation period, through integration into the genome near host proto-oncogenes (insertional mutagenesis). In addition, retroviruses and sequences thereof, such as long terminal repeat (LTR) regions, play an important role in genomic targeting strategy and tissue-specific promotion.

EFFECTS OF INFECTIOUS AGENTS ON RESEARCH

Aside from retroviruses, the numerous other agents that infect laboratory mice have diverse effects on research. Overviews of documented effects have been published (Bhatt et al., 1986b; Lindsey et al., 1991). Although many of these infections induce minimal or no discernable overt clinical signs, these reviews emphasize the protean effects of subclinical infections upon research results, especially immune responses. Immune effects are varied, with immunostimulation, immunosuppression, altered responses to other pathogens, altered allograft as well as isograft responses, and altered tumor prevalence or altered transplantable tumor kinetics. The causes of these varied immune effects are equally varied, with some agents directly infecting lymphoid cells or macrophages, some functioning as B- or T-cell mitogens, and others having undefined indirect effects. Aside from these diverse effects upon immune response, this chapter emphasizes effects that are likely to have impact upon cancer research.

A number of infectious agents have been associated with altered responses to naturally occurring, virally induced, chemically induced, and transplantable tumors. The mechanisms for these effects are largely unknown but are most likely to occur with viruses that possess polytropism for a variety of cell types, such as mouse hepatitis virus, lymphocytic choriomeningitis virus, and reovirus 3. Many strains of mouse hepatitis virus, the most common and most contagious of these agents, can have tropism for a variety of cell types and have a propensity for both direct and indirect immune modulation, as well as oncolysis. Mouse hepatitis virus has been shown to alter host responses to transplanted and induced tumors, including solid tumors, ascites tumors (Braunsteiner and Friend, 1954; Collins and Parker, 1972; Fox et al., 1977; Nelson, 1959; Manaker et al., 1961), and hybridomas (Fox et al., 1977). Mouse hepatitis virus infection of athymic nude mice has also been shown to elicit rejection of xenografts (Akimaru et al., 1981; Kyriazis et al., 1979). Lymphocytic choriomeningitis virus has an equally nefarious reputation but is less common as a contagious agent within mouse populations. It is, however, a somewhat common contaminant of tumor lines and virus stocks, including leukemia viruses (Bhatt et al., 1986a; Collins and Parker, 1972). Lymphocytic choriomeningitis virus has been shown to induce delayed allograft and tumor rejection (Guttler et al., 1975; Lehmann-Grube et al., 1972) as well as inhibit tumor formation induced by polyoma virus (Hotchin, 1962), Rauscher virus (Youn and Barski, 1966), and mammary tumor virus (Padnos and Molomut, 1973). Reovirus has a wide mammalian host range and cell tropism. It has been shown to frequently contaminate transplantable tumor lines (Collins and Parker, 1972), cause ascites tumor lysis (Bennette, 1960; Nelson and Tarnowski, 1960), and even suppress respiratory tumor formation in chemical carcinogenesis studies (Theiss et al., 1978).

Parvoviruses, which include minute virus of mice (MVM) and mouse parvovirus (MPV) serotypes, require dividing cell populations for virus replication. These viruses are common in modern mouse colonies and are common contaminants of leukemia virus stocks, transplantable tumors, hybridomas, and other cell lines (Collins and Parker, 1972; Parker et al., 1970). Infection of mice with MVM has been shown to suppress Ehrlich ascites tumors (Guetta et al., 1986). MPVs are lymphocytotropic and particularly prone to cause persistent infections and immunomodulation in adult mice. Infection with MPV potentiates rejection of tumor allografts and also induces rejection of syngeneic grafts through mechanisms believed to be mediated by perturbation of T cells (McKisic et al., 1996, 1998).

Sendai virus, a parainfluenza virus with strict tropism for respiratory epithelium, also has remarkably diverse effects on immune response and cancer biology. Despite its respiratory tropism, Sendai virus has a number of strong immunomodulatory effects and has been associated with isograft rejection and altered responses to transplanted and induced tumors (Streilein et al., 1981), including Friend leukemia (Wheelock, 1967). Sendai virus has been shown to contaminate a variety of transplantable tumor lines (Collins and Parker, 1972; Matsuya et al., 1978; Takeyama et al., 1979). As far its direct effects upon respiratory epithelium, Sendai virus can both increase and decrease responses to respiratory carcinogenesis and spontaneous prevalence of pulmonary tumors (Nettesheim et al., 1974, 1981; Peck et al., 1983).

In contrast to the above-mentioned viruses, lactate dehydrogenase–elevating virus is transmitted inefficiently, primarily through pugilistic behavior among males. Furthermore, this virus has very strict tropism for specific subpopulations of macrophages. In spite of these caveats, lactate dehydrogenase–elevating virus was shown to contaminate 50 transplantable tumor lines (Riley, 1968). More recent surveys of transplantable tumors, liver tissues, and cell lines have also revealed similar findings of widespread lactate dehydrogenase–elevating virus contamination (Goto et al., 1998; Nicklas et al., 1993). This agent can enhance or suppress transplantable Ehrlich ascites tumor growth (depending upon stage of infection) (Bailey et al., 1965), plasmacytomas (Michaelides and Schlesinger, 1974), and chemically induced fibrosarcomas (Henderson et al., 1979) as well as respiratory adenomas induced with urethan (Theiss et al., 1980) and tumors induced with vinyl chloride–vinyl acetate foreign-body implants (Brinton and Brand, 1977). Infection of mice with lactate dehydrogenase–elevating virus has been shown to alter the incidence of MMTV-induced mammary tumors (Riley, 1966) and Moloney sarcoma virus–induced tumors (McDonald, 1983; Turner et al., 1971). Other rare infections of mice, which are potentially still possible, have been shown to contaminate transplantable tumors or alter host response during passage of such tumors, including *Encephalitozoon cuniculi* (Arison et al., 1966) and *Salmonella enteritidis* (Ashley et al., 1976).

PSEUDOCANCER

Several infectious agents influence cancer research by producing lesions that can be misconstrued as cancer or preneoplastic change. Sendai virus, for example, is a highly contagious respiratory parainfluenza virus that causes necrotizing bronchiolitis and interstitial pneumonia in immunocompetent mice. When mice recover from infection, their respiratory mucosa undergoes a number of proliferative events in the repair process, including hyperplasia of respiratory epithelium, Clara cells, and alveolar type 2 pneumocytes. Such mice can manifest intense nonkeratinizing squamous metaplasia, which has been misconstrued as squamous cell carcinomas, and cuboidal metaplasia of alveoli and terminal airways, which has been misconstrued as adenomas.

Intestinal carcinogenesis studies must take into account the pathobiology of rodent mucosal proliferation, with caution advised against anthropomorphism. A number of infectious agents induce mucosal inflammation and proliferation in immunocompetent as well as immunodeficient mice, including mouse hepatitis virus, *C. rodentium*, several species of *Helicobacter*, and others. *Citrobacter rodentium* and enteric *Helicobacter* species induce marked mucosal proliferative lesions of the cecum and colon. *Citrobacter* infections are transient, with restoration of normal mucosa after recovery, whereas *Helicobacter* infections are persistent. The transient proliferative activity induced by *Citrobacter* reveals the events that can transpire with an intense but transient proliferative stimulus. The fully florid lesion reveals marked crypt distortion with cytokinetic events that mimic those of neoplasia and preneoplasia, including expansion of the crypt proliferative compartment with mitotic activity and labeling of surface (luminal) mucosa (Barthold, 1979; Barthold et al., 1978). Another feature of this disease can be aggressive growth of crypts through the muscularis mucosa. This is most apt to occur in areas of severe inflammation. Early descriptions of *Citrobacter*-related disease referred to this as colitis cystica (Brynjolfsson and Lombard, 1969).

Crypt invasiveness is a characteristic of rodent intestinal mucosa, and in the mouse, it can occur under the stimulus of intense hyperplasia induced by *Citrobacter, Helicobacter*, and other agents and is particularly prominent in traumatized rectal mucosa that has prolapsed, regardless of cause (mice have a very short and therefore poorly anchored rectum, which is prone to prolapse). These events are accentuated when the proliferative stimulus is prolonged, such as some *Helicobacter*-induced lesions. Hamsters, infected with *L. intracellularis*, are an extreme example of this rodent phenomenon (Jacoby and Johnson, 1981). *Lawsonia* induces ileal hyperplasia, but the syndrome has been called (erroneously) enzootic adenocarcinoma because of the highly invasive nature of rodent crypt epithelium, which has actually been found in mesenteric lymph nodes (Jonas et al., 1965). Although *Lawsonia* is rare in rats and mice, organisms of similar appearance have been reported in rats with invasive carcinoma of the colon (Vandenberghe et al., 1985) and it has been observed in mice with hyperplastic enteric lesions (Franklin, 2002).

Both *Citrobacter* and *Helicobacter* can induce intense mucosal hyperplasia, which can become quite dysplastic and misinterpreted as neoplasia or preneoplasia. The significance of mucosal invasiveness during hyperplasia is most pressing with current research using *Helicobacter* as a mouse model for inflammatory bowel disease (IBD) and the association of that disease with colon cancer in humans. Aside from the fact that the lesions induced by *Helicobacter* in mice do not resemble the human condition of IBD (Crohn's disease and ulcerative colitis), the hyperplastic and dysplastic changes that occur in these mice, including invasive crypt activity, can easily be misconstrued as neoplastic. For example, "nonmetastatic colon cancer" has been described in transforming growth factor beta 1 (TGFβ1) mutant mice infected with *Helicobacter* but not treated with carcinogen (Engle et al., 1999, 2002). Another report described colon cancer in *Helicobacter*-infected Rag2 knockout mice, with descriptions of cellular atypia, dysplasia, surface mucosal labeling, invasive crypt activity, and cystic crypts (Erdman et al., 2003), all similar to reversible events described in *Citrobacter* colitis (Barthold, 1979; Barthold et al., 1978). Although the conclusions in these reports may be valid, the neoplastic nature of such lesions must be carefully assessed in order to differentiate such changes from the documented effects of bacterial proliferative disease in the mouse bowel.

The invasive nature of rodent mucosal epithelium is not limited to bowel, as "adenomyosis," with invasion of endometrial glandular structures into the myometrium, is common in the female reproductive tract of the mouse and can be accentuated with opportunistic chronic bacterial infections, such as *K. oxytoca*.

PROMOTION OF CARCINOGENESIS BY INFECTIOUS DISEASE

In addition to being misconstrued as neoplastic or preneoplastic lesions, the proliferative stimuli induced by viral and bacterial infections have a number of documented effects upon carcinogenesis in a number of tissues. These effects involve both natural tumor incidence and modification of experimental carcinogenesis protocols. *Helicobacter hepaticus*, a bacterium that infects the lower bowel and seeds the liver, has been associated with an earlier onset and higher prevalence of hepatic tumors in strain A mice (Ward et al., 1994). Indeed, this abnormal incidence of neoplasia prompted the initial discovery of *Helicobacter* infections in mice and the subsequent realization that several species of this organism also induce enteric proliferative lesions. The promotional effect of *H. hepaticus* infection on *N*-nitrosodimethylamine chemical carcinogenesis of the liver has also been documented (Diwan et al., 1997). The proliferative stimulus during the repair phase of Sendai viral infection can significantly promote the development of bronchioalveolar adenomas in strains of mice that are prone to these tumors (Percy and Barthold, 2000). Strain A mice infected with Sendai virus and mouse hepatitis virus were found to develop fewer pulmonary adenomas induced by 10-chloromethyl-9-choroanthracene, and mice infected with Sendai virus developed more pulmonary adenomas induced by 7,12-dimethylbenz(*a*)anthracene compared to mice infected with mouse hepatitis virus (Peck et al., 1983).

Intestinal carcinogenesis has been shown to be significantly influenced by bacterial infections that induce proliferative changes in the mucosa. The transient and reversible hyperplastic stimulus induced by *C. rodentium* has been shown to stimulate the development of preneoplastic foci, with reduction of the latent period and increased susceptibility when mice are treated with the chemical carcinogen 1,2-dimethylhydrazine (Barthold and Jonas, 1977). The focal atypia that arise in such mice would rank as carcinomas in situ or be deemed neoplastic by many pathologists, yet these lesions are reversible when the bacteria-induced hyperplastic stimulus is removed (Barthold and Jonas, 1977). They are specific to treatment with carcinogen, but despite their morphology, they are still dependent upon mucosal proliferation to foster their progression. More recently, *Citrobacter*-induced hyperplasia has been shown to accelerate tumor development in genetically altered MIN (mucosal intestinal neoplasia) mice (Newman et al., 2001).

HUMAN SAFETY: ZOONOTIC HAZARDS

A small, but important number of mouse infectious agents infect humans. These include agents as disparate as mites (*O. bacoti*), tapeworms (*R. nana*), protozoa (*C. parvum*), bacteria (*L. ballum, Salmonella* spp. and possibly *H. bilis*), and viruses (lymphocytic choriomeningitis virus and Sendai virus) (Percy and Barthold, 2000). Lymphocytic choriomeningitis virus is particularly dangerous, as it silently infects mice, cell lines, leukemia virus stocks, and transplantable tumors, and is particularly dangerous when amplified to high titer in mice with T-cell deficiencies, such as nude, SCID, and other genetically altered mice with defined or undefined immune deficiencies (Bhatt et al., 1986a; Dykewicz et al., 1992). Sendai virus, a respiratory parainfluenza type 1 virus (PIV1), is closely related to human PIV1, and there has been a long-held debate as to whether humans are susceptible to this murine virus or if the virus was originally of human origin. Recent studies have shown that Sendai virus and human PIV1 replicate to equal titer in the upper and lower respiratory tract of both African green monkeys and chimpanzees, suggesting that Sendai virus lacks a significant host range restriction and could very well pose a zoonotic hazard (Skiadopoulos et al., 2002). *Helicobacter bilis* has been identified [by polymerase chain reaction (PCR) and 16 S recombinant DNA (rDNA) analysis] in humans with biliary disease, including cancer (Matsukura et al., 2002). Zoonotic disease control is best approached by expecting the unexpected. Xenografts of sheep origin have been found to be productively infected with *Coxiella burnetti*, a zoonotic agent that causes Q fever in humans (Criley et al., 2001). Monitoring for and controlling zoonotic agents is an institutional responsibility for protecting human health and emphasizes the importance of core institutional support for laboratory animal care programs.

CLINICAL EFFECTS OF INFECTIOUS AGENTS

The long association of these pathogens and their hosts has created a microbial détente, so that clinical disease is often not manifest, particularly when the infection is enzootic within the population, thus tempered by "herd immunity." In addition to genetic resistance, maternal transfer of antibody from the immune dam to the fetus or neonate is central to this phenomenon, so that the susceptible young are protected, and infection is maintained subclinically among the more disease resistant adults within the population. This creates a "stealth bomber" effect in laboratory mouse populations, as pathogens with significant effects on immune responses, physiology, and tumor biology, not to mention zoonotic hazard to human handlers, can exist without overt clinical signs. Given the complexities of agent strain (virulence, tropism, etc.), host genetic and age-related susceptibility or resistance, immune competence, enzootic or epizootic herd status, genetic alteration, and experimental protocol, it is nearly impossible to predict what, if any, clinical or research effects are likely to occur when mice are infected. A review of the pathology of various infectious diseases in mice is available (Percy and Barthold, 2000).

Infections in immunodeficient mice are an obvious exception, as these mice are often unable to eliminate infections and therefore develop chronic progressive disease. T-cell-deficient mice are particularly vulnerable to progressive wasting syndrome when infected with agents that are not overtly pathogenic, such as *Pneumocystis carinii*, Sendai virus, pneumonia virus of mice, avirulent strains of mouse hepatitis virus, parvoviruses, and others. *Helicobacter* infection has emerged as a major problem for immunodeficient mice, with the growing recognition of this genus, including *H. hepaticus, H. bilis, H. muridarum, H. rodentium, H. typhlonius, H. ganmani*, and other, as-yet-to-be-named species. When immunodeficient mice are infected with more virulent agents, such as some strains of mouse hepatitis virus or ectromelia virus, mortality, rather than chronic wasting disease, occurs. Immunodeficient mice can support and amplify a wide variety of infectious agents (Percy and Barthold, 2000).

As science becomes more molecular in focus, there is increasing naivete of the scientific community about the need to test for and control these agents. As a recent editorial aptly documents, there has always been a contrary opinion on disease control in mice (Gordon, 2002). This opinion advocates allowing infectious agents to establish themselves unabated in laboratory mouse populations, thereby mimicking the "wild-type" mouse. This fails to recognize the unique quality of the laboratory mouse: There is no such thing as a wild-type laboratory mouse and no such thing as wild-type microflora in this genetically and microbially unique laboratory animal. Wild-type microflora would be impossible to establish, considering the long list of agents in Table 3.1, and would be impossible to standardize among laboratories. The power of the laboratory mouse is its *homogeneity of genetic constitution*, which allows for exquisite control of genetic variables, and control of infectious variables is an equally critical requisite for modern mouse research. Thus, as with genetic constitution, microbial status does not necessarily have to be standardized but most certainly needs to be defined so that science can be reproducible.

ACKNOWLEDGMENT

This work was supported by grant U42-RR14905, University of California Mutant Mouse Regional Resource Center, from the National Center for Research Resources, National Institutes of Health.

REFERENCES

Akimaru K, Stuhlmiller GM, Siegler HF (1981): Influence of mouse hepatitis virus on the growth of human melanoma in the peritoneal cavity of the athymic nude mouse. *J Surg Oncol* 17:327–339.

Arison RN, Cassaro JA, Pruss MP (1966): Studies on murine ascites-producing agent and its effect on tumor development. *Cancer Res* 26:1915–1920.

Ashley MP, Neoh SH, Kotlarski I, Hardy D (1976): Local and systemic effects in the non-specific tumour resistance induced by attenuated Salmonella enteritidis 11RX in mice. *Aust J Exp Biol Med Sci* 54:157–168.

Bailey JM, Clough J, Lohous A (1965): Influence of LDH viruses on growth of Ehrlich ascites tumor in mice. *Proc Soc Exp Biol Med* 119:1200–1204.

Barthold S, Jonas AM (1977): Morphogenesis of early 1, 2-dimethylhydrazine-induced lesions and latent period reduction of colon carcinogenesis in mice by a variant of *Citrobacter freundii*. *Cancer Res* 37:4352–4360.

Barthold SW (1979): Autoradiographic cytokinetics of colonic mucosal hyperplasia in mice. *Cancer Res* 39:24–29.

Barthold SW, Coleman JL, Jacoby RO, Livstone EM, Jonas AM (1978): Transmissible murine colonic hyperplasia. *Vet Pathol* 15:223–236.

Beck JA, Lloyd S, Hafezparast M, Lennon-Pierce M, Eppig JT, Festing MFW, Fisher EMC (2000): Geneologies of mouse inbred strains. *Nat Genet* 24:23–25.

Bennette JG (1960): Isolation of a non-pathogenic tumour-destroying virus from mouse ascites. *Nature* 187:72–73.

Bhatt PN, Jacoby RO, Barthold SW (1986a): Contamination of transplantable murine tumors with lymphocytic choriomeningitis virus. *Lab Anim Sci* 36:136–139.

Bhatt PN, Jacoby RO, Morse HC III, New AE (1986b): *Viral and Mycoplasmal Infections of Laboratory Rodents: Effects on Biomedical Research*. Academic, Orlando, FL.

Bishop CE (1985): Most classical *Mus musculus domesticus* laboratory mouse strains carry a *Mus musculus musculus* Y chromosome. *Nature* 325:70–72.

Braunsteiner H, Friend C (1954): Viral hepatitis associated with transplantable mouse leukemias. I. Acute hepatic manifestations following treatment with urethane or methylformamide. *J Exp Med* 100:665–677.

Brinton MA, Brand I (1977): Delayed foreign-body tumorigenesis in mice infected with lactate dehydrogenase-elevating virus. *J Natl Cancer Inst* 59:1027–1029.

Brynjolfsson G, Lombard LS (1969): Colitis cystica in mice. *Cancer* 23:225–229.

Buller RML, Weinblat AC, Hamburger AW, Wallace GD (1987): Observations on the replication of ectromelia virus in mouse-derived cell lines: Implications for epidemiology of mousepox. *Lab Anim Sci* 37:28–32.

Coffin JM, Hughes SH, Varmus HE (1997): The interactions of retroviruses and their hosts. In Coffin JM, Hughes SH, Varmus HE (Eds), *Retroviruses*. Cold Spring Harbor Laboratory, Cold Spring Harbor, NY, pp 343–435.

Collins MJ, Parker JC (1972): Murine virus contamination of leukemia viruses and transplantable tumors. *J Natl Cancer Inst* 49:1139–1143.

Criley JM, Carty AJ, Besch-Williford CL, Franklin CL (2001): *Coxiella burnetti* infection in C.B-17 scid-bg mice xenotransplanted with fetal bovine tissue. *Comp Med* 51:357–360.

Diwan BA, Ward JM, Ramljak D, Anderson LM (1997): Promotion by *Helicobacter hepaticus*-induced hepatitis of hepatic tumors initiated by N-nitrosodimethylamine in male A/JCr mice. *Toxicol Pathol* 25:597–605.

Dykewicz CA, Dato VM, Fisher-Hoch SP, Howarth MV, Perez-Oronoz GI, Ostroff SM, Gary H Jr, Schonberger LB, McCormick JB (1992): Lymphocytic choriomeningitis outbreak associated with nude mice in a research institute. *J Am Med Assoc* 267:1349–1353.

Engle SJ, Boying JB, Boivin GP, Ormsby I, Gartside PS, Doetschman T (1999): Transforming growth factor beta 1 suppresses nonmetastatic colon cancer at an early stage of tumorigenesis. *Cancer Res* 59:3379–3386.

Engle SJ, Ormsby I, Pawlowski S, Boivin GP, Croft J, Balish E, Doetschman T (2002): Elimination of colon cancer in germ-free transforming growth factor beta 1-deficient mice. *Cancer Res* 62:6362–6366.

Erdman SE, Poutahidis T, Tomczak M, Rogers AB, Cormier K, Plank B, Horwitz BH, Fox JG (2003): CD4+ CD25+ regulatory T lymphocytes inhibit microbially induced colon cancer in Rag2-deficient mice. *Am J Pathol* 162:691–702.

Ferris SD, Sage RD, Wilso AC (1982): Evidence from mtDNA sequences that common laboratory strains of inbred mice are descended from a single female. *Nature* 295:163–165.

Fox JG, Murphy JC, Igras VE (1977): Adverse effects of mouse hepatitis virus on ascites myeloma passage in the BALB/cJ mouse. *Lab Anim Sci* 27:173–179.

Frankel WN, Stoye JP, Taylor BA, Coffin JM (1990): A genetic linkage map of endogenous murine leukemia viruses. *Genetics* 124:221–236.

Franklin CL (2002): personal communication.

Gaskins HR, Prochazka M, Hamaguchi K, Serreze DV, Leiter EH (1992): Beta cell expression of endogenous xenotropic retrovirus distinguishes diabetes-susceptible NOD/Lt from resistant NON/Lt mice. *J Clin Invest* 90:2220–2227.

Gordon JW (2002): Barrier facilities for transgenic rodents in academic centers—a two-edged sword. *Comp Med* 52:397–402.

Goto K, Ishihara K-I, Kuzuoka A, Ohnishi Y, Itoh T (2001): Contamination of transplantable human tumor-bearing lines by *Helicobacter hepaticus* and its elimination. *J Clin Microbiol* 39:3703–3704.

Goto K, Takakuru A, Yoshimura M, Ohnishi Y, Itoh T (1998): Detection and typing of lactate dehydrogenase-elevating virus RNA from transplantable tumors, mouse liver tissues, and cell lines, using polymerase chain reaction. *Lab Anim Sci* 48:99–102.

Gross L (1970). The development of inbred strains of mice, and its impact on experimental cancer research. In Gross L (Ed), *Oncogenic Viruses*. Pergamon, Oxford, pp. 229–237.

Guetta E, Granziani Y, Tal J (1986): Suppression of Ehrlich ascites tumors in mice by minute virus of mice. *J Natl Cancer Inst* 76:1177–1180.

Guttler F, Bro-Jorgensen K, Jorgensen PN (1975): Transient impaired cell-mediated tumor immunity after acute infection with lymphocytic choriomeningitis virus. *Scand J Immunol* 4:327–336.

Henderson DC, Chang RWS, Turk JL (1979): Infection with LDH virus alters host response to tumours. *Br J Cancer* 39:555–564.

Hill LR, Kille PS, Weiss DA, Craig TM, Coghla LG (1999): *Demodex musculi* in the skin of transgenic mice. *Comtemp Topics Lab Anim Sci* 38:13–18.

Hotchin J (1962): The biology of lymphocytic choriomeningitis infection: Virus-induced immune disease. *Cold Spring Harbor Symp Quant Biol* 27:479–499.

Izui SP, McConahey PJ, Theofilopoulous AN, Dixon FJ (1979): Association of circulating retroviral gp70- and anti-gp70 immune complexes with murine systemic lupus erythematosis. *J Exp Med* 149:1099–1116.

Jacoby RO, Johnson EA (1981): Transmissible ileal hyperplasia. *Adv Exp Med Biol* 134:267–289.

Jenkins NA, Copeland NG, Taylor BA, Lee BK (1981): Dilute (d) coat colour mutation of DBA/2J mice is associated with the site of integration of an ecotropic MuLV genome. *Nature* 293:370–374.

Jonas AM, Tomita Y, Wyand DS (1965): Enzootic intestinal adenocarcinoma in hamsters. *J Am Vet Med Assoc* 147:1102–1108.

Kozak CA, O'Neill RR (1987): Diverse wild mouse origins of xenotropic, mink cell focus-forming, and two types of ecotropic proviral genes. *J Virol* 61:3082–3088.

Kyriazis AP, DiPersio L, Michaei JG, Pesce AJ (1979): Influence of mouse hepatitis virus (MHV) infection on the growth of human tumors in the athymic mouse. *Int J Cancer* 23:402–409.

Law LW, Moloney JB (1961): Studies of congenital transmission of a leukemia virus in mice. *Proc Soc Exp Biol Med* 108:715.

Lehmann-Grube F, Niemeyer I, Lohler J (1972): Lymphocytic choriomeningitis virus of the mouse. IV. Depression of the allograft reaction. *Med Microbiol Immunol* 158:16–25.

Lindsey JR, Boorman GA, Collins MJ Jr, Hsu C-K, Van Hoosier GL Jr, Wagner JE (1991) *Infectious Diseases of Mice and Rats*. Institute of Laboratory Animal Resources, National Research Council, National Academy of Sciences, Washington, DC.

Lipman NS, Perkins S, Nguyen H, Pfeffer M (2000): Mousepox resulting from use of ectromelia virus-contaminated, imported mouse serum. *Comp Med* 50:426–435.

Manaker RA, Piczak CW, Miller AA, Stanton MF (1961): A hepatitis virus complicating studies with mouse leukemia. *J Natl Cancer Inst* 27:29–45.

Matsukura N, Yokomuro S, Yamada S, Tajiri T, Sundo T, Hadama T, Kamiya S, Naito Z, Fox JG (2002): Association between *Helicobacter bilis* in bile and biliary tract malignancies: *H. bilis* in bile from Japanese and Thai patients with benign and malignant diseases in the biliary tract. *Jpn J Cancer Res* 93:842–847.

Matsuya Y, Kusano T, Endo S, Takahashi N, Yamane I (1978): Reduced tumorigenicity by addition in vitro of Sendai virus. *Eur J Cancer* 14:837–850.

McCance DJ, Sebesteny A, Griffin BE, Balkwill F, Tilly R, Gregson NA (1983): A paralytic disease in nude mice associated with polyoma virus infection. *J Gen Virol* 64: 57–67.

McDonald TL (1983): Blocking of cell-mediated immunity to Moloney murine sarcoma virus-transformed cells by lactate dehydrogenase virus-antibody complex. *J Natl Cancer Inst* 70:493–497.

McKisic M, Macy JD Jr, Delano ML, Jacoby RO, Paturzo FX, Smith AL (1998): Mouse parvovirus infection potentiates allogeneic skin graft rejection and induces syngeneic graft rejection. *Transplantation* 65:436–446.

McKisic MD, Paturzo F, Smith AL (1996): Mouse parvovirus infection potentiates rejection of tumor allografts and modulates T cell effector function. *Transplantation* 61:292–299.

Michaelides MC, Schlesinger S (1974): Effect of acute or chronic infection with lactic dehydrogenase virus (LDV) on the susceptibility of mice to plasmacytoma MOPC-315. *J Immunol* 112:1560–1564.

Michalides RE, Wagenaar J, Hilkens B, Groner B, Hynes NE. (1982): Acquisition of proviral DNA of mouse mammary tumor virus in thymic leukemia cells from GR mice. *J Virol* 43:819–829.

Morse HC III (1981). The laboratory mouse—a historical perspective. In Foster HL, Small JD, Fox JG (Eds), *The Mouse in Biomedical Research*, Vol. 1. Academic, New York, pp 1–16.

Morse HC III, Yetter RA, Stimpling JH, Pitts OM, Fredrickson TN, Hartley JW (1985): Greying with age in mice: Relation to expression of murine leukemia virus. *Cell* 41: 439–448.

Nelson JB (1959): Emergence of hepatitis virus in mice injected with ascites tumor. *Proc Soc Exp Biol Med* 102: 357–360.

Nelson JB, Tarnowski GS (1960): An oncolytic virus recovered from Swiss mice during passage of an ascites tumour. *Nature* 188:866–867.

Nettesheim P, Schreiber H, Cresia DA, Richter CB (1974): Respiratory infections in the pathogenesis of lung cancer. *Recent Results in Cancer Res* 44:138–157.

Nettesheim P, Topping DC, Jambasi R (1981): Host and environmental factors enhancing carcinogenesis in the respiratory tract. *Ann Rev Pharmacol Toxicol* 21:133–163.

Newman JV, Kosaka T, Kosaka T, Sheppard BJ, Fox JG, Schauer DB (2001): Bacterial infection promotes colon tumorigenesis in *Apc*Min/+ mice. *J Infect Dis* 184: 227–230.

Nicklas W, Kraft V, Meyer B (1993). Contamination of transplantable tumors, cell lines, and monoclonal antibodies with rodent viruses. *Lab Anim Sci* 43:296–300.

Nicklas W, Weiss J (2000): Survey of embryonic stem cells for murine infective agents. *Comp Med* 50:410–411.

Padnos M, Molomut N (1973): Inhibition of mouse tumors and viruses by the M-P strain of LCM virus. In Lehmann-Grube F (Ed), *Lymphocytic Choriomeningitis Virus and Other Arenaviruses*, Springer, New York.

Parker JC, Collins MJ, Cross SS, Rowe WP (1970): Minute virus of mice. II. Prevalence, epidemiology, and occurrence as a contaminant of transplanted tumors. *J Natl Cancer Inst* 45:305–310.

Peck R, Eaton G, Peck E, Peck EB, Litwin S (1983): Influence of Sendai virus on carcinogenesis in strain A mice. *Lab Anim Sci* 33:154–156.

Percy DH, Barthold SW (2000): *Pathology of Laboratory Rodents and Rabbits*. Iowa State University Press, Ames, IA.

Riley V (1966): Spontaneous mammary tumors: Decrease of incidence in mice infected with enzyme-elevating virus. *Science* 153:1657–1658.

Riley V (1968): Lactate dehydrogenase in the normal and malignant state in mice and the influence of a benign enzyme-elevating virus. *Methods Cancer Res* 4:493–618.

Sebesteny A, Tilly R, Balkwill F, Trevan D (1980): Demyelination and wasting associated with polyomavirus infection in nude (nu/nu) mice. *Lab Anim* 14:337–345.

Silver LM (1995): *Mouse Genetics: Concepts and Applications*. Oxford University Press, Oxford, UK.

Skiadopoulos MH, Surman SR, Riggs JM, Elkins WR, St Claire M, Nishio M, Garcin D, Kolakofsky D, Collins PL, Murphy BR (2002): Sendai virus, a murine parainfluenza virus type 1, replicates to a level similar to human PIV1 in the upper and lower respiratory tract of African green monkeys and chimpanzees. *Virology* 297:153–160.

Steffen DL, Bird S, Weinberg RA (1980): Evidence for the Asiatic origin of endogenous AKR type murine leukemia proviruses. *J Virol* 35:824–835.

Stoye JP, Fenner S, Greenoak GE, Moran C, Coffin JM (1988): Role of endogenous retroviruses as mutagens: The hairless mutation of mice. *Cell* 54:383–391.

Streilein J, Shadduck J, Pakes SP (1981): Effects of splenectomy and Sendai virus infection on rejection of male skin isografts by pathogen free C57BL/6 female mice. *Transplantation* 32:34–37.

Takeyama H, Kawashima K, Yamada K, Ito Y (1979): Induction of tumor resistance in mice by L1210 leukemia cells persistently infected with HVJ (Sendai virus). *Gann* 70: 493–501.

Theiss JC, Shimkin MB, Stoner GD, Kniazeff AJ, Hoppenstand RD, (1980): Effect of lactate dehydrogenase virus on chemically induced mouse lung tumorigenesis. *Cancer Res* 40:64–66.

Theiss JC, Stoner GD, Kniazeff AK (1978): Effect of reovirus infection on pulmonary tumor response to urethan in strain A mice. *J Natl Cancer Inst* 61:131–134.

Tucker PK, Lee BK, Lundrigan BL, Eicher EM (1992): Geographic origin of the Y chromosomes in "old" inbred strains of mice. *Mammalian Genome* 3:254–261.

Turner W, Ebert PS, Bassin R, Spahn G, Chirigos MA (1971): Potentiation of murine sarcoma virus (Harvey) (Maloney) oncogenicity in lactic dehydrogenase-elevating virus-infected mice. *Proc Soc Exp Biol Med* 136:1314–1318.

Vandenberghe J, Verheyen A, Lauwers S, Goboes K (1985): Spontaneous adenocarcinoma of the ascending colon in Wistar rats: The intracytoplasmic presence of a Campylobacter-like bacterium. *J Comp Path* 95:45–55.

Ward JM, Fox JG, Anver MR, Haines DC, George CV, Collins MJ Jr, Gorelick PL, Nagashima K, Gonda MA, Gilden RV, Tully JG, Russell RJ, Beneveniste RE, Paster BJ (1994): Chronic active hepatitis and associated liver tumours in mice caused by persistent bacterial infection with a Helicobacter species. *J Natl Cancer Inst* 86:1222–1227.

Wheelock E (1967): Inhibitory effects of Sendai virus on Friend leukemia in mice. *J Natl Cancer Inst* 38:771–778.

Youn JK, Barski G (1966): Interference between lymphocytic choriomeningitis and Rauscher leukemia in mice. *J Natl Cancer Inst* 37:381–388.

Zhang DJ, Tsiagbe VK, Huang C, Thorbecke GJ (1996): Control of endogenous mouse mammary tumor virus superantigen expression in SJL lymphomas by a promoter within the env region. *J Immunol* 157:3510–3517.

4

GERMLINE MODIFICATION STRATEGIES

SILVIA GRISENDI AND PIER PAOLO PANDOLFI

Cancer Biology and Genetics Program and Department of Pathology, Memorial Sloan-Kettering Cancer Center, New York, New York

Modeling human disease in the mouse has become an extremely important source of knowledge in biomedical research by deepening our understanding of human tumorigenesis and generating experimental systems for the evaluation of novel therapeutic approaches. Various strategies exist for engineering gain and loss-of-function mutant alleles by means of conventional and conditional germline modification methodologies. We describe classical transgenic and gene-targeting approaches as well as their more advanced conditional versions, providing specific examples that illustrate both their utility and their disadvantages in modeling human cancer in the mouse.

INTRODUCTION

Modeling human disease in the mouse has become an extremely important source of knowledge in biomedical research by deepening our understanding of human tumorigenesis and generating experimental systems for the evaluation of novel therapeutic approaches.

The mouse shares conspicuous physiologic, anatomic, and genomic similarities with humans. During the past 20 years, an impressive arsenal of genetic tools, along with the recent availability of the mouse genome sequence, have made the laboratory mouse the key organism for modeling human genetic disease.

Historically, inbred mouse strains that develop certain types of tumors either spontaneously or following treatment with carcinogenic agents were studied to understand aspects of disease etiology (see http://tumor.informatics.jax.org, http://potency.berkeley. edu/). These earlier models were nevertheless characterized by long tumor latency and, more importantly, by the lack of defined correlation between genotype and phenotype. Therefore, the establishment of the first transgenic mouse tumor models, initially by overexpression of viral and cellular oncogenes in specific tissues (Adams et al., 1985; Brinster et al., 1984; Hanahan, 1985; Stewart et al., 1984) and then through the introduction of gene-targeting technology in mouse embryonic stem (ES) cells (Thomas and Capecchi, 1987), opened the way to a new and revolutionary conceptual framework for the creation of faithful mouse models to study tumor biology through thc manipulation of the mouse germline.

So far, a considerable number of oncogene-bearing transgenic mice or conventional tumor suppressor gene (TSG) knockout mice have been generated (for a list see http://cancermodels.nci.nih.gov). These cancer-prone mouse strains have provided considerable insights into the role of individual genes and their mutated counterparts in tumorigenesis, and of the cooperation of individual mutations in tumor development.

However, cancer is a disease characterized by the progressive accumulation of somatic mutations. Such sporadic lesions eventually lead to tumor growth inside a genetically wild-type environment, which often actively contributes to tumor progression. Therefore, although tumor models that use conventional transgenic or knockout approaches might be used to model familial forms of cancer, they do not mimic sporadic multistep tumorigenesis because the initiating mutation is present throughout the body or a particular tissue. Moreover, germline mutations of either oncogenes or TSGs often lead to embryonic lethality, which makes it unfeasible to study the effects of mutations on tumor development in adult mice. Various strategies for conditional gene expression

Mouse Models of Human Cancer, edited by Eric C. Holland
ISBN 0-471-44460-X

in the mouse have therefore been developed which allow the induction of somatic mutations in a tissue-specific and time-controlled fashion. These advancements are currently making possible the creation of sophisticated models that better mimic tumor onset and progression. Such improved models allow, for instance, the dissection of cell-specific requirements for individual genes and the evaluation of complex issues such as tumor–stroma interactions. Furthermore, mouse models with controlled gene expression allow the generation of compound mutants carrying several conditional or regulatable alleles, in which the contribution of individual mutations to the tumor phenotype can be accurately evaluated.

Tumorigenic mutations exert their effects through either dominant gain-of-function oncogene activation or recessive loss-of-function TSG inactivation. This chapter will provide an overview of the various strategies for engineering gain and loss-of-function mutant alleles by means of conventional and conditional germline modification methodologies. We will describe classical transgenic and gene-targeting approaches as well as their more advanced conditional versions, providing specific examples that illustrate both their utility and their disadvantages in modeling human cancer in the mouse.

GAIN-OF-FUNCTION GERMLINE MODIFICATION STRATEGIES

To model in the mouse the effect of deregulated oncogenes or dominant-negative TSG mutations, various gain-of-function germline modification strategies have been developed based on two fundamental approaches: (a) the transgenic approach and (b) the gene-targeting approach. The former is based on the transfer of transgenic sequences by random nonhomologous integration into the mouse genome. The second approach takes advantage of homologous recombination in ES cells to introduce exogenous genetic information into specific genetic loci and to perform site-specific recombinase-mediated chromosomal rearrangements.

Transgenic Approach

From a historical point of view, the first successful manipulation of the mouse genome was achieved by microinjection of foreign DNA into the pronuclei of fertilized zygotes (Fig. 4.1, method 1); this is still the most extensively and successfully used method of gene transfer in the mouse (for extensive reviews see Hanahan, 1989; Jaenisch, 1988; Palmiter and Brinster, 1986). Once introduced into the pronucleus, the transgene is usually integrated at random sites with variable frequency into the genome. Because it is not possible to predict the transgene integration sites or to control the number of copies of the transgene that integrated, additional systems of gene transfer in the mouse have been developed, such as viral infection of preimplantation embryos (Lois et al., 2002; Soriano et al., 1986) or, more extensively, ES cell–mediated transgenesis (Fig. 4.1, method 2). Random integration transgenesis performed in ES cells offers the valuable opportunity to select among different integration sites for the desired one, for example a single copy–single site transgene (Ding et al., 2001; Novak et al., 2000), prior to introduction of the transgene in vivo. In addition, in vitro differentiation of transgenic ES cell lines allows for screening of appropriate expression patterns of a given transgene before committing to its analysis in mice.

In a standard experimental setting, a transgene is under the control of its own regulatory sequences or, in some cases, of a ubiquitous promoter, which drives the nonspecific expression of the exogenous sequences in more than one histologic compartment (Figs. 4.1*a*,*b*). This has been considered a straightforward methodology to test in vivo the oncogenic potential of a given wild-type or aberrant gene product and the involvement of such elements in tumor pathogenesis. A successful example of this strategy was one of the first attempts at modeling leukemia in mice, where expression of the p190 *BCR-ABL* fusion product was placed under the control of the ubiquitous metallothionein 1 (MT) gene promoter (Heisterkamp et al., 1990). Mice transgenic for the MT-*BCR-ABL*^{p190} DNA construct died of acute leukemia (myeloid or lymphoid) 10–58 days after birth, thus providing crucial evidence for a causal relationship between the *BCR-ABL* fusion product and the human leukemia.

Nevertheless, the genetic alterations associated with cancer are often somatic lesions acquired in specific cell types. It is therefore important to restrict the expression of the transgene at least to the cellular compartment where the oncogenic product should exert its effect. This specificity can be achieved by constructing expression vectors in which appropriate promoters or regulatory elements direct expression of the transgene in a specific target tissue/cell or at a specific developmental stage (Fig. 4.1*c*). Tissue-specific expression of oncogenes has provided useful tumorigenesis models for several kinds of tumors; for instance, one of the earliest cancer models was generated by expressing the c-*myc* oncogene under the control of the MMTV (mouse mammary tumor virus) promoter such that transgenic mice developed spontaneous mammary adenocarcinomas by four to eight months of age (Stewart et al., 1984). Interestingly, changing the specificity of the regulatory elements controlling expression of the same oncogene leads to a different tumor phenotype, such that expression of c-*myc* under the control of an immunoglobulin enhancer induces the formation of B-cell lymphomas (Adams et al., 1985).

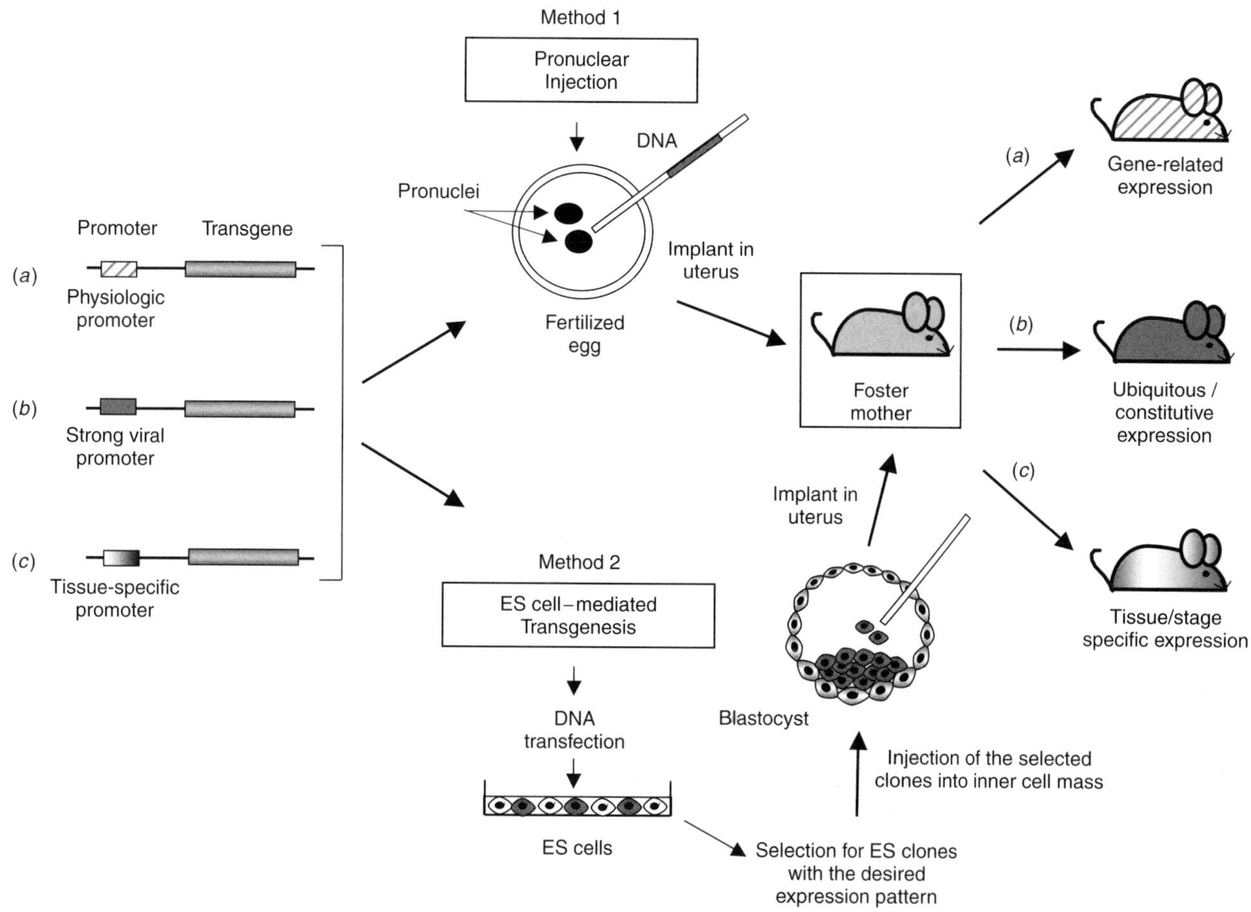

Figure 4.1. The transgenic approach: pronuclear injection (method 1) and ES cell–mediated transgenesis (method 2) of a transgene driven by (*a*) physiologic, (*b*) ubiquitous, or (*c*) tissue-specific promoter. Method 1: To generate transgenic mice, the DNA construct carrying the transgene is injected into one of the two pronuclei of a freshly fertilized egg, where it undergoes random integration into the genome. When the resulting zygote reaches the stage of a two-cell embryo, it is implanted in a pseudopregnant foster mother and the offspring is tested for the presence and appropriate expression of the transgene. Method 2: alternatively, transgenic mice can be generated through DNA transfection into ES cells. ES cells that have randomly integrated the transgene are selected using a positively selectable marker included in the vector. ES cell clones are analyzed for expression pattern and number of insertional events of the transgene and consequently injected into the inner cell mass of a blastocyst. Blastocysts are then transferred to a pseudopregnant female and the offspring is screened for pups that have germline transmission of the transgene. The expression pattern in the transgenic offspring will be different depending on the promoter used to drive the expression of the transgene. For example, in (*a*) the transgene is expressed according to the expression pattern of the related endogenous gene; in (*b*) the transgene is constitutively and ubiquitously expressed under the control of a constitutively active promoter, such as viral regulatory sequences; and in (*c*) the transgene is expressed in specific tissues and/or developmental stages under the control of a tissue-specific promoter.

A variety of tissue-specific promoters have been employed to produce constructs expressed in transgenic mice (for a list see the TBASE database for transgenic animals at http://tbase.jax.org or http://www.mshri.on.ca/nagy), and the list of oncogenes which have now been overexpressed in vivo in specific histologic compartments is extensive. In many cases, the correct pattern of gene expression has been obtained, producing an invaluable amount of information revealing important players of tumor pathogenesis in vivo.

The choice of a promoter which drives expression in an appropriate tissue and at the right time turned out to be critical in modeling transgenic mice for acute promyelocytic leukemia (APL), a distinct subtype of acute myeloid leukemia characterized by a block in myeloid differentiation at the promyelocytic stage (for a review see Merghoub et al., 2001). After several attempts, the oncogenic fusion product *PML-RAR*α, which is most frequently associated with APL, was expressed under the control of the human cathepsin G (*hGC*) promoter. This promoter becomes active during the transition from myeloblasts to myelocytes, the stage of differentiation at which the APL blasts are blocked. *hCG*-directed expression of *PML-RAR*α causes a myeloproliferative syndrome characterized by a block in myeloid differentiation at the promyelocytic stage in 100% of transgenic mice; 15–30% of these mice go on to develop fatal APL-like leukemia after 6–14 month latency period (Grisolano et al., 1997; He et al., 1997). Another faithful mouse model of APL was generated through the expression of *PML-RAR*α under the control of the human migration inhibitory factor-related protein 8 (*hMRP8*) promoter, which becomes active at the myeloblastic stage of myeloid differentiation (Brown et al., 1997). Recently, a very informative model has been reported where a human *PML-RAR*α cDNA is targeted into the genome (knockin; see also the discussion of the knockin approach in the next section) specifically at the 5′ untranslated region of the murine cathepsin G gene using homologous recombination in ES cells. In this case, the exogenous sequence is directly under the control of the desired endogenous promoter, thus allowing for a more physiologic expression of the transgene. This model produced a high-penetrance APL phenotype, with more than 90% of knockin mice developing APL between 6 and 16 months of age (Westervelt et al., 2003).

Using different promoters to evaluate the effect of the expression of one oncogene in different cell subpopulations inside the same tissue is a valid strategy to identify the subset of cells from which a certain kind of tumor arises. Following this approach, Brown and colleagues have shown that expression of mutant *ras* from a truncated keratin 5 gene promoter leads to the formation of skin tumors significantly more aggressive compared to the ones generated by overexpressing *ras* under the control of the keratin 10 or keratin 1 gene promoters. Keratin 5 gene is mainly expressed in the hair follicle cells, while keratin 10 and 1 gene promoters direct the expression mainly to suprabasal cells of the epidermis. This experimental approach allowed supporting the notion that the nature of the cell in which tumor initiation occurs is a major determinant of malignant potential (Brown et al., 1998).

It must be underlined that, to accomplish the creation of improved mouse tumor models faithfully reproducing crucial aspects of multistep tumorigenesis, more informative techniques have been developed. These take advantage of sophisticated constructs that allow regulation of the oncogene expression with high specificity in defined cellular subpopulations and, most importantly, allow the activation of the oncogenic products in a time-controlled fashion. Temporal control is achieved through the use of ligand-regulated control of gene expression or of site-specific recombinase-based methods (see following sections).

These applications have brought even greater advancements to the field when combined with the use of ES cell homologous recombination-based gene-targeting technology to introduce exogenous genetic elements and/or specific sequence modifications in the mouse genome.

Randomly integrated transgenes, obtained by pronuclear injection or by ES cell transfection, may in fact show variable patterns of expression due to at least two main shortcomings of this methodology. First, possible unwanted phenotypic effects might appear due to secondary effects related to the integration of the transgene at sensitive genomic sites. This can lead to unpredictable alterations in other genes' expression, which will be inherited by the entire progeny of the founder transgenic mouse. In addition, the transgenic approach necessarily leads to the introduction of a transgene in a genetic background where the corresponding wild-type alleles are both present and normally active. This condition is not mimicking the physiology of the tumor cell, where, in the presence of an oncogenic mutation, a single wild-type allele remains to code for the normal gene product.

Such technical limitations can be avoided by the use of homology recombination-based techniques to create knockin alleles. As illustrated in the following section, this strategy allows the sequence of interest to be inserted in the genome at a predefined locus of integration through to the replacement of the endogenous counterpart, which can control the expression of the transgene at physiologic levels.

Knockin Approach

The use of homologous recombination to replace an endogenous gene with exogenous sequences has been termed "knockin" (Hanks et al., 1995). The knockin approach has been utilized by a number of investigators to introduce and test the oncogenic potential of transgenes or mutated versions of genes of interest in the mouse genome. It allows circumventing some of the shortcomings encountered by utilizing a conventional transgenic approach, as mentioned in the previous section. In particular, since a precise genomic locus is targeted by homologous recombination, knockins prevent the expression problems related to the number and position of the transgene integration sites and conveniently place the mutant sequence under the transcriptional control of the endogenous counterpart (Fig. 4.2). In addition, this strategy can be employed to

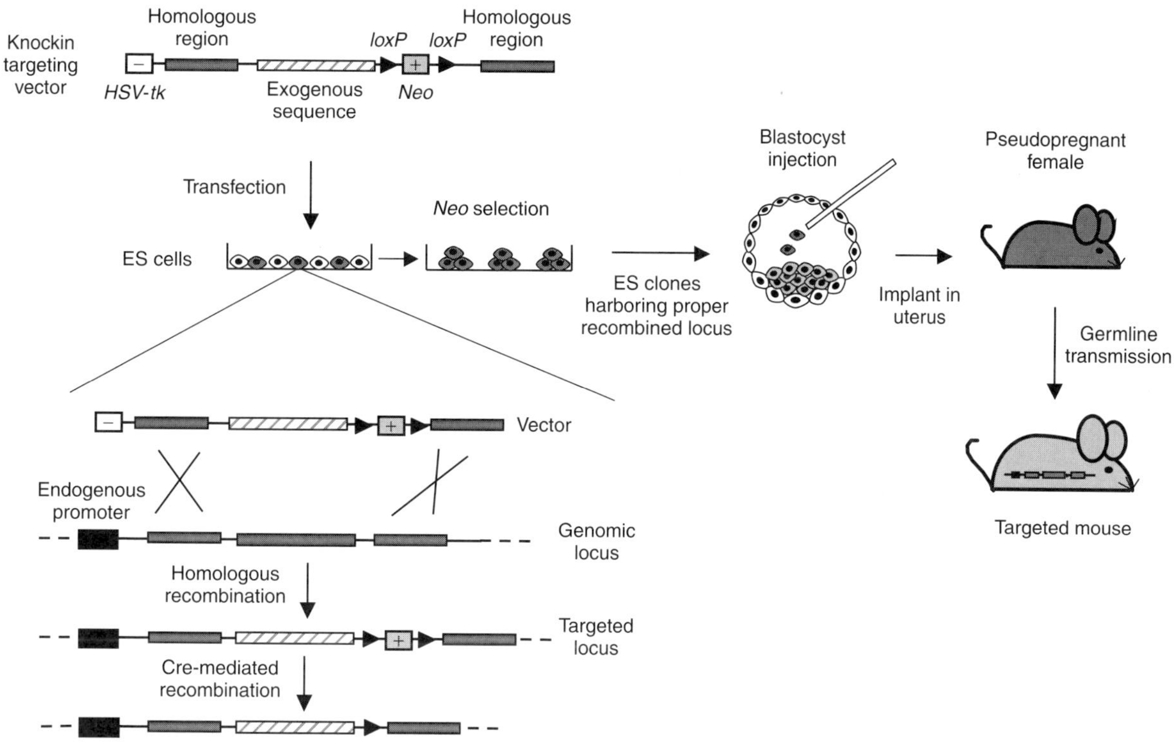

Figure 4.2. A classical knockin approach results in the replacement of an endogenous genetic segment with an exogenous/mutated counterpart at the physiologic genetic locus. The targeting vector is constructed to contain two region of homology to the desired chromosomal integration site flanking the exogenous sequence, a positive (+) and a negative (−) selectable marker. The positive selectable marker can be flanked by *loxP* sites. The targeting vector is transfected into ES cells, where homologous recombination between sequences of the vector and the genomic locus will result in the replacement of the target gene with the exogenous segment and the positive selectable marker (targeted locus). The negative marker is lost upon homologous recombination, while the positive marker will be excised through Cre-mediated recombination. Clones of ES cells selected for the presence of the proper recombined locus are independently microinjected into blastocysts, which are transferred to a pseudopregnant female. The resulting progeny is screened for germline transmission of the recombined locus in order to select heterozygous targeted mice first, which are subsequently crossed to produce progeny carrying the homozygous mutation.

introduce, at desired positions in the genome, different kinds of regulatory elements, such as site-specific recombinase target sequences, that can be used to achieve further genetic modifications (see following sections).

The basic concept in building a targeting vector to knock in a DNA sequence by insertional replacement is to use two segments of homology to the gene of interest (homology arms) that should flank the genetic region to be replaced (Fig. 4.2). Usually, a positive selectable marker is placed in the target vector between the homology arms (e.g., the neomycin phosphotransferase gene, *Neo*) along with the sequence to be inserted, and a negative selectable marker is positioned outside one of the homology arms (e.g., the herpes simplex virus 1 thymidine kinase gene, *HSV-tk*). If homologous recombination occurs at the arms, the negative selectable marker is not introduced into the genome and the cell will survive the negative selection. Optimally, knockin constructs should be designed so that no endogenous regulatory sequences are deleted and positive selectable markers can be eliminated after the targeting event. This is achieved by the use of site-specific recombinases such as the Cre-*loxP* system (see following paragraphs).

This conventional knockin vector design has largely been employed to obtain the replacement of an endogenous sequence with its mutated oncogenic counterpart under the regulation of the physiologic promoter, thereby closely mimicking what occurs, at the somatic level, in cancer cells. The replacement reduces the normal gene, if autosomal, to heterozygosity as, for instance, in the majority of chromosomal translocations associated with leukemia and lymphoma. In this combination, the relative dosage of the normal and aberrant gene is, in principle, comparable to the pathophysiologic condition in humans. This approach has been used to generate an improved model, compared to the transgenic approach, of the *BCR-ABL*-associated acute lymphoblastic leukemia (ALL). The *BCR-ABL*^{p190} fusion gene has been targeted into the *bcr* locus to generate mice that developed B cell ALL, closely reproducing the phenotype observed in human patients (Castellanos et al., 1997).

The same conceptual approach also allows for more subtle modifications of endogenous genes—for example, when the vector design includes point mutations or small insertions/deletions engineered either into one of the homology arms (de Vries et al., 2002) or, more frequently, in the replacing segment located between the homologous regions (Spring et al., 2001) (Fig. 4.3). This is an informative strategy to induce the activation of proto-oncogenes or to generate dominant-negative versions of TSGs through the introduction of mutations found in human patients at the appropriate genetic locus. For instance, knockin mice that are engineered to carry an activating point mutation (R24C) in the cyclin-dependent kinase 4 (Cdk4) gene, associated with human melanoma, expressed normal levels of the mutated protein (Rane et al.,

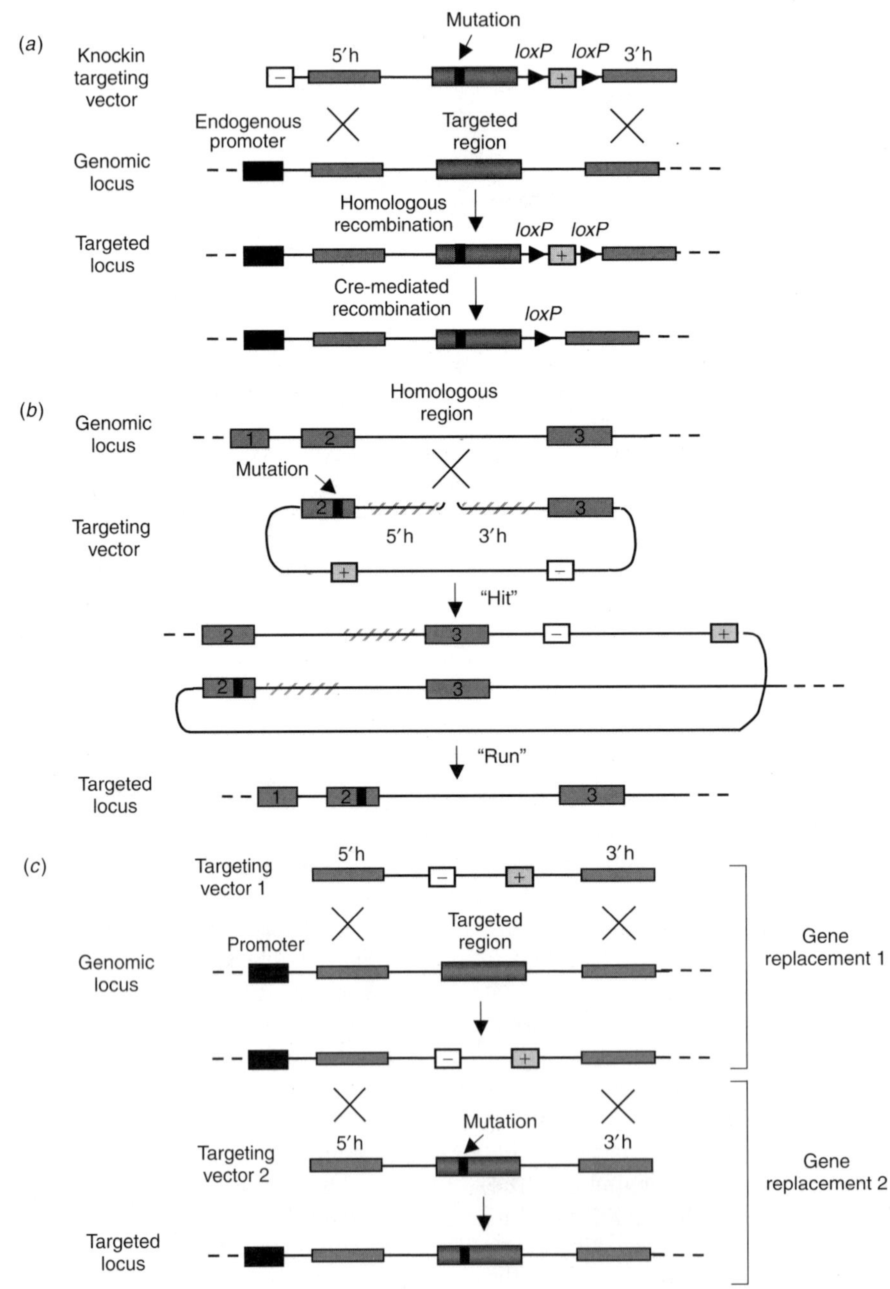

1999) and developed a wide spectrum of tumors (Sotillo et al., 2001). This provided experimental evidence of an in vivo role of Cdk4 oncogenic activation in tumorigenic processes. In addition, dominant-negative inhibition of the TSG p53 has been studied in a mouse model in which point mutations in the DNA-binding domain of p53 were knocked in at the *p53* endogenous locus, allowing for a physiologic level of expression of the mutant protein at (de Vries et al., 2002). In a similar vein, a knockin model bearing in heterozygosity an inactivating in-frame deletion at the ataxia telangiectasia mutated (ATM) genetic locus acquired an increased susceptibility to developing tumors (Spring et al., 2002). Interestingly, this mimicked the related ataxia telangiectasia human disease, characterized by radiosensitivity, genome instability, and predisposition to cancer (Lavin and Shiloh, 1997). The last two examples illustrate the peculiar case of the use of the knockin approaches to obtain the dominant-negative inhibition of a TSG by the replacement with the mutated counterparts. It must be noted that this procedure eventually leads to a loss and not to a gain of the function of the gene of interest (see also the section on loss-of-function germline modification strategies later in the chapter).

As pointed out for the transgenic methods, conventional knockin approaches have one fundamental limitation: while the genetic mutations associated with human tumorigenesis occur mostly at the somatic level, in these gain-of-function models the mutated genes are introduced in the germline as an inheritable character. Therefore, unlike somatic lesions resulting in the expression of an aberrant gene product in a specific subset of cells, the knocked-in gene is expressed throughout embryonic development at the same extent in all the tissues and follows the expression pattern of the normal counterpart. For example, in hematopoietic malignancies, many of the genes involved in cancer-associated chromosomal translocations play an important role not only in hematopoiesis but also in development. In addition, leukemia-associated fusion proteins often exert a dominant negative effect over their normal alleles. For that reason, one of the major advantages of the knockin approach sometimes turns into a substantial impediment because of the lack of tissue specificity expression, which can result in embryonic lethality. This has been the case for the *AML1-ETO* fusion gene resulting from the t(8;21) translocation observed in the AML-M2 subtype of acute myeloid leukemia (AML). When the *AML-ETO* fusion gene was knocked in the normal mouse *Aml1* locus in the attempt to reproduce the human leukemia, the replacement resulted in embryonic lethality of the heterozygous mutant due to a block in foetal hemopoiesis. The phenotype was strikingly similar to the one observed in the *Aml1* null mutants (Okuda et al., 1998; Yergeau et al., 1997). Therefore, although the leukemogenic role of *AML-ETO* was not clarified by this model, the observed phenotype was nevertheless extremely informative, since it supported the notion that the *AML-ETO* fusion gene could act in vivo in a dominant negative manner and block *AML1* function (Okuda et al., 1998).

In conclusion, due to the aforesaid limitations, the knockin approach improved and complemented but did not replace the transgenic approach. More importantly, both approaches have been recently radically revolutionized by the introduction of conditional and inducible systems to regulate exogenous gene expression, thus allowing for a more accurate reproduction of human tumorigenesis.

Figure 4.3. Introduction of subtle mutations into a specific genomic locus through three different knockin approaches. (*a*) Nonselectable mutations are introduced into the genome in a single step through homologous recombination coupled with site-specific recombination. The mutation is carried by a standard knockin targeting vector that includes two regions of homology (5′h and 3′h) flanking the exogenous segment containing the mutation, a positive selectable marker (+) flanked by *loxP* sites, and a negative selectable marker (−) located outside the 5′ homologous region. Upon homologous recombination, the mutated exogenous sequence replaces the wild-type genomic sequence and the negative marker is lost. The positive marker can be subsequently excised by the use of Cre recombinase. This method is characterized by the necessity to leave behind some extra sequences such as, in this specific example, the recognition site for the recombinase. This event could cause interference with the engineered mutation and can be avoided using pure replacement methods based on homologous recombination integration, such as the "hit-and-run" (*b*) and the "double-replacement" (*c*) approaches. (*b*) The hit-and-run strategy is based on the use of an insertion targeting vector, where the selectable marker is not placed between the arms of homology but connect them from outside. Upon homologous recombination ("hit"), the entire vector inserts into a single site, creating a duplication of the genomic sequence determined by the segments of homology. Therefore, after integration, the genome could loop, the duplicated regions could pair, and homologous recombination can occur ("run"). Generally, the positive selectable marker (+) helps the identification of the hit, while the negative selectable marker (−) is used to identify the run. If a point mutation is engineered into one of the homology arms of the vector, the run event can leave the mutation behind. (*c*) In the double-replacement strategy, two targeting steps are used. The first targeting vector inserts a positive and a negative selectable marker inside the region of homology. The positive selectable marker is used in this step to identify the targeted clone, which then becomes the subject of the second targeting event. In the second step, a different targeting vector is used that contains a sequence homologous to the genetic segment of interest. This vector is devoid of any selectable marker and carries the desired mutation. Selection against the negative selectable marker is then applied, and resistant clones are screened for the presence of the desired mutation. As with the hit-and-run method, the advantage of this strategy is the high purity; the disadvantage is generally a low efficiency.

CONDITIONAL GAIN-OF-FUNCTION STRATEGIES

As mentioned above, uncontrolled expression of an oncogene in transgenic models or mutation of a gene throughout the mouse by targeted mutagenesis of ES cells cannot model the development of sporadic cancer, which arises amid a genetically "normal" background. Furthermore, many germline oncogene-activating mutations can cause lethality, usually due to phenotypes unrelated to the cell type of interest, or can occasionally trigger compensatory pathways during embryogenesis. While the creation of chimeric mice, consisting of both normal and mutant cells, can overcome the problem of embryonic lethality and allow for development of tumor models (Cichowski et al., 1999), this strategy is limited by the inability to breed mice with the relevant genotype. To circumvent these issues, various "conditional" strategies have been developed that allow control of gene expression in both a tissue- and temporal-specific manner. Below are described different strategies that have been developed over the past decade for somatic and temporally controlled oncogene expression/activation in mouse models of cancer (for a complete review see Lewandoski, 2001).

Inducible Oncogenes

A first strategy is based on the use of chemically induced transcription factors, responsible for the regulated expression of the target transgene. In this class of approaches, the most widely used is the tetracycline-dependent regulatory system developed by Gossen and Bujard (1992). This is a binary system, in which the effector, the tetracycline transactivator (tTA), is a fusion protein containing the herpes simplex virus VP-16 transactivation domain fused to the *Escherichia coli* tetracycline repressor protein (TetR). tTA specifically binds both tetracycline and the operator sequences (*tetO*) of the *tet* operon, opportunely located upstream of the target transgene. In the original version of this system ("*tet* off"), the tTA cannot bind DNA in the presence of tetracycline (inducer), whereas in a modified version ("*tet* on") the "reverse tTA" (rtTA) binds DNA only in the presence of the inducer (Fig. 4.4*a*). Importantly, to provide tissue specificity, the tetracycline transactivator can be expressed from selected promoters, so that mice bitransgenic for the *tet* operon oncogene and rtTA will express the oncogene in specific tissues when the drinking water is supplemented with the doxycycline (less toxic version of tetracycline), whereas mice bitransgenic for the *tet* operon oncogene and tTA will constitutively express the oncogene in specific tissues unless doxycycline is provided. The major advantage of this approach is the ability to repeatedly turn on and off the expression of transgenes, making them suitable for studies on the temporal requirements for an oncogene expression at distinct stages of tumorigenesis. Reversibility of tumorigenesis was demonstrated for instance in the case of acute B-cell leukemias induced by *BCR-ABL*. Mice harboring a *BCR-ABL* transgene inducible by removal of doxycycline developed fatal B-lymphoid leukemia that completely regressed upon doxycycline readministration, showing that *BCR-ABL*-induced B-ALL is dependent on continued expression of the oncogene, even at advanced stages of leukemia (Huettner et al., 2000). This approach has been used in several other studies to assess sustained expression requirements and reversibility of tumorigenesis. Examples include a very informative inducible H-*ras-G12V* mouse model of melanoma in an *INK4a-ARF*-deficient background (Chin et al., 1999), an inducible c-*myc* model of T-cell lymphoma and acute myeloid leukemia (Felsher and Bishop, 1999), and a K-*ras-G12D* model of non-small-cell lung cancer (Fisher et al., 2001).

Another successful inducible system is based on the tissue-specific expression of a fusion product consisting of the studied oncoprotein and a mutated ligand-binding domain from the estrogen receptor (ER^T), which is activated in response to treatment with 4-hydroxytamoxifen (4-OHT). This system offers the advantage of post-transcriptional control of an inducible oncogene. In the absence of the inducer, the oncoprotein–ER^T fusion product is maintained in a latent inactive state that is released upon hormone binding, thereby activating the fusion protein (Mattioni et al., 1994) (Fig. 4.4*b*). This approach has been employed to study the role of a sustained activation of c-*myc* in cancer (Pelengaris et al., 1999). Transgene expression was directed to suprabasal keratinocytes using the involucrin promoter and caused papillomatosis upon persistent induction by 4-OHT. Withdrawal of 4-OHT resulted in loss of c-*myc* target gene expression and tumor regression, demonstrating dependence of tumor maintenance on prolonged c-*myc* activity for tumor maintenance (Pelengaris et al., 1999).

The strategies presented above have been instrumental in demonstrating that tumor viability is dependent on continuous oncogene function in vivo. However, possible disadvantages such as leakiness (unwanted expression in the absence of the inducer) or poor inducibility (insufficient expression in the presence of the inducer) sometimes favor the use of different systems to temporally control oncogene expression/activation, that is, ones based on the use of site-specific recombinases.

Conditional/Somatic Activation of Oncogenes by Site-Specific and Spontaneous Recombination

Cre- or Flp-mediated Recombination. The use of site-specific recombinases has significantly expanded the

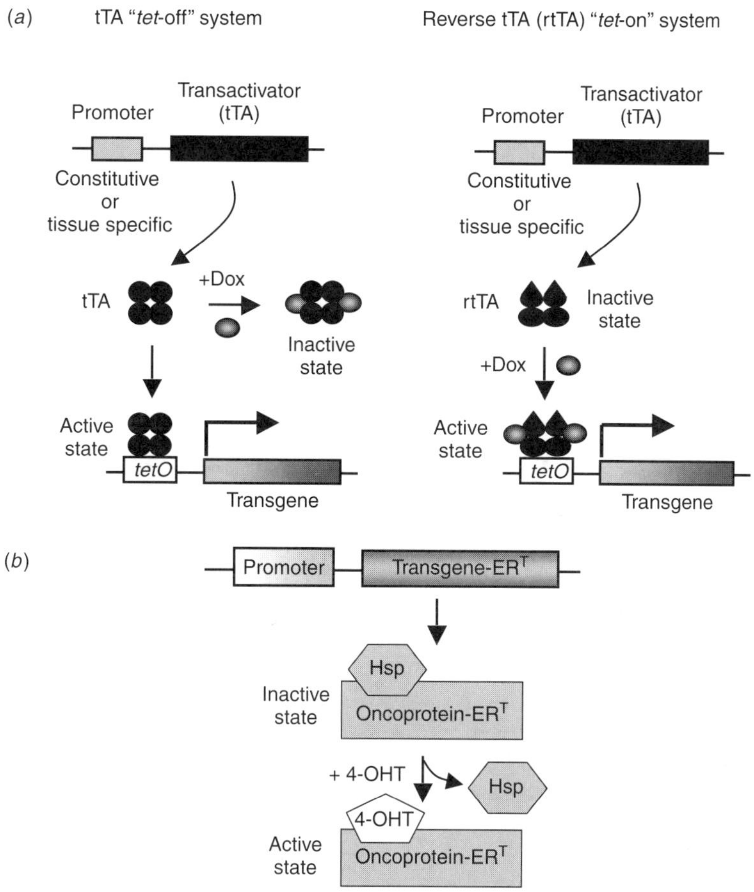

Figure 4.4. Systems that allow inducible expression of a transgene. (*a*) In the tetracycline-based transactivation system the transgene is cloned under the control of the operator sequence (*tetO*) of the *E. coli tet* operon. Tetracycline-regulatable gene expression uses either a tetracycline (*tet*)-repressible transcriptional transactivator (tTA, left) or a *tet*-inducible transactivator (rtTA, right). In the *tet*-off system, in the absence of doxycycline (Dox), tTA complex binds specifically to the *tetO* sequence and activates the transcription of the transgene. In the presence of Dox, tTA molecules undergo conformational changes that result in loss of sequence specific DNA binding. In the *tet*-on system, the rtTA complex binds to the *tetO* sequence and activate transcription only in the presence of Dox. (*b*) Tamoxifen-regulatable transgene expression is based on the tissue-specific expression of a fusion product between the transgene and a mutated ligand-binding domain from the estrogen receptor (ER^T). In the absence of the inducer 4-hydroxytamoxifen (4-OHT), the fusion protein is sequestered by heat shock proteins (HSPs), while addition of 4-OHT leads to the fusion protein release and activation.

array of genetic manipulations that can be carried out in mice in order to study the pathogenesis of cancer. Two members of the integrase family of site-specific recombinases: Cre from P1 bacteriophage (Sauer and Henderson, 1988) and Flp from *Saccharomyces cerevisiae* (O'Gorman et al., 1991) are used at present to conditionally control gene expression by site-specific DNA recombination. Cre and Flp recombinases catalyze a conservative DNA recombination event between two 34-bp recognition sites (*loxP* and *FRT*, respectively) that differ in primary sequence but share secondary structure (Fig. 4.5*a*). Because these recombinases do not require accessory proteins or high-energy cofactors for their activity, they are suitable for use in the foreign milieu of the mouse cell. Nonetheless, to improve its activity, Flp has been modified to alter its optimum temperature at 37°C, the body temperature of the mice (Buchholz et al., 1998).

The Cre-*loxP* system is the most frequently used mainly because it was the first well-developed recombination system. However, the overall strategy is the same for the Flp-*FRT* system. Although use of the Flp/*FRT* system is unlikely to be as widespread as that of the Cre-*loxP* system, it continues to be developed as a genetic tool and

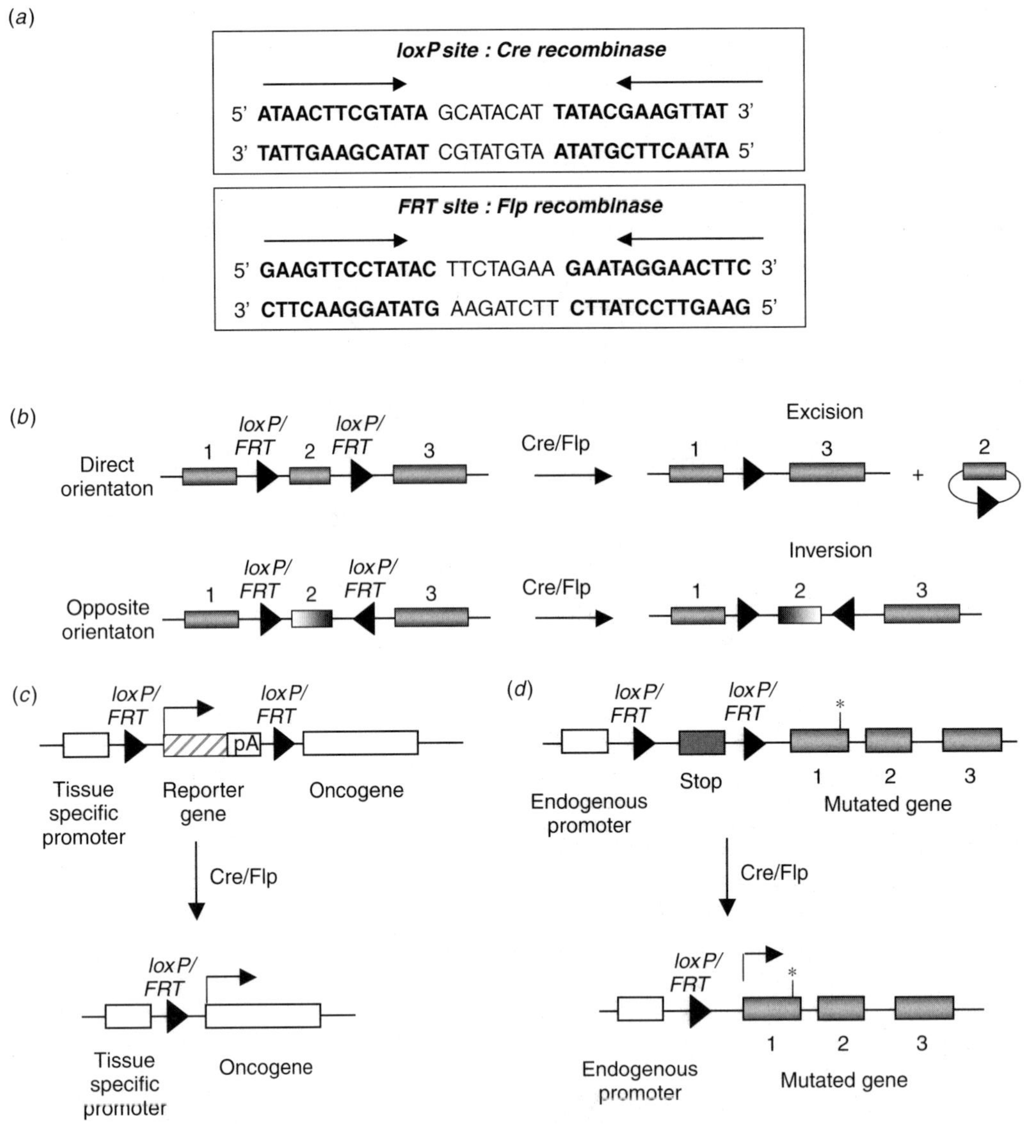

Figure 4.5. Site-specific recombinases. (*a*) Recognition sites for Cre and Flp recombinases. Both recombinases recognize unique 34-bp target sequences, *loxP* and *FRT* sites, comprised of inverted repeats separated by an 8-bp spacer sequence. This core spacer defines the orientation of the target sequence. One recombinase monomer binds each symmetrical repeated element, while the core sequence provides the site of strand cleavage, exchange, and ligation. (*b*) Cre and Flp recognize sites that can be located both in the same or in the opposite orientation; directly orientated sites lead to excision of intervening DNA, while inverted sites cause inversion of intervening DNA. (*c*) Cre/*loxP* or Flp/*FRT* recombination-mediated oncogene expression (flanked by *loxP* or *FRT* sites). Conditional oncogene activation can be achieved by using a transgene in which a floxed/flrted transcriptional silencer (e.g., a reporter gene) disconnects the oncogene from a promoter. In the absence of the recombinase enzyme, the reporter gene is expressed, and the oncogene behind the polyadenylation (pA) signal remains silent. Following recombinase-mediated excision of the silencing element, the oncogene is placed under the direct control of the promoter. (*d*) Activation of an endogenous proto-oncogene can be conditionally regulated by targeting (knocking in) the desired mutated sequence under the control of the endogenous promoter, along with a floxed/flrted transcriptional silencer (e.g., a "stop" cassette) that separates the promoter from the coding genetic region. Expression of the oncogene is then induced in the presence of a recombinase, which removes the stop cassette from the gene.

will be useful as a supplementary system. Cre/Flp catalyze the deletion of a segment of DNA that is flanked by *loxP/FRT* sites (floxed/flrted), leaving behind a single target site, or the inversion of the floxed–flrted DNA fragment if the flanking *loxP/FRT* sites are placed in opposite orientation (Fig. 4.5*b*).

Conditional oncogenes are created by the insertion of a *loxP*-flanked DNA element between the promoter and the mutant oncogene-coding sequences (Lakso et al., 1992). Conditional oncogene expression is therefore activated by recombinase-mediated removal of the transcriptional silencer, allowing tissue-specific oncogene expression (Fig. 4.5*c*). The silencer element prevents transcription and/or translation because it contains polyadenylation sequences, false translation initiation codons, and/or splice donor sequences.

Such conditional alleles can be constructed using classic transgenic technology combined with promoters of various tissue specificities. However, it must be noted that transgenic mice that harbor multiple copies of conditional transgenes may be susceptible to chromosomal breakage at *loxP* sites (and presumably *FRT* sites) (Lewandoski and Martin, 1997). An interesting conditional transgenic model of lung cancer has been recently described in which mutated K-*ras* can be sporadically activated using adenoviral-based delivery of Cre recombinase into lung epithelial cells. This enables deletion of a floxed GFP (green fluorescent protein) expression cassette placed between a ubiquitously active β-actin promoter and the K-*ras-V12* cDNA (Meuwissen et al., 2001).

Conditional mouse strains can also be created by gene targeting at the endogenous proto-oncogene loci. This alternative approach avoids potential alterations in cell biology that can result from abnormal gene expression levels as a result of transgenic promoters (Fig. 4.5*d*). Indeed, a conditional knockin version of K-*ras-G12D* was recently created by targeting the mutant allele to the endogenous K-*ras* locus. This mutant, yet silent, K-*ras* allele was activated using an adenovirus vector to deliver Cre recombinase to the lung (Jackson et al., 2001). These mice displayed a variety of hyperproliferative epithelial lesions.

The way in which to accomplish recombinase-mediated conditional gene activation is dependent on how recombinase expression is controlled. Generally, recombination is accomplished by crosses with transgenic mice expressing the recombinase in a tissue-specific manner. However, ligand-regulated inducible Cre (Indra et al., 1999) and Flp recombinases (Logie and Stewart, 1995) have also been constructed through fusion with the estrogen receptor or by placing them under the control of a *tet*-responsive promoter (Hasan et al., 2001), thus providing temporal control of recombination (for an updated list of Cre-expressing mice, see: http://www.mshri.on.ca/nagy/default.htm). In addition, both adenoviral (Akagi et al., 1997) and retroviral (Pfeifer et al., 2001; Silver and Livingston, 2001) vectors have been successfully used to introduce the recombinase activity. Delivery of Cre using viral vectors provides some interesting advantages over transgenic delivery. First of all, tumor multiplicity can be controlled by limiting the multiplicity of infection, thus allowing a closer approximation of human cancer in which single tumors form and progress through increasing stages of malignancy. Second, the introduction of Cre via a single infection serves as the initiating event, which therefore synchronizes tumor progression.

Finally, the Cre protein can be introduced directly into a system upon appropriate modifications, as in the case of the purified cell-permeable Cre-MTS (membrane translocation sequence from the Kaposi fibroblast growth factor) fusion protein, which can be transferred directly in cultured cells and in mice (Jo et al., 2001).

The most salient drawback of recombinase-mediated gene switching is its irreversibility: Once switching has occurred in a cell, all of its progeny will carry the recombined allele, even if the promoter that drives the recombinase is no longer active in these descendants. In addition, tight control of recombinase expression is crucial to prevent anomalous recombination of conditional alleles outside the tissue of interest. Ectopic recombinase expression can be particularly disturbing when it occurs during gametogenesis or in the early embryo. In this case, the floxed allele can be converted to a null allele in the germline and transmitted as such to subsequent generations of mice. To avoid these side effects, the extent of recombinase-mediated gene switching in vivo should be carefully characterized. This can be done in reporter mouse strains that carry a marker gene that is separated from a ubiquitously active promoter by a transcriptional terminator, so that the marker gene is transcribed when recombination occurs. The most widely used strain to monitor expression of the Cre recombinase is the *ROSA26-lacZ* reporter (R26R) strain, which shows uniform expression of *lacZ* upon recombination of the reporter construct in the germline (Mao et al., 1999; Soriano, 1999). Two other strains with recombinase reporter constructs in the endogenous *ROSA26* locus visualize activity of the Cre recombinase through expression of enhanced green fluorescent protein (EGFP) (Mao et al., 2001) or Flp recombinase activity through expression of placental alkaline phosphatase (PLAP) (Awatramani et al., 2001).

Spontaneous Recombination. A new exciting approach for generating oncogene-initiated murine tumor models is to create latent mutant alleles that become expressed only upon sporadic and somatic recombination in vivo. In this strategy, spontaneous recombination events occurring at low frequency in somatic cells allow random oncogene activation, thereby mimicking the sporadic occurrence of mutations as observed in human cancer.

This elegant 'hit-and-run" strategy has been recently developed to produce a mouse model for an activated K-*ras*-induced sporadic cancer (Johnson et al., 2001), in which an engineered recombination event leads to a low-frequency expression of activated K-*ras* (Fig. 4.6). To prime chromosomes for a mutagenic recombination event, two nonfunctional partial K-*ras* gene copies were knocked in in tandem, and intrachromosomal recombination between homologous sequences occurred, thus providing the initiating event. Interestingly, the mice were predisposed to a variety of tumors, most predominantly to lung adenocarcinomas. An advantage that latent alleles share with the conditional activation of oncogenes is that cells which undergo the recombination event and express the mutant oncogene are in direct contact with normal cells, thereby recapitulating an important aspect of host–tumor cell interaction found during tumor development in humans. In addition, the low but significant incidence of oncogene activation gives rise to scattered cells that, at least initially, express the mutant gene under normal physiologic control. On the other hand, it should be noted that although this strategy effectively models sporadic gene mutation, it does not allow control over gene expression.

CHROMOSOME ENGINEERING

Karyotype analysis of many human cancers often reveals gross chromosomal abnormalities such as large deletions, inversions, and translocations. Commonly observed chromosome breakpoints are usually associated with mutations that facilitate tumor growth because of tumor suppressor gene inactivation or activation of a novel oncogene produced by fusion of two normally distinct chromosomal segments (Adams, 1985; Rabbitts, 1994). As mentioned above, transgenic models have played an important role in assessing the function of some of these fusion products in cellular transformation in vivo (Adams et al., 1985; Rabbitts, 1994); however, dominant lethality of these transgenes can limit the utility of standard pronuclear injection or conventional gene-targeting approaches. In addition, these strategies generally introduce only one of the two alleles caused by a conservative recombination event such as a reciprocal translocation. Strategies have recently been developed to introduce defined chromosomal abnormalities associated with human tumors into the mouse genome by engineering them in ES cells using the Cre-*loxP* site-specific recombination system (Ramirez-Solis et al., 1995; Smith et al., 1995). The general experimental procedure takes advantage of gene-targeting technology to sequentially introduce two *loxP* sites into two loci in the ES cell genome. Subsequent expression of the gene that encodes Cre recombinase in double-targeted ES cells

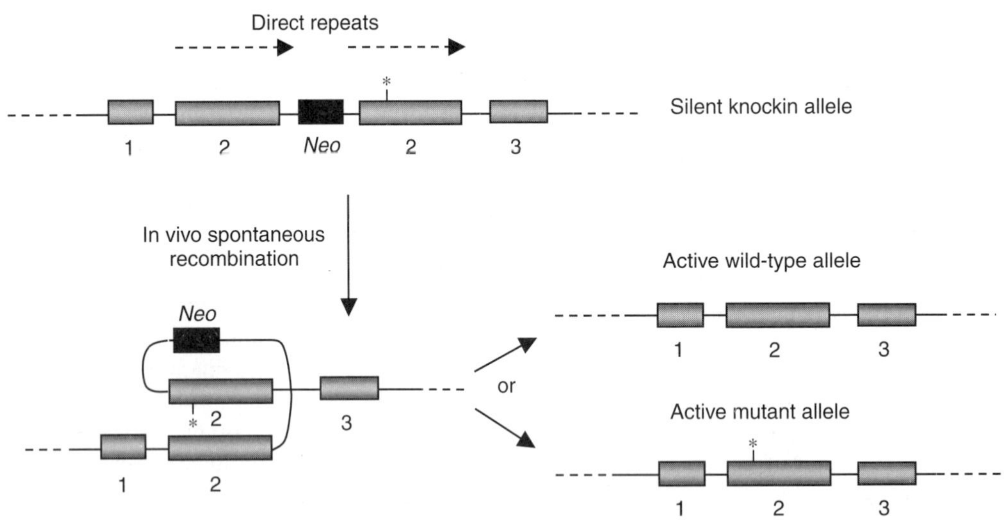

Figure 4.6. Sporadic oncogene activation through spontaneous recombination is accomplished by engineering a silent mutant allele in which a terminator (e.g., the neomycin resistance gene, *Neo*) separates two copies of an exon with an activating mutation in the latter (indicated by an asterisk). Within mouse tissues, spontaneous recombination occurs either within or between chromosomes during cell division. Occasionally, this will result in the excision of the *Neo* from the engineered locus and the concomitant production of a functional wild-type or mutant allele, depending on the exact site of recombination. Deletion events mediated through homologous repetitive sequences are frequent mutational events found in human tumors.

induces recombination between the two targeted *loxP* sites to generate the rearranged chromosome.

Deletions and inversion can be engineered by placing the recombination target sequences at desired sites within a single chromosome (Ramirez-Solis et al., 1995). Orientation of the recombination substrates determines whether the intervening sequence will be deleted or inverted. This approach has been recently applied to the study of chromosomal deletions associated with epithelial tumors. Specifically, a mouse strain has been created in which a large deletion on a distal region of chromosome 9 has been introduced to reproduce the loss of a gene-rich region at 3p21.3, which is linked to a wide range of human tumors (Smith et al., 2002). Although the deletion was embryonic lethal in homozygosity and the heterozygotes developed normally, these mice will provide the opportunity to model multistep tumorigenesis upon induction of further genetic damage. This is because deletion of 3p21.3 is believed to be a very early, preneoplastic event in the development of some types of breast and lung cancers.

Translocations can be similarly created through insertion of recombination target sequences into nonhomologous chromosomes to prime them for the recombinational event (Fig. 4.7). It must be noted though that the efficiency of Cre-*loxP* recombination between nonhomologous chromosomes is several orders of magnitude lower than that of recombination between *loxP* sites placed on the same chromosome.

To induce chromosomal rearrangements in vivo, mice that carry the double *loxP* site insertion are crossed with transgenic mice that express Cre under the control of a regulatory element with the desired tissue and temporal specificity. This approach has been applied to induce interchromosomal recombination of *Mll* and *Af9* genes or *Aml1* and *Eto* genes (Buchholz et al., 2000; Collins et al., 2000), associated with human AML. In both cases, the involved loci were successfully translocated not only ex vivo in primary cell lines but also in vivo, thus demonstrating the feasibility of inducing recombination between nonhomologous chromosomes in mice. Although Cre-mediated translocation events have been molecularly identified in mice by both groups,

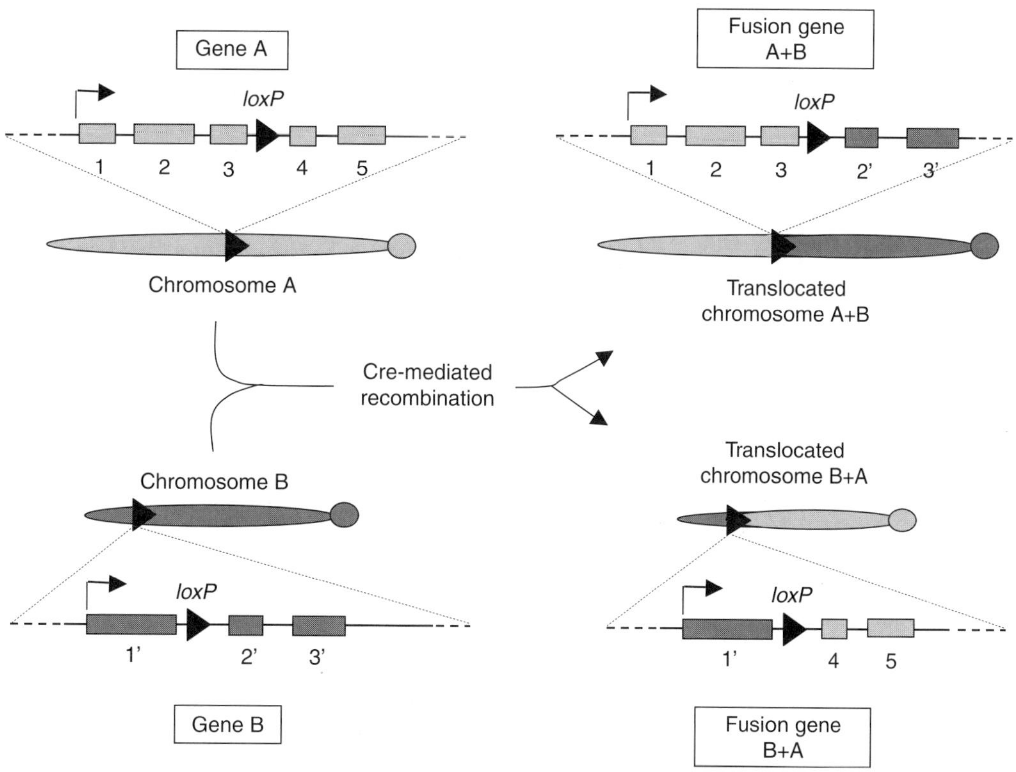

Figure 4.7. Induced translocations caused by site-specific recombination. Cre-mediated interchromosomal translocation is accomplished by targeting *loxP* sites to the desired breakpoint positions into two nonhomologous chromosomes (chromosomes A and B). Expression of Cre will allow recombination between the two *loxP* sites, thus translocating the two genes in a reciprocal manner. Fusion genes can be generated in a tissue-specific manner when the Cre transgene is appropriately expressed inside the desired cellular compartment.

neither of them have reported development of leukemia in the mice. This is possibly due to the translocations being induced in insensitive cell types, and crosses to additional Cre transgenics might be important for further characterization.

LOSS-OF-FUNCTION GERMLINE MODIFICATION STRATEGIES

Cancer arises not only as a consequence of activating gain-of-function mutations but also following loss-of-function alterations of TSGs (Knudson, 1993; Marshall, 1991). Loss-of-function genetic events that happen during tumorigenesis vary from single inactivating mutations within specific genes to deletions of entire chromosomal regions. Above we mentioned some strategies that have been recently developed to reproduce in vivo the deletion of genetic fragments by the use of site-specific recombination systems. Here we will focus on the functional inactivation of specific TSGs. In vivo modeling of TSG inactivation has been accomplished over the past two decades by utilizing the homologous recombination-mediated gene-targeting approach in ES cells (Thomas and Capecchi, 1987), initially through standard methodologies and later through more sophisticated conditional strategies.

Knockout Approach

The use of "knockout" (KO) targeting vectors to achieve null mutations in specific genetic loci has been, and still is, of extreme importance in providing critical information on gene function, in both normal development and disease processes.

The basic concept of building a targeting vector to knock out a gene is to use two segments of homologous sequences to the gene of interest that flank a part of the gene essential for function. Replacement vectors are the most widely used type of vector in gene-targeting experiments (see also the discussion of the knockin approach and Fig. 4.2). In a knockout replacement vector, a positive selectable marker is placed between the homology arms; upon homologous recombination between the arms of the vector and the corresponding genomic regions, the positive selectable marker will replace the functionally essential segment and create a "null" allele (Fig. 4.8*a*). Disruption or deletion of the coding sequence by the positive selectable marker will in most instances ablate a gene's function. However, it must be noted that in some situations a truncated protein may be generated which retains some biologic activity. Generally, since there are concerns that the presence of a selectable marker expression cassette can have an effect on the expression of neighboring genes (Olson et al., 1996), *loxP* or *FRT* sites are placed around the selectable marker in order to remove it once recombination has occurred (Fig. 4.8*b*).

Germline TSG mutants or conventional KO mice were initially developed in efforts to mimic various human inherited cancer predisposition syndromes. In these syndromes, generally a mutated TSG allele is inherited in heterozygosity, and cancer develops during the lifetime of the individual following inactivation of the second allele, according to the Knudson two-hit model (Knudson, 1996). In mouse models developed to mimic these syndromes, germline inactivation of the first allele is accomplished by standard mutagenesis through gene targeting, while inactivation of the second allele occurs in vivo in cells that undergo loss of heterozygosity (LOH). Most mouse strains heterozygous for TSG expression were found to be tumor prone and provided useful models of human cancers, demonstrating loss of the remaining wild-type allele in tumors (for a review see Ghebranious and Donehower, 1998; Jacks, 1996).

In general, while many of these TSG knockouts yielded mice with greatly increased susceptibility to cancer, only a few of them closely reproduced the actual human cancer syndromes associated with a particular TSG. One of the best models involved the *APC* gene, which is inactivated in more than 80% of human colorectal cancer and whose germline mutations are the cause of the human syndrome called familial adenomatous polyposis. *APC* heterozygous mice generated by gene targeting developed multiple polyps throughout the small intestine, and tumor analysis indicated that the wild-type *APC* gene was lost during tumorigenesis (Oshima et al., 1995). As in most cases, homozygous mice died during embryogenesis, and further analysis of *APC* genetic lesions in the mouse required the use of conditional strategies to inactivate the gene (see following section).

It is important to note that the expected outcome of conventional gene-targeting experiments is not always observed also due to species-specific differences of largely unknown origin. For example, mice heterozygous for the *Rb* gene developed pituitary adenomas rather than the retinoblastomas typically observed in children with a defective germline *Rb* allele (Hu et al., 1994; Jacks et al., 1992). Nevertheless, as with most TSG heterozygote mice, the *Rb* heterozygous mice developed early tumors, thus providing a number of important insights into the role of *Rb* in tumorigenesis. Likewise, *p53*-null mice developed cancer at a very young age (Donehower et al., 1992), but the spectrum of tumors only partially overlapped with those seen in the human Li–Fraumeni syndrome, a familial cancer predisposition associated with germline *p53* mutations (Malkin et al., 1990; Srivastava et al., 1990).

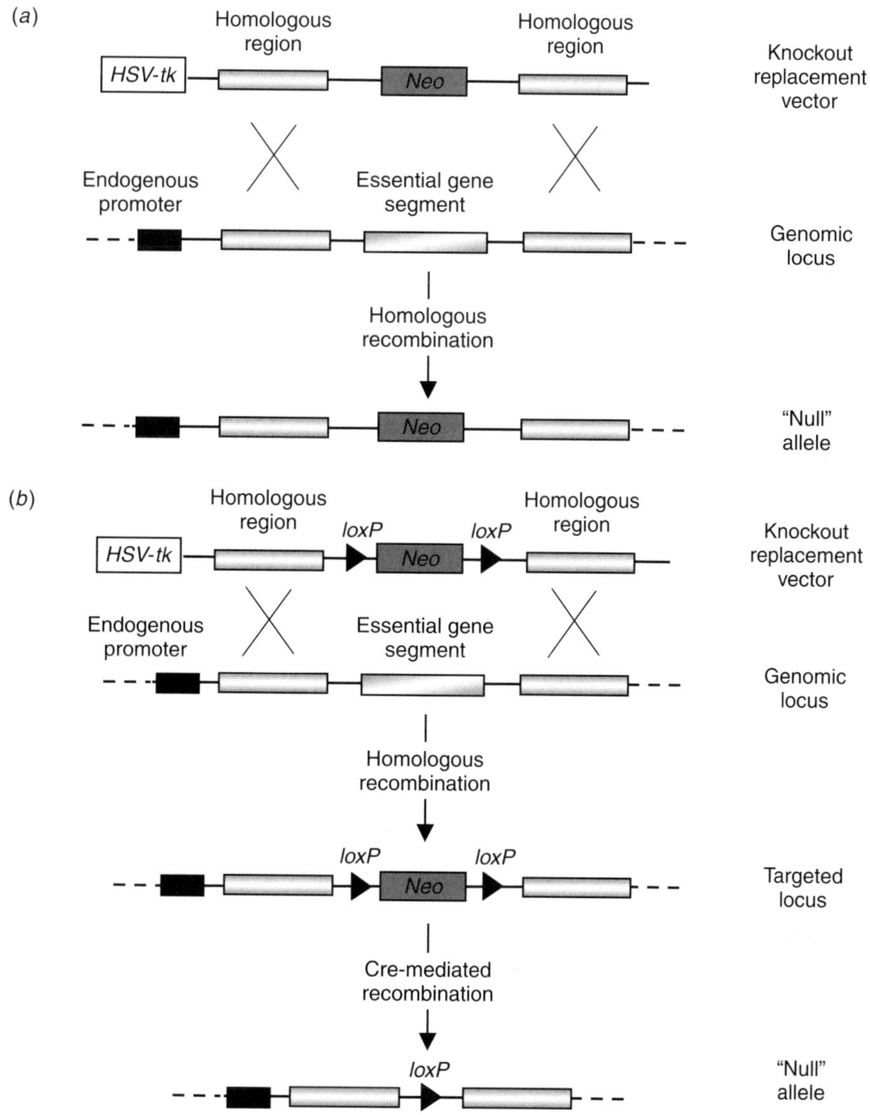

Figure 4.8. The knockout approach. (*a*) A classical knockout approach results in the replacement of an endogenous, functionally relevant genetic segment with a terminator element, generally a positive selectable marker (e.g., the neomycin resistance gene, *Neo*). The targeting vector is constructed to contain two region of homology to the desired chromosomal integration site flanking *Neo*. A negative selectable marker (e.g., the herpes simplex virus 1 thymidine kinase gene, *HSV-tk*) is placed outside one or both of the homology arms. The negative marker is lost upon homologous recombination, while the positive marker replaces the desired genetic segment. (*b*) *Neo* can be excised through Cre-mediated recombination in vitro or in vivo when the targeting vector is designed to contain recombinase-specific target sequences (*loxP* sites), flanking the positive selectable marker.

PTEN-deficient mice were also recently generated (Di Cristofano et al., 1998; Stambolic et al., 1998), and null embryos displayed an early lethal phenotype. Nevertheless, these models were able to confirm that *PTEN* mutations can predispose to cancer because *PTEN* heterozygous mice were susceptible to the early development of tumors. These mice showed a broad array of tumor types, including apparent colon adenocarcinomas, leukemia, germline tumors (Di Cristofano et al., 1998), and T-cell lymphomas (Stambolic et al., 1998). Interestingly, they also presented hyperplasias and dysplasias of the prostate and skin, which are characteristic of the human syndromes associated with *PTEN* germline mutations.

In conclusion, few TSG heterozygous KO mouse strains recapitulated faithfully the clinical features of the cognate human syndrome. In addition, it was often impossible to generate homozygous animals due to deleterious effects on normal development. Pleiotropic effects of germline mutations and the deviating phenotypes from those caused by similar mutations in human patients can be minimized by the use of conditional strategies to restrict induction of the desired mutations in a spatial and temporal fashion.

Conditional Knockout

The ability to inactivate a TSG in the mouse in a temporally and spatially controlled manner is a crucial advance in murine tumor modeling. This methodology has been called "conditional" gene targeting because it refers to a gene modification in the mouse that is restricted to certain cell types (tissue specific), to a particular stage during development (temporal specific), or to both.

Conditional gene targeting significantly expands the study of gene function in tumorigenesis in several ways. Very importantly, a widely expressed gene can be tested for loss of function in a particular cell lineage without being influenced by gene loss effects in adjacent tissues, which remain genetically wild type. In addition, by

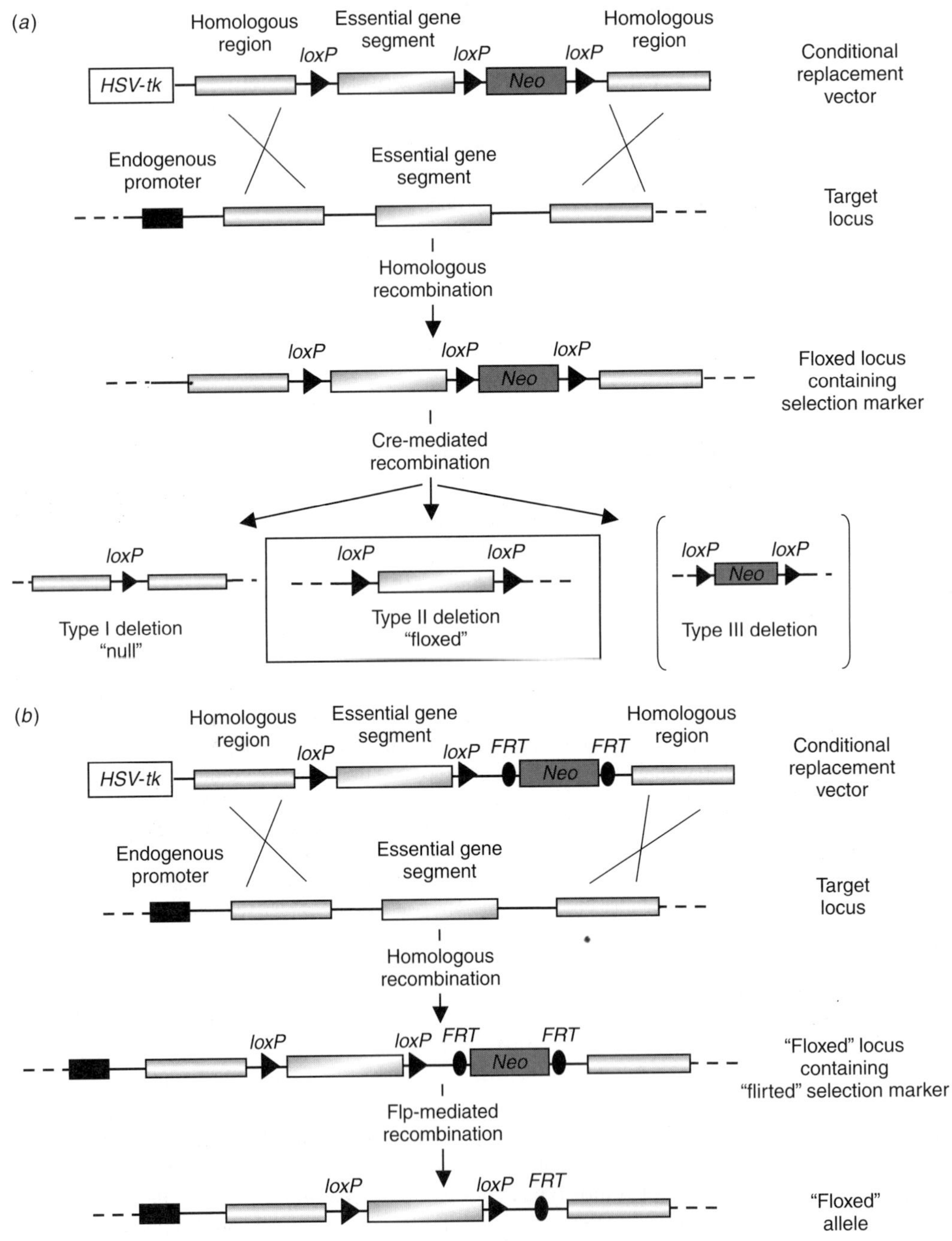

inducing the gene modification at a particular stage in development, the organism does not have an opportunity to adapt to the genetic alteration as the wild-type gene product was previously present. As a consequence, this strategy eliminates the appearance of compensatory responses during development and provides a more precise relationship between genotype and phenotype. Finally, conditional gene targeting allows the investigation of gene loss of function at late embryonic stages or in the adult when constitutively null mutations lead to a severe or lethal phenotype earlier during embryonic development.

For the conditional inactivation of a gene, standard gene-targeting techniques are used to produce a mouse strain in which an essential region of the gene of interest is floxed, so that specific Cre expression results in the inactivation of this allele (Fig. 4.9). In most cases, the strategy is based on the use of the Cre-*loxP* system (Le and Sauer, 2001), but the Flp-*FRT* recombination system is increasingly utilized as an alternative to the Cre system (Rodriguez et al., 2000).

Prior to recombination, the conditional allele is designed to have wild-type activity; thus mice homozygous for a conditional TSG allele lack an overt phenotype and are viable. Generally, *loxP* sites are placed in introns, but they can alternatively be inserted in 5′ or 3′ flanking regions without compromising gene expression. A selectable marker cassette, generally a neomycin resistance gene, is used to select for targeted ES cell clones and is therefore part of the initial floxed allele. However, the placement of *Neo* in introns or in 5′ or 3′ flanking or untranslated regions can interfere with the expression of the floxed allele, compromising its wild-type activity in the absence of Cre-mediated recombination. In some cases, it has been determined that criptic splice sites in *Neo* interfere with normal splicing and therefore reduce wild-type messenger RNA (mRNA) levels. A common solution to this problem, carried out in the first tissue-specific knockout experiments, is to use three *loxP* sites positioned such that *Neo*, as well as the essential region of the gene of interest, is floxed (Gu et al., 1994). A partial Cre-mediated recombination event then removes *Neo*, leaving the essential region floxed (Fig. 4.9*a*). This can be achieved in vitro in ES cells by transiently introducing a Cre expression cassette or in vivo by breeding mice that carry floxed alleles to mice that mosaically express Cre in the germline such that recombination occurs between two of the three *loxP* sites at a reasonable frequency (Holzenberger et al., 2000). Alternatively, the positive selectable marker can be flanked with *FRT* repeats instead of *loxP* sites, the target gene remaining flanked by direct *loxP* sites (Fig. 4.9*b*). The positive selection gene can then either be deleted directly in the targeted ES clones by transient expression of the Flp recombinase gene or it is first transmitted through the germline and the resulting mice are then crossed to a transgenic strain that expresses Flp early in embryonic development such as the *hACTB::FLP* deleter strain (Dymecki, 1996, Farley et al., 2000).

Floxed alleles where the two *loxP* sites flanking critical exons are placed in an opposite orientation have also been recently generated in order to obtain, instead of deletion, inversion of the floxed segment. When Cre expression is induced, the floxed region is inverted and, since in this case the process is reversible, it can repeat the inversion to the original state. Under sustained expression of the recombinase enzyme, inversions can happen several times, thus leading to alternate gene expression and to a general reduction of the level of the corresponding gene product (hypomorphism).

To achieve conditional deletion of the genetic region of interest, Cre recombinase can be introduced both by intercrosses with Cre transgenic mice or by somatic delivery. Transgenic mice that express either Cre or Flp

Figure 4.9. The conditional knockout approach: strategies for generating a *loxP*-flanked (floxed) gene. (*a*) A conditional replacement vector is constructed to contain two regions of homology and two *loxP* sites flanking the genetic segment to be deleted following Cre-mediated recombination. An additional *loxP* site is introduced next to the positive selectable marker *Neo* in a way that a total of three *loxP* sites flank both an essential coding segment and the selectable marker. In a targeted allele, none of the coding exons should be disturbed, and the new sequence insertion should occur only in the introns in order to preserve proper functioning of the allele. When a Cre expression vector is transiently expressed in ES cells heterozygous for the targeted allele, both total and partial excision could occur among the three *loxP* sites, resulting in three possible scenarios: one with the desired floxed gene (type II deletion), one with a null allele (type I deletion), and another with a floxed selectable marker (type III deletion). Type I and II deletions are the desired outcomes: The first one provides a null allele for the gene of interest, whereas the type II deleted allele is expected to be a wild-type equivalent functional allele. After germline transmission, the functionality of the type II allele depends on the expression of the Cre recombinase. To achieve conditional deletion of the gene of interest, Cre recombinase can be introduced by intercrosses of the floxed mice with Cre transgenic mice or by somatic delivery (see text for details). Cre will remove the *loxP*-flanked essential exon from the targeted allele, rendering it nonfunctional. (*b*) A second type of conditional replacement vector combines the use of Cre and Flp recombinases to obtain a more straightforward and safer way of generating a floxed allele. The selectable marker is flanked by *FRT* sites, while two *loxP* sites are placed around an essential genetic segment. The selectable marker can be deleted directly in the targeted ES clones by transient expression of the Flp recombinase gene or following germline transmission of the targeted allele and intercrossing with Flp transgenic mice. The conditional deletion of the floxed allele is then obtained as in (*a*).

recombinase from a variety of cellular promoters are viable and fertile (for a list see http://www.mshri.on.ca/nagy). When these are bred to mice carrying conditional TSG mutations, the TSG can be biallelically inactivated to allow the generation of organ- and cell lineage–specific tumor models. By selecting appropriate promoters or utilizing a ligand-regulatable Cre-estrogen receptor fusion protein or by taking advantage of viral systems to somatically deliver recombinase-expressing genes (discussed earlier under Cre- or Flp-Mediated Recombination), it is possible to control the timing of Cre or Flp recombinase expression in order to avoid interference with essential stages of development (Metzger and Chambon, 2001). The necessity of regulating and checking for the level and pattern of recombinase expression has also been discussed in the section on Cre- or Flp-mediated recombination; we would like to add that it has been recently reported that constitutive expression of Cre recombinase can cause karyotypic abnormalities in spermatids (Schmidt et al., 2000) and primary fibroblasts (Loonstra et al., 2001; Silver and Livingston, 2001) and should be considered as a possible confounding variable in conditional tumor models.

Several studies have used conditional strategies to circumvent embryonic lethality associated with germline mutations in a number of TSGs, including, for example, *Brca1, Brca2, Rb, Apc*, and *Nf2*. In many cases, tissue-specific inactivation of the gene of interest induced tumor phenotypes that were similar to those seen in humans. For instance, mouse mammary tumor models with conventional *Bcra1* or *Bcra2* knockouts were not successful because homozygous mutants died early in embryogenesis, and, in contrast to humans, heterozygous mice were not susceptible to developing tumors. However, a conditional mutation of either *Brca1* (Xu et al., 1999) or *Brca2* (Ludwig et al., 2001) in mouse mammary gland epithelium was found to promote mammary tumorigenesis. Alternatively, Cre recombinase was transiently delivered with recombinant adenoviruses to create an organ-specific tumor model of *Apc* deletion in the colon (Shibata et al., 1997). A conditional *Nf2* allele was used to create Schwann cell tumors similar to those seen in the cognate human disease, neurofibromatosis type 2. While *Nf2* null mice die early in embryogenesis and conventional *Nf2* KO heterozygous mice primarily develop osteosarcomas (McClatchey et al., 1998), mice carrying specific *Nf2* inactivation in Schwann cell precursors were predisposed to schwannoma and also developed additional characteristics of the human disease (Giovannini et al., 2000).

The accumulation of numerous tumor-susceptible and conditional knockout mice strains has aided investigators in the analysis of genetic interactions among different TSGs. Intercrossing of two different TSG knockout mice to generate doubly deficient offspring often revealed levels of cooperativity between two genes. Interestingly, tumor incidence rates can be accelerated and tumor spectra altered by combining tumor suppressor deficiencies. For instance, a model has been described where, in order to induce tissue-specific deletion of a conditional *Brca1* allele in adult mammary gland epithelium, Cre expression was driven by the MMTV long terminal repeat (MMTV-LTR) or by the whey acidic protein (WAP) promoter (Xu et al., 1999). These females developed mammary tumors with relatively long latency and low penetrance, and tumor formation was markedly accelerated in a *p53* heterozygous background. Interestingly, most of the resulting tumors had lost the wild-type *p53* allele. Similar synergistic effects were observed between *Brca2* and *p53*. WAP-Cre-mediated inactivation of *Brca2* caused the formation of mammary tumors with long latency (Ludwig et al., 2001), whereas keratin 14 promoter (K14)-driven Cre-mediated inactivation of *Brca2* did not predispose to the formation of epithelial tumors. However, K14-Cre-mediated inactivation of both *Brca2* and *p53* caused a high frequency of mammary tumors (Jonkers et al., 2001).

Compound mutants that harbor several conditional TSG alleles can theoretically be bred without compromising their viability and, combined with the use of a rapidly growing list of Cre tools, they can provide highly regulatable models of multiple TSG inactivations in a spatially and temporally controlled manner.

COMPOUND MODELS

An ideal mouse tumor model should faithfully reproduce the tumorigenic process in every possible aspect and, in addition, mimic the pathologic, genetic, aetiologic, and therapeutic characteristics of its human counterpart (Hann and Balmain, 2001). Models that are able to meet all these criteria are virtually impossible to generate, but it is nevertheless crucial to aim at generating a minimum set of features which will define the quality of a particular tumor model. First, the model should produce well-defined sporadic tumors with significant penetrance and within a narrow time window. This should be combined with an assessment of contribution made by individual mutations to the tumor phenotype to measure the effect of each single variable in an otherwise stable genetic background that contains a fixed number of defined complementing mutations. Consequently, the model should also allow the identification of additional mutations that cooperate with the predefined mutation in the pathogenetic process. Meeting all these criteria is virtually possible when different kinds of mutations are introduced into the same model in a tissue-specific and time-controlled fashion. Combining gain-of-function and loss-of-function events within the same cells by intercrossing independently derived mouse strains provides a powerful genetic approach for determining the specific collaborating events in tumorigenesis.

For instance, this combined approach has been used to create a very informative mouse model of melanoma in an *INK4a*-deficient background (Chin et al., 1999). In this study, expression of H-*ras-G12V* in a doxycycline-inducible manner causes the appearance of melanoma in the presence of a null mutation for the tumor suppressor *INK4a-ARF*. Use of the doxo-regulatable system was instrumental to show that, upon *INK4a-ARF* loss of function, the development and maintenance of melanoma depends on sustained expression of an activated oncogene. The withdrawal of doxycycline and consequent H-*ras-G12V* down regulation resulted in clinical and histologic regression of primary and explanted tumors. Interestingly, a detailed evaluation of this melanoma model allowed the identification of additional components involved in the tumor maintenance, such as the role of microenvironment endothelial cells, which apparently received survival signals from oncogene-expressing melanocytes (Chin et al., 1999).

Another advanced compound model has been reported by Johnson and colleagues (see also the earlier discussion of spontaneous recombination) in which sporadic human lung cancer is reproduced by means of an elegant genetic technique named "hit and run" (Johnson et al., 2001) (Fig. 4.6). The authors have produced a strain of mice in which an active, mutant K-*ras* gene is generated at random in some cells through occurrence of spontaneous recombination. This randomness echoes the sporadic incidence of K-*ras* mutations in many human cancers. Significantly, the resulting mice are prone to lung cancers that acquire metastatic features when this strain is intercrossed with *p53*-deficient mice. This is an improved model that, by introducing sporadic mutational events to a defined mutated germline background, represents a unique experimental paradigm that closely mimics the occurrence of somatic lesions in a tumor-prone background as observed in humans. In addition, this kind of approach can greatly facilitate understanding of how the molecular mechanisms that fuel tumorigenesis vary depending on the cell type and the preceding oncogenic event(s). Combining mutations in mice has provided, and will continue to provide, a major tool with which to understand these dynamics, although much more work is needed before unifying principles are fully uncovered.

RNA INTERFERENCE

In recent years, sequence-specific gene silencing has become an area of increasing focus, first of all because of its interesting biology and, concomitantly, because of its potential as an experimental tool. Homology-dependent gene-silencing events triggered by double-stranded RNA (dsRNA) are currently identified under the term *RNA interference* (RNAi) (for a review see Denli and Hannon, 2003). RNAi is a biological phenomenon that has triggered the development of powerful tools for manipulating gene expression in mammalian cells, providing a rapid loss-of-function method to assess gene function in a number of organisms. Very recently, plasmid-based systems to perform RNAi in vivo have been developed and have been used to generate germline transmitted transgenic mice expressing the inhibitory RNA sequences toward specific gene targets (see details below).

RNAi first emerged as a biologic peculiarity in *Caenorhab-ditis elegans* (Fire et al., 1998) and plants (Jorgensen et al., 1996) but was quickly harnessed as a genetic tool in these systems. It has since become clear that dsRNA-induced silencing phenomena are present in evolutionary diverse organisms including fungi and metazoans. A combination of genetic and biochemical studies suggest that many of these biologic events share a common mechanism. The general prevailing model begins with the conversion of the dsRNA silencing "trigger" into small interfering RNAs (siRNAs) by a RNase III family nuclease called Dicer (Bernstein et al., 2001). These small RNAs ($\sim$22–25 nucleotides in size) become incorporated into a multicomponent nuclease complex which uses the sequence of the siRNA as a guide to identify and destroy homologous mRNAs (Nykanen et al., 2001; Zamore et al., 2000). It has been suggested that the RNAi pathway may have evolved early in cukaryotes as a cell-based immunity against viral and genetic parasites. However, the RNAi pathway is also described as being involved in the regulation of endogenous gene targets during metazoan and plant development (for a review see Hannon, 2002).

Different methods have so far been developed in order to exogenously trigger the RNAi pathway in mammalian cells through the introduction of dsRNAs of different length and design (for a review see Paddison and Hannon, 2002). One of these strategies is based on the design of specific small RNA structures, known as short hairpin RNAs (shRNAs), which are engineered to suppress the expression of desired genes in cultured mammalian cells. Silencing by shRNAs requires the RNAi machinery and correlates with the production of siRNAs, which are the signature of RNA interference (Paddison et al., 2002).

shRNAs can be synthesized exogenously, but they can also be transcribed from RNA polymerase III promoters in vivo using plasmid-based systems. This permits the construction of continuous cell lines or transgenic animals in which RNAi enforces stable and heritable gene silencing. Germline transmission of RNAi in mice has been achieved for instance by Carmell and colleagues (2003) using shRNA expression vectors. Using standard pronuclear injection of shRNAs vectors directed against a variety of targets, they obtained transgenic animals, but virtually none of them showed a distinct or reproducible phenotype that was expected for a hypomorphic allele of the targeted genes. On the contrary, an ES cell–mediated transgenesis approach provided the opportunity of verifying the presence of the shRNA and its activity toward the target gene in cell culture: About 80% of the isolated clones showed reduction in the expression of the target gene, and chimera mice obtained from blastocyst injection gave germline transmission of the shRNA expression construct which was noted in numerous F_1 progeny (Carmell et al., 2003).

It must be noted that, while application of the RNAi approach leads to a loss of function of the gene of interest, in vivo models that use this strategy are created through the transgenic approach, mentioned here as a gain-of-function germline modification strategy (see the discussion of the transgenic approach earlier in the chapter).

This successful new methodology opens the door to using RNAi as a complement to standard knockout methodologies as a powerful means of rapidly assessing the consequences of suppressing a gene of interest in a living animal. It can be predicted that, in combination with technologically advanced tools, the use of shRNAs will ultimately provide methods for tissue-specific, inducible and reversible suppression of gene expression in mice.

REFERENCES

Adams JM (1985): Oncogene activation by fusion of chromosomes in leukaemia. *Nature* 315: 541–542.

Adams JM, Harris AW, Pinkert CA, Corcoran LM, Alexander WS, Cory S, Palmiter RD, Brinster RL (1985): The c-myc oncogene driven by immunoglobulin enhancers induces lymphoid malignancy in transgenic mice. *Nature* 318: 533–538.

Akagi K, Sandig V, Vooijs M, Van der Valk M, Giovannini M, Strauss M, Berns A (1997): Cre-mediated somatic site-specific recombination in mice. *Nucleic Acids Res* 25: 1766–1773.

Awatramani R, Soriano P, Mai JJ, Dymecki S (2001): An Flp indicator mouse expressing alkaline phosphatase from the ROSA26 locus. *Nat Genet* 29: 257–259.

Bernstein E, Caudy AA, Hammond SM, Hannon GJ (2001): Role for a bidentate ribonuclease in the initiation step of RNA interference. *Nature* 409: 363–366.

Brinster RL, Chen HY, Messing A, van Dyke T, Levine AJ, Palmiter RD (1984): Transgenic mice harboring SV40 T-antigen genes develop characteristic brain tumors. *Cell* 37: 367–379.

Brown D, Kogan S, Lagasse E, Weissman I, Alcalay M, Pelicci PG, Atwater S, Bishop JM (1997): A PMLRARalpha transgene initiates murine acute promyelocytic leukemia. *Proc Natl Acad Sci USA* 94: 2551–2556.

Brown K, Strathdee D, Bryson S, Lambie W, Balmain A (1998): The malignant capacity of skin tumours induced by expression of a mutant H-ras transgene depends on the cell type targeted. *Curr Biol* 23: 516–524.

Buchholz F, Angrand PO, Stewart AF (1998): Improved properties of FLP recombinase evolved by cycling mutagenesis. *Nat Biotechnol* 16: 657–662.

Buchholz F, Refaeli Y, Trumpp A, Bishop JM (2000): Inducible chromosomal translocation of AML1 and ETO genes through Cre/loxP-mediated recombination in the mouse. *EMBO Rep* 1: 133–139.

Carmell MA, Zhang L, Conklin DS, Hannon GJ, Rosenquist TA (2003): Germline transmission of RNAi in mice. *Nat Struct Biol* 10: 91–92.

Castellanos A, Pintado B, Weruaga E, Arevalo R, Lopez A, Orfao A, Sanchez-Garcia I (1997): A BCR-ABL (p190) fusion gene made by homologous recombination causes B-cell acute lymphoblastic leukemias in chimeric mice with independence of the endogenous bcr product. *Blood* 90: 2168–2174.

Chin L, Tam A, Pomerantz J, Wong M, Holash J, Bardeesy N, Shen Q, O'Hagan R, Pantginis J, Zhou H, Horner JW II, Cordon-Cardo C, Yancopoulos GD, DePinho RA (1999): Essential role for oncogenic Ras in tumour maintenance. *Nature* 400: 468–472.

Cichowski K, Shih TS, Schmitt E, Santiago S, Reilly K, McLaughlin ME, Bronson RT, Jacks T (1999): Mouse models of tumor development in neurofibromatosis type 1. *Science* 286: 2172–2176.

Collins EC, Pannell R, Simpson EM, Forster A, Rabbitts TH (2000): Inter-chromosomal recombination of Mll and Af9 genes mediated by cre-loxP in mouse development. *EMBO Rep* 1: 127–132.

Denli AM, Hannon GJ (2003): RNAi: An ever-growing puzzle. *Trends Biochem Sci* 28: 196–201.

de Vries A, Flores ER, Miranda B, Hsieh HM, van Oostrom CT, Sage J, Jacks T (2002): Targeted point mutations of *p53* lead to dominant-negative inhibition of wild-type *p53* function. *Proc Natl Acad Sci USA* 99: 2948–2953.

Di Cristofano A, Pesce B, Cordon-Cardo C, Pandolfi PP (1998): Pten is essential for embryonic development and tumour suppression. *Nat Genet* 19: 348–355.

Ding H, Roncari L, Shannon P, Wu X, Lau N, Karaskova J, Gutmann DH, Squire JA, Nagy A, Guha A (2001): Astrocyte-specific expression of activated p21-ras results in malignant astrocytoma formation in a transgenic mouse model of human gliomas. *Cancer Res* 61: 3826–3836.

Donehower LA, Harvey M, Slagle BL, McArthur MJ, Montgomery CA Jr, Butel JS, Bradley A (1992): Mice deficient for *p53* are developmentally normal but susceptible to spontaneous tumours. *Nature* 356: 215–221.

Dymecki SM (1996): Flp recombinase promotes site-specific DNA recombination in embryonic stem cells and transgenic mice. *Proc Natl Acad Sci USA* 93: 6191–6196.

Farley FW, Soriano P, Steffen LS, Dymecki SM (2000): Widespread recombinase expression using FLPeR (flipper) mice. *Genesis* 28: 106–110.

Felsher DW, Bishop JM (1999): Reversible tumorigenesis by MYC in hematopoietic lineages. *Mol Cell* 4: 199–207.

Fire A, Xu S, Montgomery MK, Kostas SA, Driver SE, Mello CC (1998): Potent and specific genetic interference by double-stranded RNA in *Caenorhabditis elegans*. *Nature* 391: 806–811.

Fisher GH, Wellen SL, Klimstra D, Lenczowski JM, Tichelaar JW, Lizak MJ, Whitsett JA, Koretsky A, Varmus HE (2001): Induction and apoptotic regression of lung adenocarcinomas by regulation of a K-Ras transgene in the presence and absence of tumor suppressor genes. *Genes Dev* 15: 3249–3262.

Ghebranious N, Donehower LA (1998): Mouse models in tumor suppression. *Oncogene* 17: 3385–3400.

Giovannini M, Robanus-Maandag E, van der Valk M, Niwa-Kawakita M, Abramowski V, Goutebroze L, Woodruff JM, Berns A, Thomas G (2000): Conditional biallelic Nf2 mutation in the mouse promotes manifestations of human neurofibromatosis type 2. *Genes Dev* 14: 1617–1630.

Gossen M, Bujard H (1992): Tight control of gene expression in mammalian cells by tetracycline-responsive promoters. *Proc Natl Acad Sci USA* 89: 5547–5551.

Grisolano JL, Wesselschmidt RL, Pelicci PG, Ley TJ (1997): Altered myeloid development and acute leukemia in transgenic mice expressing PML-RAR alpha under control of cathepsin G regulatory sequences. *Blood* 89: 376–387.

Gu H, Marth JD, Orban PC, Mossmann H, Rajewsky K (1994): Deletion of a DNA polymerase beta gene segment in T cells using cell type-specific gene targeting. *Science* 265: 103–106.

Hanahan D (1985): Heritable formation of pancreatic beta-cell tumours in transgenic mice expressing recombinant insulin/simian virus 40 oncogenes. *Nature* 315: 115–122.

Hanahan D (1989): Transgenic mice as probes into complex systems. *Science* 246: 1265–1275.

Hanks M, Wurst W, Anson-Cartwright L, Auerbach AB, Joyner AL (1995): Rescue of the En-1 mutant phenotype by replacement of En-1 with En-2. *Science* 269: 679–682.

Hann B, Balmain A (2001): Building "validated" mouse models of human cancer. *Curr Opin Cell Biol 13*: 778–784.

Hannon GJ (2002): RNA interference. *Nature* 418: 244–251.

Hasan MT, Schonig K, Berger S, Graewe W, Bujard H (2001): Long-term, noninvasive imaging of regulated gene expression in living mice. *Genesis* 29: 116–122.

He LZ, Tribioli C, Rivi R, Peruzzi D, Pelicci PG, Soares V, Cattoretti G, Pandolfi PP (1997): Acute leukemia with promyelocytic features in PML/RARalpha transgenic mice. *Proc Natl Acad Sci USA* 94: 5302–5307.

Heisterkamp N, Jenster G, ten Hoeve J, Zovich D, Pattengale PK, Groffen J (1990): Acute leukaemia in bcr/abl transgenic mice. *Nature* 344: 251–253.

Holzenberger M, Lenzner C, Leneuve P, Zaoui R, Hamard G, Vaulont S, Bouc YL (2000): Cre-mediated germline mosaicism: A method allowing rapid generation of several alleles of a target gene. *Nucleic Acids Res* 28: E92.

Hu N, Gutsmann A, Herbert DC, Bradley A, Lee WH, Lee EY (1994): Heterozygous Rb-1 delta 20/+mice are predisposed to tumors of the pituitary gland with a nearly complete penetrance. *Oncogene* 9: 1021–1027.

Huettner CS, Zhang P, Van Etten RA, Tenen DG (2000): Reversibility of acute B-cell leukaemia induced by BCR-ABL1. *Nat Genet* 24: 57–60.

Indra AK, Warot X, Brocard J, Bornert JM, Xiao JH, Chambon P, Metzger D (1999): Temporally-controlled site-specific mutagenesis in the basal layer of the epidermis: Comparison of the recombinase activity of the tamoxifen-inducible Cre-ER(T) and Cre-ER(T2) recombinases. *Nucleic Acids Res* 27: 4324–4327.

Jacks T (1996): Tumor suppressor gene mutations in mice. *Annu Rev Genet* 30: 603–636.

Jacks T, Fazeli A, Schmitt EM, Bronson RT, Goodell MA, Weinberg RA (1992): Effects of an Rb mutation in the mouse. *Nature* 359: 295–300.

Jackson EL, Willis N, Mercer K, Bronson RT, Crowley D, Montoya R, Jacks T, Tuveson DA (2001): Analysis of lung tumor initiation and progression using conditional expression of oncogenic K-ras. *Genes Dev* 15: 3243–3248.

Jaenisch R (1988): Transgenic animals. *Science* 240: 1468–1474.

Jo D, Nashabi A, Doxsee C, Lin Q, Unutmaz D, Chen J, Ruley HE (2001): Epigenetic regulation of gene structure and function with a cell-permeable Cre recombinase. *Nat Biotechnol* 19: 929–933.

Johnson L, Mercer K, Greenbaum D, Bronson RT, Crowley D, Tuveson DA, Jacks T (2001): Somatic activation of the K-ras oncogene causes early onset lung cancer in mice. *Nature* 410: 1111–1116.

Jonkers J, Meuwissen R, van der Gulden H, Peterse H, van der Valk M, Berns A (2001): Synergistic tumor suppressor activity of BRCA2 and *p53* in a conditional mouse model for breast cancer. *Nat Genet* 29: 418–425.

Jorgensen RA, Cluster PD, English J, Que Q, Napoli CA (1996): Chalcone synthase cosuppression phenotypes in petunia flowers: Comparison of sense vs. antisense constructs and single-copy vs. complex T-DNA sequences. *Plant Mol Biol* 31: 957–973.

Knudson AG (1993): Antioncogenes and human cancer. *Proc Natl Acad Sci USA* 90: 10,914–10,921.

Knudson AG (1996): Hereditary cancer: Two hits revisited. *J Cancer Res Clin Oncol* 122: 135–140.

Lakso M, Sauer B, Mosinger B Jr, Lee EJ, Manning RW, Yu SH, Mulder KL, Westphal H (1992): Targeted oncogene activation by site-specific recombination in transgenic mice. *Proc Natl Acad Sci USA* 15: 6232–6236.

Lavin MF, Shiloh Y (1997): The genetic defect in ataxia-telangiectasia. *Annu Rev Immunol* 15: 177–202.

Le Y, Sauer B (2001): Conditional gene knockout using Cre recombinase. *Mol Biotechnol.* 17: 269–275.

Lewandoski M (2001): Conditional control of gene expression in the mouse. *Nat Rev Genet* 2: 743–755.

Lewandoski M, Martin GR (1997): Cre-mediated chromosome loss in mice. *Nat Genet* 17: 223–225.

Logie C, Stewart AF (1995): Ligand-regulated site-specific recombination. *Proc Natl Acad Sci USA* 92: 5940–5944.

Lois C, Hong EJ, Pease S, Brown EJ, Baltimore D (2002): Germline transmission and tissue-specific expression of transgenes delivered by lentiviral vectors. *Science* 295: 868–872.

Loonstra A, Vooijs M, Beverloo HB, Allak BA, van Drunen E, Kanaar R, Berns A, Jonkers J (2001): Growth inhibition and DNA damage induced by Cre recombinase in mammalian cells. *Proc Natl Acad Sci USA* 98: 9209–9214.

Ludwig T, Fisher P, Murty V, Efstratiadis A (2001): Development of mammary adenocarcinomas by tissue-specific knockout of Brca2 in mice. *Oncogene* 20: 3937–3948.

Malkin D, et al (1990): Germ line *p53* mutations in a familial syndrome of breast cancer, sarcomas, and other neoplasms. *Science* 250: 1233–1238.

Mao X, Fujiwara Y, Chapdelaine A, Yang H, Orkin SH (2001): Activation of EGFP expression by Cre-mediated excision in a new ROSA26 reporter mouse strain. *Blood* 97: 324–326.

Mao X, Fujiwara Y, Orkin SH (1999): Improved reporter strain for monitoring Cre recombinase-mediated DNA excisions in mice. *Proc Natl Acad Sci USA* 96: 5037–5042.

Marshall CJ (1991): Tumor suppressor genes. *Cell* 64: 313–326.

Mattioni T, Louvion JF, Picard D (1994): Regulation of protein activities by fusion to steroid binding domains. *Methods Cell Biol* 43: 335–352.

McClatchey AI, Saotome I, Mercer K, Crowley D, Gusella JF, Bronson RT, Jacks T (1998): Mice heterozygous for a mutation at the Nf2 tumor suppressor locus develop a range of highly metastatic tumors. *Genes Dev* 12: 1121–1133.

Merghoub T, Gurrieri C, Piazza F, Pandolfi PP (2001): Modeling acute promyelocytic leukemia in the mouse: New insights in the pathogenesis of human leukemias. *Blood Cells Mol Dis* 27: 231–248.

Metzger D, Chambon P (2001): Site- and time-specific gene targeting in the mouse. *Methods* 24: 71–80.

Meuwissen R, Linn SC, van der Valk M, Mooi WJ, Berns A (2001): Mouse model for lung tumorigenesis through Cre/lox controlled sporadic activation of the K-Ras oncogene. *Oncogene* 20: 6551–6558.

Novak A, Guo C, Yang W, Nagy A, Lobe CG (2000): Z/EG, a double reporter mouse line that expresses enhanced green fluorescent protein upon Cre-mediated excision. *Genesis* 28: 147–155.

Nykanen A, Haley B, Zamore PD (2001): ATP requirements and small interfering RNA structure in the RNA interference pathway. *Cell* 107: 309–321.

O'Gorman S, Fox DT, Wahl GM (1991): Recombinase-mediated gene activation and site-specific integration in mammalian cells. *Science* 251: 1351–1355.

Okuda T, Cai Z, Yang S, Lenny N, Lyu CJ, van Deursen JM, Harada H, Downing JR (1998): Expression of a knocked-in AML1-ETO leukemia gene inhibits the establishment of normal definitive hematopoiesis and directly generates dysplastic hematopoietic progenitors. *Blood* 91: 3134–3143.

Olson EN, Arnold HH, Rigby PW, Wold BJ (1996): Know your neighbors: Three phenotypes in null mutants of the myogenic bHLH gene MRF4. *Cell* 85: 1–4.

Oshima M, Oshima H, Kitagawa K, Kobayashi M, Itakura C, Taketo M (1995): Loss of Apc heterozygosity and abnormal tissue building in nascent intestinal polyps in mice carrying a truncated Apc gene. *Proc Natl Acad Sci USA* 92: 4482–4486.

Paddison PJ, Caudy AA, Bernstein E, Hannon GJ, Conklin DS (2002): Short hairpin RNAs (shRNAs) induce sequence-specific silencing in mammalian cells. *Genes Dev* 16: 948–958.

Paddison PJ, Hannon GJ (2002): RNA interference: The new somatic cell genetics? *Cancer Cell* 2: 17–23.

Palmiter RD, Brinster RL (1986): Germ-line transformation of mice. *Annu Rev Genet* 20: 465–499.

Pelengaris S, Littlewood T, Khan M, Elia G, Evan G (1999): Reversible activation of c-Myc in skin: Induction of a complex neoplastic phenotype by a single oncogenic lesion. *Mol Cell* 3: 565–577.

Pfeifer A, Brandon EP, Kootstra N, Gage FH, Verma IM (2001): Delivery of the Cre recombinase by a self-deleting lentiviral vector: Efficient gene targeting in vivo . *Proc Natl Acad Sci USA* 98: 11,450–11,455.

Rabbitts TH (1994): Chromosomal translocations in human cancer. *Nature* 372: 143–149.

Ramirez-Solis R, Liu P, Bradley A (1995): Chromosome engineering in mice. *Nature* 378: 720–724.

Rane SG, Dubus P, Mettus RV, Galbreath EJ, Boden G, Reddy EP, Barbacid M (1999): Loss of Cdk4 expression causes insulin-deficient diabetes and Cdk4 activation results in beta-islet cell hyperplasia. *Nat Genet* 22: 44–52.

Rodriguez CI, Buchholz F, Galloway J, Sequerra R, Kasper J, Ayala R, Stewart AF, Dymecki SM (2000): High-efficiency deleter mice show that FLPe is an alternative to Cre-loxP. *Nat Genet* 25: 139–140.

Sauer B, Henderson N (1988): Site-specific DNA recombination in mammalian cells by the Cre recombinase of bacteriophage P1. *Proc Natl Acad Sci USA* 85: 5166–5170.

Schmidt EE, Taylor DS, Prigge JR, Barnett S, Capecchi MR (2000): Illegitimate Cre-dependent chromosome rearrangements in transgenic mouse spermatids. *Proc Natl Acad Sci USA* 97: 13,702–13,707.

Shibata H, Toyama K, Shioya H, Ito M, Hirota M, Hasegawa S, Matsumoto H, Takano H, Akiyama T, Toyoshima K, Kanamaru R, Kanegae Y, Saito I, Nakamura Y, Shiba K, Noda T (1997): Rapid colorectal adenoma formation initiated by conditional targeting of the Apc gene. *Science* 278: 120–123.

Silver DP, Livingston DM (2001): Self-excising retroviral vectors encoding the Cre recombinase overcome Cre-mediated cellular toxicity. *Mol Cell* 8: 233–243.

Smith AJ, De Sousa MA, Kwabi-Addo B, Heppell-Parton A, Impey H, Rabbitts P (1995): A site-directed chromosomal translocation induced in embryonic stem cells by Cre-loxP recombination. *Nat Genet* 9: 376–385.

Smith AJ, Xian J, Richardson M, Johnstone KA, Rabbitts PH (2002): Cre-loxP chromosome engineering of a targeted deletion in the mouse corresponding to the 3p21.3 region of homozygous loss in human tumours. *Oncogene* 21: 4521–4529.

Soriano P (1999): Generalized lacZ expression with the ROSA26 Cre reporter strain. *Nat Genet* 21: 70–71.

Soriano P, Cone RD, Mulligan RC, Jaenisch R (1986): Tissue-specific and ectopic expression of genes introduced into transgenic mice by retroviruses. *Science* 234: 1409–1413.

Sotillo R, Dubus P, Martin J, de la Cueva E, Ortega S, Malumbres M, Barbacid M (2001): Wide spectrum of tumors in knockin mice carrying a Cdk4 protein insensitive to INK4 inhibitors. *EMBO J* 20: 6637–6647.

Spring K, Ahangari F, Scott SP, Waring P, Purdie DM, Chen PC, Hourigan K, Ramsay J, McKinnoh PJ, Swift M, Lavin MF (2002): Mice heterozygous for mutation in Atm, the gene involved in ataxia-telangiectasia, have heightened susceptibility to cancer. *Nat Genet* 32: 185–190.

Spring K, Cross S, Li C, Watters D, Ben-Senior L, Waring P, Ahangari F, Lu SL, Chen P, Misko I, Paterson C, Kay G, Smorodinsky NI, Shiloh Y, Lavin MF (2001):Atm knockin mice harboring an in-frame deletion corresponding to the human ATM 7636del9 common mutation exhibit a variant phenotype. *Cancer Res* 1: 4561–4568.

Srivastava S, Zou ZQ, Pirollo K, Blattner W, Chang EH (1990): Germ-line transmission of a mutated *p53* gene in a cancer-prone family with Li-Fraumeni syndrome. *Nature* 348: 747–749.

Stambolic V, Suzuki A, de la Pompa JL, Brothers GM, Mirtsos C, Sasaki T, Ruland J, Penninger JM, Siderovski DP, Mak TW (1998): Negative regulation of PKB/Akt-dependent cell survival by the tumor suppressor PTEN. *Cell* 95: 29–39.

Stewart TA, Pattengale K, Leder P (1984): Spontaneous mammary adenocarcinomas in transgenic mice that carry and express MTV/myc fusion genes. *Cell* 38: 627–637.

Thomas KR, Capecchi MR (1987): Site-directed mutagenesis by gene targeting in mouse embryo-derived stem cells. *Cell* 51: 503–512.

Westervelt P, Lane AA, Pollock JL, Oldfather K, Holt MS, Zimonjic DB, Popescu NC, DiPersio JF, Ley TJ (2003): High-penetrance mouse model of acute promyelocytic leukemia with very low levels of PML-RAR alpha expression. *Blood* 102: 1857–1865.

Xu X, Wagner KU, Larson D, Weaver Z, Li C, Ried T, Hennighausen L, Wynshaw-Boris A, Deng CX (1999): Conditional mutation of Brca1 in mammary epithelial cells results in blunted ductal morphogenesis and tumour formation. *Nat Genet* 22: 37–43.

Yergeau DA, Hetherington CJ, Wang Q, Zhang P, Sharpe AH, Binder M, Marin-Padilla M, Tenen DG, Speck NA, Zhang DE (1997): Embryonic lethality and impairment of haematopoiesis in mice heterozygous for an AML1-ETO fusion gene. *Nat Genet* 15: 303–306.

Zamore PD, Tuschl T, Sharp PA, Bartel DP (2000): RNAi: Double-stranded RNA directs the ATP-dependent cleavage of mRNA at 21 to 23 nucleotide intervals. *Cell* 101: 25–33.

Zhu Y, Romero MI, Ghosh P, Ye Z, Charnay P, Rushing EJ, Marth JD, Parada LF (2001): Ablation of NF1 function in neurons induces abnormal development of cerebral cortex and reactive gliosis in the brain. *Genes Dev* 15: 859–876.

5

SOMATIC CELL GENE TRANSFER

LENE UHRBOM
Department of Genetics and Pathology, Rudbeck Laboratory, Uppsala University, Uppsala, Sweden
ERIC C. HOLLAND
Departments of Surgery (Neurosurgery), Neurology, and Cancer Biology and Genetics
Memorial Sloan Kettering Cancer Center, New York, New York

Experimentally induced, virally mediated oncogenesis has been used to study cancer for nearly 100 years. During this time, many of the fundamentals of cancer biology have been established. The technology differs from germline modification strategies in that the viral vectors that induce cancer in experimental animals achieve their goal by transferring genes into a limited number of somatic cells, usually after birth. New advances in these systems now support experimental designs aimed at assessing the effects of mixtures of genetic alterations, identification of the contribution of specific alterations on the histology of tumor cells, and the identification of cell-of-origin for certain cancers. These somatic cell gene transfer strategies are complementary to germline modification technologies in unraveling the complexities of cancer biology.

INTRODUCTION

Somatic cell gene transfer differs fundamentally from germline modification strategies. In the preceding chapter, germline modification strategies to generate gain-of-function and loss-of-function genetic alterations that are inherited in a Mendelian pattern have been discussed. By contrast, somatic cell genetic modifications result in an alteration within a subset of the cells in the animal and are not inherited (unless they occur in germ cells). However, a genetically modified somatic cell will pass on the mutation to its daughter cells during proliferation within a tumor. As a consequence, each animal undergoing somatic cell gene transfer is a unique event or a founder, and the altered cells are embedded within a primarily unmodified host. Observations of naturally occurring experiments of somatic cell gene transfer by DNA and RNA tumor viruses have greatly contributed to our understanding of how cancer arises and provided the basis for developing powerful molecular genetic tools.

DNA and RNA Tumor Viruses

There are two major types of viruses that can cause tumor formation in mammals. These are the DNA tumor viruses and the RNA tumor viruses. The DNA tumor viruses have two alternative life cycles. They can replicate lytically in a permissive cell and lead to killing of the infected cell. Alternatively, the viral genome may replicate episomally along with the host genome, without killing the host cell. One of the first DNA tumor virus to be isolated and studied for its tumor-forming ability was the mouse polyoma virus (PyV) (Gross, 1953). This virus belongs to the family *Papovaviridae*, which includes the two subfamilies of DNA tumor viruses, papillomaviruses and polyomaviruses. Excellent in-depth information on this subject is available in textbook form (Knipe et al., 2001). Papovaviruses have nonenveloped, double-stranded DNA genomes that replicate in the nucleus of nonpermissive host cells. The viral DNA is maintained as a stable multicopy plasmid replicated on average once per cell

Mouse Models of Human Cancer, edited by Eric C. Holland
ISBN 0-471-44460-X

cycle and is distributed equally to the daughter cells. The ability for these viruses to cause tumors is related to their expression of viral genes that give rise to proteins that disrupt the regulation of normal cellular processes.

SV40 (simian virus 40) is probably the most studied member of the polyomavirus family. All polyomaviruses have a common genome structure and express the T antigens, the large-T antigen (LT), the small-T antigen (ST), and the 17K T antigen (17KT). The mouse polyomavirus also expresses the middle-T antigen (MT). All the T antigens are produced by differential splicing of the early region transcript and were originally identified as antigens expressed in tumors induced by these viruses. The LT is necessary and often sufficient to induce viral tumorigenesis by SV40 (Saenz-Robles et al., 2001). Three regions within the LT have been shown to be required for tumorigenesis, the J domain, probably due to its interaction with hsc70, the LXCXE motif, necessary for interaction with retinoblastoma (RB) family members, and a bipartite domain in the carboxy-terminal end of the protein that binds to p53. The functional consequence of expressing LT is disruption of both p53 and RB functions that are involved in G1 cell cycle control, apoptosis, and DNA repair. The oncogenic activity of PyV is dependent on the expression of the MT. The resulting oncoprotein has the ability to simultaneously activate many signal transduction pathways, such as phosphoinositol 3-kinase (PI3K), phospholipase (PLCγ), SHC, and SRC, involved in proliferation and survival (Dilworth, 2002).

DNA tumor viruses act as dominant mutations of the cellular genome and have been very helpful as a tool to discover and study targets of human oncogenesis. Mutations of different cellular components of the pathways affected by the viral SV40 LT and PyV MT proteins are found in most human cancers. By studying the interaction between these and other viral oncoproteins and their cellular target proteins, some of the basic mechanisms of human oncogenesis have been revealed. However, as a delivery system for somatic cell gene transfer and as model system to identify genes involved in tumorigenesis a different viral family has had a much larger impact, namely the retroviruses.

Retroviruses are single-stranded RNA viruses that can be subdivided into seven groups based on sequence homologies. Five of these groups contain retroviruses with oncogenic potential. The seven groups include the avian sarcoma and leukosis viruses, the mammalian B-type viruses, the murine leukemia–related viruses, human T-cell leukemia–bovine leukemia viruses, and D-type viruses. Unlike a DNA tumor virus, a retrovirus integrates its genome into the genome of the host cell and is propagated to progeny cells in the form of proviral DNA. There are two mechanisms of retroviral oncogenesis. One is the acute transforming mechanism whereby the retrovirus carries a viral oncogene, v-*onc*. This is a cellular gene (c-*onc* or proto-oncogene) that has been incorporated into the retroviral genome (Bishop, 1983; Varmus, 1984). Appropriately regulated proto-oncogenes are not oncogenic under normal circumstances. However, when misexpressed by the retrovirus, these genes can cause transformation of cells once integrated into the host genome. A second method of oncogenesis is via retroviral insertional mutagenesis with slow-transforming retroviruses that occurs when the proviral DNA of a non-oncogene-carrying retrovirus is integrated into the genome in such manner as to either activate or disrupt the expression of cellular genes in its vicinity. Insights made from exploring these two mechanisms of retrovirus-mediated oncogenesis have led to an immense body of knowledge as well as the development of model systems that are of great importance for further advances in the field of cancer biology, and this will be discussed in more detail in the rest of this chapter.

IMPORTANT EVENTS IN THE HISTORY OF RETROVIRUSES

The history of retroviruses begins almost 100 years ago with the discovery of an agent causing neoplastic diseases in chickens. It was the Danish scientists Ellerman and Bang (1908) who showed that chicken leukosis was caused by a virus. A few years later Peyton Rous (1911) was able to demonstrate cell-free transmission of sarcomas. The experiment showed that cancer could be transferred between animals by a substance too small to see. One well-studied group of viruses causing tumor formation in birds and later figuring prominently in brain tumor modeling efforts has collectively been named avian sarcoma and leukosis virus (ASLV), and the specific virus discovered by Rous was termed Rous sarcoma virus (RSV). During the next few decades numerous retroviruses that were able to induce sarcomas, hematopoietic tumors, and carcinomas in many different animal species were isolated.

The next important step in retrovirology was the development of a method for quantitation of RSV. A first approach was done by infecting the chorioallantoic membrane of chick embryos with RSV. The small tumors that appeared could be approximately correlated with viral concentrations (Keogh, 1938). However, it was not until the use of cell culture became a robust tool that retroviruses could be precisely quantified. Temin and Rubin (1958) were able to show that RSV could cause oncogenic transformation of chick embryo fibroblasts *in vitro* and, furthermore, that each transformed focus was caused by a single viral particle. Each particle could transform a cell and its progeny and therefore appeared to alter the genetic content of the cell. The development of the focus assay was instrumental for many of the subsequent findings made in the retroviral field.

In 1964 Howard Temin presented the "provirus hypothesis" (Temin, 1964) to a skeptical scientific community. He proposed the generation of a DNA copy of the viral RNA genome that would then be integrated into the host genome. The hypothesis was accepted six years later with the discovery of the enzyme reverse transcriptase within retroviral particles (Baltimore, 1970). This seminal finding showed that retroviruses could convert RNA to DNA and caused a paradigm change in the way of thinking that extended far beyond the field of retrovirology.

The discovery of the host cell origin of the v-*src* oncogene of RSV in 1976 (Stehelin, 1976) concludes this brief overview of the important findings in the history of retrovirology. Subsequently, all retroviral oncogenes have been found to be acquisitions of cellular genes encoding components of signal transduction pathways with normal functions in cell growth control. This experiment of nature, to capture random genes from the genome and select for those that induce cancer when inappropriately expressed, emphasizes the importance of these pathways in oncogenesis.

RETROVIRUS LIFE CYCLE

A retroviral particle contains two identical single-stranded, positive-sense RNA molecules packaged together with the crucial enzymes reverse transcriptase (RT) and integrase (IN). The RNA genome contains the *gag*, *pol*, and *env* genes surrounded by a short direct repeat sequence (the R region) and U5 region at the 5′ end and U3 and R regions at the 3′ end of the RNA. Reverse transcription of the RNA followed by integration will result in a provirus which has two identical long terminal repeat (LTR) sequences at the 5′ and 3′ ends of its DNA containing the U3, R, and U5 regions (Fig. 5.1). Extensive

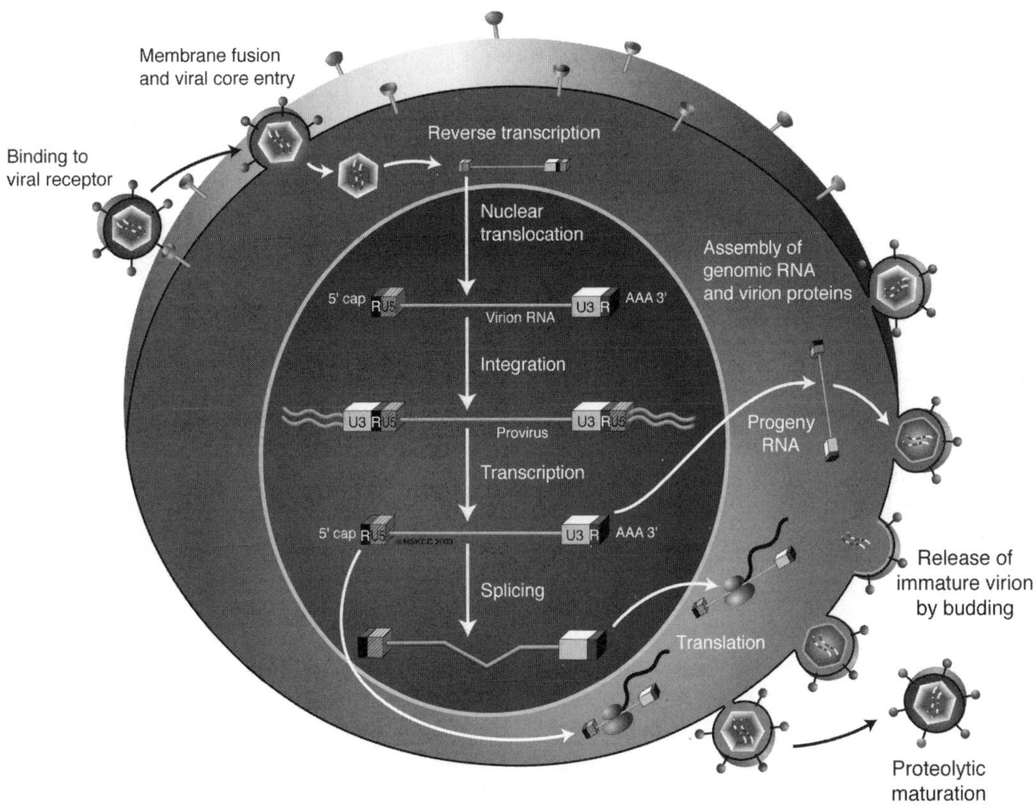

Figure 5.1. Retrovirus life cycle. Retroviral particles bind to a target cell via interaction between the viral envelope protein and the viral receptor. The virus is internalized and the viral single-strand RNA genome is converted to a double-strand DNA and subsequently inserted into the genome of the target cell, a stage referred to as a provirus. The provirus is transcribed and spliced into several RNA species that encode viral proteins and serve as the genome of the next generation of virus. These particles are constructed and bud off the infected cell and undergo proteolytical maturation to become the next generation of virus (Coffin et al., 1997).

and detailed knowledge on retroviral biology can be found in (Coffin et al., 1997).

Specificity of Viral Infection

The ability of a retrovirus to infect a host cell is determined by the expression of glycoproteins encoded by the *env* gene. These envelope glycoproteins are initially expressed on the surface of the infected cell and then incorporated into the virion membrane and bind to specific receptors of the host cell. The host range, or "tropism," of the virus is limited to those cells that express the receptors recognized by the viral envelope protein. An ecotropic virus is only infectious to cells of their natural host whereas xenotropic viruses may only infect cells other than their natural hosts. Polytropic and amphotropic viruses can infect both host cells and cells of other species. For several retroviruses the receptors have been identified. For the ASLVs there are five subgroups for the most studied viruses, A through E. The gene encoding the receptor for subgroup A ASLVs was designated *tv-a* for "tumor virus A" and has been identified as a low-density lipoprotein–related receptor (Bates et al., 1993), while the gene encoding the receptor for subgroup B was designated *tv-b* and has been identified as a tumor necrosis factor–like receptor (Brojatsch et al., 2000), etc. ASLV subgroups A and B are restricted to avian cells because mammalian cells do not express these proteins, while viruses belonging to subgroups C through E have the variable capability to also infect mammal cells (Vogt and Ishizaki, 1966). A separate example comes from the murine leukemia viruses (MLVs) that have been classified into four subgroups based on host range. In the ecotropic subgroup some of the most well-known murine retroviruses can be found, such as Moloney MLV and Friend MLV. The receptor responsible for binding the envelope glycoprotein has been mapped to the *Rec1* locus and encodes a cationic amino acid transporter (Albritton et al., 1989)

Viral Intracellular Life Cycle

After the viral particle has entered the host cell, the viral RNA is converted into a double-stranded linear DNA molecule by reverse transcriptase (Baltimore, 1970; Temin, 1970) while still in the core particle. Subsequently, the viral DNA is imported into the nucleus and inserted in the host genome in a process that requires cell division (Mulligan, 1993). The integration is mediated via the integrase gene product and results in a stably integrated provirus that is transmitted to progeny cells in a Mendelian manner. The transcription of proviral DNA is mediated by the host cell RNA polymerase II. The LTR contains all necessary control elements for retroviral transcription. The U3 region holds the enhancer and promoter elements. The R region at the 5′ LTR contains the transcription start site and at the 3′ LTR the RNA processing signal. Transcription of the provirus gives rise to one initial primary transcript. Some of these primary transcripts will be spliced to generate subgenomic alternative mRNAs. The viral transcripts will be exported out of the nucleus and translated into the various viral proteins which will be transported via the secretory pathway to the site of packaging where the virion particles will be assembled together with two copies of the unspliced viral mRNA and finally bud off the cell surface (Fig. 5.1)

PRODUCTION OF RETROVIRUSES FOR INFECTION OF MURINE CELLS

Murine Retroviruses

There are several ways to produce retroviruses for infection. One common method is to use a replication-defective retroviral vector which may contain the viral LTR sequences and packaging signal derived from a murine retrovirus but where the viral genes have been replaced with a multiple cloning site, a resistance marker gene, and an additional promoter (Fig. 5.2*a*). The inserted gene of interest can be driven either from the 5′ retroviral LTR or from the additional internal promoter. Due to a size limitation of the packaged viral genomic RNA (9–11 kb), deleting the endogenous viral genes allows the gene of interest to be larger than would be possible if the viral genes are present. A cell line producing replication-competent helper viruses or a packaging cell line is needed to produce infectious particles from such a replication-defective retroviral vector. The retroviral vector may be either transiently or stably transfected into the helper virus (Fig. 5.2*b*) producing cells or packaging cells (Fig. 5.2*c*). Alternatively, the retroviral construct may be transfected into cells followed by screening for high-expressing clones and subsequent infection of these clones with helper virus supernatant.

When rescuing the retroviral vector with helper viruses, the resulting retroviral supernatant will contain a mixture of replication-competent helper virus and replication-incompetent oncogene-carrying virus. This mixture will have the ability to spread the infection within the population of susceptible cells. If a packaging cell line that expresses the *env* and *gag-pol* in *trans* is used, only the defective viral genome will be packaged, resulting in the production of retroviral particles that are replication defective but infection competent. Such a design will allow for one round of infection but inability for the virus to spread. If an *env* gene with different tropism is used, retroviral supernatants able to infect cells of other species may be produced.

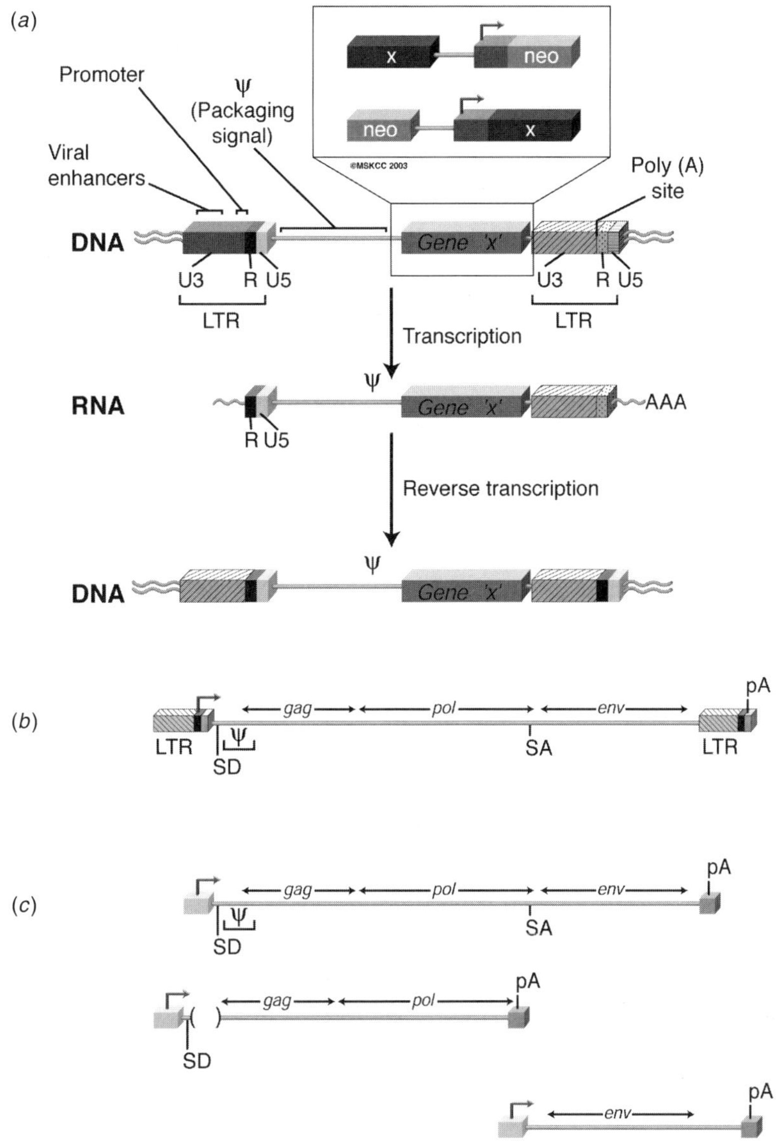

Figure 5.2. MLV-based retroviral vectors. (*a*) General structure of a retroviral vector with the viral genes replaced by a gene of interest (gene "X") and including the packaging sequence (ψ) that allows the RNA transcript to be included in the virion particles. (*a*) The absence of viral genes in the vector requires the production of these proteins in *trans* from either a wild-type helper virus or constructions that produce these proteins from exogenous promoters. (*c*) Removal of the packaging sequence and separating the viral genes in separate constructs reduce the chance of packaging the helper RNAs and recombination leading to the inadvertent generation of contaminating wild-type virus.

When using replication-competent retroviruses, a given cell can only be infected once with the same virus subgroup. This phenomenon is referred to as receptor interference or superinfection resistance (Levy, 1992). Interference occurs when the virally produced *env* gene product of an infected cell interacts with and blocks the receptor necessary for infection and thereby prevents subsequent superinfection by retroviruses of the same subgroup. Such a cell will, however, be susceptible to subsequent infection by retroviruses of different subgroups that use different receptors.

Avian Retroviruses, RCAS/TV-A

The use of avian retroviral vectors for gene transfer to mice requires that the mouse cells be genetically modified to express the receptor used by the retrovirus. The most commonly used system of this sort uses vectors based on the

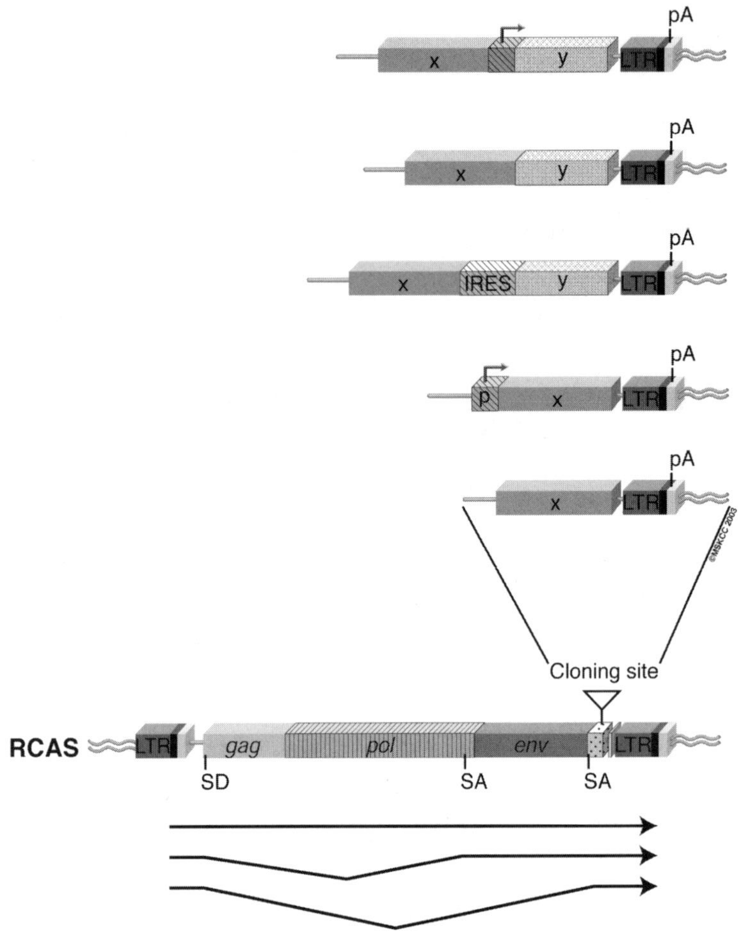

Figure 5.3. RCAS retroviral vectors. RCAS vectors contain the required viral genes allowing replication in avian cells. The gene or genes of interest are inserted 3′ to a splice acceptor site, allowing their expression from the LTR promoter. In addition, internal promoters, internal ribosome entry sites (IRESs), or protein fusions can be used to produce multiple proteins or modify gene expression from these vectors. These vectors will only infect mammalian cells if they are engineered to express the RCAS receptor, tv-a.

subgroup A avian leukosis virus (ALV) (Vogt and Ishizaki, 1965), referred to as RCAS vectors (Petropoulos and Hughes, 1991), and their receptor, tv-a (Bates et al., 1993). Since tv-a is not normally expressed on mammalian cells, the infection with RCAS is extremely low or nonexistent. However, when mammalian cells are engineered to express tv-a, they become highly susceptible to infection by RCAS vectors. A splice acceptor site in RCAS just upstream of the multiple cloning site (used to express v-*src* in RSV) leads to efficient processing of mRNA representing the inserted gene (Fig. 5.3). However, due to inappropriate splicing of primary viral mRNA and poor translation of viral proteins in mammalian cells, no viral particles are produced. As a result, these vectors do not spread from one tv-a-expressing mammalian cell to another. Furthermore, because the viral envelope protein is not produced in mammalian cells, receptor interference does not occur and there exists the possibility of infecting the same cell with several different oncogene-carrying viruses (Holland and Varmus, 1998). RCAS vectors can be engineered to express one or two genes or gene fusions or utilize internal promoters depending on the experimental needs (Fig. 5.3).

The absolute requirement for expression of *tv-a* by target cells allows the engineering of mice that limit infectability by RCAS vectors to specific cell types. If *tv-a* is expressed as a transgene from a promoter active only in specified cell types, the gene transfer achieved by RCAS vector infection will be limited to the cells that utilize the tissue-specific promoter (Fisher et al., 1999). In most cases, the transgenic expression of *tv-a* has no obvious phenotype, and so the only effect of *tv-a* expression is making specific cell types susceptible to gene transfer. Several mouse lines have been generated that express

Table 5.1. A list of transgenic mouse lines that express *TV-A* from tissue specific promoters

Promoter	Tissue Expression	Reference
α-Actin	Muscle	Federspiel et al., 1994
β-Actin	All tissues	Federspiel et al., 1996
Nestin	Brain	Holland et al., 1998a
GFAP	Brain	Holland and Varmus, 1998
GP-Ib-α	Platelets	Murphy and Leavitt, 1999
DCT	Melanoblasts	Dunn et al., 2000
K5	Epithelia	Orsulic et al., 2002
MMTV	Breast	Y. Li and H. E. Varmus, unpublished
Albumin	Liver	B. Lewis and H. E. Varmus, unpublished
Elastase	Pancreas	Lewis et al., 2003
SP-C	Lung	G. H. Fisher and H. E. Varmus, unpublished

GFAP (glial fibrillary acidic protein), GP-Ib-alpha (glycoprotein Ib), DCT (dopachrome tautomerase), K5 (keratin 5), MMTV (mouse mammary tumor virus), SP-C (surfactant protein C). Adopted from Pao et al. 2003.
HEV = Varmus, GHF = Fisher

tv-a in specific cell types (Table 5.1). The infectable, tv-a-expressing cell types represent the originating cells for tumors that arise from specified RCAS-mediated gene transfer events; comparison between two different *tv-a* transgenic lines essentially tests the effect of cell-of-origin for specific oncogenic stimuli. In some cases, *tv-a* can be expressed from promoters that are active in specific differentiation states of a particular cell lineage; comparison between such transgenic mouse lines tests the importance of the differentiation status of the cell-of-origin.

DELIVERY OF RETROVIRAL VECTORS

The main requirements for infection of cells by standard retroviral vectors are for the target cell to undergo at least one round of replication and to express the receptor for the virus. Following these requirements, there are two basic methods for the use of somatic cell gene transfer in the modeling of cancer. First, retroviral vectors can be used to transfer genes to cells in culture, and then these cells can be analyzed and secondarily transferred to mice and allowed to develop into tumors. Second, viruses can be used to directly infect cells *in vivo*.

For *in vitro* infection, the target cells are isolated and grown in culture. The cell-type specificity is provided by establishing a cell culture from specific organs prior to infection combined with using the RCAS/TV-A system to infect a subpopulation of cells limited by the promoter used to express tv-a. After infection, the cells are injected back into the same mouse or a different mouse of the same genetic strain. Alternatively, recipient mice can be severe combined immunodeficient (SCID) or nude to reduce the chance of rejection of the tumor. The advantage of this approach is the increased accessibility of cells to infection in some cases for organs/sites that are technically difficult to reach *in vivo*, for example at embryonic stages, further limiting the target cell type for infection. The main disadvantage is that the tumors are not actually induced *in vivo* and during the process of culturing cells *ex vivo* there may be a selection of a certain cell population or characteristic that is not representative of *in vivo* oncogenesis.

The strategy for *in vivo* delivery of retroviral vectors to the mouse is mainly determined by the accessibility of the target cells. In some cases, intraperitoneal injection of cell-free supernatant can result in a highly efficient infection of tv-a-expressing cells (Gaur et al., 2001). In other cases, using replication-competent systems, viral stocks are injected directly into target tissues (Uhrbom et al., 1998). Finally, for non-replication-competent systems such as RACS/TV-A, injection of producer cells directly into the target organ is required for sufficient gene transfer (Federspiel et al., 1994; Holland and Varmus, 1998) (Fig. 5.4). In the case of *tv-a* transgenic systems, it is likely that each cell type (specific *tv-a* transgenic line) may require a significant amount of effort to optimize techniques for gene transfer using vectors carrying marker genes.

As mentioned above, efficient infection with standard retroviral vectors requires at least one round of replication of the target cells. Choosing a time point for infection when target cells are mitotically active will therefore improve the efficiency of gene transfer. Alternatively, target cells may be stimulated to proliferate using hormones or other methods. Nondividing cells can be infected with vectors based on lentiviruses (Quinonez and Sutton, 2002). Most of the lentivirus vectors require

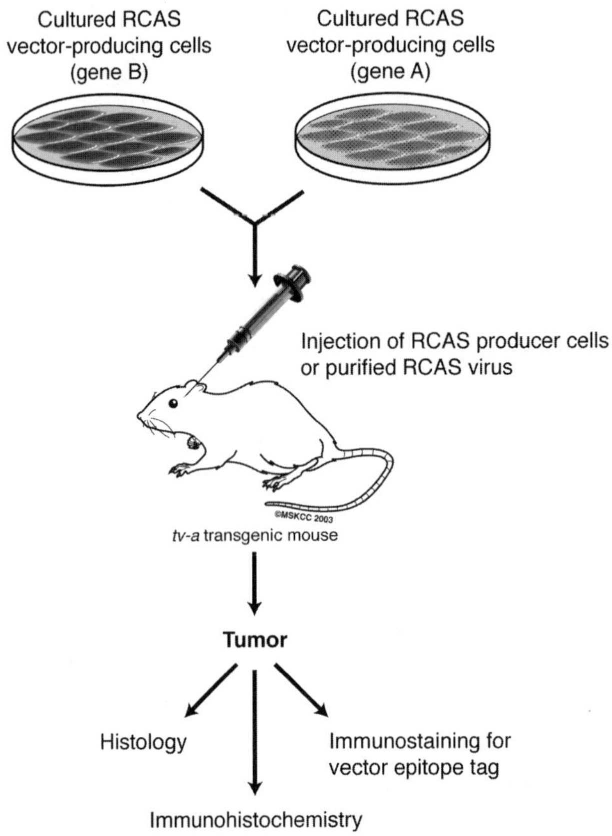

Figure 5.4. *In vivo* somatic cell gene transfer with RCAS vectors. One or more RCAS vectors are propagated in chicken fibroblasts as separate cultures. Either the cell supernatant or the producer cells themselves are injected into mice that express *tv-a* from a tissue-specific promoter allowing receptor-mediated gene transfer. Combinations of genes are transferred by mixing producer cells. Tumors that arise are analyzed for histology, gene expression, and presence of the viral vector.

pseudotyping of the vector with an envelope that targets the appropriate cell type. For MLV-based vectors lentivirus systems have been used to transfer genes *in vivo* to neurons (Naldini 1996), photoreceptors (Miyoshi, 1997), hepatocytes (Kafri, 1997), airway epithelial cells (Kobinger, 2001), and many other cell types (Quinonez and Sutton, 2002). Lentivirus vectors have also been applied to mice transgenic for *tv-a* expression. In this case lentivirus vectors have been pseudotyped with the ALV subgroup A envelope protein targeting infection to *tv-a*-expressing cells (Lewis et al., 2001). Unfortunately, the viral titer of these hybrid vectors is far lower than for MLV- or RCAS-based vectors, which currently limits their use.

INSERTIONAL MUTAGENESIS

Retroviruses that do not carry oncogenes also have the ability to cause cancer. Depending on where a retrovirus is inserted in the host cell genome, it may have different effects on the expression of surrounding cellular genes. These effects are dependent on the orientation and distance in relation to the cellular gene and can cause an upregulation or inhibition of the cellular gene function. The capacity of retroviruses to tag genes of importance for oncogenesis has been a powerful tool and led to the discovery of a long list of novel tumor loci (Jonkers and Berns, 1996; Mikkers et al., 2002; Mikkers and Berns, 2003; Neil and Cameron, 2002).

Mechanisms of Insertional Mutagenesis

Activation of cellular genes is caused by the enhancer or promoter activities of the proviral LTR sequences (Fig. 5.5). A promoter insertion occurs when the proviral DNA is inserted in the same transcriptional orientation upstream of the transcription start site of the target gene (Fig. 5.5*a*). Transcription of the cellular gene will be initiated from the viral LTR and will lead to a deregulated expression. Activation by enhancement is probably

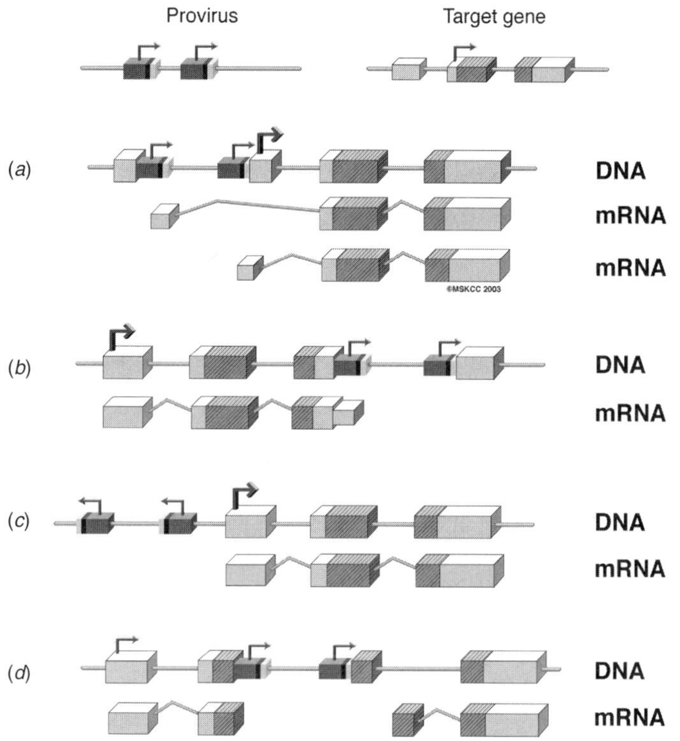

Figure 5.5. Mechanisms of insertional mutagenesis. There are multiple potential mechanisms for alteration of gene expression by the insertion of a provirus within or in the vicinity of an endogenous gene. (*a*) Increased expression can be achieved by insertion of a provirus into the promoter or 5′-untranslated region of a gene, in the same transcriptional orientation as the endogenous gene, which will cause a deregulated expression from the retroviral LTR. (*b*) A different mechanism to cause increased expression is when the retrovirus inserts in the 3′-untranslated region of the gene, causing removal of mRNA destabilization sequences. (*c*) The most common mechanisms of increased expression is when the inserted retoviral LTR causes an enhancer effect on an endogenous gene promoter. This can be achieved regardless of the orientation of the proviral DNA, and can work over long distances. (*d*) Proviral insertion can also cause interruption of gene expression by inserting its DNA into a gene, in either orientation.

the most common mechanism of insertional mutagenesis (Fig. 5.5*c*) because the orientation and distance between the proviral DNA and the target gene is more flexible. The proviral DNA may be inserted upstream or downstream of the target gene and the distance between the target gene and the provirus may be several hundred kilobase pairs. Yet another way of activating a cellular gene is by insertion of the virus into the gene, causing prematurely transcriptional termination and removal of mRNA-destabilizing sequences (Fig. 5.5*b*).

Inactivation of a tumor suppressor gene by insertional mutagenesis is not as common as insertional activation. This is probably due to the fact that in most cases both alleles have to be mutated. The inactivation of a given allele may occur when the proviral DNA is inserted into a gene, causing transcriptional termination (Fig. 5.5*d*). Such an insertion may also cause the expression of a truncated protein which may be either activating or dominant negative.

Several variations on this theme have been used recently in leukemia and lymphoma models. One method has used MLV retroviruses to infect and generate myeloid leukemias in wild-type mice and identified common integration sites (Suzuki et al., 2002). Several of these genes have been validated as causally related to tumor formation. A second variation on this approach combines loss of a tumor suppressor gene with insertional mutagenesis. *Ink4a-Arf−/−* mice were infected with MLV retroviruses and found to acquire leukemia much sooner than either loss of *Ink4a-Arf* or insertional mutagenesis alone, implying cooperative effects between the two genetic alterations. Again, many genes were identified that had multiple integrations, implying causality (Lund et al., 2002). Finally, mice transgenic for *myc* expression and deleted for *pim1* and *pim2*, genes that cooperate with *myc* in lymphoma formation, were infected with MLV viruses. The goal of this experiment was to identify genes that encoded proteins that were either downstream of *pim* kinase, or equivalent to it in the cooperation with *myc* during lymphomagenesis. Of the many genes that were identified, the *pim* family member *pim3* was tagged, supporting the utility of this experimental approach (Mikkers et al., 2002).

These approaches make the assumption that the viruses integrate randomly. If integration site bias exists, these approaches will identify many genes not related to cancer. In addition, these approaches assume that the nearest gene to the integration site is the one related to the formation of cancer, which also may not be true in all cases. It is clear that validation of the causal nature of the identified genes by germline or somatic cell modeling strategies is a required step in such a project (Neil and Cameron, 2002). Finally, the use of insertional mutagenesis strategies requires large numbers of infected cells per animal. This experimental approach is ideally suited for liquid tumors but more difficult for solid tumors unless replication-competent viruses are used. The classic story of MMTV infection identifying genes leading to breast cancer, such as *Wnt 1*, *FGF3*, and others (Nusse and Varmus, 1982), is less likely to be repeated in an RCAS/TV-A system where only a few cells are infected.

RETROVIRAL TRANSFER MODELS

Investigation of the potential roles of proto-oncogenes and tumor suppressor genes in defined animal model systems led to the development of retroviral-based mouse models of cancer. As mentioned above, there are three main strategies for using retroviruses as a means to stably transfer a specific gene into a somatic cell in a mouse. One is to infect the target cells *ex vivo* and transplant them back into the animal, the second is to inject retroviruses directly into the mouse, and the third is to use genetically modified mice where a defined cell population is susceptible to retroviral infection (RCAS/TV-A). A few examples from each of these strategies will be described in order to illuminate their different benefits.

Transplantation of Fetal Cells Infected with Engineered Retroviruses

The strategy to induce brain tumors by culturing fetal brain cells was first performed more than 10 years ago in rats (Wiestler et al., 1989). Recently this strategy was repeated to induce brain tumors in mice (Bachoo et al., 2002). Primary neural stem cells were isolated from the subventricular zone of E13.5 mouse embryos. These cells were infected *in vitro* and subsequently transplanted into the brain of SCID mice to score for tumorigenic ability. The study compared the tumorigenic ability between different combinations of mutations and the differentiation status of the infected cells. A second recent example of *ex vivo* gene transfer uses the RCAS/TV-A system to model ovarian cancer (Orsulic et al., 2002). In this work, *TV-A* is driven from the keratin 5 promoter, and the ovaries are removed, grown in culture, and infected with RCAS vectors. Although the keratin 5 promoter is active in many epithelial cells throughout the body, the *ex vivo* nature of the infection results in limiting gene transfer to ovarian epithelial cells. The cells are then replaced into the peritoneal cavity of nude mice. Specific combinations of gain of function and loss of function lead to the formation of tumors with histologic characteristics of human ovarian cancer (Orsulic et al., 2002) (see Chapter 10). A final example of *ex vivo* gene transfer combines the tetracycline induction system (described in the preceding chapter) with the RCAS/TV-A system. In this case the reverse tet transcriptional transactivator (rtTA) was encoded in an RCAS vector. The rtTA protein can bind to the tet operator (tet-o) and allows transcription of the gene cloned downstream of the tet-o in the presence of doxycycline (dox). Mouse embryo fibroblasts (MEFs) from beta-actin promoter tv-a and tet-o-Kras4bG12D transgenic mice were infected with RCAS-rtTA. The MEFs were then placed in the flanks of nude mice. Tumors formed in the presence of dox, which regressed upon removal of dox. (Pao et al., 2003).

Direct Infection in Vivo with MMLV-Based Tumor Formation Systems

There are two examples of induction of tumors by modified MMLV-based retroviruses *in vivo*. In both models the human *PDGFB* gene has been inserted into a Moloney MLV vector. In the first experiment, the virus was injected subcutaneously into newborn mice. This generated fibrosarcomas at the site of injection which were of polyclonal or oligoclonal origin (Pech et al., 1989). A gratifying parallel between human and mouse was subsequently shown when all specimens of a specific sarcoma type, dermatofibrosarcoma protuberance, have a translocation between *PDGFB* and a collagen gene *COL1A1* (Simon et al., 1997), leading to an overexpression of *PDGFB* (Greco et al., 1998). A second *PDGFB*/MMLV-induced tumor model generates brain tumors. Here, the cDNA of human *PDGFB* was inserted in a minimal retroviral vector containing essentially only the packaging sequence and LTRs of Moloney murine leukemia virus (MMLV). This vector was transfected into NIH 3T3 fibroblast cells that were subsequently infected with wild-type MMLV (helper virus). From these cells viral supernatants were collected which contained both recombinant *PDGFB*/MMLV and helper virus. Newborn mice were injected intracerebrally, and highly malignant tumors developed after a latency of 13–29 weeks (Uhrbom et al., 1998). Tumors grew invasively and diffusely, and most showed features of human glioblastomas or primitive neuroectodermal tumors (PNETs). In human gliomas platelet-derived growth factor (PDGF) ligands and receptors are often coexpressed, suggesting the presence of autocrine growth factor stimulation in these cells (Maxwell et al., 1990). The parallels between human and mouse gliomas is illustrated by the amplification of the *PDGFRα* gene found in approximately 10% of gliomas (Fleming et al., 1992). These alterations can be found in all grades of gliomas, suggesting that PDGF plays an early role in the development of these tumors.

The receptor for MMLV is expressed in many different cell types in the mouse. This fact together with the use of helper virus should allow efficient spread of both the recombinant virus and the helper virus upon infection. However, the infection appears to be relatively local as tumors mostly arise at the site of injection and because *PDGFB*-expressing cells never have been detected outside the tumors with *in situ* hybridization (Uhrbom et al., 1998).

Direct Gene Transfer in Vivo Using RCAS Vectors

Although there are many unpublished studies using RCAS vectors to induce tumors by direct gene transfer *in vivo* into various cell types, the majority of published work is from studies in the brain. The system was originally designed to model gliomas, and more time has been spent optimizing this system for brain tumor work than any other organ site to date. Several features of the potential of the RCAS system can be illustrated by the work in the brain thus far.

Determining the Cell-of-Origin for Experimental Cancers

The identity of the cell-of-origin can be defined in these systems because the expression of *tv-a* is required for gene transfer and is expressed as a transgene from promoters that are active only in specific cell types, As an example, two mouse lines have been designed that express *tv-a* from the nestin promoter and the glial fibrillary acidic protein GFAP promoter. The nestin promoter is expressed in central nervous system (CNS) progenitor cells (Lendahl et al., 1990) and the *Ntv-a* mouse supports gene transfer to these cells (Holland et al., 1998a). By contrast, in postnatal mice the GFAP promoter is active primarily in astrocytes and the *Gtv-a* mice support gene transfer to these cells (Holland and Varmus 1998). Once a cell is infected with an RCAS vector, the oncogene is expressed from the constitutively active viral LTR. The overall effect is that the gene transfer event is tissue specific but any subsequent alterations in the differentiation status of the cell do not affect the expression of the oncogene. This system allows for a lineage tracing analysis to be performed concurrently with the formation of tumors (Dai et al., 2001; Dai and Holland, 2003).

Combinations of Oncogenes

Mammalian cells infected with RCAS vectors do not produce viral envelope proteins; therefore they do not generate restriction to superinfection and will allow multiple vectors encoding different genes to infect given *tv-a*-expressing cells. Because of this feature, combining multiple genes and analysis of multiple permutations of genes is possible and significantly easier with the RCAS/TV-A system than for germline modification strategies. In germline strategies, combinations are achieved by breeding individual mouse lines together. In the RCAS/TV-A system, a single mouse line is used to support gene transfer, and the combinations are generated by mixing viruses together (Holland et al., 2000). As many as three genes have been independently demonstrated as essential for the formation of gliomas. The use of a library of vectors to simultaneously infect mice and select for the combination that leads to tumor formation has been discussed at length and would demonstrate the strength of this system at its extreme. However, to date, tumors arising from such complex mixtures of vectors have not been reported.

Effects of Specific Oncogenes on Character of Genetically Engineered Tumors

Each gene transferred in RCAS experiments is generally transferred by an individual vector; infecting with combinations of vectors will result in some cells with only infection by single vectors rather than the combination. In some cases, a single gene is sufficient to generate tumors. In such cases, the tumor arises from all cells while some of these tumor cells additionally have the second vector and others not. If cells infected with the second vector are detectable by immunostaining for an epitope tag on the virally transduced gene, such an experimental design will show what effect the second vector has on the histologic appearance of the tumor cells (Uhrbom et al., 2002). Another example of this design is the mixture of PDGF and Akt vectors in the formation of gliomas. The PDGF vector is sufficient to generate gliomas of an oligodendroglioma histology. Cells additionally infected with the Akt vector (as identified by a hemaglutinin epitope tag on the virally transduced constitutively active Akt) show an astrocytoma phenotype. These data imply that activation of the Akt pathway *in vivo* converts cells from one glioma histology to another (Dai et al., submitted).

Bicistronic Vectors

The above experimental design using multiple vectors each carrying one gene can show cooperative effects in the formation of tumors if the effects promote growth and are therefore selected for. However, if the interaction between genes inhibits tumor formation, such combined experiments are much less likely to be useful because there are always some cells that are infected by a single vector. If the double infections do not grow into tumors, the mice will develop tumors from the singly infected cells. In this case it is better to generate a single vector that encodes two genes. The first cistron encodes that oncogene and the second cistron encodes the modifying gene. The second gene can be translated either from an internal ribosome entry site (IRES) within the same mRNA or from an actual second promoter leading to a second mRNA (Fig. 5.3). One published example of this strategy is the demonstration that expression of the GTPase-activating protein (GAP) domain of p190RhoGAP blocks the formation of PDGF-driven tumors. In this case the GAP domain was used rather than the whole p190 protein because of size limitations of the viral vector (Wolf et al., 2003).

REFERENCES

Albritton LM, et al. (1989): A putative murine ecotropic retrovirus receptor gene encodes a multiple membrane-spanning protein and confers susceptibility to virus infection. *Cell* 57(4):659–666.

Bachoo RM, et al. (2002): Epidermal growth factor receptor and *Ink4a/Arf*: Convergent mechanisms governing terminal differentiation and transformation along the neural stem cell to astrocyte axis. *Cancer Cell* 1:269–277.

Baltimore D (1970): RNA-dependent DNA polymerase in virions of RNA tumour viruses. *Nature* 226(252):1209–1211.

Bates P, et al. (1993): A receptor for subgroup A Rous sarcoma virus is related to the low density lipoprotein receptor. *Cell* 74(6):1043–1051.

Bishop JM (1983): Cellular oncogenes and retroviruses. *Annu Rev Biochem* 52:301–354.

Brojatsch J, et al. (2000): TVB receptors for cytopathic and noncytopathic subgroups of avian leukosis viruses are functional death receptors. *J Virol* 74(24):11490–11494.

Coffin JM, et al. (1997): *Retroviruses*. Cold Spring Harbor Laboratory, Plainview, NY.

Dai C, et al. (2001): PDGF autocrine stimulation dedifferentiates cultured astrocytes and induces oligodendrogliomas and oligoastrocytomas from neural progenitors and astrocytes *in vivo*. *Genes Dev* 15(15):1913–1925.

Dai C, Holland EC (2003): Astrocyte differentiation states and glioma formation. *Cancer J* 9(2):72–81.

Dai C, Shih A, Lyustikman Y, Hu X, Fuller GN, Rosenblum MK, Holland EC: The characteristics of astrocytomas and oligodendrogliomas are caused by two distinct and interchangeable signaling formats, 2004 (submitted).

Dilworth SM (2002): Polyoma virus middle T antigen and its role in identifying cancer-related molecules. *Nat Rev Cancer* 2(12):951–956.

Dunn KJ, Williams BO, Li Y, Pavan WJ (2000): Neural crest-directed gene transfer demonstrates Wntl role in melanocyte expansion and differentiation during mouse development. *Proc Natl Acad Sci USA* 97:10050–10055.

Ellerman V, Bang O (1908): Experimentelle leukämie bei hühnern. *Zentralbl Bakteriol Parasitenkd Infectionskr Hyg Abt Orig* 46:595–609.

Federspiel MJ, et al. (1994): A system for tissue-specific gene targeting: Transgenic mice susceptible to subgroup A avian leukosis virus-based retroviral vectors. *Proc Natl Acad Sci USA* 91(23):11241–11245.

Federspiel MJ, Swing DA, Eagleson B, Reid SW, Hughes SH (1996): Expression of transduced genes in mice generated by infecting blastocysts with avian leukosis virus-based retroviral vectors. *Proc Natl Acad Sci USA* 93:4931–4936.

Fisher GH, Varmus HE (unpublished).

Fisher GH, et al. (1999): Development of a flexible and specific gene delivery system for production of murine tumor models. *Oncogene* 18(38):5253–5260.

Fleming TP, et al. (1992): Amplification and/or overexpression of platelet-derived growth factor receptors and epidermal growth factor receptor in human glial tumors. *Cancer Res* 52(16):4550–4553.

Gaur M, et al. (2001): Characterization of Mpl mutants using primary megakaryocyte-lineage cells from mpl(-/-) mice: A new system for Mpl structure-function studies. *Blood* 97(6):1653–1661.

Greco A, et al. (1998): Transforming activity of the chimeric sequence formed by the fusion of collagen gene COL1A1 and the platelet-derived growth factor B-chain gene in dermatofibrosarcoma protuberans. *Oncogene* 17(10):1313–1319.

Gross LA (1953): A filterable agent, recovered from Akt leukemic extracts, causing salivary gland carcinomas in C3H mice. *Proc Soc Exp Biol Med* 83:414–421.

Holland EC, et al. (2000): Combined activation of Ras and Akt in neural progenitors induces glioblastoma formation in mice. *Nat Genet* 25(1):55–57.

Holland EC, et al. (1998a): A constitutively active epidermal growth factor receptor cooperates with disruption of G1 cell-cycle arrest pathways to induce glioma-like lesions in mice. *Genes Dev* 12(23):3675–3685.

Holland EC, et al. (1998b): Modeling mutations in the G1 arrest pathway in human gliomas: Overexpression of CDK4 but not loss of Ink4a-Arf induces hyperploidy in cultured mouse astrocytes. *Genes Dev* 12(23):3644–3649.

Holland EC, Varmus HE (1998): Basic fibroblast growth factor induces cell migration and proliferation after glia-specific gene transfer in mice. *Proc Natl Acad Sci USA* 95(3):1218–1223.

Jonkers J, Berns A (1996): Retroviral insertional mutagenesis as a strategy to identify cancer genes. *Biochim Biophys Acta* 1287(1):29–57.

Kafri T, Blomer U, Peterson DA, Gage FH, Verma IM (1997): Sustained expression of genes delivered directly into liver and muscle by lentiviral vectors. *Nat Genet* 17:314–317.

Keogh EV (1938): Ectodermal lesions produced by the virus of Rous sarcoma. *Br J Exp Pathol* 19:1–9.

Knipe DM, et al. (2001): *Fundamental Virology*. Lippincott Williams & Wilkins, Philadelphia.

Kobinger GP, Weiner DJ, Yu QC, Wilson JM (2001): Filovirus-pseudotyped lentiviral vector can efficiently and stably transduce airway epithelia *in vivo*. *Nat Biotechnol* 19:225–230.

Lendahl U, et al. (1990): CNS stem cells express a new class of intermediate filament protein. *Cell* 60(4):585–595.

Levy JA (1992): *The Retroviridae*. Plenum, New York.

Lewis BC, et al. (2001): Development of an avian leukosis-sarcoma virus subgroup A pseudotyped lentiviral vector. *J Virol* 75(19):9339–9344.

Lewis BC, Klimstra DS, Varmus HE (2003): The c-myc and PyMT oncogenes induce different tumor types in a somatic mouse model for pancreatic cancer. *Genes Dev* 17:3127–3138.

Lewis B, Varmus HE (unpublished).

Li Y, Varmus HE (unpublished).

Lund AH, et al. (2002): Genome-wide retroviral insertional tagging of genes involved in cancer in Cdkn2a-deficient mice. *Nat Genet* 32(1):160–165.

Maxwell M, et al. (1990): Coexpression of platelet-derived growth factor (PDGF) and PDGF-receptor genes by primary human astrocytomas may contribute to their development and maintenance. *J Clin Invest* 86(1):131–140.

Mikkers H, et al. (2002): High-throughput retroviral tagging to identify components of specific signaling pathways in cancer. *Nat Genet* 32(1):153–159.

Mikkers H, Berns A (2003): Retroviral insertional mutagenesis: Tagging cancer pathways. *Adv Cancer Res* 88:53–99.

Miyoshi H, Takahashi M, Gage FH, Verma IM (1997): Stable and efficient gene transfer into the retina using an HIV-based lentiviral vector. *Proc Natl Acad Sci USA* 94:10319–10323.

Mulligan RC (1993): The basic science of gene therapy. *Science* 260(5110):926–932.

Murphy GJ, Leavitt AD (1999): A model for studying megakaryocyte development and biology. *Proc Natl Acad Sci USA* 96:3065–3070.

Naldini L, Blomer U, Gallay P, Ory D, Mulligan R, Gage FH, Verma IM, Trono D (1996): *In vivo* gene delivery and stable transduction of nondividing cells by a lentiviral vector. *Science* 272:263–267.

Neil JC, Cameron ER (2002): Retroviral insertion sites and cancer: Fountain of all knowledge? *Cancer Cell* 2(4):253–255.

Nusse R, Varmus HE (1982): Many tumors induced by the mouse mammary tumor virus contain a provirus integrated in the same region of the host genome. *Cell* 31(1):99–109.

Orsulic S, et al. (2002): Induction of ovarian cancer by defined multiple genetic changes in a mouse model system. *Cancer Cell* 1(1):53–62.

Pao W, et al. (2003): Use of avian retroviral vectors to introduce transcriptional regulators into mammalian cells for analyses of tumor maintenance. *Proc Natl Acad Sci USA* 100(15):8764–8769.

Pech M, et al. (1989): Generation of fibrosarcomas *in vivo* by a retrovirus that expresses the normal B chain of platelet-derived growth factor and mimics the alternative splice pattern of the v-sis oncogene. *Proc Natl Acad Sci USA* 86(8):2693–2697.

Petropoulos CJ, Hughes SH (1991): Replication-competent retrovirus vectors for the transfer and expression of gene cassettes in avian cells. *J Virol* 65(7):3728–3737.

Quinonez R, Sutton RE (2002): Lentiviral vectors for gene delivery into cells. *DNA Cell Biol* 21(12):937–951.

Rous P (1911): A sarcoma of the fowl transmissible by an agent separable from the tumor cells. *J Epx Med* 13:397–411.

Saenz-Robles MT, et al. (2001): Transforming functions of Simian Virus 40. *Oncogene* 20(54):7899–7907.

Simon MP, et al. (1997): Deregulation of the platelet-derived growth factor B-chain gene via fusion with collagen gene COL1A1 in dermatofibrosarcoma protuberans and giant-cell fibroblastoma. *Nat Genet* 15(1):95–98.

Stehelin D, et al. (1976): DNA related to the transforming gene(s) of avian sarcoma viruses is present in normal avian DNA. *Nature* 260(5547):170–173.

Suzuki T, et al. (2002): New genes involved in cancer identified by retroviral tagging. *Nat Genet* 32(1):166–174.

Temin HM (1964): Nature of the provirus of Rous sarcoma. *Natl Cancer Inst Monogr* 17:557–570.

Temin HM, Mizutani S (1970): RNA-dependent DNA polymerase in virions of Rous sarcoma virus. *Nature* 226:1211–1213.

Temin HM, Rubin H (1958): Characteristics of an assay for Rous sarcoma virus and Rous sarcoma cells in tissue culture. *Virology* 6:669–688.

Uhrbom L, et al. (1998): Induction of brain tumors in mice using a recombinant platelet-derived growth factor B-chain retrovirus. *Cancer Res* 58(23):5275–5279.

Uhrbom L, Dai C, Celestino JC, Rosenblum MK, Fuller GN, Holland EC (2002): Ink4a-Arf loss cooperates with KRas activation in astrocytes and neural progenitors to generate glioblastomas of various morphologies depending on activated Akt. *Cancer Res* 62:5551–5558.

Varmus HE (1984): The molecular genetics of cellular oncogenes. *Annu Rev Genet* 18:553–612.

Vogt PK, Ishizaki R (1965): Reciprocal patterns of genetic resistance to avian tumor viruses in two lines of chickens. *Virology* 26(4):664–672.

Vogt PK, Ishizaki R (1966): Patterns of viral interference in the avian leukosis and sarcoma complex. *Virology* 30(3):368–374.

Wiestler OD, et al. (1989): Tumour induction in fetal brain transplants exposed to the viral oncogenes polyoma middle T and v-src. *IARC Sci Publ* 96:267–274.

Wolf RM, et al. (2003): p190RhoGAP can act to inhibit PDGF-induced gliomas in mice: a putative tumor suppressor encoded on human chromosome 19q13.3. *Genes Dev* 17(4):476–487.

PART II

ORGAN SITE–SPECIFIC MOUSE MODELS OF HUMAN CANCER

6

LUNG CANCER

JONATHAN M. KURIE
Department of Thoracic and Head and Neck Medical Oncology, University of Texas M. D. Anderson Cancer Center, Houston, Texas
PARVIZ MINOO
Department of Pediatrics, Women and Children's Hospital, University of Southern California, Los Angeles, California
STEPHEN S. HECHT
Department of Laboratory Medicine and Pathology, University of Minnesota Cancer Center, Minneapolis, Minnesota
ALEXANDER NIKITIN
Department of Biomedical Sciences, Cornell University, Ithaca, New York
FRANCO J. DEMAYO
Department of Molecular and Cellular Biology, Baylor College of Medicine, Houston, Texas

Lung cancer most often stems from a single environmental cause but has diverse genetic, biochemical, and histologic characteristics. This diversity has made the basic causes of lung cancer difficult to surmise. Current mouse models support genetic alterations in lung epithelial cell transformation, but these models do not recapitulate the diversity of histologic appearance, site of origin, or natural history of lung cancer in humans. Advances in mouse model technology provide hope that the significance of the myriad genetic and biochemical alterations in human lung cancer will eventually be understood and that future models will better represent their human counterpart. Progress in this area will enhance both the basic understanding of lung tumorigenesis and the speed with which the anticancer efficacy of investigational drugs can be investigated in preclinical trials.

HUMAN LUNG CANCER

Incidence and Epidemiology

Lung cancer is the most common form of cancer worldwide and is the most common cause of cancer-related death in the United States (Greenlee et al., 2000; Parkin et al., 2001). In 2000, new cases diagnosed numbered 164,100 in the United States and 1.2 million cases worldwide and lung cancer accounted for an estimated 156,900 fatalities in the United States. Tobacco consumption is linked to 80% to 90% of all lung cancer cases. Changes in smoking trends in the United States have led to decreases in lung cancer incidence among men and increases among women. Lung cancer accounted for 25% of all cancer deaths in women in 2000 and is the primary contributor to the increase in total cancer deaths in women in recent years. Among men, lung cancer incidence decreased from 1990 to 1996 by approximately 2.6% per year. Smoking cessation campaigns waged by public health organizations in the United States have led to increases in the number of former smokers. As the number of former smokers increases, the percentage of patients with lung cancer who are former smokers has increased, now accounting for approximately 50% of all lung cancer cases (Burns, 2000).

Histopathology

Lung cancer has been classified into four major histologic subtypes: small-cell lung cancer (SCLC) and the three

Mouse Models of Human Cancer, edited by Eric C. Holland
ISBN 0-471-44460-X

main non-small-cell lung cancer (NSCLC) subtypes (squamous, large cell, and adenocarcinoma). Squamous cell tumors arise in central airways, whereas adenocarcinomas typically arise peripherally. Large-cell carcinoma is thought to be a poorly differentiated form of the other types of NSCLCs. Adenocarcinoma is at present the most common form of lung cancer.

Cells of Origin

Temporal, histologic, and genetic evidence suggests that bronchial dysplasia and carcinoma in situ are precursors of squamous cell cancer and that atypical adenomatous hyperplasia is a precursor of adenocarcinoma. Although no lung cancer progenitor cells have been identified, adenocarcinomas express proteins specific to type II alveolar cells, Clara cells, or mucin-producing cells, suggesting that they originate from progenitor cells in the epithelium of bronchioles or alveoli, and SCLC cells express neuropeptides and have histologic features that suggest a neuroendocrine origin (Betz et al., 1995; Broers et al., 1992; Gazdar et al., 1990; Jensen-Taubman et al., 1998; Kitamura et al, 1997; Rozengurt, 1999; Shijubo et al., 1995). Bronchioloalveolar carcinoma is a subtype of adenocarcinoma distinguished by its alveolar growth pattern; it too expresses type II alveolar cell markers.

Treatment and Prognosis

Treatment of lung cancer is based on disease stage at diagnosis (Minna et al., 2000; Mountain, 1997). In stage I and II disease, tumors are isolated to the lung parenchyma and hilar lymph nodes. Tumors in stage III NSCLC invade mediastinal structures or have metastasized to mediastinal lymph nodes or pleural fluid. Disease in stage IV NSCLC has metastasized to distant sites within or outside the lung. SCLC is considered to be limited if it is confined within the chest or extensive if it has spread to extrathoracic sites. Surgical resection is the treatment of choice for early-stage (i.e., stage I or II) NSCLC. Standard treatment for stage III disease is combined chemotherapy and radiotherapy, and stage IV disease is typically treated with chemotherapy alone.

The likelihood of survival is inversely related to disease stage at diagnosis (Minna et al., 2000; Mountain, 1997). In patients with NSCLC, treatment produces long-term survival in 40% to 70% of patients with stage I or II disease, 20% to 30% of those with stage III disease, and essentially none of those with stage IV disease. In patients with limited-stage SCLC, long-term survival rates range from 10% to 25%, whereas only 1% to 2% of patients with extensive-stage disease achieve long-term survival. Given the poor outcome of treatment for advanced lung cancer, efforts have moved toward prevention. Thus far, large-scale prevention trials using vitamin A derivatives to reverse histologic evidence of bronchial premalignancy or to reduce lung cancer incidence among individuals at high risk of developing lung cancer have demonstrated no therapeutic benefit (Hong and Sporn, 1997).

No improvement in prognosis for people who develop lung cancer has been seen over the past three decades. Such improvements will require a better understanding of lung cancer biology to identify the crucial events required for the transformation of human bronchial epithelial cells and the maintenance of lung cancer cell survival.

Molecular Biology

Lung cancer cells contain a vast array of genetic alterations, including DNA point mutations, DNA hypermethylation, gene deletions and amplifications, and changes in chromosome copy number (Minna et al, 2002). The presence of many such changes in bronchial premalignant lesions and their tendency to cluster around specific genes or chromosomal regions provides indirect evidence that these events have a role in lung tumorigenesis. Some of the affected genes encode proteins that are components of signal transduction pathways required for cellular proliferation, survival, and invasion.

Genes with potential oncogenic or mitogenic roles are activated in lung cancer through DNA point mutations, amplification, or overexpression (Minna et al, 2002). Examples of such genes include c-*myc*, K-*ras*, telomerase catalytic component (*hTERT*), epidermal growth factor receptor (*EGFR*), and other *ErbB* family members, gastrin-releasing peptide, bombesin, cyclin D1, cyclin-dependent kinase-4, *bcl-2*, and the *p110* catalytic component of phosphatidylinositol 3-kinase. Genes in which tumor suppressor activity is deleted, mutated, or transcriptionally silenced in lung cancer include *p53*, retinoblastoma gene, *p16* and *p27* cyclin-dependent kinase inhibitors, *PTEN* lipid phosphatase, retinoic acid receptor (*RAR*)-β, human 8-oxoguanine DNA glycosylase (*hOGG1*), insulinlike growth factor binding protein 3 (*IGFBP3*), death-associated protein kinase (*DAPK*), O6-methylguanine-DNA methyltransferase (*MGMT*), and adenomatous polyposis coli (*APC*).

Several novel genes with tumor suppressor activity in the chromosome 3p region (*FHIT*, *FUS1*, *RASSF1A*, *SEMA3B*, and *ROBO1*) were recently identified as often being deleted in lung cancer (Minna et al., 2002). In addition to deletional events, these genes are transcriptionally silenced in lung cancer cells by promoter methylation. Loss of chromosome 3p has also been detected in bronchial dysplasia, squamous metaplasia, and histologically normal areas of bronchial epithelium from smokers, making it the earliest genetic event known in lung tumorigenesis.

Implications for Mouse Modeling

Given the rapid advances in the identification of genetic events in lung cancer cells, the events that have important roles in lung tumorigenesis must be sorted out from those that form genetic "background noise." Mouse models present a unique means of addressing this issue. Although some progress has been made, more complex mouse models are needed to test the role of genetic events, in isolation and in combination, to determine whether a single event is sufficient to induce lung tumorigenesis or, alternatively, whether lung tumorigenesis is a stochastic process requiring multiple events to occur in a specific sequence or combination.

LUNG MORPHOGENESIS

Understanding lung morphogenesis is crucial to understanding lung tumorigenesis for several reasons. First, lung tumorigenesis is, in part, a disease of epithelial differentiation and as such is regulated by the genes and biochemical pathways that are activated during lung morphogenesis. Second, understanding the ontogeny of the cellular compartments within the lung is essential for clarifying which progenitor cells within the bronchial epithelium are the precursors of the major histologic subtypes of lung cancer. Third, for some situations the development of effective mouse models of lung cancer requires tissue-specific expression of genes of interest. Achieving this goal requires first identifying which genes are expressed, and in which cellular compartments, during lung development.

The lung is a highly efficient gas exchange organ. To achieve this purpose, the lung primordium, a group of endodermal cells from the ventral wall of the anterior foregut, undergoes complex morphogenetic changes that are directed by cell–cell interactions with the splanchnic mesenchyme. Lung functionality requires that the structure, distribution, and organization of highly specialized epithelial cell types along the proximodistal axis are under strict developmental control. In general, developmental processes can be viewed as an integration of "environmental signaling" into changes in cell behavior. This integration function takes place through the activity of the morphogenetic transcription factors (Ten Have-Opbroek, 1981).

In the lung, signaling is mediated by peptide growth factors such as fibroblast growth factors (FGFs) and bone morphogenetic protein 4 (BMP-4) and by morphogens such as sonic hedgehog (SHH) and the retinoids. Morphogenetic signaling originating from the cell's environment is translated into cellular behavior through changes in the activity of transcription factors, of which NKX2.1, the hepatocyte nuclear factor 3 (HNF-3), and GATA6 have been the most well-studied. The role of NKX2.1 as a transcriptional regulator of lung development is somewhat better understood than that of the others. Because transcription factors by definition modulate the activity of downstream target genes, the potential targets are presumably those that encode components of the extracellular matrix (ECM), cell–cell receptors, and cell–ECM receptors. The sum of these reactions leads to changes in cell behaviors such as proliferation, migration, and differentiation, which are at the heart of morphogenesis (Monaghan et al., 1993).

Specification of Lung Primordium

The lung primordium originates from the ventral wall of the anterior foregut (Ten Have-Opbroek, 1981). A "hierarchy" of transcriptional regulators, such as members of the HNF-3 gene family, may participate in temporal and spatial specification of the lung primordium. Later in development, HNF-3α and HNF-3β persist in fully differentiated adult bronchial epithelium (Monaghan et al., 1993) and participate in regulation of lung-specific genes such as *SpB* (Bohinski et al., 1994). Little direct evidence exists regarding the signals that specify the lung primordium, although no evidence of lung structure can be ascertained in compound homozygous *Gli2/Gli3* mutants (Motoyama et al., 1998). Three GLI transcription factors are expressed in the lung mesenchyme with distinct temporal and spatial specificities. SHH, which is known to activate the *Gli*, is expressed in the pulmonary epithelium and is thought to interact with the cellular receptor *patched* on mesenchymal cells. However, the effects of *Gli2/Gli3* compound mutation seem to be independent of SHH signaling, because *Shh*(−/−) mutant embryos are born with lung parenchyma, although that parenchyma is abnormal (Pepicelli et al., 1998).

Morphogenesis and Cellular Organization along the Trachea

In E9.5 mouse embryos, the expression pattern of a homeodomain transcription factor, NKX2.1 (also known as TTF-1 or TEBP), coincides with the dorsoventral boundary within the anterior foregut, clearly distinguishing the pulmonary primordium from that of the esophagus (Minoo et al., 1999). Little to no information is available, however, regarding the mechanisms that govern the outgrowth of the lung primordium into a tracheal tube. Even in the most severe experimentally induced phenotypes of lung abnormalities (i.e., those resulting from disruptions of the FGF-10 pathway), a tracheal tube nevertheless forms (Min et al., 1998; Sekine et al., 1999). This observation suggests that the trachea and distal lung may have different, distinct origins. In lungs of

Nkx2.1(−/−) mice, the trachea is markedly foreshortened and remains fused to the esophagus, resembling a relatively rare anatomic deformity in humans known as a complete tracheoesophageal cleft. Similar tracheoesophageal phenotypes have been described for mice with *Shh*(−/−), *Gli*2(−/−)/*Gli*3(+/−), and compound *RAR*-α1(−/−)/*RAR*-β2(−/−) mutations (Mendelsohn et al., 1994; Motoyama et al., 1998; Pepicelli et al., 1998). The relationship, if any, among retinoids, *Shh*, and *Nkx2.1* in the process of tracheoesophageal septation remains to be deciphered. *Shh* is expressed in the lungs of *Nkx2.1*(−/−) mice, and conversely *Nkx2.1* can be detected in the lungs of *Shh*(−/−) mice, suggesting independent but perhaps functionally parallel roles for the two loci in tracheal morphogenesis.

The trachea is lined with several phenotypically distinct cell types, including ciliated and nonciliated columnar epithelial cells and many secretory cells such as serous and goblet cells. Recently, Perl and colleagues (2002) identified restriction points in the lineage of progenitors of trachea and distal lung during lung development. In support of these findings, we identified a *cis*-active DNA element that can distinguish two cell lineages in transgenic mice carrying various lengths of the 5′ regulatory regions of the *Nkx2.1* gene linked to LacZ (Pan et al., 2004). These are the first indications that the fate of highly specialized cell types for distal versus proximal structures may be determined very early during embryogenesis. However, the specific signaling and transcriptional mediators necessary for differentiation and spatial organization of the various tracheal (proximal) cell types remain largely unknown. Hepatocyte nuclear factor 3/forkhead homologue 4 (HFH-4), a member of the winged-helix family of transcription factors expressed in the bronchial epithelium, seems to be necessary for differentiation of ciliated epithelial cells in the lung (Chen et al., 1998). *Hfh-4* seems to be not only necessary but also sufficient for differentiation of columnar ciliated epithelial cells, as ectopic expression of *Hfh-4* results in the appearance of a subset of columnar cells in the distal lung, where normally only cuboidal type II and squamous type I epithelial cells abound (Tichelaar et al., 1999).

Branching Morphogenesis and Differentiation of Specialized Epithelial Cells

Distal lung development and differentiation of highly specialized cell types along the proximodistal axis requires branching morphogenesis. Branching morphogenesis depends on a set of complex developmental interactions between two distinct tissues—the endodermally derived epithelium and the mesoderm. Each tissue contributes to lung development by fine tuning the spatial and temporal pattern of gene expression for a distinct array of signaling, transcriptional, and ECM-related molecules. In general, the nature of the interactions is based on the exchange of developmental cues that establish positional information at the level of individual cells.

Epithelial-Based Signaling and Transcription. The critical players in epithelial–mesenchymal interactions include several signaling molecules and transcription factors (Monaghan et al., 1993). In the lung epithelium, several of the major signaling molecules with established roles are WNT, BMP-4, SHH, and platelet-derived growth factor (PDGF). *Wnt7b* is expressed in the airway epithelium and is required for proper lung mesenchymal growth and vascular development (Shu et al., 2002). *Wnt5a* is expressed in both the lung epithelium and the mesenchyme. *Wnt5a*-mediated signaling controls lung development through regulating cell proliferation and epithelial–mesenchymal communications (Li et al., 2002). Disruption of *Wnt5a* causes increased proliferation and abnormal structural development without affecting cellular differentiation (Li et al., 2002). Ectopic overexpression of *Wnt5a* leads to changes in the expression of several key regulators of lung morphogenesis, including *Fgf-10* (Fig. 6.1) and *Bmp-4* (Li and Minoo, unpublished). Expression of *Bmp-4,* a major participant in distal lung morphogenesis, is confined to the growing tip of the branching epithelium. Targeted disruption of *Bmp-4* results in embryonic lethality before E10 and is therefore uninformative for lung morphogenesis (Lawson et al., 1999). Ectopic overexpression of *noggin*, an inhibitor of BMP-4, and *Alk-6*, a dominant-negative BMP-4 receptor, result in proximalization of the lung (Bellusci et al., 1996; Weaver et al., 1999). *Bmp-4* is thought to control epithelial cell proliferation during lung development (Weaver et al., 1999).

Another important signaling molecule originating from the lung epithelium is SHH. Disruption of SHH signaling

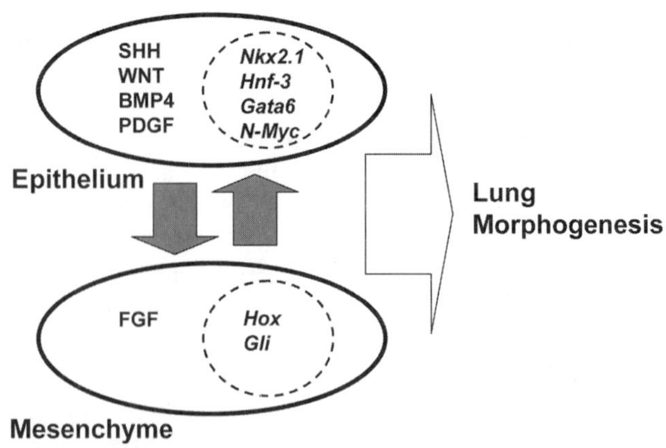

Figure 6.1. Key mediators of epithelial–mesenchymal interactions during lung morphogenesis.

perturbs proximodistal lung morphogenesis. In midgestation, *Shh* is expressed uniformly in the lung epithelium. The paracrine role of *Shh* in cell–cell communication during branching morphogenesis is supported by the localization of its receptor on the mesenchymal cells. Homozygous disruption of *Shh* causes profound structural abnormalities during lung development; however, limited distal-lung morphogenesis and cellular differentiation nevertheless can be observed, as surfactant protein–expressing cells can be detected in the lungs of *Shh*(−/−) mice (Pepicelli et al., 1998).

The identity and function of epithelial-based transcription factors are relatively well known. The transcription factors HNF-3, HFH-4, GATA6, N-MYC, and NKX2.1 are expressed in the lung epithelium, and although the evidence for HNF-3 and GATA6 is indirect, all of these factors seem to be key regulators of epithelial–mesenchymal interactions. In mutations that allow the lung to progress beyond the mainstem bronchi, disruption of *Nkx2.1* leads to the most severe defects in morphogenesis. In *Nkx2.1*(−/−) embryos, no distal lung structure develops beyond the secondary or tertiary bronchi (Minoo et al., 1999). *Nkx2.1* is a homeodomain transcriptional regulator that is expressed in the lung, thyroid, and ventral forebrain. In cell culture studies, *Nkx2.1* activates the promoters of lung-specific surfactant protein genes such as *SpB* (Bohinski et al., 1994). Suppression of NKX2.1 by antisense oligonucleotides (Minoo et al., 1995) or disruption by homologous recombination in vivo (Kimura et al., 1996) results in severely abnormal lungs and postnatal death. Morphologically, lung epithelium in *Nkx2.1*(−/−) mice is composed solely of stratified and pseudostratified columnar epithelial cells, which are characteristically found in the proximal lung. Absence of distal lung epithelial–specific differentiation markers (e.g., *SpC or SpA*), combined with evidence of ciliary differentiation and synthesis of acid mucopolysaccharides (usually found in differentiated tracheal and bronchial epithelia), further suggests that lungs of *Nkx2.1*(−/−) mice are arrested in a proximal state. Importantly, the expression pattern of regionally specific, differentially spliced forms of vascular endothelial growth factor (*Vegf*) transcripts is uniquely proximal in lungs from *Nkx2.1*(−/−) mice, providing further evidence that distal lung morphogenesis is inhibited in the absence of *Nkx2.1* (Yuan et al., 2000). Attempts to identify the downstream targets of NKX2.1 revealed that expression of *Bmp-4*, α-integrins, and collagen type IV was reduced or absent in lungs of *Nkx2.1*(−/−) mice (Yuan et al., 2000). Recently the tight junction protein claudin-18 was identified as being downstream of *Nkx2.1* (Niimi et al., 2001). Whether the abnormalities observed in lungs of *Nkx2.1*(−/−) mice can be ascribed to the absence of claudin-18 remains to be determined.

The N-*myc* proto-oncogene encodes a basic helix–loop–helix–leucine zipper transcription factor thought to be involved in cellular proliferation and differentiation. N-*myc* is expressed in the lung epithelium, and reception or interpretation of mesenchymal signaling by epithelial cells during lung branching morphogenesis depends on the normal activity of this transcription factor. Mouse embryos with targeted disruption of N-*myc* develop abnormal, dysfunctional lungs (Stanton et al., 1992).

Gata5 and *Gata6* are expressed in the lung, but expression of *Gata6* coincides with that of *Nkx2.1* and persists in adult alveolar type II cells (Shaw-White et al., 1999). Disruption of the *Gata6* locus in mice leads to early embryonic lethality before emergence of the lung primordium (Morrisey et al., 1998). Also, tagging stem cells in *Gata6*(−/−) wild-type chimeras with lacZ revealed that the tag appeared in every embryonic tissue except for the bronchial epithelium, indicating that *Gata6* plays a crucial role in the development of pulmonary epithelial cell lineages (Morrisey et al., 1998).

Mesenchyme-Based Signaling and Transcription. That the mesenchyme of the lung controls epithelial cell behavior has been known since classical embryologic approaches were applied to the study of lung morphogenesis. The lung mesenchyme is thought to arise from the splanchnic mesenchyme and possibly the neural crest. Recombination studies show that in nascent lung rudiments the mesenchyme instructs the behavior of the epithelium such that the tracheal epithelium, which normally does not branch, can be induced to undergo branching if exposed to peripheral mesenchyme (Shannon et al., 1998). In the lung mesenchyme, the most critical mediator of branching morphogenesis is FGF-10. FGFs are made by the mesenchyme but act through their cognate receptors found on epithelial cells. Targeted disruptions for *Fgf-10* results in profound interruptions in lung morphogenesis, specifically lack of lung structures distal to mainstem bronchi (Min et al., 1998; Sekine et al., 1999).

The role of FGFs in directing epithelial morphogenesis is probably related to their ability to direct epithelial cell proliferation as well as cellular differentiation. Transcriptional factors activated by FGF-induced cell proliferation such as N-MYC and C-FOS have been identified and probably play similar roles in the proliferation of lung epithelial and mesenchymal cells. The morphoregulatory transcription factors activated in the lung by FGF-10 remain elusive. In limb morphogenesis, ectopic application of FGF leads to activation of *SnR,* which encodes a vertebrate homologue of the *Drosophila* zinc finger transcription factor *Snail* (Issac et al., 2000). Other transcription factors, in particular *Tbx-4* and *Tbx-5,* are also activated in response to FGF signaling in early limb specification.*Tbx* transcription factors are expressed in the lung epithelium (Campbell et al., 1995), but their relationship to FGF signaling has not been explored. Modulation of BMP-4 levels in the lung epithelium by FGF-10 has been proposed as a mechanism affecting epithelial cell proliferation (Weaver et al., 1999).

Information on the identity and the role of mesenchymally based transcription factors in the lung is scant. In addition to members of the *Gli* family, which are downstream of SHH signaling, transcription factors in the *Hox* cluster are known to be expressed in the lung mesenchyme. *Hoxa-5* (−/−) mice die postnatally of abnormalities associated with morphogenesis of the trachea and the lung, including decreased pulmonary surfactant production (a marker of cell differentiation) and abnormal expression of epithelial transcription factors *Nkx2.1, Hnf-3*, and *Myc* (Aubin et al., 1997).

Alveologenesis and Differentiation of Other Highly Specialized Epithelial Cells. From a physiologic standpoint, alveologenesis, the process by which the lung saccular compartments are subdivided into distinct alveoli, is a key step in maturation of the pulmonary system and establishment of effective gas exchange. The FGF and PDGF signaling pathways were recently implicated in alveologenesis (Bostrom et al., 1996; Weinstein et al., 1998), but the transcription factors downstream of these pathways remain unknown. Only limited information is available regarding the signaling mechanisms and the transcriptional factors that participate in alveologenesis. It is reasonable to speculate that many of the factors involved in early lung morphogenesis that continue to be expressed in the mature lung, such as *Hnf-3, GATA6*, and *Nkx2.1,* could have roles in this process. Direct demonstration of the involvement of these factors in alveologenesis requires the generation of conditional mutants, a topic described later in this chapter.

A specific molecule may have distinctly different roles during early versus late lung development. For example, the FGF pathway, abrogation of which leads to disruption of early lung morphogenesis, also participates in alveologenesis. All four FGF receptors are expressed in neonatal lungs (Weinstein et al., 1998). Mice with mutations at both the *FgfR-3* and the *FgfR-4* loci are normal at birth but die prematurely with symptoms of pulmonary insufficiency and growth retardation. The lungs of *FgfR-3/FgfR-4* knockout mice are phenotypically consistent with a complete block in alveologenesis (Weinstein et al., 1998). Another important clue is related to PDGF signaling. PDGF ligands interact with various cell types through two types of receptor tyrosine kinases, PDGF-Rα and PDGF-Rβ (Bostrom et al., 1996). PDGF-A is expressed in the embryonic lung by epithelial cells; its receptors are found on the mesenchymal cells surrounding the branching epithelium, a configuration that establishes a paracrine signaling pathway with important implications for epithelial–mesenchymal crosstalk. Targeted disruption of PDGF-A disrupts alveolar septation and causes early neonatal death (Bostrom et al., 1996).

The cellular lineage of specialized lung epithelial cells remains unknown. Cell-specific markers are currently available for at least three cell types within the conducting airways and the alveoli. Clara cells, which are nonciliated columnar epithelial cells that populate the conducting airways, can be distinguished by the presence of Clara cell–specific protein (CCSP). CCSP is not specific to lung tissue and is thought to be required for innate immunity (Zheng et al., 1999). Alveolar type II cells, which constitute 15% of the total lung cell population but cover only about 5% of the alveolar surface area, synthesize pulmonary surfactant proteins and lipids. Alveolar type II cells can be distinguished by their synthesis of surfactant protein genes, the transcription of which in turn depends on (or is at least affected by) *Nkx2.1. Nkx2.1* and surfactant protein C (*SpC*) are expressed in nearly all epithelial cells in early lung development. Tracheal and bronchial epithelial cells gradually lose *Nkx2.1* expression later in development, as distal lung morphogenesis ensues (Yuan et al., 2000). However, *Nkx2.1* expression is selectively maintained in parallel in a subset of cells that seem to be the progenitors of alveolar type II cells. In the mature lung, both *Nkx2.1* and *SpC* are confined to alveolar type II cells. Alveolar type I cells cover most of the alveolar surface area and are critical in gas exchange. Aquaporin-5 and T1α can be used as cell-specific markers of alveolar type I cells, although neither is lung specific (Kreda et al., 2001; Williams et al., 1996).

Transgenic ablation studies suggest that alveolar type II cells are the progenitors of alveolar type I cells (Korfhagen et al., 1990). This finding is consistent with the observed dynamics of alveolar type II and type I cell populations during injury to the alveolar surface area. Under these conditions, alveolar type II cells undergo proliferation and redifferentiation into alveolar type I cells to reestablish the functional homeostasis of the alveolar surface (Evans et al., 1973). During the repair of lung injuries, rapid cell turnover occurs in response to the destruction of the alveolar epithelium, and "intermediate" or "hyperplastic" alveolar type II cells that exhibit morphologic features of alveolar type II and type I phenotypes have been observed. Similar intermediate cells can be seen in rat lungs treated with keratinocyte growth factor (Fehrenbach et al., 1999). Transitions from alveolar type II to type I phenotypes can also be observed in cell culture (Danto et al., 1995). In vitro, the transition from one phenotype to another exhibits significant plasticity in that epithelial cells that have already undergone transition toward the alveolar type I phenotype can be induced to reexpress the phenotypic markers of alveolar type II cells by the addition of peptide growth factors or selection of specific ECM proteins (Danto et al., 1995). The latter findings support the concept that the developmental mechanisms involved in determining cell lineage and cellular differentiation along specific pathways may be governed by cell–cell interactions that would include signaling molecules. Although transcriptional regulation of alveolar type I cell differentiation remains largely unknown, lungs of transgenic mice made to overexpress a GATA6-engrailed fusion protein are devoid of alveolar type I cells, implicating GATA6 in the differentiation process (Yang et al., 2002).

Conclusions

In the lung, the two compartments of mesoderm and endodermally derived epithelium, each of which has distinct sets of signaling molecules and transcription factors, interact to implement the highly complex pattern of structural growth and cellular differentiation involved in lung development. In deciphering the details of lung development, valuable insights have been gained from studies of transgenic manipulation of specific genes. Such studies have also provided intriguing clues regarding potential overlap and interactions among various morphoregulatory molecules (e.g., *Shh, RAR, Gli,* and *Nkx2.1*) in tracheal development. Neither the signaling nor the transcriptional factors that are required for lung development are unique to the lung. Future efforts therefore must be aimed at elucidating the unique combinatorial interactions of these "nonspecific" components that culminate in the highly specific pathway of lung development.

CARCINOGEN-INDUCED MODELS OF LUNG CANCER

Susceptibility to induction of lung tumors by carcinogens parallels susceptibility to the development of spontaneous lung tumors. Spontaneous lung tumor development in mice has been reviewed elsewhere (Shimkin and Stoner, 1975; Stoner, 1998). Among inbred mouse strains, the strain A mouse is the most susceptible to lung tumor development without carcinogen treatment; the next most susceptible is strain SWR (Swiss mice). Strains BALB/c, CR, O20, and DD have intermediate susceptibility, whereas strains CBA and C3H are relatively resistant and strains DBA and C57BL are the most resistant to spontaneous lung tumor development. In the most susceptible strain A mice, lung tumors may appear in some animals at age 3–4 months and the incidence rises to 100% by 18–24 months. In the most resistant strain C57BL mice, lung tumors rarely occur even in old animals (Shimkin and Stoner, 1975; Stoner, 1991). Lung tumor multiplicity is another indicator of susceptibility. Findings from Stoner's laboratory demonstrate that strain A mice have an average of 0.2–0.4 lung tumors per mouse at age 30–32 weeks (Stoner, 1991), findings consistent with our own (Hecht et al., 1989, 2002), and older mice may have as many as two or more tumors per mouse. Among strain A mice, the incidence and multiplicity of spontaneous lung tumors have remained relatively stable over the past 50–60 years (Stoner, 1998). Lung tumors in strain A mice arise in the peripheral lung and are readily identified and counted using a dissecting microscope.

Applications for Evaluating Carcinogens

The strain A mouse has been used extensively in carcinogen evaluation (Maronpot et al., 1986; Shimkin and Stoner, 1975; Stoner, 1998). Carcinogen-induced neoplasia in mice progresses from hyperplasia to adenoma to carcinoma in a well-documented sequence (Belinsky et al., 1992; Foley et al., 1991; Stoner, 1998). Tumors are readily induced by a variety of genotoxic carcinogens, including polycyclic aromatic hydrocarbons (PAHs), *N*-nitroso compounds, carbamates, nitrogen mustards, halides, metal salts, aflatoxins, and hydrazines, among others. Single doses often produce multiple tumors within several months. For example, single doses of the PAHs dibenz[*a,h*]anthracene (1–10 mg/kg) or benz[*a*]pyrene (BaP) (50–200 mg/kg) can induce up to 30 lung tumors per mouse 8 months after intraperitoneal injection (Nesnow et al., 1995). Single doses of the nitrosamine 4-(methylnitrosamino)-1-(3-pyridyl)-1-butanone (NNK) (50–100 mg/kg) induce 4–12 lung tumors per mouse 4 months after intraperitoneal injection (Hecht et al., 1989). Responses generated by particular carcinogens are highly reproducible. For example, ethyl carbamate (urethane) produces one lung tumor per milligram of carcinogen dose, and this response has not changed during the 40 years in which this compound has been tested (Stoner and Shimkin, 1982). The tumorigenic response to carcinogens should be compared against the 0.2–0.4 lung tumors that develop per untreated or vehicle control–treated strain A mouse. The range of responses to different classes of carcinogens varies widely. Strain A mice, for example, are far more sensitive to PAH and *N*-nitroso compounds than to aromatic amines (Stoner, 1998). The relative simplicity of the strain A mouse bioassay is a significant advantage in preliminary evaluations of potential carcinogenicity of test compounds. The assay is rapid, operationally simple, and relatively inexpensive. Lung tumors per mouse, the main endpoint for evaluations of tumorigenic activity, can be readily quantified, making statistical evaluations straightforward. Representative samples of lung adenomas can be confirmed by histopathologic examination. Histologic features of the tumors have been reviewed elsewhere (Stoner, 1998).

Recently, the strain A mouse bioassay was applied to evaluate the effects of environmental tobacco smoke (Witschi et al., 2000). Witschi and co-workers demonstrated that a mixture of sidestream (89%) and mainstream (11%) cigarette smoke reproducibly caused a small but significant increase in lung tumor multiplicity in mice exposed for five months and then given a four-month resting period. Tumorigenicity in this model apparently resulted from the gas-phase constituents of cigarette smoke, which conflicts with results obtained in Syrian golden hamster inhalation studies that demonstrated laryngeal tumor induction by the particulate phase of smoke (IARC Monographs, 1986). This assay nevertheless provides a convenient method of studying the lung tumorigenicity of inhaled tobacco smoke, which is difficult and expensive to do by other methods.

Although the operational simplicity and relatively low cost of the strain A mouse bioassay are distinct advantages in assessing potentially carcinogenic compounds, the results must be evaluated cautiously. Activity in this system does not necessarily signal lung tumorigenicity of a compound in other strains or species. The strain A mouse is highly susceptible to lung tumor induction, and virtually any genotoxic carcinogen will produce a response in the lung. For example, *N*-nitrosodimethylamine and aflatoxin B_1, well-established hepatocarcinogens, strongly induce lung tumors in strain A mice. Results of the strain A mouse bioassay have also been compared with those of two-year carcinogenicity studies in rats and mice conducted by the U.S. National Toxicology Program, and they do not agree well. However, approximately 50% of the compounds chosen for that comparison were aromatic amines, a class of carcinogens to which the strain A mouse is not very responsive (Maronpot et al., 1986).

The CD-1 mouse, an outbred line derived from the original colony of Swiss mice started in 1926, has been widely used for evaluation of carcinogens (Manenti et al., 2003). Spontaneous lung tumor formation in CD-1 mice averages 21.8% in animals 19–24 months old as compared with nearly 100% in strain A mice (Manenti et al., 2003). The incidence of lung tumors in CD-1 mice is similar to that of BALB/c mice and higher than that of the B6C3F1 mice (8.6% to 11.4%) used in the National Toxicology Program (Manenti et al., 2003). Susceptibility to lung tumor induction by carcinogens parallels the spontaneous incidence of lung tumors. Vinyl chloride, as an example, induces lung tumors in CD-1 and A/J mice but not in B6C3F1 mice (Manenti et al., 2003). A major determinant of susceptibility to lung tumorigenesis in strain A and CD-1 mice is the pulmonary adenoma susceptibility 1 (*Pas1*) locus (Manenti et al., 2003). Between 95% and 98% of CD-1 mice carry genetic polymorphisms of the K-*ras2* and *Pthlh* genes, which are tightly linked with the *Pas1* locus (Manenti et al., 2003). Fine mapping of *Pas1–3* quantitative trait loci has also been reported (Wang et al., 2003).

Newborn CD-1 and Swiss-Webster mice have been extensively used to evaluate the tumorigenicity of metabolites of PAHs (Conney, 1982). PAH diol epoxides, which were implicated in DNA-binding studies as ultimate carcinogens of PAHs, were first shown to have higher tumorigenic activities than their parent compounds by assays in newborn mice. Depending on stereochemical features, certain PAH diol epoxides injected on the first, eighth, or fifteenth day of life produced high incidences of lung tumors in these mice (Conney, 1982). The robust induction of lung tumors by PAH diol epoxides in newborn mice is consistent with other studies demonstrating lung tumor induction in newborn mice by various carcinogens (Anderson et al., 1991; Toth, 1968). Female Swiss mice treated topically with *N*-nitroso-bis(2-chloroethyl)urea develop tumors with squamous histology, which is the only model of squamous lung tumorigenesis developed to date (Rehm, 1991).

The U.S. National Toxicology Program uses a standard protocol for testing chemicals for carcinogenic activity in B6C3F1 mice and F344/N rats (available at http://ntp-server.niehs.nih.gov). These studies are designed to characterize and evaluate toxicologic potential, including carcinogenic activity. Chemicals are selected for testing based on human exposure, level of production, and chemical structure. Compounds judged to have clear evidence of lung carcinogenicity in B6C3F1 mice include benzene, choroprene, ethylene oxide, nitromethane, 1,3-butadiene, glycidol, benzofuran, tetranitromethane, bis(2-chloro-1-methylethyl)ether, and dichloromethane.

Mechanistic Studies of Metabolic Activation of Carcinogens

Ethyl carbamate (urethane) is a prototypic lung tumorigen in mice that has high activity in strain A and CD-1 mice. The Millers were the first to propose and demonstrate that vinyl carbamate is a proximate carcinogen of ethyl carbamate (Fig. 6.2*a*) (Dahl et al., 1978, 1980). Vinyl carbamate is more tumorigenic to the mouse lung than ethyl carbamate in strain A and CD-1 mice (Dahl et al., 1978, 1980). The oxidation of ethyl carbamate to vinyl carbamate and vinyl carbamate epoxide is catalyzed by cytochrome P450 2E1 (Guengerich and Kim, 1991). Vinyl carbamate epoxide, which is mutagenic and carcinogenic, reacts with DNA to produce a variety of adducts, including 1,N^6-etheno-dA and 3,N^4-etheno-dC (Fig. 6.2*a*) (Leithauser et al., 1990; Park et al., 1993). These adducts have miscoding properties and have been

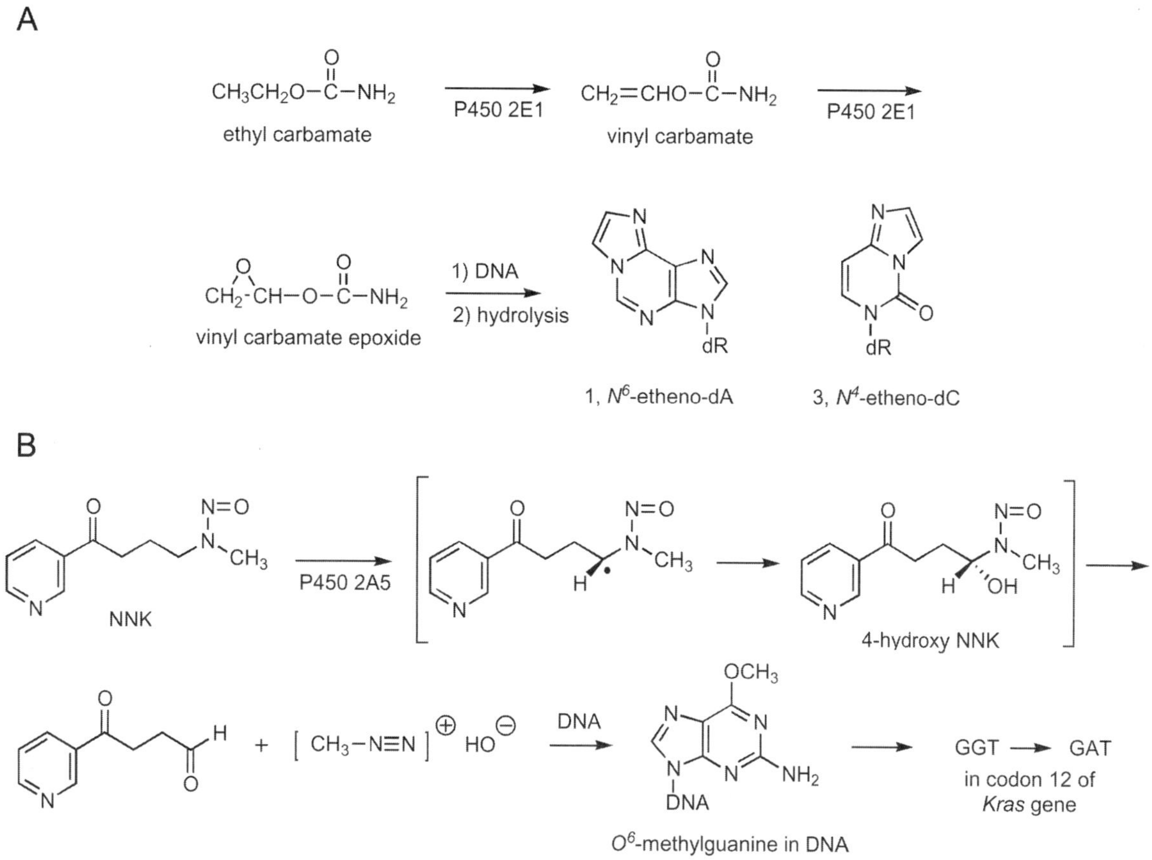

Figure 6.2. (*a*) Metabolic activation of ethyl carbamate to DNA adducts in mouse lung. (*b*) Metabolic activation of NNK in mouse lung leading to formation of O^6-methylguanine in DNA and *K*-ras mutations.

detected in the lungs of mice treated with vinyl carbamate or ethyl carbamate (Fernando et al., 1996; Titis and Forkert, 2001). They are considered to be the critical DNA lesions formed from ethyl carbamate.

NNK is a tobacco-specific lung carcinogen believed to have a significant causative role in human lung cancer (Hecht, 1999; Hecht and Hoffmann, 1988). NNK induces lung tumors in both sensitive and resistant strains of mice (Hecht, 1998). A single intraperitoneal administration of NNK to strain A mice induces multiple tumors that ultimately progress to adenocarcinoma (Belinsky et al., 1992; Hecht et al., 1989). The single-dose NNK lung tumor model is particularly useful for mechanistic studies. Abstraction of the (4*R*)-proton of NNK catalyzed by cytochrome P450 2A5 and other P450s in the mouse lung initiates a cascade of reactions that results in tumor development (Fig. 6.2*b*) (Jalas et al., 2003). The critical intermediate is 4-hydroxyNNK, which spontaneously decomposes to the methane diazonium ion (Hecht, 1998). This intermediate reacts with DNA to produce O^6-methylguanine, among other DNA adducts (Hecht, 1998), which is known to cause G-to-A transition mutations in DNA (Loechler et al., 1984). A strong correlation between persistent O^6-methylguanine levels in pulmonary DNA and lung tumors per mouse has been observed in strain A mice treated with a single dose of NNK (Jalas et al., 2003; Peterson and Hecht, 1991). Another DNA adduct–forming pathway of NNK metabolism inhibits the repair of O^6-methylguanine (Peterson et al., 2001). Moreover, GGT-to-GAT mutations are observed in codon 12 of the K-*ras* oncogene isolated from such tumors, consistent with the important role of O^6-methylguanine (Belinsky et al., 1989; Ronai et al., 1994). Clearly, O^6-methylguanine plays a critical role in strain A mouse lung tumorigenesis induced by NNK. However, the mechanism by which NNK induces lung tumors in the F-344 rat lung is considerably more complex (Hecht, 1998).

PAHs are an important class of environmental carcinogens that are likely to have a significant causative role in lung cancer in smokers and in occupationally exposed individuals (Hecht, 1999; International Agency for Research on Cancer, 1983). PAHs are effective lung carcinogens in strain A and CD-1 mice. Dose–response studies of a variety of PAHs with different structural features revealed 100-fold differences in tumor potency in strain A mice (Nesnow et al., 1998). DNA adducts were formed by means of diol epoxides and related metabolites (Nesnow et al., 1998). Time-integrated DNA adduct levels in mouse lung were found to be linearly related to dose for each PAH and determined tumor outcome in most cases (Nesnow et al., 1998). Mutations in the K-*ras* oncogene were generally consistent with adduct structures and properties, although for some PAHs the relationship was not so clear (Nesnow et al., 1998). PAH diol epoxide metabolites also form well-characterized DNA adducts in the lungs of newborn mice, in which they induce multiple tumors (Melikian et al., 1991).

In summary, the evidence is compelling that mouse lung tumors in carcinogen-treated animals result from the formation of specific DNA adducts and consequent mutations in critical genes.

Applications for Evaluating Chemopreventive Agents

The strain A mouse model has been widely used to investigate potentially chemopreventive agents against lung tumors. Advantages of this model include the reproducible induction of multiple quantifiable lung tumors by PAHs, NNK, carbamates, and other carcinogens, often after only one dose; the short time to tumor induction; and the consequently low cost. The effects of chemopreventive agents can be evaluated simply in terms of changes in numbers of lung tumors per mouse. Groups of 15–20 mice are generally sufficient for evaluation of significant effects. A variety of dosing methods, including inhalation, can be used. Moreover, effects on tumor initiation can be separated from effects on tumor promotion. Thus, "blocking agents" can be identified by their ability to decrease lung tumors per mouse when given before or concurrently with the carcinogen, whereas "suppressing agents" can be identified by their ability to decrease lung tumors per mouse when given after the carcinogen. Effects on the progression stage of carcinogenesis can also be evaluated in longer studies that allow adenomas to progress to adenocarcinomas.

Use of the strain A mouse assay for identifying chemopreventive agents against lung tumorigenesis induced by BaP or NNK has been reviewed elsewhere (Hecht, 1997, 1998). Numerous agents have been found to inhibit BaP-induced lung tumorigenesis, including benzyl isothiocyanate, β-naphthoflavone, butylated hydroxyanisole, ellagic acid, ferulic acid, chlorogenic acid, and diallyl sulfide. The single-dose NNK model has been used extensively as well (Hecht, 1998; Hecht et al., 1989), and more than 50 potential chemopreventive anti-NNK agents have been evaluated. Successful agents include 2-phenethyl isothiocyanate (PEITC), a series of arylalkyl isothiocyanates, indole-3-carbinol, tea extracts, caffeine, diallyl sulfide, ellagic acid, sulindac and other nonsteroidal anti-inflammatory drugs, D-limonene, and *myo*-inositol (Hecht, 1998). A variation of the single-dose NNK assay involves oral administration of NNK for seven weeks (Castonguay and Rioux, 1997). An advantage of this modified assay is that the chronic administration more closely mimics the human situation. However, animals treated with NNK in the chronic oral dose assay tend to lose weight, which is not a problem in the single-dose assay.

Although the strain A mouse lung tumorigenesis assay is exceptionally simple and useful for the identification of chemopreventive agents, confirmation of chemopreventive activity in another system is desirable. PEITC and 6-phenylhexyl isothiocyanate are two examples of chemopreventive agents that were first tested in a strain A mouse NNK model and were subsequently shown to effectively inhibit lung cancer induction by NNK in rats (Hecht, 1998). PEITC is being evaluated in human trials as a chemopreventive agent against lung cancer.

Another model developed for assessing chemopreventive efficacy involves oral administration of a mixture of BaP plus NNK (3 μmol of each) to strain A mice once weekly for 8 weeks. The mice are killed 26 weeks after the first dose, when about 20 lung tumors are present per mouse (Hecht et al., 1994). Advantages of this model include the use of two important lung carcinogens in tobacco smoke in combination and a dosing design that allows testing at various stages during or after carcinogen administration, both of which more closely approximate human exposure. Using this model, we have demonstrated the chemopreventive efficacy of isothiocyanates, their *N*-acetylcysteine conjugates, and *myo*-inositol, alone and in combination (Hecht et al., 1999a,b, 2000, 2002).

As noted in the section on evaluation of carcinogens, inhalational exposure to environmental tobacco smoke induces a small but significant increase in lung tumor multiplicity in strain A mice (Witschi, 2000). This model has also been used to evaluate chemopreventive agents. In the studies reported to date, only a combination of *myo*-inositol and dexamethasone, given during or after exposure to tobacco smoke, successfully inhibited lung tumor multiplicity (Witschi, 2000).

GENETIC MODELS OF LUNG CANCER

Genetic engineering of the murine genome has afforded investigators the opportunity to manipulate pulmonary gene expression in vivo to examine the molecular interactions involved in oncogenic transformation and disease progression and to test potential therapeutic interventions (Herzig and Christofori, 2002). However, modeling human lung cancer in mice is not a simple task. Human lung cancer consists of several histologically identifiable subtypes depending on which of the pulmonary epithelial cell types lining the airways are transformed (Farzan and Farzan, 1997). Examples of human lung cancer histology and mouse models of lung cancer that recapitulate them are illustrated in Figure 6.3. The success of establishing an animal model for such a purpose depends on targeting a specific genetic modification to the appropriate pulmonary epithelial cell type at the appropriate time in lung development. Currently, many tools are available to target genetic manipulations to the pulmonary epithelium. The models that have been generated thus far can be considered in the following groups: conventional transgenic models, conditional transgenic models, gene-targeted models, and conditional gene-targeted models.

Conventional Transgenic Mouse Models

Conventional transgenic mouse models are generated by constructing a fusion gene in which a cell-specific promoter is fused with the coding region of the gene of interest, the expression of which is to be targeted to the desired tissue. This transgene DNA is subsequently microinjected into fertilized murine oocytes, where it undergoes random integration into the mouse genome (Bockamp et al., 2002; Palmiter and Brinster, 1986). The first transgenic mouse model to develop lung cancer resulted from the serendipitous expression of an H-*ras* oncogene under the control of the liver-specific albumin enhancer/promoter construction (Maronpot et al., 1991; Sandgren et al., 1989). Pulmonary expression of this transgene by a liver-specific promoter was unexpected and most likely caused by integration of the transgene in some locus that allowed transcription factors common to the regulation of pulmonary and hepatic genes to direct the pulmonary expression of this transgene. Since the time of that early experiment, the specific targeting of gene expression to the lung—and the generation of mouse models for lung cancer—has been facilitated by the identification of DNA promoter elements for genes that are specific to the pulmonary epithelium (Whitsett et al., 2001).

Two DNA promoter elements that target specific pulmonary epithelial cells that are particularly useful are those for the *SpC* and *CCSP* genes. The former targets transgene expression primarily to the alveolar type II cells of the distal airways (Glasser et al., 1991), and the latter targets transgene expression to the nonciliated secretory cells of the airways (Ray et al., 1995; Stripp et al., 1992). These promoters have been used to target many regulatory proteins to the lungs to investigate the role of various proteins in lung development, physiology, and oncogenic transformation (Whitsett et al., 2001). The first oncogene to be targeted specifically to the lung was the simian

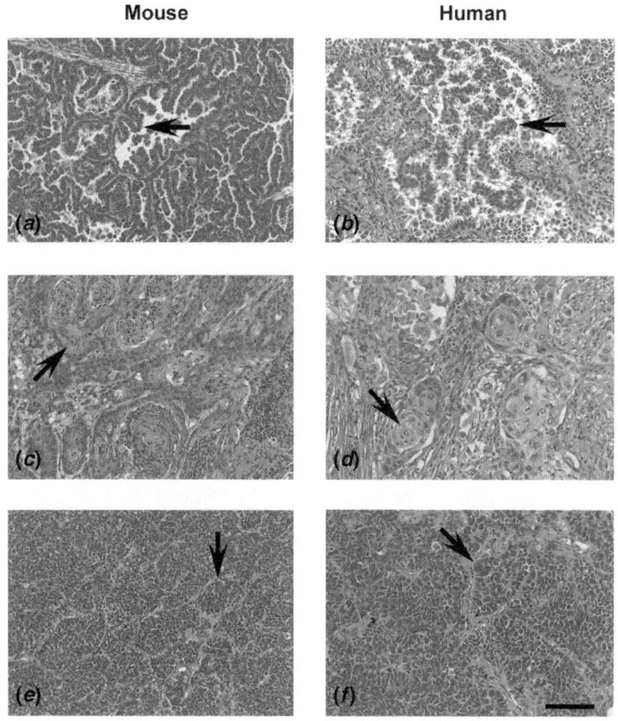

Figure 6.3. Comparative pathology of human and mouse lung neoplasms. (*a*) Papillary adenocarcinoma developed in transgenic mouse expressing the SV40 large T antigen under the control of *CC10* promoter (Magdaleno et al., 1997). (*b*) Human papillary adenocarcinoma. In both (*a*) and (*b*) tumor cells form a papillary pattern along fibrovascular cores (arrows). (*c*) Mouse keratinizing squamous cell carcinoma induced by intratracheal administration of 3-methylcholanthrene (Rehm et al, 1991). (*d*) Human keratinizing squamous cell carcinoma. In both (*a*) and (*b*) squamous differentiation is manifested by formation of keratin pearls (arrows). (*e*) Neuroendocrine carcinoma developed in the mouse carrying $p53^{floxP/floxP}$ and $Rb1^{floxP/floxP}$ conditional alleles after intratracheal instillation of recombinant adenovirus expressing Cre recombinase (Meuwissen et al., 2001). (*f*) Human small-cell lung carcinoma. In both (*e*) and (*f*) cells with small cytoplasm and granular nuclear chromatin are densely packed and arranged in nesting pattern (arrows). Hematoxylin and eosin. ScanScope images. Calibration bar: 100 μm.

virus (SV) 40 large T antigen. This antigen is a potent viral oncogene that directly disrupts the function of proteins that regulate cell cycle control; it has been used to study oncogenic transformation in numerous tissues in mouse models (Furth, 1998). Both *SpC* (Wikenheiser et al., 1992) and *CCSP* (DeMayo et al., 1991; Magdaleno et al., 1997) promoters have been used to target this oncogene to the lungs; both models resulted in adenocarcinoma of the lung, albeit of different cellular origin, with rapid transformation of the pulmonary epithelium and tumors appearing in the first few months of life. The SpC–large-T model led to transformation of alveolar type II cells and distal Clara cells (Wikenheiser et al., 1992);

the CCSP–large-T model led to transformation of Clara cells (Magdaleno et al., 1997). These models have been useful for investigating which regulatory pathways are disrupted during tumor progression and for generating specific cell lines with which to investigate the molecular regulation of gene expression (Ikeda et al., 1994; Magdaleno et al., 1997). These promoters have also been used to target other tumor-promoting genes to the cells of the airways.

The SpC promoter has been used by numerous lung cancer researchers to determine the effect of cell cycle regulatory proteins on the transformation of alveolar type II cells and distal Clara cells. This promoter has been used the most often, not because the alveolar type II cells are the ideal cell type for modeling human lung cancer, but rather because the SpC promoter is a powerful and predictable way of targeting gene expression to the lung.

Conventional transgenic models using the SpC promoter have shown that deregulated expression of c-*myc* (Ehrhardt et al., 2001), both wild-type and mutated c-*Raf-1* (Kerkhoff et al., 2000), and the receptor tyrosine kinase (RON) causes pulmonary adenocarcinomas in transgenic mice (Chen et al., 2002a,b). Combinations of these transgenic models with other genetic or epigenetic manipulations have been used to demonstrate interactions between signaling pathways and carcinogenic insult in the development of lung cancer. Epidermal growth factor (EGF) facilitated the development of adenocarcinoma in SpC-c-*myc* transgenic mice (Ehrhardt et al., 2001), and mice expressing c-*myc* and EGF in the lungs were more susceptible to the tobacco carcinogen NNK than those that did not (Ehrhardt et al., 2003). Another group showed that *Raf-1* induction of pulmonary adenocarcinoma was affected by ablation of the apoptotic regulatory gene *bcl-2* and by the cell cycle regulators *p53* and *p21*(CIP1/WAF1). Specifically, ablation of *bcl-2* delayed the appearance of tumors in SpC-*Raf-1* mice and caused both an increase in apoptosis and a decrease in cell proliferation during the development of tumors in those mice (Fedorov et al., 2002). Loss of *p53* and *p21* in the presence of deregulated *Raf-1* accelerated tumor progression, and, interestingly, loss of *p53* affected the phenotype of the transformed cells (Fedorov et al., 2003). These experiments demonstrate the power of using conventional transgenic models, in combination with gene ablation and carcinogenic insult, to study transformation of the pulmonary epithelium.

The CCSP promoter has been used, albeit to a lesser extent, to target oncogenes to the lung. Although the CCSP promoter targets a pulmonary cell type that may be more clinically relevant for the investigation of lung cancer, this promoter is not as potent as the SpC promoter in directing high levels of gene expression to the lungs and has not been as well studied. Pulmonary oncogenic transformation has been demonstrated in mice expressing c-*myc* transgenes (Geick et al, 2001) or dominant negative *p53* transgenes (Tchou-Wong et al., 2002). In the latter study, expression of a dominant negative *p53* transgene in Clara cells facilitated transformation when the mice were exposed to BaP (Tchou-Wong et al., 2002). The CCSP promoter has also been used to investigate the role of the human Achete Schuete homologue 1(*hASH-1*) gene, which regulates pulmonary neuroendocrine cell differentiation (Borges et al., 1997), in regulating Clara cell proliferation and transformation (Linnoila et al., 2000). In those studies, *hASH-1* was placed under the control of the CCSP promoter to determine whether the *hASH-1* could make Clara cells express neuroendocrine cell markers. Expression of *hASH-1* in this model did not lead to expression of neuroendocrine markers but did lead to hyperplasia of the Clara cells. Moreover, when CCSP-*hASH-1* mice were crossed with CCSP–large T mice, tumor progression was enhanced and the adenocarcinomas observed did express a neuroendocrine phenotype (Linnoila et al., 2000). These findings indicate that *hASH-1* may have a role in causing hyperplasia in neuroendocrine cell tumors and can accelerate a neuroendocrine phenotype in transformed Clara cells.

Although the SpC and CCSP promoters are powerful tools for establishing animal models of lung cancer, they do not target all of the pulmonary epithelial cell types that are transformed in the various types of lung cancer. Promoters are still needed to target genes specifically to the upper respiratory tract to generate animal models for pulmonary squamous cell carcinoma; also needed are promoters that direct transgene expression to the neuroendocrine cells of the lung to generate animal models for SCLC. In an attempt to target oncogene expression to neuroendocrine cells, Sunday and colleagues (1999) expressed the v-H-*ras* gene in the lungs of mice under the control of a calcitonin promoter. In that study, expression of v-H-*ras* induced hyperplasia of both neuroendocrine cells and nonneuroendocrine cells, and the tumors that progressed were mostly of nonneuroendocrine origin. The lack of neuroendocrine cell specificity in this model may have been due in part to the diverse biologic consequences of transgene expression in early lung development (Sunday et al., 1999).

Another limitation of these conventional transgenic approaches for generating lung cancer models is the need to target the time at which the genes are transcribed. Transcription under the control of the SpC promoter occurs early in fetal lung development (Wert et al., 1993), and that under the CCSP promoter occurs later (Ray et al., 1996). Expression of genes that will disrupt the developmental processes during branching morphogenesis and alveologenesis will result in an embryonic-lethal phenotype and be of limited value for studying the progression of lung cancer (Peters et al., 1994; Simonet et al., 1995). The ability to initiate transgene expression at specific times would allow a model to focus more effectively on identifying the sequence of events governing tumor progression. Another limitation of the conventional transgenic approach is that transgene transcription, once initiated, is irreversible. The ability to more tightly control the "window" of transgene expression would also allow more precise determination of the molecular events regulating cancer progression. Finally, the ability to terminate transgene expression would allow the stages of tumor progression to be identified independent of any growth-promoting properties that the transgene might have. Current transgenic technology has evolved to a state in which transgenes can be conditionally expressed in the lungs upon the administration of an exogenous compound. Examples of such conditional expression systems are described in the following sections.

Conditional Transgenic Mouse Models

Currently, the most effective systems for regulating the conditional expression of transgenes in mice are the ligand-inducible binary transgenic systems (DeMayo and Tsai, 2001; Lewandoski, 2001). In such systems, at least two transgene constructions, a regulator transgene and a target transgene, are used to regulate the expression of the desired gene. The regulator transgene encodes a transcription factor, the activity of which is controlled by the administration of an exogenous compound. The regulator is placed under the control of a lung-specific promoter to ensure that the transcription factor is expressed in the lung. The second transgenic construct, the target construct, contains the coding sequence of a regulatory gene of interest under the transcriptional control of *cis*-acting DNA elements that are responsive to the DNA-binding domain of the regulator transgene. The target transgene is silent until the regulator transgene is activated by the administration of the exogenous compound. Currently, two ligand-inducible binary transgenic systems have been established for the conditional regulation of gene expression in the lung: the tetracycline transactivator *(tTA)*–inducible system and the mifepristone gene switch. These systems are described further in the following paragraphs.

Tetracycline Transactivator–Inducible System. The most common ligand-inducible binary transgenic system in current use is the *tTA* regulatory system, which has been extremely useful for generating ligand-dependent and reversible transgenic mouse models (Shockett and Schatz, 1996). The *tTA* systems are based on the Tn10-specific tetracycline resistance operon in *Escherichia coli* (Hillen and Berens, 1994). This system uses the *tet* repressor gene to activate target genes that are under the control of promoter elements containing the *tet* response element—the *tet* operon. Current engineering of the *tet* gene for the ligand-inducible binary transgenic systems has generated two forms of this regulator, the *tTA* ("*tet* off") and the reverse tetracycline transactivator (*rtTA*, "*tet* on") forms. In the first form, the *tTA* acts as repressor in the presence of tetracycline; in the absence of tetracycline, the *tTA* causes the transcription of the *tet*-responsive target transgene. In the second and more commonly used form, tetracycline-dependent transactivators are generated by the reverse *tet* transactivator (*rtTA*). In the absence of ligand, the *rtTA* does not bind DNA or activate target gene transcription; however, upon administration of the ligand (tetracycline or doxycycline), the *rtTA* binds to the *tet* operon and activates transgene expression (Gossen et al., 1995). In the mouse lung, either the CCSP promoter or the SpC promoter has been used to drive expression of the *rtTA* (Ray et al., 1997; Tichelaar et al., 2000).

Conditional transgenic systems have been used to investigate the role of several regulatory proteins in the transformation of the pulmonary epithelium, namely the growth factors FGF-7 (Tichelaar et al., 2000), FGF-10 (Clark et al., 2001), and FGF-18 (Whitsett et al., 2002) and the activated K-*ras* oncogene (Fisher et al., 2001). Regulated expression of FGF-7 (Tichelaar et al., 2000) and FGF-10 (Clark et al., 2001) caused increased cell proliferation upon administration of doxycycline, and the induction of FGF-10 resulted in the development of adenomas. Both effects were reversible upon withdrawal of doxycycline.

Although the *tTA* model has generated useful findings, the most informative animal model for lung cancer has been the *rtTA* system to induce the expression of activated K-*ras* (Fisher et al., 2001). In these studies, activation of K-*ras* by using a CCSP-driven *rtTA* regulator resulted in the development of pulmonary adenocarcinoma. Upon withdrawal of doxycycline, K-*ras* expression was terminated and the tumors regressed as a result of apoptosis. These mice were then crossed with mice deficient in either *p53* or *Ink4A/Arf* to investigate the interaction of these tumor suppressor genes with the development and regression of the K-*ras*-dependent tumors. Induction of K-*ras* expression in p53(−/−) or Ink4A/ARF(−/−) mice caused the tumors to arise faster. However, the absence of these tumor suppressor genes did not prevent the tumors from regressing upon termination of K-*ras* expression. The ability to activate and terminate oncogene expression in the lungs in systems such as this allows clarification of the molecular mechanisms that regulate pulmonary epithelial cell transformation.

Mifepristone Gene Switch. The second binary transgenic system used to regulate transgene expression in mice is based on a mutated human progesterone receptor. A deletion of the carboxy-terminus ligand-binding domain of that receptor alters the specificity of ligand binding such that the receptor binds antiprogestins such as mifepristone (Vegeto et al., 1992). A chimeric receptor was then created consisting of the mutated ligand-binding domain, the DNA-binding domain of the yeast Gal4 transcription factor, and the activation domain of the nuclear factor kappa B p65 protein (Burcin et al., 1999; Wang et al., 2000). The resulting GLp65 regulator strongly activates a promoter containing the upstream activating sequences of the Gal4 gene upon the addition of mifepristone (Wang et al., 1997).

The GLp65 regulator has been targeted to the lung in transgenic mice by using the SpC promoter (Zhao et al., 2000, 2001). The SpC-GLp65 regulator was used to control the expression of FGF-3 (also known as the *int-2* oncogene) in mice. Activation of the FGF-3 transgene in response to mifepristone resulted in one of two phenotypes depending on the level of FGF-3 transgene expression: Low-level induction of FGF-3 induced massive free alveolar macrophage infiltration, whereas high-level induction of FGF-3 resulted in diffuse alveolar type II cell hyperplasia, a condition similar to adenomatous alveolar hyperplasia, which is considered a precursor of adenocarcinoma in humans. More importantly, both phenotypes were reversible after the withdrawal of mifepristone (Zhao et al., 2001). Thus, systems such as these can be used to activate or repress expression of a target transgene by using the synthetic ligand mifepristone.

Gene-Targeted Models

Aside from the ability to express genes in the pulmonary epithelium, current technology also allows gene expression to be ablated. In this model, endogenous genes are ablated by homologous recombination in embryonic stem (ES) cells (Smithies et al., 1985; Wong and Capecchi, 1986). Whereas conventional approaches to gene ablation ablate the gene in the entire animal, this homologous recombination technology has evolved to allow control of the cell type in which the gene is ablated as well as the timing of the ablation. This powerful technology has been used to investigate the role of many genes in pulmonary development and physiology.

The first use of homologous recombination to evoke development of lung cancer was the introduction of a targeting event in which somatic recombination would allow activation of a mutant K-*ras* gene (Johnson et al., 2001). In this study, mice were generated in which one K-*ras* allele contained an oncogenic mutation and the other allele was wild type. The mutant K-*ras* gene was expressed when somatic recombination occurred between the targeted allele and the wild-type allele. Mice expressing the targeted K-*ras* mutation developed pulmonary adenocarcinoma, presumably caused by the sporadic activation of the mutant allele. This approach has the advantage over transgenic approaches in that the transgene could be expressed under the control of the endogenous K-*ras* promoter, and its levels of expression better reflected those of physiologic conditions. This approach could also overcome the potential for embryonic lethality that could have occurred if the mutation was present in every cell during development. However, a limitation of this approach is that tumors formed in other organs in addition to the lung. Because the genetic modification apparently occurred in all of the cells of the mice, this approach is also limited to genetic modifications that do not affect animal viability. Nevertheless, these limitations can be overcome by the use of recombinase technology in conditional gene-targeted models such as those described below.

Conditional Gene-Targeted Models. In conditional knockout models, tissue-specific expression of recombinases is used to ablate genes in a tissue- and cell-specific fashion (Lewandoski, 2001; Ryding et al., 2001). Site-specific recombinases, as their name implies, splice DNA sequences that are flanked by specific sequences recognized by the recombinases. The most common recombinase system used to induce gene ablation in mice is Cre-*loxP*. In this system, the Cre (causes recombination) protein of bacteriophage P1 directs recombination between *loxP* (locus of crossover P1) sites and ablates gene sequences that are flanked by two 34-bp *loxP* sites (floxed genes) (Sauer and Henderson, 1988). Homologous recombination in ES cells is used to insert recombinase recognition sequences around an entire gene or around exons coding for regulatory domains. Transgenic technology is then used to express site-specific recombinase in a tissue-specific manner. This combination of the ES cell technology with the transgenic technology allows the gene of interest, flanked by the recombinase sites, to be modified in a tissue-specific fashion.

To date, these recombinases have not been used in a direct transgenic approach to ablate genes specifically in the lung, primarily because of the use of the SpC and CCSP promoters, which would activate the recombinase during embryonic development. Ablation of the desired gene under the control of these promoters may disrupt lung development and result in an embryonic-lethal phenotype. However, this limitation has been overcome by regulating the expression of activity of the recombinase. Currently, two approaches have been used to regulate recombinase activity temporally in the lung. These approaches involve using *rtTA* to regulate Cre recombinase expression in the lung and using adenoviruses to deliver the recombinase to the lung.

rtTA-Regulated Cre Recombinase. Temporal expression of Cre recombinase in the lung has been achieved by using a tetracycline-regulated system in which Cre recombinase is placed under the control of the minimal tetracycline-responsive promoter. Whitsett's laboratory has used this approach to study pulmonary epithelial cell lineage (Perl et al., 2002). In this study, *rtTA*-regulated Cre expression was used to edit *loxP* sites and to activate the expression of alkaline phosphatase and green fluorescent protein reporter genes in the lungs of mice at specified times during embryonic development, with the goal of tracing the lineage of the cells lining the airways of the lung. The next step in the use of this approach will be to activate or ablate genes proposed to be in the oncogenic pathway and investigate the influence of those genes on lung development.

Virally Delivered Cre Recombinase. Adenoviruses have also been used to deliver Cre recombinase to initiate lung tumorigenesis in mice. In one case, gene targeting was used to insert a transcription stop sequence flanked by *loxP* sites into the endogenous mutated K-*ras* gene (Jackson et al., 2001). In another set of experiments, a mutated K-*ras* transgene was placed under the transcriptional control of the β-actin promoter, and a green fluorescent protein reporter gene, flanked by *loxP* sites, was inserted between the K-*ras* coding region and the promoter (Meuwissen et al., 2001). Adenoviral delivery of Cre recombinase in both cases edited out the sequences blocking K-*ras* transcription, thereby allowing K-*ras* expression and producing adenocarcinoma in the airways of the mice. This approach has the major advantage of allowing ablation of genes at specific times without needing to breed Cre transgenic and *loxP* homozygous mice. The disadvantage of this approach is the loss of the control over the cell specificity of the gene ablation.

CONCLUSIONS

During the process of lung tumorigenesis, the bronchial epithelium undergoes a myriad of genetic and biochemical events. Which of these events is required for lung tumorigenesis is not known; thus, identifying such events is critical for developing effective preventive and therapeutic strategies for patients with lung cancer. Mouse models provide one avenue by which these events can be studied. Considered in the context of the molecular events known to occur in human lung tumors, however, current strategies of mouse lung model development are simplistic. Future strategies should incorporate more of the known oncogenic events in human lung tumors. Processes in lung cancer that affect the entire genome, such as genomic instability and alterations in DNA methylation, should also be recapitulated. Ongoing technologic advances in mouse model development will allow specific cellular compartments within the lung to be targeted. Progress in this area will enhance both the basic understanding of lung tumorigenesis and the speed with which the anticancer efficacy of investigational drugs in preclinical trials can be investigated.

ACKNOWLEDGMENT

This work was supported by grant U01 CA084306 from the National Institutes of Health.

REFERENCES

Anderson LM, Hecht SS, Kovatch RM, Amin S, Hoffmann D, Rice JM (1991): Tumorigenicity of the tobacco-specific carcinogen 4-(methylnitrosamino)-1-(3-pyridyl)-1-butanone in infant mice. *Cancer Lett* 58: 177–181.

Aubin J, Lemieux M, Tremblay M, Berard J, Jeannotte L (1997): Early postnatal lethality in Hoxa-5 mutant mice is attributable to respiratory tract defects. *Dev Biol* 192: 432–445.

Belinsky SA, Devereux TR, Foley JF, Maronpot RR, Anderson MW (1992): Role of the alveolar type II cell in the development and progression of pulmonary tumors induced by 4-(methylnitrosamino)-1-(3-pyridyl)-1-butanone in the A/J mouse. *Cancer Res* 52: 3164–3173.

Belinsky SA, Devereux TR, Maronpot RR, Stoner GD, Anderson MW (1989): The relationship between the formation of promutagenic adducts and the activation of the K-ras proto-oncogene in lung tumors from A/J mice treated with nitrosamines. *Cancer Res* 49: 5305–5311.

Bellusci S, Henderson R, Winnier G, Oikawa T, Hogan B (1996): Evidence from normal expression and targeted misexpression that bone morphogenetic protein-4 (Bmp-4) plays a role in mouse embryonic lung morphogenesis. *Development* 122: 1693–1702.

Betz C, Papadopoulos T, Buchwald J, Dammrich J, Muller-Hermelink HK (1995): Surfactant protein gene expression in metastatic and micrometastatic pulmonary adenocarcinomas and other non-small cell lung carcinomas: Detection by reverse transcriptase-polymerase chain reaction. *Cancer Res* 55: 4283–4286.

Bockamp E, Maringer M, Spangenberg C, Fees S, Fraser S, Eshkind L, Oesch F, Zabel B (2002): Of mice and models: Improved animal models for biomedical research. *Physiol Genomics* 11: 115–132.

Bohinski RJ, Di Lauro R, Whitsett JA (1994): The lung-specific surfactant protein B gene promoter is a target for thyroid transcription factor 1 and hepatocyte nuclear factor 3, indicating common factors for organ-specific gene expression along the foregut axis. *Mol Cell Biol* 9: 5671–5681.

Borges M, Linnoila RI, van de Velde HJ, Chen H, Nelkin BD, Mabry M, Baylin SB, Ball DW (1997): An achaete-scute homologue essential for neuroendocrine differentiation in the lung. *Nature* 386: 852–855.

Bostrom H, Willetts K, Pekny M, Leveen P, Lindahl P, Hedstrand H, Pekna M, Hellstrom M, Gebre-Medhin S, Schalling M, Nilsson M, Kurland S, Tornell J, Heath JK, Betsholtz C (1996): PDGF-A signaling is a critical event in lung alveolar myofibroblast development and alveogenesis. *Cell* 85: 863–873.

Broers JLV, Jensen SM, Travis WD, Pass H, Whitsett JA, Singh G, Katyal SL, Gazdar AF, Minna JD, Linnoila RI (1992): Expression of surfactant associated protein-A and clara cell 10 kilodalton mRNA in neoplastic and non-neoplastic human lung tissue as detected by in situ hybridization. *Lab Invest* 66: 337–346.

Burcin MM, Schiedner G, Kochanek S, Tsai SY, O'Malley BW (1999): Adenovirus-mediated regulable target gene expression in vivo. *Proc Natl Acad Sci USA* 96: 355–360.

Burns DM (2000): Primary prevention, smoking and smoking cessation: Implications for future trends in lung cancer prevention. *Cancer* 89: 2506–2509.

Campbell C, Goodrich K, Casey G, Beatty B (1995): Cloning and mapping of a human gene (TBX2) sharing a highly conserved protein motif with the Drosophila *omb* gene. *Genomics* 28: 255–260.

Castonguay A, Rioux N (1997): Inhibition of lung tumorigenesis by sulindac: Comparison of two protocols. *Carcinogenesis* 18: 491–496.

Chen J, Knowles HJ, Hebert JL, Hackett BP (1998): Mutation of the mouse hepatocyte nuclear factor/forkhead homologue 4 gene results in absence of cilia and random left-right asymmetry. *J Clin Invest* 102: 1077–1082.

Chen YQ, Zhou YQ, Fisher JH, Wang MH (2002a): Targeted expression of the receptor tyrosine kinase RON in distal lung epithelial cells results in multiple tumor formation: Oncogenic potential of RON in vivo. *Oncogene* 21: 6382–6386.

Chen YQ, Zhou YQ, Fu LH, Wang D, Wang MH (2002b): Multiple pulmonary adenomas in the lung of transgenic mice overexpressing the RON receptor tyrosine kinase. Recepteur d'origine nantais. *Carcinogenesis* 23: 1811–1819.

Clark JC, Tichelaar JW, Wert SE, Itoh N, Perl AK, Stahlman MT, Whitsett JA (2001): FGF-10 disrupts lung morphogenesis and causes pulmonary adenomas in vivo. *Am J Physiol Lung Cell Mol Physiol* 280: L705–L715.

Conney AH (1982): Induction of microsomal enzymes by foreign chemicals and carcinogenesis by polycyclic aromatic hydrocarbons: G.H.A. Clowes Memorial Lecture. *Cancer Res* 42: 4875–4917.

Dahl GA, Miller JA, Miller EC (1978): Vinyl carbamate as a promutagen and a more carcinogenic analog of ethyl carbamate. *Cancer Res* 38: 3793–3804.

Dahl GA, Miller EC, Miller JA (1980): Comparative carcinogenicities and mutagenicities of vinyl carbamate, ethyl carbamate, and ethyl N-hydroxycarbamate. *Cancer Res* 40: 1194–1203.

Danto SI, Shannon JM, Borok Z, Zabski SM, Crandall ED (1995): Reversible transdifferentiation of alveolar epithelial cells. *Am J Respir Cell Mol Biol* 12: 497–502.

DeMayo FJ, Finegold MJ, Hansen TN, Stanley LA, Smith B, Bullock DW (1991): Expression of SV40 T antigen under control of rabbit uteroglobin promoter in transgenic mice. *Am J Physiol* 261: L70–L76.

DeMayo FJ, Tsai SY (2001): Targeted gene regulation and gene ablation. *Trends Endocrinol Metab* 12: 348–353.

Ehrhardt A, Bartels T, Geick A, Klocke R, Paul D, Halter R (2001): Development of pulmonary bronchiolo-alveolar adenocarcinomas in transgenic mice overexpressing murine c-myc and epidermal growth factor in alveolar type II pneumocytes. *Br J Cancer* 84: 813–818.

Ehrhardt A, Bartels T, Klocke R, Paul D, Halter R (2003): Increased susceptibility to the tobacco carcinogen 4-(methylnitrosamino)-1-(3-pyridyl)-1-butanone in transgenic mice overexpressing c-myc and epidermal growth factor in alveolar type II cells. *J Cancer Res Clin Oncol* 129: 71–75.

Evans MJ, Cabral LJ, Stephens RJ, Freeman G (1973): Renewal of alveolar epithelium in the rat following exposure to NO_2. *Am J Pathol* 70: 175–198.

Farzan S, Farzan DA (Ed.). (1997). *Lung Cancer. A Concise Handbook of Respiratory Diseases*. Appleton and Lange, Stanford, CT.

Fedorov LM, Papadopoulos T, Tyrsin OY, Twardzik T, Gotz R, Rapp UR (2003): Loss of p53 in cRaf-induced transgenic lung adenoma leads to tumor acceleration and phenotypic switch. *Cancer Res* 63: 2268–2277.

Fedorov LM, Tyrsin OY, Papadopoulos T, Camarero G, Gotz R, Rapp UR (2002): Bcl-2 determines susceptibility to induction of lung cancer by oncogenic CRaf. *Cancer Res* 62: 6297–6303.

Fehrenbach H, Kasper M, Tschernig T, Pan T, Schuh D, Shannon JM, Muller M, Mason RJ (1999): Keratinocyte growth factor-induced hyperplasia of rat alveolar type II cells in vivo is resolved by differentiation into type I cells and by apoptosis. *Eur Resp J* 14: 534–544.

Fernando RC, Nair J, Barbin A, Miller JA, Bartsch H (1996): Detection of 1,N^6-ethenodeoxyadenosine and 3,N^4-ethenodeoxycytidine by immunoaffinity/^{32}P-postlabelling in liver and lung DNA of mice treated with ethyl carbamate (urethane) or its metabolites. *Carcinogenesis* 17: 1711–1718.

Fisher GH, Wellen SL, Klimstra D, Lenczowski JM, Tichelaar JW, Lizak MJ, Whitsett JA, Koretsky A, Varmus HE (2001): Induction and apoptotic regression of lung adenocarcinomas by regulation of a K-Ras transgene in the presence and absence of tumor suppressor genes. *Genes Dev* 15: 3249–3262.

Foley JF, Anderson MW, Stoner GD, Gaul BW, Hardisty JF, Maronpot RR (1991): Proliferative lesions of the mouse lung: Progression studies in strain A mice. *Exp Lung Res* 17: 157–168.

Furth PA (1998): SV40 rodent tumour models as paradigms of human disease: Transgenic mouse models. *Dev Biol Stand* 94: 281–287.

Gazdar AF, Linnoila RI, Kurita Y, Oie HK, Mulshine JL, Clark JC, Whitsett JA (1990): Peripheral airway cell differentiation in human lung cancer cell lines. *Cancer Res* 50: 5481–5487.

Geick A, Redecker P, Ehrhardt A, Klocke R, Paul D, Halter R (2001): Uteroglobin promoter-targeted c-MYC expression in transgenic mice cause hyperplasia of Clara cells and malignant transformation of T-lymphoblasts and tubular epithelial cells. *Transgenic Res* 10: 501–511.

Glasser SW, Korfhagen TR, Wert SE, Bruno MD, McWilliams KM, Vorbroker DK, Whitsett JA (1991): Genetic element from human surfactant protein SP-C gene confers bronchiolar-alveolar cell specificity in transgenic mice. *Am J Physiol* 261: L349–L356.

Gossen M, Freundlieb S, Bender G, Muller G, Hillen W, Bujard H (1995): Transcriptional activation by tetracyclines in mammalian cells. *Science* 268: 1766–1769.

Greenlee RT, Murray T, Bolden S, Wingo PA (2000): Cancer statistics, 2000. *CA Cancer J Clin* 50: 7–33.

Guengerich FP, Kim DH (1991): Enzymatic oxidation of ethyl carbamate to vinyl carbamate and its role as an intermediate in the formation of 1,N^6-ethenoadenosine. *Chem Res Toxicol* 4: 413–421.

Hecht SS (1997): Approaches to chemoprevention of lung cancer based on carcinogens in tobacco smoke. *Environ Health Perspect* 105(Suppl. 4): 955–963.

Hecht SS (1998): Biochemistry, biology, and carcinogenicity of tobacco-specific *N*-nitrosamines. *Chem Res Toxicol* 11: 559–603.

Hecht SS (1999): Tobacco smoke carcinogens and lung cancer. *J Natl Cancer Inst* 91: 1194–1210.

Hecht SS, Hoffmann D (1988): Tobacco-specific nitrosamines, an important group of carcinogens in tobacco and tobacco smoke. *Carcinogenesis* 9: 875–884.

Hecht SS, Isaacs S, Trushin N (1994): Lung tumor induction in A/J mice by the tobacco smoke carcinogens 4-(methylnitrosamino)-1-(3-pyridyl)-1-butanone and benzo[*a*]pyrene: A potentially useful model for evaluation of chemopreventive agents. *Carcinogenesis* 15: 2721–2725.

Hecht SS, Kenney PMJ, Wang M, Trushin N, Agarwal S, Rao AV, Upadhyaya P (1999a): Evaluation of butylated hydroxyanisole, *myo*-inositol, curcumin, esculetin, resveratrol, and lycopene as inhibitors of benzo[*a*]pyrene plus 4-(methylnitrosamino)-1-(3-pyridyl)-1-butanone-induced lung tumorigenesis in A/J mice. *Cancer Lett* 137: 123–130.

Hecht SS, Kenney PMJ, Wang W, Trushin N, Upadhyaya P (1999b): Effects of phenethyl isothiocyanate and benzyl isothiocyanate, individually and in combination, on lung tumorigenesis induced in A/J mice by benzo[*a*]pyrene and 4-(methylnitrosamino)-1-(3-pyridyl)-1-butanone. *Cancer Lett* 150: 49–56.

Hecht SS, Kenney PMJ, Wang M, Upadhyaya P (2000): Dose-response study of *myo*-inositol as an inhibitor of lung tumorigenesis induced in A/J mice by benzo[*a*]pyrene and 4-(methylnitrosamino)-1-(3-pyridyl)-1-butane. *Cancer Lett* 167: 1–6.

Hecht SS, Morse MA, Amin S, Stoner GD, Jordan KG, Choi CI, Chung FL (1989): Rapid single-dose model for lung tumor induction in A/J mice by 4-(methylnitrosamino)-1-(3-pyridyl)-1-butanone and the effect of diet. *Carcinogenesis* 10: 1901–1904.

Hecht SS, Upadhyaya P, Wang M, Bliss RL, McIntee EJ, Kenney PMJ (2002): Inhibition of lung tumorigenesis in A/J mice by *N*-acetyl-*S*-(*N*-2-phenethylthiocarbamoyl)-L-cysteine and *myo*-inositol, individually and in combination. *Carcinogenesis* 23: 1455–1461.

Herzig M, Christofori G (2002): Recent advances in cancer research: Mouse models of tumorigenesis. *Biochim Biophys Acta* 1602: 97–113.

Hillen W, Berens C (1994): Mechanisms underlying expression of Tn10 encoded tetracycline resistance. *Annu Rev Microbiol* 48: 345–369.

Hong WK, Sporn MB (1997): Recent advances in chemoprevention of cancer. *Science* 278: 1073–1077.

Ikeda K, et al. (1994): Immortalization of subpopulations of respiratory epithelial cells from transgenic mice bearing SV40 large T antigen. *Am J Physiol* 267: L309–317.

International Agency for Research on Cancer (1983): Polynuclear Aromatic Compounds, Part 1. Chemical, Environmental and Experimental Data. *IARC Monographs on the Evaluation of the Carcinogenic Risk of Chemicals to Humans vol. 32*. IARC, Lyon, France.

International Agency for Research on Cancer (1986): Tobacco Smoking. *IARC Monographs on the Evaluation of the Carcinogenic Risk of Chemicals to Humans vol. 38*. IARC, Lyon, France, pp 131–2.

Issac A, Cohn MJ, Ashby P, Ataliotis P, Spicer DB, Cooke J, Tickle C (2000): FGF and genes encoding transcription factors in early limb specification. *Mech Dev* 93: 41–48.

Jackson EL, Willis N, Mercer K, Bronson RT, Crowley D, Montoya R, Jacks T, Tuveson DA (2001): Analysis of lung tumor initiation and progression using conditional expression of oncogenic K-ras. *Genes Dev* 15: 3243–3248.

Jalas JR, McIntee EJ, Kenney PMJ, Upadhyaya P, Peterson LA, Hecht SS (2003): Stereospecific deuterium substitution attenuates the tumorigenicity and metabolism of the tobacco-specific nitrosamine 4-(methylnitrosamino)-1-(3-pyridyl)-1-butanone (NNK). *Chem Res Toxicol* 16: 794–806.

Jensen-Taubman SM, Steinberg SM, Linnoila RI (1998): Bronchiolization of the alveoli in lung cancer: Pathology patterns of differentiation and oncogene expression. *Int J Cancer* 75: 489–496.

Johnson L, Mercer K, Greenbaum D, Bronson RT, Crowley D, Tuveson DA, Jacks T (2001): Somatic activation of the K-ras oncogene causes early onset lung cancer in mice. *Nature* 410: 1111–1116.

Kerkhoff E, Fedorov LM, Siefken R, Walter AO, Papadopoulos T, Rapp UR (2000): Lung-targeted expression of the c-Raf-1 kinase in transgenic mice exposes a novel oncogenic character of the wild-type protein. *Cell Growth Differ* 11: 185–190.

Kimura S, Hara Y, Pineau T, Fernandez-Salguero P, Fox HC, Ward JM, Gonzalez FJ (1996): The T/ebp null mouse: Thyroid-specific enhancer-binding protein is essential for the organogenesis of the thyroid, lung, ventral forebrain, and pituitary. *Genes Dev* 10: 60–69.

Kitamura H, Kameda Y, Ito T, Hayashi H, Nakamura N, Nakatani Y, Inayma Y, Kanisawa M (1997): Cytodifferentiation of atypical adenomatous hyperplasia and bronchioloalveolar lung carcinoma: Immunohistochemical and ultrastructural studies. *Virchows Arch* 431: 415–424.

Korfhagen TR, Glasser SW, Wert SE, Bruno MD, Daugherty CC, McNeish JD, Stock JL, Potter SS, Whitsett JA (1990): Cis-acting sequences from a human surfactant protein gene confer pulmonary-specific gene expression in transgenic mice. *Proc Natl Acad Sci USA* 87: 6122–6126.

Kreda SM, Gynn MC, Fenstermacher DA, Boucher RC, Gabriel SE (2001): Expression and localization of epithelial aquaporins in the adult human lung. *Am J Respir Cell Mol Biol* 24: 224–234.

Lawson KA, Dunn NR, Roelen BA, Zeinstra LM, Davis AM, Wright CV, Korving JP, Hogan BLM (1999): *Bmp4* is required for the generation of primordial germ cells in the mouse embryo. *Genes Dev* 13: 424–436.

Leithauser MT, Liem A, Stewart BC, Miller EC, Miller JA (1990): 1,N^6-Ethenoadenosine formation, mutagenicity and murine tumor induction as indicators of the generation of an electrophilic epoxide metabolite of the closely related carcinogens ethyl carbamate (urethane) and vinyl carbamate. *Carcinogenesis* 11: 463–473.

Lewandoski M (2001): Conditional control of gene expression in the mouse. *Nat Rev Genet* 2: 743–755.

Li C, Xiao J, Hormi K, Borok Z, Minoo P (2002): Wnt5a participates in distal lung morphogenesis. *Dev Biol* 248: 68–81.

Linnoila RI, Zhao B, DeMayo JL, Nelkin BD, Baylin SB, DeMayo FJ, Ball DW (2000): Constitutive achaete-scute homologue-1 promotes airway dysplasia and lung neuroendocrine tumors in transgenic mice. *Cancer Res* 60: 4005–4009.

Loechler EL, Green CL, Essigmann JM (1984): in vivo mutagenesis by O^6-methylguanine built into a unique site in a viral genome. *Proc Natl Acad Sci USA* 81: 6271–6275.

Magdaleno SM, Wang G, Mireles VL, Ray MK, Finegold MJ, DeMayo FJ (1997): Cyclin-dependent kinase inhibitor expression in pulmonary Clara cells transformed with SV40 large T antigen in transgenic mice. *Cell Growth Differ* 8: 145–155.

Manenti G, Galbiati F, Noci S, Dragani TA (2003): Outbred CD-1 mice carry the susceptibility allele at the pulmonary adenoma susceptibility 1 (*Pas1*) locus. *Carcinogenesis* 24: 1143–1148.

Maronpot RR, Palmiter RD, Brinster RL, Sandgren EP (1991): Pulmonary carcinogenesis in transgenic mice. *Exp Lung Res* 17: 305–320.

Maronpot RR, Shimkin MB, Smith LH, Cline JM (1986): Strain A mouse pulmonary tumor test results for chemicals previously tested in the National Cancer Institute carcinogenicity tests. *J Natl Cancer Inst* 76: 1101–1112.

Melikian AA, Prahalad KA, Amin S, Hecht SS (1991): Comparative DNA binding of polynuclear aromatic hydrocarbons and their dihydrodiol and bay region diol epoxide metabolites in newborn mouse lung and liver. *Carcinogenesis* 12: 1665–1670.

Mendelsohn C, Lohnes D, Decimo D, Lufkin T, LeMeur M, Chambon P, Mark M (1994): Function of the retinoic acid receptors (RARs) during development (II). Multiple abnormalities at various stages of organogenesis in RAR double mutants. *Development* 120: 2749–2771.

Meuwissen R, Linn SC, van der Valk M, Mooi WJ, Berns A (2001): Mouse model for lung tumorigenesis through Cre/lox controlled sporadic activation of the K-Ras oncogene. *Oncogene* 20: 6551–6558.

Min H, Danilenko DM, Scully SA, Bolon B, Rihg BD, Tarpley JE, DeRose M, Simonet WS (1998): *Fgf-10* is required for both limb and lung development and exhibits striking functional similarity to Drosophila branchless. *Genes Dev* 12: 3156–161.

Minna J, Fauci AS, Braunwald E, Isselbacher KJ, Wilson JD, Martin JB, Kasper DL, Hauser SL, Longo DL (2000): Neoplasms of the lung. In *Harrison's Principles of Internal Medicine*, 15th ed. McGraw-Hill, New York, Chapter 88.

Minna J, Roth JA, Gazdar AF (2002): Focus on lung cancer. *Cancer Cell* 1: 49–52.

Minoo P, Hamdan H, Bu D, Warburton D, Stepanik P, deLemos R (1995): TTF-1 regulates lung epithelial morphogenesis. *Dev Biol* 172: 694–708.

Minoo P, Su G, Drum H, Bringas P, Kimura S (1999): Defects in tracheoesophageal and lung morphogenesis in Nkx2.1 (−/−) mouse embryos. *Dev Biol* 209: 60–71.

Monaghan AP, Kaestner KH, Grau E, Schutz G (1993): Postimplantation expression patterns indicate a role for the mouse forkhead/HNF-3α, β and γ genes in determination of the definitive endoderm, chordamesoderm and neuroectoderm. *Development* 119: 567–578.

Morrisey E, Tang Z, Sigrist K, Lu M, Jiang F, Hon S, Parmacek M (1998): GATA6 regulates HNF4 and is required for differentiation of visceral endoderm in the mouse embryo. *Genes Dev* 12: 3579–3590.

Motoyama J, Liu J, Mo R, Ding Q, Post M, Hui CC (1998): Essential function of Gli2 and Gli3 in the formation of lung, trachea and esophagus. *Nat Genet* 20: 54–57.

Mountain CF (1997): Revisions in the international system for staging lung cancer. *Chest* 111: 1710–1717.

Nesnow S, Ross JA, Mass MJ, Stoner GD (1998): Mechanistic relationships between DNA adducts, oncogene mutations, and lung tumorigenesis in strain A mice. *Exp Lung Res* 24: 395–405.

Nesnow S, Ross JA, Stoner GD, Mass MJ (1995): Mechanistic linkage between DNA adducts, mutations in oncogenes and tumorigenesis of carcinogenic environmental polycyclic aromatic hydrocarbons in strain A/J mice. *Toxicology* 103: 403–413.

Niimi T, Nagashima K, Ward JM, Minoo P, Zimonjic DB, Popescu NC, Kimura S (2001): Claudin-18, a novel downstream target gene for the T/EBP/NKX2.1 homeodomain transcription factor, encodes lung- and stomach-specific isoforms through alternative splicing. *Mol Cell Biol* 21: 7380–7390.

Palmiter RD, Brinster RL (1986): Germ-line transformation of mice. *Annu Rev Genet* 20: 465–499.

Pan Q, Li C, Xiao J, Kimura S, Minoo P (2004): In vivo characterization of Nkx2.1 promoter/enhancer elements in transgenic mice. *Gene* 331: 73–82.

Park KK, Liem A, Stewart BC, Miller JA (1993): Vinyl carbamate epoxide, a major strong electrophilic, mutagenic and carcinogenic metabolite of vinyl carbamate and ethyl carbamate (urethane). *Carcinogenesis* 14: 441–450.

Parkin DM, Bray FI, Devesa SS (2001): Cancer burden in the year 2000. The global picture. *Eur J Cancer* 37: 4–66.

Pepicelli CV, Lewis ML, McMahon P (1998): Sonic hedgehog regulates branching morphogenesis in the mammalian lung. *Curr Biol* 8: 1083–1086.

Perl AK, Wert SE, Nagy A, Lobe CG, Whitsett JA (2002): Early restriction of peripheral and proximal cell lineages during formation of the lung. *Proc Natl Acad Sci USA* 99: 10,482–10,487.

Peters K, Werner S, Liao X, Wert S, Whitsett J, Williams L (1994): Targeted expression of a dominant negative FGF receptor blocks branching morphogenesis and epithelial differentiation of the mouse lung. *EMBO J* 13: 3296–3301.

Peterson LA, Hecht SS (1991): O^6-Methylguanine is a critical determinant of 4-(methylnitrosamino)-1-(3-pyridyl)-1-butanone tumorigenesis in A/J mouse lung. *Cancer Res* 51: 5557–5564.

Peterson LA, Thomson NM, Crankshaw DL, Donaldson EE, Kenney PJ (2001): Interactions between methylating and pyridyloxobutylating agents in A/J mouse lungs: Implications for 4-(methylnitrosamino)-1-(3-pyridyl)-1- butanone-induced lung tumorigenesis. *Cancer Res* 61: 5757–5763.

Ray MK, Magdaleno SW, Finegold MJ, DeMayo FJ (1995): Cis-acting elements involved in the regulation of mouse Clara cell-specific 10-kDa protein gene. In vitro and in vivo analysis. *J Biol Chem* 270: 2689–2694.

Ray MK, Wang G, Barrish J, Finegold MJ, DeMayo FJ (1996): Immunohistochemical localization of mouse Clara cell 10-KD protein using antibodies raised against the recombinant protein. *J Histochem Cytochem* 44: 919–927.

Ray P, Tang W, Wang P, Homer R, Kuhn C 3rd, Flavell RA, Elias JA (1997): Regulated overexpression of interleukin 11 in the lung. Use to dissociate development-dependent and -independent phenotypes. *J Clin Invest* 100: 2501–2511.

Rehm S, Lijinsky W, Singh G, Katyal SL (1991): Mouse bronchial cell carcinogenesis. Histologic characterization and expression of Clara cell antigen in lesions induced by N-nitrosobis-(2-chloroethyl) ureas. *Am J Pathol* 139: 413–422.

Ronai ZA, Gradia S, El Bayoumy K, Amin S, Hecht SS (1994): Contrasting incidence of *ras* mutations in rat mammary and mouse skin tumors induced by *anti*-benzo[*c*]phenanthrene-3,4-diol-1,2-epoxide. *Carcinogenesis* 15: 2113–2116.

Rozengurt E (1999): Autocrine loops, signal transduction and cell cycle abnormalities in the molecular biology of lung cancer. *Curr Opin Oncol* 11: 116–122.

Ryding AD, Sharp MG, Mullins JJ (2001): Conditional transgenic technologies. *J Endocrinol* 171: 1–14.

Sandgren EP, Quaife CJ, Pinkert CA, Palmiter RD, Brinster RL (1989): Oncogene-induced liver neoplasia in transgenic mice. *Oncogene* 4: 715–724.

Sauer B, Henderson N (1988): Site-specific DNA recombination in mammalian cells by the Cre recombinase of bacteriophage P1. *Proc Natl Acad Sci USA* 85: 5166–5170.

Sekine K, Ohuchi H, Fujiwara M, Yamasaki M, Yoshizawa T, Sato T, Yagishita N, Matsui D, Koga Y, Itoh N, Kato S (1999): *Fgf10* is essential for limb and lung formation. *Nat Genet* 21: 138–141.

Shannon JM, Nielsen LD, Gebb SA, Randell SH (1998): Mesenchyme specifies epithelial differentiation in reciprocal recombinants of embryonic lung and trachea. *Dev Dyn* 212: 484–494.

Shaw-White JR, Bruno MD, Whitsett JA (1999): GATA-6 activates transcription of thyroid transcription factor-1. *J Biol Chem* 274: 2658–2664.

Shijubo N, Honda Y, Fujishima T, Takahashi H, Kodama T, Kuroki Y, Akino T, Abe S. (1995): Lung surfactant protein-A and carcinoembryonic antigen in pleural effusions due to lung adenocarcinoma and malignant mesothelioma. *Eur Respir J* 8: 403–406.

Shimkin MB, Stoner GD (1975): Lung tumors in mice: Application to carcinogenesis bioassay. *Adv Cancer Res* 21: 1–58.

Shockett PE, Schatz DG (1996): Diverse strategies for tetracycline-regulated inducible gene expression. *Proc Natl Acad Sci USA* 93: 5173–5176.

Shu W, Jiang YQ, Lu MM, Morrisey EE (2002): Wnt7b regulates mesenchymal proliferation and vascular development in the lung. *Development* 29: 4831–4842.

Simonet WS, DeRose ML, Bucay N, Nguyen HQ, Wert SE, Zhou L, Ulich TR, Thomason A, Danilenko DM, Whitsett JA (1995): Pulmonary malformation in transgenic mice expressing human keratinocyte growth factor in the lung. *Proc Natl Acad Sci USA* 92: 12,461–12,465.

Smithies O, Gregg RG, Boggs SS, Koralewski MA, Kucherlapati RS (1985): Insertion of DNA sequences into the human chromosomal beta-globin locus by homologous recombination. *Nature* 317: 230–234.

Stanton BR, Perkins AS, Tessarollo L, Sassoon DA, Parada LF (1992): Loss of N-myc results in embryonic lethality and failure of the epithelial component of the embryo to develop. *Genes Dev* 6: 2235–2247.

Stoner GD (1991): Lung tumors in strain A mice as a bioassay for carcinogenicity of environmental chemicals. *Exp Lung Res* 17: 405–423.

Stoner GD (1998): Introduction to mouse lung tumorigenesis. *Exp Lung Res* 24: 375–383.

Stoner GD, Shimkin MB (1982): Strain A mouse lung tumor bioassay. *J Am Coll Toxicol* 1: 145–169.

Stripp BR, Sawaya PL, Luse DS, Wikenheiser KA, Wert SE, Huffman JA, Lattier DL, Singh G, Katyal SL, Whitsett JA (1992): Cis-acting elements that confer lung epithelial cell expression of the CC10 gene. *J Biol Chem* 267: 14,703–14,712.

Sunday ME, Haley KJ, Sikorski K, Graham SA, Emanuel RL, Zhang F, Mu Q, Shahsafaei A, Hatzis D (1999): Calcitonin driven v-Ha-ras induces multilineage pulmonary epithelial hyperplasias and neoplasms. *Oncogene* 18: 4336–4347.

Tchou-Wong KM, Jiang Y, Yee H, LaRosa J, Lee TC, Pellicer A, Jagirdar J, Gordon T, Goldberg JD, Rom WN (2002): Lung-specific expression of dominant-negative mutant p53 in transgenic mice increases spontaneous and benzo(*a*)pyrene-induced lung cancer. *Am J Respir Cell Mol Biol* 27: 186–193.

Ten Have-Opbroek, AAW (1981): The development of the lung in mammals: An analysis of concepts and findings. *Am J Anat* 162: 201–219.

Tichelaar JW, Lim L, Costa RH, Whitsett JA (1999): HNF-3/forkhead homologue-4 influences lung morphogenesis and respiratory epithelial cell differentiation in vivo. *Dev Biol* 213: 405–417.

Tichelaar JW, Lu W, Whitsett JA (2000): Conditional expression of fibroblast growth factor-7 in the developing and mature lung. *J Biol Chem* 275: 11,858–11,864.

Titis AP, Forkert P-G (2001): Strain-related differences in bioactivation of vinyl carbamate and formation of DNA adducts in lungs of A/J, CD-1, and C57BL/6 mice. *Toxicol Sci* 59: 82–91.

Toth B (1968): A critical review of experiments in chemical carcinogensis using newborn animals. *Cancer Res* 28: 727–738.

Vegeto E, Allan GF, Schrader WT, Tsai MJ, McDonnell DP, O'Malley BW (1992): The mechanism of RU486 antagonism is dependent on the conformation of the carboxy-terminal tail of the human progesterone receptor. *Cell* 69: 703–713.

Wang M, Lemon WJ, Liu G, Wang Y, Iraqi FA, Malkinson AM, You M (2003): Fine mapping and identification of candidate pulmonary susceptibility 1 genes using advanced intercross lines. *Cancer Res* 63: 3317–3324.

Wang Y, DeMayo FJ, Tsai SY, O'Malley BW (1997): Ligand-inducible and liver-specific target gene expression in transgenic mice. *Nat Biotechnol* 15: 239–243.

Wang Y, Tsai SY, O'Malley BW (2000): An antiprogestin regulable gene switch for induction of gene expression in vivo. *Adv Pharmacol* 47: 343–355.

Weaver M, Yingling JM, Dunn NR, Bellusci S, Hogan BLM (1999): *Bmp* signaling regulates proximal-distal differentiation of endoderm in mouse lung development. *Development* 126: 4005–4015.

Weinstein M, Xu X, Ohyama K, Deng CX (1998): FGFR-3 and FGFR-4 function cooperatively to direct alveogenesis in the murine lung. *Development* 125: 3615–3623.

Wert SE, Glasser SW, Korfhagen TR, Whitsett JA (1993): Transcriptional elements from the human SP-C gene direct expression in the primordial respiratory epithelium of transgenic mice. *Dev Biol* 156: 426–443.

Whitsett JA, Clark JC, Picard L, Tichelaar JW, Wert SE, Itoh N, Perl AK, Stahlman MT (2002): Fibroblast growth factor 18 influences proximal programming during lung morphogenesis. *J Biol Chem* 277: 22,743–22,749.

Whitsett JA, Glasser SW, Tichelaar JW, Perl AK, Clark JC, Wert SE (2001): Transgenic models for study of lung morphogenesis and repair: Parker B. Francis lecture. *Chest* 120: 27S–30S.

Wikenheiser KA, Clark JC, Linnoila RI, Stahlman MT, Whitsett JA (1992): Simian virus 40 large T antigen directed by transcriptional elements of the human surfactant protein C gene produces pulmonary adenocarcinomas in transgenic mice. *Cancer Res* 52: 5342–5352.

Williams MC, Cao Y, Hinds A, Rishi AK, Wetterwald A (1996): T1 alpha protein is developmentally regulated and expressed by alveolar type I cells, choroids plexus, and ciliary epithelia of adult rats. *Am J Respir Cell Mol Biol* 14: 577–585.

Witschi H (2000): Successful and not so successful chemoprevention of tobacco smoke-induced lung tumors. *Exp Lung Res* 26: 743–755.

Witschi H, Espiritu I, Dance ST, Miller MS (2002): A mouse lung tumor model of tobacco smoke carcinogenesis. *Toxicol Sci* 68: 322–330.

Wong EA, Capecchi MR (1986): Analysis of homologous recombination in cultured mammalian cells in transient expression and stable transformation assays. *Somat Cell Mol Genet* 12: 63–72.

Yang H, Lu MM, Zhang L, Whitsett JA, Morrisey EE (2002): GATA6 regulates differentiation of distal lung epithelium. *Development* 129: 2233–2246.

Yuan B, Li C, Kimura S, Engelhardt R, Smith B, Minoo P (2000): Inhibition of distal lung morphogenesis in *Nkx2.1* (−/−) embryos. *Dev Dyn* 217: 180–190.

Zhao B, Chua SS, Burcin MM, Reynolds SD, Stripp BR, Edwards RA, Finegold MJ, Tsai SY, DeMayo FJ (2001): Phenotypic consequences of lung-specific inducible expression of FGF-3. *Proc Natl Acad Sci USA* 98: 5898–5903.

Zhao B, Magdaleno S, Chua S, Wang YL, Burcin M, Elberg D, Finegold M, Tsai S, DeMayo FJ (2000): Transgenic mouse models for lung cancer. *Exp Lung Res* 26: 567–579.

Zheng F, Kundu GC, Zhang Z, Ward J, DeMayo F, Mukherjee AB (1999): Uteroglobin is essential in preventing immunoglobulin A nephropathy in mice. *Nat Med* 5: 1018–1025.

7

MAMMARY GLAND CANCER

Dalit Barkan, Cristina Montagna, Thomas Ried, and Jeffrey E. Green
National Cancer Institute, National Institutes of Health, Bethesda, Maryland

It has been 20 years since the first genetically engineered mouse model of mammary cancer was published (Stewart et al., *1984) using transgenic technology. Since the first report there have been well over 100 mouse models of mammary cancer developed which have provided critical insights into the functions of numerous genes involved in mammary oncogenesis. Advances in manipulating the mouse genome have provided the technology to induce mammary cancers in mice that arise from genetic mutations that have been identified in human cancer, thus generating models based upon molecular lesions relevant to the human disease. Recent work has focused on validating these models as systems which represent, at least in part, various aspects of the human disease and which may be used for preclinical testing of novel therapies. The challenge remains to establish the predictive value of such models for chemoprevention and therapeutic response for treating human breast cancer.*

HUMAN BREAST CANCER

Breast cancer is the most common malignancy in women and constitutes 18% of all female cancers (Bange et al., 2001). Several risk factors are associated with an increased risk of developing breast cancer, including genetic factors, hormonal factors, preexisting benign lesions, and environmental factors. It has been estimated that less than one-third of breast cancers are related to genetic factors with environmental influences being the primary cause of at least two-thirds of the cases (Lichtenstein et al., 2000). Many animal models have been developed which attempt to recapitulate genetic lesions found in human breast cancer.

Genetic Factors Five to ten percent of women diagnosed with breast cancer have a defined hereditary predisposition to breast cancer (Claus et al., 1996). However, in the absence of a known inherited genetic mutation, women with a family history of breast cancer are at increased risk for developing breast cancer, possibly due to shared environmental exposures and/or unexplained genetic factors.

The most common type of cytogenetic alteration in the progression of both familial and sporadic breast cancers is the loss of heterozygosity (LOH) at multiple loci. LOH results following the deletion of a normal dominant gene at a cancer susceptibility locus, thus leaving only one, often mutated allele. LOH studies have identified several critical tumor suppressor genes related to breast cancer, including *TP53, BRCA-1, BRCA-2*, and *PTEN*. *TP53* (located on chromosome 17p13) was the first tumor suppressor gene shown to confer risk of an inherited breast cancer syndrome (Malkin et al., 1990) and is also commonly mutated in sporadic tumors (Culotta and Koshland, 1993). Mutations in *TP53* have been identified as the genetic abnormalities responsible for the Li-Fraumeni syndrome of hereditary breast cancer (Malkin et al., 1990; Srivastava et al., 1990).

P53 is a DNA-binding protein and has been referred to as the "guardian of the genome." This title reflects its critical function to trigger cellular responses to DNA damage, leading to cell cycle checkpoint arrest at the G1 phase or initiating an apoptotic response to eliminate

Mouse Models of Human Cancer, edited by Eric C. Holland
ISBN 0-471-44460-X

the damaged cell entirely (Lane, 1992). Mutations in P53 may result in loss of tumor suppressor activity, enhanced genetic instability, and gain of tumor promotion function (Culotta and Koshland, 1993; Elledge and Lee, 1995).

BRCA1 (Bluman et al., 1999; Ottman et al., 1983) and *BRCA2* (Miki et al., 1994) are two other important tumor suppressor genes associated with familial breast and ovarian cancer. Both *BRCA1* and *BRCA2* mutations are commonly inherited in an autosomal dominant manner. *BRACA1* (chromosome 17q21) contains a zinc- and DNA-binding "ring finger" motif, a carboxyl terminal *BRCT* domain, and a nuclear localization sequence (Miki et al., 1994). Mutations in *BRCA1* gene are associated with a 50% to 85% risk of developing breast cancer during a woman's lifetime, with a particularly striking predisposition to early-onset breast cancer (Ford et al., 1994; Struewing et al., 1997).

BRCA2 (chromosome 13q13) mutations are associated with familial breast cancer affecting both females and males and is also associated with ovarian cancer (Dickson and Lipmann, 2002). Women with *BRCA2* mutations are thought to be at a similar risk of developing breast cancer as those with *BRCA1* mutations (Phelan et al., 1996; Easton et al., 1997). Structure and function of the two proteins appear to be similar with both playing important roles in regulating cell cycle "check point" control mechanisms involving cell cycle arrest, cell death, and DNA repair (Dickson and Lipmann, 2002).

Breast cancer also occurs as part of other familial syndromes, including Cowden syndrome and Muir syndrome. Mutations in the *PTEN* (or *MMAC1*) gene (chromosome 10q22–23) have been described in Cowden's syndrome of heredity breast cancers and multicutaneous lesions (Liaw et al., 1997). *PTEN* encodes a 403-amino acid phosphatase that dephosphorylates phosphoinositides and phosphoamino acids which are important second messengers in signal transduction. *PTEN* functions to suppress cell survival and proliferation signaling (Ahmad et al., 1999; Datta et al., 1999; Weng et al., 1999). In Muir syndrome, mutations in the *MLH-1* and *MLH-2* genes (mismatch repair genes) have been identified (Anderson, 1980; Entius et al., 2000).

Environmental Factors Exposure to ionizing radiation increases breast cancer risk, especially if exposure occurs before the age of 40 (Preston et al., 2002). Women who received mantle irradiation for the treatment of Hodgkin's disease early in life have been shown to be at increased risk for breast cancer (Gervais-Fagnou et al., 1999). Other environmental factors have been suggested to increase breast cancer risk, but they are not conclusive. These include exposure to electromagnetic fields and organochlorine pesticides (Calle et al., 2002).

A link between diet and breast cancer risk has been proposed based on the large international variation in breast cancer incidence. Studies of immigrant groups suggested that these differences are not due solely to genetic factors (Buell, 1973; McMichael and Giles, 1988). Several studies have not found a relationship between consumption of specific types of fat and risk of breast cancer (Holmes et al., 1999; Hunter et al., 1996; Willett et al., 1987). However, a positive association between energy balance and breast cancer risk has been identified. Weight gain after age 18 was shown to be associated with increased risk in postmenopausal breast cancers (Huang et al., 1997; Ziegler et al., 1996).

The development of breast cancer in many women appears to be related to exposure to female reproductive hormones. Increased exposure to endogenous estrogens during early age at menarche, late age at first full-term pregnancy, and late age at menopause increases the risk of developing breast cancer (Gail et al., 1989; Rosner and Colditz, 1996). Elevated estrogen and estradiol levels in postmenopausal women, as a consequence of obesity or postmenopausal therapy, increase the risk of developing breast cancer (Cauley et al., 1999).

Genomic Changes The analysis of chromosomes in cancer cells has played an important role in elucidating genetic alterations of the cancer genome, and it is firmly established that chromosomal aberrations are key events during tumorigenesis (Heim and Mitelman, 1995). In hematologic malignancies, chromosomal aberrations are frequently present as balanced translocations; the genetic consequence of such translocations is the deregulation of a potential oncogene via aberrant expression or the formation of an oncogenic fusion protein (Rowley, 1990). In tumors of epithelial origin, like breast cancer, such balanced chromosomal translocations are rather the exception (Ghadimi et al., 1999, 2000; Heim and Mitelman, 1995). With the advance of methods for molecular cloning, cytogenetic analyses were complemented by techniques that are based on DNA fluorescence in situ hybridization (FISH). These molecular cytogenetic methods have greatly contributed to our understanding of the complex rearrangements observed in solid tumors, in particular in mouse models of carcinomas.

FISH analyses are based on the labeling of DNA probes with fluorescently tagged nucleotides. Following hybridization to metaphase chromosomes or interphase cells, for instance in tissue sections, entire chromosomes, chromosomal subregions, or specific genes can be visualized. Further development of FISH techniques led to the development of comparative genomic hybridization (CGH) (Kallioniemi et al., 1992) and spectral karyotyping

(SKY) (Liyanage et al., 1996; Schröck et al., 1996) or M-FISH (Speicher et al., 1996). These techniques permit visualization of the entire human or mouse tumor genomes in a single hybridization. CGH and SKY can therefore be used to survey chromosomal damage in cancer genomes without a prior knowledge of the cytogenetic abnormalities (Fig. 7.1). Numerous investigators have applied CGH to human breast carcinomas, and it is well reported that, similar to other tumors of epithelial origin, the pattern and distribution of chromosomal gains and losses are consistent in the variety of samples analyzed reviewed by Tirkkonen et al. (1998).

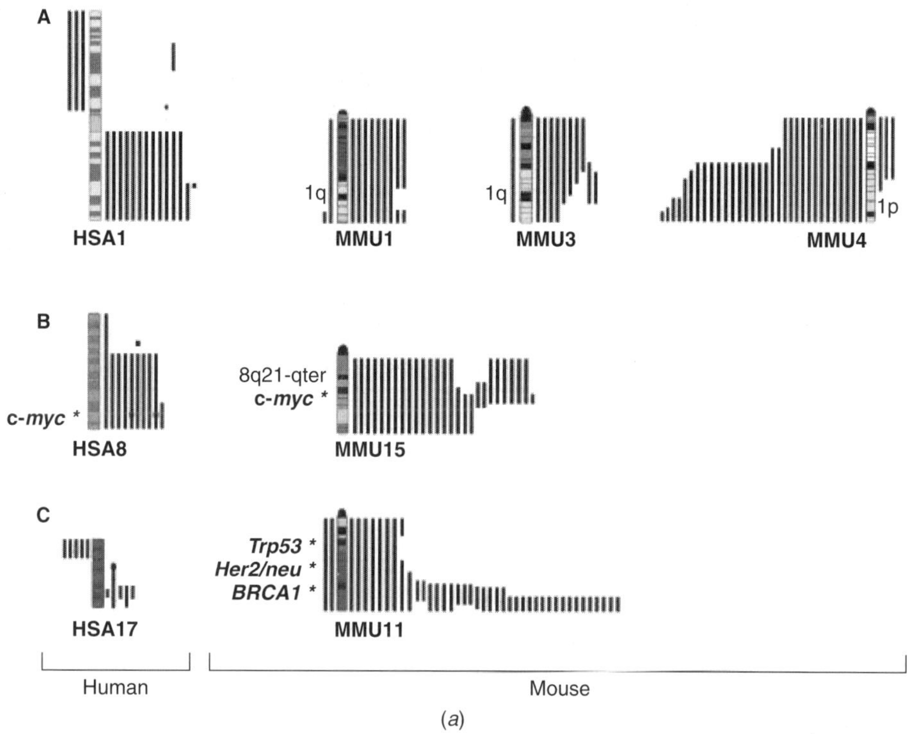

Figure 7.1. (*a*) **Summary of CGH and SKY analyses of human breast carcinomas and mouse mammary gland adenocarcinomas.** (A) In human breast cancers, chromosome arm 1p is lost, while copy number increases on chromosome arm 1q are common. The orthologous regions in the mouse genome are subject to similar copy number variations, although with different relative frequencies. These summary graphs are based on the analysis of 20 human breast carcinomas and 70 mouse tumors. (B) The comparative analysis of human chromosome 8 reveals similar results. The frequent gain of chromosome 8q and the c-myc oncogene (which maps to band 8q24) is recapitulated in murine tumor models. Note that the minimally overlapping region of genomic amplification on mouse chromosome 15 includes the locus of the c-myc oncogene. (C) The gain of the long arm of chromosome 17, which is commonly observed in human breast cancers, correlates well with genomic amplification of distal mouse chromosome 11. A divergence, however, exists. While copy number decreases that map to chromosome arm 17p are rather common in breast cancers, loss of orthologous regions in the mouse genome is an exception. The inactivation of the tumor suppressor gene TP53, which is a recurrent aberration in breast cancer, seems to be less important in mouse mammary gland tumorigenesis, or occurs by mechanisms, other than chromosome loss. (*b*) **Examples of involvement of human chromosomes 1, 8 and 17, and mouse chromosomes 1, 3, 4, 11, and 15 in rearrangements in BRCA1 deficient human breast cancer and murine tumor models that emerge in mice conditionally KO mutants for BRCA1.** In the human cell line, chromosome 1 is involved in numerous chromosomal translocations, which result in copy number increases of the long arm, yet loss of chromosome 1p. Equally, multiple segments of chromosome 8q are dispersed in this highly rearranged karyotype. Chromosome 17 appears to be normal. Orthologous chromosome in BRCA1 deficient murine tumors: Note that copy number losses of mouse chromosome 4 and gains of mouse chromosome 15 are similar to copy number changes in human BRCA1 related breast cancers. (See color insert).

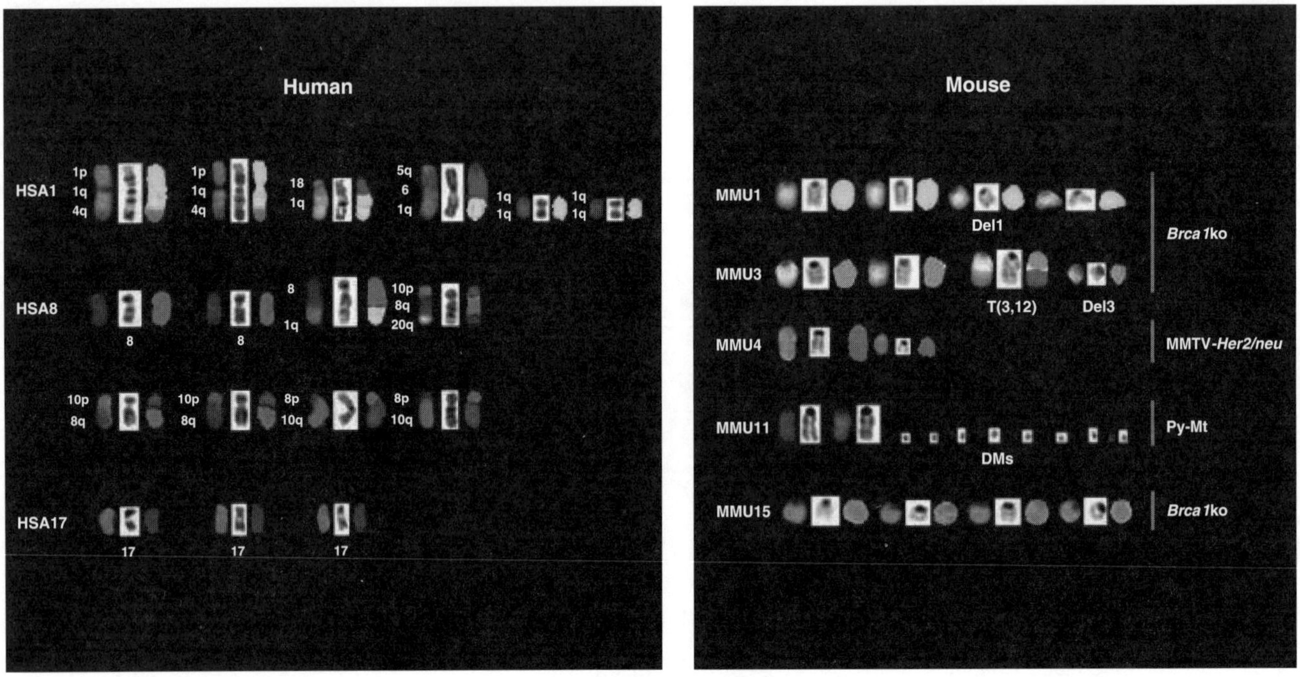

(*b*)

Figure 7.1. (*continued*).

Regions that are recurrently gained in human breast cancers include chromosome arms 1q, 8q, 11q, 17q, and 20q, whereas chromosome arms 8p and 17p are commonly lost. CGH analysis has also helped in the identification of novel regions of genomic amplification that usually point to the location of oncogenes. In breast carcinoma such amplicons map to chromosomes 17q and 20q (Kallioniemi et al. 1994; Ried et al., 1995). Based on the strictly conserved distribution of these imbalances, fluorescently labeled probe pools were devised that can be used for early detection of breast cancer cells in cytologic samples (e.g., fine-needle aspirates) with high sensitivity and specificity (Heselmeyer-Haddad et al., 2002).

Molecular Changes in Breast Cancer

Molecular Alterations in Sporadic Breast Cancer Sporadic breast cancer is the most prevalent form of breast cancer and is thought to arise from the accumulation of multiple mutations or alterations in the expression of important cell regulatory genes. Amplification of oncogenes like c-*myc* *ErbB-2/Her2*, cyclin D1, epithelial growth factor (EGF) receptor and mutations of *TP53*, and *CDH1*, the gene encoding E-cadherin, have been associated with breast cancer.

The nuclear transcription factor c-*myc* is known to mediate cell cycle progression, apoptosis, and genomic stability. In addition, c-*myc* plays a role in cellular transformation via both transcriptional up regulation and transcriptional repression of target genes (Dang, 1999; Facchini and Penn, 1998). In human breast cancer, c-*myc* is amplified in approximately 30% and overexpressed in approximately 70% of all cases, thus suggesting an important role in breast carcinogenesis (Strange et al., 1992). *ErbB-2/Her2* encodes a tyrosine kinase receptor belonging to the EGF receptor (EGFR) superfamily (Dougall et al., 1994). *ErbB-2* is overexpressed in approximately 30% of breast tumors and amplified in 20% of human breast cancer (Slamon et al., 1987). Elevated *ErbB-2* expression is detected in many in situ and invasive human ductal carcinomas and rarely observed in hyperplasias and dysplasias (Allred et al., 1992; Mansour et al., 1994). Elevated *ErbB-2* is associated with aggressive tumor phenotype and poor prognosis in invasive breast cancer. It correlates to tumor grade, size, relapse rate, and metastases (Hynes and Stern, 1994; Mansour et al., 1994). The *CCND1* gene encodes cyclin D1, an important regulatory molecule of the cell cycle. The overexpression of cyclin D1 results in abnormal cell proliferation. Cyclin D1 amplification occurs in approximately 30% of breast cancer cases and its prognostic significance is under investigation. The EGFR is aberrantly expressed in 40% of human breast cancers and occasionally overexpressed in those tumors with bad prognosis (Humphreys and Hennighausen, 2000; Normanno et al., 1994). This tyrosine kinase receptor plays a crucial role in the development and progression of human breast cancers.

Mutations in tumor suppressor genes are found in sporadic tumors. Fifteen to thirty-four percent of breast cancers have mutations in the *TP53* gene (Hartmann et al., 1997). The majority of the mutations are missense and some result in dominant-negative function. Failure of normal *P53* function normally results in an increased risk of genomic instability and mutations in additional genes (Donehower, 1997). The *CDH1* gene encodes the E-cadherin protein. This gene is located on chromosome 16q22.1, a region frequently deleted in breast tumors. E-cadherin has been shown to be mutated in a large portion of lobular breast tumors (Berx et al., 1995, 1996; Vos et al., 1997), where one copy of the gene is deleted and the other copy mutated. Reduced expression of E-cadherin has been detected in lobular breast cancers as well as in ductal breast cancers. E-cadherin is a calcium-dependent cell adhesion molecule, involving homophilic cell–cell interactions. The loss of function of E-cadherin is associated with enhanced malignant invasive growth of breast cancer cells (Ingvarsson, 1999). However E-cadherin, has been found to be overexpressed in inflammatory breast cancer, challenging our current understanding of the role of E-cadherin in the metastatic process (Colpaert et al., 2003).

Though mutations in *BRCA-1* have not been found in sporadic tumors, inactivation of the normal function of the protein may occur in sporadic tumors, including mislocalization of the protein (Chen et al., 1995) to the cytoplasm or decreased expression of *BRCA-1* during tumor progression (Thompson et al., 1995)).

Loss of heterozygosity of chromosomes 1, 3p, 6q, 7q, 8p, 9p, 10q, 11, 13q, 16q, 17, 18q, 22q, and X are frequently observed in breast cancer and may result in the inactivation or loss of tumor suppressor genes. However, it is still unclear what gene losses are critical in the pathogenesis of breast cancer. Further studies will resolve these questions.

Gene Expression Profiling of Human Breast Cancer

The application of gene expression profiling using microarray technology is a major advance in attempting to understand and categorize the molecular complexity of breast cancer. Several studies using complementary DNA (cDNA) arrays have established molecular profiles for tumor classification and have identified novel biomarkers of oncogenesis and potential candidate targets for therapeutic intervention (Desai et al., 2002a; Perou et al., 1999; Sorlie et al., 2001; van't Veer et al., 2002).

Several studies have begun to explore classifying sporadic breast tumors based upon alterations in gene expression. Perou et al. (1999) studied 42 patients with locally advanced disease. A subset of 476 genes were identified which could classify the tumors into four major groups: (1) *ErbB2*-positive cluster, (2) normal breastlike, (3) luminal cell-like, and (4) a novel subgroup of basal cell–like breast tumors. This classification was further validated by the study of Sorlie et al. (2001) in which the same intrinsic gene set was used to classify 78 breast carcinomas, 3 fibroadenomas, and 4 normal breast samples in which additional luminal subgroups were identified. In addition, the classification system was associated with clinical outcome (Perou et al., 1999; Sorlie et al., 2001).

Using a different microarray approach, Hedenfalk et al. (2001) found unique identifiers that could distinguish between tumors with *BRCA1*-positive mutations versus those with *BRCA2*-positive mutations. *BRCA1* mutations appeared to affect genes involved in DNA repair and apoptosis pathways. Furthermore, the mutations in the *BRCA1/2* gene resulted in gene expression profiles, which are distinct from sporadic tumors.

Other microarray studies have shown that breast cancers tend to segregate into clusters depending on the status of estrogen receptor (ER) expression. Gruvberger et al. (2001) and van't Veer et al (2002) demonstrated that the ER status of tumors was associated with distinctive gene expression patterns. This set of genes from the study of Gruvberger et al. were also used in another study (Khan et al., 2001) to train the artificial neural network to define a subset of signature genes whose expression patterns are proven to correctly predict whether a tumor is ER positive or ER negative.

Gene profiling has also been used to predict response to alternative therapies and clinical outcome. Another study using 98 tumors with known clinical outcome was performed using an artificial neural network approach to determine whether initial gene expression signatures were predictive of prognosis. This approach segregated tumors into two dominant ER-positive and ER-negative groups and identified a set of 231 genes that could predict clinical outcome (van't Veer, 2002). Ramaswamy et al. (2003) have identified a set of 17 genes in the primary adenocarcinomas that can serve as a molecular signature of metastasis in primary solid tumors.

All of this microarray work requires further validation using much larger numbers of patients but holds the promise of identifying important new features of oncogenesis and new molecular targets and individualizing patient therapy.

Progression of Breast Cancer Lesions and Relevance of Diagnostic and Prognostic Markers

The development and progression of breast cancer is thought to occur over many years and evolve through multiple stages (Fig. 7.2). There has been a dramatic increase in the incidence of invasive breast cancers over the past several decades. More than 80% of all mammary

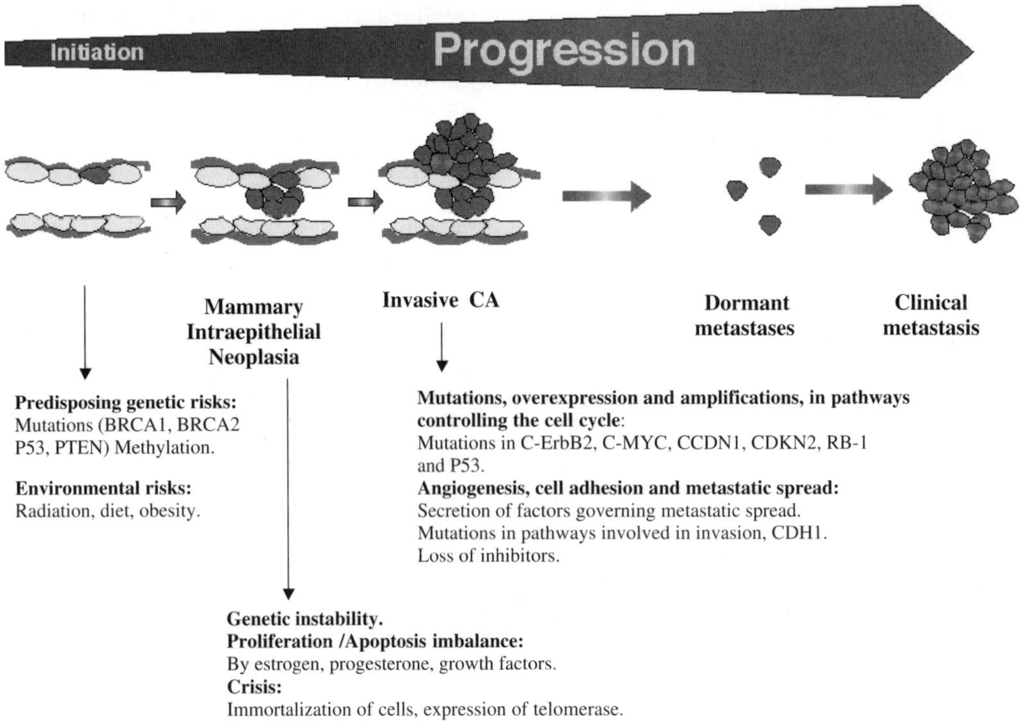

Figure 7.2. Biologic and molecular changes commonly associated with human breast cancer progression.

carcinomas are ductal carcinomas which are thought to originate from the terminal ductal lobular units. Ductal carcinoma in situ (DCIS) represents an early preinvasive stage in the development of invasive breast carcinoma. It is characterized pathologically by a proliferation of malignant epithelial cells within the mammary ductal–lobular system. However, there are several distinct histologic patterns of DCIS which are associated with different clinical outcomes (Schulze-Garg et al., 2000). Several systems have been proposed to classify DCIS lesions based upon morphologic criteria (Silverstein, 1997; Silverstein et al., 1995; Winer et al., 2002).

Hormone Responsiveness The status of ER-α expression in breast tumors is an important prognostic variable and generally identifies which patients can benefit from antihormonal therapy. In humans more then 60% of human breast cancers at initial presentation are ER positive. However, only two-thirds of these ER-positive tumors respond to endocrine therapy (Winer et al., 2002), whereas 5% to 10% of the cases that are diagnosed as ER negative respond initially to endocrine therapy (Allegra and Lippman, 1980; DeSombre et al., 1978; Elledge and Fuqua 2002; Osborne et al., 1980; Paridaens et al., 1980). The expression of a progesterone receptor (PR) is being utilized as another prognostic variable (Lamy et al., 2002).

ER and PR expression appear to be upregulated in DCIS and in hormone-dependent breast cancer compared to normal mammary epithelium. The determination of PR status improves the predictability of the hormone dependency of the tumor; however, this relationship remains imperfect and calls for future development of better assays for determining the functionality of ER and PR.

Unfortunately, tumors which are initially hormone responsive become hormone unresponsive during treatment or at relapse and are generally resistant to hormone therapy (Dowsett, 1996). Alterations in ER responsiveness may occur through several mechanisms, including promoter methylation (Ottaviano et al., 1994), variant splice forms (Elledge and Fuqua, 2002), or mutation (Fuqua, 1994).

The transition of mammary cancer from hormonal dependence to hormone independence is a key aspect of this disease, which is not understood. Debate continues as to whether hormone-responsive tumors actually become hormone independent or whether two separate cells of origin are responsible for these biologically diverse tumor types (Nandi et al., 1995). This has important implications for designing mouse mammary cancer models which to date have been primarily ER negative.

Metastases

The metastatic spread of breast cancer is the primary cause of patient morbidity and mortality. Although it has generally been assumed that metastases are a continuation of the process of tumor progression, recent microarray studies have suggested that the metastatic potential may be conferred very early in the development of breast cancer lesions (van de Vijver et al., 2002). Additional studies suggest that genetic predisposition may play an important role in determining the likelihood of mestastatic spread (Hunter et al., 2001). The process of metastases is reviewed elsewhere (Chambers et al., 2002).

Breast cancer cells will frequently metastasize to bone, liver, brain, and lungs. More then a century ago Stephan Paget (1889) tried to explain this phenomenon by suggesting the tumor cells ("seed") will grow only when cultivated in an appropriate organ ("soil"). James Ewing's (1928) theory contradicted the Paget theory. He suggested that circulatory patterns between the primary tumor and the specific secondary site are sufficient to explain this phenomenon. Current evidence supports a role for both theories (Chambers et al., 2002). Recent studies have shown that breast cancer cells as well as breast primary tumors express chemokine receptors CXCR4 and CCR7 at high levels. Interestingly their specific homing ligands *CXCL12* and *CCL21* are found at elevated levels at sites where breast tumors usually metastesize (Muller et al., 2001). Though physical factors also take part in determining the site where the cancer cells arrest, other factors such as chemokine receptor ligand interactions and changes in intracellular signaling may result in several alterations in the motility and invasion properties of the cells (Chambers et al., 2002). Once the cancer cells have reached the secondary site, they face several fate options. Cancer cells can exist in a secondary site as solitary cells, small preangiogenic metastases, or larger vascularized metastases. At each step, only a portion of the cells will proceed to the next phase and the remainder arrest in state of dormancy or die. Dormant solitary cells are cells that neither proliferate nor undergo apoptosis (Naumov et al., 2002). The "dormant" preangiogenic micromates are those cells in which active proliferation is balanced by active apoptosis (Folkman, 2002). Tumor growth will occur when this small metastasis acquires the ability to become vascularized.

Recurrence of breast cancer occurs after a long latent period, and metastasis may become manifested a decade or more after primary treatment. The dormant status of cancer cells may account for this clinical state and is further supported by studies that have modeled dormancy mathematically. These studies emphasize that continuous slow growth is unlikely and cannot account for the clinical latency (Demicheli, 2001). Therefore, for future therapeutic approaches, it is important to investigate the mechanisms leading to the maintenance and escape of the dormant cells.

Imaging

Early detection of breast cancer by mammographic imaging has become a critical approach in the battle against breast cancer and can substantially reduce mortality, although this has recently been questioned (Gotzsche and Olsen, 2000; Olsen and Gotzsche, 2001). A more recent study seems to support the use of imaging for improving survival from breast cancer (Fletcher and Elmore, 2003).

There are several imaging modalities under investigation for the early detection of breast cancer. Both anatomic imaging [i.e., mammography, computed tomography (CT), and magnetic resonance imaging (MRI)] and molecular imaging [i.e., positron emission tomography (PET)] are being investigated. Anatomic imaging detects structural abnormalities with high accuracy. Most experts consider routine mammography effective in preventing breast cancer deaths, at least in women over the age of 50 (Roque and Andre, 2002). The sensitivity of mammography for the detection of breast cancer approaches 85%. Mammography can detect smaller, nonpalpable breast cancers and DCIS (Moody-Ayers et al., 2000). However, mammography is partially effective in the detection of malignant tumors. The detection of malignant tumors depends on a number of factors, including breast tissue density, lesion size, patient age and hormone status, overall image quality, and experience and skills of the radiologists (Roque and Andre, 2002). A second limitation of mammography is its low specificity; 70% to 90% of breast biopsies reveal benign pathology (Winer et al., 2002). Thus, while mammographic screening reduces breast cancer mortality, there are women who cannot be screened reliably with mammography. Mammographic MRI in many studies demonstrates greater than 90% sensitivity for detecting invasive breast carcinomas (Arthur and Bluemke, 2002) and may be used for detecting malignancy in high-risk patients when mammography is compromised owing to radiographically dense breasts, scarring, or silicon augmentation. MRI is also a very useful tool to search for a primary tumor in patients who exhibit positive axillary nodes but have no evidence of breast cancer based on conventional imaging or physical examination (Arthur and Bluemke, 2002). There are also several studies suggesting the use of MRI in the preoperative staging and posttreatment monitoring of breast cancer.

Ultrasound (US) examination of the breast is most often used to differentiate a cystic from a solid mass detected on palpation or mammography. The use of US was also shown to differentiate benign from malignant masses, therefore limiting the need to perform a biopsy on lesions with benign features (Arthur and Bluemke, 2002). Diagnosis and staging of the axillary nodes is possible also by using U.S.-guided FNAB (fine-needle aspiration biopsy) of pathologic nodes, which increase the specificity. Another radiologic tool that is used for preoperative diagnosis of metastatic nodes is CT imaging. CT is a cross-sectional method that produces high-quality images of the axilla. The advantage of CT scanning is that there is a large field of view while a significant disadvantage is that small involved nodes may not be recognized (Lernevall, 2000).

The imaging agent 2-deoxy-2 (2) fluoro-D-glucose (FDG) measures elevated metabolism. Breast cancer cells exhibit markedly increased glucose metabolic activity and are therefore likely to be detected by FDG PET scanning. PET scanning together with the imaging agent 2-deoxy-2 (2) FDG diagnoses, stages, and restages cancer with high diagnostic accuracy. PET scanning detects lesions over 1 cm in size. Therefore, the size of most cancers at the time of detection is well within the special resolution of dedicated PET scanners. Furthermore, dense breasts do not exhibit increased glycolitic activity and, therefore, do not affect the cancer detection. However, PET scanners cannot reliably detect DCIS and should not be used as a general screening tool because of its limited spatial resolution after image reconstruction (Czernin, 2002). FDG PET does, however, stage axillary node involvement with an accuracy of about 85% (Czernin, 2002), since small involved nodes may be missed. Whole-body staging is also very important for deciding certain treatments for breast cancer patients, and PET whole-body staging tends to be superior to the combined conventional imaging. It was reported that PET could detect distant metastatic disease with an accuracy of 100% (Schirrmeister et al., 2001).

Chemoprevention

Chemoprevention holds the greatest promise of substantially reducing the morbidity and mortality of breast cancer. The success of inhibiting estrogen signaling in ER-positive breast cancer has clearly demonstrated the value of the chemopreventive approach. The major groups of chemopreventive agents can be classified according to their function: selective estrogen receptor modulators (SERMs); selective estrogen receptor downregulators (SERDs), which lack uterine agonist activity (Cummings et al., 1999; Dauvois et al., 1993; Gibson et al., 1991; Osborne et al., 2000; Wakeling et al., 1991); aromatase inhibitors; isoflavones, which alter estrogen metabolism; and gonadotropin-releasing hormone agonists. Agents which are more active in ER-positive tissues include retinoids, rexinoids, and deltanoids. Other agents are thought to be active regardless of the hormone receptor status and include polyamine synthesis inhibitors, tyrosine kinase inhibitors, cyclooxygenase 2 (COX-2) inhibitors, combined demethylating agents and histone deacetylase inhibitors, metalloprotease inhibitors, and angiogenesis inhibitors (Gibson et al., 1991). Since human cancer prevention trials are difficult to perform, testing such compounds in animal models may lead to important insights in a relatively short period of time.

Inhibitors of Estrogen Signaling Interference of estrogen signaling has proved to be quite beneficial in treating hormone-responsive breast cancers and preventing recurrence (Gajdos and Jordan, 2002). As a chemoprevention drug, tamoxifen has been shown to decrease the incidence of invasive and noninvasive breast cancer by 50%. Greater reductions were seen in women with a prior diagnosis of atypical ductal hyperplasia or lobular carcinoma in situ (Fisher et al., 1998). Tamoxifen was the first agent approved for this purpose and acts by inhibiting translocation and nuclear binding of estrogen receptors, thereby blocking estrogen stimulation of breast cancer (Jaiyesimi et al., 1995; Tonetti and Jordan, 1995). However, tamoxifen is associated with an increased risk of uterine cancer. Another potential clinical problem that may emerge upon treatment with tamoxifen is the development of resistance to the drug (Wiebe et al., 1993). Several mechanisms have been proposed for the emergence of resistance (Erlichman and Loprinzi, 2002).

To reduce adverse side effects of an estrogen-signaling blockade, SERMs have been developed. Newer generation SERMs such as toremifen and raloxifene have become available which may have fewer side effects. Raloxifene, like tamoxifen, reduces the incidence of ER-positive tumors but has no growth-promoting effect on the uterine epithelium (Rosenbaum Smith and Osborne 2000)

Aromatase Inhibitors Aromatase inhibitors block the synthesis of estrogen from androgens in the peripheral tissues. The peripheral tissue is the main source of estrogen synthesis in postmenopausal women. Aromatase inhibitors fall into two classes: type I agents, which bind to the androgen substrate binding site of the enzyme (irreversible inhibition), and type II agents, which are nonsteroidal drugs, usually azoles, which have a reversible mechanism of action (Brodie et al., 1999). It was shown that the aromatase inhibitors can decrease proliferation and increase apoptosis in breast tissues (Sasano et al., 1999). There are new and highly potent and selective aromatase inhibitor agents: letrozole, anastrozole, and exemestane, which are being clinically tested. They have been shown to be effective in patients whose cancer progresses on tamoxifen therapy. Another major benefit of these inhibitors is the relative lack of toxicity (Brodie et al., 1999).

Retinoids Preclinical experimental studies have shown that retinoids inhibit cancer cell proliferation and show substantial antitumor effects. Retinoids regulate proliferation and apoptosis and interfere with activator protein 1 (AP1) signaling. Interestingly, AP1 signaling has been shown to have enhanced activity in certain tamoxifen-resistant breast cancer cells (Schiff et al., 2000). These observations make retinoids an attractive candidate for treatment of hormone-insensitive breast cancer. Fenretinoid, a synthetic retinoid, is now in clinical trial (Torrisi and Decensi, 2000)

Polyamine Synthesis Inhibitors Difluoromethylornithine (DFMO) is an irreversible inhibitor of ornithine decarboxylase, the rate-limiting enzyme of polyamine synthesis that is often up regulated in breast cancer. It is attractive as a potential chemopreventive agent because it has the potential to disrupt biologic processes important for both ER-negative and ER-positive tumor formation (Glikman et al., 1989; Hobbs et al., 2002) and has minimal side effects at low doses. DFMO is currently being used in a double-blind randomized phase II chemoprevention trial conducted in a group of women at high risk for the development of breast cancer. However, DFMO administered for six months does not modulate breast risk biomarkers tested in this study (Fabian et al., 2002).

Tyrosine Kinase Inhibitors Up regulation of cyclin D1 has been demonstrated in 40% of all breast cancers, emphasizing its importance in tumor formation. Cyclin D1 forms a holoenzyme complex with cyclin-dependent kinase and thus regulates the transition through the cell cycle. Several agents such as staurosporine were designed to inhibit the function of the cyclin-dependent kinases to suppress tumor growth (Sausville et al., 2000). *her-2/ERbB2* is also found to be amplified and overexpressed in 60% of ductal carcinoma in situ (Hynes and Stern, 1994). Herceptin is a monoclonal antibody to *her-2/ERbB2* developed to block the tyrosine kinase activity of this receptor. In combination with chemotherapy, it was shown to increase the overall response rate as well as to increase mean survival time by 25% (Slamon et al., 2001). However, it may lead to heart failure in combination with doxorubicin. (Cook-Bruns, 2001).

COX-2 Inhibitors Prostaglandins are overproduced in many cancers and induce cell proliferation, motility, and tumor-associated angiogenesis (Nathanson et al., 2001). COX-1 and COX-2 are the enzymes which control the rate-limiting step in the production of prostaglandins. Epidemiologic studies suggest that nonsteroidal anti-inflammatory drugs may reduce the relative risk of breast cancer (Harris et al., 1999; Sharpe et al., 2000). This is now under further investigation.

Histone Deacetylase Inhibitors Changes in chromatin structure may silence gene expression. The histone deacetylase enzyme inhibitor blocks histone deacetylase activity, thus leading to a less compact chromatin structure and reactivation of genes, including tumor suppressor genes (Collingwood et al., 1999). The human acetylase *EP300* was found to be mutated in some epithelial cancers and was shown to be a tumor suppressor (Gayther et al., 2000). The deacetylase inhibitor suberoylanilide hydroxamic acid was shown to lead to cell cycle arrest and growth inhibition of breast cancer cells (Huang and Pardee, 2000; Richon et al., 2000). This drug is now being evaluated in phase I clinical trials.

Metalloprotease Inhibitors and Angiogenesis Inhibitors Tumor expansion and metastasis often involve the induction of matrix metalloprotease (MMP) enzymes and induction of angiogenic factors like vascular endothelial growth factor (VEGF) (Chambers et al., 2002; Folkman, 2002). MMP enzymes have been implicated in tumor invasion and metastasis. MMP enzymes enable malignant cells to break down basement membranes and subsequently invade into other cellular compartments. Some inhibitors of MMP enzymes are already in clinical trials and some in preclinical development. Antiangiogenic compounds, which prevent proliferation of endothelial cells directly or inhibit VEGF and fibroblast growth factor (FGF), are also in clinical trials. The SU5416 is one of the inhibitors directed against the receptor tyrosine kinase for VEGF receptor (FLK-1/VEGF-2) and is the most advanced in phase III trials for the treatment of a variety of solid tumors. Inhibitors of FLK-1 activity are in early clinical trials (Bange et al., 2001). These antiangiogenic drugs may have additional value in treating breast cancer since the endothelial cell is the primary target of action. These cells are genetically stable and, therefore, are not likely to develop resistance to treatments (Boehm et al., 1997). The use of antiangiogenic compounds may be particularly useful in combination with other therapies.

MODELING MAMMARY CANCER IN THE MOUSE

Genetics and Methods

The first inbred mouse strain having a high incidence of mammary tumors was developed more than 60 years ago (Callahan and Smith, 2000). The finding of the "milk factor" in mouse strains having high incidence of mammary tumors led to the discovery of the mouse mammary tumor virus (MMTV) and cellular proto-oncogenes. (Bittner, 1936; Hennighausen, 2000). *Wnt-1* was the first gene discovered, to be activated by viral insertion in mouse mammary tumors (Nusse and Varmus, 1982). Furthermore, mouse mammary tumor virus long terminal repeat (MMTV-LTR) could direct expression of oncogenes to the mammary gland (Stewart et al., 1984). These findings led to a new era in designing genetically engineered mouse models of mammary cancer.

Genetically engineered mouse models of mammary cancer have become important model systems to study breast cancer. Many models have been designed to mimic human breast cancer using several approaches: (1) overexpression of genes associated with the human cancer development, (2) knockout or knockin of target genes, and (3) generation of dominant-negative proteins to disrupt normal protein function.

The majority of transgenic models use the MMTV-LTR promoter (Stewart et al., 1984) or the mammary-specific whey acidic protein (WAP) promoter to overexpress oncogenes (Schoenenberger et al., 1988). Both of these promoters are hormonally regulated, and pregnancy has often been used to stimulate transgene expression and a phenotype in the mice. The MMTV-LTR may be expressed throughout mammary gland development, with increased expression during pregnancy, while the WAP promoter is generally activated during pregnancy and lobuloalveolar differentiation (Hutchinson and Muller, 2000). Since early pregnancy is a protective factor in humans for reducing the risk of breast cancer (Kampert et al., 1988), the stimulation of a transgene by pregnancy is a disadvantage of these models for studying mammary cancer progression as it relates to human breast cancer biology. Furthermore, expression patterns of either the MMTV-LTR or WAP vary among transgenic mouse lines due to variability in the site of transgene integration sites. A β-lactoglobulin (BLG) promoter (Barash et al., 1994) is another mammary-specific promoter that has also been utilized for transgenic studies. Other promoters have been used to target the mammary epithelium, including the rat prostatein C3(1) (Green et al., 2000) or cytokeratin K14 promoters (Jonkers et al., 2001). However, these promoters are not mammary specific and are active in epithelial cells of multiple tissues. More recently, the RCAS system of gene delivery (Fults et al., 2002) has been applied to mammary cancer modeling (Yi et al., unpublished data) and will provide an important means of studying the interactions between multiple genes in mammary oncogenesis.

While well over 100 mammary cancer models have been developed, this review will only focus on several representative examples of commonly used models that have contributed to our understanding of the disease. A more comprehensive listing of available mammary cancer models is provided in Table 7.1. The mouse models that overexpress the oncogenes *myc, Her2/neu*, polyomavirus middle T (*PyMT*), and *Wnt-1* will be discussed, as will models that inactivate tumor suppressor genes, including SV40 TAg, p53, BRACA, and PTEN knockout mice and mice with certain compound genetic mutation. The reader is referred to the National Cancer Institute (NCI) Cancer Models Database (http://emice.nci.nih.gov) and the Jackson Labs Mutant Mouse Database for additional and updated listings of mammary cancer mouse models.

The validation of transgenic mammary cancers model as systems to study aspects of human breast cancer must include such parameters as the natural history of disease progression, histopathology, gene and/or protein expression patterns, hormone responsiveness and metastasis to other organs, and response to therapeutic agents.

Histologic Comparison to Human Disease and Correlation to Human Tumor Gene Expression

In 1999, a landmark classification system was developed by veterinary and human pathologists to establish a unified nomenclature to describe the various types of histopathologic mammary lesions of genetically engineered mice (GEM), based upon an examination of 39 mouse models of mammary cancer (Cardiff et al., 2000). The Annapolis consensus identified distinct categories of mammary gland lesions: (1) lesions that resemble those found in spontaneous mouse mammary tumorigenesis, (2) lesions that have a unique "signature" tumor phenotype that was specific for the transgene, and (3) lesions that resemble those found in human breast diseases (Fig. 7.3).

The classification scheme and nomenclature are outlined in Table 7.2.

Mouse Models which Overexpress Oncogenes Associated with Human Breast Cancer

Myc. c-*myc* was the first human oncogene to be overexpressed in the mouse mammary epithelium (Stewart et al., 1984). Several different c-*myc* transgenic mammary models have been developed (Jamerson et al., 2000) that overexpress c-*myc* in a mammary cells under the MMTV-LTR (Stewart et al., 1984) or WAP promoters (Schoenenberger et al., 1988). Some of the *myc* tumors develop areas with densely sclerotic stroma, which is a typical feature of human breast cancer (Cardiff and Wellings, 1999; Cardiff et al., 2000). A more aggressive tumor phenotype is observed when additional oncogenes are coexpressed in the same animal, such as *Ras* (Sinn et al., 1987; Jamerson et al., 2000). TGFα can act to suppress c-*myc*-induced apoptosis and cooperate with c-*myc* in promoting a synergistic cell cycle progression (Amundadottir et al., 1995). *TGFα/myc* bitransgenic mice develop multiple mammary adenocarcinomas with a much reduced latency compared to c-*myc* transgenic mice and do so without requiring pregnancy, ovarian hormone stimulation, or estrogen (Jamerson et al., 2000; Sandgren et al., 1995).

TGFα. Members of the EGF family, including *TGFα*, can function as autocrine and/or juxtacrine growth factors in mammary epithelial cells (Normanno et al., 1994).

Table 7.1. Mammary Cancer Models

Model	Promoter	Phenotype	References
		Transgene Models	
FGF-3 (int-2)	MMTV-LTR	Hyperplasia, adenocarcinoma	Muller et al., 1990
FGF7 (KGF)	MMTV-LTR	Hyperplasia, adenocarcinoma, metastasis	Kitsberg and Leder, 1996
Heregulin (NDF)	MMTV-LTR	Hyperplasia, adenocarcinoma	Krane and Leder, 1996; Weinstein and Leder, 2000
Transforming growth factor (TGFα)	MMTV-LTR	Abnormal mammary gland development, hyperplasia, adenoma, adenocarcinoma	Jhappan et al., 1990; Matsui et al., 1990; Sandgren et al., 1990, 1995
ErB2 (neu)	MMTV-LTR	Hyperplasia, adenocarcinoma, lung metastasis	Guy et al., 1992; Muller et al., 1998
RET-1	MMTV-LTR	Hyperplasia, adenocarcinoma	Iwamoto et al., 1990
Polyoma virus middle T (PyV-mT)	MMTV-LTR	Hyperplasia, adenocarcinoma, lung metastasis	Guy et al., 1992
RAS	MMTV-LTR	Hyperplasia, adenocarcinoma	Sinn et al., 1987
Ck2	MMTV-LTR	Hyperplasia, dysplasia, adenocarcinoma	Landesman-Bollag et al., 2001
WNT 1	MMTV-LTR	Hyperplasia, adenocarcinoma	Tsukamoto et al., 1988
NOTCH4 (INT3)	MMTV-LTR, WAP	Dysplasia, adenocarcinoma, lung metastasis	Smith et al., 1995; Gallahan et al., 1996
WNT10b	MMTV-LTR	Hyperplasia, adenocarcinoma	Lane and Leder, 1997
MTS1	MMTV-LTR	Adenocarcinoma and lung metastasis	Ambartsumian et al., 1996
β-Catenin	MMTV-LTR	Hyperplasia, adenocarcinoma	Michaelson and Leder, 2001
COX-2	MMTV-LTR	Hyperplasia, dysplasia, adenocarcinoma and lymph node metastasis	Liu et al., 2001a
CDC37	MMTV-LTR	Hyperplasia, adenocarcinoma, Adenosquamous carcinoma	Stepanova et al., 2000
MT-MMP-1	MMTV-LTR	Hyperplasia, dysplasia, periductal fibrosis, adenocarcinoma, lung metastasis	Ha et al., 2001
Stromelysin	WAP	Hyperplasia, adenocarcinoma, carcinoma, lung and kidney metastasis	Sternlicht et al., 1999
SV40 TAg	WAP; C3(1)	Hyperplasia, adenocarcinoma, soft-tissue sarcoma, osteosarcoma, lung metastasis	Husler et al., 1998; Maroulakou et al., 1994; Tzeng et al., 1993
HGF	MT	Abnormal mammary gland development, adenocarcinoma, adenosquamous carcinoma	Takayama et al., 1997
IGFII	BGL, H19	Hyperplasia, adenoma, adenocarcinoma, adenosquamous carcinoma, metastasis to spleen	Bates et al., 1995; Pravtcheva and Wise, 1998
TGFα	MMTV-LTR	Hyperplasia, adenomas, adenocarcinomas	Matsui et al., 1990
		Gene-Targeting Models	
PTEN*		Fibroadenoma, adenocarcinoma metastasis to lung and lymph node in *PTEN* (+/−) mice	Stambolic et al., 2000
PTEN*	MMTV-LTR	Hyperplasia, fibroadenoma, adenocarcinoma	Li et al., 2002

(*continued overleaf*)

Table 7.1. (*continued*)

Model	Promoter	Phenotype	References
BRCA1* (conditional D of exon 11 or constitutively truncation)	MMTV-Cre; WAP-Cre	Hyperplasia and adenocarcinoma in *Brca1*$^{ko/co}$ mice; adenocarcinoma in mice homozygous for truncated Brca1 mice (*Brca1*$^{tr/tr}$)	Ludwig et al., 2001; Xu et al., 1999
BRCA2* (conditional D of exons 3 and 4; constitutively D of exon 27)	WAP-Cre	Hyperplastic alveolar nodule (HAN); invasive adenocarcinoma adenosquamous carcinoma, spindle cell carcinoma	Ludwig et al., 2001; McAllister et al., 2002
P-cadherin*		Hyperplasia, dysplasia	Radice et al., 1997
APCmin**		Hyperplasia and adenocarcinoma in heterozygote	Moser et al., 1993
		Compound Genetic Alterations	
FGF3(int-2) × WNT1	MMTV-LTR	Accelerated tumorigenesis and higher incidence in bitransgenic mice	Kwan et al., 1992
TGFα × MYC	WAP	Shortened tumor latency; tumor in nulliparous female and male in bitransgenic mice	Sandgren et al., 1995
ErbB2(neu) × p53-172 H	MMTV-LTR, WAP	Shortened tumor latency	Li et al., 1997
PyV-mT × SHCor GRB2	MMTV-LTR	Accelerated tumor progression	Rauh et al., 1999
RAS × MYC	MMTV-LTR	Accelerated tumorigenesis	Sinn et al., 1987
RAS × p21*	MMTV-LTR	Accelerated tumorigenesis; metastasis in lung and abdomen in *RAS*;*p21*$^{-/-}$mice	Adnane et al., 2000
Akt × Mutant pyv-middleT (Mty315/322F)	MMTV-LTR	Shortened tumor latency in bitransgenic mice	Hutchinson et al., 2001
p53* × Brca1*		Increased tumor incidence in *Brca1*$^{+/-}$; *p53*$^{-/-}$ mice	Cressman et al., 1999
MYC × rtTA	MMTV-LTR, TetO	Hyperplasia, adenocarcinoma, invasive adenocarcinoma	D'Cruz et al., 2001; Stewart et al., 1984
MYC × BCL-2	MMTV-LTR; WAP	Accelerated tumorigenesis	Jager et al., 1997
SV40TAg × BCL-2	WAP	Inhibition of apoptosis, shortened tumor latency	Furth et al., 1999
SV40TAg × bax	C3(1)	Accelerated tumorigenesis, decreased survival rates	Shibata et al., 1999
SV40TAg × Ki-Ras*	C3(1)	Decreased tumor incidence in $SV40 - TAg$; $Ki - Ras^{+/-}$ mice	Liu et al., 2001b
SV40TAg × maspin	WAP	Reduced tumor growth angiogenesis and metastasis	Shi et al., 2003
BRCA1* × p53*	MMTV-Cre; WAP-Cre	Invasive carcinoma in Brcalko/co; MMTV-Cre*Trp53*$^{+/-}$ mice	Xu et al., 1999
CDC37 × CYCLIN D1	MMTV-LTR	Shortened latency, lung metastasis	Stepanova et al., 2000
BRCA2* (conditional D of exon 11) × p53* (conditional inactivation)	K14-Cre	Dysplasia, adenocarcinoma, adenosquamous carcinoma	Jonkers et al., 2001

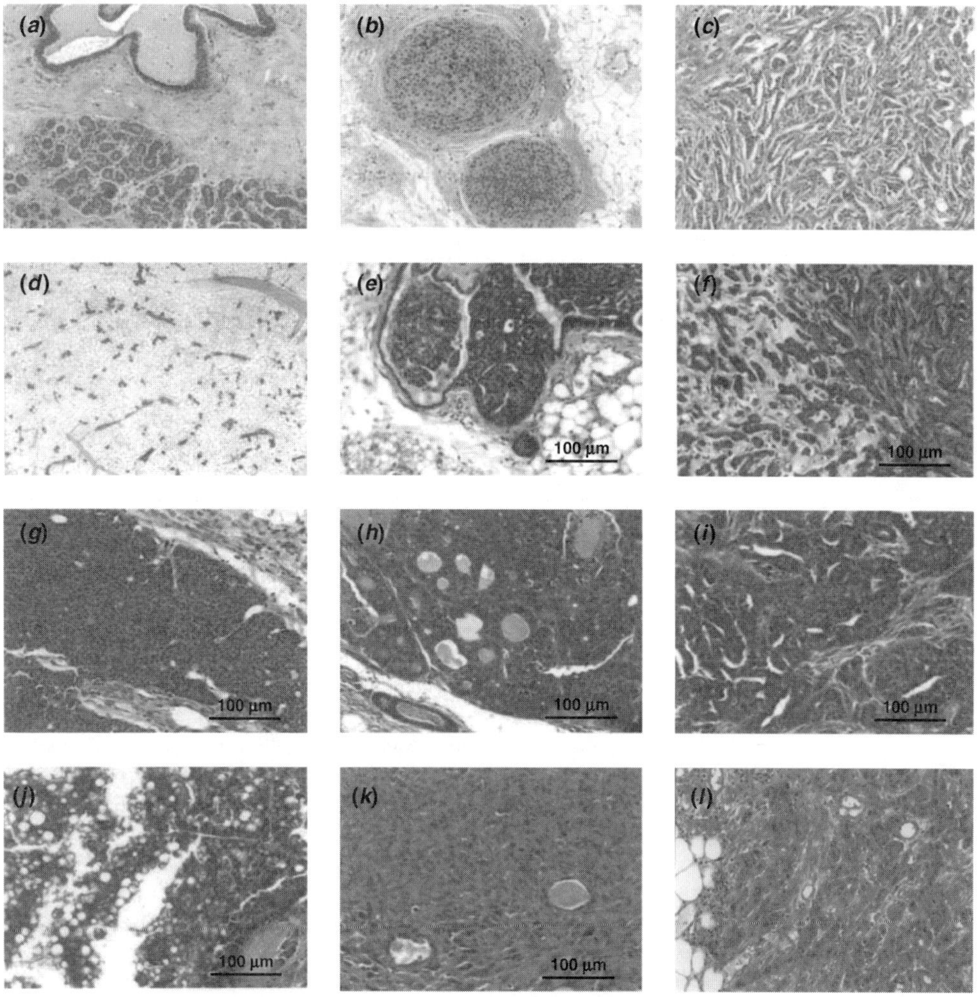

Figure 7.3. Histologic comparisons of human and mouse mammary tumors: (*a*) normal human non-lactating mammary gland; (*b*) human DCIS (20X); (*c*) human poorly differentiated infiltrating breast carcinoma. Nuclei are highly pleomorphic with tumor cells surrounded by dense connective tissue; (*d*) normal mouse adult virgin FVB/N mammary gland (20X); (*e*) mammary intraepithelial neoplasia in C3(1)/TAg mouse resembling human DCIS.; (*f*) invasive adenocarcinoma in C3(1)/TAg mouse resembling human tumor in C; (*g*) MMTV-her2/neu tumor with typical undifferentiated, solid, nodular appearance. Tumor is highly vascularized with uniform nuclei and red/pink cytoplasm; (*h*) MMTV-myc tumor with cystic and necrotic features. Large blue cells with prominent nucleoli are typical of myc-induced tumors; (*i*) MMTV-PyMT tumor with well differentiated and poorly differentiated characteristics. Papillary and invasive features with significant fibrosis are visible; (*j*) MMTV-TGFα adenosquamous carcinoma with secretory and papillary characteristics; (*k*) poorly differentiated adenocarcinoma from p53−/− transplant model; (*l*) tumor from BRCA1−/− mouse model. Images A-K and diagnoses courtesy of Robert Cardiff, UC Davis (Mammary Cancer in Humans and Mice; a tutorial for comparative pathology CD ROM) and image L courtesy of Chuxia Deng, NIDDK, NIH. (See color insert).

TGFα overexpression under the control of the MMTV-LTR leads to hyperplasia of alveoli and terminal ducts in female transgenic mice. Multiple histologic abnormalities develop, including lobular hyperplasias, cystic hyperplasias, adenomas, and adenocarcinomas (Matsui et al., 1990).

Her-2/neu. Several strains of mice overexpressing *Her-2/neu* under the MMTV-LTR promoter have been derived. These mice exhibit rapid mammary tumor development (Guy et al., 1996; Muller and Illiger, 2000) and have histologic features of human DCIS and comedo-carcinomas (Cardiff and Wellings, 1999). These

Table 7.2. Nomenclature for Describing Histopathology in GEM Mammary Cancer Models

Descriptor	Definition
Glandular	Tumor is composed of glands.
Acinar	Tumor is composed of small glandular clusters with small lumens. While this is a subclass of glandular, it is very characteristic of MMTV-induced tumors.
Cribriform	Tumor is composed of sheets or nests of cells forming lumens with round, punched-out spaces.
Papillary	Tumor has fingerlike projections composed of epithelium covering a central cerebrovascular core.
Solid	Tumor is composed of solid sheets of epithelial cells with little or no glandular differentiation.
Squamous	Tumors are composed solely of squamous cells with or without keratinization; absence of glandular pattern.
Fibroadenoma	Tumor is composed of a proliferation of both myxoid and fibrous stroma and glands.
Adenomyoepithelioma	Tumor is composed of myoepithelium and glands.
Adenosquamous	Tumor has both glandular and squamous differentiation.
Not otherwise specified (NOS)	Tumor does not have any of the other common descriptor patterns.
Modifiers	**Definition**
Biologic potential	
Carcinoma	Neoplasm originating from epithelium with proven malignant biologic behavior
Adenocarcinoma	Neoplasm originating from glandular epithelium with proven malignant biologic behavior
Adenoma	Neoplasm originating from glandular epithelium without proven malignant biologic behavior
Mammary intraepithelial neoplasia (MIN)	Spectrum of intraluminal epithelial proliferations with cytologic atypia, including in situ carcinomas
Hyperplasia	Any increase in cell number without cytologic atypia
Tumor	Any space-occupying mass with unknown biologic potential
Property	
Atypia	Cells with abnormal nuclear morphology
Necrosis	Cell death generally not applied to programmed cell death (apoptosis)
Fibrosis	Increased or abnormal deposition of connective tissue
Secretory	Tissues or glands producing and exporting lipid or protein
Metaplasia	A change from one adult cell type to another adult cell type
Topographic	
Diffuse	All of the mammary gland involved in the process
Focal	One area of the mammary gland involved in the process
Multifocal	Multiple foci in mammary gland
Inducer (etiology)	
Gene induced	Tumors that have morphologic or cytologic patterns characteristic of specific transgenes or specific mutations (*myc* type, *ras* type, *erbB2* type)
MMTV induced	Tumors known to be induced by (MMTV)
Chemically induced	Tumors known to be induced by chemical carcinogen
Hormone induced	Tumors known to be induced by exogenous hormones
Biological/experimental context	
Biologic	Parity, pregnancy, lactation, involution, hormones
Experimental	Promoter, exogenous hormones or chemicals

Source: Cardiff et al. (2000).

models have elevated expression of ErbB-2 protein in both normal and tumor epithelium. However, in the epithelia of the tumors there is elevation in the phosphorylated form of ErbB-2 as a consequence of the acquisition of activating mutations in the *erb*B-2 transgene (Siegel et al., 1994). Although such somatic mutations are generally not found in the human cases, an alternative splice form of ErbB-2 with oncogenic activity has been described. This alternative splice form possesses a deletion in the juxta-transmembrane domain of ErbB-2 that bears remarkable similarity to the sporadic mutations in the MMTV/*erb*B-2 strains (Siegel et al., 1999; Dankort and Muller, 2000). Thus, this novel splice variant may play a critical role in the induction of human breast cancer.

While transgenic overexpression of oncogenes has been a valuable tool in modeling cancer in mice, it generally does not recapitulate molecular and physiologic changes which occur in human tumors. To address this point, several *Her-2/neu* models have been developed using the *Her-2/neu* promoter (Andrechek et al., 2000; Weinstein et al., 2000). These models develop mammary tumors and resemble more closely the molecular changes which occur in some forms of human breast cancer.

Wnt-1. *Wnt-1* was identified as the gene often dysregulated and etiologically associated with mammary cancer produced by infection of the MMTV (Nusse and Varmus, 1982). *Wnt-1* is an extracellular glycoprotein which signals through the Notch family and β-cathenin. β-Cathenin is known to modulate several genes that are involved in cellular growth, proliferation, and development (Li et al., 2000b). Overexpression of *Wnt-1* in transgenic mice using the MMTV-LTR has confirmed the oncogenicity of *Wnt-1* (Li et al., 2000b). The MMTV-*Wnt* transgenic mice develop extensive ductal hyperplasia early in life and mammary adenocarcinomas in approximately 50% of the female transgenic mice by six months of age (Li et al., 2000b). This is one of the few mammary cancer models where tumors appear to be ER positive (Li et al., 2000b).

Cyclin D1. Overexpression of cyclin D1 is a common feature of human breast cancer. Transgenic mice overexpressing cyclin D1 develop tumors relatively late (at about 16 months of age) (Wang et al., 1994). Overexpression of cyclin D1 resulted in abnormal mammary cell proliferation, including the development of mammary adenocarcinomas.

Polyoma Middle T. Unlike the other models, expression of the polyomavirus middle T (PyMT) oncogene, driven by the MMTV promoter (Wagner and Risau, 1994), leads to rapid mammary tumor formation. The PyMT model produces atypical mammary hyperplasia within three weeks of birth, which leads to rapid production of multifocal mammary adenocarcinomas that metastasize with high frequency to the lungs. Some tumors exhibit regions with densely sclerotic stroma (Cardiff et al., 2000). PyMT activates phosphoinositol 3-kinase (PI3K) cascade and interacts with c-*src* kinase and, therefore, affects downstream signaling of this pathway (Guy et al., 1992).

Conditional Expression of Oncogenes Inducible systems to express genes in the mammary gland under temporal control have begun to be established (Gunther et al., 2002). Using a tetracycline-inducible system, the *myc* oncogene has been expressed and turned off at different stages of mammary lesion development. These studies have demonstrated that deregulated c-*myc* expression in this inducible system resulted in the formation of invasive mammary adenocarcinomas (D'Cruz et al., 2001). Inducible systems will provide a very powerful means of determining functional interactions between oncogenes, suppressor genes, and other cellular machinery.

Models with Loss of Tumor Suppressor Genes

SV40-TAg. The early region of the simian virus 40 (SV40) containing large T and small t antigens (*SV40-TAg*) can induce mammary carcinogenesis when expressed using the rat C3(1) or WAP promoters. The mechanism by which *TAg* induces transformation and tumorigenesis involves its ability to bind to and functionally inactivate two important tumor suppressor genes, *p53* and *Rb* (Dyson et al., 1989; Mietz et al., 1992). Although the expression of *SV40* has not been correlated to human breast cancer, the inactivation of *p53* and *Rb* have been correlated in many human cancers, including breast cancer. Mutations or deletions in these tumor suppressor genes lead to abnormalities in cell cycle regulation, genomic instability increased cellular proliferation, and defects in the activation of apoptosis pathways. Thus, induction of *TAg* in a transgenic model may utilize mechanistic pathways that are also involved in the development of human breast cancer (Green et al., 2000; Li et al., 2000a).

The regulatory region of the C3(1) prostate steroid binding protein gene has been shown to drive the expression of *TAg* in the mammary epithelium and results in progressive mammary lesions and tumor development in all female mice (Green et al., 2000; Maroulakou et al., 1994). At 8 weeks of age, atypia of mammary ductal epithelial cells is observed and represents the earliest stage of low-grade MIN. The atypical lesions progressed to high-grade MIN by about 12 weeks of age and appear similar to human DCIS. By 18 weeks of age, invasive carcinomas are observed, at which time tumors may be grossly palpable (Green et al., 2000). A significant reactive stromal component is often observed in the tumors which have a varying metastatic rate, depending upon the strain background (Maroulakou et al., 1997). The *TAg* tumors also demonstrate a high degree of aneuploidy.

TAg has been expressed in mammary epithelial cells under the WAP promoter leading mammary hyperplasia and adenocarcinoma formation (Li et al., 2000a). *TAg* expression is triggered by hormonal stimulation, during either estrus or pregnancy. Breast adenocarcinomas developed in 100% of the female mice by approximately 8–9 months of age.

p53. *p53* is a tumor suppressor gene frequently disrupted in human breast cancers and responsible for the Li-Fraumeni syndrome. $p53^{null}$ mice rarely develop mammary tumors due to the formation of lymphomas and sarcomas within the first few months of life (Donehower et al., 1995b). However, a conditional knockout of the *p53* gene in the mammary gland has been developed which does generate mammary tumors which metastasize (E. Lee, personal communication). Another approach has transplanted mammary tissue from $p53^{-/-}$ mice into wild-type mammary fat pads which subsequently develop adenocarcinomas in about two-thirds of the transplants with frequent metastases (Jerry et al., 2000). These tumors are characterized by high rates of aneuploidy which mimic the human advanced-stage breast cancers.

BRCA1 and BRCA2. Since the homozygous null mutation of the *BRCA1* gene leads to embryonic lethality (Liu et al., 1996), conditional mouse mutants have been developed in which the *BRCA1* gene may be disrupted in specific tissues. The deletion of *BRCA1* in the mammary gland via the MMTV-Cre or WAP-Cre promoter have resulted in transgenic models that appear to develop diverse mammary tumors via the inactivation of *p53* (Xu et al., 1999).

By 10–13 months $Brca1^{ko/co}$; WAP-Cre and MMTV-Cre mouse developed highly invasive adenocarcinomas. Mammary tumors develop after a long latency period and were associated with genomic instability, including aneuploidy and chromosomal rearrangements. Conditional knockout of the *BRCA2* gene in mammary epithelial cells has also been developed. These tumors appeared after long latency of approximately 15 months (Ludwig et al., 2001a,b). These invasive adenocarcinomas are histologically quite uniform, exhibiting predominantly a solid, nodular tumor pattern with very few variants, in striking contrast to the morphologic heterogeneity of analogous *Brca1*-associated tumors (Ludwig et al., 2001a,b).

PTEN. The phosphatase and tensin homologue deleted on chromosome 10 (*PTEN*) gene has been identified as a tumor suppressor gene which is mutated in several types of human cancers, including breast cancer, as well as three inherited autosomal dominant inherited disorders (Cowden syndrome, Lhermitte-Duclos disease, and Bannayan-Zonana syndrome). Mice with a conditional loss of *PTEN* in the mammary gland exhibit developmental abnormalities of the mammary gland and mammary gland tumors (Li et al., 2002). The histologic features of these tumors are pleomorphic, ranging from benign fibroadenomas to adenocarcinomas with mammary glands containing squamous epithelium.

Genomic Analysis of Mouse Mammary Cancer Models: Correlation to Human Disease

The presence of specific chromosomal imbalances in human breast cancers suggest that tumorigenesis in the breast requires the acquisition and maintenance of these genomic aberrations. In an effort to determine whether similar aberrations occur in mouse models of human breast cancer, extensive cytogenetic analyses of murine tumors were performed using SKY and CGH (Liu et al., 1998; McCormack et al., 1998; Montagna et al., 2002; Weaver et al., 1999, 2002). These analyses included the mapping of genomic imbalances in some 90 cases of experimentally induced mammary gland adenocarcinomas, extending from models where oncogenes such as c-*myc* or *her2/neu* are overexpressed under the control of mammary gland specific promoters, and tumors that arise when the function of tumor suppressor genes is impaired by either the expression of viral genes or the conditional, tissue-specific knockout (KO). Complete karyotypes can also be retrieved from the following website: http://www.ncbi.nlm.nih.gov/sky/skyweb.cgi.

As in human breast cancer, mammary gland adenocarcinomas in the mouse display a specific distribution of genomic imbalances (Fig. 7.1). For instance, in many tumor models chromosome 4 is partially or completely deleted and lost, whereas mouse chromosomes 11 and 15 are commonly gained. Mouse chromosome 4 contains regions that are orthologous to human chromosome 1p, a region that has been described to be subject to allelic loss in human breast carcinomas (see http://www.ncbi.nlm.nih.gov/genome/guide/mouse/index.html and http://www.ensembl.org/Mus_musculus/). Mouse chromosome 15 is orthologous predominantly to human chromosome arm 8q, a region that is very frequently gained in human cancers. In humans molecular analyses have confirmed that c-*myc* is the target for the recurrent gains in epithelial cancers. The analysis of mouse chromosome 11 is particularly interesting. This chromosome contains the entire sequence orthologous to human chromosome 17. Numerous CGH analyses have shown that human 17p is frequently lost, while the long arm is frequently gained. The tumor suppressor gene *TP53* maps to 17p and several known or presumed oncogenes are located on 17q, including *ERBB2* on 17q13, several candidate genes on 17q21–23 (Monni et al., 2001), and *MSF* on 17q25 (Montagna et al., 2003). Unlike mouse chromosome 15, for which we detect copy number increases that include the entire chromosome in 50% of the analyzed cases, mouse chromosome 11 therefore contains genes whose deletion or amplification, respectively, is observed in human breast cancer. A pattern of chromosomal aberrations that is consistent with this requirement is achieved by complex chromosomal aberrations, that is, jumping translocations that we detected in breast cancer models in both

BRCA1-conditional KO (Weaver et al., 2002) and *PyV-mT*-overexpressing mice (Montagna et al., 2003).

In general, the pattern of genomic imbalances in experimentally induced mouse mammary gland adenocarcinomas is similar to human breast cancer. However, significant differences exist. Human breast carcinomas have a certain degree of chromosomal instability, as measured by the number of chromosomal copy number alterations divided by the respective numbers of cases (ANCA) (Ried et al., 1999). This number varies from 8.2 to 13.5. In mouse mammary gland adenocarcinomas, however, in the majority of models we observed lower ANCA values. While in tumors induced by strong oncogenic stimuli, such as overexpression of *Her2/neu* or c-*myc* under the control of the MMTV promoter, only relatively few chromosomal imbalances occur (ANCA = 3.4–5.7), this number is considerably higher in tumors that arise as a consequence of the tissue-specific inactivation of *Brca1*. In the KO mouse model the ANCA index increased to 14.2, a number similar to the one observed in human breast cancer. It seems possible that strong oncogenic stimuli, such as vastly increased copy numbers for c-*myc*, circumvent the need to acquire secondary genetic insults, which, however, would be necessary for the development of tumors in the absence of just one copy of *Brca1*.

Gene Expression Profiling The advent of microarray technology provides a means of capturing and comparing the expression patterns of the entire genome of multiple samples in a high-throughput manner. Tens of thousands of genes in a single tumor can be analyzed for their expression patterns and compared with expression profiles from histologically similar tumors to define molecular classification schemes. This technology can identify important genes and pathways of oncogenesis (Desai et al., 2002a; Perou et al., 1999; Sorlie et al., 2001; van't Veer et al., 2002) which can then be used to develop diagnostic, preventative, and therapeutic agents. Microarray technologies can also be used as one means of validating mouse models by comparing expression patterns in mouse tumors with those of human breast cancer. In addition, mammary cancer models can be appropriately selected for experimental use based upon their gene expression profiles and experimental question. Recent studies have identified gene expression signatures in mouse mammary tumors which are dependent upon the initiating oncogenic event (Desai et al., 2002b) (http://dc.nci.nih.gov) (Fig. 7.4).

Six different mammary tumor models initiated by five different oncogenes were analyzed in the first microarray study which revealed that the *TAg* tumors had the greatest number of specific gene changes compared to the other models which affected numerous regulatory pathways. The *myc* model showed many similarities to the *TAg* model but demonstrated alterations in a significant number of genes related to transcription factors and ribosomal RNA. Models induced by *ras*, *erbB2/her-2/neu*, and *PyMT* clustered together with many common alterations in genes related to G proteins and the *ras* signaling pathway. Many genes with unknown function have been identified using this approach and several are under investigation. Recent studies have demonstrated that models which inactivate the *p53* or *BRCA1* tumor suppressor genes cluster separately from the transgenic models, although significant similarities exist with the *TAg* model (Fig. 7.4). Further insights will be gained as additional mammary cancer models using other oncogenic pathways are analyzed using microarray analyses.

Ideally, a direct comparison between the gene expression profiles of mouse mammary cancer and human breast cancers will provide an important means for model validation. However, at present, this cannot be accomplished since there is no common reference to compare gene expression profiles between species. The normal composition of mammary tissue of the mouse and human is strikingly different and thus cannot be used as a reference from which to compare changes in gene expression from tumors arising in each species. New informatics approaches will need to be developed to overcome this obstacle.

Hormone Responsiveness

ERα expression and tumor sensitivity to antiestrogen therapy is an important characteristic of human breast cancer. A major deficiency in transgenic mouse mammary models is that the tumors are largely ER negative and do not respond directly to hormonal therapy, whereas half of all human breast lesions are ER positive and hormone responsive (Maglione et al., 2001; Yoshidome et al., 2000). This problem is further confounded by the fact that the two most commonly used promoters, MMTV and WAP, are hormone responsive. The lack of hormone-responsive mouse models is a disadvantage of using mouse models to study breast cancer. Unlike most mouse models, chemically induced mammary cancer models in the rat are *ERα* positive and are hormone responsive (Tominaga et al., 1985).

Very early mammary lesions in the C3(1)/*TAg* model express *ERα*, but expression is lost as the tumors progress to high-grade MIN and invasive carcinomas (Yoshidome et al., 2000). Mice lacking *ERα* expression (ERKO mice) and carrying the C3(1)/*TAg* transgene do not develop mammary tumors but develop tumors in other organ sites (Yoshidome et al., 2000). However, mice carrying MMTV-*Wnt-1* or MMTV-*neu* in an ERKO background do develop mammary tumors, but with a much longer latency (Bocchinfuso et al., 1999; Hewitt et al., 2002). These results support a model suggesting that some breast cancers may directly evolve from *ERα*-negative cells (Nandi et al., 1995). One study suggests

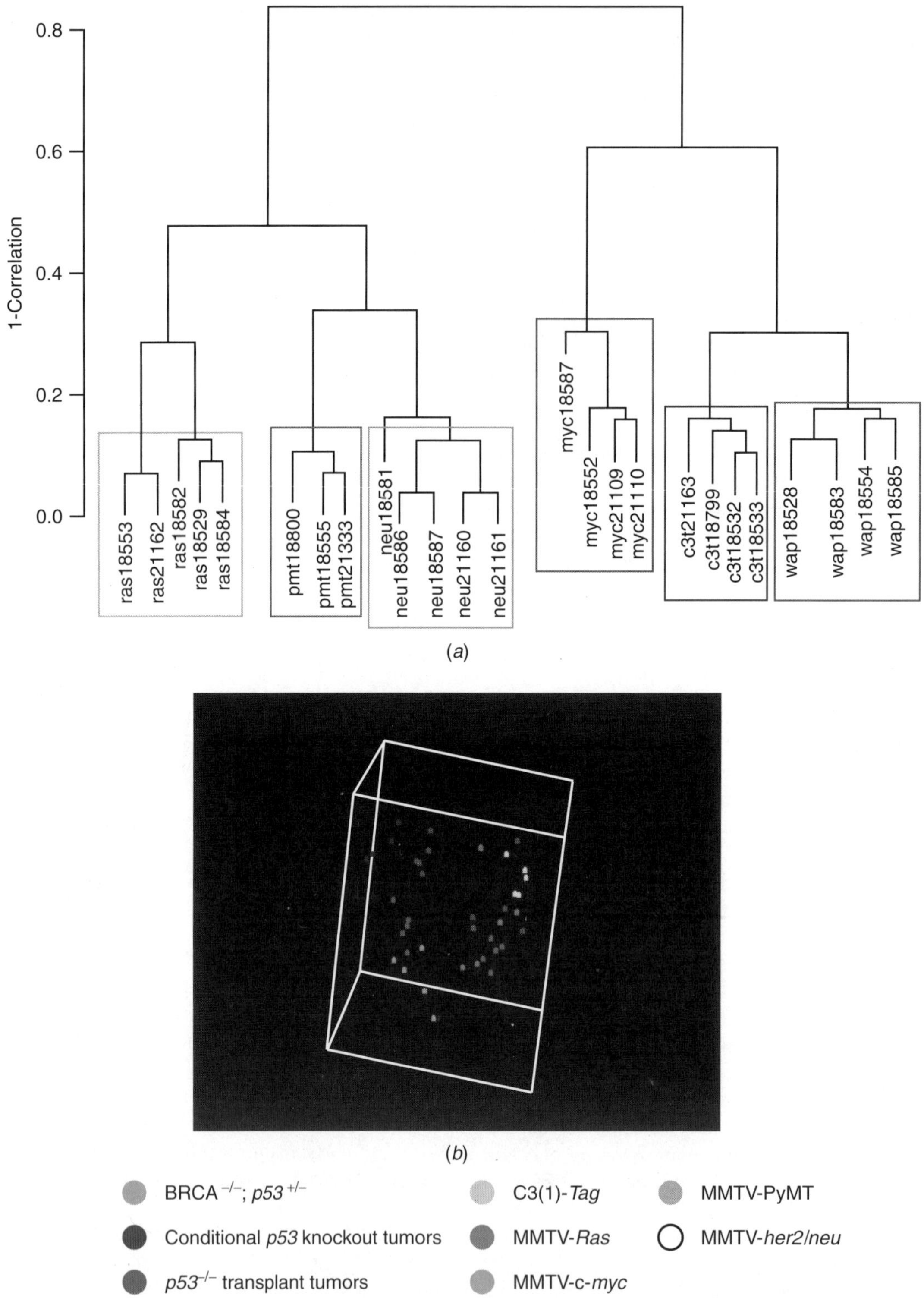

Figure 7.4. (*a*) **Comparative cDNA microarray analysis of mouse models of human breast cancer**. The tumors fall into three distinct groups: (1) MMTV-*neu*, MMTV-*ras*, and MMTV-PyMT; (2) MMTV-*myc*; and (3) the T antigen group (Desai et al, 2002b). (*b*) **Graphical distinctions between gene expression profiles of mouse models of human breast cancer represented in three dimensions by multidimensional scaling**. (See color insert).

that MMTV-PyMT tumors are ER positive (Maglione et al., 2001) as are MMTV-*Wnt-1* tumors. However, hyperplastic growth in the *Wnt*-transgenic mice persisted despite ovariectomy (Bocchinfuso et al., 1999), indicating that the tumors may not be hormone responsive. *p53*-null mammary transplants have also introduced an interesting insight on molecular events involved in hormonal response of mammary tumors. Hormonal stimulation of *p53*-null mammary transplants dramatically increase the tumor incidence in this model, suggesting that *p53* also plays a pivotal role in regulating hormone-induced molecular processes (Jerry et al., 2000). Furthermore, Medina et al. (2003) demonstrated that *p53*-null mammary cells generate a significant percentage of ER- negative tumors in mice, but the premalignant cells are ER positive and highly responsive to ovarian hormones during the normal and premalignant stages. A conditional deletion of *p53* in the mammary epithelium also appears to produce a low percentage of ER-positive tumors (E. Lee, personal communication). These models may be particularly useful for understanding the origin of ER-negative tumor cells as well as the effects of chemopreventive agents on the development of both ER-positive and ER-negative mammary tumors.

Metastasis

Modeling of metastatic disease in transgenic mammary cancer models has been extremely limited to date. Metastases that do occur are primarily found in the lung and occasionally in the liver. The frequencies of tumor metastasis in many mouse models are low, although this may be significantly impacted by the background strain (Lifsted et al., 1998). Overexpression of wild-type ErbB-2 by MMTV results in mammary tumors, which frequently metastasize to the lungs (Guy et al., 1992). MMTV-PyMT transgenic mice display the highest frequency of metastatic lesions (100%) to the lungs, whereas the *Wnt-1* model develop metastases predominantly to the lungs in about 50% of the mice (Guy et al., 1992). The C3(1)-*TAg* transgenic mice also develop lung metastases in 15–50% of the mice, depending upon the background strain (Maroulakou et al., 1997), although metastases were also occasionally observed in the liver, adrenal, and heart. Human breast cancer often metastasizes to the bone, a characteristic which has not been recapitulated in any genetically engineered model to date. Since this is a critical issue and remains a major deficiency of mouse models for human breast cancer, approaches which help define the actual metastatic potential of existing models or identify potential molecular mechanisms to aid in the development of metastatic mouse models need to be more fully explored.

Several studies have used GEM animals to identify genes which modify metastatic behavior. For instance, alteration in glycosylation of cell surface receptors resulted in a dramatic decrease in MMTV-PyMT metastases (Granovsky et al., 2000). Transgenic mice expressing a *TGF*β antagonist of the soluble type II TGFβ receptor were resistant to the development of metastases when crossed with the MMTV-*neu* transgenic mouse model of metastatic breast cancer (Yang et al., 2002). Other studies have shown the importance of MMPs in mammary tumor metastases in transgenic mice (Sternlicht et al., 2000).

In vivo Imaging Techniques

Advanced imaging techniques have recently been developed for rodents (see Part IV in this book). It is now possible to image mice using CT, US, MRI, and positron enhanced imaging to obtain important structural and anatomic information regarding tumors and metastases in living animals. An example of an MRI image of primary and metastatic mammary tumors is shown in Fig. 7.5. New emerging techniques of molecular and functional imaging allow visualization of the physiology and cellular or molecular processes as they take place in living cells and tissues. Single-photon emission CT and PET are molecular imaging techniques that use radiolabeled molecules to image molecular interactions in biologic processes in vivo. PET technology currently provide tomographic images of the entire body and can be used in animal models for diagnosis and localizing tumor tissues as well as monitoring therapy (Schwaiger, 2002). in vivo fluorescent (Hoffman, 2002) and bioluminescent imaging (luciferase imaging) (Vooijs et al., 2002)

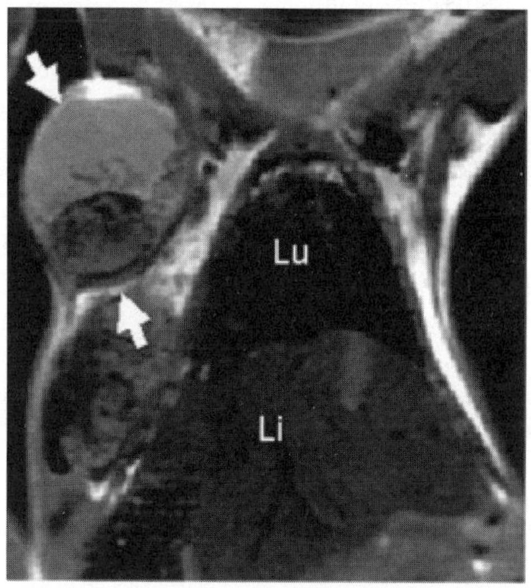

Figure 7.5. MRI visualization of mouse MMTV-LTR-*her2/neu* mammary tumor in axillary region (arrowheads): 1-mm-thick slices at 312 μ m in-plane resolution in 6 min; Lu, lung; Li, liver.

are other emerging tools for imaging tumor progression in real time.

Real-time imaging of gene expression in the whole animal is also possible (Bhaumik and Gambhir, 2002; Gratton and Fabiani, 2001; Ray et al., 2001, 2003; Wu et al., 2002a) and allows the monitoring of gene therapy. This emerging technology holds great promise for coupling phenotypic alterations with gene expression.

Prevention/Therapeutic Trials

Although much has been learned about the molecular biology of GEM models of mammary cancer, few studies have addressed the predictive value of such models for preclinical testing. These models offer an important opportunity to assess the efficacy of selected compounds for chemoprevention and novel therapeutic strategies. Attention is now also drawn to finding agents which may prevent or treat ER-negative breast cancer. Existing mouse models of human breast cancer may be ideally suited for testing such compounds since most tumors which develop in these models are ER negative. Indeed, few models have been tested and validated for this purpose. Drugs which are commonly used to treat human breast cancer were not developed using GEM models as preclinical test systems. Xenograft system have, instead, been widely used because they are relatively inexpensive and can produce results quickly. However, attention is now being focused on how to utilize GEM models for preclinical testing of compounds which target specific pathways.

Retinoids Retinoids and their analogues have been investigated for their chemopreventive properties in several transgenic mammary cancer models. MMTV/c-*neu* transgenic mice were continuously administered Arotinoid Ro 40-8757, resulting in an increased number of mammary ducts, decreased mammary ductal branching, and decreased tumor incidence and multiplicity (Rao et al., 1999). Another study using the C3(1)-*SV40-TAg* model found that 9-*cis*-retinoic acid [a ligand for retinoic acid receptor (RAR) and retinoic X recept (RXR)] delayed tumor incidence and significantly reduced tumor multiplicity (Wu et al., 2000). An RXR-selective retinoid, LGD1069, has also been shown to be effective in inhibiting mammary cancer development in transgenic mice with fewer side effects than 9-*cis*-retinoic acid (Wu et al., 2002b). These chemopreventive effects may be attributed to the induction of cellular differentiation and/or the modulation of gene expression through retinoic acid receptors and are currently in clinical trials.

Polyamine Synthesis Inhibitors The C3(1)-*SV40-TAg* mice have also been used in trials to determine the efficacy of DFMO in mammary cancer prevention (Green et al., 2001). DFMO is an inhibitor of ornithine decarboxylase, an essential enzyme in polyamine synthesis. C3(1)-*TAg* mice administered DFMO orally for 12 weeks had decreased tumor incidence and tumor multiplicity in a dose-dependent manner. DFMO appeared to inhibit tumor progression rather than to inhibit tumor initiation (Green et al., 2001). Bis(ethyl)polyamine analogues also inhibit polyamine synthesis and were evaluated for their preventive effects in the MMTV-*neu* transgenic breast cancer mouse model (Shah et al., 1999). Mice given this analogue exhibited a 40% reduction in tumor number compared to controls, but overall survival was not increased.

Cyclin-Dependent Kinase Inhibitors Cyclin-dependent kinase inhibitors such as UCN-01 and flavopiridol have been developed to inhibit cell cycle progression and induce apoptosis in cancer cells of patients. However, further research is needed to credential these types of drugs for preclinical testing in transgenic models (Kavanaugh et al., 2002; Senderowicz, 2000).

Farnesyl Transferase Inhibitors (FTIs) FTI prevents the oncogenic proteins N and H-ras from localizing to the plasma membrane and becoming activated via posttranslational modification. FTIs also have been shown to work through alternative molecular pathways by inhibiting cell cycle progression and increasing levels of apoptosis (Barrington et al., 1998). These agents have been tested and shown effective in different models of breast cancer, which overexpress *ras* (Barrington et al., 1998; Mangues et al., 1998; Omer et al., 1997; Shi et al., 2000). However FTIs are not effective in blocking K-Ras4B processing and signaling (Lerner et al., 1997).

Cox-2 Inhibitors Overexpression of *Cox-2* in the mouse mammary gland was shown to be sufficient to induce mammary gland tumorigenesis (Liu et al., 2001a). Indeed, Celecoxib, a selective cyclooxygenase 2 inhibitor, was shown to significantly reduce the incidence of mammary tumors in MMTV/*neu* mice (Howe et al., 2002) and C3(1)-*TAg* mice (Kavanaugh, unpublished). Therefore, inhibition of *Cox-2* may represent a future target for chemoprevention therapy for breast cancer.

Immunotherapy Augmentation of an immune response against tumors has become a promising approach for anti-cancer therapy through the use of vaccines, cytokines, and costimulatory molecules. Since transgenic mammary cancer models appear to have functioning immune systems, they are attractive models in which to test immunotherapeutic approaches. Several studies indicate that IL-12 may impede tumor angiogenesis. Interleukin-12 (IL-12) injected into *erb*B-2/*her2*/*neu* transgenic mice decreased tumor progression (Boggio et al., 2000; Cifaldi et al., 2001). Combining IL-12 with IL-2 treatment inhibited tumor formation and actually induced regression of established tumors in the C3(1)-*TAg* mammary model (Wigginton et al., 2001). This combination therapy is currently in clinical trials. IL-12 was also introduced together with the angiogenesis inhibitor angiostatin to established PyMT transplantable tumors in FVB mice. The combination of angiostatin and IL-12 resulted in a transient regression of 96% of tumors and a sustained regression in 54% of the tumors (Gyorffy et al., 2001).

The *PyMT* and *erb*B-2/*her2*/*neu* transgenic models have also been used to test the efficiency of B7 costimulating molecules for enhancing the antitumor effect of IL-2 (Emtage et al., 1998). A complete regression of tumor formation in the PyMT and *erb*B-2/*her2*/*neu* models was observed.

Tumor Vaccines The use of vaccination strategies that generate T-cell responses with or without accompanying antibody responses may serve to mitigate the problem of developing resistance to immune therapy. Breast cancer vaccines include sialyl-Tn-keyhole limpet hemocyanin (KLH) vaccine (Theratope), MUC1 mucin peptides, and *her2*/*neu* peptide vaccines (Mitchell, 2002). Several studies have demonstrated that vaccination of MMTV-*neu* mice with a *neu* expression plasmid protected them from tumor formation but also resulted in a variety of autoimmune responses (Cefai et al., 1999; Esserman et al., 1999; Kiessling et al., 2002; Pupa et al., 2001).

Antiangiogenic Compounds Antiangiogenic compounds inhibit tumor growth by preventing the generation of new blood vessels to support tumor expansion and metastases. Testing these compounds in transgenic mammary models will provide insight to their efficacy and mechanisms of action, especially in the early premalignant stage, where the "angiogenic switch" occurs (Hanahan et al., 1996). MMTV/*neu* transgenic mice treated with liposome-mediated angiostatin exhibited reduced tumor size and metastases, but tumor incidence was not decreased (Sacco et al., 2001). A different antiangiogenic compound, endostatin, has been shown to inhibit tumor formation in the C3(1)-*TAg* mammary cancer mouse model by several approaches. The administration of recombinant mouse endostatin (Yokoyama et al., 2000), a mutated form of human endostatin, or expression of endostatin using an adenoviral gene therapy approach has demonstrated significant inhibition of tumor formation (Calvo et al., 2002a, b).

Genetically altered mice which model human breast cancer have been used in relatively few prevention and preclinical studies. However, the use of mouse models of mammary cancer offers a valuable means of testing multiple types of preventive and therapeutic approaches, alone or in combination, which may be applicable to human breast cancer.

CONCLUSIONS ABOUT DISEASE GENERATED FROM MOUSE MODELING DATA

Limitations of Existing Mammary Cancer Models

Although there are similarities between human breast cancer and the cancers that arise in transgenic mouse models of mammary cancer, there are distinct differences which can limit the use of GEM models for studying the biology of the disease and testing therapeutic agents. The basic architecture of the mouse mammary gland is quite different from that of the human mammary gland and important physiologic differences exist as well. GEM models only approximate important features of human breast cancer. Most mouse mammary tumors lack significant stromal components, which are often characteristic of human tumors, suggesting that important stromal–epithelial interactions may differ between the species. Half of all human breast cancers are hormone dependent and respond to hormonal therapy, whereas most mouse tumors are hormone independent (Cardiff et al., 2000). The lack of tumor metastasis to bone and other common sites of breast cancer dissemination is a major drawback of existing models. The few GEM models which metastasize do so primarily to the lung, whereas most human mammary tumors metastasize to the lymph nodes or bone (Pattengale et al., 1989). Also, there is less of an inflammatory response in mouse mammary tumors compared to human breast cancer.

The genetic background of mice has a significant impact on the biology, onset, and growth of the tumor (Lifsted et al., 1998) and provides an important means to find genes which modify the cancer phenotype. The mouse strain can also influence tumor latency and even the type of tumor caused by the transgene (Lifsted et al., 1998). This information and potential limitations must be considered when interpreting the phenotype of a GEM model.

While our knowledge of breast cancer has been greatly enhanced through the study of GEM models, existing GEM models of mammary cancer only represent specific aspects of human breast cancer. New promoters which target the mammary epithelium but are not hormone responsive will be identified. Given the increasing knowledge of cancer genomics and proteomics, it is likely that models with metastases to the bone will be developed allowing for the study of ways to inhibit this lethal form of tumor progression. Additionally, it is conceivable that techniques will be advanced which will allow for gene targeting to stromal cells that will enable the more formal study of epithelial–stromal interactions. In conclusion, we can look forward to the development of models with additional characteristics of human breast cancer that will greatly aid in our endeavor to prevent and treat this disease.

ACKNOWLEDGMENT

This work was supported in part by the National Cancer Institute (NCI) Intramural Mouse Mammary Cancer Models Collective Component of the NCI Mouse Models of Human Cancer Consortium.

REFERENCES

Adnane J, et al. (2000): Loss of p21WAF1/CIP1 accelerates Ras oncogenesis in a transgenic/knockout mammary cancer model. *Oncogene* 19(47):5338–5347.

Ahmad S, et al. (1999): Role of AKT1 in 17beta-estradiol- and insulin-like growth factor I (IGF-I)-dependent proliferation and prevention of apoptosis in MCF-7 breast carcinoma cells. *Biochem Pharmacol* 58(3):425–430.

Allegra JC, Lippman ME (1980): Estrogen receptor status and the disease-free interval in breast cancer. *Recent Results Cancer Res* 71:20–25.

Allred DC, et al. (1992): Overexpression of HER-2/neu and its relationship with other prognostic factors change during the progression of in situ to invasive breast cancer. *Hum Pathol* 23(9):974–979.

Ambartsumian NS, et al. (1996): Metastasis of mammary carcinomas in GRS/A hybrid mice transgenic for the mts1 gene. *Oncogene* 13(8):1621–130.

Amundadottir LT, et al. (1995): Synergistic interaction of transforming growth factor alpha and c-myc in mouse mammary and salivary gland tumorigenesis. *Cell Growth Differ* 6(6):737–748.

Anderson DE (1980): An inherited form of large bowel cancer: Muir's syndrome. *Cancer* 45(5 Suppl):1103–1107.

Andrechek ER, et al. (2000): Amplification of the neu/erbB-2 oncogene in a mouse model of mammary tumorigenesis. *Proc Natl Acad Sci USA* 97(7):3444–3449.

Arthur EL, Bluemke DA (2002): Magnetic resonance imaging. In Devita VT Jr., Hellman S, Rosenberg SA (Eds), *Cancer Principles and Practice of Oncology*, Vol. 1. Lippincott-Williams & Wilkins, Philadelphia, 673.

Bange J, et al. (2001): Molecular targets for breast cancer therapy and prevention. *Nat Med* 7(5):548–552.

Barash I, et al. (1994): Ectopic expression of beta-lactoglobulin/human serum albumin fusion genes in transgenic mice: hormonal regulation and in situ localization. *Transgenic Res* 3(3):141–151.

Barrington RE, et al. (1998): A farnesyltransferase inhibitor induces tumor regression in transgenic mice harboring multiple oncogenic mutations by mediating alterations in both cell cycle control and apoptosis. *Mol Cell Biol* 18(1):85–92.

Bates P, et al. (1995): Mammary cancer in transgenic mice expressing insulin-like growth factor II (IGF-II). *Br J Cancer* 72(5):1189–1193.

Berx G, et al. (1995): E-cadherin is a tumour/invasion suppressor gene mutated in human lobular breast cancers. *EMBO J* 14(24):6107–6115.

Berx G, et al. (1996): E-cadherin is inactivated in a majority of invasive human lobular breast cancers by truncation mutations throughout its extracellular domain. *Oncogene* 13(9):1919–1925.

Bhaumik S, Gambhir SS (2002): Optical imaging of Renilla luciferase reporter gene expression in living mice. *Proc Natl Acad Sci USA* 99(1):377–382.

Bittner JJ (1936): *Science* 84:162.

Bluman LG, et al. (1999): Attitudes, knowledge, and risk perceptions of women with breast and/or ovarian cancer considering testing for BRCA1 and BRCA2. *J Clin Oncol* 17(3):1040–1046.

Bocchinfuso WP, et al. (1999): A mouse mammary tumor virus-Wnt-1 transgene induces mammary gland hyperplasia and tumorigenesis in mice lacking estrogen receptor-alpha. *Cancer Res* 59(8):1869–1876.

Boehm T, et al. (1997): Antiangiogenic therapy of experimental cancer does not induce acquired drug resistance. *Nature* 390(6658):404–407.

Boggio K, et al. (2000): Ability of systemic interleukin-12 to hamper progressive stages of mammary carcinogenesis in HER2/neu transgenic mice. *Cancer Res* 60(2):359–3564.

Brodie A, et al. (1999): Aromatase and its inhibitors. *J Steroid Biochem Mol Biol* 69(1–6):205–210.

Buell P (1973): Changing incidence of breast cancer in Japanese-American women. *J Natl Cancer Inst* 51(5):1479–1483.

Callahan R, Smith GH (2000): MMTV-induced mammary tumorigenesis: Gene discovery, progression to malignancy and cellular pathways. *Oncogene* 19(8):992–1001.

Calle EE, et al. (2002): Organochlorines and breast cancer risk. *CA Cancer J Clin* 52(5):301–309.

Calvo A, et al. (2002a): Adenovirus-mediated endostatin delivery results in inhibition of mammary gland tumor growth in C3(1)/SV40 T-antigen transgenic mice. *Cancer Res* 62(14):3934–3938.

Calvo A, et al. (2002b): Inhibition of the mammary carcinoma angiogenic switch in C3(1)/SV40 transgenic mice by a mutated form of human endostatin. *Int J Cancer* 101(3):224–234.

Cardiff RD, et al. (2000): The mammary pathology of genetically engineered mice: the consensus report and recommendations from the Annapolis meeting. *Oncogene* 19(8):968–988.

Cardiff RD, Wellings SR (1999): The comparative pathology of human and mouse mammary glands. *J Mammary Gland Biol Neoplasia* 4(1):105–122.

Cauley JA, et al. (1999): Elevated serum estradiol and testosterone concentrations are associated with a high risk for breast cancer. Study of Osteoporotic Fractures Research Group. *Ann Intern Med* 130(4 Pt 1):270–277.

Cefai D, et al. (1999): Targeting HER-2/neu for active-specific immunotherapy in a mouse model of spontaneous breast cancer. *Int J Cancer* 83(3):393–400.

Chambers AF, et al. (2002): Dissemination and growth of cancer cells in metastatic sites. *Nat Rev Cancer* 2(8):563–572.

Chen Y, et al. (1995): Aberrant subcellular localization of BRCA1 in breast cancer. *Science* 270(5237):789–791.

Cifaldi L, et al. (2001): A light, nontoxic interleukin 12 protocol inhibits HER-2/neu mammary carcinogenesis in BALB/c transgenic mice with established hyperplasia. *Cancer Res* 61(7):2809–2812.

Claus EB, et al. (1996): The genetic attributable risk of breast and ovarian cancer. *Cancer* 77(11):2318–2324.

Collingwood TN, et al. (1999): Nuclear receptors: Coactivators, corepressors and chromatin remodeling in the control of transcription. *J Mol Endocrinol* 23(3):255–275.

Colpaert CG, et al. (2003): Inflammatory breast cancer shows angiogenesis with high endothelial proliferation rate and strong E-cadherin expression. *Br J Cancer* 88(5):718–725.

Cook-Bruns N (2001): Retrospective analysis of the safety of Herceptin immunotherapy in metastatic breast cancer. *Oncology* 61 (Suppl 2):58–66.

Cressman VL, et al. (1999): Mammary tumor formation in p53- and BRCA1-deficient mice. *Cell Growth Differ* 10(1):1–10.

Culotta E, Koshland Jr DE (1993): p53 sweeps through cancer research. *Science* 262(5142):1958–1961.

Cummings SR, et al. (1999): The effect of raloxifene on risk of breast cancer in postmenopausal women: Results from the MORE randomized trial. Multiple Outcomes of Raloxifene Evaluation. *JAMA* 281(23):2189–2197.

Czernin J (2002): Clinical applications of FDG-PET in oncology. *Acta Med Austriaca* 29(5):162–170.

Dang CV (1999): c-Myc target genes involved in cell growth, apoptosis, and metabolism. *Mol Cell Biol* 19(1):1–11.

Dankort DL, Muller WJ (2000): Signal transduction in mammary tumorigenesis: A transgenic perspective. *Oncogene* 19(8):1038–1044.

Datta SR, et al. (1999): Cellular survival: A play in three Akts. *Genes Dev* 13(22):2905–2927.

Dauvois S, et al. (1993): The antiestrogen ICI 182780 disrupts estrogen receptor nucleocytoplasmic shuttling. *J Cell Sci* 106(Pt 4):1377–1388.

D'Cruz CM, et al. (2001): c-MYC induces mammary tumorigenesis by means of a preferred pathway involving spontaneous Kras2 mutations. *Nat Med* 7(2):235–239.

Demicheli R (2001): Tumour dormancy: Findings and hypotheses from clinical research on breast cancer. *Semin Cancer Biol* 11(4):297–306.

Desai KV, et al. (2002a): Chipping away at breast cancer: Insights from microarray studies of human and mouse mammary cancer. *Endocr Relat Cancer* 9(4):207–220.

Desai KV, et al. (2002b): Initiating oncogenic event determines gene-expression patterns of human breast cancer models. *Proc Natl Acad Sci USA* 99(10):6967–6972.

DeSombre E, et al. (1978): *Estrophilin and Endocrine Responsive of Breast cancer*. Ravenn, New York.

Dickson RB, Lipmann ME (2002): Molecular biology of breast cancer. In Devita VT Jr., Hellman S, Rosenberg SA (Eds), *Cancer Principles and Practice of Oncology*, Vol. 2, Lippincott-Williams & Wilkins, Philadelphia, 1633–1645.

Donehower LA, (1997): Genetic instability in animal tumorigenesis models. *Cancer Surv* 29:329–352.

Donehower LA, et al. (1995b): Effects of genetic background on tumorigenesis in p53-deficient mice. *Mol Carcinog* 14(1):16–22.

Dougall WC, et al. (1994): The neu-oncogene: Signal transduction pathways, transformation mechanisms and evolving therapies. *Oncogene* 9(8):2109–2123.

Dowsett M (1996): Endocrine resistance in advanced breast cancer. *Acta Oncol* 35(Suppl 5):91–95.

Dyson N, et al. (1989): The human papilloma virus-16 E7 oncoprotein is able to bind to the retinoblastoma gene product. *Science* 243(4893):934–937.

Easton DF, et al. (1997): Cancer risks in two large breast cancer families linked to BRCA2 on chromosome 13q12–13. *Am J Hum Genet* 61(1):120–128.

Elledge RM, Fuqua SAW (2002): Estrogen and progesterone receptors. In Harris JR, Lippman ME, Morrow M, Osborne CK (Eds), *Diseases of the Breast*, 2nd Ed. Lippincott-Williams & Wilkins, Philadelphia, 471.

Elledge RM, Lee WH (1995): Life and death by p53. *Bioessays* 17(11):923–930.

Emtage PC, et al. (1998): Enhanced interleukin-2 gene transfer immunotherapy of breast cancer by coexpression of B7–1 and B7–2. *J Interferon Cytokine Res* 18(11):927–937.

Entius MM, et al. (2000): Microsatellite instability and expression of hMLH-1 and hMSH-2 in sebaceous gland carcinomas as markers for Muir-Torre syndrome. *Clin Cancer Res* 6(5):1784–1789.

Erlichman C, Loprinzi CL (2002): Hormonal therapies. In Devita VT Jr., Hellman S, Rosenberg SA (Eds), *Cancer Principles and Practice of Oncology*, Vol. 2. Lippincott-Williams & Wilkins, Philadelphia, 478–486.

Esserman LJ, et al. (1999): Vaccination with the extracellular domain of p185neu prevents mammary tumor development in neu transgenic mice. *Cancer Immunol Immunother* 47(6):337–342.

Ewing J (1928): *A Treatise on Tumors*. W.B. Saunders, Philadelphia.

Fabian CJ, et al. (2002): A phase II breast cancer chemoprevention trial of oral alpha-difluoromethylornithine: Breast tissue, imaging, and serum and urine biomarkers. *Clin Cancer Res* 8(10):3105–3117.

Facchini LM, Penn LZ (1998): The molecular role of Myc in growth and transformation: Recent discoveries lead to new insights. *Faseb J* 12(9):633–651.

Fisher B, et al. (1998): Tamoxifen for prevention of breast cancer: Report of the National Surgical Adjuvant Breast and Bowel Project P-1 Study. *J Natl Cancer Inst* 90(18):1371–1388.

Fletcher SW, Elmore JG (2003): Clinical practice. Mammographic screening for breast cancer. *N Engl J Med* 348(17):1672–1680.

Folkman J (2002): Role of angiogenesis in tumor growth and metastasis. *Semin Oncol* 29(6 Suppl 16):15–18.

Ford D, et al. (1994): Risks of cancer in BRCA1-mutation carriers. Breast Cancer Linkage Consortium. *Lancet* 343(8899): 692–695.

Fults D, et al. (2002): MYC expression promotes the proliferation of neural progenitor cells in culture and in vivo. *Neoplasia* 4(1):32–39.

Fuqua SA (1994): Estrogen receptor mutagenesis and hormone resistance. *Cancer* 74(3 Suppl):1026–1029.

Furth PA, et al. (1999): Loss of anti-mitotic effects of Bcl-2 with retention of anti-apoptotic activity during tumor progression in a mouse model. *Oncogene* 18(47):6589–6596.

Gail MH, et al. (1989): Projecting individualized probabilities of developing breast cancer for white females who are being examined annually. *J Natl Cancer Inst* 81(24):1879–1886.

Gajdos C, Jordan VC (2002): Selective estrogen receptor modulators as a new therapeutic drug group: concept to reality in a decade. *Clin Breast Cancer* 2(4):272–281.

Gallahan D, et al. (1996): Expression of a truncated Int3 gene in developing secretory mammary epithelium specifically retards lobular differentiation resulting in tumorigenesis. *Cancer Res* 56(8):1775–1785.

Gayther SA, et al. (2000): Mutations truncating the EP300 acetylase in human cancers. *Nat Genet* 24(3):300–303.

Gervais-Fagnou DD, et al. (1999): Breast cancer in women following supradiaphragmatic irradiation for Hodgkin's disease. *Oncology* 57(3):224–231.

Ghadimi BM, et al. (1999): Interphase cytogenetics: At the interface of genetics and morphology. *Anal Cell Pathol* 19(1):3–6.

Ghadimi BM, et al. (2000): Centrosome amplification and instability occurs exclusively in aneuploid, but not in diploid colorectal cancer cell lines, and correlates with numerical chromosomal aberrations. *Genes Chromosomes Cancer* 27(2):183–190.

Gibson MK, et al. (1991): The mechanism of ICI 164,384 antiestrogenicity involves rapid loss of estrogen receptor in uterine tissue. *Endocrinology* 129(4):2000–2010.

Glikman P, et al. (1989): Polyamine involvement in the growth of hormone-responsive and -resistant human breast cancer cells in culture. *Cancer Res* 49(6):1371–1376.

Gotzsche PC, Olsen O (2000): Is screening for breast cancer with mammography justifiable? *Lancet* 355(9198):129–134.

Granovsky M, et al. (2000): Suppression of tumor growth and metastasis in Mgat5-deficient mice. *Nat Med* 6(3):306–312.

Gratton G, Fabiani M (2001): Shedding light on brain function: The event-related optical signal. *Trends Cogn Sci* 5(8):357–363.

Green JE, et al. (2000): The C3(1)/SV40 T-antigen transgenic mouse model of mammary cancer: ductal epithelial cell targeting with multistage progression to carcinoma. *Oncogene* 19(8):1020–1027.

Green JE, et al. (2001): 2-Difluoromethylornithine and dehydroepiandrosterone inhibit mammary tumor progression but not mammary or prostate tumor initiation in C3(1)/SV40 T/t-antigen transgenic mice. *Cancer Res* 61(20):7449–7455.

Gruvberger S, et al. (2001): Estrogen receptor status in breast cancer is associated with remarkably distinct gene expression patterns. *Cancer Res* 61(16):5979–5984.

Gunther EJ, et al. (2002): A novel doxycycline-inducible system for the transgenic analysis of mammary gland biology. *Faseb J* 16(3):283–292.

Guy CT, et al. (1992): Induction of mammary tumors by expression of polyomavirus middle T oncogene: A transgenic mouse model for metastatic disease. *Mol Cell Biol* 12(3):954–961.

Guy CT, et al. (1996): Activated neu induces rapid tumor progression. *J Biol Chem* 271(13):7673–7678.

Guy CT, et al. (1992): Expression of the neu protooncogene in the mammary epithelium of transgenic mice induces metastatic disease. *Proc Natl Acad Sci USA* 89(22):10578–10582.

Gyorffy S, et al. (2001): Combined treatment of a murine breast cancer model with type 5 adenovirus vectors expressing murine angiostatin and IL-12: A role for combined anti-angiogenesis and immunotherapy. *J Immunol* 166(10): 6212–6217.

Ha HY, et al. (2001): Overexpression of membrane-type matrix metalloproteinase-1 gene induces mammary gland abnormalities and adenocarcinoma in transgenic mice. *Cancer Res* 61(3):984–990.

Hanahan D, et al. (1996): Transgenic mouse models of tumour angiogenesis: The angiogenic switch, its molecular controls, and prospects for preclinical therapeutic models. *Eur J Cancer* 32A(14):2386–2393.

Harris RE, et al. (1999): Genetic induction and upregulation of cyclooxygenase (COX) and aromatase (CYP19): An extension of the dietary fat hypothesis of breast cancer. *Med Hypotheses* 52(4):291–292.

Hartmann A, et al. (1997): The molecular epidemiology of p53 gene mutations in human breast cancer. *Trends Genet* 13(1):27–33.

Hedenfalk I, et al. (2001): Gene-expression profiles in hereditary breast cancer. *N Engl J Med* 344(8):539–548.

Heim S, Mitelman F (1995): *Cancer Cytogenetics*, Wiley-Liss, New York.

Hennighausen L. (2002). "Mouse models for breast cancer." *Oncogene* 19(8): 966–967.

Heselmeyer-Haddad K, et al. (2002): Detection of chromosomal aneuploidies and gene copy number changes in fine needle aspirates is a specific, sensitive, and objective genetic test for the diagnosis of breast cancer. *Cancer Res* 62(8):2365–2369.

Hewitt SC, et al. (2002): Lack of ductal development in the absence of functional estrogen receptor alpha delays mammary tumor formation induced by transgenic expression of ErbB2/neu. *Cancer Res* 62(10):2798–2805.

Hobbs CA, et al. (2002): Deregulation of polyamine biosynthesis alters intrinsic histone acetyltransferase and deacetylase activities in murine skin and tumors. *Cancer Res* 62(1):67–74.

Hoffman RM (2002): Green fluorescent protein imaging of tumor cells in mice. *Lab Anim (NY)* 31(4):34–41.

Holmes MD, et al. (1999): Association of dietary intake of fat and fatty acids with risk of breast cancer. *JAMA* 281(10):914–920.

Howe LR, et al. (2002): Celecoxib, a selective cyclooxygenase 2 inhibitor, protects against human epidermal growth factor receptor 2 (HER-2)/neu-induced breast cancer. *Cancer Res* 62(19):5405–5407.

Huang L, Pardee AB (2000): Suberoylanilide hydroxamic acid as a potential therapeutic agent for human breast cancer treatment. *Mol Med* 6(10):849–866.

Huang Z, et al. (1997): Dual effects of weight and weight gain on breast cancer risk. *JAMA* 278(17):1407–1411.

Humphreys RC, Hennighausen L (2000): Transforming growth factor alpha and mouse models of human breast cancer. *Oncogene* 19(8):1085–1091.

Hunter DJ, et al. (1996): Cohort studies of fat intake and the risk of breast cancer—a pooled analysis. *N Engl J Med* 334(6):356–361.

Hunter KW, et al. (2001): Predisposition to efficient mammary tumor metastatic progression is linked to the breast cancer metastasis suppressor gene Brms1. *Cancer Res* 61(24):8866–8872.

Husler MR, et al. (1998): Lactation-induced WAP-SV40 TAg transgene expression in C57BL/6 J mice leads to mammary carcinoma. *Transgenic Res* 7(4):253–263.

Hutchinson J, et al. (2001): Activation of Akt (protein kinase B) in mammary epithelium provides a critical cell survival signal required for tumor progression. *Mol Cell Biol* 21(6):2203–12.

Hutchinson JN, Muller WJ (2000): Transgenic mouse models of human breast cancer. *Oncogene* 19(53):6130–6137.

Hynes NE, Stern DF (1994): The biology of erbB-2/neu/HER-2 and its role in cancer. *Biochim Biophys Acta* 1198(2–3): 165–184.

Ingvarsson S (1999): Molecular genetics of breast cancer progression. *Semin Cancer Biol* 9(4):277–288.

Iwamoto T, et al. (1990): Oncogenicity of the ret transforming gene in MMTV/ret transgenic mice. *Oncogene* 5(4):535–542.

Jager R, et al. (1997): Overexpression of Bcl-2 inhibits alveolar cell apoptosis during involution and accelerates c-myc-induced tumorigenesis of the mammary gland in transgenic mice. *Oncogene* 15(15):1787–1795.

Jaiyesimi IA, et al. (1995): Use of tamoxifen for breast cancer: Twenty-eight years later. *J Clin Oncol* 13(2):513–529.

Jamerson MH, et al. (2000): Dual regulation of proliferation and apoptosis: c-myc in bitransgenic murine mammary tumor models. *Oncogene* 19(8):1065–1071.

Jerry DJ, et al. (2000): A mammary-specific model demonstrates the role of the p53 tumor suppressor gene in tumor development. *Oncogene* 19(8):1052–1058.

Jhappan C., C. Stahle, et al. (1990). "TGF alpha overexpression in transgenic mice induces liver neoplasia and abnormal development of the mammary gland and pancreas." *Cell* 61(6):1137–46.

Jonkers J, et al. (2001): Synergistic tumor suppressor activity of BRCA2 and p53 in a conditional mouse model for breast cancer. *Nat Genet* 29(4):418–425.

Kallioniemi A, et al. (1992): Comparative genomic hybridization for molecular cytogenetic analysis of solid tumors. *Science* 258:818–821.

Kallioniemi A, et al. (1994): Detection and mapping of amplified DNA sequences in breast cancer by comparative genomic hybridization. *Proc Natl Acad Sci USA* 91:2156–2160.

Kampert JB, et al. (1988): Combined effect of childbearing, menstrual events, and body size on age-specific breast cancer risk. *Am J Epidemiol* 128(5):962–979.

Kavanaugh CJ, et al. (2002): Pre-clinical applications of transgenic mouse mammary cancer models. *Transgenic Res* 11(6):617–633.

Khan J, et al. (2001): Classification and diagnostic prediction of cancers using gene expression profiling and artificial neural networks. *Nat Med* 7(6):673–679.

Kiessling R, et al. (2002): Cellular immunity to the Her-2/neu protooncogene. *Adv Cancer Res* 85:101–144.

Kitsberg DI, Leder P (1996): Keratinocyte growth factor induces mammary and prostatic hyperplasia and mammary adenocarcinoma in transgenic mice. *Oncogene* 13(12):2507–2515.

Krane IM, Leder P (1996): NDF/heregulin induces persistence of terminal end buds and adenocarcinomas in the mammary glands of transgenic mice. *Oncogene* 12(8):1781–1788.

Kwan H, et al. (1992): Transgenes expressing the Wnt-1 and int-2 proto-oncogenes cooperate during mammary carcinogenesis in doubly transgenic mice. *Mol Cell Biol* 12(1):147–154.

Lamy PJ, et al. (2002): Progesterone receptor quantification as a strong prognostic determinant in postmenopausal breast cancer women under tamoxifen therapy. *Breast Cancer Res Treat* 76(1):65–71.

Landesman-Bollag E, et al. (2001): Protein kinase CK2 in mammary gland tumorigenesis. *Oncogene* 20(25):3247–3257.

Lane DP (1992): Cancer. p53, guardian of the genome. *Nature* 358(6381):15–16.

Lane TF, Leder P (1997): Wnt-10b directs hypermorphic development and transformation in mammary glands of male and female mice. *Oncogene* 15(18):2133–2144.

Lerner EC, et al. (1997): Inhibition of the prenylation of K-Ras, but not H- or N-Ras, is highly resistant to CAAX peptidomimetics and requires both a farnesyltransferase and a geranylgeranyltransferase I inhibitor in human tumor cell lines. *Oncogene* 15(11):1283–1288.

Lernevall A (2000): Imaging of axillary lymph nodes. *Acta Oncol* 39(3):277–281.

Li B, et al. (1997): neu/ERBB2 cooperates with p53–172 H during mammary tumorigenesis in transgenic mice. *Mol Cell Biol* 17(6):3155–3163.

Li G, et al. (2002): Conditional loss of PTEN leads to precocious development and neoplasia in the mammary gland. *Development* 129(17):4159–4170.

Li M, et al. (2000a): WAP-TAg transgenic mice and the study of dysregulated cell survival, proliferation, and mutation during breast carcinogenesis. *Oncogene* 19(8):1010–1019.

Li Y, et al. (2000b): Use of MMTV-Wnt-1 transgenic mice for studying the genetic basis of breast cancer. *Oncogene* 19(8):1002–1009.

Liaw D, et al. (1997): Germline mutations of the PTEN gene in Cowden disease, an inherited breast and thyroid cancer syndrome. *Nat Genet* 16(1):64–67.

Lichtenstein P, et al. (2000): Environmental and heritable factors in the causation of cancer–analyses of cohorts of twins from Sweden, Denmark, and Finland. *N Engl J Med* 343(2):78–85.

Lifsted T, et al. (1998): Identification of inbred mouse strains harboring genetic modifiers of mammary tumor age of onset and metastatic progression. *Int J Cancer* 77(4):640–644.

Liu CH, et al. (2001a): Overexpression of cyclooxygenase-2 is sufficient to induce tumorigenesis in transgenic mice. *J Biol Chem* 276(21):18563–18569.

Liu CY, et al. (1996): Inactivation of the mouse Brca1 gene leads to failure in the morphogenesis of the egg cylinder in early postimplantation development. *Genes Dev* 10(14):1835–1843.

Liu SC, et al. (1998): Markers of cell proliferation in normal epithelia and dysplastic leukoplakias of the oral cavity. *Cancer Epidemiol Biomarkers Prev* 7(7):597–603.

Liyanage M, et al. (1996): Multicolour spectral karyotyping of mouse chromosomes. *Nat Genet* 14(3):312–315.

Ludwig T, et al. (2001a): Tumorigenesis in mice carrying a truncating Brca1 mutation. *Genes Dev* 15(10):1188–1193.

Ludwig T, et al. (2001b): Development of mammary adenocarcinomas by tissue-specific knockout of Brca2 in mice. *Oncogene* 20(30):3937–3948.

Maglione JE, et al. (2001): Transgenic polyoma middle-T mice model premalignant mammary disease. *Cancer Res* 61(22):8298–8305.

Malkin D, et al. (1990): Germ line p53 mutations in a familial syndrome of breast cancer, sarcomas, and other neoplasms. *Science* 250(4985):1233–1238.

Mangues R, et al. (1998): Antitumor effect of a farnesyl protein transferase inhibitor in mammary and lymphoid tumors overexpressing N-ras in transgenic mice. *Cancer Res* 58(6):1253–1259.

Mansour EG, et al. (1994): Prognostic factors in early breast carcinoma. *Cancer* 74(1 Suppl):381–400.

Maroulakou IG, et al. (1994): Prostate and mammary adenocarcinoma in transgenic mice carrying a rat C3(1) simian virus 40 large tumor antigen fusion gene. *Proc Natl Acad Sci USA* 91(23):11236–11240.

Maroulakou IG, et al. (1997): Reduced p53 dosage associated with mammary tumor metastases in C3(1)/TAG transgenic mice. *Mol Carcinog* 20(2):168–174.

Matsui Y, et al. (1990): Development of mammary hyperplasia and neoplasia in MMTV-TGF alpha transgenic mice. *Cell* 61(6):1147–1155.

McAllister KA, et al. (2002): Cancer susceptibility of mice with a homozygous deletion in the COOH-terminal domain of the Brca2 gene. *Cancer Res* 62(4):990–994.

McCormack SJ, et al. (1998): Myc/p53 interactions in transgenic mouse mammary development, tumorigenesis and chromosomal instability. *Oncogene* 16(21):2755–2766.

McMichael AJ, Giles GG (1988): Cancer in migrants to Australia: Extending the descriptive epidemiological data. *Cancer Res* 48(3):751–756.

Medina D, et al. (2003): Hormone dependence in premalignant mammary progression. *Cancer Res* 63(5):1067–1072.

Michaelson JS, Leder P (2001): Beta-catenin is a downstream effector of Wnt-mediated tumorigenesis in the mammary gland. *Oncogene* 20(37):5093–5099.

Mietz JA, et al. (1992): The transcriptional transactivation function of wild-type p53 is inhibited by SV40 large T-antigen and by HPV-16 E6 oncoprotein. *EMBO J* 11(13):5013–5020.

Miki Y, et al. (1994): A strong candidate for the breast and ovarian cancer susceptibility gene BRCA1. *Science* 266(5182):66–71.

Mitchell MS (2002): Cancer vaccines, a critical review—Part II. *Curr Opin Investig Drugs* 3(1):150–1508.

Mogavero GT, et al. (1992): Inflammatory breast cancer: CT evaluation. *Clin Imaging* 16(3):183–186.

Monni O, et al. (2001): Comprehensive copy number and gene expression profiling of the 17q23 amplicon in human breast cancer. *Proc Natl Acad Sci USA* 98(10):5711–5716.

Montagna C, et al. (2002): Centrosome abnormalities, recurring deletions of chromosome 4, and genomic amplification of HER2/neu define mouse mammary gland adenocarcinomas induced by mutant HER2/neu. *Oncogene* 21(6):890–898.

Montagna C, et al. (2003): The Septin 9 (MSF) gene is amplified and overexpressed in mouse mammary gland adenocarcinomas and human breast cancer cell lines. *Cancer Res* 63(9):2179–2187.

Moody-Ayers SY, et al. (2000): "Benign" tumors and "early detection" in mammography-screened patients of a natural cohort with breast cancer. *Arch Intern Med* 160(8):1109–1115.

Moser AR, et al. (1993): ApcMin, a mutation in the murine Apc gene, predisposes to mammary carcinomas and focal alveolar hyperplasias. *Proc Natl Acad Sci USA* 90(19):8977–8981.

Muller A, et al. (2001): Involvement of chemokine receptors in breast cancer metastasis. *Nature* 410(6824):50–56.

Muller L, Illiger HJ (2000): Determination of oncogenes c-erbB-2 or Her-2/neu for prognosis in breast cancer. *Strahlenther Onkol* 176(1):44–46.

Muller WJ, et al. (1998): Oncogenic activation of Neu/ErbB-2 in a transgenic mouse model for breast cancer. *Biochem Soc Symp* 63:149–157.

Muller WJ, et al. (1990): The int-2 gene product acts as an epithelial growth factor in transgenic mice. *EMBO J* 9(3):907–913.

Nandi S, et al. (1995): Hormones and mammary carcinogenesis in mice, rats, and humans: a unifying hypothesis. *Proc Natl Acad Sci USA* 92(9):3650–3657.

Nathanson KL, et al. (2001): Breast cancer genetics: What we know and what we need. *Nat Med* 7(5):552–556.

Naumov GN, et al. (2002): Persistence of solitary mammary carcinoma cells in a secondary site: A possible contributor to dormancy. *Cancer Res* 62(7):2162–2168.

Normanno N, et al. (1994): Epidermal growth factor-related peptides in the pathogenesis of human breast cancer. *Breast Cancer Res Treat* 29(1):11–27.

Nusse R, Varmus HE (1982): Many tumors induced by the mouse mammary tumor virus contain a provirus integrated in the same region of the host genome. *Cell* 31(1):99–109.

Olsen O, Gotzsche PC (2001): Cochrane review on screening for breast cancer with mammography. *Lancet* 358(9290): 1340–1342.

Omer CA, et al. (1997): Farnesyl: Proteintransferase inhibitors as agents to inhibit tumor growth. *Biofactors* 6(3):359–366.

Osborne CK, et al. (1980): Epidermal growth factor stimulation of human breast cancer cells in culture. *Cancer Res* 40(7):2361–2366.

Osborne CK, et al. (2000): Selective estrogen receptor modulators: Structure, function, and clinical use. *J Clin Oncol* 18(17):3172–3186.

Ottaviano YL, et al. (1994): Methylation of the estrogen receptor gene CpG island marks loss of estrogen receptor expression in human breast cancer cells. *Cancer Res* 54(10):2552–2555.

Ottman R, et al. (1983): Practical guide for estimating risk for familial breast cancer. *Lancet* 2(8349):556–558.

Paget S (1889): The distribution of secondary growths in cancer of the breast. *Lancet* 1:99–101.

Paridaens R, et al. (1980): Clinical significance of the quantitative assessment of estrogen receptors in advanced breast cancer. *Cancer* 46(12 Suppl):2889–2895.

Pattengale PK, et al. (1989): Animal models of human disease. Pathology and molecular biology of spontaneous neoplasms occurring in transgenic mice carrying and expressing activated cellular oncogenes. *Am J Pathol* 135(1):39–61.

Perou CM, et al. (1999): Distinctive gene expression patterns in human mammary epithelial cells and breast cancers. *Proc Natl Acad Sci USA* 96(16):9212–9217.

Phelan CM, et al. (1996): Mutation analysis of the BRCA2 gene in 49 site-specific breast cancer families. *Nat Genet* 13(1):120–122.

Pravtcheva DD, Wise TL (1998): Metastasizing mammary carcinomas in H19 enhancers-Igf2 transgenic mice. *J Exp Zool* 281(1):43–57.

Preston DL, et al. (2002): Radiation effects on breast cancer risk: A pooled analysis of eight cohorts. *Radiat Res* 158(2):220–235.

Pupa SM, et al. (2001): Prevention of spontaneous neu-expressing mammary tumor development in mice transgenic for rat proto-neu by DNA vaccination. *Gene Ther* 8(1):75–79.

Radice GL, et al. (1997): Precocious mammary gland development in P-cadherin-deficient mice. *J Cell Biol* 139(4): 1025–1032.

Ramaswamy S, et al. (2003): A molecular signature of metastasis in primary solid tumors. *Nat Gen* 33:49–54.

Rao GN, et al. (1999): Changes associated with delay of mammary cancer by retinoid analogues in transgenic mice bearing c-neu oncogene. *Breast Cancer Res Treat* 58(3):241–254.

Rauh MJ, et al. (1999): Accelerated mammary tumor development in mutant polyomavirus middle T transgenic mice expressing elevated levels of either the Shc or Grb2 adapter protein. *Mol Cell Biol* 19(12):8169–8179.

Ray P, et al. (2001): Monitoring gene therapy with reporter gene imaging. *Semin Nucl Med* 31(4):312–320.

Ray P, et al. (2003): Optical bioluminescence and positron emission tomography imaging of a novel fusion reporter gene in tumor xenografts of living mice. *Cancer Res* 63(6):1160–1165.

Richon VM, et al. (2000): Histone deacetylase inhibitor selectively induces p21WAF1 expression and gene-associated histone acetylation. *Proc Natl Acad Sci USA* 97(18):10014–10019.

Ried T, et al. (1999): Genomic changes defining the genesis, progression, and malignancy potential in solid human tumors: A phenotype/genotype correlation. *Genes Chromosomes Cancer* 25(3):195–204.

Ried T, et al. (1995): Comparative genomic hybridization of formalin-fixed, paraffin-embedded breast tumors reveals different patterns of chromosomal gains and losses in fibroadenomas and diploid and aneuploid carcinomas. *Cancer Res* 55(22):5415–5423.

Roque AC, Andre TC (2002): Mammography and computerized decision systems: A review. *Ann NY Acad Sci* 980:83–94.

Rosenbaum Smith SM, Osborne MP (2000): Breast cancer chemoprevention. *Am J Surg* 180(4):249–251.

Rosner B, Colditz GA (1996): Nurses' health study: Log-incidence mathematical model of breast cancer incidence. *J Natl Cancer Inst* 88(6):359–364.

Rowley JD (1990): The Philadelphia chromosome translocation. A paradigm for understanding leukemia. *Cancer* 65:2178–2184.

Sacco MG, et al. (2001): Systemic gene therapy with anti-angiogenic factors inhibits spontaneous breast tumor growth and metastasis in MMTVneu transgenic mice. *Gene Ther* 8(1):67–70.

Sandgren E.P., N.C. Luetteke, et al. (1990). "Overexpression of TGF alpha in transgenic mice: induction of epithelial hyperplasia, pancreatic metaplasia, and carcinoma of the breast." *Cell* 61(6):1121–35.

Sandgren EP, et al. (1995): Inhibition of mammary gland involution is associated with transforming growth factor alpha but not c-myc-induced tumorigenesis in transgenic mice. *Cancer Res* 55(17):3915–3927.

Sasano H, et al. (1999): Effects of aromatase inhibitors on the pathobiology of the human breast, endometrial and ovarian carcinoma. *Endocr Relat Cancer* 6(2):197–204.

Sausville EA, et al. (2000): Inhibition of CDKs as a therapeutic modality. *Ann N Y Acad Sci* 910:207–221; discussion 221–222.

Schiff R, et al. (2000): Oxidative stress and AP-1 activity in tamoxifen-resistant breast tumors in vivo. *J Natl Cancer Inst* 92(23):1926–1934.

Schirrmeister H, et al. (2001): Fluorine-18 2-deoxy-2-fluoro-D-glucose PET in the preoperative staging of breast cancer: Comparison with the standard staging procedures. *Eur J Nucl Med* 28(3):351–358.

Schoenenberger CA, et al. (1988): Targeted c-myc gene expression in mammary glands of transgenic mice induces mammary tumours with constitutive milk protein gene transcription. *EMBO J* 7(1):169–175.

Schröck E, et al. (1996): Multicolor spectral karyotyping of human chromosomes. *Science* 273(5274):494–497.

Schulze-Garg C, et al. (2000): A transgenic mouse model for the ductal carcinoma in situ (DCIS) of the mammary gland. *Oncogene* 19(8):1028–1037.

Schwaiger M (2002): Functional imaging for assessment of therapy. *Br J Radiol* 75 **Spec No**:S67–73.

Senderowicz AM (2000): Small molecule modulators of cyclin-dependent kinases for cancer therapy. *Oncogene* 19(56):6600–6606.

Shah N, et al. (1999): Antitumor effects of bis(ethyl)polyamine analogs on mammary tumor development in FVB/NTgN (MMTVneu) transgenic mice. *Cancer Lett* 146(1):15–23.

Sharpe CR, et al. (2000): Nested case-control study of the effects of non-steroidal anti-inflammatory drugs on breast cancer risk and stage. *Br J Cancer* 83(1):112–120.

Shi B, et al. (2000): The farnesyl protein transferase inhibitor SCH66336 synergizes with taxanes in vitro and enhances their antitumor activity in vivo. *Cancer Chemother Pharmacol* 46(5):387–393.

Shi HY, et al. (2003): Modeling human breast cancer metastasis in mice: Maspin as a paradigm. *Histol Histopathol* 18(1):201–206.

Shibata MA, et al. (1999): Haploid loss of bax leads to accelerated mammary tumor development in C3(1)/SV40-TAg transgenic mice: reduction in protective apoptotic response at the preneoplastic stage. *EMBO J* 18(10):2692–2701.

Siegel PM, et al. (1994): Novel activating mutations in the neu proto-oncogene involved in induction of mammary tumors. *Mol Cell Biol* 14(11):7068–7077.

Siegel PM, et al. (1999): Elevated expression of activated forms of Neu/ErbB-2 and ErbB-3 are involved in the induction of mammary tumors in transgenic mice: Implications for human breast cancer. *EMBO J* 18(8):2149–2164.

Silverstein MJ (1997): *Ductal Carcinoma In situ of the Breast*. Williams & Wilkins, Baltimore.

Silverstein MJ, et al. (1995): Prognostic classification of breast ductal carcinoma-in-situ. *Lancet* 345(8958):1154–1157.

Sinn E, et al. (1987): Coexpression of MMTV/v-Ha-ras and MMTV/c-myc genes in transgenic mice: Synergistic action of oncogenes in vivo. *Cell* 49(4):465–475.

Slamon DJ, et al. (1987): Human breast cancer: Correlation of relapse and survival with amplification of the HER-2/neu oncogene. *Science* 235(4785):177–182.

Slamon DJ, et al. (2001): Use of chemotherapy plus a monoclonal antibody against HER2 for metastatic breast cancer that overexpresses HER2. *N Engl J Med* 344(11):783–792.

Smith GH, et al. (1995): Constitutive expression of a truncated INT3 gene in mouse mammary epithelium impairs differentiation and functional development. *Cell Growth Differ* 6(5):563–577.

Sorlie T, et al. (2001): Gene expression patterns of breast carcinomas distinguish tumor subclasses with clinical implications. *Proc Natl Acad Sci USA* 98(19):10869–10874.

Speicher MR, et al. (1996): Karyotyping human chromosomes by combinatorial multi-fluor FISH. *Nat Genet* 12(4):368–375.

Srivastava S, et al. (1990): Germ-line transmission of a mutated p53 gene in a cancer-prone family with Li-Fraumeni syndrome. *Nature* 348(6303):747–749.

Stambolic V, et al. (2000): High incidence of breast and endometrial neoplasia resembling human Cowden syndrome in pten+/− mice. *Cancer Res* 60(13):3605–3611.

Stepanova L, et al. (2000): The oncoprotein kinase chaperone CDC37 functions as an oncogene in mice and collaborates with both c-myc and cyclin D1 in transformation of multiple tissues. *Mol Cell Biol* 20(12):4462–4473.

Sternlicht MD, et al. (2000): The matrix metalloproteinase stromelysin-1 acts as a natural mammary tumor promoter. *Oncogene* 19(8):1102–1113.

Sternlicht MD, et al. (1999): The stromal proteinase MMP3/stromelysin-1 promotes mammary carcinogenesis. *Cell* 98(2):137–146.

Stewart TA, et al. (1984): Spontaneous mammary adenocarcinomas in transgenic mice that carry and express MTV/myc fusion genes. *Cell* 38(3):627–637.

Strange R, et al. (1992): Apoptotic cell death and tissue remodelling during mouse mammary gland involution. *Development* 115(1):49–58.

Struewing JP, et al. (1997): The risk of cancer associated with specific mutations of BRCA1 and BRCA2 among Ashkenazi Jews. *N Engl J Med* 336(20):1401–1408.

Takayama H, et al. (1997): Diverse tumorigenesis associated with aberrant development in mice overexpressing hepatocyte growth factor/scatter factor. *Proc Natl Acad Sci USA* 94(2):701–706.

Thompson ME, et al. (1995): Decreased expression of BRCA1 accelerates growth and is often present during sporadic breast cancer progression. *Nat Genet* 9(4):444–450.

Tirkkonen M, et al. (1998): Molecular cytogenetics of primary breast cancer by CGH. *Genes Chromosomes Cancer* 21(3):177–184.

Tominaga T, et al. (1985): Effective sequential administration of tamoxifen and medroxyprogesterone acetate for 7,12-dimethylbenz[a]-anthracene-induced rat mammary tumors in relation to hormone receptors. *Jpn J Cancer Res* 76(11):1120–1125.

Tonetti DA, Jordan VC (1995): Possible mechanisms in the emergence of tamoxifen-resistant breast cancer. *Anticancer Drugs* 6(4):498–507.

Torrisi R, Decensi A (2000): Fenretinide and cancer prevention. *Curr Oncol Rep* 2(3):263–270.

Tsukamoto AS, et al. (1988): Expression of the int-1 gene in transgenic mice is associated with mammary gland hyperplasia and adenocarcinomas in male and female mice. *Cell* 55(4):619–625.

Tzeng YJ, et al. (1993): Breast cancer formation in transgenic animals induced by the whey acidic protein SV40 T antigen (WAP-SV-T) hybrid gene. *Oncogene* 8(7):1965–1971.

van de Vijver MJ, et al. (2002): A gene-expression signature as a predictor of survival in breast cancer. *N Engl J Med* 347(25):1999–2009.

van't Veer LJ, et al. (2002): Gene expression profiling predicts clinical outcome of breast cancer. *Nature* 415(6871):530–536.

Vooijs M, et al. (2002): Noninvasive imaging of spontaneous retinoblastoma pathway-dependent tumors in mice. *Cancer Res* 62(6):1862–1867.

Vos CB, et al. (1997): E-cadherin inactivation in lobular carcinoma in situ of the breast: An early event in tumorigenesis. *Br J Cancer* 76(9):1131–1133.

Wagner EF, Risau W (1994): Oncogenes in the study of endothelial cell growth and differentiation. *Semin Cancer Biol* 5(2):137–145.

Wakeling AE, et al. (1991): A potent specific pure antiestrogen with clinical potential. *Cancer Res* 51(15):3867–3873.

Wang TC, et al. (1994): Mammary hyperplasia and carcinoma in MMTV-cyclin D1 transgenic mice. *Nature* 369(6482):669–671.

Weaver Z, et al. (2002): Mammary tumors in mice conditionally mutant for Brca1 exhibit gross genomic instability and centrosome amplification yet display a recurring distribution of genomic imbalances that is similar to human breast cancer. *Oncogene* 21(33):5097–5107.

Weaver ZA, et al. (1999): A recurring pattern of chromosomal aberrations in mammary gland tumors of MMTV-cmyc transgenic mice. *Genes Chromosomes Cancer* 25(3):251–260.

Weinstein EJ, et al. (2000): A mouse model for breast cancer induced by amplification and overexpression of the neu promoter and transgene. *Mol Med* 6(1):4–16.

Weinstein EJ, Leder P (2000): The extracellular region of heregulin is sufficient to promote mammary gland proliferation and tumorigenesis but not apoptosis. *Cancer Res* 60(14):3856–3861.

Weng LP, et al. (1999): PTEN suppresses breast cancer cell growth by phosphatase activity-dependent G1 arrest followed by cell death. *Cancer Res* 59(22):5808–5814.

Wiebe VJ, et al. (1993): Tamoxifen resistance in breast cancer. *Crit Rev Oncol Hematol* 14(3):173–188.

Wigginton JM, et al. (2001): Complete regression of established spontaneous mammary carcinoma and the therapeutic prevention of genetically programmed neoplastic transition by IL-12/pulse IL-2: Induction of local T cell infiltration, Fas/Fas ligand gene expression, and mammary epithelial apoptosis. *J Immunol* 166(2):1156–1168.

Willett WC, et al. (1987): Dietary fat and the risk of breast cancer. *N Engl J Med* 316(1):22–28.

Winer EP, Morrow M, Osborne CK, Harris JR (2002): Malignant tumors of the breast. In Devita VT Jr., Hellman S, Rosenberg SA (Eds), *Cancer Principles and Practice of Oncology*, Vol. 2. Lippincott-Williams & Wilkins, Philadelphia.

Wu JC, et al. (2002a): Optical imaging of cardiac reporter gene expression in living rats. *Circulation* 105(14):1631–1634.

Wu K, et al. (2000): 9-cis-Retinoic acid suppresses mammary tumorigenesis in C3(1)-simian virus 40 T antigen-transgenic mice. *Clin Cancer Res* 6(9):3696–3704.

Wu K, et al. (2002b): Suppression of mammary tumorigenesis in transgenic mice by the RXR-selective retinoid, LGD1069. *Cancer Epidemiol Biomarkers Prev* 11(5):467–474.

Xu X, et al. (1999): Conditional mutation of Brca1 in mammary epithelial cells results in blunted ductal morphogenesis and tumour formation. *Nat Genet* 22(1):37–43.

Yang YA, et al. (2002): Lifetime exposure to a soluble TGF-beta antagonist protects mice against metastasis without adverse side effects. *J Clin Invest* 109(12):1607–1615.

Yokoyama Y, et al. (2000): Effect of endostatin on spontaneous tumorigenesis of mammary adenocarcinoma in a transgenic mouse model. *Cancer Res* 60(16):4362–4365.

Yoshidome K, et al. (2000): Estrogen promotes mammary tumor development in C3(1)/SV40 large T-antigen transgenic mice: Paradoxical loss of estrogen receptoralpha expression during tumor progression. *Cancer Res* 60(24):6901–6910.

Ziegler RG, et al. (1996): Relative weight, weight change, height, and breast cancer risk in Asian-American women. *J Natl Cancer Inst* 88(10):650–660.

8

PROSTATE CANCER

Katharine Ellwood-Yen and Charles Sawyers
Department of Hematology and Oncology, University of California—Los Angeles, Los Angeles, California

Prostate cancer is one of the leading causes of cancer death among men in the United States. Development of model systems to study the disease is imperative. Both transgenic and knockout mice have been generated and offer great hope for future studies. While this type of research is still in its infancy, several promising models have been generated which share similar features to the human disease. Many experiments have been conducted using these models to identify proteins and pathways important to the initiation and progression of prostate cancer. Perhaps the most significant are microarray experiments performed on mouse tissues that have identified expression changes in several genes such as Nkx3.1, Pim-1, and prostate stem cell antigen (PSCA). These genes have already been implicated in human prostate cancer and therefore strengthen the argument for use of mouse model systems to tease out the molecular mechanisms of disease progression. It is only a matter of time before researchers are able to interpret data from the mouse and translate the results to human therapeutics.

OVERVIEW OF HUMAN PROSTATE CANCER

Prostate cancer is the most common noncutaneous cancer in U.S. men, and according to the American Cancer Society, it is second only to lung cancer as the leading cause of cancer death [1]. While the cause of prostate cancer is unknown, the best-documented risk factors associated with the disease include age, ethnicity, and family history. In the United States, the average age of patients at the time of diagnosis is 70 and the risk of developing the disease is much higher in men if their father or brother is also afflicted. Furthermore, prostate cancer is much more prevalent in African American men and much less common in men of Asian or American Indian descent. Some studies suggest that the prevalence of prostate cancer among some races may be due to differences in diet, adding it as a fourth risk factor to the disease.

CLINICAL DIAGNOSIS OF HUMAN PROSTATE CANCER

The prostate is an androgen-regulated gland in the man's reproductive system that makes and stores seminal fluid. It is approximately the size of a walnut and is located below the bladder and in front of the rectum. It surrounds the upper portion of the urethra, and if the prostate grows too large, as in the case of benign prostatic hyperplasia (BPH) or a tumor, the flow of urine can be impeded. Other symptoms that indicate a problem with the prostate include a need to urinate frequently, painful or burning urination, difficulty in having an erection, and frequent pain or stiffness in the lower back, hips, or thighs [2]. Men presenting with any of these symptoms will undergo such tests as a digital rectal examination, a urinalysis to determine if there is blood or an infection present, and a blood test to measure serum prostate-specific antigen (PSA) levels. PSA is the most important clinical marker for adenocarcinoma of the prostate. This protein is secreted normally by prostatic epithelium into the lumina or prostatic acini and ducts, and blood serum concentrations are normally less than 4.0 ng/mL. Serum elevations of PSA (between 4 and 10 ng/mL) may be

Mouse Models of Human Cancer, edited by Eric C. Holland
ISBN 0-471-44460-X

Table 8.1. Transgenic Mouse Models of Prostate Cancer

Model	Strain	PIN	Cancer	Neuroendocrine	Metastasis	Lobes	Androgen Independent (AI)
1. TRAMP	C57BL/6, C57BL/6 × FVB	8 weeks	12 weeks	Yes	Yes	DL	Yes
2. LADY	CD1	10–15 weeks	6–7 months	Yes	Yes	DLVA	Yes
3. C(3)-SV40 TAg	FVB	2–3 months	7–11 months	No	Yes	DLV	No
4. Cryptidin-2 SV40 TAg	FVB	8–10 weeks	24 weeks	Yes	Yes	DLVA	Yes
5. Gγ/T-15 SV40 TAg	C57BL/6 × CBA/J	+	16–24 weeks	Yes	Yes	DV	Yes
6. ARR_2Pb-*Myc*	FVB	2 weeks	6 months	No	No	DLVA	No
7. MPAKT	FVB	+	No	No	No	LV	No
8. Pb-mAR	FVB	12 months	No	No	No	DLV	No
9. C3(1)-*Myc*	C57BL/6	8 months	No	No	No	V	No
10. Bk5-IGF-1	FVB × ICR	>6 months	>6 months	Yes	No	DLVA	Yes

Note: A = anterior lobe; D = dorsal lobe, V = ventral lobe, L = lateral lobe.

Table 8.2. Knockout Mouse Models of Prostate Cancer

Model	Strain	PIN	Cancer	Neuroendocrine	Metastasis	Lobes	Androgen Independent (AI)
1. *Pten*(+/−)	C57BL/6	+	No	No	No	N/A	No
2. *Pten*(+/−)/*Cdkn1b*(−/−)	C57BL/6	N/A	3 months	No	No	DLV	No
3. Conditional *Pten*(−/−)	129/Balb/c	6 weeks	1.5 months	No	Yes	DLVA	Yes
4. *Nkx3.1* (−/−)	129/SvImJ, C57BL/6J	>1 year	No	No	No	N/A	No
5. Conditional *Nkx3.1* (−/−)	C57BL/6	15 weeks	No	No	No	DLVA	No
6. *Nkx3.1* (−/−)/*Pten*(+/−)	C57BL/6J-129/SvJ	N/A	6 months	No	No	DLA	No

Note: N/A represents data not available.

caused by either benign or malignant prostate disease and can indicate an increase in prostate tissue mass caused by tumor formation or BPH [3]. If test results suggest the presence of disease, a biopsy will be performed to determine if cancer cells are present and, if so, the grade of tumor will be determined.

Staging is performed to determine the extent of the disease. In stage I prostate cancer, there is no evidence that the disease has spread outside of the prostate, and it is usually found incidentally during surgery for BPH or in response to increasingly elevated serum PSA [2]. In stage II, although the tumor has not spread from the prostate, the patient has an elevated serum PSA with a corresponding larger tumor that can be felt during a rectal examination. Stage III cancer has spread outside the prostate and invades nearby tissue, and in stage IV, prostate cells have infiltrated the lymph nodes and other parts of the body. Various imaging techniques are used by physicians to stage tumors in hopes of assessing the grade of the tumor which will ultimately aid in clinical evaluation and treatment.

CLINICAL IMAGING OF HUMAN PROSTATE CANCER

While clinical imaging of many types of human tumors has proven successful, imaging of the prostate is not very effective in the diagnosis of prostate cancer. Listed below are some of the standard and newer technologies that are being implemented by clinicians and researchers to help aid in the diagnosis and treatment of the disease, yet many of these procedures do not give accurate enough information to warrant the cost. More research is needed to improve the imaging capability of the prostate, which will ultimately provide a powerful means to assess the extent of the disease in humans.

Transrectal Ultrasonography

Because prostate cancer is multifocal in nature, many false-negative biopsies can make this method of disease assessment somewhat misleading. Therefore, as diagnostic imaging becomes more sensitive, more physicians are relying on these procedures to more accurately diagnose and treat patients with prostate cancer [4]. Transrectal ultrasonography (TRUS) is currently the standard method of imaging for detecting prostate cancer in the United States, and it is used to guide biopsies in patients with abnormal digital rectal examinations or elevated serum PSA levels [4,5]. Because the zones of the prostate are not distinguishable in normal healthy men by sonography, the appearance of various zones in the images would indicate the development of BPH or cancer. Therefore, the sonogram images that are created by TRUS allow physicians to look at the whole prostate for gross changes in shape and morphology; however, only cancer localized to the peripheral zone can be reliably detected [4]. Such a tool is helpful in ruling out false negatives obtained by biopsy alone but also has its limitations. Recent work to improve this technology has shown some real advances. For example, two- and three-dimensional ultrasound has provided real-time color coding of the prostate which can also help to distinguish cancerous from noncancerous prostate tissue [5].

Magnetic Resonance Imaging/Three-Dimensional Proton Magnetic Resonance Spectroscopic Imaging

Magnetic resonance imaging (MRI) combined with three-dimensional proton magnetic resonance spectroscopic imaging (3D-^{1}H-MRSI) can also be used to define therapeutic strategies for patients suffering from prostate cancer [6]. MRI has been shown to be 70% effective in diagnosing cancerous lesions, and when combined with 3D-MRSI, it can be a useful tool in staging the extent of the disease in the local region of the prostate [4]. Recent studies have shown that patients who underwent these procedures preprostatectomy gave surgeons a clearer idea of where the cancer was located, the extent of extracapsular spread, and the aggressiveness of the cancer [4]. In MRI scans, prostate cancer usually appears as an area of abnormal signal intensity against the homogeneous high-signal-intensity background of the peripheral zone. Low-signal lesions in the peripheral zone can indicate cancer but can just as easily represent benign conditions such as prostatitis, hemorrhage or dystrophic changes related to radiation, or androgen withdrawal therapies. Therefore, because of the high cost and the lack of a sufficiently high positive predictive value for cancer detection, MRI alone is not routinely used in cancer screening.

Proton 3D-MRSI is a newly developed technique that provides metabolic information about the prostate and is done in conjunction with traditional MRI scans. The primary reason patients would be referred for MRI/3D-MRSI would be to help define treatment strategies based on the localization, staging, and aggressiveness of the cancer [4]. During this procedure, levels of prostatic metabolites are measured. In areas of cancer, high levels of choline are found while significantly lower levels of citrate are observed in the peripheral zone [4]. Similarly, ratios of choline plus creatine to citrate are greatly elevated in cancerous regions and can be detected by this technique. Furthermore, studies have shown that MRI/3D-MRSI can discriminate between residual or recurrent prostate cancer from normal and necrotic tissue after cryosurgery [4]. Therefore, the morphologic data obtained by MRI combined with the detection of metabolic changes obtained by 3D-MRSI can greatly improve the specificity of cancer detection and localization. New studies have also demonstrated the potential of this procedure in guiding brachytherapy for localized prostate cancer [7]. 3D-MRSI allows for precise delineation of cancer lesions within the prostate gland that allows for delivery of localized ionizing radiation [7,8].

Computed Tomography

Computed tomography (CT) and bone scintigraphy are two techniques mainly used to detect metastatic disease. By CT scan, the prostate gland appears as a soft-tissue structure with uniform homogenous attenuation, and it is difficult to distinguish between surrounding muscles, vessels, and perineal structures [4]. In fact, gross extracapsular extension can only be diagnosed when there is extensive soft-tissue expansion into the periprostatic fat. Because CT lacks the soft-tissue contrast resolution that is required for depicting prostate cancer, it is not a very useful tool in the diagnosis of the disease. Instead, it is used to identify spreading of the disease into adjacent tissues, detect lymph node metastasis, and help doctors develop a strategy for radiation therapy.

Radionuclide Bone Scintigraphy

The most common site of metastasis for patients with advanced stages of prostate cancer is the bone. In fact, bone metastasis can be observed in approximately 85% of patients dying from the disease [4]. Radionuclide bone scintigraphy is used to detect these metastatic lesions because it is more sensitive that plain X-ray films, which can only detect bone metastases that have caused a change in bone density by at least 50% [4]. During this procedure, the entire skeleton is scanned and small metastatic lesions can be accurately defined. The results of such scans can then be used by the practitioner to develop patient-specific treatments. For example, studies have shown that early intervention with bone-specific treatments such as radionuclides and biophosphonates along with high-dose chemotherapy may prolong the survival of the patient [9]. Similarly, classifying bone scans according to the site of metastasis, axial (torso) versus appendicular (appendages) bone, has proven to be a good indicator of patient outcome. Patients who had metastatic lesions found in the axial region had a far better prognosis than those suffering from appendicular metastases [10]. However, while bone scans are more sensitive than traditional radiographs, they have a much lower specificity and should be correlated with traditional films of suspect areas in order to make a more definitive diagnosis.

Positron Emission Tomography

Positron emission tomography (PET) is based on pharmaceuticals combined with positron-emitting isotopes that accumulate in both normal and abnormal cells [11]. The production of these radiopharmaceuticals relies heavily on the ability to combine a radioisotope with an appropriate compound so the compound is not greatly altered in its biologic properties and yet remains easy to image. The most commonly used PET radiopharmaceutical is fluoro-D-glucose (FDG). This compound is taken in by cells undergoing a high rate of metabolism and undergoes only the first step of glycolysis. Because many cancer cells express an increased number of glucose transporters, they rapidly take up the drug where it accumulates, resulting in elevated FDG in these cells relative to normal cells [11]. This technique has proven far less effective in the clinical diagnosis of prostate cancer for several reasons. First, well-differentiated prostate cancer has a lower level of glucose metabolism when compared to other types of cancers. Also, because the prostate is located in close proximity to the bladder, which takes up high levels of FDG, imaging of the primary cancer is often impeded by high levels of background. Also, PET images are not able to differentiate between primary tumors and BPH [4,12]. Therefore, new studies are being conducted to try and identify better tracers that would be more applicable to prostate cancer. Recently, a new radiopharmaceutical, ^{11}C-acetate, was used as a tracer for PET imaging and a high sensitivity was observed in prostate cancer lesions [13]. This tracer was also better at detecting lymph node metastases and demonstrated an enhanced intrapelvic uptake [14]. Similarly, intense research is being focused on citrate since high levels of this compound are secreted by normal prostate tissue but are lost in cancerous lesions. This differential is currently being used for MRI/3D-MRSI and offers hope for success in PET imaging as well [11]. At this point, until better tracers have been developed, PET imaging is most effective in the detection of lymph node and other organ metastases [4].

RADIOLABELED MONOCLONAL ANTIBODIES

Radiolabeled monoclonal antibodies against PSAs are currently being exploited in an attempt to improve imaging of prostate cancer. The most well known monoclonal antibody that has been developed is ProstaScint and has recently received Food and Drug Administration (FDA) approval for imaging of prostate cancer on soft tissue. This murine monoclonal antibody (mAb) is labeled with indium-111 and is directed against a glycoprotein expressed in prostate epithelium known as prostate-specific membrane antigen (PSMA) [4]. Interestingly, the mAb is directed against an intracellular epitope. This protein is expressed in nearly all prostate cancers and its expression increases in poorly differentiated, metastatic, and hormone-refractory disease [15]. Early results from the ProstaScint study group suggest that imaging may help to distinguish between local or distant recurrence of the cancer in patients who have a rising PSA following radical prostatectomy. Such information would aid physicians in determining if such a patient would benefit from salvage radiation therapy after a failed surgery.

DIETARY EFFECTS ON PROSTATE CANCER

Recently, a great deal of research has been conducted to ascertain the effects of diet on prostate cancer initiation and progression. This research stems from epidemiologic studies of immigrants from Asia into the United States suggesting that diet can play a major contributing role to the onset of prostate cancer. For example, a dramatic increase in breast, colon, and prostate cancer was observed in Japanese immigrants when compared to native Japanese people, and this increase may be attributed to diets higher in animal fat in the United States [15]. For example, charcoal-cooked red meat and fish containing heterocyclic amines have been shown to induce cancers of the mammary glands, colon, and prostate in rats [15,16]. Similarly, just as some foods have been shown to increase the risk of prostate cancers, others have demonstrated some beneficial effects against the disease. There is a dramatic difference in mortality from prostate cancer between Japan and the United States, which may be linked to the difference in dietary intake of soybeans rich in isoflavones [17]. Other protective dietary factors include micronutrients such as selenium, zinc, carotenoids, lycopenes, and vitamins E and D [17]. Therefore, intense research is being conducted to determine what role diet can have on the onset, progression, and survival of prostate cancer patients.

ANATOMY OF THE PROSTATE

Human Anatomy

The human prostate is made up of both glandular and nonglandular components that are tightly fused together in a common capsule [18]. There are three histologically distinctive glandular zones located within the fibromuscular stroma [19]. The peripheral zone makes up approximately 70% to 75% of the prostate, the central zone makes up about 20% to 25% of the tissue, and the small transition zone comprises 5% to 10% of the organ (Fig. 8.1) [18]. The large peripheral zone is characterized by ducts, which radiate laterally from the distal prostatic urethra. This zone also has typically smaller acini with less complex ductal branching. Approximately 75% of human prostate carcinomas arise from this region, and due to the distal location of this zone, by the time patients present

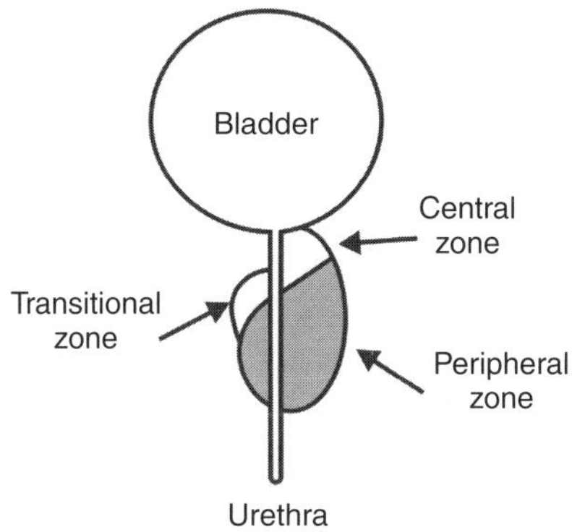

Figure 8.1. Human prostate gland. The three zones that make up the single gland are noted.

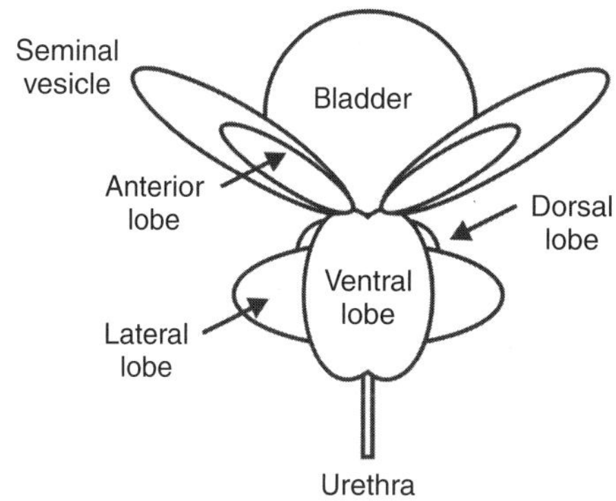

Figure 8.2. Mouse prostate. The four lobes are labeled in relation to the seminal vesicles, bladder, and urethra.

with symptoms, the tumors are usually quite large. The central zone is a proximal wedge-shaped glandular region that surrounds the ejaculatory ducts and is quite broad at the base of the prostate. The central zone contains ducts and acini, which are much larger than both the peripheral or transition zones; however, it has much less stroma when compared to the other two regions [18]. The transition zone, which makes up only a very small portion of the prostate, is composed of two separate gland groups located bilaterally to the urethra. The ducts of the transition zone arise from the urethra and branch entirely anteriorly and anterolaterally away from the other two zones [18]. Tumors from this region occur in approximately 25% of prostate cancer patients and are usually found at an early stage due to obstruction of the urethra. The stroma of the transition zone is comprised of compact interlacing smooth-muscle bundles and is much more compact than the adjacent loose peripheral zone stroma. Furthermore, the majority of clinically significant BPH arises from this region of the prostate [19].

Mouse Anatomy

There are several similarities between the human and the mouse prostate, which validate this model system to study the molecular changes that lead to the onset and progression of prostate cancer. However, there are also some striking differences in the gross and microanatomy of the two organs which may impact the pathologic analysis of the disease and complicate the interpretation of the data into therapeutically useful information for clinicians [20]. First, unlike the human prostate, the mouse organ is divided into anatomically distinct lobes which are not encased by stroma into a capsule forming a single gland (Fig. 8.2). The mouse prostate is made up of four different and discrete lobes called the anterior (also called the coagulating gland), ventral, dorsal, and lateral prostate. The latter are often grouped together and referred to as the dorsolateral prostate. The anterior lobe lies adjacent to the seminal vesicles and demonstrates a more papillary and cribiform growth pattern compared to the other lobes [20]. The architecture of the anterior prostate and the relationship to the seminal vesicle are reminiscent of the human central zone [20].

The dorsal prostate, which is composed of branching ducts and glands, is lined with columnar epithelium and displays a moderate degree of infolding. The lateral prostate contains only sparse regions of infoldings and is lined by cuboidal and short columnar epithelium. The ventral prostate has only focal epithelial tufting or infolding but has abundant luminal spaces which contain homogenous pale serous secretions [20]. The lobes of the mouse prostate gland have histologic similarities to the human organ. They contain luminal secretory cells, basal cell layers, and a small population (less than 1%) of neuroendocrine cells. Therefore, while the gross anatomy of the mouse prostate differs quite dramatically from the human, the histologic similarities allow scientists to utilize this animal for basic research. Because of the ease of genetic manipulation in the mouse, a vast number of animal models have been developed to study prostate cancer. Due to the differences between human and mouse, a great deal of effort has been made in developing guidelines that will allow researchers to grade murine prostate cancer and ultimately provide helpful insight into the mechanism of this disease [21].

HYPERPLASIA IN THE MOUSE

Just as in humans, prostates of elderly wild-type mice also display regions of increased epithelial cell number, which is referred to as hyperplasia. This increase in cell number may also arise as a consequence of transgene expression and therefore warrants definition. Hyperplasia can be either focal, involving one or several glands, or diffuse, involving the majority (>50%) of the glands. These cells do not have abundant cytoplasm or other atypical changes that are typically found in prostate intraepithelial neoplasia (PIN). At this time there is no way to determine if hyperplasia occurs as a result of aging or as a result of transgene expression in the mouse.

PROSTATE INTRAEPITHELIAL NEOPLASIA IN A MOUSE

The precursor of prostate cancer in humans is felt to be PIN, a lesion characterized by cellular proliferation within existing glands accompanied by atypical cytology. PIN in the mouse can be similarly defined and several attempts have been made to classify these lesions. Park et al. have subdivided murine PIN (mPIN) into four types [21]. PIN I lesions are characterized by having relatively small foci with one to two layers of atypical cells. The stroma surrounding the ducts is intact, and the cells are generally more columnar, larger, and taller than adjacent normal cells. They also display an abundant amount of cytoplasm with hyperchromatic nuclei. Areas that fall under the classification of PIN II have larger foci that contain two or more layers of atypical cells; however, these cells still do not fill the lumen. Cribiform and tufting patterns of these cells may be clearly discernable as well as an abundant pale pink cytoplasm with increasing nuclear pleomorphism and hyperchromasia [21]. The foci found in PIN III lesions contain atypical cells which either do or almost completely fill the lumen of the gland. These atypical cells are very often poorly oriented with relatively pale cytoplasm along with increasingly severe nuclear pleomorphism and hyperchromasia. The nuclear-to-cytoplasmic ratio is reversed and mitotic figures are often present. In PIN IV, the foci of atypical cells completely fill the lumen of the ducts, which are no longer smooth but instead look distorted with irregular and bulging shapes. Recently, pathologists associated with the Mouse Models of Human Cancer Consortium (MMHCC) have redefined the criteria for mPIN in an attempt to simplify this classification into a single group. Mouse PIN is now characterized histologically as having multifocal proliferative lesions of epithelial cells affecting several ductules within individual lobes (Fig. 8.3). Cribiform and tufting patterns of growth are also observed, but these lesions must also exhibit progressive nuclear atypia (Fig. 8.3). Atypical cells are usually characterized by the presence of large irregular nuclei, hyperchromatic or vesicular chromatin patterns, and prominent nucleoli with amphophilic cytoplasm (Fig. 8.3). Finally, in earlier studies of mouse models of prostate cancer dysplasia was often used to describe these lesions; however, today it is felt that this term is synonymous with mPIN.

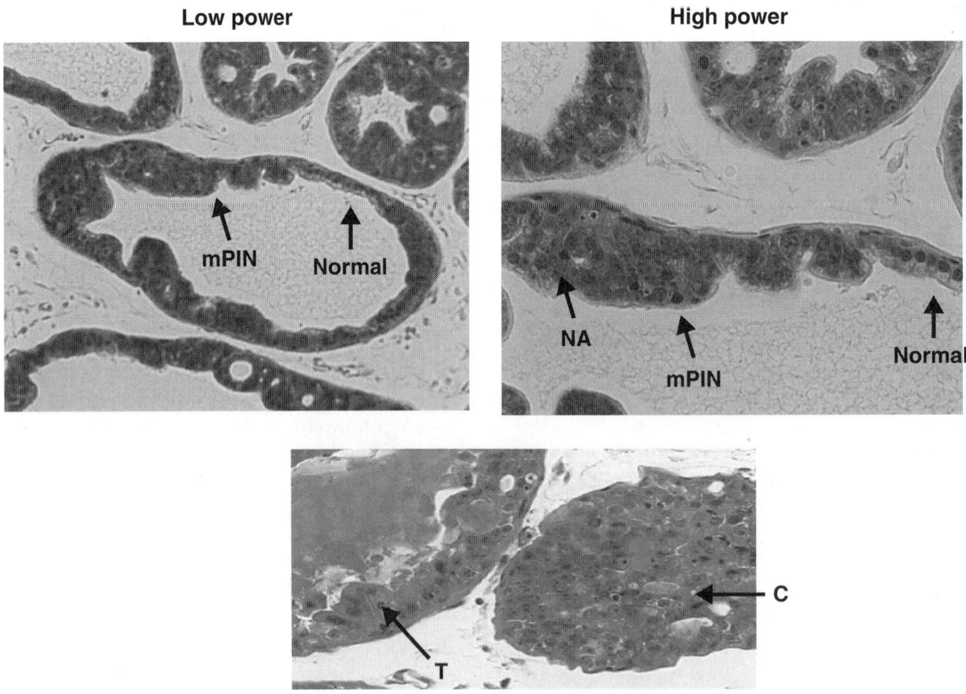

Figure 8.3. Histology showing mouse PIN lesions. Early stages of PIN shown by the piling up of cells adjacent to normal tissue are noted. As these lesions progress, they take on a cribiform (C) or tufting (T) pattern of growth. Nuclear atypia (NA) is also noted by the presence of large irregular nuclei containing prominent nucleoli.

PROSTATIC ADENOCARCINOMA IN MICE

Invasive cancer is characterized by the extension of malignant cells (usually as numerous nests of acini) through the basement membrane extending into the prostatic stroma and periprostatic adipose tissue surrounding the gland (Fig. 8.4). An increased stromal response (desmoplasia), penetration through the basement membrane (confirmed by immunohistochemistry using antilaminin antibodies), lymphovascular invasion, and metastases are also hallmarks of the disease. Prostatic malignancies can arise from either epithelial or neuroendocrine cells. Neuroendocrine differentiation, although very rare (1% to 5% of human cancers), occurs in several forms, including small-cell neuroendocrine carcinoma, carcinoid-like tumors, and adenocarcinoma with focal neuroendocrine differentiation. Histologically, these tumors have distinguishing architectural features such as rosette formation and "salt-and-pepper" chromatin, which differs them from adenocarcinomas (Fig. 8.5). Immunohistochemistry can also be used to help classify these types of tumors. Anticytokeratin antibodies are used to identify epithelial tumors, whereas positive staining with chromogranin A or synaptophysin would indicate neuroendocrine differentiation. Epithelial tumors exhibit positive staining with anticytokeratin antibodies and negative staining for neuroendocrine markers (chromogranin A and synaptophysin). Neuroendocrine tumors exhibit positive staining for chromogranin A and synaptophysin and either negative or dotlike positivity for cytokeratin markers. Because great advances have been made in the genetic manipulation of the mouse and because prostate-specific promoters such as probasin have been identified, many different models for prostate cancer have been developed using both transgenic and knockout strategies [22].

TRANSGENIC MOUSE MODELS OF PROSTATE CANCER

TAg-Based Models

TRAMP. The TRAMP (transgenic adenocarcinoma mouse prostate) mouse was generated by overexpression of the simian virus 40 (SV40) large-T and small-t tumor antigens from the minimal prostate-specific probasin promoter (−426 to +28 bp) in a pure C57BL/6 background [23]. The large-T antigen transforms cells by binding to and inactivating the tumor suppressors p53 and Rb [24,25]. Expression of the small-t antigen has been shown to activate cyclin D1 and inactivate protein phosphatase 2A, a protein that activates multiple signaling cascades [26–28]. Since p53, Rb, and cyclin D1 have all been implicated in human prostate cancer, it is reasonable to target these proteins in the mouse prostate using both the large-T and small-t antigens [29,30].

Recently, a comprehensive study of the pathobiology of TRAMP mice was conducted using [C57BL/6 TRAMP × FVB] F1 mice [31]. PIN lesions were observed in all

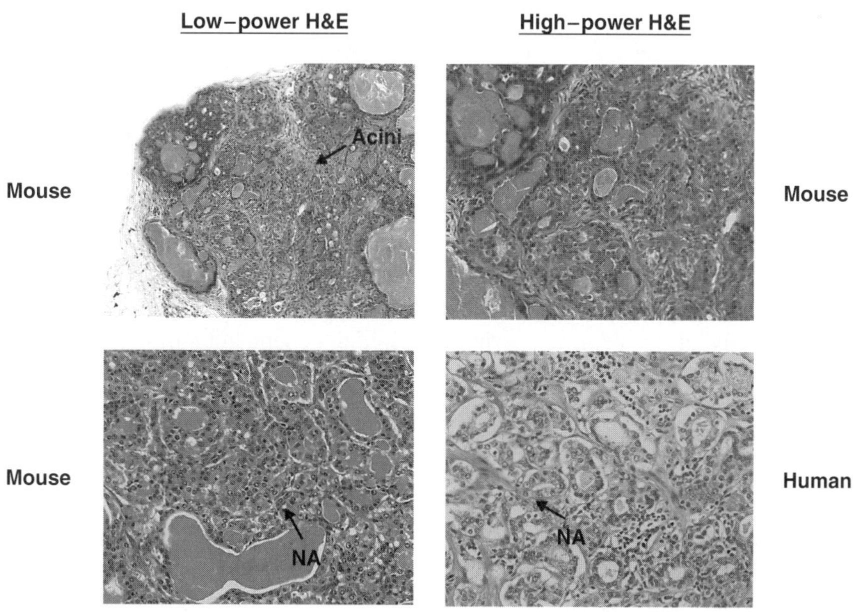

Figure 8.4. Histology showing adenocarcinoma of mouse prostate. An increased number of tightly packed small acini are found in a haphazard arrangement invading into the surrounding stroma. A picture of human prostate cancer is also shown for comparison. Nuclear atypia (NA) is noted.

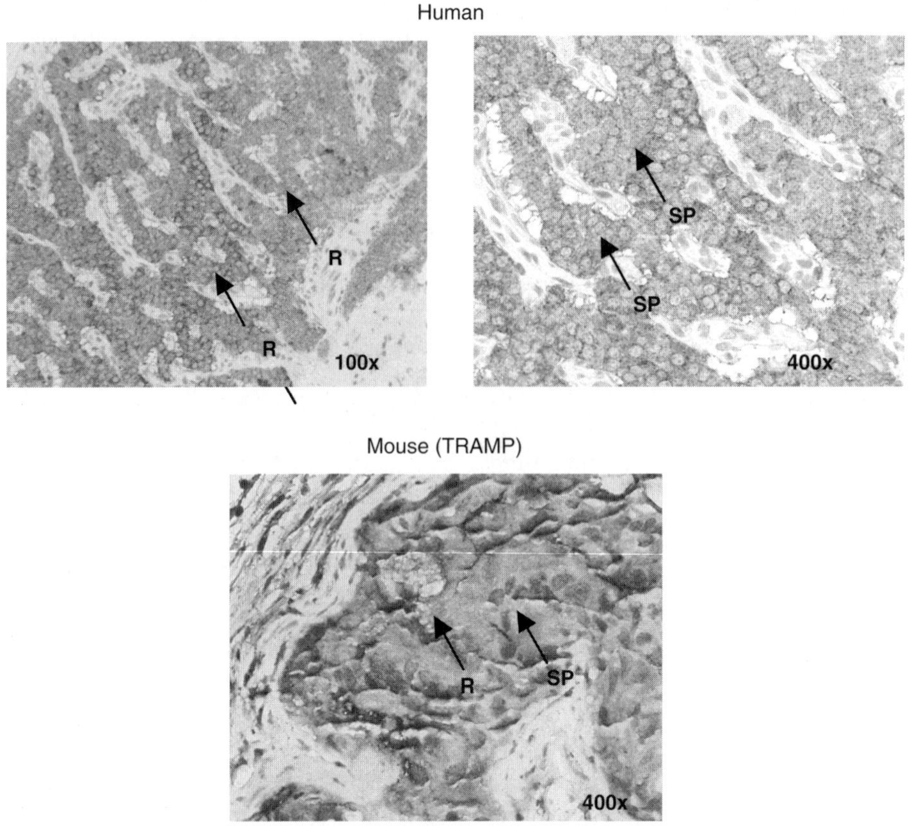

Figure 8.5. Histology showing neuroendocrine tumors of mouse prostate from LADY model. The distinct architecture of these cells is characterized by rosette formation (R) and salt-and-pepper chromatin (SP). Shown are characteristic human prostate tumors that do not share these features.

lobes of the mice by 8 weeks of age, and focal regions of cancer were evident in 80% of these mice [31]. By 12 weeks, 100% of the mice developed cancer, which progressed at different rates depending on which lobe was affected [31]. The dorsal and lateral lobes developed PIN more rapidly than the ventral lobe, and the anterior lobe remained predominately normal. As the tumors became more advanced and poorly differentiated, they displayed neuroendocrine features, whereas early PIN lesions failed to stain positive for synaptophysin (a neuroendocrine marker), suggesting that the tumors did not arise from neuroendocrine precursors or that the cells of the PIN lesions had lost the ability to differentiate. This result demonstrates the stochastic nature of the neuroendocrine phenotype as the mice reached end-stage disease [31]. TRAMP mice also displayed metastatic lesions to the pelvis and lymph nodes in 44% of the mice by 16 weeks of age. Metastasis was also observed in the lung and salivary glands, although to a lesser extent [31].

To date, the only clinical treatment for end-stage prostate cancer is hormone ablation therapy. To study the effects of androgen withdrawal in TRAMP mice, the animals were castrated and tumor burden was measured over time. Results showed that castration led to a decrease in tumor incidence that was followed by the outgrowth of poorly differentiated androgen-independent tumors [31]. This phenomenon mimics the outgrowth of tumors in patients undergoing hormone ablation therapy. One caveat to these experiments is that castration results in silencing of the transgene because the probasin promoter is androgen regulated. Therefore, upon castration, the initial genetic alteration [i.e., expression of the large-T antigen (TAg)] is lost, and in order for relapse to occur, secondary genetic events must have occurred. Further analysis of the tumors following castration showed that they had all undergone neuroendocrine differentiation, a feature that is rare in human prostate cancer [32]. There was also an increase in the number of metastatic lesions (44% to 71%) following surgery. These results demonstrate the aggressiveness of the tumors once they have become androgen independent. Therefore, because the TRAMP mice do develop prostate cancer, they are useful for studying potential therapeutics that could be used in patients suffering from end-stage disease.

However, the model is limited by how quickly the animals succumb to the disease. Because tumors arise so rapidly, it makes it difficult to study the genetic events that occur during the onset and progression of the disease. Furthermore, the fact that tumors progress with a neuroendocrine phenotype points out the differences these mice have when compared to human prostate cancer [33].

LADY. The LADY transgenic mouse was generated by overexpression of the large-T antigen from the large (−11,000- to +28-bp) rat probasin promoter in the CD1 strain [34]. These mice show a much less aggressive phenotype when compared to the TRAMP animals. They initially display multifocal low-grade PIN lesions, which first develop in the dorsolateral prostate and are characterized by glandular proliferation and nuclear atypia [35]. These lesions go on to develop adenocarcinoma and neuroendocrine prostate cancer that commonly metastasize to the lymph nodes, liver, and the lung [35]. Careful analysis of the LADY mice revealed that the rate of disease progression varied among individual sublines, most likely due to varying degrees of transgene expression. In general, by 5–10 weeks of age, all transgenic male mice develop multifocal hyperplastic lesions in all lobes of the mouse prostate; however, the dorsolateral and anterior lobes showed the most dramatic phenotype [36]. The rate at which high-grade dysplasia appeared varied between 10–15 weeks in some sublines and as late as 25–30 weeks in others [36]. The ventral lobe in all cases showed only hyperplasia and sometimes low-grade dysplasia, suggesting that TAg expression was not sufficient for tumor formation in the ventral lobe of the mouse prostate [36]. There was also a marked increase in stromal hyperplasia, a feature that is not common in human prostate cancer [35]. During the progression to cancer, the lesions are characterized by nuclear stratification, nuclear enlargement, nuclear hyperchromasia, and increased mitosis and apoptosis [33]. These nuclear changes are more pronounced and contain more hyperchromasia, pleomorphism, and chromatin clumping and less conspicuous nucleolar enlargement when compared to human prostate cancer [33].

Castration experiments were performed to determine what role androgen played on tumor formation and growth. Mice were castrated at 7 weeks when they displayed clear histologic evidence of hyperplasia and were followed for up to 50 weeks postcastration. No observable regrowth of tumor was evident following castration at this early time point, suggesting that early-stage disease progression is dependent upon androgen [37]. When mice were castrated at 11 weeks of age in sublines that display clear adenocarcinomas, castration induced a dramatic decrease in tumor size. Eventually, between 2 and 6 months postcastration, tumors were able to regrow, demonstrating that in the LADY prostate cancer model there is an emergence of androgen-insensitive disease [33].

In summary, the LADY model for mouse prostate cancer is capable of inducing PIN in male mice with 100% penetrance. These lesions spontaneously progress to adenocarcinomas as well as neuroendocrine prostate cancers and commonly undergo metastases. Furthermore, even though there is a variable rate of disease progression depending on which subline is used, in general the kinetics of disease progression is rapid.

C3(1) SV40 Large-T Antigen. The C3(1) SV40 FVB transgenic mice were generated by flanking the 5′ promoter region of the rat prostatic steroid binding protein [38] to the early region of the SV40 TAg [39]. This regulatory region is capable of targeting expression to the prostate as well as to the mammary gland. In male mice, hyperplastic changes were seen in the dorsal and ventral lobes of the prostate, and by 2–3 months of age, PIN can be observed [40]. Adenocarcinoma developed in approximately one-third of the animals between 7 and 11 months of age, and one of the mice did display metastasis to the lung, but this phenotype was not prominent. The cancerous lesions had poorly defined borders and contained cells with hyperchromatic pleomorphic nuclei with a large number of atypical mitotic figures. These lesions were locally invasive as demonstrated by hematoxylin and eosin (H&E) staining and contained cribiform glandular structures with disorganized hyperchromatic cells [39,41]. Therefore, histologic analysis showed that the tumors share similarities to the more poorly differentiated variants of human prostate cancer [40].

One caveat of using this promoter is that it is not specific to the prostate and it is also capable of driving expression of the transgene in mammary tissue. Therefore, upon analysis of female mice, two-thirds of these animals showed atypical hyperplasia in mammary ducts and acini by three months of age [39]. By six months, these mice succumbed to multifocal mammary adenocarcinoma with occasional metastasis to the lung [39]. Furthermore, the development of these tumors was accelerated if the animals had become pregnant, demonstrating the hormonal regulation of this promoter.

Because these mice show a consistent development of PIN which progresses to adenocarcinoma, it makes them a nice model to study the multistage progression of tumors in both the mammary gland and the prostate. They can be used to study hormone response elements in vivo and to define factors of androgen- and estrogen-specific regulation. However, because the promoter is not specific to the prostate (also expressed in the mammary gland, thyroid, salivary gland, and cartilage), it makes it difficult to follow progression of the disease.

Cryptidin-2 SV40 TAg. The cryptidin promoter (nucleotides −6500 to +34) was used to drive expression of the SV40 TAg in a subset of neuroendocrine cells in all lobes of the FVB mouse prostate [42]. Foci, similar to human PIN, were distinguishable in all lobes of the prostate by 8–10 weeks of age, which coincides with the expression of the transgene in prostatic epithelium. The lesions had increased cellularity and multiple mitotic and apoptotic bodies, and the cells had enlarged nuclei with variable shape, increased chromatin density, and multiple enlarged nucleoli [42]. By 12 weeks all transgenic animals had PIN lesions and 70% of these mice demonstrated local invasion of neoplastic cells through the basement membrane into the surrounding stroma [42]. Histologic examination of prostates from mice at 24 weeks of age showed large tumor nodules composed of solid masses of neoplastic cells [42]. By this time, metastasis to the lymph nodes, liver, lung, and bone marrow was evident in 40% of the mice. Male mice expressing the transgene died between 5 and 7 months of age [42].

The H&E-stained sections showed cellular architecture consistent with neuroendocrine differentiation, suggesting that in these mice prostate cancer initiated from neuroendocrine cell lineages. This result was confirmed by immunohistochemistry using two well-characterized neuroendocrine markers, synaptophysin and chromogranin A [42]. Finally, castration of animals at both early and late time points revealed that androgen ablation was not capable of inhibiting the PIN and tumor phenotypes in the prostate. This result indicates that progression of prostate cancer in neuroendocrine cells is not dependent upon androgen and that the promoter is not androgen regulated, consistent with the fact that these cells do not express the androgen receptor [42]. The highly reproducible kinetics of these mice makes them a nice model to study neuroendocrine cancer, which can be studied from initiation all the way through to advanced metastatic disease. These mice can also be used to identify gene products that could act as genetic markers for advanced-stage disease. Identification of such markers would provide a valuable tool to clinicians when deciding what course of action to take while treating patients with prostate cancer. The limitation of these mice is that they only develop neuroendocrine tumors, which account for only a very low number of human prostate cancers and therefore are not useful in studying less aggressive forms of the disease.

Gγ/T-15 SV40 TAg. When the fetal globin promoter (Gγ/T-15) was linked to the SV40 TAg viral oncogene and microinjected into the pronuclei of (C57BL/6J × CBA/J) F2 mouse zygotes, appropriate expression of the transgene was seen in embryonic erythroid tissue; however, male transgenic mice developed prostate cancer with both epithelial and neuroendocrine features [43]. Transient transfection assays into human prostate cell lines showed that the Gγ-globin promoter sequence between −201 and −140 is highly active in androgen-independent human prostate DU-145 and PPC-1 cell lines and less responsive in the androgen-dependent LNCaP cells. The authors concluded that this region of the promoter contains cryptic *cis*-acting DNA elements that allow expression in the prostate [43]. Therefore, this model system is now being used to study highly aggressive androgen-independent end-stage prostate cancer.

Neoplastic epithelial cells consistent with PIN lesions were seen in the ventral prostate of mice at early time points. Histologic analysis (H&E) showed that these foci remained trapped in the basement membrane and do not invade into the surrounding stroma. Palpable prostate tumors developed in both the dorsal and ventral lobes between 16 and 24 weeks of age with 100% penetrance in homozygous transgenic males [44]. Visible metastatic lesions were evident in the renal lymph node, adrenal glands, and kidneys and micrometastasis was detectable by ribonuclease (RNase) protection in the lung and bone. Castration experiments showed that hormone ablation failed to impede tumor development, demonstrating the androgen-independent nature of these tumors. Finally, immunohistochemistry revealed that tumors contained both epithelial and neuroendocrine features as demonstrated by the presence of chromogranin A, a neuroendocrine marker, as well as cytokeratin 8, which marks cells of epithelial lineage. Therefore, because these mice are an androgen-independent neuroendocrine prostate cancer that metastasizes to the lymph nodes, adrenal glands, lungs, and bone, it is a good model to study rare and aggressive forms of the disease. However, because the fetal globin promoter is not prostate specific and causes tumors in many sites, including the adrenal gland and brown adipose, it limits the reproducibility of the data obtained by these mice. Furthermore, this model was derived from a single transgenic founder.

Non-TAg-Based Models

ARR$_2$Pb-c-Myc. Amplification of c-*MYC* at both the genomic and transcript level can be seen in human prostate cancer [45–47]. To understand the role of c-*Myc* in the progression of the disease, a transgenic mouse was generated using the small composite probasin promoter ARR$_2$Pb [48] driving expression of human c-*Myc* in FVB mice [49]. Consistent with published data, transgene expression was seen in mice by two weeks. Surprisingly, histologic analysis revealed that the prostates of these young mice already displayed PIN lesions, suggesting that c-*Myc* is sufficient to initiate early stages of the disease. PIN lesions were still evident in mice at three months of age, whereas adenocarcinoma developed by six months in the dorsolateral and ventral lobes and, to a lesser extent, in the anterior lobe with 100% penetrance [49]. Similar results were obtained when c-*Myc* was driven by the canonical Pb promoter (−426 to +28); however, the kinetics of disease progression were slightly delayed. PIN lesions developed in the Pb-*Myc* mice by four weeks; however, cancer was not evident until approximately one year. Metastasis was not clearly evident in either the ARR_2Pb-*Myc* or the Pb-*Myc* mice, but some lung nodules where evident in older transgenic animals. Finally, castration of these mice at three and six months caused complete regression of the PIN and cancer lesions, suggesting that androgen may be the survival signal that inhibits c-*Myc*-induced apoptosis. However, because the transgene is driven by androgen signaling, it is not possible to test this hypothesis in this model system.

Because these mice provide a nice model for prostate cancer progression, microarray analyses were performed to identify gene expression changes that were important as the prostates progressed from PIN to cancer. Murine genechips were purchased from Affymetrix and biotin-labeled cRNA from multiple PIN and cancer samples were hybridized to the chips. Analysis of the data revealed that unsupervised hierarchical clustering was capable of generating a gene list that distinguishes between wild-type and transgenic mice. When genes were ranked by degree of differential expression using Student's *t*-test, several genes relevant to prostate cancer were observed [49]. Among the most interesting were *Pim-1* and *Nkx3.1*, both of which have been previously implicated in human prostate cancer [50, 51]. Finally, cross-species bioinformatics comparisons between mouse and human datasets identified gene expression patterns common to both the c-*Myc* model and human prostate cancer. The most striking result was again the presence of *Pim-1* near the top of the list. This result is important because *Pim-1* has been shown to cooperate with c-*Myc* in lymphomagenesis [52, 53] and suggests that this type of analysis could be used to identify complementing oncogenes in other transgenic model systems. Therefore, the ARR_2Pb-*Myc* model is a powerful tool that can be used to study the progression of prostate cancer.

C3(1)-c-Myc. In this transgenic model approximately 4.1 kb of the 5′ upstream promoter region of the rat C3(1) gene was fused to exons 2 and 3 of mouse c-*Myc* and microinjected into C57BLC mice [54]. RNase protection and in situ hybridization demonstrated the presence of transcript in the ventral prostate and testes of male transgenic mice. Prostates from these animals appeared slightly larger than normal, and the epithelial cells had a much more basophilic appearance, enlarged and overlapping nuclei, and a high nuclear-to-cytoplasmic ratio similar to human PIN in mice over 8 months of age [54]. Testis showed no obvious phenotype despite the fact that c-*Myc* message was detectable. Upon aging (up to 18 months), no overt prostate cancer phenotype was observable, making this a poor model system to study the progression of the disease. Since c-*Myc* can induce prostate cancer under the ARR_2Pb promoter (see above), this model most likely failed due to low levels of transgene protein expression [54].

MPAKT. The loss of *PTEN*, or somatic mutations in the protein, is extremely common in human prostate cancer [55]. Such mutations lead to deregulation of the phosphoinositol 3-kinase (PI3K) signaling pathway resulting in constitutive activation of downstream targets such as the Akt/PKB kinase family. Primary prostate tumors also show an increase in Akt kinase activity, suggesting that activation of Akt may be important in the progression of the disease [56]. To address this question, the minimal rat probasin promoter (−426 to +28) was used to drive expression of myr-HA-Akt1 (constitutively active form of *Akt*) specifically in the mouse prostate in an FVB background [57]. A single line, MAPKT, was found to have specific expression of myr-HA-Akt1 in the ventral prostate and in 10% of the mice; a low level of expression was also seen in the lateral lobe. Expression and Akt activation was detected as early as two days postnatal [57].

Histologic analysis of the ventral prostate demonstrated hyperplastic and dysplastic epithelium that contained disorganized multicell layers, intraepithelial lumen formation, loss of cell polarity, nuclear atypia, and apoptotic bodies consistent with the definition of mouse PIN. No phenotype was identifiable in the dorsal or anterior, lobes but in mice where expression was present in the lateral lobe, PIN was also noted. The MPAKT mice also developed PIN lesions similar to the *Pten*(+/−) mice, consistent with the fact that activation of *Akt* occurs as a result of *PTEN* loss. These mice were followed for over 78 weeks but never developed invasive cancer, whereas the *Pten*(+/−) mice do develop prostate adenocarcinomas [57]. Therefore, activation of *AKT* is sufficient to induce PIN, but other events associated with *Pten* loss are necessary to drive tumor formation.

Microarray analysis was performed using Affymetrix murine genechips to study the patterns of gene expression in MPAKT versus nontransgenic animals. Unsupervised hierarchical clustering was used to separate FVB wild-type mice from the MPAKT mice, and a ranked gene list was generated. Among the most significantly upregulated genes were *PSCA* and angiogenin-3 [57]. *PSCA* is a gene that is expressed during prostate development and is upregulated in human prostate tumors and angiogenin-3 is a member of secreted proteins that induce angiogenesis [57]. *PSCA* was also found to be upregulated by microarray analysis in the conditional *Pten*(−/−) knockout described below, indicating that these model systems are capable of identifying important markers for prostate cancer. This model system provides the opportunity to study mutagenic, environmental, dietary, and cooperative events associated with early stages of the disease. However, it is limited by the fact that only a single founder line developed PIN and these mice do not progress to cancer.

Pb-mAR. To study the role of the androgen receptor in prostate cancer, a transgenic mouse expressing AR from the rat probasin promoter (−426 to +28) was generated in FVB mice [58]. The transgene was expressed in all lobes of the prostate and could be found in the nuclei of secretory epithelial cells and in scattered stroma cells of the ventral and dorsal lateral prostate. Upon histologic analysis, dysplastic lesions could be found in mice 12 months and older and displayed characteristics similar to human PIN. For example, the epithelial cells showed pronounced crowding and grew in a cribiform pattern. They also displayed enlarged vesicular nuclei, which contained one or more prominent nucleoli and apoptotic cells [58]. The periodic acid/Schiff (PAS) marker showed early signs of microinvasion of AR-positive cells into the basement membrane but no clear breakthrough. Also, Ki67 staining revealed a high level of proliferation in PIN lesions, which was balanced by an increase in apoptosis [58]. Therefore, since the Pb-mAR mice never develop prostate cancer, they are not useful for studying the progression of the disease. However, because this model is biologically relevant and mice do develop PIN lesions with the potential to progress to cancer, they can be used to study second genetic events that might be necessary to drive tumor formation. The model can also be used to test various drugs and the effect of diet on early stages of the disease.

BK5-IGF-1. Serum insulinlike growth factor 1 (IGF-1) levels have been identified as an important risk factor for prostate cancer, and the IGF-1 receptor plays an important role in neoplastic transformation [59,60]. Therefore, the bovine keratin 5 (BK5) promoter was used to drive expression of IGF-1 in the basal layer of multiple epithelia including the prostate. Following microinjection of DNA into the pronuclei of FVB × ICR F_1 embryos, injected embryos were transferred to the oviducts of pseudopregnant ICR females for term maturation [61]. Between two and three months of age, expression of the transgene had no effect on prostate tissue, which still appeared normal. However, prostate tissue from older mice (six months and older) presented with atypical hyperplastic changes in all lobes of the gland. The overall size of the gland was increased when compared to nontransgenic controls, and lesions in the ventral and dorsolateral lobes consisted of crowded acini with irregular shapes and contours [62].

Focal regions of adenocarcimona were evident as well as vascularization of the surrounding stroma. Immunohistochemistry using the keratin-5 antibody showed that the basal layer of cells was either disrupted or totally absent in the more dysplastic regions, as demonstrated by the absence of endogenous keratin-5 staining [62]. This result highlights the local invasive properties of these lesions. Approximately 50% of transgenic animals older that six months developed both adenocarcinomas and neuroendocrine tumors (small-cell carcinomas), indicating a relatively high incidence of tumor formation. Castration of these mice caused no significant difference in the rate of apoptosis upon androgen withdrawal [62]. Therefore, these data suggest that IFG-1 alone is not able to compensate for the lack of androgen and prevent apoptosis following castration. This model is useful in that it allows one to follow the development of prostate cancer from the early stages of hyperplasia to PIN and finally to adenocarcinoma; however, because the promoter is not specific to the prostate, tumors in other organs can complicate any findings made using this model system.

KNOCKOUT MODELS OF PROSTATE CANCER

Pten(+/−)

The *PTEN* (phosphatase and tensin homologue deleted from chromosome 10) gene encodes a dual-specificity protein phosphatase that can dephosphorylate serine, threonine, and tyrosine residues. It is located on chromosome 10p23, a genomic region that is frequently deleted in many human cancers, including human glioblastomas and advanced prostate cancer. Somatic mutations and deletions have also been identified in prostate, brain, breast, endometrial, skin, and kidney tumors. A *PTEN* loss-of-function mutation was created by homologous recombination in mouse embryonic stem (ES) cells [63]. A targeting vector was designed to replace exons 4 and 5, which encode for the entire *PTEN*–phosphatase domain plus part of the two α-helix motifs flanking the catalytic core, with the neomycin resistance gene. Positive ES clones were expanded and microinjected into C57BL/6 blastocysts. *Pten* heterozygotes were intercrossed to produce *Pten*(−/−) mice; however, none of the homozygous progeny were viable past embryonic day 7.5 [63]. Analysis of the *Pten*(+/−) mice revealed tumors in the colon and testes as well as the development of teratomas, thyroid papillary adenocarcinomas, and leukemia. Analysis of the prostate gland showed enlarged and complex acini, increased cellularity, and frequent mitotic figures consistent with human PIN lesions [63,64]. However, these mice failed to progress to adenocarcinoma of the prostate, indicating that secondary genetic alterations are required to drive tumor formation. Therefore, these mice are useful to study the early stages of prostate cancer but are not useful to study the progression of the disease.

Pten(+/−)/*Cdkn1b*(−/−)

The $p27^{KIP1}$ cyclin-dependent kinase inhibitor is a negative regulator of cell cycle progression. Expression studies of $p27^{KIP1}$ have shown it to be low or absent in advanced-stage human prostate cancer, and it is therefore a good prognostic marker for clinicians [65,66]. Targeted deletion of *Cdkn1b* (the gene that encodes $p27^{KIP1}$) results in only mild hyperplasia of the mouse prostate even at advanced age [67]. *PTEN* has been proposed to regulate $p27^{KIP1}$ protein stability [68]; therefore, in an attempt to exacerbate the mild phenotype, $Pten^{+/-}$ mice were crossed with $Cdkn1b^{-/-}$ mice in a C57BL/6 background [69]. Reduction in the copy number of *Cdkn1b* caused a decrease in survival in the compound *Pten*(+/−)/*Cdkn1b*(+/−) mutants from 51 weeks to 35 weeks, and the *Pten*(+/−)/*Cdkn1b*(−/−) mice showed an even more severe phenotype with a mean survival of 15 weeks [69]. Histologic analysis revealed multiple tumors in many organs, and all of the double mutant male mice analyzed developed prostate cancer within 3 months in the dorsolateral and ventral lobes. An increase in proliferation was also evident in both *Pten*(+/−)/*Cdkn1b*(+/−) and *Pten*(+/−)/*Cdkn1b*(−/−) mice as demonstrated by Ki67 staining [69]. Therefore, this model illustrates the cooperative effect of *PTEN* and $p27^{KIP1}$ as tumor suppressors in epithelial tissue. Because these two genes have both been implicated in the human disease, it illustrates the strength of this model in dissecting the signaling pathways involved in cancer progression. However, these mice develop tumors very quickly and in multiple organs complicating the interpretation of data.

Conditional *Pten*(−/−)

Because *PTEN* alteration can be observed in 30% of primary prostate cancers and in 63% of metastatic disease and because loss of *PTEN* function causes embryonic lethality, $PTEN^{loxp/loxp}$; ARR_2Probasin-Cre mice were generated in order to achieve prostate-specific *Pten* deletion [70]. To generate the $Pten^{L/L}$;C+ mice, ARR_2Pb-Cre mice were crossed to $Pten^{L/L}$ mice on a 129/Balb/c background. Deletion of both alleles of *Pten* caused the prostates of these animals to become progressively larger than wild-type, and histologic analysis of these mice revealed that mPIN lesions were found in 100% of the animals by 6 weeks of age. These mice showed foci of hyperplasia containing two to three layers of cells; however, the duct profiles were still intact. Therefore, this homozygous deletion of *Pten* significantly reduced the latency of mPIN formation from 8–10 months, which is seen in $Pten^{+/-}$ (see above), to 1.5 months in the conditional knockouts [70]. This result suggests that *PTEN* loss is sufficient to initiate prostate cancer.

Homozygous deletion of *PTEN* also led to invasive adenocarcinoma in all lobes of the mouse prostate by 9–10 weeks of age. Neoplastic cells filled and expanded the lumens of the ducts while the surrounding tunica was dissolved and the branches were fused together. These lesions also showed signs of severe atypia, cells outside the fibromuscular tunica and a distinct host response to invasive cells [70]. These tumors failed to show any histological or immunohistochemical evidence of neuroendocrine differentiation. Metastasis was seen to the lymph nodes and lung.

The *Pten* conditional knockouts are a good model system to test the effects of hormone ablation because loss of *Pten* in this model system is not dependent upon the androgen. Castration experiments carried out on 14-week-old mice revealed that *Pten*-null prostate cancer cells are less sensitive to androgen ablation therapy when compared to wild-type controls and, in fact, tumors still remained 7–10 months postcastration. Therefore, loss of *Pten* can lead to androgen-independent tumor formation. Finally gene expression profiles were studied using microarray analysis. Comparison of 29-week-old *Pten*-null prostates with age-matched wild-type controls demonstrated gene expression changes known to be associated with human prostate cancer. In particular, loss of *Nkx3.1* (frequently lost in human disease, see above) and an increase in *PSCA*, a prostate cell surface antigen that is overexpressed in local and metastatic human disease, were observed. The loss of *Nkx3.1* is interesting because it was also noted in the ARR_2Pb-*Myc* model [49], suggesting that there could be common pathways for disease progression that are illuminated through the use of transgenic mice. Overall, this model mimics the course of human prostate cancer progression and can be used to dissect the mechanisms of resistance to hormone ablation therapy and to identify molecular signatures important for drug design. The one drawback is the speed at which the animals develop cancer. The transition from mPIN to adenocarcinoma is quite rapid, and therefore, studies of progression can be difficult to interpret.

Nkx3.1(−/−)

During embryogenesis, the *NKX3.1* homeobox gene is the earliest known marker for prostate epithelium and is expressed in vivo during all stages of prostate differentiation [71]. Human *NKX3.1* maps to chromosome 8p21, a region that frequently undergoes loss of heterozygosity during prostate carcinogenesis. Although this protein is not mutated in prostate cancer, it does undergo epigenetic inactivation through loss of protein expression, making it a potential prostate-specific tumor suppressor gene. To test this hypothesis, targeted disruption of this gene via homologous recombination was performed in mice. The targeting vector effectively deleted the homeodomain as well as the carboxy-terminal protein sequences, thus generating a null mutation that was maintained in a hybrid 129/SvImJ and C57Bl/6J strain background [71].

Histologic analysis of both homozygous and heterozygous *Nkx3.1* mice revealed that these animals developed hyperplasia and dysplastic lesions by one year of age [71]. Prostate glands from older mice have shown a more severe phenotype in both the epithelium and stroma [72]. In fact, 70% of homozygous mutants developed features resembling human PIN. There was a cribiform architecture of the epithelial cells, and the epithelial nuclei were larger with open chromatin and prominent nucleoli. The smooth muscle of the fibromuscular layer also became attenuated and sometimes discontinuous, indicating the potential to develop into adenocarcinoma. The stromal layer was also attenuated, leaving an increased epithelial–stromal ratio, which may reflect a decreased dependence of epithelial cells on the stromal layer [72]. Finally, there was no increase in the number of neuroendocrine cells, consistent with the fact that the number of these cells is not normally increased in PIN lesions. However, despite the histopathologic changes that occur as these mice age, they never develop prostate carcinoma even at the oldest time points (i.e., 18 months).

Because *Nkx3.1* is important for prostate development, the $Nkx3.1^{-/-}$ knockout mice show developmental defects, which complicate the interpretation of the phenotype described above [71]. For this reason, a conditional knockout model using Cre-mediated prostate-specific recombination was generated in C57BL/6J mice [73]. These mice also displayed focal epithelial hyperplasia and PIN with enlarged nuclei with prominent nucleoli [73]. Therefore, because all the *Nkx3.1* mutant mice do develop prostate-specific alterations, these models are valuable for studying the initiating events associated with prostate cancer. These mice can also be used to dissect the signaling pathways associated with the progression of the disease by generating compound mutant mice; however, on their own, they are only useful for studying early stages of the disease.

Nkx3.1(−/−)/Pten(+/−)

To determine if loss of *Nkx3.1* and *Pten* can cooperate in prostate cancer progression, compound heterozyogotes were intercrossed ($Nkx3.1^{+/-} \times Pten^{+/-}$) in a mixed C57BL/6J-129/SvJ strain background [74]. Analysis of the compound mutants showed a synergistic effect on cancer development in the anterior and dorsolateral lobes in 60% of the mice by 6 months of age. Both the $Nkx3.1^{+/-} : Pten^{+/-}$ and $Nkx3.1^{-/-} : Pten^{+/-}$ developed extensive multifocal lesions made of poorly differentiated cells with atypical nuclei and frequent mitotic figures. These lesions also displayed a high proliferative index, as demonstrated by the abundance of Ki67 staining, and are classified as PIN/early carcinomas [74]. By 12 months, 100% of the prostates from these compound mice displayed these lesions.

Immunohistochemical analysis showed the loss of *Nkx3.1* staining; however, LOH studies revealed that the allele was retained. This is in complete contrast to *Pten*, which did suffer allelic loss in 9 of the 10 compound heterozygotes. Finally, in an attempt to understand the mechanism of cancer progression in the compound mice, levels of phospho-Akt were analyzed. Immunohistochemistry showed strong staining of phospho-Akt in compound mutant mice, which was also present in the single mutants, albeit to a lesser extent [74]. These data show the synergistic activation of Akt in the compound mutants and shed some light on a possible mechanism for the progression of the disease. However, although this model does illuminate some of the genetic components of prostate cancer progression, it is not possible to determine the sequential order of events (i.e., *Nkx3.1* loss followed by *Pten* loss or vice versa) that ultimately leads to cancer. Therefore, new models with inducible genes must be made to truly dissect the mechanism of prostate cancer progression.

FUTURE DIRECTIONS

Over the past five years, as researchers have developed a better understanding of the signaling pathways involved in prostate cancer initiation and progression, there has been an increase in the number of mouse models developed to study the disease. As one can see, each model offers its own unique contribution. The first models that were developed relied on expression of the TAg. While these mice do develop prostate cancer, interpretation of the data is complicated by the fact that so many pathways are disrupted by expression of the transgene. Also, these mice undergo neuroendocrine differentiation, a rare event in human prostate cancer patients. More recently, models have been derived from some genes implicated in human disease, such as c-*Myc* and *PTEN*. In both cases, the expression of the transgene or the conditional knockout of *PTEN* elicited PIN lesions that progressed to adenocarcinoma without neuroendocrine differentiation, much like the human disease. Finally, gene expression studies using these mice has identified several genes that have already been implicated in human prostate cancer. Therefore, microarray analysis provides a unique opportunity to study global gene expression profiles that can then be compared to human tumors. This translational impact not only validates the mouse as a model system for prostate cancer but also can offer great insight into the molecular mechanism of the disease.

Hormone ablation therapy is the standard treatment for patients suffering from prostate cancer. Probably the greatest limitation to many of the mouse models is that, in order to achieve prostate-specific transgene expression, they rely on the androgen-regulated probasin promoter. Therefore, interpretation of androgen withdrawal in the mouse is complicated by the fact that castration shuts off the transgene. The conditional knockout of *PTEN* provides a unique opportunity to study the effects of castration without shutting off the initiating genetic event, but these mice succumb to disease rather rapidly. To study the effects of hormone ablation in the mouse, new ways of driving transgene expression, independent of androgen, must be achieved. However, several of the models that exist today provide an excellent tool for the study of prostate cancer, which will hopefully lead to new and improved therapeutics for patients suffering from the disease.

ACKNOWLEDGMENTS

The authors acknowledge grant support from The Damon Runyon Cancer Research Foundation, CaP CURE, The Linda and Arthur Gelb Center for Translational Research, and the National Cancer Institute (RO1 CA85912, PO1 CA89021, P50 CA09038).

REFERENCES

1. Harris R, Lohr KN (2002): Screening for prostate cancer: An update of the evidence for the U.S. Preventive Services Task Force. *Ann Intern Med* 137(11):917–929.
2. Anonymous (2002): Information about detection, symptoms, diagnosis, and treatment of prostate cancer. NIH Publication No. 00–1576.
3. Vollmer PAHART (1998): Relationships between serum prostate-specific antigen and histopathologic appearances of prostate carcinoma. In Virginia MD, LiVolsi A. (Eds.), *Pathology of the Prostate*, Vol. 34. W. B. Saunders, Philadelphia, pp 253–281.
4. Coakley FV, et al. (2002): Prostate cancer tumor volume: Measurement with endorectal MR and MR spectroscopic imaging. *Radiology* 223(1):91–97.
5. Balaji KC, et al. (2002): Role of advanced 2 and 3-dimensional ultrasound for detecting prostate cancer. *J Urol* 168(6):2422–2425.
6. Adusumilli S, Pretorius ES (2002): Magnetic resonance imaging of prostate cancer. *Semin Urol Oncol* 20(3): 192–210.
7. DiBiase SJ, et al. (2002): Magnetic resonance spectroscopic imaging-guided brachytherapy for localized prostate cancer. *Int J Radiat Oncol Biol Phys* 52(2):429–438.
8. Mizowaki T, et al. (2002): Towards integrating functional imaging in the treatment of prostate cancer with radiation: the registration of the MR spectroscopy imaging to ultrasound/CT images and its implementation in treatment planning. *Int J Radiat Oncol Biol Phys* 54(5):1558–1564.
9. O'Sullivan JM, Cook GJ (2002): A review of the efficacy of bone scanning in prostate and breast cancer. *Q J Nucl Med* 46(2):152–159.
10. Rigaud J, et al. (2002): Prognostic value of bone scan in patients with metastatic prostate cancer treated initially with androgen deprivation therapy. *J Urol* 168(4 Pt 1):1423–1426.
11. Mathews D, Oz OK (2002): Positron emission tomography in prostate and renal cell carcinoma. *Curr Opin Urol* 12(5):381–385.
12. Salminen E, et al. (2002): Investigations with FDG-PET scanning in prostate cancer show limited value for clinical practice. *Acta Oncol* 41(5):425–429.
13. Oyama N, et al. (2003): 11C-acetate PET imaging of prostate cancer: Detection of recurrent disease at PSA relapse. *J Nucl Med* 44(4):549–555.
14. Dimitrakopoulou-Strauss A, Strauss LG (2003): PET imaging of prostate cancer with 11C-acetate. *J Nucl Med* 44(4):556–558.
15. Yao D, et al. (2002): The utility of monoclonal antibodies in the imaging of prostate cancer. *Semin Urol Oncol* 20(3):211–218.
16. Sinha R (2002): An epidemiologic approach to studying heterocyclic amines. *Mutat Res* 506–507:197–204.
17. Shirai T, et al. (2002): Diet and prostate cancer. *Toxicology* 181–182:89–94.
18. Myers RP (2002): Gross and applied anatomy of the prostate, Vol. 3. In Kantoff JTIAPW (Ed), *Prostate Cancer Principles and Practices*, Lippincott Williams and Wilkins, Philadelphia, pp 3–15.
19. McNeal JE (1998): Anatomy and normal histology of the human prostate. In *Pathology of the Prostate*, Vol 34. Virginia MD, LiVolsi A W.B. Saunders, Philadelphia, pp 19–34.
20. Shappell S (2003): Anatomic Considerations: Man vs. Mouse. National Cancer Institute, http://emice/nci.nih.gov/emice/mouse_models/orgen_models/prostate_models/man_vs_mouse/anatomic_considerations
21. Park JH, et al. (2002): Prostatic intraepithelial neoplasia in genetically engineered mice. *Am J Pathol* 161(2):727–735.
22. Tuveson DA, Jacks T (2002): Technologically advanced cancer modeling in mice. *Curr Opin Genet Dev* 12(1): 105–110.
23. Greenberg NM, et al. (1995): Prostate cancer in a transgenic mouse. *Proc Natl Acad Sci USA* 92(8):3439–3443.

24. Levine AJ (1990): Tumor suppressor genes. *Bioessays* 12(2):60–66.
25. Levine AJ, Momand J (1990): Tumor suppressor genes: The p53 and retinoblastoma sensitivity genes and gene products. *Biochim Biophys Acta* 1032(1):119–136.
26. Han EK, et al. (1998): Cyclin D1 expression in human prostate carcinoma cell lines and primary tumors. *Prostate* 35(2):95–101.
27. Pallas DC, et al. (1990): Polyoma small and middle T antigens and SV40 small t antigen form stable complexes with protein phosphatase 2A. *Cell* 60(1):167–176.
28. Sontag E, et al. (1993): The interaction of SV40 small tumor antigen with protein phosphatase 2A stimulates the map kinase pathway and induces cell proliferation. *Cell* 75(5):887–897.
29. Brooks JD, et al. (1996): An uncertain role for p53 gene alterations in human prostate cancers. *Cancer Res* 56(16):3814–3822.
30. Brooks JD, et al. (1995): Allelic loss of the retinoblastoma gene in primary human prostatic adenocarcinomas. *Prostate* 26(1):35–39.
31. Kaplan-Lefko PJ, et al. (2003): Pathobiology of autochthonous prostate cancer in a pre-clinical transgenic mouse model. *Prostate* 55(3):219–237.
32. Gingrich JR, et al. (1997): Androgen-independent prostate cancer progression in the TRAMP model. *Cancer Res* 57(21):4687–4691.
33. Scott B, et al. (2001): Transgenic mouse models of prostate carcinoma: Anatomic, histopathologic, and molecular considerations. In Abel E-NLAPD (Ed). *Prostate Cancer: Scientific and Clinical Aspects: Bridging the Gap*, Imperial College Press, London.
34. Kasper S, et al. (1998): Development, progression, and androgen-dependence of prostate tumors in probasin-large T antigen transgenic mice: a model for prostate cancer. *Lab Invest* 78(6):i–xv.
35. Masumori N, et al. (2001): A probasin-large T antigen transgenic mouse line develops prostate adenocarcinoma and neuroendocrine carcinoma with metastatic potential. *Cancer Res* 61(5):2239–2249.
36. Yan Y, et al. (1997): Large fragment of the probasin promoter targets high levels of transgene expression to the prostate of transgenic mice. *Prostate* 32(2):129–139.
37. Kasper S, et al. (1998): Development, progression, and androgen-dependence of prostate tumors in probasin-large T antigen transgenic mice: A model for prostate cancer [corrected and republished in Lab Invest 1998 Jun;78(6):i–xv]. *Lab Invest* 78(3):319–333.
38. Kundu SD, et al. (2000): Absence of proximal duct apoptosis in the ventral prostate of transgenic mice carrying the C3(1)-TGF-beta type II dominant negative receptor. *Prostate* 43(2):118–124.
39. Maroulakou IG, et al. (1994): Prostate and mammary adenocarcinoma in transgenic mice carrying a rat C3(1) simian virus 40 large tumor antigen fusion gene. *Proc Natl Acad Sci USA* 91(23):11236–11240.
40. Shibata MA, et al. (1998): The C3(1)/SV40 T antigen transgenic mouse model of prostate and mammary cancer. *Toxicol Pathol* 26(1):177–182.
41. Shibata MA, et al. (1996): Progression of prostatic intraepithelial neoplasia to invasive carcinoma in C3(1)/SV40 large T antigen transgenic mice: Histopathological and molecular biological alterations. *Cancer Res* 56(21):4894–4903.
42. Garabedian EM, et al. (1998): A transgenic mouse model of metastatic prostate cancer originating from neuroendocrine cells. *Proc Natl Acad Sci USA* 95(26):15382–15387.
43. Perez-Stable C, et al. (1996): Prostate, adrenocortical, and brown adipose tumors in fetal globin/T antigen transgenic mice. *Lab Invest* 74(2):363–373.
44. Perez-Stable C, et al. (1997): Prostate cancer progression, metastasis, and gene expression in transgenic mice. *Cancer Res* 57(5):900–906.
45. Jenkins RB, et al. (1997): Detection of c-myc oncogene amplification and chromosomal anomalies in metastatic prostatic carcinoma by fluorescence in situ hybridization. *Cancer Res* 57(3):524–531.
46. Kaltz-Wittmer C, et al. (2000): FISH analysis of gene aberrations (MYC, CCND1, ERBB2, RB, and AR) in advanced prostatic carcinomas before and after androgen deprivation therapy. *Lab Invest* 80(9):1455–1464.
47. Fleming WH, et al. (1986): Expression of the c-myc protooncogene in human prostatic carcinoma and benign prostatic hyperplasia. *Cancer Res* 46(3):1535–1538.
48. Zhang J, et al. (2000): A small composite probasin promoter confers high levels of prostate-specific gene expression through regulation by androgens and glucocorticoids in vitro and in vivo. *Endocrinology* 141(12):4698–4710.
49. Ellwood-Yen K, et al. (2003): Myc-driven murine prostate cancer shares molecular features with human prostate tumors. *Cancer Cell* 4(3):223–238.
50. Dhanasekaran SM, et al. (2001): Delineation of prognostic biomarkers in prostate cancer. *Nature* 412(6849):822–826.
51. Bowen C, et al. (2000): Loss of NKX3.1 expression in human prostate cancers correlates with tumor progression. *Cancer Res* 60(21):6111–6115.
52. van Lohuizen M, et al. (1989): Predisposition to lymphomagenesis in pim-1 transgenic mice: Cooperation with c-myc and N-myc in murine leukemia virus-induced tumors. *Cell* 56(4):673–682.
53. van Lohuizen M, et al. (1991): Identification of cooperating oncogenes in E mu-myc transgenic mice by provirus tagging. *Cell* 65(5):737–752.
54. Zhang X, et al. (2000): Prostatic neoplasia in transgenic mice with prostate-directed overexpression of the c-myc oncoprotein. *Prostate* 43(4):278–285.
55. Cairns P, et al. (1997): Frequent inactivation of PTEN/MMAC1 in primary prostate cancer. *Cancer Res* 57(22):4997–5000.

56. Sun M, et al. (2001): AKT1/PKBalpha kinase is frequently elevated in human cancers and its constitutive activation is required for oncogenic transformation in NIH3T3 cells. *Am J Pathol* 159(2):431–437.
57. Pradip K, et al. (2003): Prostate intraepithelial neoplasia induced by prostate restricted Akt activation: The MPAKT model. *Proc Natl Acad Sci USA* 100(13):7841–7846.
58. Stanbrough M, et al. (2001): Prostatic intraepithelial neoplasia in mice expressing an androgen receptor transgene in prostate epithelium. *Proc Natl Acad Sci USA* 98(19):10823–10828.
59. Baserga R (1995): The insulin-like growth factor I receptor: A key to tumor growth? *Cancer Res* 55(2):249–252.
60. Chan JM, et al. (1998): Plasma insulin-like growth factor-I and prostate cancer risk: A prospective study. *Science* 279(5350):563–566.
61. DiGiovanni J, et al. (2000): Constitutive expression of insulin-like growth factor-1 in epidermal basal cells of transgenic mice leads to spontaneous tumor promotion. *Cancer Res* 60(6):1561–1570.
62. DiGiovanni J, et al. (2000): Deregulated expression of insulin-like growth factor 1 in prostate epithelium leads to neoplasia in transgenic mice. *Proc Natl Acad Sci USA* 97(7):3455–3460.
63. Di Cristofano A, et al. (1998): Pten is essential for embryonic development and tumour suppression. *Nat Genet* 19(4):348–355.
64. Podsypanina K, et al. (1999): Mutation of Pten/Mmac1 in mice causes neoplasia in multiple organ systems. *Proc Natl Acad Sci USA* 96(4):1563–1568.
65. Thomas GV, et al. (2000): Preoperative prostate needle biopsy p27 correlates with subsequent radical prostatectomy p27, Gleason grade and pathological stage. *J Urol* 164(6):1987–1991.
66. Yang RM, et al. (1998): Low p27 expression predicts poor disease-free survival in patients with prostate cancer. *J Urol* 159(3):941–945.
67. Cordon-Cardo C, et al. (1998): Distinct altered patterns of p27KIP1 gene expression in benign prostatic hyperplasia and prostatic carcinoma. *J Natl Cancer Inst* 90(17): 1284–1291.
68. Mamillapalli R, et al. (2001): PTEN regulates the ubiquitin-dependent degradation of the CDK inhibitor p27(KIP1) through the ubiquitin E3 ligase SCF(SKP2). *Curr Biol* 11(4):263–267.
69. Di Cristofano A, et al. (2001): Pten and p27KIP1 cooperate in prostate cancer tumor suppression in the mouse. *Nat Genet* 27(2):222–224.
70. Wang S, et al. (2003): Prostate-specific deletion of the murine Pten tumor suppressor gene leads to metastatic prostate cancer. *Cancer Cell* 4(3):209–221.
71. Bhatia-Gaur R, et al. (1999): Roles for Nkx3.1 in prostate development and cancer. *Genes Dev* 13(8):966–977.
72. Kim MJ, et al. (2002): Nkx3.1 mutant mice recapitulate early stages of prostate carcinogenesis. *Cancer Res* 62(11):2999–3004.
73. Abdulkadir SA, et al. (2002): Conditional loss of Nkx3.1 in adult mice induces prostatic intraepithelial neoplasia. *Mol Cell Biol* 22(5):1495–1503.
74. Kim MJ, et al. (2002): Cooperativity of Nkx3.1 and Pten loss of function in a mouse model of prostate carcinogenesis. *Proc Natl Acad Sci USA* 99(5):2884–2889.

9

SKIN CANCER

Marcus Bosenberg
Department of Pathology, University of Vermont, Burlington, Vermont

Skin cancer is the most common form of human cancer and causes significant morbidity and mortality. A wide variety of techniques have been used to model skin cancer in mice, including genetically engineered mice, exposure to chemical carcinogens or ultraviolet light, and xenograft studies of human skin and human skin reconstructs. These models have been used to validate human cancer genetics, identify new oncogenes and tumor suppressors, and to develop novel therapies for use in human skin cancer.

INTRODUCTION

Skin cancer is the most prevalent form of human cancer. Over 1 million new cases of skin cancer occur in the United States every year, most of which are either basal cell carcinomas (900,000 cases/year) or squamous cell carcinomas (200,000 cases/year) (Miller and Weinstock, 1994). Although the incidence of melanoma is lower (53,600 cases/year), it is the leading cause of skin cancer deaths, resulting in an estimated 7400 of the 9600 skin cancer deaths/per year.

The skin forms a relatively impermeable barrier against fluid loss, chemical exposure, infectious agents, mechanical damage, and ultraviolet radiation. In addition, skin plays important roles in thermoregulation and the sensation of touch, temperature, and pain. Several cell types are present in the skin to carry out these diverse functions. The epidermis; hair follicles; sebaceous, apocrine, and eccrine glands; and nails arise from the embryonic ectoderm. The neural crest gives rise to melanocytes, nerves, and sensory receptors, while pilar muscles, fibroblasts, adipocytes, blood and lymph vessels, and cells of the immune system are derived from the mesoderm. Skin varies dramatically at different body sites with increased numbers of sebaceous glands on the face, numerous eccrine glands and ducts on the palms and soles, and thickened dermis of skin of the back.

Mouse and human skin share many characteristics, but several important differences exist (Peckham and Heider, 1999). Melanocytes in human skin are predominantly located at the junction of epidermis and dermis with some melanocytes present in hair follicles. In contrast, murine melanocytes are predominantly within hair follicles or the interfollicular dermis but rarely are present at the dermal–epidermal junction. Human epidermis is generally about twice as thick as mouse epidermis. The proportion of skin covered by terminal hair follicles is much higher in mice relative to humans, and eccrine glands are restricted to glabrous skin in mice (footpads) but have a widespread distribution in humans.

A diverse set of over 100 distinct tumors arises from the many cell types that comprise human skin (McKee, 1996). The biologic behavior and clinical course of different cutaneous malignancies vary markedly. Basal cell carcinomas are locally invasive but only very rarely metastasize or cause death. Melanomas, on the other hand, are exceptionally metastatic at low tumor burdens, and nearly all of the mortality due to melanoma is secondary to metastatic disease. Squamous cell carcinomas occasionally metastasize but more often are locally invasive. Several other less common forms of cutaneous malignancies result in significant mortality, including angiosarcoma and primary cutaneous neuroendocrine carcinoma (Merkel cell carcinoma).

Several advances in the understanding of tumor invasion and metastasis have occurred in recent years;

Mouse Models of Human Cancer, edited by Eric C. Holland
ISBN 0-471-44460-X

however, many basic aspects of these processes are still poorly understood (Hanahan and Weinberg, 2000; Yoshida et al., 2000). Tissue culture–based experiments have helped elucidate many of the cell autonomous changes that result in cancer; however, it is not possible to accurately model the complex process of tumor formation, invasion, and metastasis entirely in vitro. Mouse models of cancer are often good approximations of human cancer counterparts and recapitulate the tumor–stromal interactions, angiogenesis, and multistep nature of tumor progression of cancers (Van Dyke and Jacks, 2002). These models will likely aid in the development and testing of novel therapeutic approaches.

Several approaches of generating and studying mouse models of human skin cancers have been developed. The accessibility of the skin and ease of evaluating skin cancer phenotypes are major experimental assets and led to the study of topically applied carcinogens nearly 100 years ago. These approaches of chemical carcinogenesis produced a multistep model of squamous cell carcinoma that recapitulated many aspects of the human disease (Klein-Szanto, 1997; Yuspa and Poirer, 1988). Exposure to ultraviolet light, an important exposure risk for human melanoma, squamous cell carcinoma, and basal cell carcinoma, has also been used to model these tumors in mice. Organotypic models of skin cancer and xenografting of human tissue onto immunodeficient mice have been developed and refined (Atillasoy et al., 1998; Berking and Herlyn, 2001; Maas-Szabowski et al., 2001). Transgenic and knockout mouse strains have been used to model human skin cancer in part because these models have the advantage of examining the effect of genetic changes observed in human malignancy on tumor formation in the mouse (Bardeesy et al., 2000; Brown and Balmain, 1995; Chin et al., 1998a; Walker and Hayward, 2002).

MALIGNANT MELANOMA

Clinical Features

The incidence of melanoma has been rapidly rising over the last 50 years in the United States, with the lifetime risk increasing from 1 in 1500 in 1935 to 1 in 75 at present (Rigel et al., 1996). In contrast to many other solid tumors, melanoma may affect a younger population and is the leading cause of death due to malignancy among females between the ages of 20 and 30 (Johnson et al., 1998). Fortunately the mortality due to melanoma has risen at a lower rate. This may be due to the earlier detection and treatment of melanomas today. Risk factors for melanoma include having a previous melanoma, a family history of melanoma, light skin color, and the presence of increased numbers of melanocytic nevi.

Malignant melanoma is classified into several subtypes based on clinical characteristics and histopathologic features (Langley et al., 1999). Superficial spreading melanoma (70% of cases) are typically flat or slightly elevated brown/black lesions that vary in color and have irregular outlines. Nodular melanomas (15% of cases) appear as small but rapidly enlarging nodules that tend to ulcerate. Acral lentiginous melanoma (8% of cases) occurs on the hands or feet and resembles superficial spreading melanoma. Lentigo maligna melanoma (5% of cases) tends to arise from pigmented lesions on the sun-damaged skin of elderly patients.

Melanoma prognosis is affected by several variables, including patient age, gender, primary tumor site, tumor thickness, tumor ulceration, mitotic rate of tumor cells, the presence of lymphovascular or perineural invasion, the presence of tumor infiltrating lymphocytes, and the presence of metastasis to lymph nodes, the skin, or visceral organs (Langley et al., 1999). Of these, the tumor thickness (Breslow thickness), tumor ulceration, and presence of metastasis have been the most useful prognostic indicators (Balch et al., 2001b). Visceral metastases and the number of lymph nodes positive for metastatic melanoma are also important prognostic indicators and are incorporated in the American Joint Committee on Cancer (AJCC) melanoma staging classification (Balch et al., 2001a). The prognosis for thin nonulcerated melanomas (less than 1 mm in thickness) is uniformly good, with a 90% 15-year overall survival. For melanomas greater than 4 mm in thickness with ulceration, prognosis is usually poor (30% 15-year overall survival). However, the ability to predict outcome in patients with intermediate-thickness melanomas (1–4 mm) is poor, with the 15-year survival ranging from 50% to 70% (Balch et al., 2001b). Improved prognostic indicators for intermediate-thickness melanomas would be of great clinical value.

Metastatic melanoma usually involves draining lymph nodes and occasionally adjacent skin first but eventually metastasizes to distant visceral sites. The lung is most commonly involved followed by brain, liver, bone marrow, and intestine. A wide variety of other sites are involved less frequently. In about 5% of cases, metastatic melanoma is detected in the absence of an identifiable primary tumor (Chang and Knapper, 1982). In these cases it is generally believed that the primary tumor has regressed.

Treatment of melanomas initially consists of wide excision of the primary tumor with 1–2-cm margins, usually after a diagnostic biopsy has already been performed. For most thin melanomas, wide excision is curative; however, for intermediate and thick melanomas, sentinel lymph node biopsy is performed to better predict outcome. This procedure involves injecting a small amount of radioactive tracer at the site of the primary tumor and monitoring the radioactive signal until it accumulates in the first draining lymph node and excising it. Completion lymphadenectomy is frequently performed in cases with a positive sentinel lymph node.

Treatment of metastatic melanoma is difficult. High-dose interferon alpha has been shown to have statistically significant efficacy in stage III melanoma (Kirkwood et al., 1996); however, the increase in overall survival is minimal (Lens and Dawes, 2002). Melanoma is notoriously resistant to conventional chemotherapy regimens. Radiation therapy is occasionally used for palliation. Interestingly, a small percentage of high-stage melanoma patients have spontaneous remissions by unexplained mechanisms (Wang et al., 1998a). This observation and the occasional finding of immune-mediated regression of primary melanomas have driven substantial effort to develop a cancer vaccine for melanoma (Jager et al., 2001).

Pathology

Evaluation of the histopathology of melanoma biopsies and excisions is of vital importance to determine prognostic indices such as tumor thickness and ulceration, adequacy of margins, whether sentinel lymph node biopsy is indicated, and whether metastatic tumor is present.

Most melanomas appear to arise de novo, without an identifiable precursor. Roughly 15% to 40% of melanomas arise in the context of dysplastic nevi, which share some clinical and histologic features of melanoma (Rhodes, 1999). Occasionally melanoma arises within a giant congenital nevus or adjacent to a common dermal or compound nevus. The great majority of melanomas have an in situ (epidermal) component, which is typically composed of severely atypical melanocytes that migrate upward into the epidermis with a proliferation of single and nested atypical melanocytes at the dermal–epidermal junction (Fig. 9.1*a*). This component of melanomas is sometimes referred to as the radial growth phase, as the lesions tend to grow laterally rather than invade the underlying dermis. The vertical growth phase is the portion of the melanoma that invades the dermis and occasionally the subcutis. The histopathologic features of melanoma subtypes can vary dramatically (Fig. 9.1). The invasive component of superficial spreading and acral lentiginous melanomas consist of nests and individual epithelioid to somewhat spindled cells with prominent nucleoli. Nodular melanoma has an exophytic growth pattern and tends to be composed of epithelioid cells with a high mitotic rate. Lentigo maligna melanoma arises in the context of a linear proliferation of atypical melanocytes at the epidermal–dermal junction in chronically sun-damaged skin. Desmoplastic/neurotropic melanoma also can be found underlying a confluent junctional proliferation of atypical melanocytes but is very spindled and usually present in collagenous stroma. Desmoplastic melanoma is frequently mistaken for a scar both clinically and histologically; however, the presence of an overlying in situ component and the tendency to see nodular lymphocytic aggregates in the deep dermis/subcutis may indicate involvement by desmoplastic melanoma. This melanoma subtype has a propensity for perineural invasion and recurs locally more frequently than other subtypes. Advanced melanoma is frequently amelanotic, and the cells can mimic a wide range of poorly differentiated carcinomas, sarcomas, and even some lymphomas.

Biology

The study of familial melanoma has provided several insights into the genetics of both inherited predisposition to melanoma and sporadic melanoma (Pollock and Trent, 2000). Familial melanoma accounts for about 10% of melanoma and the *CDKN2A* locus on chromosome 9p21 is mutated roughly 20% of familial melanoma kindreds (Greene, 1999). This complex locus encodes two gene products, $p16^{INK4A}$ and $p14^{ARF}$, that are unrelated at the protein level, but both function as tumor suppressors (Chin et al., 1998b). The $p16^{INK4A}$ functions in the retinoblastoma (RB) pathway, binding to and reducing the kinase activity of CDK4 or CDK6 bound to cyclin D. Most germline mutations in *CDKN2A* involve exon 2, which is shared between both gene products in alternative reading frames, or involve exon 1α, which is $p16^{INK4A}$ specific (Fargnoli et al., 1998; Pollock et al., 1996). Activating germline mutations in *CDK4* have been demonstrated in a subset of familial melanoma kindreds. These mutations prevent $p16^{INK4A}$ from binding to CDK4/cyclin D complexes (Zuo et al., 1996). Retinoblastoma patients also have an increased risk for developing melanoma, further emphasizing the importance of the RB pathway in melanoma.

The *TP53* is infrequently mutated in human melanoma; however, several alterations in the p53 pathway have been demonstrated. Germline mutations in exon 1β of *CDKN2A* have been demonstrated in a few cases of patients with familial or multiple primary melanomas. These mutations specifically inactivate $p14^{ARF}$ and do not alter $p16^{INK4}$ (Hewitt et al., 2002; Randerson-Moor et al., 2001; Rizos et al., 2001). $p14^{ARF}$ normally functions in the p53 pathway, binding to murine double minute chromosomes (MDM2) and preventing it from inactivating p53. Increased MDM2 protein has been demonstrated in a large proportion of sporadic melanomas, which would be predicted to result in reduced p53 levels (Polsky et al., 2001). Apoptic protease activating factor-1 (APAF-1) functions downstream of p53 as an initiator of apoptosis and is inactivated in a subset of melanomas by combination of hemizygous deletion and promoter hypermethylation (Soengas et al., 2001). Mouse models of melanoma have highlighted the importance of the p53 pathway in melanoma

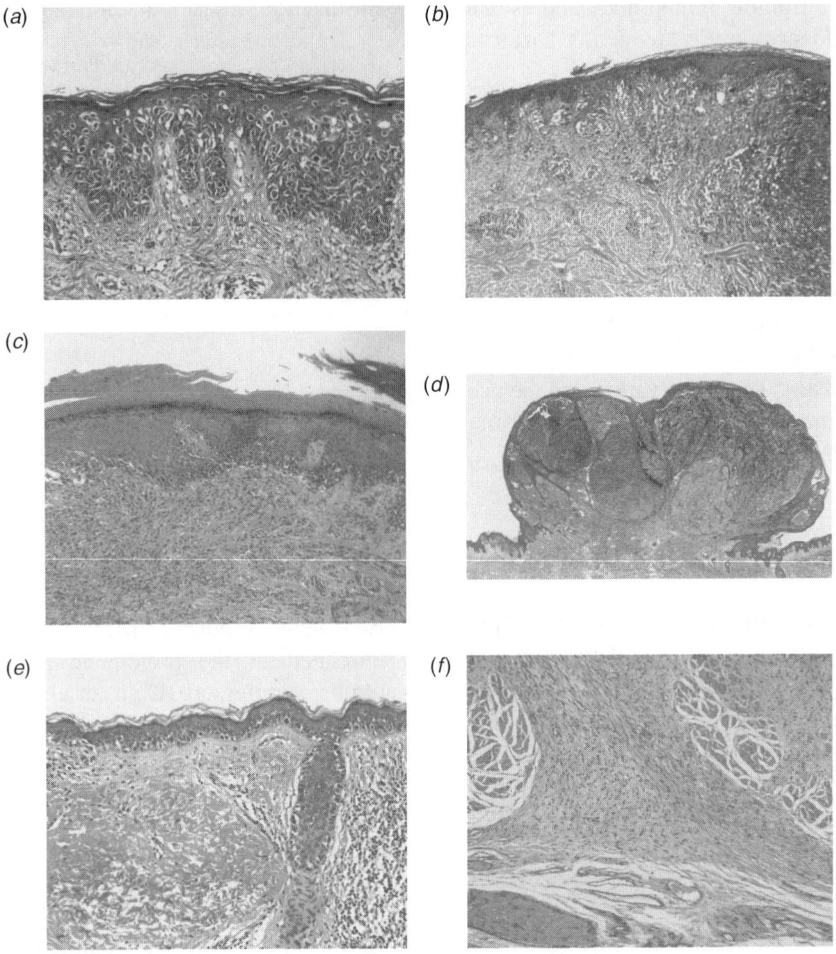

Figure 9.1. Human melanoma. (*a*) Melanoma in situ. Severely atypical melanocytes (with relatively clear cytoplasm) are seen migrating upward with the epidermis. Intraepidermal spread (Pagetoid spread) is a common feature in early human melanoma. (*b*) Invasive melanoma arising in a dysplastic nevus. Invasive melanoma (right portion of image, full thickness) is present adjacent to residual dysplastic nevus (left, nested proliferation of melanocytes at dermal–epidermal junction). (*c*) The invasive component of superficial spreading and acral lentiginous melanomas consist of nests and individual epithelioid to somewhat spindled cells with prominent nucleoli. (*d*) Nodular melanoma has an exophytic growth pattern and tends to be composed of epithelioid cells with a high mitotic rate. (*e*) Lentigo maligna melanoma arises in the context of a linear proliferation of atypical melanocytes that extends down hair follicles in sun-damaged skin. (*f*) Desmoplastic/neurotropic melanoma also can be found underlying a lentiginous confluent proliferation of atypical melanocytes but is very spindled and usually present in collagenous stroma.

as well. Mice lacking p53 and expressing an activated allele of H-*ras* have an increased rate of melanoma formation (Bardeesy et al., 2001), and p53 pathway alterations occur in nearly all examined cases of melanoma arising in $p16^{INK4}$-deficient mice expressing an activated H-*ras* allele in melanocytes (Kannan et al., 2003).

Ultraviolet radiation is an important etiologic factor in skin cancer. Xeroderma pigmentosum patients have defects in repairing ultraviolet-(UV-)induced DNA damage and have a markedly increased incidence of melanoma, squamous cell carcinoma, and basal cell carcinoma (Cleaver and Crowley, 2002). Epidemiologic evidence suggests that severe sunburns during childhood in particular are associated with an increased subsequent risk for melanoma formation (Whiteman et al., 2001). In contrast, chronic ultraviolet exposure appears to predominantly increase rates of basal cell carcinoma and squamous cell carcinoma. Several mouse models of skin cancer are induced or accelerated by UV irradiation, including several models based on knockout mice lacking gene products of one of the xeroderma pigmentosum complementation groups (de Vries et al., 1995; Nakane et al., 1995; Sands et al., 1995).

Activation of the RAS-RAF–mitogen activated protein kinase (MAPK) pathway is emerging as a central feature of melanocytic neoplasia. Activating mutations in *RAS* genes occur in 15% to 25% of cutaneous melanomas (Herlyn and Satyamoorthy, 1996) and commonly involve codon 61 of *NRAS*. Transgenic expression of an activated allele of H-*ras* induces melanoma in *Ink4a/Arf*-deficient mice, supporting a role for RAS in melanoma formation. Activating mutations of *BRAF* have been identified in about 66% of malignant melanomas and 80% of melanoma cell lines (Davies et al., 2002) with activating mutations in *RAS* genes found in roughly half of the remaining melanoma lines that lack *BRAF* alterations, suggesting that activation of RAS or BRAF occurs in most cases of melanoma. Identification of activating mutations of *BRAF* in 82% of melanocytic nevi (Pollock et al., 2003b) and the presence of *HRAS* mutations and amplifications in a subset of Spitz nevi (Bastian et al., 2000b) suggest that activation of the RAS-RAF-MAPK pathway is insufficient to drive melanoma formation but may play a more general role in melanocytic proliferations.

Tyrosine kinases appear to play a role in both the survival and migration of melanocytes and in melanoma formation and progression (Easty and Bennett, 2000). Inactivating mutations in the *KIT* tyrosine kinase or kit ligand result in piebaldism in humans and mice (White and Steel murine loci). Mouse models of melanoma have been produced that involve activation of either the Ret or Met tyrosine kinases (Iwamoto et al., 1991; Noonan et al., 2001; Otsuka et al., 1998). Metastases have occurred more frequently in some of these models than by overexpression of activated H-*ras*, suggesting that non-RAS signaling by tyrosine kinases may play a role in progression to metastasis.

The chromosomal alterations that occur during the progression of malignant melanoma have been extensively characterized. Karyotyping of melanoma cell lines and comparative genomic hybridization of tumors have demonstrated recurrent loss of chromosome 1p, 6q, 9p, 10q, 11q, and 12q and gains of chromosome 1q, 6p, 7, 8q, 11q, 17q, and 22q (Bastian et al., 1998; Pollock and Trent, 2000). It is likely that each of these regions contains one or more genes involved in melanoma biology. Some histologic subtypes of melanoma appear to have a higher incidence of alterations in specific loci. Cyclin D1 (*CCND1*; chromosome 11q13) has been shown to be amplified more frequently in acral lentiginous melanoma (Bastian et al., 2000a) and the neurofibromatosis 1 gene (*NF1*) undergoes loss of heterozygosity in a relatively high proportion of desmoplastic melanomas (Gutzmer et al., 2000).

Mouse Models of Melanoma

Early human melanoma is characterized by upward spread of atypical melanocytes within the epidermis (Pagetoid spread) (Fig. 9.1*a*). Pagetoid spread is uncommon in mouse melanoma models and may be a reflection of the follicular/dermal location of murine melanocytes. Transgenic expression of mitogenic or survival signals by keratinocytes has resulted in increased numbers of melanocytes at the epidermal/dermal junction (Kunisada et al., 1998) and combined with perinatal UV irradiation has resulted in a mouse model with striking Pagetoid spread (Noonan et al., 2001) (Fig. 9.2*a*).

Several mouse models of melanoma studied to date have involved expression of a transgene under the control of the melanocyte-specific tyrosinase promoter including the simian virus 40 (SV40) early region (Bradl et al., 1991) and activated alleles of H-*ras* (Broome Powell et al., 1999; Chin et al., 1997). Alternatively, melanomas have developed in the context of more ubiquitous transgene expression, such as the expression of Ret (Iwamoto et al., 1991) or hepatocyte growth factor (Otsuka et al., 1998) under the control of the metallothionein promoter. In one model, histologically benign appearing melanocytic lesions spontaneously arise with very short latency and high frequency in a transgenic strain of mice (Pollock et al., 2003a; Zhu et al., 1998) (Figs. 9.2*c*,*d*). These lesions frequently progress to invasive lesions and/or metastasize. This highly penetrant phenotype appears to be integration site specific and the relevant gene has been demonstrated to encode the metabotropic glutamate receptor (Grm1). Transgenic mice expressing Grm1 under the control of the melanocyte-specific dopachrome tautomerase promoter develop generalized melanocytic hyperplasia. A subset of mice progress to malignant melanoma and metastases. Murine melanomas have also arisen at increased rates in knockout mice in the context of chemical carcinogenesis protocols, typically involving 7,12-dimethylbenz(*a*)anthracene (DMBA) as an initiator and 12-*O*-tetradecanoylphorbol-13-acetate (TPA) as a promoter.

Several mouse models have been constructed to recapitulate the genetic changes observed in human melanoma (Fig. 9.2, Table 9.1). Knockout mice lacking both p16^{INK4A} and p19^{ARF} (*Ink4a/Arf*−/−) develop melanomas at a high rate in the setting of tyrosinase-driven activated H-*ras* expression but rarely develop melanomas in the absence of a second genetically engineered change. Inducible expression of *H*-ras in *Ink4a/Arf*−/− mice resulted in melanoma formation that was dependent on *H*-ras expression for tumor maintenance, as cessation of H-*ras* expression resulted in regression of tumorous lesions to microscopic deposits within 12 days (Chin et al., 1999). Interestingly, specific inactivation of either p16^{INK4A} or p19^{ARF} results in increased

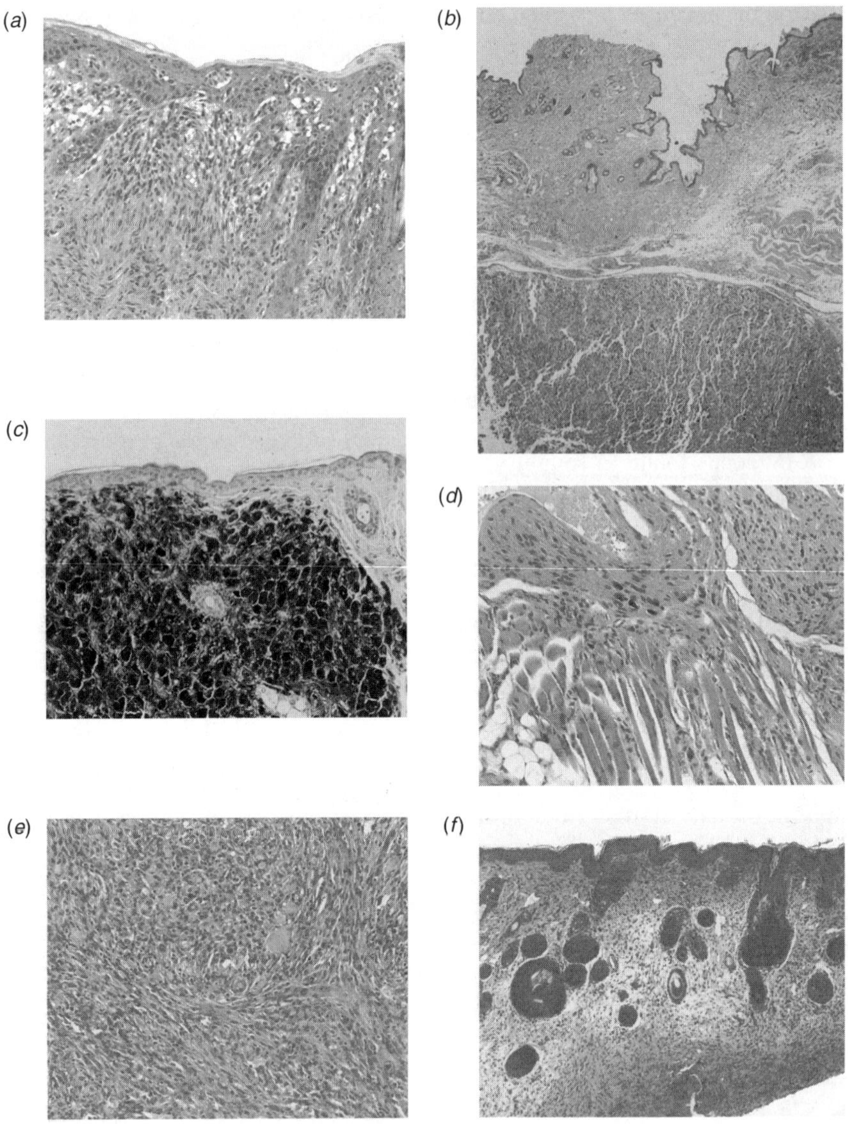

Figure 9.2. Mouse models of melanoma. (*a*) Melanoma arising in Mt-*Hgf* transgenic mice with perinatal UV irradiation (Noonan et al., 2001). Note the extensive intraepidermal spread of melanoma cells overlying the invasive portion of the tumor (*b*) Melanoma arising in Mt-*Hgf* transgenic mice without perinatal UV irradiation. In the absence of perinatal UV exposure, melanomas arising in Mt-*Hgf* mice are dermal nodules that lack intraepidermal spread. (*c*) Melanocytoma (epithelioid blue nevus). Heavily pigmented melanocytic tumor present on the ear composed of large melanocytes with copious cytoplasm filled with melanin (Zhu et al., 1998). No mitoses or significant atypia are present on bleached sections. (*d*) Invasive melanoma. Amelanotic melanocytes can be seen invading underlying skeletal muscle. This lesion arose in the same genetic context that produced the heavily pigmented lesion in (*c*) (Zhu et al., 1998). (*e*) Amelanotic melanoma arising in a Tyr-*rtTA* tetOp-H-ras^{G12V} *Ink4a*−/− mouse that had been fed doxycycline (to induce H-ras^{G12V} expression). Melanoma cells have spindled and epithelioid cytology with numerous mitoses. (*f*) Regression of an amelanotic melanoma from a Tyr-*rtTA* tetOp-H-ras^{G12V} *Ink4a*−/− mouse that had withdrawn from doxycycline for a period of 12 days, thereby shutting off H-ras^{G12V} expression. Grossly, the tumor has regressed completely at this point, but an altered dermis is present with scattered spindle cells with small amounts of cytoplasm separated by edematous stroma. Trp-1 immunohistochemistry was positive in several of the spindled cells.

Table 9.1. Mouse Models of Melanoma

Transgene	Inactivated Gene	Treatment	Phenotype	References
Tyr-*SV40E*			Predominantly ocular but also cutaneous melanomas with a high rate of metastasis	Bradl et al., 1991
Tyr-*SV40E*		Perinatal UV	UV exposure induces melanoma with subsequent metastatic lesions	Kelsall and Mintz, 1998; Klein-Szanto et al., 1994
Mt-*Ret*			Cutaneous hyper-pigmentation and melanocytic tumors	Iwamoto et al., 1991
Mt-*Ret*		UV	Increased rates of melanocytic tumors with UV treatment	Kato et al., 2000
Mt-*Hgf*			Melanomas develop after long latency with metastases occurring in a subset of cases	Otsuka et al., 1998
Mt-*Hgf*		Chronic UV	Increased rates of nonmelanoma skin tumors	Noonan et al., 2000
Mt-*Hgf*		Perinatal UV	Melanomas with striking intraepidermal spread	Noonan et al., 2001
Mt-*Hgf*	*Ink4a/Arf−/−*	Perinatal UV	Melanomas develop with decreased latency in UV-treated mice on *Ink4a/Arf−/−* background	Recio et al., 2002
TP-*ras*		DMBA, UV	Cutaneous melanomas develop after topical DMBA application; few melanomas arise after UV treatment; untreated animals are hyperpigmented	Broome Powell et al., 1999
Tyr-H-ras^{G12V}	*Ink4a/Arf−/−*		Cutaneous and ocular melanomas without evidence of metastasis	Chin et al., 1997
Tyr-*rTTA*; tetOp-H-ras^{G12V}	*Ink4a/Arf−/−*		Cutaneous melanomas dependent on doxycycline-induced *H*-ras expression	Chin et al., 1999
	$Ink4a^{\Delta 2/3/*}$	DMBA	Cutaneous melanomas develop after DMBA treatment, a subset of which metastasize	Krimpenfort et al., 2001
	Ink4a−/−	DMBA	Cutaneous melanomas develop after DMBA treatment at a low rate	Sharpless et al., 2001
Tyr-H-ras^{G12V}	*Ink4a−/−*	Perinatal UV	Cutaneous melanomas in 50% of mice by 40 weeks	Kannan et al., 2003; Sharpless et al., 2003
Tyr-H-ras^{G12V}	*Arf−/−*	Perinatal UV	Incidence and penetrance of cutaneous melanomas increased by UV treatment	Kannan et al., 2003; Sharpless et al., 2003

(*continued overleaf*)

Table 9.1. (*continued*)

Transgene	Inactivated Gene	Treatment	Phenotype	References
Tyr-H-*ras*^{G12V}	*p53−/−*		Cutaneous and ocular melanomas without evidence of metastasis	Bardeesy et al., 2001
	Cdk4^{R24C} knockin	DMBA	Cutaneous melanomas develop after DMBA treatment, a subset of which metastasize	Sotillo et al., 2001
	Pten+/−, *Ink4a/Arf−/−*		Cutaneous melanomas develop spontaneously at a low rate in these highly tumor-prone mice	You et al., 2002
TG3			Highly penetrant melanocytic proliferations with short latency; a subset of lesions progress to invasive tumors and metastases	Zhu et al., 1998
Dct-*Grm1*			Generalized melanocytic hyperplasia with limited transformation to malignant melanoma, a subset of which exhibit metastases	Pollock et al., 2003a
K14-*stem cell factor*			Hyperpigmentation with increased melanin production and increased epidermal melanocytes	Kunisada et al., 1998

Note: See also Walker and Hayward, 2002.

rates of melanoma in the setting of activated *H*-ras expression (Kannan et al., 2003; Sharpless et al., 2003), underscoring the importance of both components of the Ink/Arf locus (*CDKN2A*). Mice containing a knockin activating mutation of the *CDK4* gene develop melanomas at a high frequency after a two-step chemical carcinogenesis protocol (Sotillo et al., 2001).

Several mouse models have used UV irradiation to enhance melanoma formation (Kannan et al., 2003; Kelsall and Mintz, 1998; Noonan et al., 2001). The dosage and timing of UV irradiation is important, as clearly demonstrated in the metallothionein–hepatocyte growth factor (MT-HGF) model. Chronic UV irradiation resulted in increased rates of squamous tumors without significant increasing rates of melanoma (Noonan et al., 2000); however, short-term perinatal UV irradiation resulted in dramatically increased rates of melanoma formation and alteration of the tumor morphology with intraepidermal spread of tumor cells (Noonan et al., 2001) (Fig. 9.2*a*), a feature not observed in sporadic or chronic UV-irradiated MT–HGF mice.

SQUAMOUS CELL CARCINOMA

Clinical Features

Approximately 200,000 cases of cutaneous squamous cell carcinoma occur in the United States per year, resulting in ~2000 deaths (Miller and Weinstock, 1994). Excessive UV radiation exposure, immunosuppression, human papilloma virus infection, certain chronic dermatoses, and topical arsenic exposure are all risk factors for developing squamous cell carcinoma (Schwartz and Stoll, 1999). Solid organ transplant recipients commonly develop squamous cell carcinoma in the setting of immunosuppression and in some instances develop several hundred separate primary tumors over the span of a several years (Jensen et al., 1999).

Most squamous cell carcinomas arise on the sun-damaged skin of the head and neck, with fewer lesions arising on the extremities and occasional tumors occurring on the trunk. Early lesions frequently present as a red, scaly spot. Later lesions may form nodules or firm plaques, either of which can ulcerate. Diagnosis is established by biopsy and histopathologic confirmation. Complete excision is curative in the vast majority of cases. Occasionally squamous cell carcinoma will invade along the perineural layer of peripheral nerves and will extend well beyond the clinically apparent mass. Local recurrence is more common in these instances, and when present on the head, direct intracranial extension may occur. Metastases occur in a minority of cases and disseminated disease is the cause of most squamous cell carcinoma–related deaths. Higher rates of metastasis ($\sim$15%) are observed with primary lesions of the lips or ears (Rowe et al., 1992). Radiation therapy is helpful in some cases of locally recurrent disease in which complete resection is difficult to achieve and in cases of limited metastatic disease.

Pathology

The development of squamous cell carcinoma is frequently a multistep process. Early lesions tend to be either actinic keratoses with atypia of the basal keratinocytic layer of the epidermis or squamous cell carcinoma in situ, in which keratinocytic atypia spans the full-thickness epidermis (Fig. 9.3). These precursors are frequently present adjacent to invasive squamous cell carcinomas which invade the dermis as nests, islands, cords, or individual cells. Several grading schemes have been developed for squamous cell carcinoma and incorporate the extent of keratinization and nuclear atypia (Broders, 1932). A widely used scheme divides tumors into well, moderately, or poorly differentiated. Although poorly differentiated tumors tend to behave more aggressively, well-differentiated tumors can also give rise to metastasis and result in death. Several histologic variants of squamous cell carcinoma have been documented, including verrucous, spindle cell, and pseudoangiosarcomatous (Fig. 9.3). Occasionally squamous tumors arise rapidly, have a crater-form morphology, and spontaneously regress. These tumors are known as keratoacanthomas. Some crater-form squamous lesions do not regress but continue to invade and grow and represent invasive squamous cell carcinomas. It is occasionally difficult to definitively determine whether a crater-form squamous lesion is a keratoacanthoma or represents a well-differentiated invasive squamous cell carcinoma on the basis of histologic examination alone.

Biology

Chronic UV radiation plays a central role in the development of the majority of squamous cell carcinomas, as demonstrated by the association with sites of sun exposure, the histologic presence of sun damage adjacent to squamous cell carcinomas, and the increased risk of squamous cell carcinoma in xeroderma pigmentosum patients.

Human papilloma virus (HPV) infects squamous epithelia at various sites, including the cervix, vagina, vulva, anorectal junction, skin, oral mucosa, and pharynx. Different HPV types tend to cause different clinical manifestations; for example, HPV types 1–4 are associated with cutaneous warts; types 6, 11, 16, and 18 are associated with genital–mucosal lesions; and types 5 and 8 are found in lesions from persons with epidermodysplasia verruciformis (EV), a rare inherited susceptibility to HPV infection (Ramoz et al., 2002). Many of the resultant lesions are benign; however, some are precursors for invasive squamous cell carcinoma. Types 16 and 18 are most often present in squamous cell carcinomas arising in the cervix and at other genital–mucosal sites. Two gene products, E6 and E7, have been shown to play key roles in transformation by HPV. E6 inhibits the function of p53 by directing its ubiquitination and degradation, while E7 inhibits the RB pathway (Munger, 2002).

Some of the genetic changes that play a role in squamous cell carcinoma formation and progression have been defined. p53 is mutated in the majority of both early lesions (actinic keratoses) and invasive squamous cell carcinomas (Ortonne, 2002). The p110α subunit of the PI3-kinase (PIK3CA) is amplified in about 30% of squamous cell carcinomas (Singh et al., 2002). Chromosomal aberrations are common in squamous cell carcinoma with chromosomes 8, 1, 11, 3, 13, 5, and 7 most frequently altered (Jin et al., 1995). Squamous cell carcinoma may occur as a manifestation of certain inherited conditions, including xeroderma pigmentosum and X-linked dyskeratosis congenita.

Mouse Models of Squamous Cell Carcinoma

The use of chemical carcinogens to induce squamous lesions in rodents has been extensively characterized (Yuspa and Poirer, 1988). Most commonly, a two-step process is used with DMBA as the initiator and TPA as the promoter. Typically papillary/polypoid exophytic squamous lesions arise first that bear some resemblance to human keratoacanthoma-like lesions (Klein-Szanto, 1997). These lesions either persist or may progress to carcinoma, with the invasive portion invading the polyp stalk and the underlying tissue. Many studies have examined suppression or enhancement of cutaneous carcinogenesis by applying topical carcinogens to transgenic or knockout mice. The Tg:AC transgenic strain, in which v-H-*Ras* is expressed under the control of the zeta globin

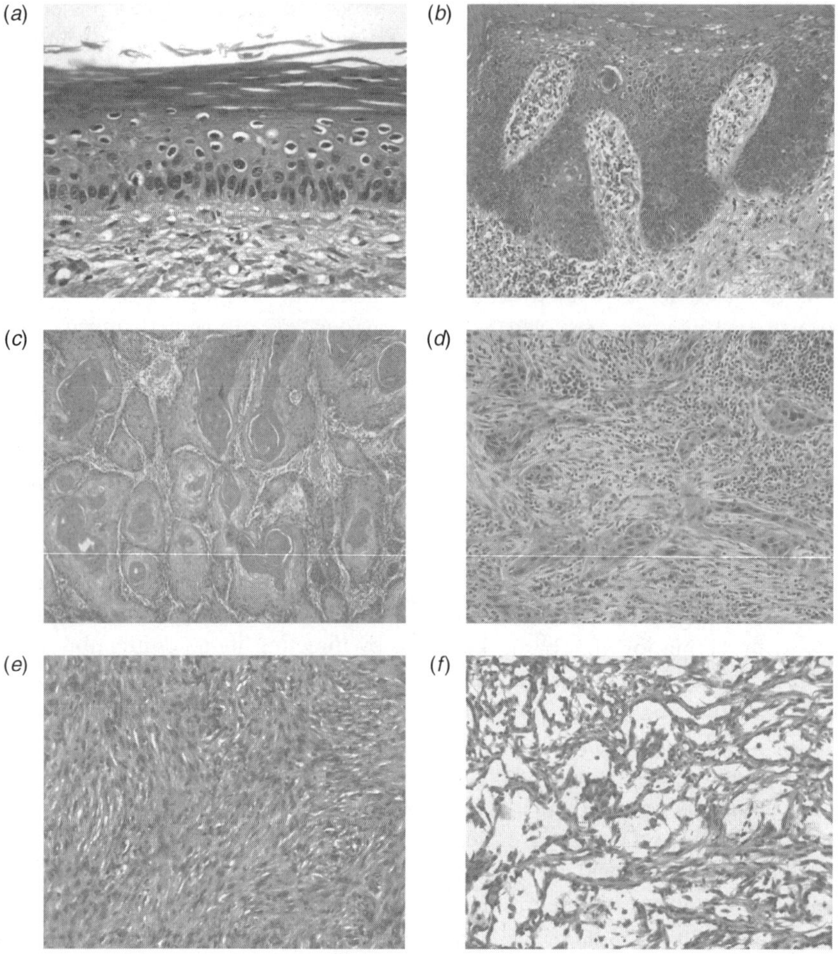

Figure 9.3. Human squamous cell carcinoma. (*a*) Actinic keratosis. Note the atypia and crowding of the basal keratinocytic layer. (*b*) Squamous cell carcinoma in situ is characterized by full-thickness keratinocytic atypia. (*c*) Well-differentiated invasive squamous cell carcinoma. Keratinizing squamous nests and islands are invading the dermis. (*d*) Poorly differentiated invasive squamous cell carcinoma. Squamous cell carcinoma cells are invading the dermis as nests and small cords with a prominent inflammatory response and desmoplasia (fibrosis). Keratinization is minimal. (*e*) Spindle cell squamous cell carcinoma resembles sarcomas and spindle cell melanomas. Identification of an origin contiguous with dysplastic overlying squamous epithelium and immunohistochemical staining for cytokeratin are helpful for definitive diagnosis. (*f*) Pseudoangiosarcomatous squamous cell carcinoma. Prominent artifactual spaces that mimic vascular channels are present in this cytokeratin-positive, endothelial-marker negative tumor. (Case kindly provided by P. H. McKee.)

promoter, is widely used to test the carcinogenicity of topically applied chemical compounds (Sistare et al., 2002).

Chronic UV irradiation has been used to induce skin cancer as well. In this setting poorly differentiated spindle cell squamous cell carcinomas are more common. Diagnosis of these lesions can be difficult, and sections should be carefully examined to determine if the dermal spindle cell component is contiguous with dysplastic overlying epithelium (Morison et al., 1986).

Transgenic mice have been constructed that have a predisposition for forming squamous cell carcinoma (Table 9.2, Fig. 9.4). Many models have used cytokeratin promoters active in the basal layer of the epidermis such as K5 or K14. One inducible model of precancerous squamous lesions has been developed by suprabasal keratinocytic expression of the c-*myc*/estrogen receptor fusion protein. Induction of *myc* activity by topical administration of 4-hydroxytamoxifen resulted in squamous proliferation with atypia, morphologically resembling squamous cell carcinoma in situ (Figs. 9.4*c*, 9.*d*). These lesions resolved in the absence of tamoxifen application. A second model involves expression of HPV-16

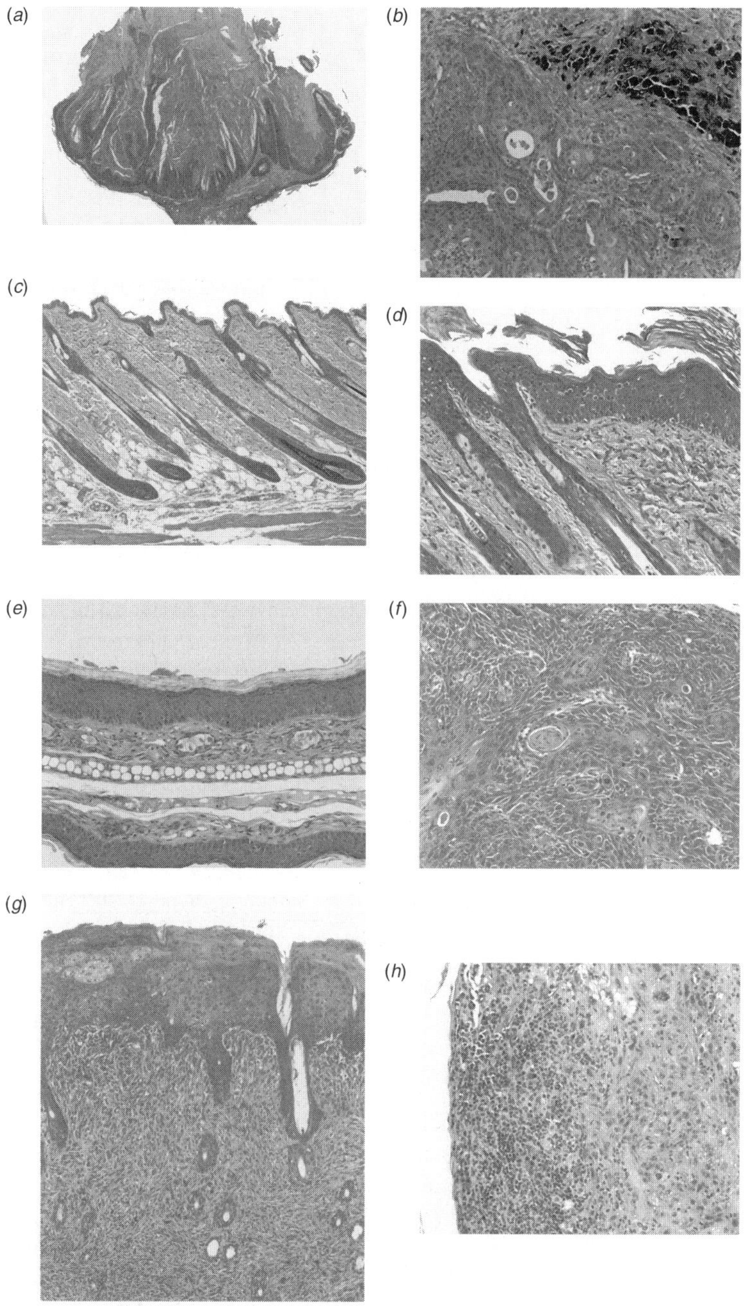

Figure 9.4. Mouse models of squamous cell carcinoma. (*a*) Squamous papilloma arising in a mouse treated with a two-step (DMBA/TPA) chemical carcinogenesis protocol. Squamous papillomas with a craterlike growth pattern and a polypoid stalk are very common in mice treated with chemical carcinogenesis. (*b*) Well-differentiated squamous cell carcinoma arising after DMBA/TPA treatment with an adjacent benign melanocytic proliferation. (*c*) Back skin from an untreated Involucrin-Myc^{ER} mouse, demonstrating normal mouse skin. (*d*) Tamoxifen treatment of an Involucrin-Myc^{ER} mouse results in squamous cell carcinoma in situ. The lesions revert to normal skin upon withdrawal of tamoxifen treatment (Pelengaris et al., 1999). (*e*) Ear skin showing squamous hyperplasia with atypia arising in a K14-*HPV16(early region)* mouse (Arbeit et al., 1994). (*f*) Invasive squamous cell carcinoma arising in a K14-*HPV16(early region)* mouse. (*g*) Spindle cell squamous cell carcinoma arising in a UV-irradiated C3H mouse. Identification of a contiguous origin in overlying dysplastic squamous epithelium is critical to differentiate some squamous cell carcinomas from other spindle cell malignancies. (*h*) Metastatic squamous cell carcinoma (right half of image) present in a lymph node in a HK1-$p53^{R172H}$ mouse treated with a two-stage chemical carcinogenesis protocol (Wang et al., 1998b).

Table 9.2. Mouse Models of Squamous Cell Carcinoma

Transgene	Inactivated Gene	Treatment	Phenotype	Reference
		DMBA/TPA	Squamous papilloma formation with progression to invasive cancer	Hennings et al., 1993; many others
Various transgenes		DMBA/TPA	Modifying effects on squamous papilloma and carcinoma incidence	Numerous studies
	p53−/−	UV irradiation	Actinic keratosis–like lesions and squamous cell carcinomas	Jiang et al., 1999
	Xpa−/−	UVB, DMBA	Squamous cell carcinomas	de Vries et al., 1995; Nakane et al., 1995
	Xpc−/−	UV	Squamous cell carcinomas	Sands et al., 1995
Tg:AC (v-H-*ras*)		Various chemical carcinogens	Squamous neoplasia with squamous cell carcinoma	Sistare et al., 2002
K14-*HPV16* early region			Squamous hyperplasia with atypia with progression to invasive carcinoma of varying grades	Arbeit et al., 1994
HK1-*TP53R172H*			Invasive squamous cell carcinoma after two-stage chemical carcinogenesis protocol	Wang et al., 1998b
K5-*IkB*			Basal keratinocytic proliferation with atypia and progression to squamous cell carcinoma	Van Hogerlinden et al., 1999
Involucrin-c-*MycER*		Tamoxifen	Squamous hyperplasia with atypia that is dependent on induction of *myc* activity by topical tamoxifen administration	Pelengaris et al., 1999

early region genes under the control of the cytokeratin 14 promoter (Arbeit et al., 1994). Progression of squamous neoplasia from hyperplasia with atypia to in situ carcinoma and invasive squamous cell carcinoma is evident in this model (Figs. 9.4*e*,*f*).

BASAL CELL CARCINOMA

Clinical Features

Basal cell carcinoma is by far the most common cancer, with nearly 1 million new cases per year in the United States. Despite the high incidence, the number of deaths attributed to basal cell carcinoma is very low. Examples of metastasizing basal cell carcinoma are extraordinarily rare (Tavin et al., 1995). Despite the low metastatic rate, basal cell carcinoma can cause significant morbidity secondary to direct invasion of adjacent tissues, especially when the primary tumor is on the face near an eye.

Basal cell carcinomas can have a variety of clinical appearances, including flat red spots, a shiny papule with telangiectatic superficial blood vessels, a slightly depressed scar, or an ulcerated mass. Definitive diagnosis is established by histopathologic examination of a skin biopsy. Several treatment modalities are available that result in removal or destruction of basal cell carcinomas, including excision, electrodesiccation and curettage, or treatment with liquid nitrogen (Martinez and Otley, 2001). For lesions adjacent to sensitive structures on the face, Moh's micrographic surgery is frequently performed. In addition, topical application of the immunomodulator Imiquimod appears to be effective in the some basal cell carcinomas, particularly superficial variants (Marks et al., 2001).

(a) (b) (c) (d)

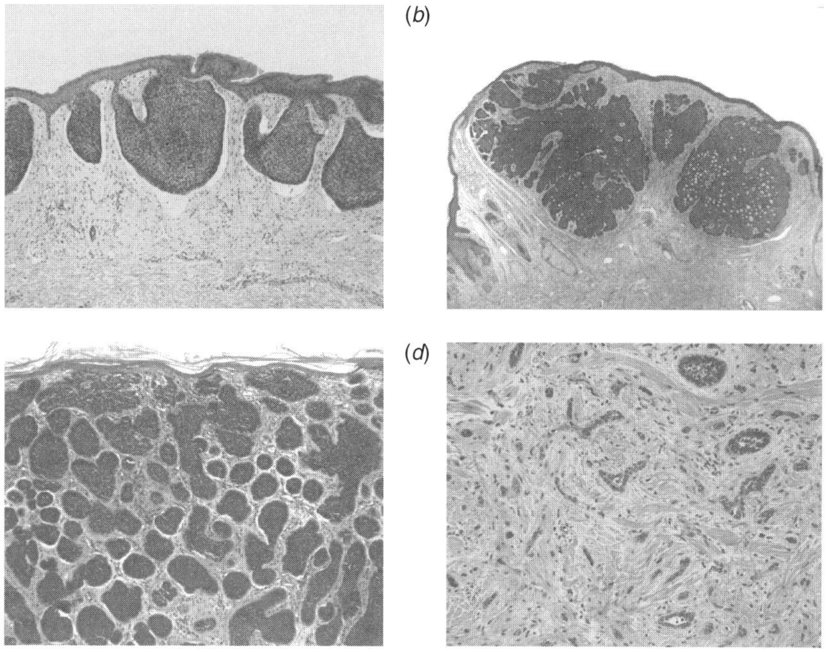

Figure 9.5. Human basal cell carcinoma. (*a*) Superficial basal cell carcinoma. Continuity with several points of the overlying epidermis is common. (*b*) Nodular basal cell carcinoma. (*c*) Micronodular basal cell carcinoma. (*d*) Sclerosing (morpheaform) basal cell carcinoma is characterized by cords, small nests, and individual basal cell carcinoma cells set in a fibrotic stroma.

Pathology

Basal cell carcinomas have a wide variety of histologic appearances (Fig. 9.5). Most commonly, superficial and/or nodular variants are observed. Characteristic features include a dermal proliferation of basaloid cells in islands, strands, or nests separated by fibromyxoid tumor stroma and frequently exhibiting an artifactual space between epithelial nests and adjacent stroma. The peripheral row of cells typically is aligned (palisades) and mitotic figures and apoptotic cells are common. Variants include superficial, nodular, micronodular, invasive, sclerosing, metatypical, and infundibulocystic, nodulocystic, adenoid, and clear cell. Follicular or sebaceous differentiation may also be observed. Occasionally, perineural invasion is present which is associated with a higher rate of local recurrence. Basal cell carcinomas with follicular differentiation may resemble other follicular neoplasms, including trichoepithelioma and trichoblastoma.

Biology

Familial predisposition to basal cell carcinoma is a characteristic of basal cell nevus/Gorlin's syndrome. Other features of basal cell nevus syndrome include palmoplantar pitting, odontogenic keratocysts of the jaw, increased incidence of medulloblastoma and rhabdomyosarcoma, calcification of the falx cerebri, and several bony/skeletal changes, including frontal bossing, kyphoscoliosis, and rib abnormalities. Linkage to 9q22 was established and mutations in the patched gene were demonstrated in several kindreds (Hahn et al., 1996). Patched is homologous to the *Drosophila* segment polarity gene by the same name that acts in the hedgehog pathway. The wealth of information generated by analysis of the hedgehog signaling pathway has greatly advanced the understanding of the genetic changes resulting in human basal cell carcinoma. Signaling is initiated by binding of the hedgehog ligand to patched, inactivating the ability of patched to inhibit the smoothened cell surface protein. Uninhibited smoothened activity results in activation of the GLI transcription factors which are frequently overexpressed in basal cell carcinoma (Ghali et al., 1999). Patched-1 mutations or deletions have been demonstrated in both familial and sporadic basal cell carcinoma. Activating Smoothened mutations have been documented in about 20% of sporadic basal cell carcinoma cases (Lam et al., 1999).

Table 9.3. Mouse Models of Basal Cell Carcinoma/Adnexal Neoplasms

Transgene	Inactivated Gene	Treatment	Phenotype	References
K5-*hedgehog*			Basaloid follicular neoplasms, some of which are superficial basal cell carcinoma type	Oro et al., 1997
	Patched+/−	Arsenic, UV, ionizing radiation	Basaloid follicular neoplasms with diverse morphologies; most are trichoblastoma type, others are infundibulocystic basal cell carcinoma-type	Aszterbaum et al., 1999
K5-*Gli1*			Basaloid follicular neoplasms with diverse morphologies; many are trichoblastoma type or cylindroma type, others are superficial basal cell carcinoma type	Nilsson et al., 2000
K5-*Gli2*			Basaloid follicular neoplasms	Grachtchouk et al., 2000;
K14-*Myc2*			Squamous and sebaceous hyperplasia with stem cell depletion	Waikel et al., 2001
K5-*MycER*			Squamous and sebaceous hyperplasia with stem cell depletion	Arnold and Watt, 2001
K14-Δ*N*87-β *catenin*			De novo hair follicle morphogenesis with occasional pilomatrixomas and trichofolliculomas	Gat et al., 1998
K14-*CreER(tam)*	β *catenin (N-term loxP)*	Tamoxifen	Conditional activation of β-catenin results in pilomatrixoma formation	Vasioukhin et al., 1999

Mouse Models of Basal Cell Carcinoma and Adnexal Neoplasms

Basal cell carcinoma has a variety of clinical and pathologic appearances, many of which are characteristic. Less common forms such as the infundibulocystic type and basal cell carcinomas with follicular differentiation morphologically overlap with a wide variety of follicular/adnexal neoplasms. Alteration of the hedgehog pathway in genetically engineered mice has resulted in a diverse array of adnexal neoplasms, some of which bear some resemblance to certain forms of human basal cell carcinoma (Table 9.3, Fig. 9.6). Examination of a subset of the tumors generated in *Ptch+/−* mice and *K5-Gli-1* mice by a panel of dermatopathologists and veterinary pathologists convened on behalf of the National Cancer Institute's Mouse Models of Human Cancer Consortium in San Juan, Puerto Rico, in 2001 resulted in the consensus opinion that the tumors were complex follicular germinative neoplasms and, although in some instances shared some features of basal cell carcinoma, are best classified as "basaloid follicular neoplasms" with further descriptors appropriate for the morphology of the particular case (see also LeBoit, 2000) (Table 9.4).

Determination of whether a particular murine adnexal tumor is benign or malignant is frequently a difficult undertaking, as the tumors tend to be circumscribed with a pushing border and do not metastasize. Other criteria, including tumor necrosis, cellular pleomorphism, high

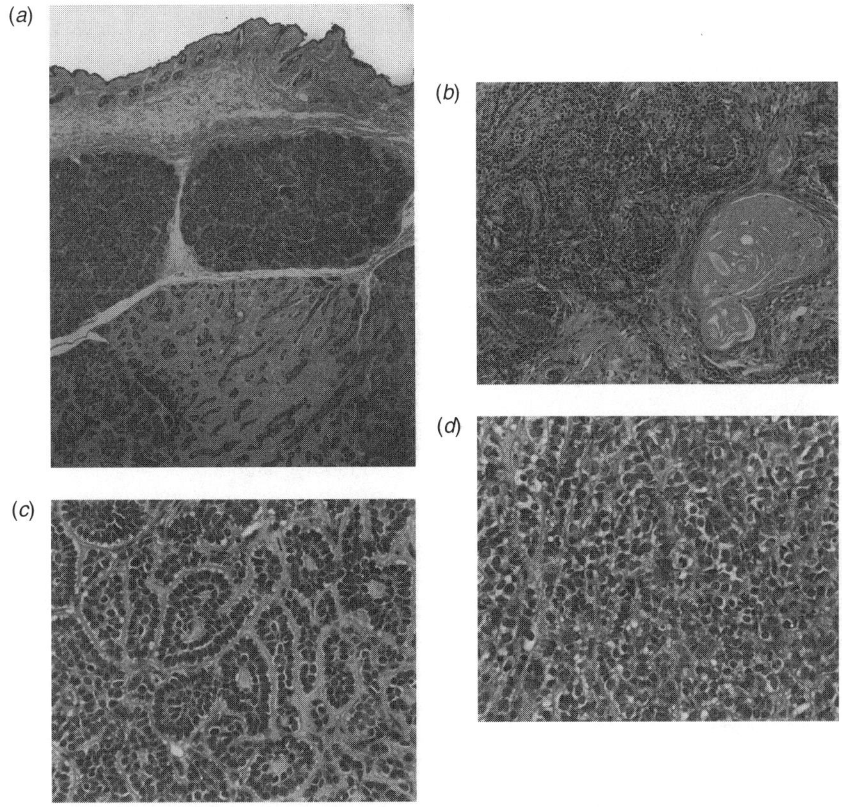

Figure 9.6. Mouse models of basal cell carcinoma/adnexal neoplasms. (*a*) Large nodular basaloid follicular neoplasm arising in a *Patched*+/− mouse that was treated with ionizing radiation and oral arsenic. The tumor lobules contain cords and nests of basaloid cells separated by eosinophilic stroma (Aszterbaum et al., 1999). (*b*) Basaloid follicular neoplasm arising in a *Patched*+/− mouse treated with ionizing radiation. The tumor is composed of cords of basaloid cells, foci resembling germinative follicles, and keratocysts that contain several hair fragments. The overall features are similar to human trichoblastomas; however, keratocysts and mature hair fragments are typically seen in more differentiated follicular neoplasms, illustrating the complexity of morphologic patterns seen in mouse follicular/adnexal neoplasms. (*c,d*) Basaloid follicular neoplasm resembling spiradenoma/cylindroma arising in a K5-*Gli1* mouse. This portion of the tumor is islands or nests of tumor cells that have a jigsaw puzzle–like arrangement, separated by dense, eosinophilic basement membrane–like material. The nests have a prominent peripheral layer of basaloid cells surrounding slightly less basophilic cells with central deposits of the eosinophilic material. (*d*) A more poorly differentiated portion of the same tumor is present and lacks the striking architecture of islands of tumor separated by dense eosinophilic material. These changes coupled with numerous mitoses and foci of necrosis are likely to represent malignant transformation.

mitotic rates, and destruction of underlying tissue, are features suggestive of a malignant adnexal tumor.

SUMMARY

A wide variety of techniques have been used to model skin cancer in mice, ranging from exposure to chemical carcinogens or ultraviolet light, the use of genetically engineered mice, and xenograft studies of human skin and human skin reconstructs. In models involving genetically engineered mice, the modifying effects of secondary genetic changes or exposure to environmental carcinogens have also been examined. These models are being further refined to more closely approximate the genetic changes present in the human cancer counterparts and contributing environmental etiologic factors.

ACKNOWLEDGMENT

The author acknowledges the support of the Mouse Models of Cancer Consortium of the National Cancer Institute and the help and guidance of the MMHCC Skin Pathology Panel, which also included David E. Elder, Andres J. Klein-Szanto, Donna F. Kusewitt, B. Jack Longley, Martin C. Mihm, and Miriam R. Anver.

Table 9.4. Classification of Neoplasms and Proliferations of Mouse Skin

Melanocytic neoplasms and proliferations
- Melanocytic proliferative disorders
  - Junctional melanocytic hyperplasia
  - Dermal melanocytic hyperplasia
  - Compound melanocytic hyperplasia
- Benign melanocytic neoplasms
  - Melanocytoma
- Melanocytic neoplasms of uncertain malignant potential
- Malignant melanocytic neoplasms
  - Melanoma: epithelioid, spindle cell, spindle and epithelioid, other
  - Metastatic melanoma

Squamous neoplasms and proliferations
- Squamous proliferative disorders
  - Squamous hyperplasia
  - Squamous hyperplasia with atypia
- Benign squamous neoplasms
  - Papilloma
- Benign basal cell tumor
- Squamous cell carcinomas in situ
- Squamous cell carcinoma
  - Well-differentiated squamous cell carcinoma
  - Moderately differentiated squamous cell carcinoma
  - Poorly differentiated squamous cell carcinoma: spindle cell squamous cell carcinoma

Adnexal neoplasms and proliferations
- Follicular neoplasms
  - Basaloid follicular neoplasm
    - Basal cell carcinoma: superficial basal cell carcinoma type, nodular basal cell carcinoma type, infundibulocystic basal cell carcinoma type
    - Pilomatrixoma type
    - Trichoblastoma type
    - Trichoepithelioma type
    - Trichofolliculoma type
    - Cylindroma type
    - Spiradenoma type
    - Other variants
- Sebaceous neoplasms and hyperplasias
  - Sebaceous hyperplasia
  - Sebaceous adenoma
  - Sebaceous carcinoma
- Eccrine neoplasms
- Apocrine neoplasms
- Mixed tumor
- Other adnexal neoplasms

Table 9.4. (*continued*)

Benign mesenchymal neoplasms
- Fibroma
- Hemangioma
- Lipoma
- Hibernoma
- Schwannoma
- Neurofibroma
- Myxoma
- Granular cell tumor

Sarcomas of the skin and subcutis
- Hemangiosarcoma
- Lymphangiosarcoma
- Malignant peripheral nerve sheath tumor
  - Malignant schwannoma
- Leiomyosarcoma
- Rhabdomyosarcoma
- Liposarcoma
- Fibrosarcoma
- Malignant spindle cell neoplasm, unclassified

Cutaneous tumors of hematompoietic origin
- Cutaneous lymphomas
  - B-cell lymphomas
  - T-cell lymphomas
  - Other lymphomas
- Plasma cell neoplasms and proliferations
  - Benign plasma cell proliferations
  - Malignant plasmacytoma
- Histiocytic neoplasms
  - Benign histiocytic neoplasms
  - Malignant histiocytic neoplasms
- Mast cell neoplasms
  - Mastocytoma
- Other cutaneous neoplasms of hematopoietic origin

Note: Prepared by Marcus W. Bosenberg, Miriam R. Anver, David E. Elder, Andres J. Klein-Szanto, Donna F. Kusewitt, B. Jack Longley, and Martin C. Mihm, Jr., as part of the Cutaneous Oncology session sponsored by the Mouse Models of Human Cancer Consortium of the National Cancer Institute (United States), held in San Juan, Puerto Rico in January 2001.

REFERENCES

Arbeit JM, Munger K, Howley PM, Hanahan D (1994): Progressive squamous epithelial neoplasia in K14-human papillomavirus type 16 transgenic mice. *J Virol* 68:4358–4368.

Arnold I, Watt FM (2001): c-Myc activation in transgenic mouse epidermis results in mobilization of stem cells and differentiation of their progeny. *Curr Biol* 11:558–568.

Aszterbaum M, Epstein J, Oro A, Douglas V, LeBoit PE, Scott MP, Epstein EH Jr (1999): Ultraviolet and ionizing radiation enhance the growth of BCCs and trichoblastomas in patched heterozygous knockout mice. *Nat Med* 5:1285–1291.

Atillasoy ES, Seykora JT, Soballe PW, Elenitsas R, Nesbit M, Elder DE, Montone KT, Sauter E, Herlyn M (1998): UVB induces atypical melanocytic lesions and melanoma in human skin. *Am J Pathol* 152:1179–1186.

Balch CM, Buzaid AC, Soong SJ, Atkins MB, Cascinelli N, Coit DG, Fleming ID, Gershenwald JE, Houghton A Jr, Kirkwood JM, McMasters KM, Mihm MF, Morton DL, Reintgen DS, Ross MI, Sober A, Thompson JA, Thompson JF (2001a): Final version of the American Joint Committee on Cancer staging system for cutaneous melanoma. *J Clin Oncol* 19:3635–3648.

Balch CM, Soong SJ, Gershenwald JE, Thompson JF, Reintgen DS, Cascinelli N, Urist M, McMasters KM, Ross MI, Kirkwood JM, Atkins MB, Thompson JA, Coit DG, Byrd D, Desmond R, Zhang Y, Liu PY, Lyman GH, Morabito A (2001b): Prognostic factors analysis of 17,600 melanoma patients: Validation of the American Joint Committee on Cancer melanoma staging system. *J Clin Oncol* 19:3622–3634.

Bardeesy N, Bastian BC, Hezel A, Pinkel D, DePinho RA, Chin L (2001): Dual inactivation of RB and p53 pathways in RAS-induced melanomas. *Mol Cell Biol* 21:2144–2153.

Bardeesy N, Wong KK, DePinho RA, Chin L (2000): Animal models of melanoma: Recent advances and future prospects. *Adv Cancer Res* 79:123–156.

Bastian BC, Kashani-Sabet M, Hamm H, Godfrey T, Moore DH II, Brocker EB, LeBoit PE, Pinkel D (2000a): Gene amplifications characterize acral melanoma and permit the detection of occult tumor cells in the surrounding skin. *Cancer Res* 60:1968–1973.

Bastian BC, LeBoit PE, Hamm H, Brocker EB, Pinkel D (1998): Chromosomal gains and losses in primary cutaneous melanomas detected by comparative genomic hybridization. *Cancer Res* 58:2170–2175.

Bastian BC, LeBoit PE, Pinkel D (2000b): Mutations and copy number increase of HRAS in Spitz nevi with distinctive histopathological features. *Am J Pathol* 157:967–972.

Berking C, Herlyn M (2001): Human skin reconstruct models: A new application for studies of melanocyte and melanoma biology. *Histol Histopathol* 16:669–674.

Bradl M, Klein-Szanto A, Porter S, Mintz B (1991): Malignant melanoma in transgenic mice. *Proc Natl Acad Sci USA* 88:164–168.

Broders AC (1932): Practical points on the microscopic grading of carcinoma. *NY State J Med* 32:667.

Broome Powell M, Gause PR, Hyman P, Gregus J, Lluria-Prevatt M, Nagle R, Bowden GT (1999): Induction of melanoma in TPras transgenic mice. *Carcinogenesis* 20:1747–1753.

Brown K, Balmain A (1995): Transgenic mice and squamous multistage skin carcinogenesis. *Cancer Metastasis Rev* 14:113–124.

Chang P, Knapper WH (1982): Metastatic melanoma of unknown primary. *Cancer* 49:1106–1111.

Chin L, Merlino G, DePinho RA (1998a): Malignant melanoma: Modern black plague and genetic black box. *Genes Dev* 12:3467–3481.

Chin L, Pomerantz J, DePinho RA (1998b): The INK4a/ARF tumor suppressor: One gene—two products—two pathways. *Trends Biochem Sci* 23:291–296.

Chin L, Pomerantz J, Polsky D, Jacobson M, Cohen C, Cordon-Cardo C, Horner JW II, DePinho RA (1997): Cooperative effects of INK4a and ras in melanoma susceptibility in vivo. *Genes Dev* 11:2822–2834.

Chin L, Tam A, Pomerantz J, Wong M, Holash J, Bardeesy N, Shen Q, O'Hagan R, Pantginis J, Zhou H, Horner JW II, Cordon-Cardo C, Yancopoulos GD, DePinho RA (1999): Essential role for oncogenic Ras in tumour maintenance. *Nature* 400:468–472.

Cleaver JE, Crowley E (2002): UV damage, DNA repair and skin carcinogenesis. *Front Biosci* 7:d1024–1043.

Davies H, Bignell GR, Cox C, Stephens P, Edkins S, Clegg S, Teague J, Woffendin H, Garnett MJ, Bottomley W, Davis N, Dicks E, Ewing R, Floyd Y, Gray K, Hall S, Hawes R, Hughes J, Kosmidou V, Menzies A, Mould C, Parker A, Stevens C, Watt S, Hooper S, Wilson R, Jayatilake H, Gusterson BA, Cooper C, Shipley J, Hargrave D, Pritchard-Jones K, Maitland N, Chenevix-Trench G, Riggins GJ, Bigner DD, Palmieri G, Cossu A, Flanagan A, Nicholson A, Ho JW, Leung SY, Yuen ST, Weber BL, Seigler HF, Darrow TL, Paterson H, Marais R, Marshall CJ, Wooster R, Stratton MR, Futreal PA (2002): Mutations of the BRAF gene in human cancer. *Nature* 417:949–954.

de Vries A, van Oostrom CT, Hofhuis FM, Dortant PM, Berg RJ, de Gruijl FR, Wester PW, van Kreijl CF, Capel PJ, van Steeg H, Verbeek SJ (1995): Increased susceptibility to ultraviolet-B and carcinogens of mice lacking the DNA excision repair gene XPA. *Nature* 377:169–173.

Easty DJ, Bennett DC (2000): Protein tyrosine kinases in malignant melanoma. *Melanoma Res* 10:401–411.

Fargnoli MC, Chimenti S, Keller G, Soyer HP, Dal Pozzo V, Hofler H, Peris K (1998): CDKN2a/p16INK4a mutations and lack of p19ARF involvement in familial melanoma kindreds. *J Invest Dermatol* 111:1202–1206.

Gat U, DasGupta R, Degenstein L, Fuchs E (1998): De Novo hair follicle morphogenesis and hair tumors in mice expressing a truncated beta-catenin in skin. *Cell* 95:605–614.

Ghali L, Wong ST, Green J, Tidman N, Quinn AG (1999): Gli1 protein is expressed in basal cell carcinomas, outer root sheath keratinocytes and a subpopulation of mesenchymal cells in normal human skin. *J Invest Dermatol* 113:595–599.

Grachtchouk M, Mo R, Yu S, Zhang X, Sasaki H, Hui CC, Dlugosz AA (2000): Basal cell carcinomas in mice overexpressing Gli2 in skin. *Nat Genet* 24:216–217.

Greene MH (1999): The genetics of hereditary melanoma and nevi. 1998 update. *Cancer* 86:2464–2477.

Gutzmer R, Herbst RA, Mommert S, Kiehl P, Matiaske F, Rutten A, Kapp A, Weiss J (2000): Allelic loss at the neurofibromatosis type 1 (NF1) gene locus is frequent in desmoplastic neurotropic melanoma. *Hum Genet* 107:357–361.

Hahn H, Wicking C, Zaphiropoulos PG, Gailani MR, Shanley S, Chidambaram A, Vorechovsky I, Holmberg E, Unden AB, Gillies S, Negus K, Smyth I, Pressman C, Leffell DJ,

Gerrard B, Goldstein AM, Dean M, Toftgard R, Chenevix-Trench G, Wainwright B, Bale AE (1996): Mutations of the human homolog of Drosophila patched in the nevoid basal cell carcinoma syndrome. *Cell* 85:841–851.

Hanahan D, Weinberg RA (2000): The hallmarks of cancer. *Cell* 100:57–70.

Helmbach H, Rossmann E, Kern MA, Schadendorf D (2001): Drug-resistance in human melanoma. *Int J Cancer* 93:617–622.

Hennings H, Glick AB, Lowry DT, Krsmanovic LS, Sly LM, Yuspa SH (1993): FVB/N mice: An inbred strain sensitive to the chemical induction of squamous cell carcinomas in the skin. *Carcinogenesis* 14:2353–2358.

Herlyn M, Satyamoorthy K (1996): Activated ras. Yet another player in melanoma? *Am J Pathol* 149:739–744.

Hewitt C, Lee Wu C, Evans G, Howell A, Elles RG, Jordan R, Sloan P, Read AP, Thakker N (2002): Germline mutation of ARF in a melanoma kindred. *Hum Mol Genet* 11:1273–1279.

Iwamoto T, Takahashi M, Ito M, Hamatani K, Ohbayashi M, Wajjwalku W, Isobe K, Nakashima I (1991): Aberrant melanogenesis and melanocytic tumour development in transgenic mice that carry a metallothionein/ret fusion gene. *EMBO J* 10:3167–3175.

Jager D, Jager E, Knuth A (2001): Vaccination for malignant melanoma: Recent developments. *Oncology* 60:1–7.

Jensen P, Hansen S, Moller B, Leivestad T, Pfeffer P, Geiran O, Fauchald P, Simonsen S (1999): Skin cancer in kidney and heart transplant recipients and different long-term immunosuppressive therapy regimens. *J Am Acad Dermatol* 40:177–186.

Jiang W, Ananthaswamy HN, Muller HK, Kripke ML (1999): p53 protects against skin cancer induction by UV-B radiation. *Oncogene* 18:4247–4253.

Jin Y, Mertens F, Jin C, Akervall J, Wennerberg J, Gorunova L, Mandahl N, Heim S, Mitelman F (1995): Nonrandom chromosome abnormalities in short-term cultured primary squamous cell carcinomas of the head and neck. *Cancer Res* 55:3204–3210.

Johnson TM, Dolan OM, Hamilton TA, Lu MC, Swanson NA, Lowe L (1998): Clinical and histologic trends of melanoma. *J Am Acad Dermatol* 38:681–686.

Kannan K, Sharpless NE, Xu J, O'Hagan RC, Bosenberg M, Chin L (2003): Components of the Rb pathway are critical targets of UV mutagenesis in a murine melanoma model. *Proc Natl Acad Sci USA* 100:1221–1225.

Kato M, Liu W, Akhand AA, Hossain K, Takeda K, Takahashi M, Nakashima I (2000): Ultraviolet radiation induces both full activation of ret kinase and malignant melanocytic tumor promotion in RFP-RET-transgenic mice. *J Invest Dermatol* 115:1157–1158.

Kelsall SR, Mintz B (1998): Metastatic cutaneous melanoma promoted by ultraviolet radiation in mice with transgene-initiated low melanoma susceptibility. *Cancer Res* 58:4061–4065.

Kirkwood JM, Strawderman MH, Ernstoff MS, Smith TJ, Borden EC, Blum RH (1996): Interferon alfa-2b adjuvant therapy of high-risk resected cutaneous melanoma: The Eastern Cooperative Oncology Group Trial EST 1684. *J Clin Oncol* 14:7–17.

Klein-Szanto AJ (1997). Neoplastic and preneoplastic lesions of the mouse skin. In Bannasch P, Gössner W (Eds). *Pathology of Neoplasia and Preneoplasia in Rodents*. Schattauer, Stuttgart, pp 1–18.

Klein-Szanto AJ, Silvers WK, Mintz B (1994): Ultraviolet radiation-induced malignant skin melanoma in melanoma-susceptible transgenic mice. *Cancer Res* 54:4569–4572.

Krimpenfort P, Quon KC, Mooi WJ, Loonstra A, Berns A (2001): Loss of p16Ink4a confers susceptibility to metastatic melanoma in mice. *Nature* 413:83–86.

Kunisada T, Yoshida H, Yamazaki H, Miyamoto A, Hemmi H, Nishimura E, Shultz LD, Nishikawa S, Hayashi S (1998): Transgene expression of steel factor in the basal layer of epidermis promotes survival, proliferation, differentiation and migration of melanocyte precursors. *Development* 125:2915–2923.

Kusewitt DF, Ley RD (1996): Animal models of melanoma. *Cancer Surv* 26:35–70.

Lam CW, Xie J, To KF, Ng HK, Lee KC, Yuen NW, Lim PL, Chan LY, Tong SF, McCormick F (1999): A frequent activated smoothened mutation in sporadic basal cell carcinomas. *Oncogene* 18:833–836.

Langley RGB, Barnhill RL, Mihm MC, Fitzpatrick TB, Sober AJ (1999). Neoplasms: cutaneous melanoma. In Freedberg IM, Eisen AZ, Wolff K, Austen KF, Goldsmith LA, Katz SI, Fitzpatrick TB (Eds). *Fitzpatrick's Dermatology in General Medicine*, 5th ed. McGraw-Hill, New York, pp 1080–1116.

LeBoit PE (2000): Puzzling follicular germinative proliferations. *Am J Dermatopathol* 22:87–88.

Lens MB, Dawes M (2002): Interferon alfa therapy for malignant melanoma: A systematic review of randomized controlled trials. *J Clin Oncol* 20:1818–1825.

Maas-Szabowski N, Szabowski A, Stark HJ, Andrecht S, Kolbus A, Schorpp-Kistner M, Angel P, Fusenig NE (2001): Organotypic cocultures with genetically modified mouse fibroblasts as a tool to dissect molecular mechanisms regulating keratinocyte growth and differentiation. *J Invest Dermatol* 116:816–820.

Marks R, Gebauer K, Shumack S, Amies M, Bryden J, Fox TL, Owens ML (2001): Imiquimod 5% cream in the treatment of superficial basal cell carcinoma: Results of a multicenter 6-week dose-response trial. *J Am Acad Dermatol* 44:807–813.

Martinez JC, Otley CC (2001): The management of melanoma and nonmelanoma skin cancer: A review for the primary care physician. *Mayo Clin Proc* 76:1253–1265.

McKee PH (1996) *Pathology of the Skin*, 2nd ed. Mosby, London.

Miller DL, Weinstock MA (1994): Nonmelanoma skin cancer in the United States: Incidence. *J Am Acad Dermatol* 30:774–778.

Morison WL, Jerdan MS, Hoover TL, Farmer ER (1986): UV radiation-induced tumors in haired mice: Identification as squamous cell carcinomas. *J Natl Cancer Inst* 77:1155–1162.

Munger K (2002): The role of human papillomaviruses in human cancers. *Front Biosci* 7:d641–649.

Nakane H, Takeuchi S, Yuba S, Saijo M, Nakatsu Y, Murai H, Nakatsuru Y, Ishikawa T, Hirota S, Kitamura Y, Kato Y, Tsunoda Y, Miyauchi H, Horio T, Tokunaga T, Matsunaga T, Nikaido O, Nishimune Y, Okada Y, Tanaka K (1995): High incidence of ultraviolet-B- or chemical-carcinogen-induced skin tumours in mice lacking the xeroderma pigmentosum group A gene. *Nature* 377:165–168.

Nilsson M, Unden AB, Krause D, Malmqwist U, Raza K, Zaphiropoulos PG, Toftgard R (2000): Induction of basal cell carcinomas and trichoepitheliomas in mice overexpressing GLI-1. *Proc Natl Acad Sci USA* 97:3438–3443.

Noonan FP, Otsuka T, Bang S, Anver MR, Merlino G (2000): Accelerated ultraviolet radiation-induced carcinogenesis in hepatocyte growth factor/scatter factor transgenic mice. *Cancer Res* 60:3738–3743.

Noonan FP, Recio JA, Takayama H, Duray P, Anver MR, Rush WL, De Fabo EC, Merlino G (2001): Neonatal sunburn and melanoma in mice. *Nature* 413:271–272.

Oro AE, Higgins KM, Hu Z, Bonifas JM, Epstein EH Jr, Scott MP (1997): Basal cell carcinomas in mice overexpressing sonic hedgehog. *Science* 276:817–821.

Ortone JP (2002): From actinic keratosis to squamous cell carcinoma. *Br J Dermatol* 146:20–23.

Otsuka T, Takayama H, Sharp R, Celli G, LaRochelle WJ, Bottaro DP, Ellmore N, Vieira W, Owens JW, Anver M, Merlino G (1998): c-Met autocrine activation induces development of malignant melanoma and acquisition of the metastatic phenotype. *Cancer Res* 58:5157–5167.

Peckham JC, Heider K (1999): Skin and subcutis. In Maronpot RR (Ed). *Pathology of the Mouse*. Cache River Press, Vienna, IL, pp 555–611.

Pelengaris S, Littlewood T, Khan M, Elia G, Evan G (1999): Reversible activation of c-Myc in skin: Induction of a complex neoplastic phenotype by a single oncogenic lesion. *Mol Cell* 3:565–577.

Pollock PM, Cohen-Solal K, Sood R, Namkoong J, Martino JJ, Koganti A, Zhu H, Robbins C, Makalowska I, Shin SS, Marin Y, Roberts KG, Yudt LM, Chen A, Cheng J, Incao A, Pinkett HW, Graham CL, Dunn K, Crespo-Carbone SM, Mackason KR, Ryan KB, Sinsimer D, Goydos J, Reuhl KR, Eckhaus M, Meltzer PS, Pavan WJ, Trent JM, Chen S (2003a): Melanoma mouse model implicates metabotropic glutamate signaling in melanocytic neoplasia. *Nat Genet* 34:108–112.

Pollock PM, Harper UL, Hansen KS, Yudt LM, Stark M, Robbins CM, Moses TY, Hostetter G, Wagner U, Kakareka J, Salem G, Pohida T, Heenan P, Duray P, Kallioniemi O, Hayward NK, Trent JM, Meltzer PS (2003b): High frequency of BRAF mutations in nevi. *Nat Genet* 33:19–20.

Pollock PM, Pearson JV, Hayward NK (1996): Compilation of somatic mutations of the CDKN2 gene in human cancers: non-random distribution of base substitutions. *Genes Chromosomes Cancer* 15:77–88.

Pollock PM, Trent JM (2000): The genetics of cutaneous melanoma. *Clin Lab Med* 20:667–690.

Polsky D, Bastian BC, Hazan C, Melzer K, Pack J, Houghton A, Busam K, Cordon-Cardo C, Osman I (2001): HDM2 protein overexpression, but not gene amplification, is related to tumorigenesis of cutaneous melanoma. *Cancer Res* 61:7642–7646.

Ramoz N, Rueda LA, Bouadjar B, Montoya LS, Orth G, Favre M (2002): Mutations in two adjacent novel genes are associated with epidermodysplasia verruciformis. *Nat Genet* 32:579–581.

Randerson-Moor JA, Harland M, Williams S, Cuthbert-Heavens D, Sheridan E, Aveyard J, Sibley K, Whitaker L, Knowles M, Bishop JN, Bishop DT (2001): A germline deletion of p14(ARF) but not CDKN2A in a melanoma-neural system tumour syndrome family. *Hum Mol Genet* 10:55–62.

Recio JA, Noonan FP, Takayama H, Anver MR, Duray P, Rush WL, Lindner G, De Fabo EC, DePinho RA, Merlino G (2002): Ink4a/arf deficiency promotes ultraviolet radiation-induced melanomagenesis. *Cancer Res* 62:6724–6730.

Rhodes AR (1999): Dysplastic nevi. In Freedberg IM, Eisen AZ, Wolff K, Austen KF, Goldsmith LA, Katz SI, Fitzpatrick TB (Eds). *Fitzpatrick's Dermatology in General Medicine*, 5th ed. McGraw-Hill, New York, pp 1060–1079.

Rigel DS, Friedman RJ, Kopf AW (1996): The incidence of malignant melanoma in the United States: Issues as we approach the 21st century. *J Am Acad Dermatol* 34:839–847.

Rizos H, Puig S, Badenas C, Malvehy J, Darmanian AP, Jimenez L, Mila M, Kefford RF (2001): A melanoma-associated germline mutation in exon 1beta/check inactivates p14ARF. *Oncogene* 20:5543–5547.

Rowe DE, Carroll RJ, Day CL Jr (1992): Prognostic factors for local recurrence, metastasis, and survival rates in squamous cell carcinoma of the skin, ear, and lip. Implications for treatment modality selection. *J Am Acad Dermatol* 26:976–990.

Sands AT, Abuin A, Sanchez A, Conti CJ, Bradley A (1995): High susceptibility to ultraviolet-induced carcinogenesis in mice lacking XPC. *Nature* 377:162–165.

Schwartz RA, Stoll HLJ (1999): Squamous cell carcinoma. In Freedberg IM, Eisen AZ, Wolff K, Austen KF, Goldsmith LA, Katz SI, Fitzpatrick TB (Eds). *Fitzpatrick's Dermatology in General Medicine*, 5th ed. McGraw-Hill, New York, pp 840–856.

Sharpless NE, Bardeesy N, Lee KH, Carrasco D, Castrillon DH, Aguirre AJ, Wu EA, Horner JW, DePinho RA (2001): Loss of p16Ink4a with retention of p19Arf predisposes mice to tumorigenesis. *Nature* 413:86–91.

Sharpless NE, Kannan K, Xu J, Bosenberg MW, Chin L (2003): Both products of the mouse Ink4a/Arf locus suppress melanoma formation in vivo. *Oncogene* 22:5055–5059.

Singh B, Reddy PG, Goberdhan A, Walsh C, Dao S, Ngai I, Chou TC, P OC, Levine AJ, Rao PH, Stoffel A (2002): p53 regulates cell survival by inhibiting PIK3CA in squamous cell carcinomas. *Genes Dev* 16:984–993.

Sistare FD, Thompson KL, Honchel R, DeGeorge J (2002): Evaluation of the Tg.AC transgenic mouse assay for testing the human carcinogenic potential of pharmaceuticals—practical pointers, mechanistic clues, and new questions. *Int J Toxicol* 21:65–79.

Soengas MS, Capodieci P, Polsky D, Mora J, Esteller M, Opitz-Araya X, McCombie R, Herman JG, Gerald WL, Lazebnik YA, Cordon-Cardo C, Lowe SW (2001): Inactivation of the apoptosis effector Apaf-1 in malignant melanoma. *Nature* 409:207–211.

Sotillo R, Garcia JF, Ortega S, Martin J, Dubus P, Barbacid M, Malumbres M (2001): Invasive melanoma in Cdk4-targeted mice. *Proc Natl Acad Sci USA* 98:13312–13317.

Tavin E, Persky MS, Jacobs J (1995): Metastatic basal cell carcinoma of the head and neck. *Laryngoscope* 105: 814–817.

Van Dyke T, Jacks T (2002): Cancer modeling in the modern era: progress and challenges. *Cell* 108:135–144.

van Hogerlinden M, Rozell BL, Ahrlund-Richter L, Toftgard R (1999): Squamous cell carcinomas and increased apoptosis in skin with inhibited Rel/nuclear factor-kappaB signaling. *Cancer Res* 59:3299–3303.

Vasioukhin V, Degenstein L, Wise B, Fuchs E (1999): The magical touch: Genome targeting in epidermal stem cells induced by tamoxifen application to mouse skin. *Proc Natl Acad Sci USA* 96:8551–8556.

Waikel RL, Kawachi Y, Waikel PA, Wang XJ, Roop DR (2001): Deregulated expression of c-Myc depletes epidermal stem cells. *Nat Genet* 28:165–168.

Walker GJ, Hayward NK (2002): Pathways to melanoma development: Lessons from the mouse. *J Invest Dermatol* 119:783–792.

Wang TS, Lowe L, Smith JW II, Francis IR, Sondak VK, Dworzanian L, Finkelstein S, Slingluff CL, Jr., Johnson TM (1998a): Complete spontaneous regression of pulmonary metastatic melanoma. *Dermatol Surg* 24:915–919.

Wang XJ, Greenhalgh DA, Jiang A, He D, Zhong L, Medina D, Brinkley BR, Roop DR (1998b): Expression of a p53 mutant in the epidermis of transgenic mice accelerates chemical carcinogenesis. *Oncogene* 17:35–45.

Whiteman DC, Whiteman CA, Green AC (2001): Childhood sun exposure as a risk factor for melanoma: A systematic review of epidemiologic studies. *Cancer Causes Control* 12:69–82.

Yoshida BA, Sokoloff MM, Welch DR, Rinker-Schaeffer CW (2000): Metastasis-suppressor genes: A review and perspective on an emerging field. *J Natl Cancer Inst* 92:1717–1730.

You MJ, Castrillon DH, Bastian BC, O'Hagan RC, Bosenberg MW, Parsons R, Chin L, DePinho RA (2002): Genetic analysis of Pten and Ink4a/Arf interactions in the suppression of tumorigenesis in mice. *Proc Natl Acad Sci USA* 99:1455–1460.

Yuspa SH, Poirer MC (1988): Chemical carcinogenesis: From animal models to molecular models in one decade. *Adv Cancer Res* 50:25–70.

Zhu H, Reuhl K, Zhang X, Botha R, Ryan K, Wei J, Chen S (1998): Development of heritable melanoma in transgenic mice. *J Invest Dermatol* 110:247–252.

Zuo L, Weger J, Yang Q, Goldstein AM, Tucker MA, Walker GJ, Hayward N, Dracopoli NC (1996): Germline mutations in the p16INK4a binding domain of CDK4 in familial melanoma. *Nat Genet* 12:97–99.

10

OVARIAN CANCER

Sandra Orsulic
Molecular Pathology Unit, Massachusetts General Hospital Cancer Center, Charlestown, Massachusetts

Mice have been used in ovarian cancer research mainly as hosts for cell lines derived from human ovarian tumors and ascites. Such models provided valuable information into the nature of metastatic ovarian cancer and possible treatment strategies. However, the complexity of genetic aberrations in human ovarian cancer cell lines precluded understanding of the initiating events responsible for ovarian cancer induction. Since the majority of ovarian cancer patients present at an advanced stage of the disease, it has been difficult to identify the precursor lesions that could be used to study the early morphologic and genetic changes in ovarian cancer. It is thought that the development of animal models in which ovarian cancer can be induced and studied during its early stages will enable better understanding of early ovarian cancer lesions and elucidate molecular events that support ovarian cancer progression. The difficulties in generating such models include the lack of an adequate ovary-specific promoter, the uncertainty about the tissue of origin for different histologic types of ovarian cancer, and the deficiency in understanding the genetic aberrations responsible for ovarian cancer induction. In spite of these difficulties, the first advances toward generating mouse models for ovarian cancer have been made.

INTRODUCTION

Ovarian cancer is the fifth leading cause of cancer death among women in the United States and has the highest mortality rate of all gynecologic cancers (Jemal et al., 2002). The etiology of ovarian cancer is not well understood. The progress in basic ovarian cancer research has been slow, mainly because of the lack of appropriate animal models. Attempts to develop animal models that recapitulate the development and pathophysiologic manifestations of human ovarian cancer have primarily resulted in the development of rare germ cell and sex cord–stromal tumors, but not epithelial tumors, which are the prevalent ovarian cancer type in women. This chapter provides an overview of the different types of ovarian cancer in women and the recent efforts to model the disease in mice in order to understand the molecular events that are responsible for ovarian cancer initiation.

OVARY DEVELOPMENT

The ovaries are paired reproductive organs located on either side of the uterus and adjacent to the lateral wall of the pelvis. The main functions of the ovary are the production of oocytes and the steroid hormones, estrogen and progesterone. The ovaries consist of several different cell types that can be generally grouped into somatic and germ cells. While the lineage of germ cells is well established, less is known about the origin of various somatic cell types in the ovary (McLaren, 2000). Precursors of primordial germ cells form in the epiblast at the beginning of gastrulation and migrate from the epiblast to the extraembryonic mesoderm (Ginsburg et al., 1990). They then migrate back into the embryo through the hindgut mesentery toward the gonadal ridges (Fig. 10.1*a*). The gonadal ridges form by proliferation of the coelomic epithelium and condensation of the underlying mesenchyme of the urogenital ridge (Fig. 10.1*a*). Induced by the underlying mesenchyme, the basement

Mouse Models of Human Cancer, edited by Eric C. Holland
ISBN 0-471-44460-X

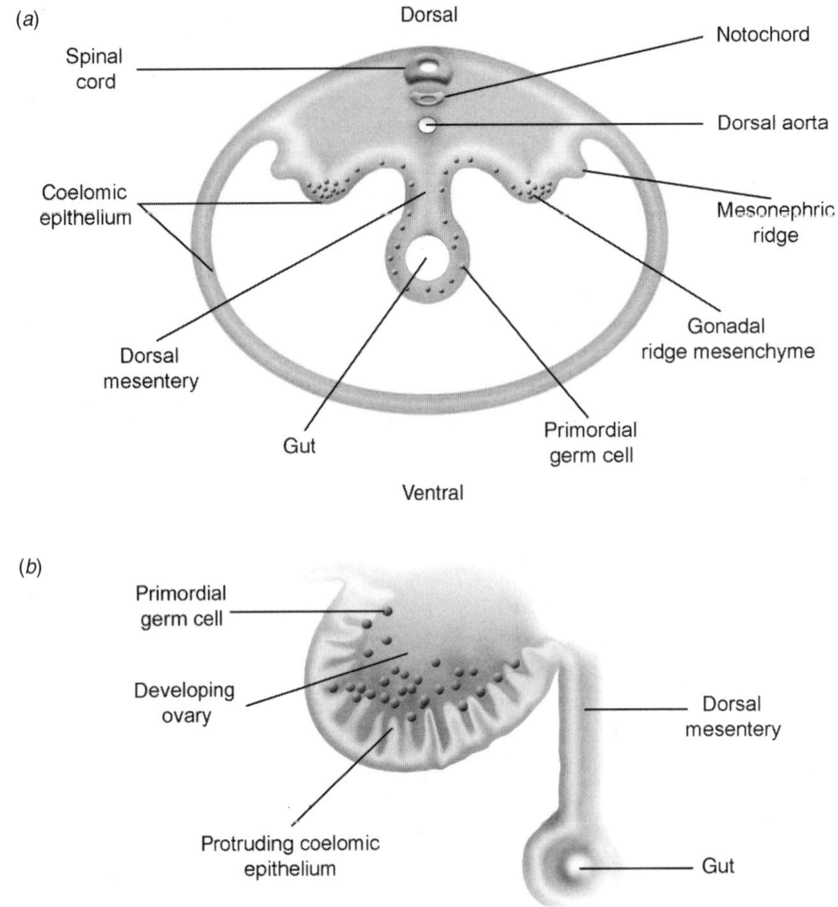

Figure 10.1. Lineages of somatic and germ cells in the ovary. (*a*) The primordial germ cells migrate through the gut mesentery and populate the gonadal ridges that form on the medial surfaces of the urogenital ridges. The gonadal ridge mesenchyme is surrounded by proliferating coelomic epithelia. (*b*) As the ovary develops, derivates from the coelomic epithelia invade the gonadal ridges. These are believed to give rise to several different types of somatic cells in the developed ovary, including the monolayered epithelial cells that cover the surface of the ovary.

membrane of the coelomic epithelium breaks down (Karl and Capel, 1998) and epithelial derivatives enter the ridge (Fig. 10.1*b*). Thus the gonadal ridges are populated by germ cells, the mesenchyme of the urogenital ridge, and the coelomic epithelium. Various tissues in the adult ovary (Fig. 10.2) are believed to arise from these three precursor cell types.

NEOPLASMS OF THE OVARY

Ovarian cancer is a broad term used for a wide range of neoplasms that originate in the ovary. Based on the putative cell of origin, ovarian neoplasms are classified into three broad groups: germ cell, sex cord–stromal cell, and surface epithelial-stromal tumors. Each group contains several histologic ovarian tumor subtypes (Table 10.1). Since epithelial ovarian tumors are the most common and the most lethal tumors of the ovaries, the majority of the text will be devoted to epithelial ovarian cancers.

Germ Cell Tumors

Ovarian germ cell tumors develop from an oocyte and account for less than 5% of ovarian neoplasms. They typically occur in teenagers and young women. Mature cystic teratomas (dermoid cysts) comprise 95% of these tumors and often contain differentiated tissues such as skin, teeth, and hair (Fig. 10.3*a*). Approximately one-third of germ cell tumors are malignant.

Sex Cord–Stromal Cell Tumors

Sex cord–stromal cell tumors account for less than 5% of ovarian neoplasms. They originate from granulosa cells,

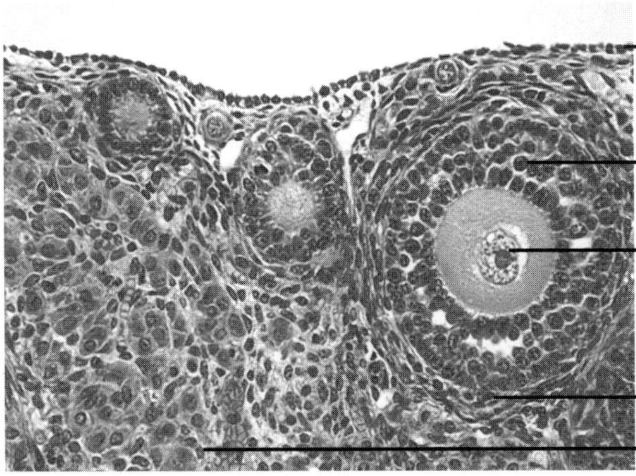

Figure 10.2. Adult mouse ovary section. The section is stained with hematoxylin and eosin (H&E), which stain nuclei and cell cytoplasm, respectively.

Table 10.1. Histologic Classification of Ovarian Tumors

Surface Epithelial–Stromal Tumors	Sex Cord–Stromal Tumors	Germ Cell Tumors
Serous	Fibroma-thecoma	Mature teratoma
Mucinous	Granulosa cell	Immature teratoma
Endometrioid	Sertoli-Leydig cell	Dysgerminoma
Clear cell	Sertoli cell	Yolk sac
Transitional	Mixed	Embryonal carcinoma
Squamous	Steroid cell	Choriocarcinoma
Mixed epithelial		Mixed germ cell
Undifferentiated carcinoma		Gonadoblastoma

theca cells and their luteinized derivatives, Sertoli cells, Leydig cells, and fibroblasts of stromal origin. Most clinically malignant sex cord–stromal tumors are of the granulosa cell type (Fig. 10.3*b*).

Surface Epithelial–Stromal Tumors

More than 90% of human ovarian cancers are epithelial and believed to originate from the ovarian surface epithelium (Scully, 1977). Epithelial ovarian cancers are predominantly a disease of perimenopausal and postmenopausal women, with 80% to 90% of ovarian cancer cases occurring after the age of 40. The peak incidence of invasive epithelial ovarian cancer occurs at age 63. Hereditary ovarian cancers occur approximately 10 years earlier (Boyd and Rubin, 1997). The most significant risk factor, other than age, is family history of ovarian cancer. The risk increases with infertility and nulliparity and decreases with multiparity. A striking discovery within the last 20 years has been the protective effect of oral contraceptives against ovarian cancer (Franceschi et al., 1991). After 5 years of oral contraceptive use, women reduce their risk of ovarian cancer by 50% (Holschneider and Berek, 2000; Purdie et al., 2003). It is believed that hormonal changes and suppression of ovulation are responsible for the protective effect of multiparity and oral contraceptives. Surface epithelial stromal tumors can be divided into four main different histologic subtypes: serous, mucinous, endometrioid, and clear cell (Figs. 10.3*c*–*f*). In some cases, ovarian epithelial cancers consist of more than one distinct histologic subtype (Table 10.1).

SCREENING AND DETECTION

The inaccessible anatomic location of the ovaries and the asymptomatic nature of the disease hinder the detection of cancer while it is still confined to the ovary. Even in later stages, the disease is usually associated with subtle symptoms such as abdominal bloating, discomfort, or changes in bowel or bladder habits. These common symptoms are often misdiagnosed or simply dismissed. As a result, over 75% of ovarian cancer patients are diagnosed at an advanced stage when peritoneal dissemination has already taken place (Holschneider and Berek, 2000). Unlike many other epithelial tumors that progress gradually and display identifiable preneoplastic lesions, the majority of ovarian epithelial cancers display characteristics of invasive carcinoma without any evidence of intermediate phases of benign and/or borderline neoplastic lesions. Ovarian cancers also appear to have a different pattern of invasion and metastasis than the majority of solid epithelial tumors. The most common route of spread is by direct extension to adjacent organs or by exfoliation of tumor cells

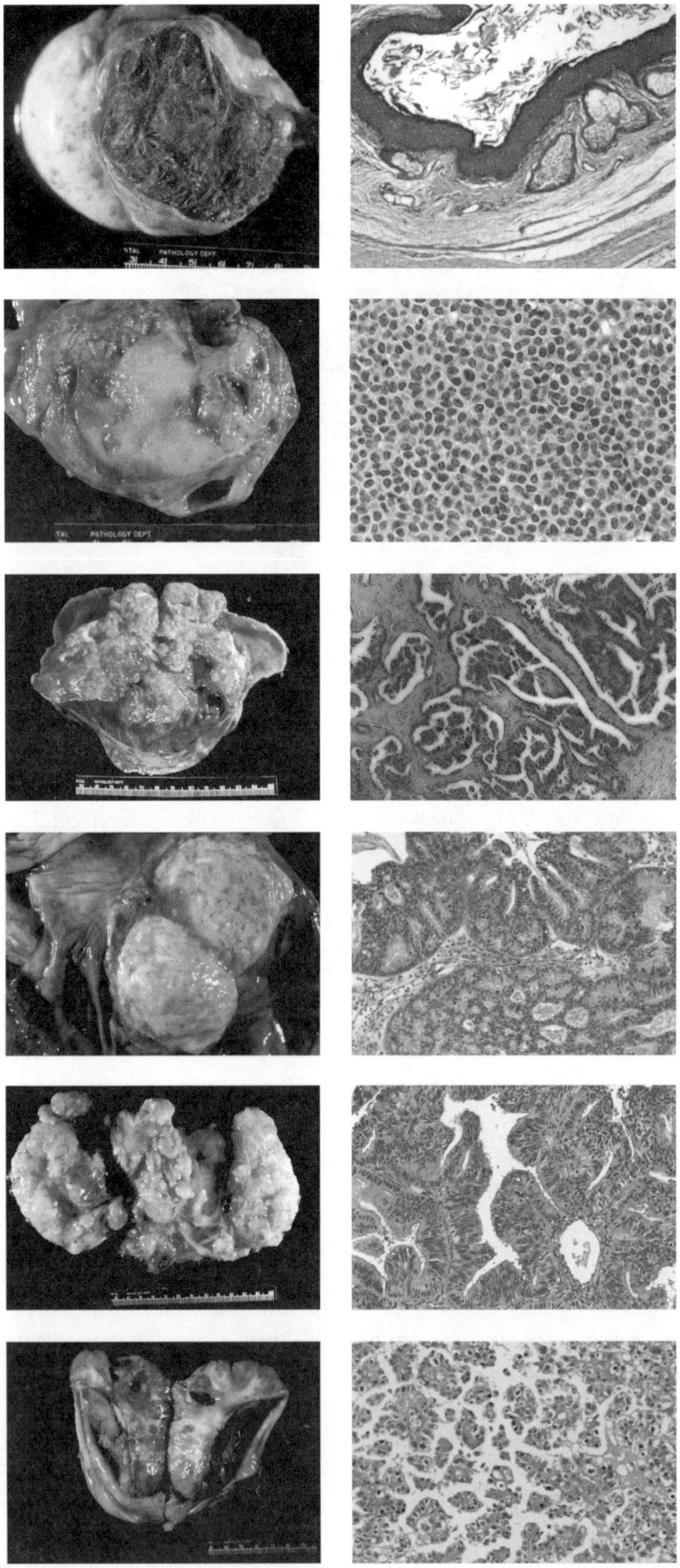

Figure 10.3. Histologic types of ovarian cancer. (*a*) Mature cystic teratoma, the most common subtype in the category of germ cell tumors. Hair is grossly seen. (*b*) Granulosa cell tumor (subtype of sex cord–stromal cell tumor). Four main human epithelial ovarian cancer subtypes: (*c*) serous, (*d*) mucinous, (*e*) endometrioid, and (*f*) clear-cell carcinoma. (Courtesy of E. Oliva.) (See color insert.)

from the ovary, followed by intraperitoneal dissemination, implantation, and growth of metastases on mesothelial surfaces. The tumor cells are often transported into the upper abdomen with the continuous clockwise movement of the peritoneal fluid resulting from bowel peristalsis and respiratory motion. In addition, ovarian cancer can spread to regional lymph nodes. Blood-borne metastases are rare.

Because the five-year survival rate for patients with early-stage ovarian cancer is significantly better than that for patients with advanced disease, there have been continued attempts to improve techniques for screening and early detection. A screening test must have sufficient sensitivity and specificity to be effective. The sensitivity is necessary in order to detect the disease at an early stage, while the specificity is needed because the incidence of ovarian cancer is relatively low in the average population. Serum tumor marker CA125 (Bast et al., 1983) and imaging using transvaginal sonography (Fig. 10.4) are the most widely used detection methods for ovarian cancer. However, neither of these methods has sufficient sensitivity and specificity to be useful for screening of asymptomatic women in the general population (Bast et al., 2002; Gallion and Bast, 1993; Helzlsouer et al., 1993; Schwartz, 2002; Schwartz et al., 1995; Schwartz and Taylor, 1995). Improvements in early diagnosis may be achieved by using a panel of markers that are differentially expressed in the serum of women with ovarian cancer. High-throughput molecular technologies, such as cDNA microarray expression profiling, comparative genomic hybridization screening, and proteomic pattern analysis with learning algorithms, have recently been used to identify a discriminatory pattern that can distinguish ovarian cancer patients from healthy women (Bayani et al., 2002; Hough et al., 2001; Ismail et al., 2000; Matei et al., 2002; Mok et al., 2001; Ono et al., 2000; Sawiris et al., 2002; Schummer et al., 1999; Schwartz et al., 2002; Shridhar et al., 2001; Tapper et al., 2001; Tonin et al., 2001; Wang et al., 1999; Welsh et al., 2001). Recent proteomic data demonstrated that it is possible to use this nonbiased global approach to extract a discriminatory fingerprint that recognizes stage I ovarian cancer with sensitivity and specificity that exceeds that of CA125 (Petricoin et al., 2002). This could be an important development in ovarian cancer screening if the method proves to be specific in a more general population.

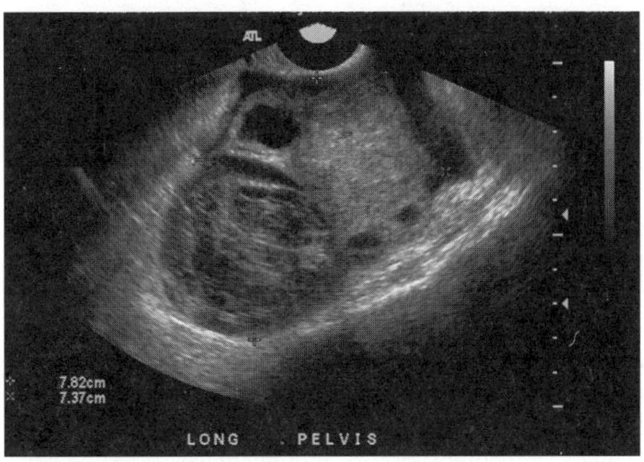

Figure 10.4. Ultrasonography image of ovarian cancer. (Courtesy of S. Mironov and H. Hricak).

TREATMENT

Since there are many types of tumors that can arise in the ovaries, the treatment depends on the type of cancer and the extent of its spread. Benign tumors of the ovary can be removed surgically without the need for further treatment. Management of malignant ovarian cancers generally requires a multimodal approach, which commonly includes surgery and combination chemotherapy. The current chemotherapy regimens typically consist of combinations of microtubule inhibitors and platinum agents. The two principal prognostic factors are stage at diagnosis and maximum residual disease following cytoreductive surgery. The stage is determined at surgery and depends upon the extent of ovarian cancer spread (Table 10.2). In most cases, the tumor has spread throughout the abdomen and cannot be completely removed by surgery. Thus the goal of cytoreductive surgery is to remove as much cancer as possible, preferably without leaving nodules larger than 1 cm in size. Most patients with advanced ovarian cancer will respond well to the initial treatment and

Table 10.2. Surgical Stages of Ovarian Cancer

Stage I:	Limited to the ovaries IA: One ovary involved IB: Both ovaries involved IC: One or both ovaries involved but with cancer on the surface of an ovary, rupture of an ovarian cyst, malignant ascites, or positive abdominal washings
Stage II:	Spread to adjacent pelvic structures IIA: Spread to uterus or fallopian tubes IIB: Spread to pelvic peritoneum IIC: Confined to the pelvis but with malignant ascites or positive abdominal washings
Stage III:	Spread to the upper abdomen IIIA: Microscopic spread to the upper abdomen IIIB: Cancer nodules less than 2 cm in the abdomen IIIC: Cancer nodules more than 2 cm in the abdomen or positive pelvic or aortic lymph nodes
Stage IV:	Distant spread beyond the abdominal cavity or visceral metastases

some will have a complete remission. However, recurrent cancers are often resistant to further treatment regimens. After aggressive cytoreductive surgery and combination chemotherapy, five-year survival rates are as follows: stage I (93%), stage II (70%), stage III (37%), and stage IV (25%) (Holschneider and Berek, 2000). Unfortunately, less than one-third of patients are diagnosed at stage I.

ETIOLOGY

Our ability to screen for early-stage ovarian cancer is hampered by deficiencies in the understanding of the molecular and morphologic steps involved in ovarian carcinogenesis. Epidemiologic studies suggest a direct correlation between the number of ovulatory cycles and the risk of ovarian cancer (Bernal et al., 1995; Perez et al., 1991; Purdie et al., 2003). The first theory about the role that ovulation could play in ovarian carcinogenesis was put forward by Fathalla (1971). He speculated that the rupture of a follicle increases the risk of ovarian cancer by causing trauma and exposing the ovarian surface epithelium to high levels of steroid hormones and gonadotropins. Also, repair of the ovulatory wound in the ovarian surface likely results in the rapid proliferation of epithelial cells, which may increase the frequency and accumulation of spontaneous mutations. Additionally, ovulation may lead to the entrapment of epithelial cells in the underlying stroma with the subsequent formation of inclusion cysts. These inclusion cysts could be the precursor ovarian cancer lesions in which the surrounding stromal environment facilitates neoplastic transformation (Fig. 10.5*a*).

Another cancer-contributing factor may result from the pluripotential nature of ovarian surface epithelial cells. Embryologically, the ovarian surface epithelium is derived from the coelomic epithelium (Fig. 10.1), which overlies the presumptive gonadal ridge (Barber, 1988; McLaren, 2000). The coelomic epithelium also gives rise to the Müllerian ducts, which are the primordia for the epithelia of the Fallopian tubes, uterine

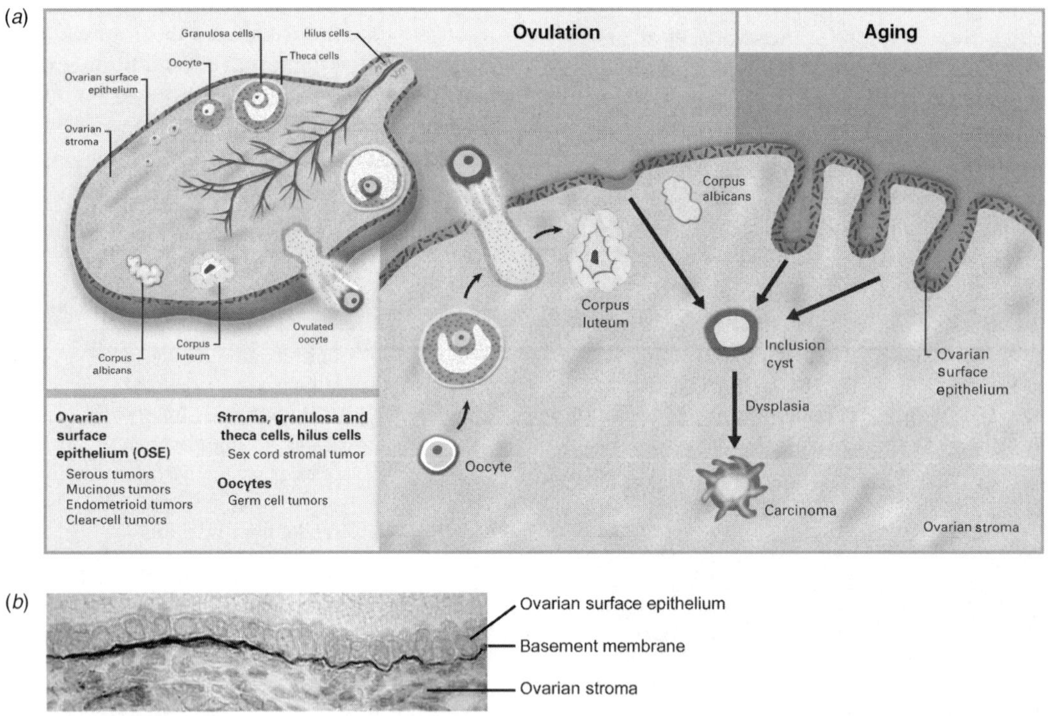

Figure 10.5. Model of ovarian carcinogenesis. (*a*) Epithelial carcinomas are thought to arise from the surface epithelium or cortical inclusion cysts. Cortical inclusion cysts form as a result of surface complexity that is associated with aging and/or as part of the surface repair process following ovulation. These cysts, under the hormonal effects of the ovarian stroma, are more susceptible to malignant change. (Reproduced with permission from Drapkin et al., 2002.) (*b*) Ovarian surface epithelium consists of a single sheet of cells that secrete a basement membrane on their basal side. The basement membrane is thought to serve both as a barrier from direct contact between epithelial and stromal cells and as a medium for directing polarized signaling between the two cell types. In this figure, the basement membrane of the mouse ovarian epithelium was detected by immunohistochemical staining of laminin.

corpus, and endocervix. Developmentally, the ovarian surface epithelium has retained properties of uncommitted pluripotential cells and, similar to the embryonic coelomic epithelium, is competent to differentiate along several different pathways. This pluripotential state may be necessary to provide the ovarian surface epithelium with the phenotypic plasticity required for its functions in ovulatory wound repair. However, it might also contribute to the susceptibility of the ovarian surface epithelium to undergo neoplastic transformation. It is thought that an early step in epithelial ovarian neoplasia involves aberrant epithelial differentiation of the cells on the ovarian surface. The ability of the ovarian surface epithelial cells to undergo Müllerian differentiation was demonstrated experimentally by ectopic expression of the homeobox gene *HOXA7* in immortalized ovarian surface epithelial cells (Naora et al., 2001). Consistent with the ability of ovarian surface epithelial cells to differentiate along Müllerian lines, malignancies of the ovary display a remarkable range of histologic features, which generally recapitulate those of the Fallopian tube, endocervix, and endometrium (Fox, 1993). Based on cellular and structural features reminiscent of normal adult tissues of Müllerian origin, malignant epithelial neoplasms are classified into serous, mucinous, clear-cell, and endometrioid carcinomas (Table 10.1, Figs. 10.3*c–f*). While substantial evidence supports the theory that serous ovarian tumors originate from the ovarian surface epithelium (Auersperg et al., 1998; Berchuck and Carney, 1997; Feeley and Wells, 2001; Ghahremani et al., 1999), the origin of mucinous, clear-cell and endometrioid ovarian tumors is less clear.

GENETIC ABERRATIONS IN SPORADIC AND HEREDITARY OVARIAN CANCERS

Cytogenetic analyses of human ovarian cancers have revealed numerous chromosomal aberrations, but they have failed to identify a consistent chromosomal aberration that is ubiquitously present in ovarian cancers (Gallion et al., 1990; Gray et al., 2003; Pejovic et al., 1992; Whang-Peng et al., 1984). Similarly, although many genetic alterations have been found to be associated with ovarian cancers (Table 10.3), a single individual genetic

Table 10.3. Alterations in Oncogenes and Tumor Suppressor Genes That Have Been Observed in Human Ovarian Carcinomas

Gene	Chromosome Location	Type of Alteration	Function
Oncogenes			
PIK3CA	3q26.3	Amplification, overexpression	Phosphatidylinositol 3-kinase activity
c-*FMS*	5q33-q35	Overexpression	Tyrosine kinase, cell growth, proliferation
EGFR	7p12	Unregulated expression	Tyrosine kinase, signaling
c-*myc*	8q24.12-q24.13	Amplification, overexpression	Proliferation, transcription, cell cycle regulation
K-*ras*	12p12.1	Mutation (codons 12,13, and 61)	Signal transduction, cell cycle regulation
Akt1	14q32.3	Amplification, overexpression	Protein kinase
HER-2/neu	17q11.2-q12	Amplification, overexpression	Receptor signaling, cell proliferation
Akt2	19q13.1-q13.2	Amplification, overexpression	Protein kinase
EEF1A2	20q13.3	Amplification, overexpression	Protein elongation factor
Tumor suppressor genes			
NOEY2 (ARHI)	1p31	Loss of expression	Induces p21, inhibits cyclin D1
MSH2	2p21	Mutation	DNA mismatch repair
FHIT	3p14.2	Altered transcripts	Unknown
SPARC	5q31.3-q32	Loss of expression	Extracellular matrix protein, cell adhesion
LOT-1	6q25	Loss of expression	Zinc-finger protein
p16 (INK4A)	9p21	Loss of expression	Cell cycle checkpoint
PTEN	10q23.3	Mutations	Phosphatase
WT1	11p13	Mutations	Transcription factor
$p27^{KIP1}$	12p13	Loss of expression	Cyclin-dependent kinase inhibitor
BRCA2	13q12.3	Mutations	DNA repair, cell cycle regulation
Rb1	13q14.2	Loss of expression, mutations	Cell cycle regulation
TP53	17p13.1	Mutations	Apoptosis, cell cycle regulation, transcription
OVCA1 and *OVCA2*	17p13.3	Loss of expression	Unknown
BRCA1	17q21	Mutations	DNA repair, cell cycle regulation, transcription

aberration that is present in all ovarian carcinomas has not been identified. It is likely that the identified genetic aberrations represent only a fraction of the genes that are involved in ovarian cancer initiation and progression (Gray et al., 2003). It is unclear whether ovarian carcinomas of different histologic subtypes develop via distinct molecular pathways. Aberrations in genes such as *p53*, c-*myc*, K-*ras*, *Akt*, and *HER-2* have been observed in all four histologic subtypes of ovarian carcinoma. However, some genetic aberrations are more prevalent in certain ovarian cancer subtypes. For example, K-*ras* mutations are more frequently found in mucinous than in serous tumors (Pieretti et al., 1995), and c-*myc* overexpression is more common in serous tumors (Baker et al., 1990; Wang et al., 1999). Mutations in *PTEN* and *ß-catenin* typically occur in endometrioid tumors but not in other ovarian cancer types (Obata et al., 1998; Wu et al., 2001; Zhai et al., 2002). Recent findings that gene expression patterns in ovarian carcinomas reflect both the morphologic features and biologic behavior of tumors (Schwartz et al., 2002; Tonin et al., 2001) indicate that the underlying genetic alterations may be the foundation of tumor heterogeneity.

The majority of ovarian cancers are sporadic, without any known familial history of ovarian cancer. Approximately 10% of ovarian cancers are hereditary (Randall et al., 1998). Most hereditary cancers can be attributed to germline mutations in the breast/ovarian cancer susceptibility gene *Brca1* (Miki et al., 1994) or *Brca2* (Wooster et al., 1995). Sporadic and hereditary cancers are similar in many respects; however, patients with hereditary cancers develop the disease earlier, display a longer recurrence-free interval following chemotherapy, and have a longer overall survival rate (Zweemer et al., 1999b). This may be related to the increased sensitivity of *Brca*-deficient tumor cells to therapeutic DNA damaging agents that produce double-stranded breaks (Scully and Livingston, 2000). Studies in mouse models have demonstrated that *p53* is highly cooperative with *Brca1* in promoting mammary and ovarian tumor development (Xu et al., 2001). It is thought that the absence of *p53* decreases apoptosis and relaxes the cell cycle control, thus preventing the *Brca1*-induced apoptosis and senescence (Cao et al., 2003). Consistently, *p53* aberrations are commonly found in ovarian and breast tumors from women heterozygous for *Brca1* (Zweemer et al., 1999a). Recent analysis of prophylactically removed ovaries from *Brca1*-heterozygous women demonstrated that *Brca1* loss of heterozygosity and inactivation of the *p53* function are the early events in the induction of hereditary ovarian cancer (Werness et al., 2000).

TISSUE CULTURE MODELS

The basic biology of the ovarian surface epithelium and the molecular mechanisms underlying the acquisition of an invasive phenotype in this tissue are not well understood. This lack of understanding is due to the difficulties in establishing an appropriate model system for epithelial ovarian carcinoma. The development of culture systems posed problems because the ovarian surface epithelium constitutes a very small fraction of the whole ovary and is difficult to separate from other ovarian cell types by physical or enzymatic means. However, in the 1980s, the first tissue culture systems for ovarian surface epithelium from different species, including human, were developed (Adams and Auersperg, 1983; Auersperg et al., 1984; Dubeau et al., 1990; Nicosia et al., 1984). Mouse and rat ovarian epithelial cells subjected to repetitious growth in culture occasionally undergo malignant transformation. Independent cell lines derived from such spontaneously transformed cultures display a range of cytogenetic changes (Testa et al., 1994).

The main difficulty in establishing cultures of human ovarian surface epithelial cells is their limited growth potential and early senescence. Unlike rodent cells, human ovarian surface epithelial cells do not spontaneously transform in culture. Even expression of the simian virus 40 T antigen (*SV40 TAg*) is not sufficient to immortalize human ovarian surface epithelial cells, although it significantly increases their growth potential (Maines-Bandiera et al., 1992). Such ovarian surface epithelial cells are nontumorigenic, but they acquire several other properties of neoplastic cells, including reduced dependence on serum and genetic instability. SV40 TAg-transformed ovarian epithelial cell lines are used in many laboratories as representative of normal human ovarian surface epithelium. However, these cell lines have frequently undergone genetic alterations following establishment in tissue culture and thus are not truly representative of normal ovarian surface epithelium. Another difficulty in modeling ovarian cancer in culture is that the ovarian stromal cells may play a crucial role in ovarian cancer induction (Ghahremani et al., 1999). In a normal ovary, the monolayered ovarian surface epithelium lies adjacent to the basement membrane (Nicosia et al., 1989) that separates the epithelium from the ovarian stroma (Fig. 10.5*b*). Thus, organ culture systems in which epithelial cells are in constant communication with the underlying ovarian stroma may be necessary to truly model the normal biologic functions of this tissue and its changes in neoplasia.

Nevertheless, the use of ovarian surface epithelial cultures, primary and immortalized, has provided the opportunity to investigate the underlying genetic changes that induce a tumorigenic state in this tissue. The first demonstrations that the immortalized ovarian surface epithelium can be transformed by introducing defined genetic elements were achieved by transfection of rodent ovarian surface epithelial cells with K-*ras* (Adams and Auersperg, 1981), H-*ras* (Hoffman et al., 1993), and

HER-2/neu (Davies et al., 1998). Recently, the transformation capability of cultured human ovarian surface epithelium was demonstrated by the introduction of *SV40 TAg*, the catalytic subunit of human telomerase (*hTERT*) and H-*ras* (Liu et al., 2004).

MOUSE MODELS

Over the last 20 years, the bulk of research on animal models for ovarian cancer has involved the xenografting of human ovarian tumors and established ovarian cancer cell lines into immunodeficient mice (Hamilton et al., 1984). The use of immunodeficient mice as surrogate hosts for human tumors has provided a valuable and reproducible system that represents the best approximation to the original human tumor. However, the majority of ovarian cancer cell lines have been established from tumors or ascites from patients with advanced ovarian cancers that have already accumulated numerous genetic changes. The complexity of genetic events in such tumors and cell lines has made it difficult to correlate the tumor phenotype to the primary genetic events that trigger tumor formation. Additionally, xenografting human cells into immunodeficient mice cannot simulate the interaction of the immune system in the development and progression of ovarian cancer, which may prove to be critical in understanding the disease.

Unlike human ovarian tumor cells that can only be introduced into immunodeficient mice, transformed mouse ovarian cells can be introduced into syngeneic mice with intact immune systems, thus making mouse models suitable for the investigation of tumor–host interactions and antitumor immune mechanisms (Roby et al., 2000). Additionally, new genes suspected to play a role in tumorigenesis could be introduced into the transformed mouse cell line. The utility of this approach was elegantly demonstrated in examining the multifaceted functions of vascular endothelial growth factor (VEGF) in modulating the tumor microenvironment and affecting the complex interactions in angiogenesis and antitumor immune mechanisms (Zhang et al., 2002).

The lack of common inbred laboratory animals that develop epithelial ovarian cancer remains one of the major obstacles to ovarian cancer research. It is unclear why spontaneous epithelial ovarian cancers are rare in laboratory animals. One possible explanation is that the ovarian surface epithelium in most laboratory animals is structurally and functionally different and thus lacks specific characteristics which predispose the human ovarian surface epithelium to neoplastic progression. It is also possible that the life span of most laboratory animals is not long enough for the development of these neoplasms. The highest incidence of ovarian epithelial tumors occurs in the postmenopausal period, which is characterized by the following changes: (1) the pool of germ cells (oocytes) is depleted from the ovary; (2) the loss of germ cell–dependent follicle development results in a reduced level of circulating estrogen; (3) the reduced estrogen production is accompanied by higher production of gonadotropins, luteinizing hormone (LH), and follicle-stimulating hormone (FSH); and (4) the structural aberrations in the ovarian surface epithelium that results from numerous ovulations and ruptures of the once smooth ovarian surface. It is possible that the accumulation of genetic aberrations in aged ovaries and postmenopausal conditions cooperate in the predisposition to ovarian cancer.

Several attempts have been made to generate mouse models relevant to human ovarian tumors, largely by trying to simulate the events that occur in the ovaries of postmenopausal women. The strategies have included depletion of oocytes, inhibition or overproduction of estrogen and gonadotropins, carcinogen-induced transformation, X-ray irradiation, neonatal thymectomy, and aging. The most common neoplasms induced by these methods were of stromal origin. Several mouse models of germ cell tumors have also been developed. Since stromal and germ cell neoplasms are very rare in women, such models did not find a niche in human ovarian cancer research. Very recently, efforts have been made to design mouse models of epithelial ovarian cancer, which is the prevalent cancer type in women. This was achieved by direct introduction of oncogenes into the ovarian surface epithelial cells or by generating genetically modified mice that are predisposed to tumor development. Currently existing mouse models for germ cell, sex cord–stromal, and epithelial tumors are described below.

Models of Germ Cell Tumors

A transgenic mouse line predisposed to the development of germ cell tumors was created by insertion of an imprinted transgene *TG.KD* (Fafalios et al., 1996). Germ cell tumors develop in 15% to 20% of hemizygous female carriers of the transgene. These tumors consist of a mixture of immature embryonal carcinoma cells and mature embryonic cells. They are frequently metastatic and, in some instances, result in death of the mouse. Genetic analyses demonstrated that the tumors in these mice were associated with the transgene integration site and did not occur in other transgenic lines with the same transgene.

Development of benign cystic germ cell tumors occurs in aging transgenic mice that overexpress the apoptosis suppression protein Bcl-2 under the ovary-specific *inhibin* gene promoter (Hsu et al., 1996). Overexpression of Bcl-2 protein in the ovary leads to decreased ovarian somatic cell apoptosis and enhanced folliculogenesis. The *bcl-2* transgene in these mice is overexpressed in somatic cells, but not in oocytes, suggesting that enhanced survival of selected somatic cells can lead to germ cell tumorigenesis.

Models of Sex Cord–Stromal Tumors

The importance of functional interaction between stromal and germ cells was demonstrated in female mice homozygous for the germ cell–deficient (*gcd*) mutation (Duncan et al., 1993; Duncan and Chada, 1993). These mice enter premature reproductive senescence due to death of the germ cells during embryonic development. Ovaries of young *gcd*-null mice are atrophic and mostly consist of a connective tissue matrix with some stromal cells. Half of these mice develop tubulostromal adenomas by one year of age. Similarly, Wx/Wv mice which contain 1% of the normal oocytes at birth rapidly lose the follicular apparatus and develop complex tubular adenomas from the surface germinal epithelium (Blaakaer et al., 1995).

It is thought that the loss of germ cells in the ovaries of postmenopausal women diminishes the number of Graafian follicles and sex hormone secretion, leading to compensatory over production of the pituitary gonadotropins LH and FSH. It has been suggested that the increase in gonadotropins contributes to the development of ovarian tumors (Capen et al., 1995). The potential involvement of the pituitary gonadotropins LH and FSH in ovarian tumorigenesis has been extensively investigated, since the production of these hormones is elevated in postmenopausal women. LH and FSH control ovary growth, differentiation, and steroidogenesis. The absence of these hormones results in infertile individuals who maintain a prepubescent state into adulthood with infantile gonads (Kendall et al., 1995). Both LH and FSH are members of the glycoprotein hormone family and are heterodimers that contain an α subunit common to each hormone and a unique β subunit that dictates biologic specificity (Pierce and Parsons, 1981). Expression of the glycoprotein α subunit and the hormone-specific β subunit is regulated by the gonadotropin-releasing hormone (GnRH), steroids, and the ovarian and pituitary peptides activins and inhibins (Matzuk et al., 1996; Pierce and Parsons, 1981). The importance of gonadotropins in ovarian tumorigenesis was elegantly demonstrated by specific suppression of gonadotropins in Wx/Wv mice (Blaakaer et al., 1995). Injection of Wx/Wv mice with GnRH agonist completely suppresses ovarian tumor development. The requirement for gonadotropins in induction of ovarian stromal tumors was demonstrated in hypogonadotropic (*hpg/hpg*) mice deficient in GnRH and lacking LH and FSH. Irradiation-induced oocyte depletion or prolonged treatment with a high dose of gonadotropins results in mesothelial adenomas and granulosa cell tumors in *hpg/+* mice but not in *hpg/hpg* mice (Tennent and Beamer, 1986).

To address the role of overproduction of LH in ovarian tumor development, several transgenic mouse models with chronic LH hypersecretion were developed (Keri et al., 2000; Nilson et al., 2000; Risma et al., 1995). The common characteristics of these mice include infrequent ovulation, a prolonged luteal phase, and development of pathologic ovarian changes such as cyst formation and enlargement of ovaries with reduced numbers of primordial follicles. Depending on the mouse strain, the aged female mice with chronically elevated LH develop luteoma or granulosa cell tumors (Nilson et al., 2000). Overproduction of FSH has also been studied in ovarian tumorigenesis. For example, mice with homozygous deletion of a member of the transforming growth factor β (TGFβ) superfamily, inhibin, develop mixed or incompletely differentiated sex cord–stromal tumors as early as four weeks with 100% penetrance (Matzuk et al., 1992). Consistent with the role of inhibin to suppress pituitary FSH synthesis and secretion, inhibin-deficient mice demonstrate an elevated concentration of FSH. Thus, besides a possible tumor suppressor role for inhibin, the accompanying rise in FSH levels in the circulation may contribute to tumor formation. Female mice deficient for the FSH receptor are infertile and have high levels of circulating FSH (Danilovich et al., 2001; Kumar et al., 1996). They also have small ovaries resulting from a blockage in folliculogenesis at the preantral stage and develop stromal tumors after 12 months of age. In addition to the high levels of circulating FSH, the elevated levels of LH in the FSH receptor mutant mice could contribute to ovarian tumor formation, which was shown to be the case in LH-overexpressing mice (Keri et al., 2000; Nilson et al., 2000; Risma et al., 1995).

To delineate the biologic role of FSH in ovarian growth and tumorigenesis, double-homozygous-mutant mice that are deficient in both inhibin and FSH were generated (Kumar et al., 1999). Double-mutant mice show a significant delay in ovarian tumor development compared with mice deficient in inhibin alone. Mice deficient in inhibin and FSH have suppressed levels of FSH, but LH is still present and could contribute to ovarian tumorigenesis. Consistent with this hypothesis, mice deficient in inhibin and GnRH, which have suppressed levels of both FSH and LH, develop only premalignant lesions in the ovary (Kumar et al., 1996). Isolated LH deficiency with normal production of FSH can also lead to ovarian tumorigenesis. Mice null for the *Lats1* gene display isolated LH hypogonadotropic hypogonadism and develop ovarian stromal cell tumors by the age of three months (Maie et al., 1999). The role of FSH in ovarian tumor development was further explored by generating gain-of-function transgenic mice that overexpress human FSH (Kumar et al., 1999). Female transgenic mice expressing high levels of FSH are infertile and develop hemorrhagic and cystic ovaries but have no signs of tumors. Together, these results suggest that prolonged exposure to elevated FSH levels does not directly cause ovarian tumorigenesis; however, FSH significantly influences the tumor progression in inhibin-deficient mice.

Transgenic mice that express the powerful viral oncogene *SV40 TAg* under regulation of the mouse inhibin-promoter develop metastatic ovarian granulosa and theca cell tumors with 100% penetrance at the age of five to six months (Kananen et al., 1995). The tumors are gonadotropin dependent and do not develop when the transgenic mice are rendered gonadotropin deficient by crossbreeding them into the *hpg/hpg* background. The suppression of gonadotropins by treating the mice with the GnRH antagonist SB-75 also results in the inhibition of tumor growth (Kananen et al., 1997).

Expression of *SV40 TAg* under the regulation of the Müllerian-inhibiting substance (MIS) promoter in transgenic mice induces development of granulosa cell tumors, which in advanced stages invade neighboring organs and develop metastases to the liver and lungs (Dutertre et al., 1997, 2001; Peschon et al., 1992). MIS binds to the Müllerian inhibitory substance type II receptor (*MISIIR*), which is specifically localized to ovarian granulosa cells (di Clemente et al., 1994; Takahashi et al., 1986) and ovarian surface epithelium (Connolly et al., 2003). Frequently, the *MISIIR* can be detected in human ovarian tumors derived from granulosa cells (Gustafson et al., 1992; Imbeaud et al., 1995) and the ovarian epithelial cells (Masiakos et al., 1999). MIS treatment of *MISIIR*-positive tumors and cell lines exhibits growth-inhibitory effects (Kim et al., 1992; Masiakos et al., 1999; Segev et al., 2000; Stephen et al., 2001).

Models of Epithelial Tumors

Epithelial tumors are the most common, and also the most deadly, ovarian tumors in women. Therefore, mouse models of epithelial ovarian tumors are highly sought after. One difficulty in establishing genetically modified mice in which gene expression is altered specifically in the ovarian surface epithelial cells is that these cells lack specialized features that could be exploited as a source of tissue-specific promoters. Researchers resorted to using promoters of genes that are expressed in several tissues, including the ovarian surface epithelium. Hamilton and colleagues used the upstream regulatory sequences of the mouse *MISIIR* gene to target expression of *SV40 TAg* to the precursor cells that generate several tissues of the female mouse reproductive tract, including the ovarian surface epithelium (Connolly et al., 2003). By 6–13 weeks of age, 50% of the transgenic female mice develop bilateral ovarian masses. Histologically, the ovarian tumors are poorly differentiated carcinomas with occasional cysts and papillary structures that resemble human ovarian serous carcinoma (Fig. 10.6*a*). The tumors are often associated with the production of bloody ascites, extensive tumor cell dissemination and invasion to the omentum, mesentery and parietal and visceral serosa. Consistent with the presence of *MISIIR* in several tissue types, other gynecologic tumors develop, albeit less frequently. Unfortunately, it may be difficult to establish stable transgenic lines of *MISIIR-TAg* mice because the rapid onset of tumor initiation renders female mice infertile. Consistent with the expression of *MISIIR* in the tubular and follicular structures of the fetal male gonads and in Sertoli and Leydig cells of adult testes, 28% of the male *MISIIR-TAg* mice develop Sertoli cell tumors but remain fertile. Thus, it may be possible to transmit the transgene and the ovarian phenotype to female offspring through the male transgenic mice. This heritable transgenic ovarian cancer model is currently the most promising model in terms of understanding how ovarian cancer is initiated. The utility of the model in studying the genesis of ovarian cancer will significantly increase with the development of conditionally controlled expression of *SV40 TAg* and with a better understanding of the individual biochemical pathways that are altered by this potent viral oncogene.

Although the ovarian tumors in the aforementioned mouse models histologically resemble human ovarian neoplasms, they may not accurately represent genetic changes that occur during tumor development. It is thought that most human cancers develop as a result of the accumulation of multiple genetic events. Thus, to dissect the multigenetic etiology of cancers, it is necessary to find technical means by which to sequentially introduce multiple genetic modifications into mammalian cells. Furthermore, the majority of human cancers arise in somatic cells, initiating neoplasia in the adult, unlike most transgenic mouse models that carry germline genetic modifications during embryonic development. Recently, a new technique for the introduction of multiple genes into somatic cells of adult mice was developed (Federspiel et al., 1996). This system is based on avian RCAS virus delivery to the cells that are programmed to express the avian TVA receptor.

In this system, viral infection can be restricted to a specific tissue of interest by placing TVA under the control of a tissue-specific promoter. The RCAS-TVA system has been used to generate mouse models for several human cancers, including ovarian cancer (reviewed in Orsulic, 2002). Due to the lack of an ovarian epithelium-specific promoter, transgenic mice that express TVA from the keratin 5 promoter were used. The ovary-specific gene delivery was ensured by isolating the ovaries from mice and infecting them ex vivo with RCAS vectors. Since ovarian surface epithelial cells are the only ovarian cells that express the TVA receptor (Fig. 10.6*b*), these are the only cells susceptible to the RCAS virus infection. RCAS

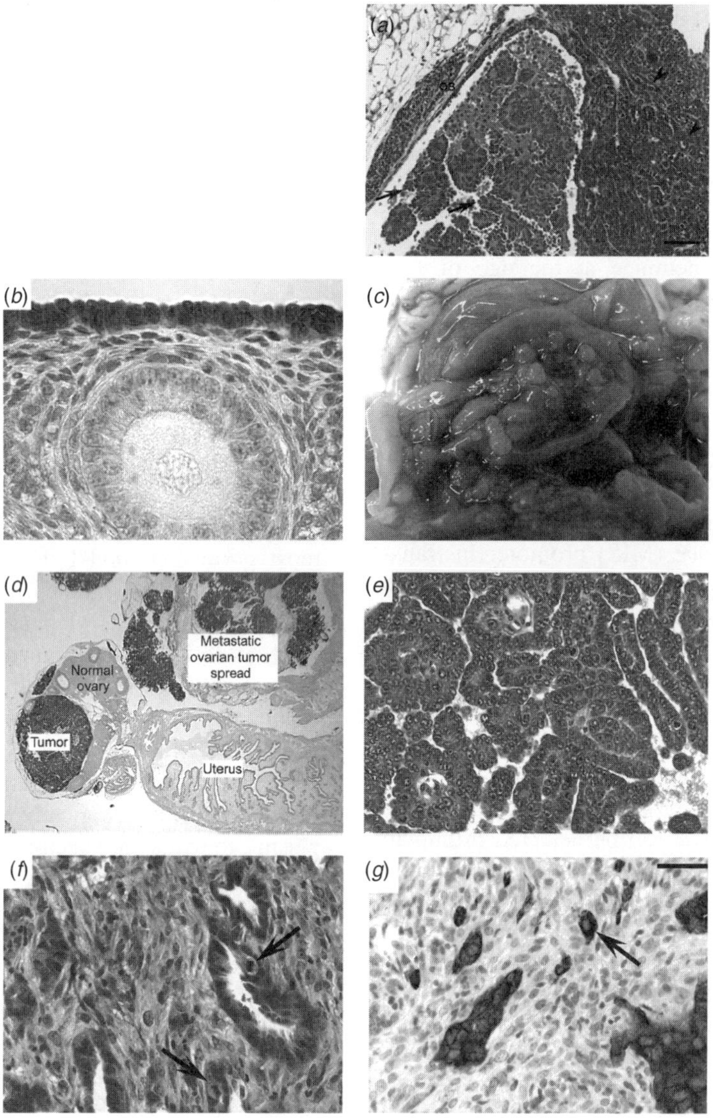

Figure 10.6. Mouse models of epithelial ovarian cancer. (*a*) The ovarian carcinoma in the mouse chimeric for expression of the simian virus 40 T antigen (*SV40 TAg*) under control of the Müllerian inhibitory substance type II receptor (*MISIIR*) promoter. Neoplastic cells form tubular (arrowheads) and papillary (arrows) structures in the ovary and in the intrabursal space, respectively, and invade the ovarian bursa (OB). (*b–e*) An RCAS-TVA mouse model for ovarian epithelial cancer. (*b*) Ovary section from a keratin 5-TVA transgenic mouse stained with antibody against the avian retroviral TVA receptor. Ectopic expression of the TVA receptor renders the cells susceptible to infection with avian RCAS viruses, which are used as vehicles for gene transfer. Cell-specific expression of the TVA receptor restricts the infection to the cells of the ovarian surface epithelium, which is the presumptive precursor tissue for ovarian carcinoma. (*c*) Intraperitoneal carcinomatosis in a mouse model for epithelial ovarian cancer resembles human ovarian cancer spread. (*d*) Metastatic ovarian tumor spread. An influenza hemaglutinin (HA)-tagged *Akt* oncogene is detected with an HA-probe in the primary ovarian tumor and in ovarian metastases. (*e*) H&E staining of an ovarian tumor induced in the RCAS-TVA mouse model. Ovarian tumors induced in the mouse model histologically resemble human ovarian papillary serous carcinoma. (*f,g*) Serous adenocarcinoma induced by selective AdCre-LoxP mediated inactivation of *p53* and *Rb1* in the ovarian surface epithelium of $p53^{loxP/loxP}Rb^{loxP/loxP}$ mice. (*f*) H&E staining representing mitotic (arrows) carcinoma cells that form glandular structures in dense fibrous tissue. (*g*) Immunohistochemical detection of keratin 8 in invasive neoplastic cells (arrow). (Courtesy of A. Y. Nikitin). (See color insert.)

vectors can be designed to carry oncogenes, dominant-negative tumor suppressor genes, and various marker genes. Thus, genetically defined aberrations can be introduced into the ovarian epithelial cells of adult female mice. This provides a very efficient means to evaluate the collaboration of candidate genes in ovarian oncogenesis. For example, the minimal genetic requirements for induction of a tumorigenic state in primary mouse ovarian epithelial cells were determined by introducing combinations of c-*myc*, K-*ras*, and *Akt* into ovarian cells from *p53*-null mice. It was demonstrated that a loss of the *p53* gene and the addition of any two of the c-*myc*, K-*ras*, and *Akt* oncogenes are sufficient to induce transformation of mouse primary ovarian epithelial cells (Orsulic et al., 2002). These genetic aberrations are commonly present in human ovarian carcinomas, although it is not known whether they act in combination to induce ovarian tumors. Orthotopic implantation of ex vivo infected ovarian cells results in metastatic ovarian tumors in four to eight weeks, depending upon the combination of oncogenic aberrations. The initial tumorigenic growth is confined to the implanted ovary followed by spread to adjacent tissues and finally metastatic growth on the surfaces of intraperitoneal organs with a special affinity for the omentum and the mesentery (Figs. 10.6*c,d*). The metastatic tumor development in this model closely resembles human ovarian tumor development and metastatic spread, including the production of ascites and tumor spread throughout the peritoneal cavity. Similar to human ovarian metastatic cancer, the tumor burden remains confined to the peritoneal cavity. Histologically, these tumors resemble human ovarian serous papillary carcinomas (Fig. 10.6*e*). Therefore, this model provides direct experimental proof that the ovarian surface epithelium is the precursor tissue for serous ovarian carcinoma.

Perhaps the greatest limitation of the RCAS-TVA system is that the initiating genetic manipulation does not accurately model the sporadic molecular events that occur in vivo. However, the genetically defined nature of the model allows for the study of genotype–phenotype correlation and thus may lead to a clearer understanding of the contributions of individual genetic aberrations to the process of tumor progression and metastasis. Understanding the collaboration of biochemical pathways in the induction of a tumorigenic state may set the stage for testing novel molecule-based therapies in the future. The similarities in metastatic behavior make this model particularly attractive for developing and testing therapeutic approaches aimed at the advanced stages of ovarian epithelial cancer in humans.

The need for an ovary-specific promoter and/or ex vivo infection and subsequent orthotopic implantation of ovarian cells can be circumvented by the direct administration of viruses into the intrabursal space of surgically exposed mouse ovaries. The ovarian bursa envelopes the mouse ovary to form an anatomically closed space that can be filled with viral supernatant and thus expose the ovarian surface to the infectious viral particles. This approach was used to introduce recombinant adenovirus with cytomegalovirus (CMV) driven Cre recombinase into the ovaries of $p53^{loxP/loxP}/Rb^{loxP/loxP}$ mice in order to concurrently inactivate *p53* and *Rb* (Flesken-Nikitin et al., 2003). The mice develop tumors that contain epithelial cells (Figs 10.6f,g), which are thought to be derived from the adenovirus-infected ovarian surface epithelial cells. The majority of mice, with both genes inactivated, develop ovarian tumors at a median of 32 weeks, while mice with either the *p53* or *Rb* gene inactivated develop ovarian tumors less frequently. These results demonstrate the collaborative effect of concurrent inactivation of *p53* and *Rb* in inducing tumorigenesis in ovarian cells. However, the long tumor latency in these mice indicates that additional genetic aberrations are required for tumor development.

AREAS IN WHICH MODELS ARE NEEDED

None of the aforementioned mouse models for ovarian cancer recapitulate the entire course of the human disease. However, most of the models successfully recapitulate the later stages of the disease and the metastatic spread in the peritoneal cavity. Probably the least understood aspect of ovarian cancer is its initiation and tissue of origin. Precursor lesions are almost never detected in patients and it is unclear whether ovarian carcinomas follow the gradual progression from benign to metastatic cancer or occur de novo, without any identifiable premalignant stages (Bell and Scully, 1994). Currently available mouse models fail to accurately model the early steps in transformation of the ovarian cells. Modeling ovarian cancer initiation in the mouse would significantly contribute to our understanding of precursor lesions and identifying recognizable histologic or molecular markers that could be used for early ovarian cancer detection. Another aspect of ovarian cancer that is poorly understood is the complexity of genetic aberrations and their role in ovarian cancer initiation and progression, and the development of drug resistance. Mouse models in which distinct genetic aberrations can be correlated with the tumor phenotype would not only help to delineate the biochemical pathways responsible for the different ovarian cancers but also provide valuable systems in which to test pathway-targeted therapy. The use of mouse models for drug testing will require concurrent development of imaging techniques for monitoring the tumor response to the drug. Development of imaging modalities in the mouse may contribute to the development of urgently needed detection methods for the early stages of ovarian cancer.

SUMMARY

The general lack of understanding of the biology and genetics of ovarian cancer is a stumbling block in creating mouse models for this disease. However, several models that approximate certain aspects of ovarian cancer development and metastatic spread have been generated. Experiments on animal cell lines and mouse models have greatly contributed to our understanding of the biology of the ovarian surface epithelium and clearly demonstrated that this is the precursor tissue for epithelial ovarian cancer. The mouse models have provided insight into the genetic and hormonal requirements for ovarian cancer progression and the understanding of the biochemical pathways that govern the development of ovarian carcinomas. Technological and informational advances, and new capabilities of manipulating gene expression in the mouse, continue to contribute to the development of more sophisticated mouse models for ovarian cancer. The next challenge is to correlate the information obtained from the mouse ovarian cancer models with human ovarian cancers.

ACKNOWLEDGMENTS

The author acknowledges the Varmus Lab and the Susan G. Komen Foundation for their support, and Kristy Daniels for assistance in preparation of the book chapter.

REFERENCES

Adams AT, Auersperg N (1981): Transformation of cultured rat ovarian surface epithelial cells by Kirsten murine sarcoma virus. *Cancer Res* 41:2063–2072.

Adams AT, Auersperg N (1983): Autoradiographic investigation of estrogen binding in cultured rat ovarian surface epithelial cells. *J Histochem Cytochem* 31:1321–1325.

Auersperg N, Edelson MI, Mok SC, Johnson SW, Hamilton TC (1998): The biology of ovarian cancer. *Semin Oncol* 25:281–304.

Auersperg N, Siemens CH, Myrdal SE (1984): Human ovarian surface epithelium in primary culture. *In Vitro* 20:743–755.

Baker VV, Borst MP, Dixon D, Hatch KD, Shingleton HM, Miller D (1990): c-myc amplification in ovarian cancer. *Gynecol Oncol* 38:340–342.

Barber HR (1988): Embryology of the gonad with reference to special tumors of the ovary and testis. *J Pediatr Surg* 23:967–972.

Bast RC Jr, Klug TL, St John E, Jenison E, Niloff JM, Lazarus H, Berkowitz RS, Leavitt T, Griffiths CT, Parker L, Zurawski VR Jr, Knapp RC (1983): A radioimmunoassay using a monoclonal antibody to monitor the course of epithelial ovarian cancer. *N Engl J Med* 309:883–887.

Bast RC Jr, Urban N, Shridhar V, Smith D, Zhang Z, Skates S, Lu K, Liu J, Fishman D, Mills G (2002): Early detection of ovarian cancer: Promise and reality. *Cancer Treat Res* 107:61–97.

Bayani J, Brenton JD, Macgregor PF, Beheshti B, Albert M, Nallainathan D, Karaskova J, Rosen B, Murphy J, Laframboise S, Zanke B, Squire JA (2002): Parallel analysis of sporadic primary ovarian carcinomas by spectral karyotyping, comparative genomic hybridization, and expression microarrays. *Cancer Res* 62:3466–3476.

Bell DA, Scully RE (1994): Early de novo ovarian carcinoma. A study of fourteen cases. *Cancer* 73:1859–1864.

Berchuck A, Carney M (1997): Human ovarian cancer of the surface epithelium. *Biochem Pharmacol* 54:541–544.

Bernal A, Mendez-Moran L, Fajardo-Gutierrez A, Gonzalez-Lira G, Escudero P, Ortiz H (1995): Univariate and multivariate analysis of risk factors for ovarian cancer: Case-control study, Mexico City. *Arch Med Res* 26:245–249.

Blaakaer J, Baeksted M, Micic S, Albrectsen P, Rygaard J, Bock J (1995): Gonadotropin-releasing hormone agonist suppression of ovarian tumorigenesis in mice of the Wx/Wv genotype. *Biol Reprod* 53:775–779.

Boyd J, Rubin SC (1997): Hereditary ovarian cancer: Molecular genetics and clinical implications. *Gynecol Oncol* 64:196–206.

Cao L, Li W, Kim S, Brodie SG, Deng CX (2003): Senescence, aging, and malignant transformation mediated by p53 in mice lacking the Brca1 full-length isoform. *Genes Dev* 17:201–213.

Capen CC, Beamer WG, Tennent BJ, Stitzel KA (1995): Mechanisms of hormone-mediated carcinogenesis of the ovary in mice. *Mutat Res* 333:143–151.

Connolly DC, Bao R, Nikitin AY, Stephens KC, Poole TW, Hua X, Harris SS, Vanderhyden BC, Hamilton TC (2003): Female mice chimeric for expression of the simian virus 40 TAg under control of the MISIIR promoter develop epithelial ovarian cancer. *Cancer Res* 63:1389–1397.

Danilovich N, Roy I, Sairam MR (2001): Ovarian pathology and high incidence of sex cord tumors in follitropin receptor knockout (FORKO) mice. *Endocrinology* 142:3673–3684.

Davies BR, Auersperg N, Worsley SD, Ponder BA (1998): Transfection of rat ovarian surface epithelium with erb-B2/neu induces transformed phenotypes in vitro and the tumorigenic phenotype in vivo. *Am J Pathol* 152:297–306.

di Clemente N, Wilson C, Faure E, Boussin L, Carmillo P, Tizard R, Picard JY, Vigier B, Josso N, Cate R (1994): Cloning, expression, and alternative splicing of the receptor for anti-Mullerian hormone. *Mol Endocrinol* 8:1006–1020.

Drapkin R, Hecht J (2002): The origins of ovarian cancer: hurdles and progress. *Women's Oncol Rev* 2: 261–268.

Dubeau L, Velicescu M, Sherrod AE, Schreiber G, Holt G (1990): Culture of human fetal ovarian epithelium in a chemically-defined, serum-free medium: A model for ovarian carcinogenesis. *Anticancer Res* 10:1233–1240.

Duncan M, Cummings L, Chada K (1993): Germ cell deficient (gcd) mouse as a model of premature ovarian failure. *Biol Reprod* 49:221–227.

Duncan MK, Chada KK (1993): Incidence of tubulostromal adenoma of the ovary in aged germ cell-deficient mice. *J Comp Pathol* 109:13–19.

Dutertre M, Gouedard L, Xavier F, Long WQ, di Clemente N, Picard JY, Rey R (2001): Ovarian granulosa cell tumors express a functional membrane receptor for anti-Mullerian hormone in transgenic mice. *Endocrinology* 142:4040–4046.

Dutertre M, Rey R, Porteu A, Josso N, Picard JY (1997): A mouse Sertoli cell line expressing anti-Mullerian hormone and its type II receptor. *Mol Cell Endocrinol* 136:57–65.

Fafalios MK, Olander EA, Melhem MF, Chaillet JR (1996): Ovarian teratomas associated with the insertion of an imprinted transgene. *Mamm Genome* 7:188–193.

Fathalla MF (1971): Incessant ovulation—a factor in ovarian neoplasia? *Lancet* 2:163.

Federspiel MJ, Swing DA, Eagleson B, Reid SW, Hughes SH (1996): Expression of transduced genes in mice generated by infecting blastocysts with avian leukosis virus-based retroviral vectors. *Proc Natl Acad Sci USA* 93:4931–4936.

Feeley KM, Wells M (2001): Precursor lesions of ovarian epithelial malignancy. *Histopathology* 38:87–95.

Flesken-Nikitin A, Choi KC, Eng JP, Shmidt EN, Nikitin AY (2003): Induction of carcinogenesis by concurrent inactivation of p53 and Rb1 in the mouse ovarian surface epithelium. *Cancer Res* 63:3459–3463.

Fox H (1993): Pathology of early malignant change in the ovary. *Int J Gynecol Pathol* 12:153–155.

Franceschi S, Parazzini F, Negri E, Booth M, La Vecchia C, Beral V, Tzonou A, Trichopoulos D (1991): Pooled analysis of 3 European case-control studies of epithelial ovarian cancer: III. Oral contraceptive use. *Int J Cancer* 49:61–65.

Gallion HH, Bast RC Jr (1993): National Cancer Institute Conference on Investigational Strategies for Detection and Intervention in Early Ovarian Cancer. *Cancer Res* 53:3839–3842.

Gallion HH, Powell DE, Smith LW, Morrow JK, Martin AW, van Nagell JR, Donaldson ES (1990): Chromosome abnormalities in human epithelial ovarian malignancies. *Gynecol Oncol* 38:473–477.

Ghahremani M, Foghi A, Dorrington JH (1999): Etiology of ovarian cancer: A proposed mechanism. *Med Hypotheses* 52:23–26.

Ginsburg M, Snow MH, McLaren A (1990): Primordial germ cells in the mouse embryo during gastrulation. *Development* 110:521–528.

Gray JW, Suzuki S, Kuo WL, Polikoff D, Deavers M, Smith-McCune K, Berchuck A, Pinkel D, Albertson D, Mills GB (2003): Specific keynote: Genome copy number abnormalities in ovarian cancer. *Gynecol Oncol* 88:S16–21; discussion S22–14.

Gustafson ML, Lee MM, Scully RE, Moncure AC, Hirakawa T, Goodman A, Muntz HG, Donahoe PK, MacLaughlin DT, Fuller AF Jr (1992): Mullerian inhibiting substance as a marker for ovarian sex-cord tumor. *N Engl J Med* 326:466–471.

Hamilton TC, Young RC, Ozols RF (1984): Experimental model systems of ovarian cancer: Applications to the design and evaluation of new treatment approaches. *Semin Oncol* 11:285–298.

Helzlsouer KJ, Bush TL, Alberg AJ, Bass KM, Zacur H, Comstock GW (1993): Prospective study of serum CA-125 levels as markers of ovarian cancer. *JAMA* 269:1123–1126.

Hoffman AG, Burghardt RC, Tilley R, Auersperg N (1993): An in vitro model of ovarian epithelial carcinogenesis: Changes in cell-cell communication and adhesion occurring during neoplastic progression. *Int J Cancer* 54:828–838.

Holschneider CH, Berek JS (2000): Ovarian cancer: Epidemiology, biology, and prognostic factors. *Semin Surg Oncol* 19:3–10.

Hough CD, Cho KR, Zonderman AB, Schwartz DR, Morin PJ (2001): Coordinately up-regulated genes in ovarian cancer. *Cancer Res* 61:3869–3876.

Hsu SY, Lai RJ, Finegold M, Hsueh AJ (1996): Targeted overexpression of Bcl-2 in ovaries of transgenic mice leads to decreased follicle apoptosis, enhanced folliculogenesis, and increased germ cell tumorigenesis. *Endocrinology* 137:4837–4843.

Imbeaud S, Faure E, Lamarre I, Mattei MG, di Clemente N, Tizard R, Carre-Eusebe D, Belville C, Tragethon L, Tonkin C, Nelson J, McAuliffe M, Bidart JM, Lababidi A, Josso N, Cate RL, Picard JY (1995): Insensitivity to anti-mullerian hormone due to a mutation in the human anti-mullerian hormone receptor. *Nat Genet* 11:382–388.

Ismail RS, Baldwin RL, Fang J, Browning D, Karlan BY, Gasson JC, Chang DD (2000): Differential gene expression between normal and tumor-derived ovarian epithelial cells. *Cancer Res* 60:6744–6749.

Jemal A, Thomas A, Murray T, Thun M (2002): Cancer statistics, 2002. *CA Cancer J Clin* 52:23–47.

Kananen K, Markkula M, Rainio E, Su JG, Hsueh AJ, Huhtaniemi IT (1995): Gonadal tumorigenesis in transgenic mice bearing the mouse inhibin alpha-subunit promoter/simian virus T-antigen fusion gene: Characterization of ovarian tumors and establishment of gonadotropin-responsive granulosa cell lines. *Mol Endocrinol* 9:616–627.

Kananen K, Rilianawati, Paukku T, Markkula M, Rainio EM, Huhtanemi I (1997): Suppression of gonadotropins inhibits gonadal tumorigenesis in mice transgenic for the mouse inhibin alpha-subunit promoter/simian virus 40 T-antigen fusion gene. *Endocrinology* 138:3521–3531.

Karl J, Capel B (1998): Sertoli cells of the mouse testis originate from the coelomic epithelium. *Dev Biol* 203:323–333.

Kendall SK, Samuelson LC, Saunders TL, Wood RI, Camper SA (1995): Targeted disruption of the pituitary glycoprotein hormone alpha-subunit produces hypogonadal and hypothyroid mice. *Genes Dev* 9:2007–2019.

Keri RA, Lozada KL, Abdul-Karim FW, Nadeau JH, Nilson JH (2000): Luteinizing hormone induction of ovarian tumors: Oligogenic differences between mouse strains dictates tumor disposition. *Proc Natl Acad Sci USA* 97:383–387.

Kim JH, Seibel MM, MacLaughlin DT, Donahoe PK, Ransil BJ, Hametz PA, Richards CJ (1992): The inhibitory effects of mullerian-inhibiting substance on epidermal growth factor induced proliferation and progesterone production of human granulosa-luteal cells. *J Clin Endocrinol Metab* 75:911–917.

Kumar TR, Palapattu G, Wang P, Woodruff TK, Boime I, Byrne MC, Matzuk MM (1999): Transgenic models to study gonadotropin function: The role of follicle-stimulating hormone in gonadal growth and tumorigenesis. *Mol Endocrinol* 13:851–865.

Kumar TR, Wang Y, Matzuk MM (1996): Gonadotropins are essential modifier factors for gonadal tumor development in inhibin-deficient mice. *Endocrinology* 137:4210–4216.

Liu J, Yang G, Thompson-Lanza JA, Glassman A, Hayes K, Patterson A, Marquez RT, Auersperg N, Yu Y, Hahn WC, Mills GB, Bast RC Jr. (2004): A genetically defined model for human ovarian cancer. *Cancer Res* 64:1655–1663.

Maines-Bandiera SL, Kruk PA, Auersperg N (1992): Simian virus 40-transformed human ovarian surface epithelial cells escape normal growth controls but retain morphogenetic responses to extracellular matrix. *Am J Obstet Gynecol* 167:729–735.

Masiakos PT, MacLaughlin DT, Maheswaran S, Teixeira J, Fuller AF Jr, Shah PC, Kehas DJ, Kenneally MK, Dombkowski DM, Ha TU, Preffer FI, Donahoe PK (1999): Human ovarian cancer, cell lines, and primary ascites cells express the human Mullerian inhibiting substance (MIS) type II receptor, bind, and are responsive to MIS. *Clin Cancer Res* 5:3488–3499.

Matei D, Graeber TG, Baldwin RL, Karlan BY, Rao J, Chang DD (2002): Gene expression in epithelial ovarian carcinoma. *Oncogene* 21:6289–6298.

Matzuk MM, Finegold MJ, Su JG, Hsueh AJ, Bradley A (1992): Alpha-inhibin is a tumour-suppressor gene with gonadal specificity in mice. *Nature* 360:313–319.

Matzuk MM, Kumar TR, Shou W, Coerver KA, Lau AL, Behringer RR, Finegold MJ (1996): Transgenic models to study the roles of inhibins and activins in reproduction, oncogenesis, and development. *Recent Prog Horm Res* 51:123–154; discussion 155–127.

McLaren A (2000): Germ and somatic cell lineages in the developing gonad. *Mol Cell Endocrinol* 163:3–9.

Miki Y, Swensen J, Shattuck-Eidens D, Futreal PA, Harshman K, Tavtigian S, Liu Q, Cochran C, Bennett LM, Ding W, Bell R, Rosenthal J, Hussey C, Tran T, McClure M, Frue C, Hattier T, Phelps R, Haugen-Strano A, Katcher H, Yakumo K, Gholami Z, Shaffer D, Stone S, Bayer S, Wray C, Bogden R, Dayananth P, Ward J, Tonin P, Narod S, Bristow PK, Norris PH, Helvering L, Morrison P, Rosteck P, Lai M, Barrett JC, Lewis C, Neuhausen S, Cannon-Albright L, Goldgar D, Wiseman R, Kamb A, Skolnick MH (1994): A strong candidate for the breast and ovarian cancer susceptibility gene BRCA1. *Science* 266:66–71.

Mok SC, Chao J, Skates S, Wong K, Yiu GK, Muto MG, Berkowitz RS, Cramer DW (2001): Prostasin, a potential serum marker for ovarian cancer: Identification through microarray technology. *J Natl Cancer Inst* 93:1458–1464.

Naora H, Montz FJ, Chai CY, Roden RB (2001): Aberrant expression of homeobox gene HOXA7 is associated with mullerian-like differentiation of epithelial ovarian tumors and the generation of a specific autologous antibody response. *Proc Natl Acad Sci USA* 98:15209–15214.

Nicosia SV, Johnson JH, Streibel EJ (1984): Isolation and ultrastructure of rabbit ovarian mesothelium (surface epithelium). *Int J Gynecol Pathol* 3:348–360.

Nicosia SV, Narconis RJ, Saunders BO (1989): Regulation and temporal sequence of surface epithelium morphogenesis in the postovulatory rabbit ovary. *Prog Clin Biol Res* 296:111–119.

Nilson JH, Abbud RA, Keri RA, Quirk CC (2000): Chronic hypersecretion of luteinizing hormone in transgenic mice disrupts both ovarian and pituitary function, with some effects modified by the genetic background. *Recent Prog Horm Res* 55:69–89; discussion 89–91.

Obata K, Morland SJ, Watson RH, Hitchcock A, Chenevix-Trench G, Thomas EJ, Campbell IG (1998): Frequent PTEN/MMAC mutations in endometrioid but not serous or mucinous epithelial ovarian tumors. *Cancer Res* 58:2095–2097.

Ono K, Tanaka T, Tsunoda T, Kitahara O, Kihara C, Okamoto A, Ochiai K, Takagi T, Nakamura Y (2000): Identification by cDNA microarray of genes involved in ovarian carcinogenesis. *Cancer Res* 60:5007–5011.

Orsulic S (2002): An RCAS-TVA-based approach to designer mouse models. *Mamm Genome* 13:543–547.

Orsulic S, Li Y, Soslow RA, Vitale-Cross LA, Gutkind JS, Varmus HE (2002): Induction of ovarian cancer by defined multiple genetic changes in a mouse model system. *Cancer Cell* 1:53–62.

Pejovic T, Heim S, Mandahl N, Baldetorp B, Elmfors B, Floderus UM, Furgyik S, Helm G, Himmelmann A, Willen H, Mitelman F (1992): Chromosome aberrations in 35 primary ovarian carcinomas. *Genes Chromosom Cancer* 4:58–68.

Perez RP, Godwin AK, Hamilton TC, Ozols RF (1991): Ovarian cancer biology. *Semin Oncol* 18:186–204.

Peschon JJ, Behringer RR, Cate RL, Harwood KA, Idzerda RL, Brinster RL, Palmiter RD (1992): Directed expression of an oncogene to Sertoli cells in transgenic mice using mullerian inhibiting substance regulatory sequences. *Mol Endocrinol* 6:1403–1411.

Petricoin EF, Ardekani AM, Hitt BA, Levine PJ, Fusaro VA, Steinberg SM, Mills GB, Simone C, Fishman DA, Kohn EC, Liotta LA (2002): Use of proteomic patterns in serum to identify ovarian cancer. *Lancet* 359:572–577.

Pierce JG, Parsons TF (1981): Glycoprotein hormones: Structure and function. *Annu Rev Biochem* 50:465–495.

Pieretti M, Cavalieri C, Conway PS, Gallion HH, Powell DE, Turker MS (1995): Genetic alterations distinguish different types of ovarian tumors. *Int J Cancer* 64:434–440.

Purdie DM, Bain CJ, Siskind V, Webb PM, Green AC (2003): Ovulation and risk of epithelial ovarian cancer. *Int J Cancer* 104:228–232.

Randall TC, Bell KA, Rebane BA, Rubin SC, Boyd J (1998): Germline mutations of the BRCA1 and BRCA2 genes in a breast and ovarian cancer patient. *Gynecol Oncol* 70:432–434.

Risma KA, Clay CM, Nett TM, Wagner T, Yun J, Nilson JH (1995): Targeted overexpression of luteinizing hormone in transgenic mice leads to infertility, polycystic ovaries, and ovarian tumors. *Proc Natl Acad Sci USA* 92: 1322–1326.

Roby KF, Taylor CC, Sweetwood JP, Cheng Y, Pace JL, Tawfik O, Persons DL, Smith PG, Terranova PF (2000): Development of a syngeneic mouse model for events related to ovarian cancer. *Carcinogenesis* 21:585–591.

Sawiris GP, Sherman-Baust CA, Becker KG, Cheadle C, Teichberg D, Morin PJ (2002): Development of a highly specialized cDNA array for the study and diagnosis of epithelial ovarian cancer. *Cancer Res* 62:2923–2928.

Schummer M, Ng WV, Bumgarner RE, Nelson PS, Schummer B, Bednarski DW, Hassell L, Baldwin RL, Karlan BY, Hood L (1999): Comparative hybridization of an array of 21,500 ovarian cDNAs for the discovery of genes overexpressed in ovarian carcinomas. *Gene* 238:375–385.

Schwartz DR, Kardia SL, Shedden KA, Kuick R, Michailidis G, Taylor JM, Misek DE, Wu R, Zhai Y, Darrah DM, Reed H, Ellenson LH, Giordano TJ, Fearon ER, Hanash SM, Cho KR (2002): Gene expression in ovarian cancer reflects both morphology and biological behavior, distinguishing clear cell from other poor-prognosis ovarian carcinomas. *Cancer Res* 62:4722–4729.

Schwartz PE (2002): Current diagnosis and treatment modalities for ovarian cancer. *Cancer Treat Res* 107:99–118.

Schwartz PE, Chambers JT, Taylor KJ (1995): Early detection and screening for ovarian cancer. *J Cell Biochem Suppl* 23:233–237.

Schwartz PE, Taylor KJ (1995): Is early detection of ovarian cancer possible? *Ann Med* 27:519–528.

Scully R, Livingston DM (2000): In search of the tumour-suppressor functions of BRCA1 and BRCA2. *Nature* 408: 429–432.

Scully RE (1977): Ovarian tumors. A review. *Am J Pathol* 87:686–720.

Segev DL, Ha TU, Tran TT, Kenneally M, Harkin P, Jung M, MacLaughlin DT, Donahoe PK, Maheswaran S (2000): Mullerian inhibiting substance inhibits breast cancer cell growth through an NFkappa B-mediated pathway. *J Biol Chem* 275:28371–28379.

Shridhar V, Lee J, Pandita A, Iturria S, Avula R, Staub J, Morrissey M, Calhoun E, Sen A, Kalli K, Keeney G, Roche P, Cliby W, Lu K, Schmandt R, Mills GB, Bast RC, Jr., James CD, Couch FJ, Hartmann LC, Lillie J, Smith DI (2001): Genetic analysis of early-versus late-stage ovarian tumors. *Cancer Res* 61:5895–5904.

Stephen AE, Masiakos PT, Segev DL, Vacanti JP, Donahoe PK, MacLaughlin DT (2001): Tissue-engineered cells producing complex recombinant proteins inhibit ovarian cancer in vivo. *Proc Natl Acad Sci USA* 98:3214–3219.

Takahashi M, Hayashi M, Manganaro TF, Donahoe PK (1986): The ontogeny of mullerian inhibiting substance in granulosa cells of the bovine ovarian follicle. *Biol Reprod* 35:447–453.

Tapper J, Kettunen E, El-Rifai W, Seppala M, Andersson LC, Knuutila S (2001): Changes in gene expression during progression of ovarian carcinoma. *Cancer Genet Cytogenet* 128:1–6.

Tennent BJ, Beamer WG (1986): Ovarian tumors not induced by irradiation and gonadotropins in hypogonadal (hpg) mice. *Biol Reprod* 34:751–760.

Testa JR, Getts LA, Salazar H, Liu Z, Handel LM, Godwin AK, Hamilton TC (1994): Spontaneous transformation of rat ovarian surface epithelial cells results in well to poorly differentiated tumors with a parallel range of cytogenetic complexity. *Cancer Res* 54:2778–2784.

Tonin PN, Hudson TJ, Rodier F, Bossolasco M, Lee PD, Novak J, Manderson EN, Provencher D, Mes-Masson AM (2001): Microarray analysis of gene expression mirrors the biology of an ovarian cancer model. *Oncogene* 20:6617–6626.

Wang K, Gan L, Jeffery E, Gayle M, Gown AM, Skelly M, Nelson PS, Ng WV, Schummer M, Hood L, Mulligan J (1999): Monitoring gene expression profile changes in ovarian carcinomas using cDNA microarray. *Gene* 229:101–108.

Welsh JB, Zarrinkar PP, Sapinoso LM, Kern SG, Behling CA, Monk BJ, Lockhart DJ, Burger RA, Hampton GM (2001): Analysis of gene expression profiles in normal and neoplastic ovarian tissue samples identifies candidate molecular markers of epithelial ovarian cancer. *Proc Natl Acad Sci USA* 98:1176–1181.

Werness BA, Parvatiyar P, Ramus SJ, Whittemore AS, Garlinghouse-Jones K, Oakley-Girvan I, DiCioccio RA, Wiest J, Tsukada Y, Ponder BA, Piver MS (2000): Ovarian carcinoma in situ with germline BRCA1 mutation and loss of heterozygosity at BRCA1 and TP53. *J Natl Cancer Inst* 92:1088–1091.

Whang-Peng J, Knutsen T, Douglass EC, Chu E, Ozols RF, Hogan WM, Young RC (1984): Cytogenetic studies in ovarian cancer. *Cancer Genet Cytogenet* 11:91–106.

Wooster R, Bignell G, Lancaster J, Swift S, Seal S, Mangion J, Collins N, Gregory S, Gumbs C, Micklem G (1995): Identification of the breast cancer susceptibility gene BRCA2. *Nature* 378:789–792.

Wu R, Zhai Y, Fearon ER, Cho KR (2001): Diverse mechanisms of beta-catenin deregulation in ovarian endometrioid adenocarcinomas. *Cancer Res* 61:8247–8255.

Xu X, Qiao W, Linke SP, Cao L, Li WM, Furth PA, Harris CC, Deng CX (2001): Genetic interactions between tumor suppressors Brca1 and p53 in apoptosis, cell cycle and tumorigenesis. *Nat Genet* 28:266–271.

Zhai Y, Wu R, Schwartz DR, Darrah D, Reed H, Kolligs FT, Nieman MT, Fearon ER, Cho KR (2002): Role of beta-catenin/T-cell factor-regulated genes in ovarian endometrioid adenocarcinomas. *Am J Pathol* 160:1229–1238.

Zhang L, Yang N, Garcia JR, Mohamed A, Benencia F, Rubin SC, Allman D, Coukos G (2002): Generation of a syngeneic mouse model to study the effects of vascular endothelial growth factor in ovarian carcinoma. *Am J Pathol* 161:2295–2309.

Zweemer RP, Shaw PA, Verheijen RM, Ryan A, Berchuck A, Ponder BA, Risch H, McLaughlin JR, Narod SA, Menko FH, Kenemans P, Jacobs IJ (1999a): Accumulation of p53 protein is frequent in ovarian cancers associated with BRCA1 and BRCA2 germline mutations. *J Clin Pathol* 52:372–375.

Zweemer RP, Verheijen RH, Menko FH, Gille JJ, van Diest PJ, Coebergh JW, Shaw PA, Jacobs IJ, Kenemans P (1999b): Differences between hereditary and sporadic ovarian cancer. *Eur J Obstet Gynecol Reprod Biol* 82:151–153.

11

PERIPHERAL NERVOUS SYSTEM TUMORS

DAVID H. GUTMANN
Department of Neurology, Washington University School of Medicine, St. Louis, Missouri
ARIE PERRY
Division of Neuropathology, Washington University School of Medicine, St. Louis, Missouri
RESHMA RANGWALA AND LARRY S. SHERMAN
Division of Neuroscience, Oregon National Primate Research Center, Oregon Health and Science University, Beaverton, Oregon

Tumors of the peripheral nervous system arise from Schwann cells and include neurofibromas, schwannomas, and malignant peripheral nerve sheath tumors (MPNSTs). Individuals with the inherited tumor predisposition syndromes, neurofibromatosis 1 (NF1) and neurofibromatosis 2 (NF2), develop neurofibromas and schwannomas, respectively, suggesting that the NF1 and NF2 tumor suppressor genes are critical regulators of Schwann cell growth. Mouse models of these tumors have focused on the Nf1 and Nf2 genes, while MPNSTs in mice and humans require additional cooperating genetic alterations.

INTRODUCTION

The most common tumors affecting the peripheral nervous system (PNS) are schwannomas and neurofibromas, benign tumors that arise from nerve sheaths (Scheithauer et al., 1999). Less common tumors include perineuroma, a benign tumor composed entirely of perineurial cells, peripheral nerve sheath myxomas, neurothekeoma, and granule cell tumors. These latter tumors have not been modeled in mice to date and will not be discussed further in this chapter. Here, we will focus on a number of mouse models of schwannomas and neurofibromas as well as tumors that mimic many of the characteristics of a rare, malignant PNS tumor, the malignant peripheral nerve sheath tumor (MPNST). These mouse models reveal numerous putative molecular mechanisms underlying the initiation and growth of these tumors and have the potential to serve as preclinical models for the evaluation of putative antitumor therapies.

Peripheral Nerves Are Organized into Fascicles Comprised of Axons and Schwann Cells

The tumors discussed in this chapter are believed to be derived, at least in part, from transformed Schwann cells. Peripheral nerves, including spinal and cranial nerves, are typically organized into fascicles of axons associated with Schwann cells. The Schwann cells can be nonmyelinating cells that interact with small-diameter axons or myelinating cells that wrap around large-diameter axons (Fig. 11.1). The myelin sheaths of these larger diameter axons promote increased conduction velocities through saltatory conduction.

Each fascicle within a peripheral nerve is surrounded by a perineurium, comprised of concentric layers of flattened fibroblast-like cells (*perineurial fibroblasts*) that have basal laminae between each layer. Within the perineurium is the endoneurium, which is composed of elongated fibroblasts surrounded by reticular and collagenous fibers that surround individual nerve fibers and contain a plexus of capillaries. A sheath of connective

Mouse Models of Human Cancer, edited by Eric C. Holland
ISBN 0-471-44460-X

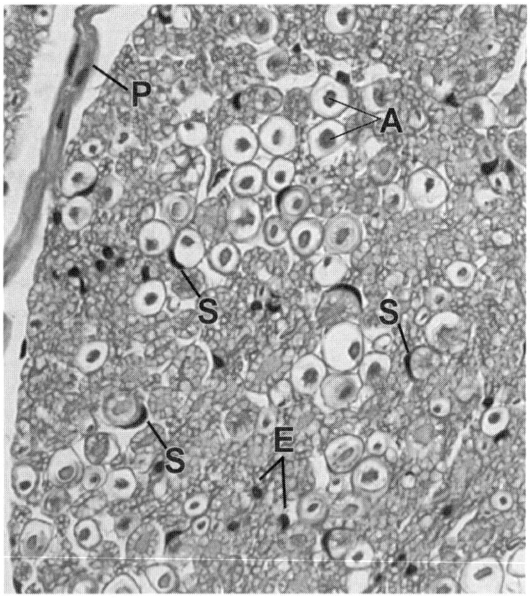

Figure 11.1. Anatomy of the peripheral nerve. Portion of a single fascicle from an adult mouse sciatic nerve stained with hematoxylin and eosin (H&E). Both myelinating (with myelin evident in white areas surrounding axons) and nonmyelinating Schwann cells (S; examples show nuclei) envelop axons (A) coursing through a nerve fascicle. The fascicle is separated from an adjacent fascicle (in the upper left corner) by perineurium (P; detached in this section) consisting of perineurial fibroblasts, whose nuclei are flattened and elongated. Within the fascicle, individual nerve fibers are separated by endoneurium that contains endoneurial fibroblasts (E). Multiple fascicles in larger nerves are surrounded by an epineurium (not shown). Transformed Schwann cells therefore interact with numerous other cell types and extracellular matrix as they form tumors.

tissue, the epineurium, envelops collections of fascicles in large nerves and is composed of fibroblasts, collagenous fibers, and a few elastic fibers. Blood vessels enter the epineurium and branch into smaller vessels that supply the perineurium and endoneurium. Thus, peripheral nerves are comprised of many cell types and different extracellular environments. Tumors derived from Schwann cells can therefore impact on the functions of these different cells types and, consequently, each of these cells and their associated extracellular matrix have the potential to influence tumor progression.

TUMORS OF PERIPHERAL NERVOUS SYSTEM

Schwannoma

Schwannomas are benign low-grade tumors classified by the World Health Organization (WHO) criteria as grade I (Kleihues and Cavenee, 2000). As the name implies, these tumors are comprised entirely of neoplastic Schwann cells. They may arise anywhere within the peripheral nervous system and are associated with peripheral or cranial nerves (Ferner and O'Doherty, 2002). The most commonly involved sites include the vestibular portion of cranial nerve VIII (vestibular schwannoma or *acoustic neuroma*) and paraspinal nerve roots. When they occur as solitary tumors, the peak incidence is between the fourth and sixth decade of life. In contrast, multiple schwannomas are frequently encountered in the context of the inherited tumor predisposition syndrome neurofibromatosis 2 (NF2) and develop one to two decades earlier.

Macroscopically, these tumors typically arise eccentrically to the involved nerve as tightly encapsulated globoid masses composed of spindle-shaped neoplastic Schwann cells (Fig. 11.2). Microscopically, schwannomas may contain areas of tightly compacted spindle-shaped cells (Antoni A pattern) with occasional nuclear palisading (Verocay bodies) or loosely packed, less cellular areas with cystic degeneration, vascular hyalinization, and/or hemorrhage (Antoni B pattern). Immunostaining reveals diffuse S100 protein and collagen IV (i.e., pericellular basement membrane) expression, with occasional glial fibrillary acidic protein (GFAP) immunoreactivity. In contrast to neurofibroma, immunostains for neurofilament protein often highlight compressed axons from the parent nerve at the periphery, rather than within the substance of the tumor.

Neurofibroma

Neurofibromas are also benign low-grade tumors classified by the WHO as grade I (Kleihues and Cavenee, 2000). Solitary neurofibromas are common and affect all ages and both genders equivalently. When multiple neurofibromas are present, the diagnosis of neurofibromatosis 1 (NF1) should be considered. In contrast to schwannomas, neurofibromas are composed not only of neoplastic Schwann cells but also of multiple other cell types. They are infiltrative intraneural and sometimes extraneural tumors that contain entrapped axons from the parent nerve. Two major types of neurofibromas have been described, including those that are more circumscribed and associated with only one nerve fascicle (discrete neurofibroma, either deep intraneural or cutaneous) (Woodruff, 1999) as well as neurofibromas that involve multiple nerve fascicles (plexiform neurofibroma) (Korf, 1999; Packer et al., 2002). Plexiform neurofibromas, unlike their discrete counterparts, may encompass large segments of the body, cause bony destruction, and transform into malignant and metastatic cancers (MPNSTs; see below) (Waggoner et al., 2000). Less commonly, variants may infiltrate large regions of skin and other soft-tissue elements and are thus referred to as *diffuse neurofibromas*.

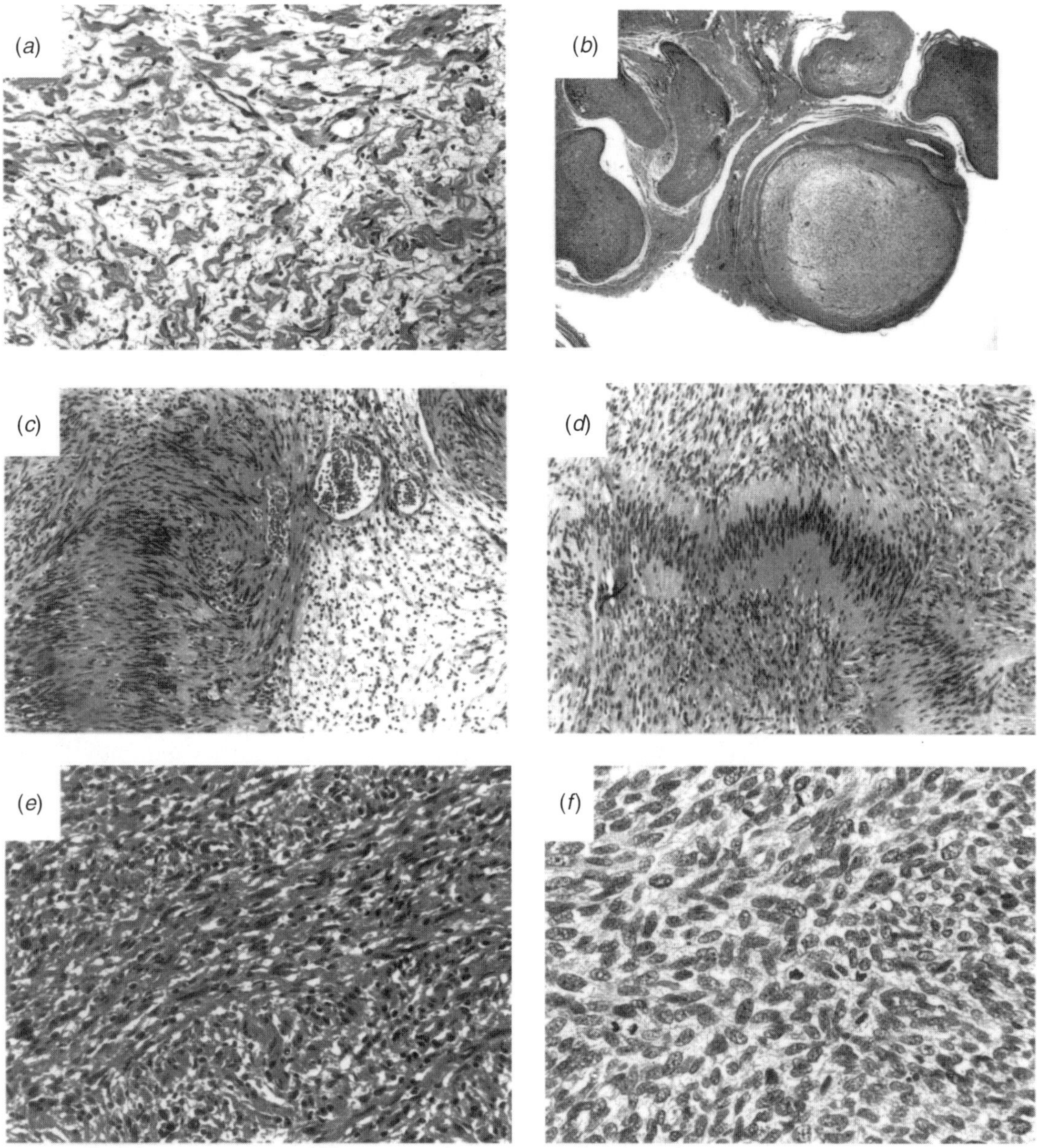

Figure 11.2. Human peripheral nerve sheath tumors. (*a*) Neurofibroma (200×) showing typical loose, myxoid appearance with thin wavy cells, small spindled nuclei, lack of mitotic activity, and patchy collagen deposition. (*b*) Plexiform neurofibroma (20×) showing enlargement of multiple adjacent nerve fascicles, thus imparting a "bag-of-worms" appearance. (*c*) Schwannoma (200×). Verocay bodies consist of palisaded (i.e., lined-up) nuclei, typically in regions of Antoni A tissue. (*d*) Schwannoma (200×) with classic biphasic appearance. The dense Antoni A tissue is on the left, whereas the loose, microcystic Antoni B tissue is on the right. Cases with predominance of the latter are sometimes difficult to distinguish from neurofibroma without special studies. (*e*) Low-grade MPNST. This lesion arose from a plexiform neurofibroma in a patient with NF1. Though the minimal criteria for malignant transformation are not clearly established, the presence of increased cell size, hypercellularity, and mitotic activity were used to establish the diagnosis in this case. (*f*) High-grade MPNST (400×). The majority of MPNSTs are high-grade sarcomas with necrosis and high mitotic index. In the absence of a preexisting neurofibroma, NF1, or an intraneural origin, they are often hard to distinguish from high-grade fibrosarcomas and other sarcoma subtypes without special studies.

Microscopically, neurofibromas contain neoplastic Schwann cells as well as nonneoplastic Schwann cells, fibroblasts, perineurial-like cells and entrapped neural elements, such as neurofilament-immunopositive axons (Fig. 11.2). In addition, mast cells are frequently encountered. The cellular elements are embedded in a rich matrix of collagen fibers and mucosubstances. Immunostaining reveals a patchy S100-positive population of neoplastic Schwann cells but no reactivity with epithelial membrane antigen (EMA).

Malignant Peripheral Nerve Sheath Tumors

MPNSTs are typically high-grade, often fatal, malignancies classified by the WHO as grade III/IV tumors (Ferner et al., 2002; Kleihues and Cavenee, 2000). These tumors have also been termed neurogenic sarcoma, neurofibrosarcoma, and malignant schwannoma. Fifty percent of MPNST are associated with NF1 and an additional 10% are associated with prior radiation. Compared to the benign Schwann cell tumors, MPNSTs are uncommon neoplasms. Nearly 70% arise from preexisting plexiform neurofibromas. Typically, MPNSTs develop in adults during the third through sixth decades and are slightly more common in women. In the context of NF1, these tumors develop one to two decades earlier (Evans et al., 2002; King et al., 2000).

MPNSTs frequently involve the buttocks, thigh, brachial plexus, and paraspinal nerves, with the sciatic nerve representing the most commonly affected peripheral nerve. Cranial nerves are not usually involved, although rare MPNSTs may arise in the fifth or eighth cranial nerve. MPNSTs are highly aggressive, frequently metastasizing cancers and often have associated intratumoral hemorrhage and necrosis.

Microscopically, MPNSTs are composed of tightly packed hyperchromatic spindle-shaped cells with frequent mitotic figures (Fig. 11.2). Minimal criteria for malignant transformation of a plexiform neurofibroma have yet to be established, but increased cellularity, cell size, and mitotic activity are generally regarded as most important in the diagnosis of low-grade MPNST. However, the majority represent high-grade sarcomas that are clearly malignant, and therefore the more difficult differential diagnosis is often with other poorly differentiated spindle cell sarcomas, particularly fibrosarcoma. Fifteen percent of tumors will also exhibit unusual histologic features suggestive of divergent cellular differentiation. Variants have been reported with evidence of muscle differentiation, epithelioid features, glandular elements, and mesenchymal differentiation ("Triton" tumor). Roughly half of MPNSTs are S100 immunoreactive, though it is often focal and/or weak. Therefore, it is sometimes difficult to prove the Schwann cell derivation on histology alone, and clinical features such as history of NF1, a preexisting plexiform neurofibroma, or a clear neural origin may be particularly useful in such cases. Ultrastructural studies are helpful for demonstrating Schwann cell derivation in a subset but, similar to immunohistochemistry, may yield nonspecific findings in poorly differentiated examples. There is some suggestion that genetic studies may also be useful in selected cases (see below).

MOLECULAR GENETICS OF PERIPHERAL NERVOUS SYSTEM TUMORS

Schwannomas

As would be predicted from the increased frequency in individuals affected with NF2, most schwannomas harbor inactivating *NF2* gene mutations. One hundred percent of NF2-associated and 60% to 80% of sporadic schwannomas exhibit bi-allelic inactivation of the *NF2* gene on chromosome 22q, associated with loss of expression of the *NF2* gene product, merlin or schwannomin (Gutmann et al., 1997; Stemmer-Rachamimov et al., 1997, 1998; Twist et al., 1994). Merlin is a 595-amino-acid protein that exhibits sequence similarity to members of the Protein 4.1 family of molecules. Protein 4.1 molecules link cell surface glycoproteins to the actin cytoskeleton and are involved in modulating cell motility (Gautreau et al., 2002; Sun et al., 2002). In keeping with merlin's inclusion in the Protein 4.1 family of molecules, merlin loss in *NF2*-deficient Schwann cells is associated with abnormalities in actin cytoskeleton-mediated processes (Gutmann et al., 1999). Similarly, merlin-deficient tumor cells are highly motile (McClatchey et al., 1998). Reexpression of wild-type merlin, but not mutant merlin molecules, results in abrogation of these abnormalities and suppression of both cell growth and motility (Morrison et al., 2001; Schulze et al., 2002).

Neurofibromas

Similar to the critical role of the *NF2* gene in schwannoma pathogenesis, loss of *NF1* gene expression is observed in all NF1-associated neurofibromas and many sporadic neurofibromas (Perry et al., 2001; Rutkowski et al., 2000). The *NF1* gene is located on chromosome 17q and codes for a 220–250-kDa protein, termed neurofibromin. Analysis of the predicted *NF1* amino sequence reveals sequence similarity between 400 residues in the central portion of neurofibromin and members of a family of proteins that negatively regulate the p21-RAS protooncogene (Cichowski and Jacks, 2001; Gutmann, 2001; Zhu and Parada, 2001). These GTPase-activating proteins (GAPs) function to inhibit RAS activity by accelerating the conversion of RAS from its active, GTP-bound conformation to an inactive GDP-bound form. Active GTP-bound RAS provides a strong mitogenic signal for a number of diverse cell types, such that loss of neurofibromin in tumor cells would be predicted to result in high levels of activated RAS and increased cell proliferation (Fig. 11.3).

Neurofibromas contain a variety of cellular elements, including Schwann cells, perineurial-like cells, fibroblasts, and mast cells. Elegant studies over the past several years have demonstrated that the neoplastic cell type is the Schwann cell. Loss of *NF1* gene expression has been identified in the neoplastic Schwann cell, but not the fibroblasts cultured from neurofibromas (Rutkowski et al., 2000). Moreover, NF1-derived Schwann cells lacking neurofibromin expression exhibit high levels of RAS-GTP, consistent with the notion that *NF1* inactivation in Schwann cells results in loss of RAS-GAP function and increased Schwann cell proliferation (Sherman et al., 2000).

Other molecules may also be involved in neurofibroma formation, including growth factors produced by nonneoplastic mast cells or fibroblasts. In this respect,

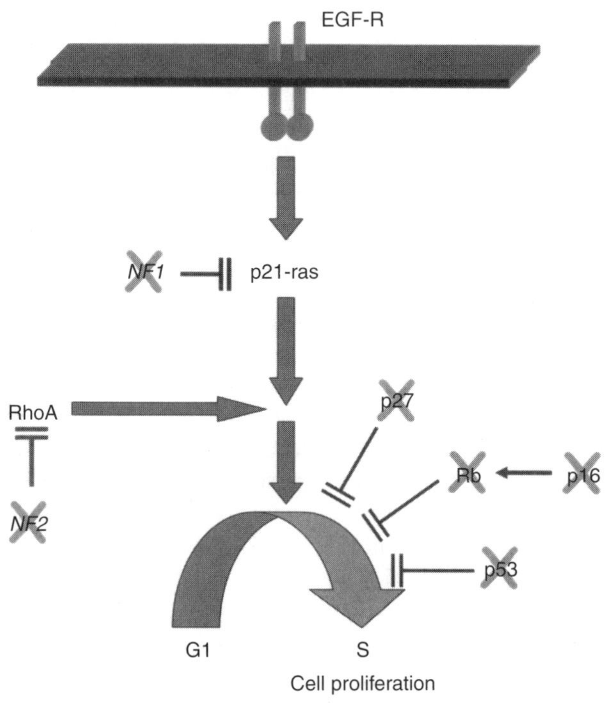

Figure 11.3. Molecular pathways involved in peripheral nerve sheath tumorigenesis. Multiple signaling pathways are affected by genetic changes in benign and malignant peripheral nerve sheath tumors. Benign neurofibromas involve inactivation of the *NF1* gene while *NF2* inactivation underlies the molecular pathogenesis of schwannomas. MPNSTs have been reported to exhibit amplification or overexpression of the EGFR as well as inactivation of the *CDKN2A/p16*, *p27*, and *TP53* tumor suppressor genes.

neurofibromas overexpress growth factors and receptors, such as epidermal growth factor receptor (EGFR) and vascular endothelial growth factor (Arbiser et al., 1998) that potentially modulate cell growth, cell invasion, and angiogenesis. Lastly, matrix metalloproteinase protein overexpression, important for tumor cell infiltration, has also been reported in neurofibromas (Muir, 1995).

Malignant Peripheral Nerve Sheath Tumors

MPNSTs likely arise from a precursor lesion (plexiform neurofibroma) that already harbors bi-allelic inactivation of *NF1* and loss of neurofibromin expression (Perry et al., 2002). Although one-third of MPNSTs do not have a clearly identified associated plexiform neurofibroma, all NF1-associated and most sporadic MPNSTs examined exhibit *NF1* inactivation and loss of neurofibromin expression. As suggested above, loss of neurofibromin expression alone is probably insufficient for MPNST formation, but *NF1* inactivation cooperates with other genetic changes to facilitate malignant transformation. Genetic studies of sporadic and NF1-associated MPNSTs have identified a number of these cooperating events, including *p53* gene inactivation (Drobnjak et al., 1994; Halling et al., 1996; Lothe et al., 2001), *p27* loss (Kourea et al., 1999a), *p16* inactivation (Kourea et al., 1999b; Nielsen et al., 1999; Perry et al., 2001), and *EGFR* overexpression (DeClue et al., 2000; Perry et al., 2001). Whereas *p16* inactivation is seen in both NF1-associated and sporadic MPNSTs, one study suggested that *p53* inactivation may be more common in sporadic MPNSTs (Birindelli et al., 2001). In comparison to other spindle cell sarcomas, *NF1* deletions and *p16* homozygous deletions are much more common in MPNSTs (Perry et al., 2002). Therefore, ancillary genetic testing may be useful in selected high-grade sarcomas where a definitive diagnosis cannot be reached by routine techniques. One interesting consistent genetic loss involves chromosome 22q, in the region of the *NF2* gene (Jhanwar et al., 1994); however, direct evidence of *NF2* inactivation or merlin loss has not been reported in MPNSTs.

MOUSE MODELS OF PERIPHERAL NERVOUS SYSTEM TUMORS

Chemically Induced Tumors

One of the earliest rodent models of malignant Schwann cell tumors came from studies of rats treated with *N*-ethyl-*N*-nitrosourea (ENU). Certain strains of rats, hamsters, and mice exposed to a single dose of ENU, either transplacentally or at early postnatal stages, develop malignant schwannomas (Nakamura et al., 1991; Nikitin et al., 1991; Perantoni et al., 1987). Interestingly, these tumors arise with the highest frequency in the trigeminal nerve. ENU-induced schwannomas have activating point mutations in the *neu/erbB2/HER-2* receptor tyrosine kinase (Nikitin et al., 1991; Perantoni et al., 1987). Although mutations in other genes, including N-*ras*, have also been identified in ENU-induced schwannomas (Perantoni et al., 1987), expressing a constitutively active *erbB2* gene in immature Schwann cells causes malignant transformation, suggesting that the ENU-induced *erbB2* point mutations are sufficient to transform Schwann cells (Sherman et al., 1999).

A problem with the ENU-induced malignant schwannomas is that they do not appear to model human disease. For one, such aggressive malignant schwannomas do not occur in humans. Furthermore, there have been no documented human Schwann cell tumors bearing *erbB2* mutations, although *erbB2* amplification is a characteristic of several human cancers, including mammary carcinomas (for review, see Yarden, 2001). This tumor model is therefore problematic for studies aimed at understanding the molecular and cellular processes underlying human schwannoma, neurofibroma, or MPNST tumorigenesis.

Models of Neurofibromas

As the majority of human Schwann cell tumors are linked to mutations in either the *NF1* or *NF2* genes, several groups have used a number of different approaches to disrupt these genes in mice (Table 11.1) (Gutmann and Giovannini, 2002). The first attempts along these lines involved mice with mutations in the *Nf1* gene. However, mice that are homozygous for a targeted *Nf1* allele die in utero between embryonic days 12.5 and 13.5 due to cardiac vessel defects (Brannan et al., 1994; Jacks et al., 1994). Mice carrying a single mutant *Nf1* allele develop some neoplasms associated with NF1 disease in humans, including pheochromocytomas and myeloid leukemia (Jacks et al., 1994). These animals also develop cognitive defects that may model some features of similar defects in NF1 patients (Silva et al., 1997). These animals do not, however, develop either neurofibromas or MPNSTs.

An alternative approach to testing whether loss of both *Nf1* alleles could induce neurofibromas was taken by Cichowski and co-workers (1999), who generated chimeric *Nf1* −/−;*Nf1* +/+ mice. Remarkably, a large proportion of the chimeras that survived developed multiple tumors (between 10 and 100) that resembled plexiform neurofibromas (Fig. 11.4). These tumors arose in peripheral nerves along the limbs, in dorsal root ganglia, along trigeminal ganglia, and in the tongue. However, it was unclear from these studies whether Schwann cells or other cell types were critical for initiating tumorigenesis.

To determine whether Schwann cell–specific loss of *Nf1* is sufficient to induce neurofibromas, Zhu and co-workers (2001, 2002) generated mice with Schwann cell–specific *Nf1* inactivation (Krox20-Cre; $Nf1^{flox/flox}$ mice) that were also heterozygous for a null *Nf1* mutation in all cells. In this way, it was possible to generate mice lacking *Nf1* expression in Schwann cells within an *Nf1* +/− cellular environment, similar to the human condition. These mice developed enlarged peripheral nerves characterized by Schwann cell hyperplasia and, occasionally, lesions that histologically resembled plexiform neurofibromas (Fig. 11.4; Zhu et al., 2002). Numerous mast cells were recruited to these tumors, consistent with the histopathology of human neurofibromas. Interestingly, mice with Schwann cell–specific loss of *Nf1* in a wild-type background also developed Schwann cell hyperplasia but not tumors (Zhu et al., 2002), suggesting that other cell types contribute to neurofibroma tumorigenesis.

MODELS OF MPNST

As discussed above, other genes in addition to *NF1* are involved in the molecular pathogenesis of MPNST in NF1 patients. To model this malignancy in mice, therefore, one might expect that it would be necessary to target both *Nf1* and other relevant tumor suppressor genes. This approach has proven to be fruitful in the case of mice carrying heterozygous mutations in both the *Nf1* and *p53* genes (Cichowski et al., 1999; Vogel et al., 1999). The *Nf1* and *p53* loci are closely linked

Table 11.1. Mouse Models of Peripheral Nervous System Tumors

Tumor Type	Genotype	Reference
Neurofibroma	Schwann cell–specific *Nf1* conditional knockout mice	Zhu et al., 2002
Plexiform neurofibroma	Chimeric *Nf1*−/− mice	Cichowski et al., 1999
Schwannoma	Schwann cell–specific *Nf2* conditional knockout mice	Giovannini et al., 1999
MPNST	*Nf1* +/−; *p53*+/− mice	Cichowski et al., 1999; Vogel et al., 1999

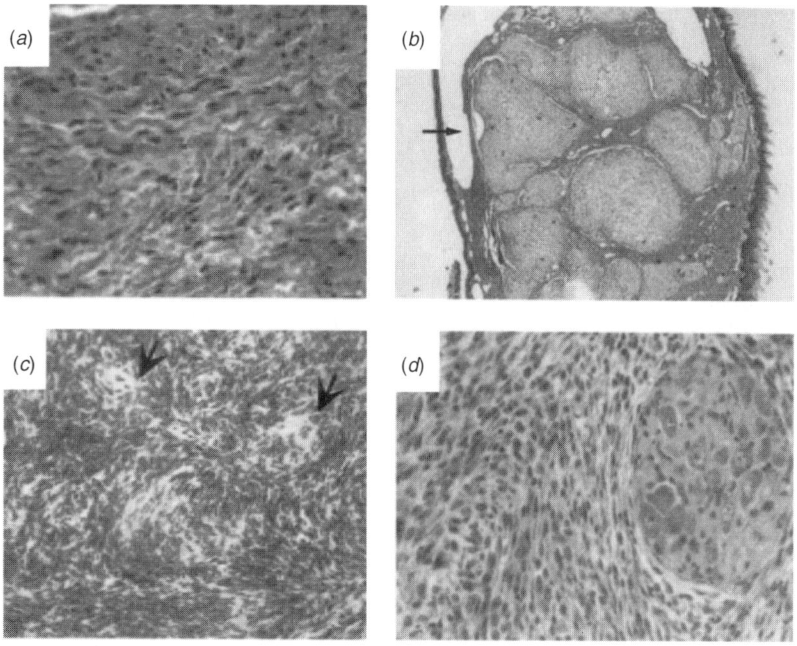

Figure 11.4. Genetically engineered mouse peripheral nerve tumors. (*a*) Representative neurofibroma exhibiting typical thin wavy cells with small spindle-shaped nuclei arose in an *Nf1* +/− mouse lacking *Nf1* expression in Schwann cells (Zhu et al., 2002). (*b*) A representative plexiform neurofibroma with multiple discrete nodular masses containing a loose matrix of collagen deposition arose in an *Nf1* +/−; *Nf1* +/+ chimeric mouse (Cichowski et al., 1999). (*c*) A representative schwannoma with Antoni A–type palisading (arrows) fibers arose in a Schwann cell–specific *Nf2* conditional knockout mouse (Giovannini et al., 2000). (*d*) A representative MPNST resembling a sarcoma and expressing the S100 Schwann cell marker arose in an *Nf1* +/−; *p53* +/− mouse (Cichowski et al., 1999). (Reproduced by permission of Cold Spring Harbor Laboratory Press and Science.).

on mouse chromosome 11. Animals bearing mutations in both genes on the same chromosome (in *cis*) but not on different chromosomes (in *trans*) developed sarcomas that had histopathologic characteristics that resemble human MPNST, indicating that the combined loss of neurofibromin and *p53* can facilitate the formation of these tumors (Fig. 11.4). Whether cooperation between loss of neurofibromin and other tumor suppressor genes altered in human MPNSTs can similarly induce MPNSTs in mice remains to be tested.

MODELS OF SCHWANNOMAS

Similar approaches to those discussed above have been utilized to model NF2-associated tumors in mice. Like *Nf1*-null mice, mice with mutations in both *Nf2* alleles die during early embryonic development, although at a much earlier stage (McClatchey et al., 1997). Mice with only one mutant *Nf2* allele are prone to the development of metastatic osteosarcomas, fibrosarcomas, and hepatocellular carcinomas, all of which exhibit loss of the wild-type *Nf2* allele (McClatchey et al., 1998). These animals do not, however, develop schwannomas and meningiomas characteristic of NF2.

An alternative approach, using the Cre-lox system, was utilized to test the possibility that biallelic loss of *Nf2* in Schwann cells is sufficient to cause schwannomas (Fig. 11.4). Giovannini and co-workers (2000) developed mice in which *Nf2* is conditionally inactivated in Schwann cells using the myelin P0 protein promoter. Interestingly, mice with mutations in the region encoding the amino-terminal domain of merlin developed Schwann cell hyperplasia followed by the formation of tumors that share many of the histologic features of human schwannoma (Giovannini et al., 1999). Mice with truncating mutations in the carboxyl-terminal region, however, appeared to be normal. Thus, loss of *Nf2* appears to be sufficient to initiate schwannoma tumorigenesis.

CONTRIBUTION OF CELLULAR ENVIRONMENT

One of the unresolved issues concerns the limited growth potential of neurofibromas. It has been long recognized that these benign tumors exhibit accelerated growth during puberty and pregnancy, suggesting hormonal influences on neoplastic Schwann cell growth. Similarly, most individuals with NF1 note that their neurofibromas appear to grow for defined periods of time and then remain unchanged for decades. These observations suggest that other factors, besides loss of *NF1* and dysregulated RAS activity in neoplastic Schwann cells, contribute to the growth of neurofibromas. The exciting results by Parada and associates with the Schwann cell–specific conditional knockout mouse suggest that *Nf1*+/− cells in the neurofibroma profoundly influence its growth (Zhu et al., 2002). It is possible that *Nf1*+/− fibroblasts or mast cells elaborate growth factors that promote *Nf1*−/− Schwann cell growth. Alternatively, *Nf1*+/− fibroblasts or perineurial-like cells in contact with *Nf1*−/− Schwann cells might respond aberrantly to growth factors produced by the neoplastic Schwann cells and facilitate tumorigenesis and neurofibroma growth. Future studies on the contribution of the tumor microenvironment to tumor formation and growth may provide critical insights into these processes and reveal additional rational targets for developmental therapeutics.

ACKNOWLEDGMENT

This work was supported by the National Institutes of Health, grants 36996 and 35848.

REFERENCES

Arbiser JL, Flynn E, Barnhill RL (1998): Analysis of vascularity of human neurofibromas. *J Am Acad Dermatol* 38: 950–954.

Birindelli S, Perrone F, Oggionni M, Lavarino C, Pasini B, Vergani B, Ranzani GN, Pierotti MA, Pilotti S (2001): Rb and TP53 pathway alterations in sporadic and NF1-related malignant peripheral nerve sheath tumors. *Lab Invest* 81: 833–844.

Brannan CI, Perkins AS, Vogel KS, Ratner N, Nordlund ML, Reid SW, Buchberg AM, Jenkins NA, Parada LF, Copeland NG (1994): Targeted disruption of the neurofibromatosis type-1 gene leads to developmental abnormalities in heart and various neural crest-derived tissues. *Genes Dev* 8: 1019–1029.

Cichowski K, Jacks T (2001): NF1 tumor suppressor gene function: Narrowing the GAP. *Cell* 104: 593–604.

Cichowski K, Shih TS, Schmitt E, Santiago S, Reilly K, McLaughlin ME, Bronson RT, Jacks T (1999): Mouse models of tumor development in neurofibromatosis type 1. *Science* 286: 2172–2176.

DeClue JE, Heffelfinger S, Benvenuto G, Ling B, Li S, Rui W, Vass WC, Viskochil D, Ratner N (2000): Epidermal growth factor receptor expression in neurofibromatosis type 1-related tumors and NF1 animal models. *J Clin Invest* 105: 1233–1241.

Drobnjak M, Latres E, Pollack D, Karpeh M, Dudas M, Woodruff JM, Brennan MF, Cordon-Cardo C (1994): Prognostic implications of p53 nuclear overexpression and high proliferation index of Ki-67 in adult soft-tissue sarcomas. *J Natl Cancer Inst* 86: 549–554.

Evans DG, Baser ME, McGaughran J, Sharif S, Howard E, Moran A (2002): Malignant peripheral nerve sheath tumours in neurofibromatosis 1. *J Med Genet* 39: 311–314.

Ferner RE, Gutmann DH, Coffin C, Grimer R, Guha A, Judson I, Sandison A, Smith M, Viskochil D (2002): International consensus statement on malignant peripheral nerve sheath tumors in neurofibromatosis. *Cancer Res* 62: 1573–1577.

Ferner RE, O'Doherty MJ (2002): Neurofibroma and schwannoma. *Curr Opin Neurol* 15: 679–684.

Gautreau A, Louvard D, Arpin M (2002): ERM proteins and NF2 tumor suppressor: The Yin and Yang of cortical actin organization and cell growth signaling. *Curr Opin Cell Biol* 14: 104–109.

Giovannini M, Robanus-Maandag E, Niwa-Kawakita M, van der Valk M, Woodruff JM, Goutebroze L, Merel P, Berns A, Thomas G (1999): Schwann cell hyperplasia and tumors in transgenic mice expressing a naturally occurring mutant NF2 protein. *Genes Dev* 13: 978–986.

Giovannini M, Robanus-Maandag E, van der Valk M, Niwa-Kawakita M, Abramowski V, Goutebroze L, Woodruff JM, Berns A, Thomas G (2000): Conditional biallelic Nf2 mutation in the mouse promotes manifestations of human neurofibromatosis type 2. *Genes Dev* 14: 1617–1630.

Gutmann DH (2001): The neurofibromatoses: When less is more. *Hum Mol Genet* 10: 747–755.

Gutmann DH, Giordano MJ, Fishback AS, Guha A (1997): Loss of merlin expression in sporadic meningiomas, ependymomas and schwannomas. *Neurology* 49: 267–270.

Gutmann DH, Giovannini M (2002): Mouse models of neurofibromatosis 1 and 2. *Neoplasia* 4: 279–290.

Gutmann DH, Sherman L, Seftor L, Haipek C, Hoang Lu K, Hendrix M (1999): Increased expression of the NF2 tumor suppressor gene product, merlin, impairs cell motility, adhesion and spreading. *Hum Mol Genet* 8: 267–275.

Halling KC, Scheithauer BW, Halling AC, Nascimento AG, Ziesmer SC, Roche PC, Wollan PC (1996): p53 expression in neurofibroma and malignant peripheral nerve sheath tumor. An immunohistochemical study of sporadic and NF1-associated tumors. *Am J Clin Pathol* 106: 282–288.

Jacks T, Shih TS, Schmitt EM, Bronson RT, Bernards A, Weinberg RA (1994): Tumour predisposition in mice heterozygous for a targeted mutation in Nf1. *Nat Genet* 7: 353–361.

Jhanwar SC, Chen Q, Li FP, Brennan MF, Woodruff JM (1994): Cytogenetic analysis of soft tissue sarcomas. Recurrent chromosome abnormalities in malignant peripheral nerve sheath tumors (MPNST). *Cancer Genet Cytogenet* 78: 138–144.

King AA, Debaun MR, Riccardi VM, Gutmann DH (2000): Malignant peripheral nerve sheath tumors in neurofibromatosis 1. *Am J Med Genet* 93: 388–392.

Kleihues P, Cavenee WK (2000): *Tumours of the Nervous System* IARC Press, Lyon.

Korf BR (1999): Plexiform neurofibromas. *Am J Med Genet* 89: 31–37.

Kourea HP, Cordon-Cardo C, Dudas M, Leung D, Woodruff JM (1999a): Expression of p27(kip) and other cell cycle regulators in malignant peripheral nerve sheath tumors and neurofibromas: The emerging role of p27(kip) in malignant transformation of neurofibromas. *Am J Pathol* 155: 1885–1891.

Kourea HP, Orlow I, Scheithauer BW, Cordon-Cardo C, Woodruff JM (1999b): Deletions of the INK4A gene occur in malignant peripheral nerve sheath tumors but not in neurofibromas. *Am J Pathol* 155: 1855–1860.

Lothe RA, Smith-Sorensen B, Hektoen M, Stenwig AE, Mandahl N, Saeter G, Mertens F. (2001): Biallelic inactivation of TP53 rarely contributes to the development of malignant peripheral nerve sheath tumors. *Genes Chromosomes Cancer* 30: 202–206.

McClatchey AI, Saotome I, Mercer K, Crowley D, Gusella JF, Bronson RT, Jacks T (1998): Mice heterozygous for a mutation at the Nf2 tumor suppressor locus develop a range of highly metastatic tumors. *Genes Dev* 12: 1121–1133.

McClatchey AI, Saotome I, Ramesh V, Gusella JF, Jacks T (1997): The Nf2 tumor suppressor gene product is essential for extraembryonic development immediately prior to gastrulation. *Genes Dev* 11: 1253–1265.

Morrison H, Sherman LS, Legg J, Banine F, Isacke C, Haipek CA, Gutmann DH, Ponta H, Herrlich P (2001): The NF2 tumor suppressor gene product, merlin, mediates contact inhibition of growth through interactions with CD44. *Genes Dev* 15: 968–980.

Muir D (1995): Differences in proliferation and invasion by normal, transformed and NF1 Schwann cell cultures are influenced by matrix metalloproteinase expression. *Clin Exp Metastasis* 13: 303–314.

Nakamura T, Hara M, Kasuga T (1991): Early lesion of *N*-nitroso-*N*-ethylurea-induced hamster neurofibromatosis model. *Cancer Res* 51: 434–438.

Nielsen GP, Stemmer-Rachamimov AO, Ino Y, Moller MB, Rosenberg AE, Louis DN (1999): Malignant transformation of neurofibromas in neurofibromatosis 1 is associated with CDKN2A/p16 inactivation. *Am J Pathol* 155: 1879–1884.

Nikitin AYu, Ballering LA, Lyons J, Rajewsky MF (1991): Early mutation of the neu (erbB-2) gene during ethylnitrosourea-induced oncogenesis in the rat Schwann cell lineage. *Proc Natl Acad Sci USA* 88: 9939–9943.

Packer RJ, Gutmann DH, Rubenstein A, Viskochil D, Zimmerman RA, Vezina G, Small J, Korf B (2002): Plexiform neurofibromas in NF1: Toward biologic-based therapy. *Neurology* 58: 1461–1470.

Perantoni AO, Rice JM, Reed CD, Watatani M, Wenk ML (1987): Activated neu oncogene sequences in primary tumors of the peripheral nervous system induced in rats by transplacental exposure to ethylnitrosourea. *Proc Natl Acad Sci USA* 84: 6317–6321.

Perry A, Kunz SN, Fuller CE, Banerjee R, Marley EF, Liapis H, Watson MA, Gutmann DH (2002): Differential NF1, p16, and EGFR patterns by interphase cytogenetics (FISH) in malignant peripheral nerve sheath tumor (MPNST) and morphologically similar spindle cell neoplasms. *J Neuropathol Exp Neurol* 61: 702–709.

Perry A, Roth KA, Banerjee R, Fuller CE, Gutmann DH (2001): NF1 deletions in S-100 protein-positive and negative cells of sporadic and neurofibromatosis 1 (NF1)-associated plexiform neurofibromas and malignant peripheral nerve sheath tumors. *Am J Pathol* 159: 57–61.

Rutkowski JL, Wu K, Gutmann DH, Boyer PJ, Legius E (2000): Genetic and cellular defects contributing to benign tumor formation in neurofibromatosis type 1. *Hum Mol Genet* 9: 1059–1066.

Scheithauer BW, Woodruff JM, Erlandson RA (1999): *Tumors of the Peripheral Nervous System* AFIP, Washington DC.

Schulze KM, Hanemann CO, Muller HW, Hanenberg H (2002): Transduction of wild-type merlin into human schwannoma cells decreases schwannoma cell growth and induces apoptosis. *Hum Mol Genet* 11: 69–76.

Sherman L, Atit R, Rosenbaum T, Cox AD, Ratner N (2000): Single cell Ras-GTP analysis reveals altered Ras activity in a subpopulation of neurofibroma Schwann cells but not fibroblasts. *J Biol Chem* 275: 30740–30745.

Sherman L, Sleeman JP, Hennigan RF, Herrlich P, Ratner N (1999): Overexpression of activated neu/erbB2 initiates immortalization and malignant transformation of immature Schwann cells in vitro. *Oncogene* 18: 6692–6699.

Silva AJ, Frankland PW, Marowitz Z, Friedman E, Laszlo GS, Cioffi D, Jacks T, Bourtchuladze R, Lazlo G (1997): A mouse model for the learning and memory deficits associated with neurofibromatosis type I. *Nat Genet* 15: 281–284.

Stemmer-Rachamimov AO, Ino Y, Lim ZY, Jacoby LB, MacCollin M, Gusella JF, Ramesh V, Louis DN (1998): Loss of the NF2 gene and merlin occur by the tumorlet stage of schwannoma development in neurofibromatosis 2. *J Neuropathol Exp Neurol* 57: 1164–1167.

Stemmer-Rachamimov AO, Xu L, Gonzalez-Agosti C, Burwick JA, Pinney D, Beauchamp R, Jacoby LB, Gusella JF, Ramesh V, Louis DN (1997): Universal absence of merlin, but not other ERM family members, in schwannomas. *Am J Pathol* 151: 1649–1654.

Sun CX, Robb VA, Gutmann DH (2002): Protein 4.1 tumor suppressors: Getting a FERM grip on growth regulation. *J Cell Sci* 115: 3991–4000.

Twist EC, Ruttledge MH, Rousseau M, Sanson M, Papi L, Merel P, Delattre O, Thomas G, Rouleau GA (1994): The neurofibromatosis type 2 gene is inactivated in schwannomas. *Hum Mol Genet* 3: 147–151.

Vogel KS, Klesse LJ, Velasco-Miguel S, Meyers K, Rushing EJ, Parada LF (1999): Mouse tumor model for neurofibromatosis type 1. *Science* 286: 2176–2179.

Waggoner DJ, Towbin J, Gottesman G, Gutmann DH (2000): Clinic-based study of plexiform neurofibromas in neurofibromatosis 1. *Am J Med Genet* 92: 132–135.

Woodruff JM (1999): Pathology of tumors of the peripheral nerve sheath in type 1 neurofibromatosis. *Am J Med Genet* 89: 23–30.

Yarden Y (2001): The EGFR family and its ligands in human cancer. signalling mechanisms and therapeutic opportunities. *Eur J Cancer* 37: S3–8.

Zhu Y, Ghosh P, Charnay P, Burns DK, Parada LF (2002): Neurofibromas in NF1: Schwann cell origin and role of tumor environment. *Science* 296: 920–922.

Zhu Y, Parada LF (2001): Neurofibromin, a tumor suppressor in the nervous system. *Exp Cell Res* 264: 19–28.

Zhu Y, Romero MI, Ghosh P, Ye Z, Charnay P, Rushing EJ, Marth JD, Parada LF (2001): Ablation of NF1 function in neurons induces abnormal development of cerebral cortex and reactive gliosis in the brain. *Genes Dev* 15: 859–876.

12

CENTRAL NERVOUS SYSTEM TUMORS

ANDREW B. LASSMAN
Department of Neurology, Memorial Sloan-Kettering Cancer Center, New York, New York
ERIC C. HOLLAND
Departments of Surgery, Neurology, and Cancer Biology and Genetics, Memorial Sloan-Kettering Cancer Center, New York, New York

Within the last several decades, a plethora of mutations and signaling abnormalities in human brain tumors have been discovered. This represents an exciting development that holds the potential for new therapies. However, conquering such tumors will also require the successful differentiation of oncogenic abnormalities from epiphenomena. Modeling of brain tumors in mice is an important technique in this endeavor; transgenic, knockout, conditional tissue expression and somatic cell gene transfer methods have all been used to create these models, both separately and in combination. In addition, use of these models will become an important modality to test newly developed small-molecule pathway inhibitors. Developed models include those of the most common primary brain tumors: the gliomas and medulloblastomas. These mouse tumors have substantial similarities to their human counterparts with respect to both histologic appearance and noninvasive imaging. In this chapter, the pathways important in driving both gliomagenesis and medulloblastomagenesis are described, along with the key features of individual tumor models presently in use.

INTRODUCTION

Of all tumor types, brain tumors are often the most difficult to treat and potentially devastating. Almost a century ago, Harvey Cushing (1908, p. 231), one of the founding fathers of modern neurosurgery, wrote:

> Aside from such simple measures as serve to palliate the severity of given symptoms ... we have long stood helpless in the presence of a brain tumor.

While tremendous advances have occurred in the treatment for other common cancers such as those of lung and breast, the prognosis for the most frequent and aggressive primary brain tumor, the glioblastoma multiforme (GBM), has remained essentially unchanged for the last 50 years with a median survival of approximately 12 months (CBTRUS, 2002; Jelsma and Bucy 1967).

One reason why primary brain tumors remain so difficult to treat has been the limited understanding of the underlying biology. As described below, most primary brain tumors are defined histologically rather than molecularly. For example, it is difficult, if not impossible, to distinguish either radiographically or histologically a de novo GBM from one that begins as a lower grade tumor. As further evidence of the early stage of translating bench discoveries to the bedside, these tumors are presently treated identically even though recent discoveries in several laboratories have clearly demonstrated that they arise from an overlapping yet distinct set of molecular abnormalities (DeAngelis, 2001; Kleihues and Ohgaki, 1999).

However, hope is on the horizon. In the last several decades, the molecular bases of brain tumor growth are coming to light (Louis et al., 2001). Brain tumor specimens and cell lines exhibit many abnormalities of both signal transduction cascades and cell cycle control.

Mouse Models of Human Cancer, edited by Eric C. Holland
ISBN 0-471-44460-X

However, the existence of such abnormalities in brain tumors does not prove they are causal. As in most epic endeavors, the lessons of history provide a framework to analyze this problem.

As tuberculosis ravaged the landscape in the late 1800s, Robert Koch published his "postulates" stating that in order to prove that a pathogen caused an infectious disease, the pathogen had to be isolated from an infected individual, be grown in culture, produce the same symptoms when used to infect a naïve host, and then be reisolated from the second host (Koch, 1884). The concepts involved in Koch's postulates also apply to brain tumors in that modeling in animals allows us to distinguish oncogenic abnormalities from epiphenomena. As we better understand tumor biology, we envision the development of new pathway modulators tailored to the specific molecular abnormalities driving individual tumors, with the hope of far more effective therapy than available treatment currently.

CLINICAL ASPECTS OF HUMAN BRAIN TUMORS

The most common primary brain tumors in adults are the gliomas. These parenchymal brain tumors usually occur within the cerebral hemispheres. Most of these are astrocytomas (including glioblastomas) and oligodendrogliomas. The majority of gliomas occur *supratentorially*, that is, above the membrane called the *tentorium*. In contrast, childhood brain tumors typically occur in the cerebellum, which is located below the tentorium (*infrantentorial*). The most common such tumor is the childhood medulloblastoma. Cerebellar tumors are also called posterior fossa tumors because the cerebellum is in the posterior fossa of the skull. Intracranial tumors also arise from the coverings of the brain (meningiomas), and the most common "brain" tumors are actually metastases from other cancers such as lung, breast, or colon (Posner, 1996; Walker et al., 1985). However, while they are important clinical problems, neither meningeal nor metastatic (or *secondary*) brain tumors arise from the brain parenchyma, and the most common and refractory primary brain tumors of adulthood and childhood are gliomas and medulloblastomas. Therefore, this chapter focuses on the biology and models of these two tumor types.

Gliomas

Several cell types comprise the normal human brain, including glia, neurons, vascular epithelium, and meningeal cells. By far, the most numerous are the glia. There are four subtypes of glia: astrocytes, oligodendrocytes, microglia, and ependymal cells. Astrocytes provide supporting structure within the brain and likely participate in homeostatic and communicative mechanisms. Oligodendrocytes produce the myelin sheaths that insulate neurons and speed action potential conduction down axons. Microglia are related to macrophages and provide immune defense while ependymal cells line the ventricles.

There are two main histologic subtypes of gliomas as mentioned briefly above: astrocytomas and oligodendrogliomas. The cell of origin for these tumors may be astrocyte and oligodendrocyte precursors, respectively, or, alternatively, these two tumors could arise from a common glioneuronal precursor. The existence of such a precursor would explain the overlapping features sometimes seen in mixed oligoastrocytomas and mixed glioneuronal tumors. This is a topic of important and ongoing research (Dai and Holland, 2003; Holland, 2001b). Suffice it to say that astrocytomas express markers of and share morphologic features with astrocytes, and oligodendrogliomas express markers of and share morphologic features with oligodendrocytes.

Astrocytomas and Glioblastoma Multiforme. Astrocytomas are composed of tumor cells with astrocytic character. These cells typically have angulated nuclei and prominent fibrillary cytoplasm and express the astrocyte-specific marker glial fibrillary acidic protein (GFAP). There are four grades of astrocytomas based on histologic criteria named grades I–IV by the World Health Organization (WHO) (Kleihues and Cavenee, 2000). Although grade I tumors are called "astrocytomas," such tumors represent a different biologic and histologic entity that typically does not infiltrate surrounding brain and arises most frequently in the posterior fossa of children. It is essentially synonymous with juvenile pilocytic astrocytoma and is surgically curable in many cases. Grade I tumors have not been the focus of mouse models and will not be discussed further here.

Grade II tumors are low-grade diffuse astrocytomas, grade III tumors are anaplastic, and grade IV are glioblastomas or GBMs. A detailed description of the pathologic criteria that define each grade is beyond the scope of this text, and the interested reader is referred to an in-depth reference (Bigner et al., 1998). In brief, criteria associated with increasing grade include pleomorphism, nuclear atypia, mitotic activity, vascular endothelial neoplasia, and necrosis with pseudopalisades. In addition, the often diffuse nature of these tumors with infiltration into surrounding brain represents an important clinical feature that renders surgical cure impossible (Fig. 12.1).

Increased grade usually implies more rapid growth and portends a worse prognosis. Median survival from the diagnosis of a low-grade astrocytoma is approximately 5–10 years, from an anaplastic astrocytoma is approximately 3 years, and from GBM is approximately

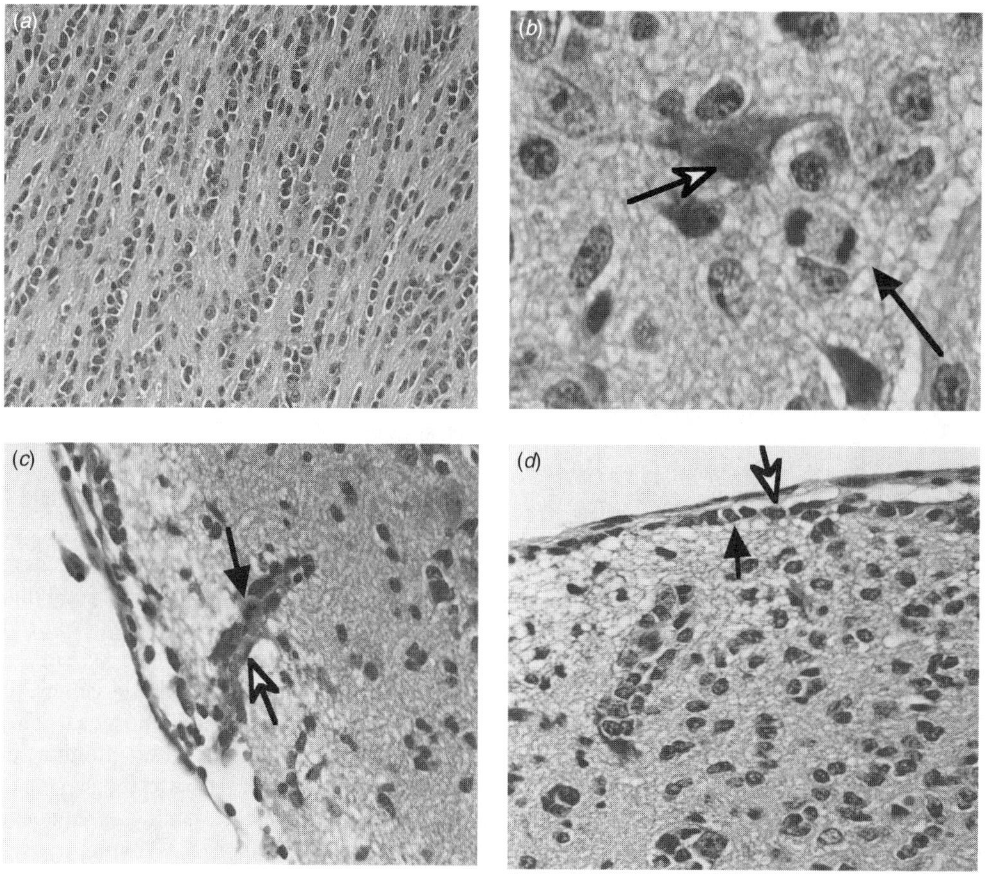

Figure 12.1. Glioma cells invading normal brain structures. Photomicrographs of mouse glioma models showing invasive features indistinguishable from the characteristic found in human gliomas. Glioma cells are seen migrating, along white matter tracts (*a*), glioma cells (closed arrow) surrounding neurons (open arrow) (*b*), glioma cells (closed arrow) lined up along a blood vessel (open arrow) (*c*), and glioma cells (closed arrow) accumulating in the subpial space (open arrow) (*d*). (See color insert).

1 year (DeAngelis, 2001). Although somewhat controversial (Peterson and DeAngelis, 2001), most patients with low-grade tumors are conservatively followed without treatment until features of increasing grade appear. In contrast, patients with high-grade astrocytomas (grades III or IV, also called "malignant" astrocytomas, although they almost never metastasize outside the brain) are usually treated aggressively with radiotherapy and chemotherapy.

In addition, GBMs are divided into two categories. Primary glioblastomas arise *de novo* without a preexisting lower grade tumor. Primary tumors typically occur in older patients and are frequently associated with activating mutations of epidermal growth factor receptor (*EGFR*), loss of the tumor suppressor phosphatase tensin homologue on chromosome 10 (*PTEN*), and the inhibitor of Cdk4 (*INK4A*) and its alternate reading frame (*ARF*). In contrast, secondary glioblastomas begin as a lower grade tumor and over time accumulate mutations and pathway abnormalities that cause progression to grade IV. Secondary GBMs occur more commonly in younger patients and have inactivating mutations of the *p53* gene but rarely demonstrate *EGFR* overexpression or activating mutations (Kleihues and Ohgaki, 1999). However, once a patient is diagnosed with a GBM, the treatment and prognosis do not differ regardless of the primary or secondary nature.

Oligodendrogliomas. Oligodendrogliomas are comprised of relatively uniform cells having regular round nuclei. The cytoplasm of these tumor cells contracts during tissue processing, giving an overall "fried egg" appearance (Fig. 12.2). In contrast to astrocytomas, oligodendroglioma tumor cells do not express GFAP. However, like astrocytomas, oligodendrogliomas invade the normal structures of the brain, migrating along white matter tracts and encircling neurons and blood vessels (Fig. 12.1). It has been proposed that this migratory behavior is a recapitulation of that exhibited by glial progenitors early

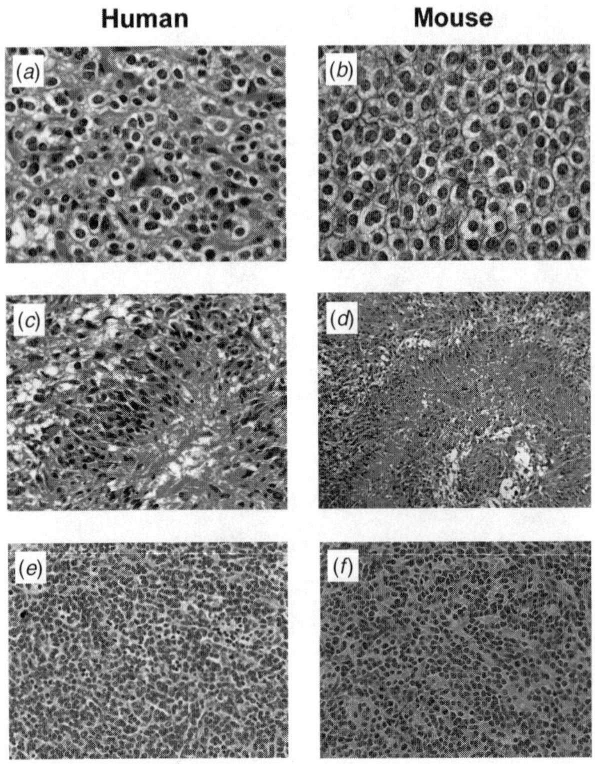

Figure 12.2. Comparative histology of (*a, c, e*) human and (*b, d, f*) mouse brain tumors. (*a*) Human oligodendroglioma with characteristic oval cells with nuclei surrounded by a white halo. (*b*) Mouse oligodendroglioma from *Pdgf* somatic cell gene transfer showing similar histologic features. (*c*) Human glioblastoma with characteristic pseudopalisading necrosis. (*d*) Pseudopalisading necrosis in a mouse glioblastoma induced by combined activation of *Ras* and *Akt* using somatic cell gene transfer. (*e*) Human medulloblastoma with characteristic small blue pleomorphic hypercellularity. (*f*) Similar histology in a mouse model of medulloblastoma induced by combined *Shh* and *Igf2* using somatic cell gene transfer. (*a, c, e*) Courtesy of Marc Rosenblum, MD, Department of Pathology, Memorial Sloan-Kettering Cancer Center, New York, NY. (*f*) Courtesy of Daniel W. Fults, MD, University of Utah School of Medicine, Salt Lake City, Utah. (See color insert.)

during development which migrate from the subventricular zone to the location that they will occupy as a differentiated glial cell.

Oligodendrogliomas are considered either low grade (WHO II) or anaplastic (WHO III). WHO IV grade gliomas with pure oligodendroglioma character are rare, implying that as these tumors develop histologic features of GBMs, they frequently also acquire astrocytic characteristics. Consistent with this view, many glioblastomas have oligodendrogliomatous features (Kleihues and Cavenee, 2000). The prognosis for patients with oligodendrogliomas is somewhat better than for patients with astrocytomas. For example, the median survival from the diagnosis of a low-grade oligodendroglioma is approximately 10 years. Like astrocytomas, many patients with low-grade oligodendrogliomas are followed conservatively without treatment unless features of anaplasia appear, at which point treatment is initiated. In contrast with astrocytomas, many patients with anaplastic oligodendrogliomas are typically treated first with multiagent chemotherapy followed by radiation therapy although this is an active area of clinic research and controversy. However, as we understand more about these tumors, including specific chromosomal abnormalities such as loss of heterozygosity on 1p and 19q that correlate with a better response to therapy, these regimens are changing (Ino et al., 2001).

Medulloblastomas

Medulloblastomas are "blue-cell tumors" comprised of sheets of small cells with little cytoplasm (Fig. 12.2). The desmoplastic variant is composed of islands of cells with more cytoplasm in a sea of smaller more typical medulloblastoma cells. In the large-cell variant, the cells are larger and more pleomorphic. These tumors are invasive at their edges, although they do not infiltrate throughout the brain tissue to the same extent as gliomas. Immunohistochemical staining is frequently positive for neuronal markers such as synaptophysin and NeuN and may be regionally positive for GFAP, indicating astrocytic differentiation in these areas.

Medulloblastomas typically occur in the cerebellum of children and young adults. Like patients with gliomas, patients with medulloblastomas are currently treated with surgical resection of the tumor followed by radiation and chemotherapy depending on tumor grade. Complete surgical resections can sometimes be achieved because of the limited spread of the tumor cells into the brain parenchyma in some patients. However, medulloblastomas have a higher propensity than gliomas to spread through the cerebrospinal fluid pathways, requiring a vigilant search for "drop" metastases elsewhere in the central nervous system and often postsurgical radiation of the entire brain and spinal cord. Unfortunately, such widespread radiation to the entire neuroaxis frequently results in substantial adverse effects on intellect that become apparent as children go through school.

Diagnosis

Neurologic symptoms such as headache, vomiting, seizures, imbalance, or hemiparesis usually alert the patient and physician to the presence of a glioma or medulloblastoma. The diagnosis is then suggested by brain image with either computerized axial tomography (CAT or CT) scanning or magnetic resonance imaging (MRI). In almost all cases except for tumors involving the skull, MRI is far superior to CT scanning and is the imaging modality of choice. The use of intravenous contrast material in either CT or MRI demonstrates areas of disrupted blood–brain barrier. Although there are exceptions, contrast enhancement is generally a characteristic of high-grade (III and IV) gliomas rather than of low-grade (II) gliomas. Positron emission tomography (PET) images the relative metabolic activity and can assist in the determination of tumor grade through noninvasive means. PET may also distinguish areas of recurrent tumor from radiation necrosis. In spite of the astonishingly high resolution of these newer noninvasive imaging techniques, the gold standard for diagnosis of these lesions remains surgical biopsy or resection and histologic analysis by a skilled neuropathologist.

HUMAN GLIOMA GENETICS: GROWTH FACTOR SIGNALING AND PROLIFERATION CONTROL

Mutations and Oncogenic Pathways in Human Brain Tumors

Over the past decade a substantial amount of work from many laboratories has documented myriad somatic mutations and pathway abnormalities in brain tumors. Many of these abnormalities are found frequently, implying an important role in the initiation or progression of these tumors. Mouse modeling has proven the causal link.

The proteins encoded by these genes generally participate in one of two important pathways. The first pathway transduces mitotic signals from outside the cell. The second pathway directly controls cell division. These are described below.

Extracellular Growth Factor Signaling Cascades Several growth factors stimulate tumor cell growth and cell division. Among these, abnormalities of platelet-derived growth factor (PDGF) and epidermal growth factor (EGF) are the best characterized in gliomas and also the growth factors most often used in brain tumor models. Therefore, this discussion will focus on these two growth factors and their receptors, PDGFR and EGFR.

Many methods of increasing growth factor signaling in human tumors have been documented. One type of abnormality is the endogenous production of these growth factors by cells that express surface receptors, creating autocrine loops. In addition, increasing the number of receptors on the cell surface by either gene amplification or overexpression causes abnormally elevated signaling in the presence of normal extracellular concentrations of growth factors. Finally, receptor mutations cause constitutive activation, rendering the receptors ligand independent. In one particular mutation (*EGFRvIII*), deletion of several exons fuses amino acids together, creating a unique epitope that allows antibodies to distinguish tumor cells from normal cells containing wild-type *EGFR*. Activating mutations of *EGFR* are far more common in primary than secondary glioblastomas, and this may represent an important therapeutic consideration in the future (Ekstrand et al., 1991, 1992; Fleming et al., 1992; Guha et al., 1995; Hermanson et al., 1992, 1996; Hurtt et al., 1992; Humphrey et al., 1990; Kleihues and Ohgaki 1999; Kuan et al., 2001; Kumabe et al., 1992; Nister et al., 1988; Robinson et al., 2001; Schlegel et al., 1994; Shoshan et al., 1999; Smith et al., 2000; Watanabe et al., 1996; Wikstrand et al., 1995).

PDGF is a well-known mitogen for glial cells in culture. Previous studies have shown that PDGF acts to maintain glial precursors in an undifferentiated and proliferating state (Noble et al., 1988) and promotes the proliferation of central nervous system (CNS) progenitors in vivo (Williams et al., 1997). Like EGF/EGFR, gliomas frequently overexpress both the PDGF ligand and its receptor, suggesting that an autocrine loop may in part drive the proliferation of these tumors in vivo (Guha et al., 1995; Nister et al., 1988). Amplification of *PDGFR* has been demonstrated in malignant gliomas, implying a selective advantage to high levels of signaling through this receptor (Hermanson et al., 1992, 1996; Smith et al., 2000). Finally, *PDGFR* overexpression appears partly limited to a subset of GBMs that may be predominantly nonoverlapping with those that signal through EGFR (Hesselager et al., 2003),

Both EGFR and PDGFR transmit their signals through the tyrosine kinase activity of their cytoplasmic domain. Hence, they are called receptor tyrosine kinases (RTKs). Following ligand binding, the receptors dimerize, inducing a conformational change that activates the tyrosine kinase activity in the cytoplasmic domain. The activated kinase then autophosphorylates the receptor and phosphorylates additional downstream activators either directly or through adapter molecules, including SRC, RAS, and AKT, phosphoinositol 3-kinase (PI3K), and phospholipase C (PLC).

Although RAS activates many pathways, one of the most important consequential signaling cascades is initiated by RAF (RAS, MEK, ERK cascade activated factor 1). RAF then phosphorylates and activates MEK (mitogen-associated protein (MAP) kinase or ERK kinase), which then phosphorylates and activates ERK (MAP/extracellular signal-regulated kinase). Some of the important targets of ERK include transcription factors, antiapoptotic proteins, and protein synthesis participants (Blume-Jensen and Hunter, 2001).

RTKs also activate PI3K, which phosphorylates phosphotidyl inositol 4-phosphate (PIP), creating phosphatidyl inositol 4,5-bisphosphate (PIP2). PI3K also phosphorylates PIP2, creating PIP3. The phospholipids are required cofactors for the activation of AKT by RTK and phosphoinositide-dependent kinase 1 (PDK1) (Alessi et al., 1997; Datta et al., 1996).

Following activation, AKT stimulates protein translation largely through mTOR (mammalian target of rapamycin), participates in cellular metabolism, and inhibits apoptosis. In addition, both RAS and AKT modulate the cell cycle regulatory system. PTEN reverses AKT activation by dephosphorylating PIP3 to PIP2 and PIP2 to PIP, thereby preventing the activation of AKT by PDK1 (Knobbe et al., 2002).

In addition, through a separate but related pathway, the phospholipid products of PI3K activation, PIP2 and PIP3, also coordinate with RTKs to activate PLC. The phospholipase hydrolyzes PIP3 into the degradation products 1,4,5-triphosphate (IP3) and diacylglycerol (DAG). These second messengers activate additional signaling pathways, including protein kinase C (PKC).

Cell Cycle Control Abnormalities of cell cycle control comprise the second broad category of pathway misdeeds that occur frequently in human brain tumors (Biernat et al., 1997a, b; Ichimura et al., 1996, 2000; Newcomb et al., 1998; Ueki et al., 1996). Normally, two major arms cooperate to maintain cells in G1 arrest. These are the p16INK4A (Inhibitor of Cdk4) pathway and the p14ARF (Alternate Reading Frame) pathway. Under senescent conditions, p16INK4A inhibits Cyclin–cyclin-dependent kinase (Cdk) heterodimers from stimulating progression through the cell cycle. Without active Cyclin-Cdks, the retinoblastoma (RB) gene product remains minimally phosphorylated and bound to the transcription factor E2F. While bound to RB, E2F is inactive. However, when p16INK4A activity is lost, Cyclin-Cdks phosphorylate RB, releasing E2F, which then elevates its own transcription and activates the cyclins and CDKs forming a positive feedback loop that pushes cells through the cell cycle. (Chellappan et al., 1991; Johnson et al., 1993; Pardee, 1989)

The p14ARF protein is formed by use of an alternate reading frame of the same gene that encodes p16INK4A. Like p16INK4A, p14ARF also suppresses proliferation. However, the players are different in this arm. Instead of acting directly on cyclins, p14ARF forms an inhibitory complex with a molecule originally isolated from murine double minute chromosomes (Mdm2 in mice, HDM2 in humans) (Fakharzadeh et al., 1991; Kamijo et al., 1998; Momand et al., 1992; Pomerantz et al., 1998; Tao and Levine, 1999). In the presence of p14ARF, HDM2 cannot inhibit the function of the p53 gene product; therefore, the p53 gene product becomes phosphorylated and thereby protected from degradation (Ashcroft et al., 1999). Although the molecular events are complex, phosphorylated p53 inhibits progression through the cell cycle mainly by stimulating expression of the Cdk inhibitor protein p21CIP1, which inhibits Cyclin-Cdks (el-Deiry et al., 1993).

As the above discussion implies, many different abnormalities of cell cycle regulation could cause increased proliferation. In one arm, such abnormalities include *p16INK4A* loss, *CDK4* amplification, or *RB* loss. In the other arm, these include *p14ARF* loss, *HDM2* amplification, or *p53* inactivation. The majority of gliomas exhibit abnormalities leading to disruption of both pathways. In addition, tumors with wild-type *INK4A-ARF* often show inactivating methylation of the gene (Biernat et al., 1997a; Costello et al., 1996; Fueyo et al., 1996; Fulci et al., 2000; He et al., 1994; Ichimura et al., 1996, 2000; Moulton et al., 1995; Reifenberger et al., 1993; Schmidt et al., 1994, 1997; Ueki et al., 1996; Wolter et al., 2001).

Human Medulloblastoma Genetics and Signaling

The genetic profile of medulloblastomas has not been documented as extensively as that of gliomas. Although a common change is loss of heterozygosity for part of chromosome 17p, this loss occurs distal to *p53*, and the specific tumor suppressor gene remains anonymous. A very aggressive subset of these tumors show amplification and overexpression of *MYC*, the gene for an important transcription factor with myriad functions, including effects on cell cycle control and cell differentiation (Fults et al., 2002; Levens, 2002).

The observations that have led to some of our greatest insight in this disease come from basal cell nevus syndrome (BCNS), an inherited disorder that frequently presents with medulloblastoma caused by germline mutations in the gene patched (*PTCH*) (Hahn et al., 1996; Johnson et al., 1996). Patched is the receptor for sonic hedgehog (SHH), a transmembrane signaling protein (Hahn et al., 1996; Johnson et al., 1996). In the absence of SHH, patched likely inhibits signaling to the effector molecule smoothened (SMO) (Stone et al., 1996). However, binding of SHH to patched releases this inhibition, resulting in the activation by SMO of GLI1, a transcriptional activator (Hui et al., 1994).

In the developing cerebellum, patched is expressed by neuronal precursors in the external granule cell layer (Dahmane and Ruiz-i-Altaba, 1999; Ruiz i Altaba et al., 2002; Wechsler-Reya and Scott, 1999) and SHH stimulates division of these cells (Wechsler-Reya and Scott, 1999). Unregulated activity of this signaling pathway may contribute to medulloblastoma formation by promoting the continued proliferation of granule cell precursors and the medulloblastoma cells they resemble. Further evidence of the importance of patched and SHH are the observations that defects in this pathway can be detected in sporadic medulloblastomas (Pietsch et al., 1997; Raffel et al., 1997; Taylor et al., 2002).

Mouse Models of Gliomas and Medulloblastomas

Chief reasons to model tumors in animals are to improve our understanding of tumor biology and to provide an experimental system to test treatments. As an equally important corollary, many such potential treatments may arise from pathway abnormalities discovered from successful modeling.

Early models used DNA alkylating agents such as nitrosoureas. When animals were exposed to these mutagens either directly or transplacentally, they often developed brain tumors (Kleihues et al., 1976; Koestner et al., 1971; Kokkinakis et al., 2001). These tumors resembled the human counterparts in many ways. However, mutagens create multiple random genetic changes, so these models in some ways also reflect the difficulties of human tumors in that they do not distinguish epiphenomena from oncogenic mutations.

In addition to chemical mutagens, the other major type of modeling system developed in prior decades and still used today is grafting of human tumors or tumor cell lines into nude mice (xenografts) or rodent tumors or tumor cell lines into animals of the same species (allografts). Reproducibility of tumor growth rate and location is one of the key strengths of graft models. However, in subcutaneous grafts into the flank, the tumors obviously cannot recapitulate the normal intracranial condition, and when grafted intracranially, these tumors do not typically infiltrate into surrounding brain. Glioma models that do not reproduce the invasive character of these tumors limit their utility (Finkelstein et al., 1994). Furthermore, despite their promising theory, xenograft models have not been predictive in the development of novel therapeutics for brain tumors.

Several genetically defined mouse models for tumors of the CNS have been published recently. They have used germline modification strategies such as transgenic and knockout techniques, somatic cell gene transfer approaches with viral vectors, or combinations of these methods. In-depth discussion of these methodologies are found in Chapters 4 and 5 of this text. Briefly, germline modification strategies provide either gain-of-function or loss-of-function mutations that are integrated into the germline and inherited in a Mendelian manner. Typically, these models create a line of mice that are at risk for the development of tumors from some organ site and identify genetic alterations that contribute to the formation of specific tumors. Secondary mutations are acquired in the cell population at risk, leading to tumor formation. These strategies more accurately reflect syndromes of inherited susceptibility to tumors such as Li–Fraumani (*p53* mutation) or retinoblastoma (*RB* mutation) than the majority of brain tumors that arise from sporadic mutations in somatic cells. In contrast, somatic cell gene transfer techniques use viral vectors to transfer genes to a limited number of somatic cells postnatally and mimic the acquisition of mutations in somatic cells that occurs in sporadic cancers. The two techniques complement each other. In the case of CNS tumors, results from models combining the two approaches provide strong evidence that the abnormalities in signal transduction and cell cycle arrest found in these tumors are causally related to their formation. Major models of gliomas and medulloblastomas currently in use are listed in Table 12.1.

Models of Astrocytoma and Glioblastoma Multiforme

Several investigators have developed models of astrocytomas using different techniques. Early experiments with the Rous sarcoma virus (RSV) showed that infection with certain subgroups of this virus into dog brains caused gliomas with astrocytic character (Bigner et al., 1969; Haguenau, 1981; Haguenau et al., 1971; Rabotti et al., 1966). Subsequently, the oncogene encoded by RSV was identified as an activated form of *Src*. Investigators observed the presence of the *SRC* gene product in human astrocytomas (Takenaka et al., 1985). Transgenic expression of v-*src* from the GFAP promoter led to the formation of brain tumors in 15% of mice with low-copy transgene and almost 50% of mice with high-copy transgene. However, it is also likely that additional signaling or cell cycle abnormalities spontaneously occur in these tumors. The histologic features of these tumors varied from low-grade astrocytoma to GBM (Weissenberger et al., 1997).

Although activating mutations of the *RAS* gene have not been documented in human gliomas (Bos, 1989; Guha et al., 1997), activation of RAS signaling occurs at high frequency (Feldkamp et al., 1999; Guha et al., 1997). In addition, the germline abnormality in patients with neurofibromatosis type 1 (NF1) creates unregulated RAS activity (Ballester et al., 1990; Xu et al., 1990), and patients with NF1 develop gliomas among other tumor types (Cichowski and Jacks, 2001). In another model, mice transgenic for an activated form of Ras driven by the GFAP promoter developed astrocytomas. Mice with high transgene copy number developed tumors with histologic features of glioblastoma, including vascular neoplasia and pseudopalisading necrosis. All of these mice died from their tumors before transmitting the transgene in the germline. Mice with lower transgene copy number also developed astrocytomas but of lower grade and longer incubation time, allowing establishment of a stable line of germline transgenics. Molecular analysis showed that many of the tumors also exhibited concurrent alterations of other genes of known importance in glioma biology, including *Ink4a-Arf, Cdk4, Mdm2, Akt*, and *Pten* (Ding et al., 2001), consistent with the "two-hit" model of oncogenesis (Knudson, 1971, 2001).

Disruption of cell cycle control by genetic mutation is common in high-grade gliomas (Henson et al., 1994; Ueki et al., 1996). By contrast, although mice with deletions of these tumor suppressors develop lymphomas and sarcomas, they generally do not form gliomas (Donehower et al., 1992; Holland, 2001a; Williams et al., 1994). For the most part, these mutations elevate the grade and shorten the latency for gliomas generated by signal transduction abnormalities (Bachoo et al., 2002; Hesselager et al., 2003; Holland et al., 1998; Reilly et al., 2000). However, there are a few notable exceptions to this general trend. First, although *Ink4a-Arf* loss does not lead to tumor formation, *Arf* loss leads to oligodendrogliomas in about 10% of mice (Kamijo et al., 1999). The reason for the discrepancy is unknown. In addition, transgenic expression of a portion of the simian virus 40 (SV40) large T antigen that blocks the function of Rb and its family members, T121, leads to the formation of diffuse astrocytic gliomas (Xiao et al., 2002). By contrast, deletion of *Rb* alone is lethal. This discrepancy is potentially due to the fact that experimental deletion of *Rb* can be compensated for by the other family members p107 and p130, proteins that T121 inhibits.

Elevated activity of the RAS pathway is found in essentially all GBMs while elevated activity of AKT signaling is found in 70% of GBMs. The activation of AKT in these tumors appears to be achieved by primarily inactivating mutations of *PTEN* rather than mutations in *AKT* itself (Choe et al., 2003; Davies et al., 1999; Duerr et al., 1998; Holland et al., 2000; Li et al., 1997; Rasheed et al., 1997). In order to model the majority of GBMs having combined RAS and AKT overactivity, both the Ras and Akt pathways were activated simultaneously in a mouse model combining transgenic and somatic cell gene transfer approaches. By infection of Tv-a transgenic mice with two RCAS vectors, one encoding activated Ras and the other activated Akt, 26% of mice developed gliomas with histologic characteristics of GBMs (Holland et al., 2000). However, only mice expressing Tv-a from the nestin promoter developed tumors, while mice expressing Tv-a from the GFAP promoter did not. Glial progenitors typically express nestin but not GFAP while the opposite is true for mature astrocytes. Therefore, the difference in tumor formation between the two lines of mice implies that less differentiated cells are more sensitive to the oncogenic effects of these signaling pathways. Furthermore, the combination of Ras and Akt signaling appeared to be critical, as neither activated Ras nor activated Akt alone caused tumors. At first glance these data appear to contrast with the transgenic model of astrocytomas where activated Ras is driven by the GFAP promoter without other intentional mutations (Ding et al., 2001) described above. However, the astrocytomas in mice transgenic for GFAP-Ras also developed additional spontaneous mutations in signaling or cell cycle control genes, including *Akt, Pten*, and *Ink4a-Arf*. Therefore, although Ras clearly contributes to gliomagenesis, additional alterations, such as Akt activation, appear equally important. The observation that Ras activation in cooperation with *Ink4a-Arf* loss also causes astrocytomas in mice bolsters this view further (Uhrbom et al., 2002).

Another model employed loss of multiple tumor suppressor genes instead of activated oncogenes. Mice have been generated with targeted deletions of both the *Nf1* and *p53* genes adjacent to each other on the same chromosome. Neurofibromin, the gene product of *NF1*, functions as an inhibitor of Ras activity. Loss of the unaffected chromosome in these mice results in simultaneous loss of the remaining alleles of both *Nf1* and *p53*, causing activation of Ras signaling and disruption of cell cycle control, respectively. Such double-knockout mice develop astrocytomas and GBMs. Not surprisingly, these gliomas have lost the remaining *Nf1* and *p53* alleles (Reilly et al., 2000).

Models of Oligodendrogliomas

Signaling through PDGFR is one of the most studied signaling cascades in both oligodendrocyte development and oligodendrogliomas (Di Rocco et al., 1998; Nister et al., 1988; Robinson et al., 2001; Shoshan et al., 1999; Smith et al., 2000). Chronic PDGF stimulation maintains glial precursors in an undifferentiated state (Guha et al., 1995; Noble et al., 1988). Modeling systems with somatic cell gene transfer techniques that force chronic PDGF stimulation *in vivo* have shown that PDGF signaling can cause the formation of gliomas with characteristics of oligodendrogliomas in mice. In one model, injection of Maloney murine leukemia virus (MMLV) carrying the gene encoding PDGF-B caused tumors with various morphologies, including both low-grade and anaplastic oligodendrogliomas (Uhrbom et al., 1998). *PDGF-B* is the cellular homology of v-*sis*, the transforming oncogene encoded by simian sarcoma virus (SSV) (Besmer et al., 1983; Betsholtz et al., 1983; Doolittle et al., 1983; Robbins et al., 1983; Waterfield et al., 1983). Therefore, this experiment parallels the formation of various glial tumors in newborn marmosets following intracranial SSV injections (Deinhardt, 1980). In another model, cell-type specific transfer of *PDGF-B* to mature astrocytes using the RCAS-TVA system (Dai et al., 2001) caused oligodendrogliomas or mixed oligoastrocytomas depending on the cell of origin. When GFAP-expressing cells were infected, approximately 40% of mice developed mixed gliomas; when nestin-expressing cells were infected, nearly 70% of mice developed gliomas that were essentially pure oligodendrogliomas in character. This further implicates PDGF signaling in dedifferentiation of mature astrocytes, as evidenced from cell culture work using the same vector, and also demonstrates that astrocytes and oligodendrocytes have overlapping features and can derive from the same cell of origin. Consistent with the loss of the tumor suppressor *INK4A-ARF* in the majority of high-grade gliomas in humans, mice with targeted deletions of the *Ink4a-Arf* locus infected with the PDGF vector developed tumors with shorter latency and higher grade than wild-type mice.

Signaling through EGFR is also abnormal in human oligodendrogliomas (Reifenberger et al., 1996). The addition of activated *EGFR* to the GFAP-Ras transgenic model for diffuse astrocytomas described above led to a shorter latency for tumor formation and conversion of the tumor histology to an oligodendroglioma or mixed oligoastrocytoma phenotype (Ding et al. 2003). Finally, mice transgenic for v-*erbB* (the constitutively active viral homologue of *EGFR)* driven by the *S100*β promoter (another gene expressed in glial progenitors) developed oligodendrogliomas (Weiss et al., 2003). These findings together implicate chronic signaling through EGFR in either dedifferentiation of mature astrocytes into transformable glial progenitors or maintenance of an undifferentiated and transformable state. Furthermore, these results also illustrate the ability of glial progenitors to become either oligodendrocytic or astrocytic tumor cells.

Unlike chronic activation of *PDGFR*, forced expression of constitutively activated *EGFR* as a single abnormality has not resulted in the formation of gliomas (Bachoo et al., 2002; Ding et al., 2003; Holland et al., 1998). The chronic signaling generated by activated forms of EGFR only results in tumor formation when combined with specific activation of other signaling pathways, such as the Ras pathway as described above, or loss of cell cycle control by simultaneous loss of *p53* and overexpression of *Cdk4* (Holland et al., 1998), or loss of *Ink4a-Arf* (Bachoo et al., 2002; Holland et al., 1998). It is not clear why signaling through the PDGFR is sufficient for tumor formation while signaling through the EGFR is not. However, in this context it is interesting that PDGFR signaling is found elevated in gliomas of all grades whereas *EGFR* mutations occur predominantly in high-grade tumors associated with other genetic and pathway abnormalities.

Models of Medulloblastomas

As noted above, the external granule cell layer (EGL) of the cerebellum responds to the proliferative effects of the SHH signaling pathway by blocking differentiation and promoting proliferation. Several mouse models of medulloblastoma have been created using SHH activation and disruption of cell cycle arrest. As noted earlier, in the absence of SHH, the receptor patched (Ptch) inhibits

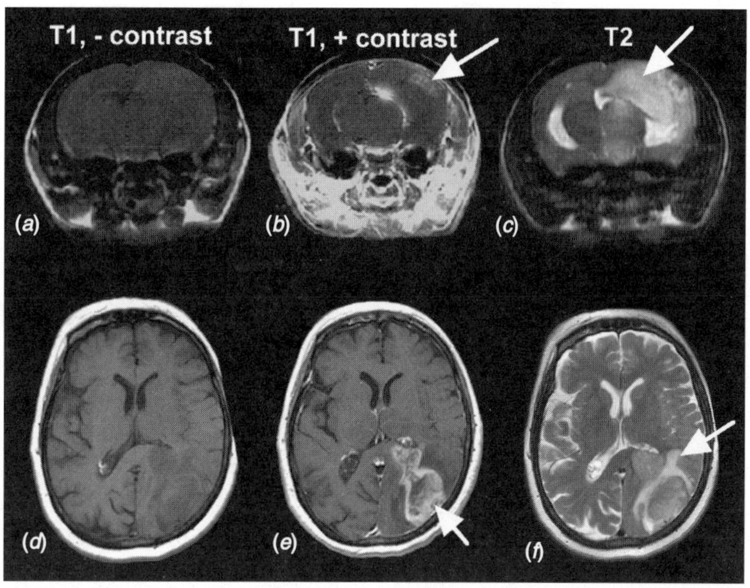

Figure 12.3. MRI scans from of GBMs in human and mouse. T1-weighted images of (*a*) mouse and (*d*) human GBMs demonstrate mild mass effect with distortion of the ventricular anatomy. Intravenous contrast leaks out of blood vessels in areas of vascular abnormalities and breakdown of the blood–brain barrier as seen in both (*b*) mouse and (*e*) human versions of the disease. (*c*, *f*) Bright areas on T2-weighted images demonstrate infiltration of nonenhancing tumor mixed with surrounding edema.

Table 12.1. Various Primary Brain Neoplasms Generated in Mice by Genetic Technologies That Appear Similar to Their Human Counterparts

Histology	Gene or Pathway	Tissue-specific Promoter[a]	References
Mixed glioma (oligoastrocytoma)	*Ink4a-Arf* + *PDGFB*	GFAP	Dai et al. 2001
Low-grade (WHO II) astrocytoma (WHO II)	*Ras*	GFAP	Ding et al. 2001
	Nfl + *p53*	All cells	Reilly et al. 2000
	src	GFAP	Weissenberger et al. 1997
Low-grade (WHO II) oligodendroglioma	*Arf*	All cells	Kamijo et al. 1999
	PDGFB	Nestin	Dai et al. 2001
	v-*erbB*	S100β	Weiss et al. 2003
	Ras + *EGFR VIII*	GFAP	Ding et al. 2003
Anaplastic (WHO III) astrocytoma	Ras	GFAP	Ding et al. 2001
	Nfl + *p53*	All cells	Reilly et al. 2000
	src	GFAP	Weissenberger et al. 1997
	Rb pathway inactivated by mutant SV40 T antigen	GFAP	Xiao et al. 2002
Anaplastic (WHO III) oligodendroglioma	*Ink4a-Arf* + *PDGFB*	Nestin	Dai et al. 2001
GBM (WHO IV astrocytoma)	*Ras* + *Akt*	Nestin	Holland et al. 2000
	PDGFB	Mixed population	Uhrbom et al. 1998
	Nfl + *p53*	All cells	Reilly et al. 2000
	Ink4a-Arf + *Ras*	GFAP or nestin	Uhrbom et al. 2002
Medulloblastoma	*Ptch*	All cells	Goodrich et al. 1997, Wetmore et al. 2000, Zurawel et al. 2000
	Ptch + *p53*	All cells	Wetmore et al. 2000
	Rb + *p53*	GFAP	Marino et al. 2000
	Shh	Mixed cerebellar progenitor cells	Weiner et al. 2002
	Shh + *myc*	Nestin	Rao et al. 2003
	Lig4 + *p53*	All cells	Lee et al. 2002
	Parp + *p53*	All cells	Tong et al. 2003

Note: In the case of knockout strains the cell of origin can only be guessed. In the case of transgenic mice or somatic cell gene transfer, the promoter used to limit expression or gene transfer is listed.

[a] GFAP, astrocytes; nestin, glioneuronal progenitors; S100β; glioneuronal progenitors.

proliferative signals through the SHH/GLI signaling pathway; therefore, *PTCH* loss is equivalent to *SHH* overexpression. As proof of the oncogenicity of this signaling mechanism, approximately 15% of mice heterozygous for *Ptch* (with a functional remaining allele) developed medulloblastomas within one year (Goodrich et al., 1997; Wetmore et al., 2000; Zurawel et al., 2000). Further evidence of the oncogenic potential of this pathway is derived from somatic cell gene transfer experiments showing that viral vectors encoding SHH cause medulloblastomas, either with MLV injected prenatally (Weiner et al., 2002) or RCAS-TVA injected postnatally (Rao et al., 2003).

Although *p53* and *RB* are not inactivated in human medulloblastomas, there is ample evidence that experimental disruption of these pathways in the appropriate cells in mouse models clearly promotes the formation of tumors with medulloblastoma histology. When $Ptch^{+/-}$ mice are crossed with mice deficient for *p53*, the tumor rate increases from less than 20% at 1 year to more than 95% by 12 weeks (Wetmore et al., 2001). Furthermore, when *p53* and *Rb* were conditionally deleted in EGL cells by Cre-*lox*, all of the mice developed tumors of the cerebellum with histologic characteristics of medulloblastomas. In contrast, when only one of these genes alone was inactivated, tumors did not form (Marino et al., 2000).

Two recent publications identify models that implicate chromosomal breakage and loss of *p53* in the formation of medulloblastomas. Poly(ADP-ribose) polymerase (PARP-1) repairs DNA and thereby suppresses tumor formation in cooperation with p53 (Tong et al., 2001).

Half of mice lacking both *p53* and *Parp-1* developed cerebellar medulloblastomas with abnormalities of signaling through the Shh-Ptch-Gli pathway (Tong et al., 2003). DNA ligase IV (LIG4) is another enzyme that participates in DNA repair (Robins and Lindahl, 1996). While *Lig4* loss was embryonically lethal, mice with combined loss of *Lig4* and *p53* survived for several months, but all developed medulloblastomas in addition to other systemic tumors (Lee and McKinnon, 2002). These findings further link *p53* loss to mouse medulloblastomagenesis and imply that *p53* loss likely re-creates a signaling abnormality arising through a different mechanism in humans that remains unidentified but with presumed therapeutic potential.

Imaging of Brain Tumor Models

One major goal of modeling is to provide an experimental system with which to test pathway inhibitors as potential therapies for brain tumors. To that end, it is important to monitor the presence of tumors and the effect of pathway modulators in living animals. Magnetic resonance imaging (MRI) is one such method. New techniques have been developed to obtain MRI images from mice that share many of the features of images from humans, including enhancement demonstrating disruption of the blood–brain barrier that correlates with high-grade tumors (Koutcher et al., 2002; Xu et al., 2003) (Fig. 12.3). Furthermore, stereotactic methods of tumor localization in humans are critical for successful neurosurgical procedures. Similar methods are now available for mice (Tada et al., 2002). Finally, bioluminescent techniques discussed at length in Chapter 21 for the detection and characterization of intracranial tumors will represent an important tool in the near future (Vooijs et al., 2002). These imaging techniques may have significant impact on the use of these models for preclinical trials, as discussed in Chapter 22.

CONCLUSION

Improved knowledge of the molecular events contributing to tumor growth is essential for the design of rational therapies for human brain tumors. The models described in this chapter have begun to distinguish oncogenic changes from epiphenomena. However, much work remains. For although many elements of both growth factor signaling and cell cycle control have been identified, the details of how these signals interact remain obscure. Understanding such molecular cross-talk will be critical for successful therapy requiring multiple-pathway modulators. In addition, the tumor models described in this chapter re-create the human counterparts in many ways (Weiss et al., 2002), and simple models with one or two genetic alterations are a useful first step. However, the human diseases are clearly more complex, as evidenced by the myriad gene mutations within single human tumor specimens. The ultimate utility of models resides in the effective translation of research into effective therapy for patients. Hopefully, that day may be near.

ACKNOWLEDGMENTS

The authors would like to thank Christina Glaster for her expertise in editing and organizing this chapter and Bill Carey for copying numerous articles. The work was funded by the Seroussi and Bressler Foundations, UO1 CA894314-1 supporting the Mouse Models of Human Cancer Consortium (ECH), and T32 CA009512 from the National Cancer Institute and the America Brain Tumor Association (ABL).

REFERENCES

Alessi DR, et al. (1997): Characterization of a 3-phosphoinositide-dependent protein kinase which phosphorylates and activates protein kinase Balpha. *Curr Biol* 7(4):261–269.

Ashcroft M, et al. (1999): Regulation of p53 function and stability by phosphorylation. *Mol Cell Biol* 19(3):1751–1758.

Bachoo RM, et al. (2002): Epidermal growth factor receptor and Ink4a/Arf: Convergent mechanisms governing terminal differentiation and transformation along the neural stem cell to astrocyte axis. *Cancer Cell* 1(3):269–277.

Ballester R, et al. (1990): The NF1 locus encodes a protein functionally related to mammalian GAP and yeast IRA proteins. *Cell* 63(4):851–859.

Besmer P, et al. (1983): The Parodi-Irgens feline sarcoma virus and simian sarcoma virus have homologous oncogenes, but in different contexts of the viral genomes. *J Virol* 46(2):606–613.

Betsholtz C, et al. (1983): Synthesis of a PDGF-like growth factor in human glioma and sarcoma cells suggests the expression of the cellular homologue to the transforming protein of simian sarcoma virus. *Biochem Biophys Res Commun* 117(1):176–182.

Biernat W, et al. (1997a): Amplification and overexpression of MDM2 in primary (de novo) glioblastomas. *J Neuropathol Exp Neurol* 56(2):180–185.

Biernat W, et al. (1997b): Alterations of cell cycle regulatory genes in primary (de novo) and secondary glioblastomas. *Acta Neuropathol (Berl)* 94(4):303–309.

Bigner DD, et al. (1998): *Russell and Rubinstein's Pathology of Tumors of the Nervous System.* New York, Oxford University Press.

Bigner DD, et al. (1969): Brain tumors induced in dogs by the Schmidt-Ruppin strain of Rous sarcoma virus. Neuropathological and immunological observations. *J Neuropathol Exp Neurol* 28(4):648–680.

Blume-Jensen P, Hunter T (2001): Oncogenic kinase signalling. *Nature* 411(6835):355–365.

Bos JL (1989): ras oncogenes in human cancer: A review. *Cancer Res* 49(17):4682–4689.

CBTRUS (2002): Statistical Report: Primary Brain Tumors in the United States, 1995–1999. Central Brain Tumor Registry of the United States, Chicago (www.cbtrus.org).

Chellappan SP, et al. (1991): The E2F transcription factor is a cellular target for the RB protein. *Cell* 65(6):1053–1061.

Choe G, et al. (2003): Analysis of the phosphatidylinositol 3′-kinase signaling pathway in glioblastoma patients in vivo. *Cancer Res* 63(11):2742–2746.

Cichowski K, Jacks T (2001): NF1 tumor suppressor gene function: Narrowing the GAP. *Cell* 104(4):593–604.

Costello JF, et al. (1996): Silencing of p16/CDKN2 expression in human gliomas by methylation and chromatin condensation. *Cancer Res* 56(10):2405–2410.

Cushing H (1908): *Surgery of the Head.* L. B. Adams, Division of Gryphon Editions, Birmingham, p 231 (reprinted in *Classics of Neurology and Neurosurgery Library*, 1983).

Dahmane N, Ruiz-i-Altaba A (1999): Sonic hedgehog regulates the growth and patterning of the cerebellum. *Development* 126(14):3089–3100.

Dai C, et al. (2001): PDGF autocrine stimulation dedifferentiates cultured astrocytes and induces oligodendrogliomas and oligoastrocytomas from neural progenitors and astrocytes in vivo. *Genes Dev* 15(15):1913–1925.

Dai C, Holland EC (2003): Astrocyte differentiation states and glioma formation. *Cancer J* 9(2):72–81.

Datta K, et al. (1996): Akt is a direct target of the phosphatidylinositol 3-kinase. Activation by growth factors, v-src and v-Ha-ras, in Sf9 and mammalian cells. *J Biol Chem* 271(48):30,835–30,839.

Davies MP, et al. (1999): Mutation in the PTEN/MMAC1 gene in archival low grade and high grade gliomas. *Br J Cancer* 79(9–10):1542–1548.

DeAngelis LM (2001): Brain tumors. *N Engl J Med* 344(2): 114–123.

Deinhardt F (1980): Biology of primate retroviruses. In Klein G (Ed), *Viral Oncology*. Raven, New York, pp 357–398.

Ding H, et al. (2001): Astrocyte-specific expression of activated p21-ras results in malignant astrocytoma formation in a transgenic mouse model of human gliomas. *Cancer Res* 61(9):3826–3836.

Ding H, et al. (2003): Oligodendrogliomas result from the expression of an activated mutant epidermal growth factor receptor in a RAS transgenic mouse astrocytoma model. *Cancer Res* 63(5):1106–1113.

Di Rocco F, et al. (1998): Platelet-derived growth factor and its receptor expression in human oligodendrogliomas. *Neurosurgery* 42(2):341–346.

Donehower L, et al. (1992): Mice deficient for p53 are developmentally normal but susceptible to spontaneous tumours. *Nature* 356(6366):215–221.

Doolittle RF, et al. (1983): Simian sarcoma virus onc gene, v-sis, is derived from the gene (or genes) encoding a platelet-derived growth factor. *Science* 221(4607):275–277.

Duerr EM, et al. (1998): PTEN mutations in gliomas and glioneuronal tumors. *Oncogene* 16(17):2259–2264.

Ekstrand AJ, et al. (1991): Genes for epidermal growth factor receptor, transforming growth factor alpha, and epidermal growth factor and their expression in human gliomas in vivo. *Cancer Res* 51(8):2164–2172.

Ekstrand AJ, et al. (1992): Amplified and rearranged epidermal growth factor receptor genes in human glioblastomas reveal deletions of sequences encoding portions of the N- and/or C-terminal tails. *Proc Natl Acad Sci USA* 89(10):4309–4313.

el-Deiry WS, et al. (1993): WAF1, a potential mediator of p53 tumor suppression. *Cell* 75(4):817–825.

Fakharzadeh SS, et al. (1991): Tumorigenic potential associated with enhanced expression of a gene that is amplified in a mouse tumor cell line. *EMBO J* 10(6):1565–1569.

Feldkamp MM, et al. (1999): Expression of activated epidermal growth factor receptors, Ras-guanosine triphosphate, and mitogen-activated protein kinase in human glioblastoma multiforme specimens. *Neurosurgery* 45(6):1442–1453.

Finkelstein SD, et al. (1994): Histological characteristics and expression of acidic and basic fibroblast growth factor genes in intracerebral xenogeneic transplants of human glioma cells. *Neurosurgery* 34(1):136–143.

Fleming TP, et al. (1992): Amplification and/or overexpression of platelet-derived growth factor receptors and epidermal growth factor receptor in human glial tumors. *Cancer Res* 52(16):4550–4553.

Fueyo J, et al. (1996): Hypermethylation of the CpG island of p16/CDKN2 correlates with gene inactivation in gliomas. *Oncogene* 13(8):1615–1619.

Fulci G, et al. (2000): p53 gene mutation and ink4a-arf deletion appear to be two mutually exclusive events in human glioblastoma. *Oncogene* 19(33):3816–3822.

Fults D, et al. (2002): MYC expression promotes the proliferation of neural progenitor cells in culture and in vivo. *Neoplasia* 4(1):32–39.

Goodrich LV, et al. (1997): Altered neural cell fates and medulloblastoma in mouse patched mutants. *Science* 277(5329): 1109–1113.

Guha A, et al. (1995): Expression of PDGF and PDGF receptors in human astrocytoma operation specimens supports the existence of an autocrine loop. *Int J Cancer* 60(2):168–173.

Guha A, et al. (1997): Proliferation of human malignant astrocytomas is dependent on Ras activation. *Oncogene* 15(23):2755–2765.

Haguenau F (1981): Comparative ultrastructure of human gliomas and experimental gliomas induced by Rous sarcoma virus (RSV) (author's transl.). *Neurochirurgie* 27(5):251–253.

Haguenau F, et al. (1971): Gliomas induced by rous sarcoma virus in the dog—an ultrastructural study. *J Natl Cancer Inst* 46(3):539–545 passim.

Hahn H, et al. (1996): Mutations of the human homolog of Drosophila patched in the nevoid basal cell carcinoma syndrome. *Cell* 85(6):841–851.

He J, et al. (1994): Analysis of glioma cell lines for amplification and overexpression of MDM2. *Genes Chromosom Cancer* 11(2):91–96.

Henson JW, et al. (1994): The retinoblastoma gene is involved in malignant progression of astrocytomas. *Ann Neurol* 36(5):714–721.

Hermanson M, et al. (1992): Platelet-derived growth factor and its receptors in human glioma tissue: Expression of messenger RNA and protein suggests the presence of autocrine and paracrine loops. *Cancer Res* 52(11):3213–3219.

Hermanson M, et al. (1996): Association of loss of heterozygosity on chromosome 17p with high platelet-derived growth factor alpha receptor expression in human malignant gliomas. *Cancer Res* 56(1):164–171.

Hesselager G, et al. (2003): Complementary effects of platelet-derived growth factor autocrine stimulation and p53 or Ink4a-Arf deletion in a mouse glioma model. *Cancer Res* 63(15):4305–4309.

Holland EC (2001a): Animal models of cell cycle dysregulation and the pathogenesis of gliomas. *J Neurooncol* 51(3):265–276.

Holland EC (2001b): Progenitor cells and glioma formation. *Curr Opin Neurol* 14(6):683–688.

Holland EC, et al. (2000): Combined activation of Ras and Akt in neural progenitors induces glioblastoma formation in mice. *Nat Genet* 25(1):55–57.

Holland EC, et al. (1998): A constitutively active epidermal growth factor receptor cooperates with disruption of G1 cell-cycle arrest pathways to induce glioma-like lesions in mice. *Genes Dev* 12(23):3675–3685.

Hui CC, et al. (1994): Expression of three mouse homologs of the Drosophila segment polarity gene cubitus interruptus, Gli, Gli-2, and Gli-3, in ectoderm- and mesoderm-derived tissues suggests multiple roles during postimplantation development. *Dev Biol* 162(2):402–413.

Humphrey PA, et al. (1990): Anti-synthetic peptide antibody reacting at the fusion junction of deletion-mutant epidermal growth factor receptors in human glioblastoma. *Proc Natl Acad Sci USA* 87(11):4207–4211.

Hurtt MR, et al. (1992): Amplification of epidermal growth factor receptor gene in gliomas: Histopathology and prognosis. *J Neuropathol Exp Neurol* 51(1):84–90.

Ichimura K, et al. (2000): Deregulation of the p14ARF/MDM2/p53 pathway is a prerequisite for human astrocytic gliomas with G1-S transition control gene abnormalities. *Cancer Res* 60(2):417–424.

Ichimura K, et al. (1996): Human glioblastomas with no alterations of the CDKN2A (p16INK4A, MTS1) and CDK4 genes have frequent mutations of the retinoblastoma gene. *Oncogene* 13(5):1065–1072.

Ino Y, et al. (2001): Molecular subtypes of anaplastic oligodendroglioma: Implications for patient management at diagnosis. *Clin Cancer Res* 7(4):839–845.

Jelsma RK, Bucy PC (1967): The treatment of glioblastoma multiforme. *Trans Am Neurol Assoc* 92:90–93.

Johnson DG, et al. (1993): Expression of transcription factor E2F1 induces quiescent cells to enter S phase. *Nature* 365(6444):349–352.

Johnson RL, et al. (1996): Human homolog of patched, a candidate gene for the basal cell nevus syndrome. *Science* 272(5268):1668–1671.

Kamijo T, et al. (1999): Tumor spectrum in ARF-deficient mice. *Cancer Res* 59(9):2217–2222.

Kamijo T, et al. (1998): Functional and physical interactions of the ARF tumor suppressor with p53 and Mdm2. *Proc Natl Acad Sci USA* 95(14):8292–8297.

Kleihues P, Cavenee W (2000): *Pathology and Genetics of Tumours of the Nervous System.* International Agency for Research on Cancer, Lyon.

Kleihues P, et al. (1976): Chemical carcinogenesis in the nervous system. *Int Rev Exp Pathol* 15:153–232.

Kleihues P, Ohgaki H (1999): Primary and secondary glioblastomas: From concept to clinical diagnosis. *Neuro-oncol* 1(1):44–51.

Knobbe CB, et al. (2002): Pten signaling in gliomas. *Neuro-oncol* 4(3):196–211.

Knudson AG (2001): Two genetic hits (more or less) to cancer. *Nat Rev Cancer* 1(2):157–162.

Knudson AG Jr (1971): Mutation and cancer: Statistical study of retinoblastoma. *Proc Natl Acad Sci USA* 68(4):820–823.

Koch R (1884): Die Aetiologie der Tuberkulose. *Mittheilungen aus dem Kaiserlichen Gesundheitsampte.* 2:1–88.

Koestner A, et al. (1971): Transplacental production with ethylnitrosourea of neoplasms of the nervous system in Sprague-Dawley rats. *Am J Pathol* 63(1):37–56.

Kokkinakis DM, et al. (2001): Characterization of initiated cells in N-methylnitrosourea-induced carcinogenesis of the CNS in the adult rat. *Neuro-oncol* 3(2):99–112.

Koutcher JA, et al. (2002): MRI of mouse models for gliomas shows similarities to humans and can be used to identify mice for preclinical trials. *Neoplasia* 4(6):480–485.

Kuan CT, et al. (2001): EGF mutant receptor vIII as a molecular target in cancer therapy. *Endocr Relat Cancer* 8(2):83–96.

Kumabe T, et al. (1992): Amplification of alpha-platelet-derived growth factor receptor gene lacking an exon coding for a portion of the extracellular region in a primary brain tumor of glial origin. *Oncogene* 7(4):627–633.

Lee Y, McKinnon PJ (2002): DNA ligase IV suppresses medulloblastoma formation. *Cancer Res* 62(22):6395–6399.

Levens D (2002): Disentangling the MYC web. *Proc Natl Acad Sci USA* 99(9):5757–5759.

Li J, et al. (1997): PTEN, a putative protein tyrosine phosphatase gene mutated in human brain, breast, and prostate cancer. *Science* 275(5308):1943–1947.

Louis DN, et al. (2001): Glioma classification: A molecular reappraisal. *Am J Pathol* 159(3):779–786.

Marino S, et al. (2000): Induction of medulloblastomas in p53-null mutant mice by somatic inactivation of Rb in the external granular layer cells of the cerebellum. *Genes Dev* 14(8):994–1004.

Momand J, et al. (1992): The mdm-2 oncogene product forms a complex with the p53 protein and inhibits p53-mediated transactivation. *Cell* 69(7):1237–1245.

Moulton T, et al. (1995): MTS1/p16/CDKN2 lesions in primary glioblastoma multiforme. *Am J Pathol* 146(3):613–619.

Newcomb EW, et al. (1998): Survival of patients with glioblastoma multiforme is not influenced by altered expression of p16, p53, EGFR, MDM2 or Bcl-2 genes. *Brain Pathol* 8(4):655–667.

Nister M, et al. (1988): Expression of messenger RNAs for platelet-derived growth factor and transforming growth factor-alpha and their receptors in human malignant glioma cell lines. *Cancer Res* 48(14):3910–3918.

Noble M, et al. (1988): Platelet-derived growth factor promotes division and motility and inhibits premature differentiation of the oligodendrocyte/type-2 astrocyte progenitor cell. *Nature* 333(6173):560–562.

Pardee AB (1989): G1 events and regulation of cell proliferation. *Science* 246(4930):603–608.

Peterson K, DeAngelis LM (2001): Weighing the benefits and risks of radiation therapy for low-grade glioma. *Neurology* 56(10):1255–1256.

Pietsch T, et al. (1997): Medulloblastomas of the desmoplastic variant carry mutations of the human homologue of Drosophila patched. *Cancer Res* 57(11):2085–2088.

Pomerantz J, et al. (1998): The Ink4a tumor suppressor gene product, p19Arf, interacts with MDM2 and neutralizes MDM2's inhibition of p53. *Cell* 92(6):713–723.

Posner JB (1996): Brain metastases: 1995. A brief review. *J Neurooncol* 27(3):287–293.

Rabotti GF, et al. (1966): Induction of multiple brain tumours (gliomata and leptomeningeal sarcomata) in dogs by Rous sarcoma virus. *Nature* 209(26):884–886.

Raffel C, et al. (1997): Sporadic medulloblastomas contain PTCH mutations. *Cancer Res* 57(5):842–845.

Rao G, et al. (2003): c-Myc enhances sonic hedgehog-induced medulloblastoma formation from nestin-expressing neural progenitors in mice. *Neoplasia* 5(3):198–204.

Rasheed BK, et al. (1997): PTEN gene mutations are seen in high-grade but not in low-grade gliomas. *Cancer Res* 57(19):4187–4190.

Reifenberger G, et al. (1993): Amplification and overexpression of the MDM2 gene in a subset of human malignant gliomas without p53 mutations. *Cancer Res* 53(12):2736–2739.

Reifenberger J, et al. (1996): Epidermal growth factor receptor expression in oligodendroglial tumors. *Am J Pathol* 149(1):29–35.

Reilly KM, et al. (2000): Nf1;Trp53 mutant mice develop glioblastoma with evidence of strain-specific effects. *Nat Genet* 26(1):109–113.

Robbins KC, et al. (1983): Structural and immunological similarities between simian sarcoma virus gene product(s) and human platelet-derived growth factor. *Nature* 305(5935):605–608.

Robins P, Lindahl T (1996): DNA Ligase IV from HeLa Cell Nuclei. *J Biol Chem* 271(39):24,257–24,261.

Robinson S, et al. (2001): Constitutive expression of growth-related oncogene and its receptor in oligodendrogliomas. *Neurosurgery* 48(4):864–873; discussion 873–874.

Ruiz i Altaba A, et al. (2002): Gli and hedgehog in cancer: Tumours, embryos and stem cells. *Nat Rev Cancer* 2(5):361–372.

Schlegel J, et al. (1994): Amplification and differential expression of members of the erbB-gene family in human glioblastoma. *J Neurooncol* 22(3):201–207.

Schmidt EE, et al. (1994): CDKN2 (p16/MTS1) gene deletion or CDK4 amplification occurs in the majority of glioblastomas. *Cancer Res* 54(24):6321–6324.

Schmidt EE, et al. (1997): Infrequent methylation of CDKN2A (MTS1/p16) and rare mutation of both CDKN2A and CDKN2B(MTS2/p15) in primary astrocytic tumours. *Br J Cancer* 75(1):2–8.

Shoshan Y, et al. (1999): Expression of oligodendrocyte progenitor cell antigens by gliomas: implications for the histogenesis of brain tumors. *Proc Natl Acad Sci USA* 96(18):10,361–10,366.

Smith JS, et al. (2000): Amplification of the platelet-derived growth factor receptor-A (PDGFRA) gene occurs in oligodendrogliomas with grade IV anaplastic features. *J Neuropathol Exp Neurol* 59(6):495–503.

Stone DM, et al. (1996): The tumour-suppressor gene patched encodes a candidate receptor for Sonic hedgehog. *Nature* 384(6605):129–134.

Tada T, et al. (2002): A head holder for magnetic resonance imaging that allows the stereotaxic alignment of spontaneously occurring intracranial mouse tumors. *J Neurosci Methods* 116(1):1–7.

Takenaka N, et al. (1985): Immunohistochemical detection of the gene product of Rous sarcoma virus in human brain tumors. *Brain Res* 337(2):201–207.

Tao W, Levine AJ (1999): P19(ARF) stabilizes p53 by blocking nucleocytoplasmic shuttling of Mdm2. *Proc Natl Acad Sci USA* 96(12):6937–6941.

Taylor MD, et al. (2002): Mutations in SUFU predispose to medulloblastoma. *Nat Genet* 31(3):306–310.

Tong WM, et al. (2001): Poly(ADP-ribose) polymerase: A guardian angel protecting the genome and suppressing tumorigenesis. *Biochim Biophys Acta* 1552(1):27–37.

Tong WM, et al. (2003): Null mutation of DNA strand break-binding molecule poly(ADP-ribose) polymerase causes medulloblastomas in p53(−/−) mice. *Am J Pathol* 162(1):343–352.

Ueki K, et al. (1996): CDKN2/p16 or RB alterations occur in the majority of glioblastomas and are inversely correlated. *Cancer Res* 56(1):150–153.

Uhrbom L, et al. (2002): Ink4a-Arf loss cooperates with KRas activation in astrocytes and neural progenitors to generate glioblastomas of various morphologies depending on activated Akt. *Cancer Res* 62(19):5551–5558.

Uhrbom L, et al. (1998): Induction of brain tumors in mice using a recombinant platelet-derived growth factor B-chain retrovirus. *Cancer Res* 58(23):5275–5279.

Vooijs M, et al. (2002): Noninvasive imaging of spontaneous retinoblastoma pathway-dependent tumors in mice. *Cancer Res* 62(6):1862–1867.

Walker AE, et al. (1985): Epidemiology of brain tumors: The national survey of intracranial neoplasms. *Neurology* 35(2):219–226.

Watanabe K, et al. (1996): Overexpression of the EGF receptor and p53 mutations are mutually exclusive in the evolution of primary and secondary glioblastomas. *Brain Pathol* 6(3):217–223; discussion 23–24.

Waterfield MD, et al. (1983): Platelet-derived growth factor is structurally related to the putative transforming protein p28sis of simian sarcoma virus. *Nature* 304(5921):35–39.

Wechsler-Reya RJ, Scott MP (1999): Control of neuronal precursor proliferation in the cerebellum by Sonic Hedgehog. *Neuron* 22(1):103–114.

Weiner HL, et al. (2002): Induction of medulloblastomas in mice by sonic hedgehog, independent of Gli1. *Cancer Res* 62(22):6385–6389.

Weiss WA, et al. (2003): Genetic determinants of malignancy in a mouse model for oligodendroglioma. *Cancer Res* 63(7):1589–1595.

Weiss WA, et al. (2002): Neuropathology of genetically engineered mice: Consensus report and recommendations from an international forum. *Oncogene* 21(49):7453–7463.

Weissenberger J, et al. (1997): Development and malignant progression of astrocytomas in GFAP-v-src transgenic mice. *Oncogene* 14(17):2005–2013.

Wetmore C, et al. (2000): The normal patched allele is expressed in medulloblastomas from mice with heterozygous germ-line mutation of patched. *Cancer Res* 60(8):2239–2246.

Wetmore C, et al. (2001): Loss of p53 but not ARF accelerates medulloblastoma in mice heterozygous for patched. *Cancer Res* 61(2):513–516.

Wikstrand CJ, et al. (1995): Monoclonal antibodies against EGFRvIII are tumor specific and react with breast and lung carcinomas and malignant gliomas. *Cancer Res* 55(14): 3140–3148.

Williams BO, et al. (1994): Extensive contribution of Rb-deficient cells to adult chimeric mice with limited histopathological consequences. *EMBO J* 13(18):4251–4259.

Williams BP, et al. (1997): A PDGF-regulated immediate early gene response initiates neuronal differentiation in ventricular zone progenitor cells. *Neuron* 18(4):553–562.

Wolter M, et al. (2001): Oligodendroglial tumors frequently demonstrate hypermethylation of the CDKN2A (MTS1, p16INK4a), p14ARF, and CDKN2B (MTS2, p15INK4b) tumor suppressor genes. *J Neuropathol Exp Neurol* 60(12): 1170–1180.

Xiao A, et al. (2002): Astrocyte inactivation of the pRb pathway predisposes mice to malignant astrocytoma development that is accelerated by PTEN mutation. *Cancer Cell* 1(2): 157–168.

Xu GF, et al. (1990): The catalytic domain of the neurofibromatosis type 1 gene product stimulates ras GTPase and complements ira mutants of S. cerevisiae. *Cell* 63(4): 835–841.

Xu S, et al. (2003): In vivo multiple-mouse imaging at 1.5 T. *Magn Reson Med* 49(3):551–557.

Zurawel RH, et al. (2000): Evidence that haploinsufficiency of Ptch leads to medulloblastoma in mice. *Genes Chromosom Cancer* 28(1):77–81.

13

MYELOID MALIGNANCIES

Scott C. Kogan

Comprehensive Cancer Center and Department of Laboratory Medicine, University of California—San Francisco, San Francisco, California

A wealth of information is available regarding genetic abnormalities that contribute to human myeloid malignancies. Modeling these mutations in mice has been facilitated by the ability to manipulate the hematopoietic system not only through germline modification but also by retroviral transduction of bone marrow. Chimeric fusion proteins resulting from balanced chromosomal aberrations as well as intragenic mutations have been introduced into mice to generate models of many types of human myeloid leukemias. These models recapitulate important features of the human diseases and have been utilized to elucidate mechanisms through which particular genetic changes contribute to myeloid neoplasia.

The study of human myeloid neoplasms has led to many advances in our understanding of these disorders and of other human malignancies. The association of specific chromosomal changes in neoplastic cells with particular malignancies was begun by the identification of the Philadelphia chromosome in chronic myelogenous leukemia (CML) (Nowell, 1960). CML has remained a leading source of insight into the pathogenesis and treatment of human neoplasms: from identifying that the Philadelphia chromosome creates an oncogenic BCR/ABL fusion protein to the development of imatinib mesylate as a molecularly targeted therapeutic agent. In addition to the t(9;22) of CML, numerous recurrent balanced chromosomal aberrations have been identified in human myeloid neoplasms. In fact, the genes altered by all of the major (and many of the minor) balanced aberrations of myeloid leukemia have been identified. Most of these aberrations result in the production of abnormal fusion proteins, while others result in dysregulated gene expression. In addition, a substantial number of recurrent point mutations and small intragenic changes have also been associated with human leukemias and related disorders. Because there is a wealth of information regarding the genetic changes that are present in these human diseases, it has been possible to model numerous such changes in mice.

HEMATOPOIESIS, NORMAL AND NEOPLASTIC

To appreciate the issues and challenges that modeling of myeloid neoplasms entails, it is necessary to understand the contexts in which these diseases arise in humans and in mice.

Development

The basic principles of hematopoietic development for mice and humans have been elucidated in a number of experimental systems, including mice. Embryonic hematopoiesis is followed by the development of the definitive hematopoietic system that functions throughout life and from which neoplasms may arise. In the mouse, at embryonic day 7.5 primitive erythroid cells appear as blood islands in the yolk sac (Russell and Bernstein, 1966). These primitive erythroid cells retain their nuclei, unlike those of definitive hematopoiesis. Hematopoietic stem cells (HSCs), defined as cells that can give rise to all blood cell lineages and that have the potential to persistently contribute to hematopoiesis, can be identified in the middle of mouse embryonic day 10 in the

Mouse Models of Human Cancer, edited by Eric C. Holland
ISBN 0-471-44460-X

aorta–gonad–mesonephros region (Muller et al., 1994). At approximately this point in embryogenesis HSCs can also be identified in the yolk sac and fetal liver. It appears that the first HSCs arise from endothelial cells, referred to as hemangioblasts. HSCs have been identified as arising in the ventral surface of the dorsal aorta (reviewed in Marshall and Thrasher, 2001; North et al., 1999).

The liver is a major site of hematopoiesis from day 11 onward, with decreasing intrahepatic hematopoiesis in the few days prior to birth (Fredrickson and Harris, 2000). During days 16 and 17 of mouse development, myelopoiesis becomes apparent in the bone marrow and spleen. In postnatal life the bone marrow is the major site of hematopoiesis in mice and humans. However, it is important to note that whereas in humans splenic hematopoiesis disappears, hematopoiesis in the spleen continues throughout life in mice.

Neoplasms of the lymphoid system are discussed in detail in Chapter 14. Lymphocytes are derived from HSCs. Important times in the development of lymphoid organs of mice include the appearance of B cells in the liver at embryonic day 12, the colonization of the thymus with lymphoid cells at embryonic day 15, and the maturation of lymphoid areas within the spleen and lymph nodes during the first two weeks of postnatal life (Fredrickson and Harris, 2000).

Differentiation

HSCs can give rise to all the hematopoietic lineages. Elegant studies have established a hierarchy of hematopoietic progenitors that have been defined by immunophenotype, that is, by the expression of particular molecules on the surface of these cells (Akashi et al., 2000; Kondo et al., 1997). HSCs that contribute to all hematopoietic lineages can be divided into long term (LT-HSC) and short term (ST-HSC). Whereas LT-HSCs can contribute to hematopoiesis for the life of the organism, ST-HSC have limited capacity for self-renewal contributing to hematopoiesis for up to about eight weeks (Morrison and Weissman, 1994). The steps through which HSCs differentiate into progenitor and differentiated cells is illustrated in Figure 13.1. In humans, the two major arms of the hematopoietic system are referred to as lymphoid (including B cells, T cells, and NK cells) and myeloid [including granulocytic, monocytic, erythroid (red blood cell), and megakaryocytic (platelet) lineages]. In mice, however, the term *myeloid* has been used by rodent pathologists in a more restricted sense to specify granulocytic and/or monocytic cells, with *nonlymphoid hematopoietic cells* being proposed to encompass all hematologic cells other than lymphocytes (Kogan et al., 2002).

Neoplasms of the nonlymphoid (myeloid) hematopoietic system are primarily characterized by the lineage(s) of the neoplastic cells as well as by their differentiation state [degree of maturation, presence of abnormal maturation (dysplasia)]. Additional important characteristics include clinical (history of exposure to carcinogenic therapies, history of a preceding disorder, persistence of unexplained abnormalities), biologic (manifestations of disease in peripheral blood, location and spread of disease in the organism, aggressiveness, transplantability), and genetic features. The particular manner in which these characteristics are utilized in the classification of nonlymphoid (myeloid) hematopoietic neoplasms of humans and mice is detailed, respectively, in the *World Health Organization Classification of Tumours. Pathology and Genetic of Haematopoietic and Lymphoid Tissues* (Jaffe et al., 2001) and in the "Bethesda Proposals for the Classification of Nonlymphoid Hematopoietic Neoplasms in Mice" (Kogan et al., 2002).

Although a broad spectrum of neoplasms involving nonlymphoid (myeloid) hematopoietic cells has been observed in both humans and mice, there are important differences in the types of disease that are commonly observed. Tumors of macrophages (histiocytes) are rare in humans but frequent in mice. Similarly, leukemias comprised primarily of erythroid cells are seen more often in mice than in humans. In contrast, whereas a number of human neoplasms caused by specific recurrent genetic abnormalities [e.g., chronic myelogenous leukemia (CML)] have been delineated, spontaneous murine diseases with homologous morphologic and genetic features have not been described.

Sites of Hematopoiesis

Neoplasms generally arise in locations in which cells normally proliferate. Hence, although lymphoid neoplasms can arise in the bone marrow, such neoplasms also frequently arise in peripheral lymphoid tissues (i.e., thymus, spleen, lymph nodes, Peyer's patches) but can develop in any location in the body (skin and the central nervous system being particularly important sites for human lymphomas). The location in which various lymphoid neoplasms appear in humans and mice is also discussed in Chapter 14.

Similarly, myeloid neoplasms usually develop in sites of normal myelopoiesis. Hence, most myeloid neoplasms in humans arise in the bone marrow, whereas in mice such neoplasms commonly arise in the spleen as well. Specifically, human acute myeloid leukemia (AML), chronic myeloproliferative disease (MPD), myelodysplastic syndrome (MDS), and myelodysplastic/myeloproliferative disease (MD/MPD) almost always arise in the bone marrow. Mouse nonlymphoid leukemia, myeloid proliferation (nonreactive), and myeloid dysplasia correspondingly arise in the marrow or spleen. In contrast to humans in whom solid tumors of myeloid cells are rare, nonlymphoid hematopoietic sarcomas are often observed in

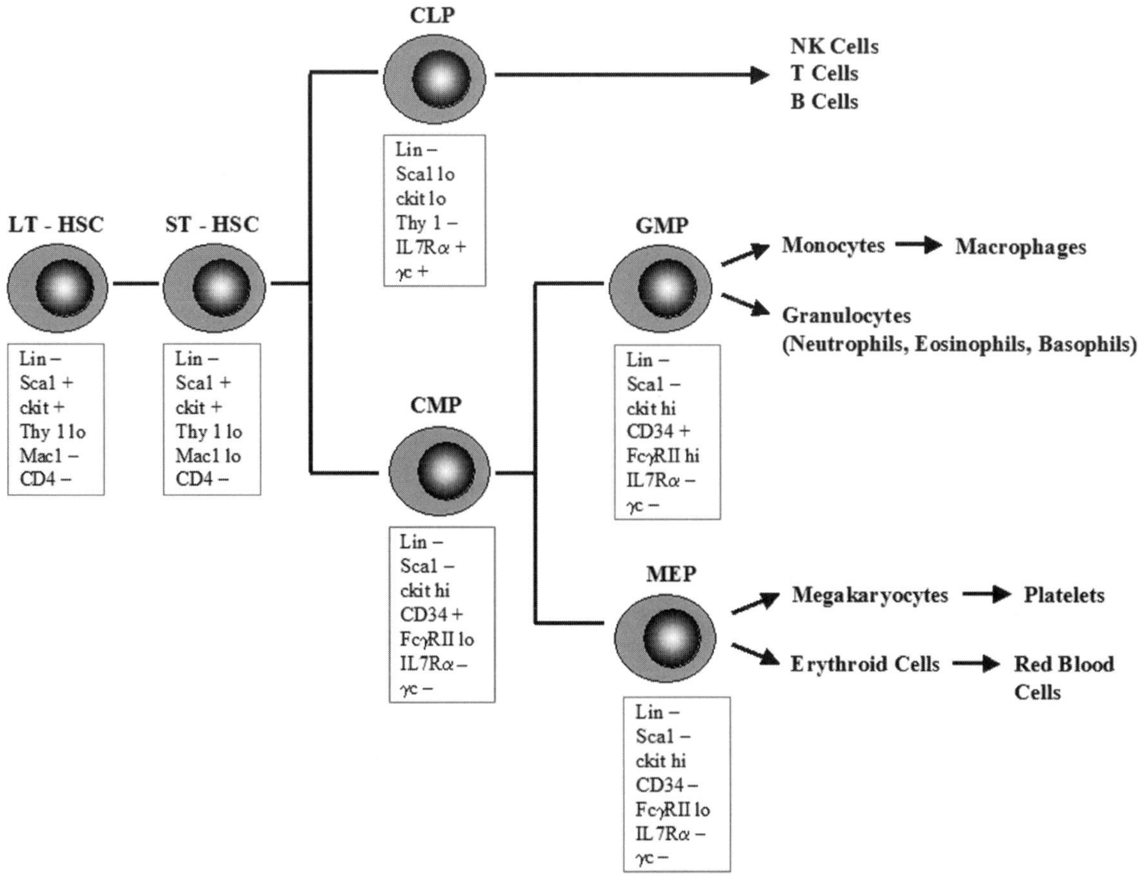

Figure 13.1. Hematopoiesis. Hematopoietic stem cells give rise to all hematopoietic lineages. LT-HSCs can contribute to hematopoiesis for the life of the organism. ST-HSCs have more limited capacity for self-renewal. Common lymphoid progenitors (CLPs) and common myeloid progenitors (CMPs) give rise to the cells of lymphoid and myeloid lineages, respectively. Lymphocyte development is detailed in Chapter 14. Granulocyte/macrophage progenitors (GMPs) and megakaryocyte/erythrocyte progenitors (MEPs) are two branches of the myeloid differentiation pathway. The immunophenotypes of these progenitors in mice are shown (Akashi et al., 2000; Kondo et al., 1997; Morrison and Weissman, 1994).

mice: Histiocytic sarcomas frequently develop from tissue macrophages of the liver, spleen, or uterus (Fredrickson and Harris, 2000).

HUMAN MYELOID NEOPLASMS

Human myeloid neoplasms encompass AML, MPD, MDS, and MD/MPD. Although in the World Health Organization (WHO) classification other nonlymphoid hematopoietic neoplasms of humans are separated from these four categories, histiocytic and mast cell neoplasms are discussed in this chapter whereas neoplasms of dendritic cells are mentioned in Chapter 14. The entities delineated in the WHO classification are listed in Table 13.1. Representative cytology of human myeloid neoplasms is shown in Figure 13.2.

Acute Myeloid Leukemias

AML is defined by the presence of greater than or equal to 20% myeloid blasts in the blood and/or bone marrow (Jaffe et al., 2001). AML is also occasionally diagnosed despite lower percentages of blasts when a myeloid neoplasm is shown to harbor *AML1/ETO* or *CBFB/MYH11* fusion genes or rarely when a solid tumor is derived from the myeloid lineage. Under the WHO classification, AML is subdivided into four major groupings. The first group is defined genetically by the presence of *AML1/ETO, CBFB/MYH11*, or *PML/RARA* (or variant *RARA*) fusion genes or by abnormalities of the *MLL* gene at 11q23. The second group is defined by the presence of abnormal (dysplastic) differentiation in greater than or equal to 50% of cells of at least two cell lineages (of neutrophilic, erythroid, and megakaryocytic lineages). The third group is defined by a clinical history

Table 13.1. World Health Organization Classification of Myeloid Malignancies

Acute Myeloid Leukemia

AML with recurrent genetic abnormalities
- Acute promyelocytic leukemia [AML with t(15;17)(q22;q12), *PML/RARa* and variants]
- AML with t(8;21)(q22;q22), *AML1/ETO*
- AML with inv(16)(p13q22) or t(16;16)(p13;q22), *CBFb/MYH11*
- AML with 11q23 (*MLL*) abnormalities

AML with multilineage dysplasia
- With prior MDS
- Without prior MDS

AML and MDS, therapy related
- Alkylating agent related
- Topoisomerase II inhibitor related

Acute myeloid leukemia not otherwise categorized
- AML minimally differentiated
- AML without maturation
- AML with maturation
- Acute myelomonocytic leukemia
- Acute monoblastic and monocytic leukemia
- Acute erythroid leukemia
- Acute megakaryoblastic leukemia
- Acute basophilic leukemia
- Acute panmyelosis with myelofibrosis
- Myeloid sarcoma

Acute leukemia of ambiguous lineage
- Undifferentiated acute leukemia
- Bilineal acute leukemia
- Biphenotypic acute leukemia

Chronic Myeloproliferative Disease

- Chronic myelogenous leukemia [Philadelphia chromosome, t(9;22)(q34;q11), *BCR/ABL* positive]
- Chronic neutrophilic leukemia
- Chronic eosinophilic leukemia (and the hypereosinophilic syndrome)
- Polycythemia vera
- Chronic idiopathic myelofibrosis (with extramedullary hematopoiesis)
- Essential thrombocythemia
- Chronic myeloproliferative disease, unclassifiable

Myelodysplastic Syndromes

- Refractory anemia
- Refractory anemia with ringed sideroblasts
- Refractory cytopenia with multilineage dysplasia
- Refractory anemia with excess blasts
- Myelodysplastic syndrome associated with isolated del(5q) chromosome abnormality
- Myelodysplastic syndrome, unclassifiable

Myelodysplastic/Myeloproliferative Diseases

- Chronic myelomonocytic leukemia
- Atypical chronic myeloid leukemia
- Juvenile myelomonocytic leukemia
- MD/MPD, unclassifiable

Histiocytic Neoplasms

- Histiocytic sarcoma
- Langerhans cell histiocytosis
- Langerhans cell sarcoma

(*continued overleaf*)

Table 13.1. (*continued*)

Mastocytosis
Cutaneous mastocytosis
Systemic mastocytosis
Indolent systemic mastocytosis
Systemic mastocytosis with associated clonal hematologic non–mast cell lineage disease
Aggressive systemic mastocytosis
Mast cell leukemia
Mast cell sarcoma
Extracutaneous mastocytoma

Source: Jaffe et al., 2001.

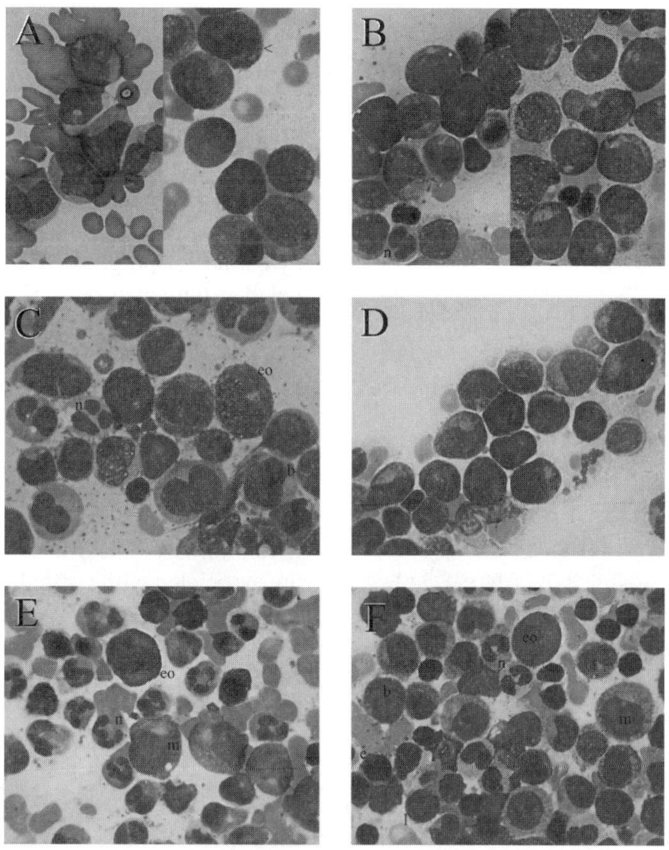

Figure 13.2. Cytology of human myeloid neoplasms. Wright's Giemsa stain. Bone marrow unless otherwise noted. Original magnification 1000×. Selected cells and components are identified: b, blast; e, erythroid cell; eo, eosinophil; l, lymphocyte; m, myelocyte; n, neutrophil. (*a*) Two examples of human APL. Left image is blood (microgranular APL) and right image is marrow aspirate (typical hypergranular APL). An abnormal promyelocyte with multiple Auer rods is indicated by an arrowhead. (*b*) Two examples of human AML with t(8;21). Left image is from a patient whose leukemic cells were heterogeneous, including dysplastic neutrophils. Right image is from a patient whose leukemic cells were more homogeneous. Auer rod is indicated by an arrowhead. (*c*) Human AML with inv(16). Note eosinophils with basophilic granulation, blasts with indented monocytic nuclei, and neutrophils. (*d*) Human AML with 11q23 abnormality. Marrow is filled with monoblasts. (*e*) Human chronic myelogenous leukemia. Marrow is filled with maturing myeloid elements. (*f*) Human juvenile myelomonocytic leukemia. Note heterogeneous mix of cells, including increased myeloid elements.

of treatment with cytotoxic chemotherapy or radiation. Finally, the fourth group is defined by traditional criteria of cell lineage (defined by morphology, cytochemistry, and immunophenotype), degree of maturation (defined by morphology), and histologic features.

Clinical Features. AMLs are aggressive diseases that if untreated are fatal in days to weeks. The lethality of these leukemias is primarily due to a loss of normal hematopoiesis with neutropenia leading to uncontrolled infection or thrombocytopenia resulting in fatal hemorrhage. Occasionally, the leukemic cells may directly initiate a fatal dysregulation of hemostasis known as disseminated intravascular coagulation.

Genetic Features. Numerous genetic abnormalities have been identified in AML. Cytogenetic analysis has important roles in classification and in defining prognosis. Table 13.2 depicts risk stratification of AML by karyotype (Grimwade et al., 1998). Whereas a great deal of progress has been made in identifying genes altered by balanced chromosomal aberrations in AML, genes that contribute to AML through unbalanced rearrangements remain largely unidentified. Many genes implicated in regulating cellular proliferation, survival, and differentiation have also been examined in AML. Table 13.3 summarizes many important genetic changes associated with human AML.

Therapy. The main treatment modality of AML is chemotherapy, generally including an anthracycline and cytosine arabinoside. However, AML is one area in which molecularly targeted therapeutics have had a significant impact. Both all-*trans* retinoic acid (ATRA) and arsenic trioxide target the PML/RARα fusion protein that underlies acute promyelocytic leukemia (APL); these two targeted therapeutics have dramatically enhanced long-term survival for APL patients. Drugs that target RAS and FLT3 have been in clinical trials and may be useful in some AML patients. Nevertheless, AML with a poor-risk karyotype remains a disease in which five-year survival is <20% (Grimwade et al., 1998).

Table 13.2. Cytogenetic Risk Groups in Human Acute Myeloid Leukemia

	Favorable
t(15;17)	t(8;21)
inv(16)	
	Intermediate
Normal	del(7q)
+8	del(9q)
+21	Abnormal 11q23
+22	
All other structural/numerical abnormalities	
	Adverse
−5	del(5q)
−7	Abnormal 3q
Complex (at least five unrelated cytogenetic abnormalities)	

Source: Grimwade et al., 1998.

Chronic Myeloproliferative Diseases

MPDs encompass myeloid neoplasms characterized by increased numbers of myeloid cells in the peripheral blood but with relatively normal differentiation (Jaffe et al., 2001). The most common of these diseases is CML, which is genetically defined by the presence of the *BCR/ABL* fusion gene. The other disorders represent a spectrum of neoplasms in which there is excessive expansion of granulocytic, erythroid, and/or megakaryocytic cells.

Clinical Features. MPDs can be indolent disorders with a modest impact on survival (e.g., essential thrombocythemia) or progressive disorders that inevitably lead to death if untreated (e.g., CML). These disorders are often associated with organomegaly (enlargement of the spleen and liver). Patients may also experience symptoms related to their increased blood cell counts (e.g., thrombosis).

Genetic Features. Whereas CML is defined by the presence of the *BCR/ABL* fusion gene, the other common MPDs [polycythemia vera (PV), chronic idiopathic myelofibrosis (CIMF), essential thrombocythemia (ET)] are associated with normal karyotypes or unbalanced chromosomal aberrations. Among the chromosomal abnormalities associated with these MPDs are del(1p) (PV), partial trisomy 1q (CIMF), del(5q)/−5 (CIMF), del(7q)/−7 (CIMF), +8 (PV/CIMF/ET), +9 (PV/CIMF/ET), del(13q) (PV/CIMF/ET), and del(20q) (PV/CIMF) (Jaffe et al., 2001).

Therapy. Therapy of MPDs is in part directed at ameliorating symptoms caused by the elevated blood counts and organomegaly. Allogeneic bone marrow transplantation can be curative. The most common MPD, CML, is treated with a variety of agents, including imatinib mesylate. Imatinib mesylate directly inhibits the abnormal BCR/ABL tyrosine kinase and can lead to prolonged remissions (Druker et al., 2001).

Myelodysplastic Syndromes

Myelodysplastic syndromes are a group of disorders characterized by cytopenias (low blood counts) and abnormal differentiation (Jaffe et al., 2001). Both cytopenias and abnormal differentiation can be manifest in

Table 13.3. Selected Genetic Abnormalities of Human Acute Myeloid Leukemia

Recurrent Balanced Aberrations (Selected of 142)	
t(15;17)(q22;q12) and variants; *PML/RARA, PLZF/RARA, NPM/RARA, NuMA/RARA, STAT5B/RARA*	
t(8;21)(q22;q22); *AML1 (RUNX1)/ETO*	
inv(16)(p13q22), t(16;16)(p13;q22); *CBFB/MYH11*	
11q23 translocations; *MLL* and multiple partners	
t(8;16)(p11;p13); *MOZ/CBP*	
t(6;9)(p23;q34); *DEK/CAN (NUP214)*	
inv(3)(q21q26), t(3;3)(q21;q26); *EVI1*	
t(3;21)(q24;q22); *AML1/MDS1/EVI1*	
t(7;11)(p15;p15); *NUP98/HOXA*	
t(1;22)(p13;q13); *RBM15/MKL1*	
Recurrent Unbalanced Aberrations (Selected of 735)	
+4, del(5q)/−5, del(7q)/−7, +8, +9, del(9q), +11, del(11q), del(12p), +13, −18, +19, del(20q), +21, +22	
Gene Amplification (Double Minute Chromosomes, Homogeneously Staining Regions)	
MYC	*MLL*
	Intragenic Mutations (Activating or Inactivating)
RAS	*CEBPA*
FLT3	*GATA1*
KIT	*PU.1 (SPI1)*
GCSFR (CSF3R)	*RB1*
AML1 (RUNX1)	*TP53*

Source: Ariyama et al., 1998; Cline, 1994; Dong et al., 1995; Grimwade et al., 1998; Jaffe et al., 2001; Kelly and Gilliland, 2002; Le Beau and Larson, 2000; Look, 1997; Mueller et al., 2002; Pabst et al., 2001; Tanaka et al., 1993; Wechsler et al., 2002.

one or more myeloid lineages (neutrophilic, erythroid, megakaryocytic). The abnormal differentiation can be morphologic dysplasia (dysgranulopoiesis, dyserythropoiesis, dysmegakaryocytopoiesis), increased myeloid blasts, or both.

Clinical Features. Patients with MDS can have stable or progressive disease in which they may experience symptoms of anemia and may be at risk of infection or hemorrhage due to neutropenia or thrombocytopenia, respectively. Although some patients do progress to acute leukemia, many MDS patients die as a result of their cytopenias.

Genetic Features. MDS has been associated with a spectrum of genetic changes, including activation of the *KRAS* or *NRAS* genes, mutation or loss of *TP53*, t(3;3)/inv(3), del(5q)/−5, del(7q)/−7, +8, +9, +11, del(11q), del(12p), −18, +19, del(20q), and +21 (Jaffe et al., 2001).

Therapy. The mainstay of MDS therapy is efforts to decrease cytopenias. A number of approaches are used, including transfusion, cytokines, and 5-azacytidine. Some patients may undergo allogeneic bone marrow transplantation with intent to cure. If patients progress to AML, they are often treated with AML-type chemotherapy, but prognosis in these patients remains poor.

Myelodysplastic/Myeloproliferative Diseases

MD/MPD is a subcategory of myeloid neoplasia that was incorporated into the WHO classification. It was created to encompass a group of heterogeneous diseases characterized by increased blood counts of certain myeloid lineage cells (usually accompanied by abnormal differentiation) and, in some patients, cytopenias of other myeloid lineage cells (Jaffe et al., 2001).

Clinical Features. Clinical characteristics of the MD/MPDs vary depending on the extent to which the increased or decreased blood counts result in clinical illness. The relative contribution of expanded hematopoiesis and cytopenias to a patient's clinical condition can shift over time.

Genetic Features. Interestingly, although the MD/MPDs are quite heterogeneous, a number of genetic abnormalities have been identified in subgroups or in individual patients (Jaffe et al., 2001). Activating mutations of a *RAS* gene are common in chronic myelomonocytic leukemia (CMML) and juvenile myelomonocytic leukemia (JMML); RAS can also be activated by loss of *NF1* in JMML patients. del(7q)/−7 is another common feature of CMML and JMML. +8, del(12p), del(20q), and i(17q) (with loss of *TP53*) are observed in CMML and atypical CML (aCML). Interestingly, balanced chromosomal aberrations involving the tyrosine kinases *PDGFRB* on chromosome 5 and *FGFR1* on chromosome 12 have been observed in MD/MPDs as well as in patients with other hematopoietic neoplasms.

Therapy. Treatment may be similar to that of MPD or MDS, depending on the abnormalities observed in an individual. Cases in which an activated tyrosine kinase is identified can benefit from treatment with tyrosine kinase inhibitors.

Other Nonlymphoid Hematopoietic Neoplasms

Histiocytic sarcomas, which are solid tumors of macrophages, are very rare in humans. They must be carefully distinguished from T- and B-cell neoplasms by excluding the presence of lymphoid features, including demonstrating the absence of clonal T- or B-cell receptor gene rearrangements.

Systemic mastocytosis is an uncommon human disorder. Systemic mastocytosis can have a variety of forms, ranging from an indolent disorder associated with cutaneous mast cell proliferations to expansion of mast cells accompanying acute leukemias to aggressive leukemias or sarcomas composed of mast cells. In humans, mast cell neoplasms are associated with activating mutations of the *KIT* receptor tyrosine kinase (Valent et al., 2001).

MOUSE MODELS OF HUMAN MYELOID LEUKEMIAS AND RELATED DISORDERS

Mouse models of human diseases can represent powerful systems for elucidating pathogenesis and for assessing novel therapies. An issue central to the use of animal models is validation. Validation is the process of delineating those attributes of the model that accurately match the attributes of the human disease (Cardiff, 2001). To facilitate the validation process, the hematopathology subcommittee of the Mouse Models of Human Cancers Consortium developed proposals for the characterization and classification of lymphoid and nonlymphoid hematopoietic neoplasms of mice (Kogan et al., 2002; Morse et al., 2002). It was hoped that these proposals will "allow investigators to diagnose lesions as well-defined entities according to accepted criteria" and that "a uniform classification will enhance the value of the scientific literature and will make it possible to meaningfully compare and contrast murine lesions with human lesions" (Kogan et al., 2002, p. 238). The entities delineated in the Bethesda Proposals for the Classification of Nonlymphoid Hematopoietic Neoplasms in Mice are listed in Table 13.4. Representative cytology of mouse nonlymphoid hematopoietic neoplasms is shown in Figure 13.3, with representative histopathology shown in Figure 13.4.

These proposals are designed to facilitate comparisons to human disease while also utilizing nomenclature for mice that is distinct from that used for humans, when necessary to highlight differences between the two species. Because the spectrum of diseases described in mice is more limited than the diseases defined in the human WHO classification, quite a few diseases noted in Table 13.1 do not have corresponding entries in Table 13.4. These differences between the classifications have been discussed in more detail elsewhere (Kogan et al., 2002).

What follows is a selected overview of mouse models of human myeloid malignancies. The focus is on models in which neoplasms arise with a penetrance that facilitates studies of pathogenesis and therapy, particularly those in which a genetic alteration found in humans induces an analogous disease in mice. Also discussed are selected cooperating events most closely associated with human disease. Hence, this chapter is not a comprehensive review of (i) all known murine nonlymphoid hematopoietic neoplasms, (ii) hematologic effects of genetic alterations in mice, (iii) the growth of abnormal cell lines in mice, or (iv) leukemia pathogenesis. The reader may also be interested in materials regarding myeloid malignancies in mice available at the website of the Mouse Models of Human Cancers Consortium.

Models of Human AML

Nonlymphoid leukemias of mice parallel the AMLs of humans with two important exceptions. First, leukemias of mice that share many common attributes (i.e., expanded myelopoiesis, suppression of normal hematopoiesis reflected as peripheral blood cytopenias, dissemination of neoplastic cells outside of normal hematopoietic tissues, and rapid lethality) can show a range of morphologic differentiation from poorly differentiated leukemias composed entirely of immature forms/blasts to well-differentiated leukemias in which there is no apparent block to maturation. Hence, in contrast to human AMLs, the mouse classification includes a category (MPD-like myeloid leukemia) that encompasses well-differentiated clinically aggressive neoplasms. Second,

Table 13.4. Bethesda Proposals for the Classification of Nonlymphoid Hematopoietic Neoplasms in Mice

Nonlymphoid Leukemia	
Features: (1) diffuse involvement of hematopoietic tissue, (2) peripheral blood cytopenias, (3) dissemination of leukemic cells, (4) additional aspect of malignancy in appearance or behavior (i.e., at least 20% immature forms/blasts and/or rapidly fatal)	
Myeloid leukemia	
Myeloid leukemia without maturation	Myelomonocytic leukemia
Myeloid leukemia with maturation	Monocytic leukemia
MPD-like myeloid leukemia	
Erythroid leukemia	
Megakaryocytic leukemia	
Biphenotypic leukemia	
Additional descriptor: acute, with multilineage dysplasia, therapy related	
Myeloid Proliferation (Nonreactive)	
Features: (1) diffuse involvement of hematopoietic tissue, (2) increased number of nonlymphoid hematopoietic cells, (3) not leukemia or myeloid dysplasia	
Myeloproliferation (genetic)	Myeloproliferative disease
Additional descriptor: with features of a named human chronic myeloproliferative disease or myelodysplastic/myeloproliferative disease	
Myeloid Dysplasia	
Features: (1) diffuse involvement of hematopoietic tissue, (2) defective maturation of nonlymphoid hematopoietic cells (3) peripheral blood cytopenias, (4) not leukemia	
Cytopenia with increased blasts	Myelodysplastic syndrome
Additional descriptor: with features of a named human myelodysplastic syndrome	
NonLymphoid Hematopoietic Sarcoma	
Features: (1) solid tumor mass without diffuse involvement of hematopoietic tissue, (2) invasive and/or metastatic	
Granulocytic sarcoma	Mast cell sarcoma
Histiocytic sarcoma	

Source: Kogan et al., 2002

myeloid sarcomas are classified as AML in the WHO classification but are included with other nonlymphoid hematopoietic sarcomas in the Bethesda Proposals (along with histiocytic sarcoma and mast cell sarcoma).

Models of AML with Recurrent Genetic Abnormalities. Mouse models have been developed by expression of *PML/RARA* (and variants), *AML1/ETO*, and *CBFB/MYH11* as well as a number of *MLL* (11q23) fusions.

Models of Acute Promyelocytic Leukemia. A number of models of human APL have been created in mice, and these models have been used for studies of pathogenesis, cooperating genetic events, and therapy. Close to 99% of human APL is associated with expression of the *PML/RARA* fusion gene. This fusion brings together coding sequences for PML, a nuclear body protein that can suppress cell growth and stimulate apoptosis, with the sequences encoding a nuclear hormone receptor for retinoic acid (RARα). In response to the RAR ligand ATRA, APL cells will undergo morphologic differentiation; the combination of ATRA with chemotherapy has greatly enhanced cure for this disease.

APL models were generated by expressing *PML/RARA* as a transgene or by reconstituting irradiated normal recipient animals with mouse bone marrow that had been transduced with a retrovirus designed to express *PML/RARA* (Brown et al., 1997; Grisolano et al., 1997; He et al., 1997; Minucci et al., 2002). The cathepsin G promoter drives transgene expression primarily at the promyelocyte stage of differentiation. Mice that express *PML/RARA* under the regulation of the cathepsin G promoter initially show mild abnormalities of myelopoiesis but go on to develop an AML (Grisolano et al., 1997; He et al., 1997). Human APL has a characteristic cytology, including cells with multiple Auer rods (abnormal elongated primary granules) and cells with primary granules that are so numerous they obscure the nuclear border. Another feature of human APL is that very few cells show maturation beyond the promyelocyte stage. The leukemias

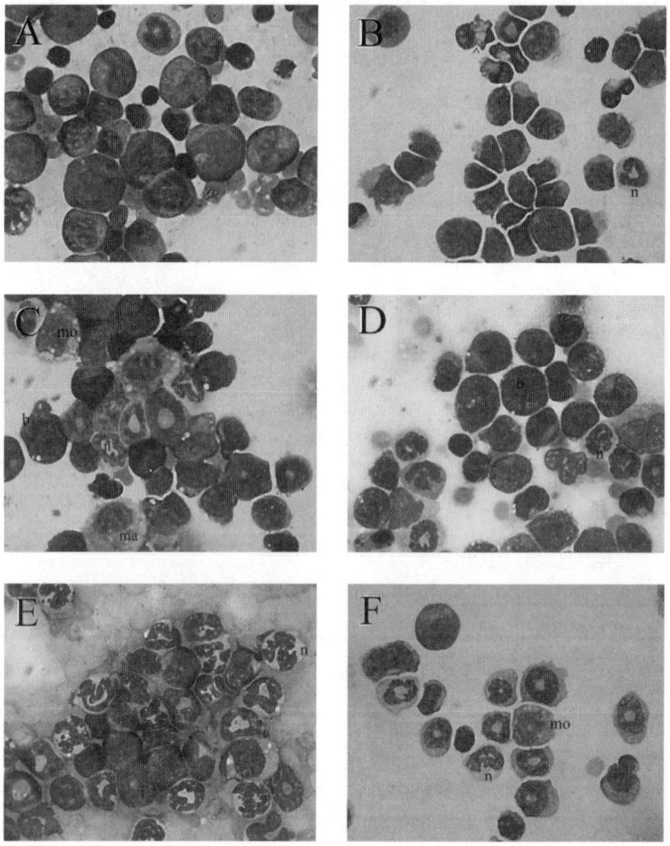

Figure 13.3. Cytology of mouse nonlymphoid hematopoietic neoplasms. Wright's Giemsa stain. Bone marrow unless otherwise noted. Original magnification 1000×. Selected cells are identified: b, immature form/blast; i, intermediate cell; ma, macrophage; mo, monocyte; n, neutrophil. (*a*) Myeloid leukemia in a *PML/RARA* transgenic mouse. Note granular cytoplasm of immature forms/blasts. (*b*) Myeloid leukemia in an *Aml1/ETO* conditional knockin mouse in the BXH-2 strain (which carries an endogenous retrovirus). A dysplastic cell with a bilobed nucleus is indicated by an arrowhead. (*c*) Myeloid leukemia in an ethylnitrosourea-treated *Cbfb/MYH11* knockin chimeric mouse. Note heterogeneous myeloid cells. (*d*) Myeloid leukemia in a recipient of mouse bone marrow transduced with an *MLL/CBP* retrovirus. Note indented nuclei of monocytic immature forms/blasts. (*e*) Myeloid leukemia in a recipient of mouse bone marrow transduced with a *BCR/ABL* retrovirus. Blood exhibits marked leukocytosis of maturing myeloid cells. (*f*) Myeloproliferative disease in an $Nf1^{-/-}$ mouse. Marrow contains a mix of neutrophilic and monocytic cells.

of cathepsin G *PML/RARA* transgenic mice exhibit some features of human APL: There is an increased number of cells with numerous primary granules, occasional Auer rods have been observed, and the disease is responsive to retinoic acid. A difference from human APL is that the leukemias in these mice show significant differentiation beyond the promyelocyte stage. Hence, in the proposals for classification in mice, most cathepsin G *PML/RARA* leukemias are considered to be AML with maturation.

The MRP8 promoter drives transgene expression broadly in myelopoiesis, being expressed at some level from the granulocyte macrophage progenitor (GMP) stage (and possibly earlier) through to mature neutrophils. Like cathepsin G *PML/RARA* mice, MRP8 *PML/RARA* mice also show mild preleukemic abnormalities (Brown et al., 1997). The MRP8 *PML/RARA* transgenic mice develop leukemias that are predominantly composed of cells with features of human promyelocytes, including large numbers of primary granules. An interesting feature of the gross pathology of the leukemias of MRP8 *PML/RARA* transgenic mice is that infiltration of lymph nodes with leukemic cells expressing high levels of myeloperoxidase imparts a green color (chloromatous appearance) to these nodes. In the mouse classification, most MRP8 *PML/RARA* leukemias are classified as AML without maturation.

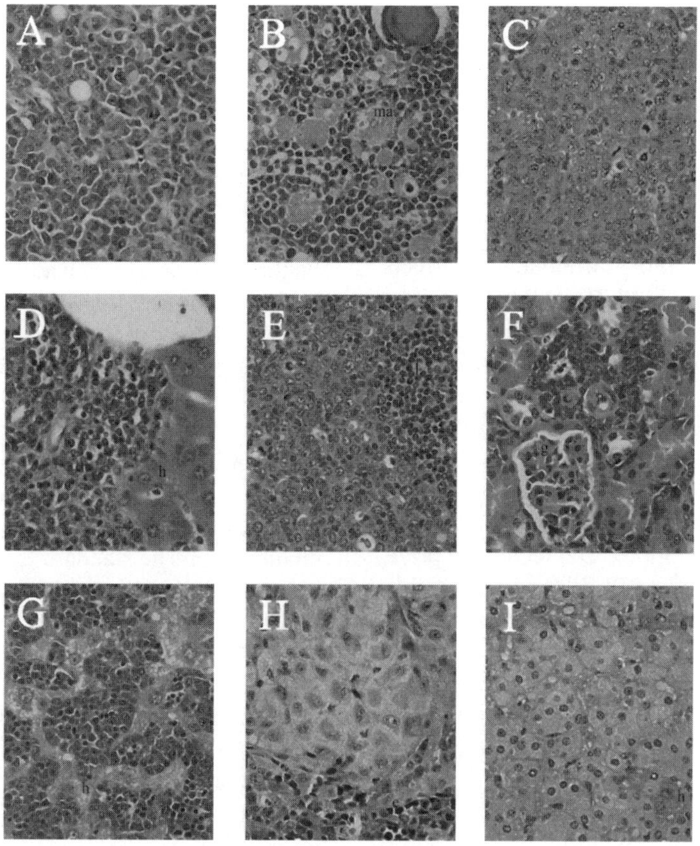

Figure 13.4. Histopathology of mouse nonlymphoid hematopoietic neoplasms. Hematoxylin and eosin stain. Original magnification 400×. Selected cells and components are identified: e, erythroid cells; g, glomerulus; h, hepatocytes; l, lymphocytes; ma, macrophages. (*a*) Bone marrow. Myeloid leukemia in a recipient of mouse bone marrow transduced with a *BCR/ABL* retrovirus. Note maturing myeloid elements. (*b*) Bone marrow. Leukemia in a BXH-2 mouse. Note immature forms/blasts accompanied by macrophages with crystalline material (pseudo-Gaucher cells). (*c*) Spleen. Myeloid leukemia in a *PML/RARA* transgenic mouse. Note immature forms/blasts with abundant cytoplasm. (*d*) Liver. Myeloid leukemia in an ethylnitrosourea-treated *Cbfb/MYH11* knockin chimeric mouse. Note that infiltration of periportal areas extends into the parenchyma as well. (*e*) Lymph node. Myeloid leukemia in a *PML/RARA* transgenic mouse. Immature forms/blasts surround a residual lymphoid follicle. (*f*) Kidney. Myeloid leukemia in a recipient of mouse bone marrow transduced with an *MLL/CBP* retrovirus. Note leukemic cells adjacent to an intact glomerulus. (*g*) Liver. Erythroid leukemia (from the study set of T. Fredrickson). Immature and maturing erythroid cells surround residual hepatocytes. (*h*) Spleen. Histiocytic sarcoma in a recipient of mouse bone marrow transduced with a *BCR/ABL* retrovirus. Note large neoplastic macrophages adjacent to residual erythroid cells of splenic red pulp. (*i*) Liver. Mast cell sarcoma (from the study set of T. Fredrickson). Neoplastic mast cells have abundant pale cytoplasm.

Efficient transduction of bone marrow with a retrovirus driving expression of *PML/RARA* has provided the basis for another mouse model for human APL (Minucci et al., 2002). In this model, preleukemic abnormalities of myelopoiesis are not apparent in vivo, but the mice nevertheless develop leukemias comprised predominantly of immature forms/blasts that have features of promyelocytes. These leukemias appear to border between AML without maturation and AML with maturation.

In mice it is more difficult than in humans to distinguish the various stages of neutrophilic and monocytic maturation. Hence, it has been proposed to call the less differentiated myeloid cells of mice "immature forms/blasts" without specifying the division of blasts from promyelocytes (Kogan et al., 2002). In regard to the leukemias of MRP8 *PML/RARA* transgenic mice and leukemias arising in marrow transduced with *PML/RARA*, it has nevertheless been suggested that fairly homogeneous strong expression of primary granule proteins may permit such leukemias to be delineated from AML without maturation and to be defined as mouse APL.

In rare cases of human APL, other *RARA* fusion genes contribute to the disease, including *PLZF/RARA, NPM/RARA, NUMA1/RARA*, and *STAT5B/RARA*. The most common of these is the *PLZF/RARA* fusion, which is usually expressed with the reciprocal fusion *RARA/PLZF*; these APLs exhibit resistance to the differentiating effects of retinoic acid. Cathepsin G *PLZF/RARA* transgenic mice develop an MPD-like myeloid leukemia in which there is no significant block to neutrophilic differentiation (He et al., 1998). To obtain a more accurate model of APL with *PLZF/RARA* fusion, cathepsin G *PLZF/RARA* transgenics were crossed to transgenic mice expressing the reciprocal *RARA/PLZF* fusion (He et al., 2000). These mice developed leukemias with increased immature forms/blasts (AML with maturation). The reciprocal product inhibits the normal activity of PLZF, and remarkably similar AMLs arise when the cathepsin G *PLZF/RARA* transgene is expressed in $Plzf^{-/-}$ animals (He et al., 2000). Cathepsin G *NPM/RARA* mice have also been reported to develop myeloid leukemia with or without impaired neutrophilic differentiation (Cheng et al., 1999).

Studies of *PML-RARA* and *PLZF-RARA* leukemia models have contributed to our understanding of APL pathogenesis and response to therapy. The differential response of these mouse leukemias to retinoic acid contributed to the recognition that dissociation of corepressor complexes, including histone deacetylases, is a critical step in retinoic acid–induced differentiation of APL cells (He et al., 1998). Additional studies investigating pathogenesis have shown that (i) other genetic changes associated with t(15;17) including the *RARA-PML* fusion and loss of one *PML* allele may contribute to the development of leukemia (Pollock et al., 1999; Rego et al., 2001); (ii) the PML domain is critical for leukemia development, doing more than simply making RARα a constitutive repressor (Kogan et al., 2000); and (iii) an activated FLT3 tyrosine kinase (which is a common additional genetic change in AML in general and in APL in particular) cooperates with *PML-RARA* to induce leukemia (Kelly et al., 2002; Sohal et al., 2003).

The APL models have also been used for preclinical studies of therapy. Arsenic trioxide is another agent with powerful antileukemia effects for APL. Work in the APL models showed that arsenic trioxide and retinoic acid cooperate in their antileukemic effects (Lallemand-Breitenbach et al., 1999; Rego et al., 2000). In addition, it was observed that the combination of histone deacetylase inhibitors or cyclic adenosine monophosphate (AMP) with retinoic acid can have antileukemic effects in leukemias resistant to retinoic acid (Guillemin et al., 2002; He et al., 2001). In addition, the *PML/RARA* plus activated FLT3 model has been used to investigate the efficacy of FLT3 kinase inhibitors in AML (Sohal et al., 2003).

Models of Human AML with t(8;21), AML1/ETO and inv(16), CBFB/MYH11. The balanced chromosomal aberrations that result in *AML1/ETO* and *CBFB/MYH11* fusions alter different subunits of the same transcription factor. *AML1* (*RUNX1*) encodes the α subunit of the core-binding factor (CBF, also known as PEBP2), which binds DNA and is able to activate transcription. *CBFB* encodes the β subunit of this factor, which associates with α and is required for function. The fusion of the α subunit of CBF to the transcriptional repressor ETO or the β subunit of CBF to a myosin smooth-muscle heavy chain disrupts the normal function of CBF. Such disruption of either CBF subunit results in marked deficits in definitive hematopoiesis.

Generating a mouse model of AML with t(8;21) proved challenging. The standard transgenic, retroviral bone marrow transduction, and knockin approaches that succeeded with *PML/RARA* or *MLL* fusions (see next section) did not yield robust models with *AML1/ETO* [transgenic and retroviral models did not readily give rise to leukemias (de Guzman et al., 2002; Rhoades et al., 2000; Schwieger et al., 2002; Yuan et al., 2001); the knockin was embryonic lethal (Yergeau et al., 1997)]. However, a conditional knockin of *ETO* into the *Aml1* locus has proved valuable (Higuchi et al., 2002). Mice with the *Aml1/ETO* fusion downstream of a lox–stop–lox cassette (see Chapter 4) were crossed with mice that expressed the Cre recombinase under the control of the interferon responsive Mx1 promoter. Activation of Cre permitted the expression of *AML1/ETO* in mouse bone marrow, but this on its own did not have much impact on hematopoiesis in vivo and did not generate an appreciable number of neoplasms. To induce disease, mice with the conditionally activated *Aml1/ETO* allele were treated with the mutagen ethylnitrosourea (ENU). As a result of this treatment, these mice developed solid neoplasms of immature myeloid cells; these lesions are classified as granulocytic sarcomas.

Human *AML1/ETO* leukemias often have a typical morphology. Blasts have fairly abundant cytoplasm with granules, including Auer rods and a prominent Golgi zone. The leukemic cells demonstrate differentiation beyond the blast stage, forming dysplastic myelocytes and more mature neutrophilic elements. Eosinophilia is common. Although the granulocytic sarcomas of ENU-treated *Aml1/ETO* conditional knockin mice do not precisely mimic the cytology of the human leukemia, it is remarkable that 22% of human patients with t(8;21) develop a solid neoplasm of myeloid cells some time during their disease course (Higuchi et al., 2002). The combination of a conditionally activated *Aml1/ETO* allele with retroviral insertional mutagenesis, rather than ENU mutagenesis, can result in AML with maturation; this model complements the ENU model by paralleling the differentiation seen in human leukemias with t(8;21) (D. Largaespada, 2002, unpublished observations).

Similar to the *Aml1/ETO* nonconditional knockin, mice with *MYH11* recombined into the *Cbfb* locus demonstrated embryonic lethality (Castilla et al., 1996). Nevertheless, mice derived from chimeras composed of normal cells and *Cbfb/MYH11* knockin embryonic stem cells proved useful (Castilla et al., 1999). Treatment of these *Cbfb/MYH11* knockin mice with ENU resulted in myeloid leukemias. Subsequently, conditional *Cbfb/MYH11* knockin mice have been used to generate leukemias following ENU or retroviral insertional mutagenesis (Landrette et al., 2002).

The *CBFB/MYH11* fusion can be seen in human leukemias with a number of morphologies but is closely associated with acute myelomonocytic leukemia with eosinophilia. Such leukemias are characterized by blasts with monocytic character accompanied by dysplastic eosinophils and decreased numbers of neutrophilic cells, which can also be dysplastic. The leukemias of *Cbfb/MYH11* mice have been described to parallel features of the human disease, including a myelomonocytic character and occasional dysplastic eosinophils.

Models of Human AML with 11q23 (MLL) Abnormalities. Bone marrow retroviral transduction and knockin approaches have been used to generate mouse models of these human leukemias. MLL is a large protein that regulates patterning during development by modulating transcription. Fusion of the *MLL* gene on human chromosome 11 with a large number of other loci is observed in AML, acute lymphoblastic leukemia (ALL), and biphenotypic acute leukemia. *MLL* can also be mutated by partial tandem duplication. Human AMLs with 11q23 abnormalities usually show monocytic differentiation and are often composed primarily of monoblasts. 11q23 abnormalities are common in patients who develop secondary AML after treatment with topoisomerase II inhibitors.

Mice with a nonconditional knockin of the *AF9* gene into the *Mll* locus demonstrated the ability of this approach to generate AML (Corral et al., 1996). Genetic manipulation of the mouse *Mll* gene has been further elaborated upon to demonstrate the potency of knockin and conditional knockin approaches, the fact that *Mll* fused to a bacterial protein can initiate leukemia, and that recapitulating a leukemia-associated chromosomal translocation in mice can generate myeloid leukemia (Collins et al., 2000; Dobson et al., 1999, 2000; T. Rabbitts, 2003, unpublished observations). Retroviral transduction of bone marrow with a number of *MLL* fusions, including *MLL/ENL, MLL/CBP, MLL/ELL, MLL/AF10, MLL/AFX, MLL/FKHRL1*, and *MLL/GAS7*, has generated additional models of leukemia with 11q23 abnormalities (DiMartino et al., 2000, 2002; Lavau et al., 1997, 2000a,b; Luo et al., 2001; So and Cleary, 2002, 2003; So et al., 2003). Studies in these retroviral models have provided evidence that interactions with molecules that effect transcription play an important role in the ability of MLL fusions to induce leukemia. These MLL models give rise to leukemias with a variety of morphologies, including leukemias that parallel the monocytic differentiation observed in many human MLL leukemias as well as biphenotypic leukemias that are congruent with an uncommon human phenotype.

Other Models of Human AML. Treatment of human patients with chemotherapy and radiotherapy can result in secondary AML. The agents that most commonly result in therapy related AML (tAML) are alkylators and topoisomerase II inhibitors. The murine leukemias that arise in MLL fusion mice serve to model a subset of topoisomerase II inhibitor–related AML. In addition, because human tAMLs induced by radiation exposure are similar to alkylator-induced tAML, murine myeloid leukemias induced by radiation represent a model of alkylating agent–related AML. Radiation exposure can induce AML in a number of strains of recombinant inbred mice, including CBA (Major and Mole, 1978). These leukemias, like human tAML, are heterogeneous (Hayata et al., 1983). Radiation-induced mouse leukemias have been used to localize a number of putative myeloid tumor suppressor loci, including a region of mouse chromosome 2 in an area syntenic to human chromosome 11p (Silver et al., 1999).

Retroviral infection has been another approach used to generate myeloid leukemias in mice through mutagenesis. Recombinant inbred mice of the BXH-2 strain have an endogenous retrovirus that inevitably causes myeloid leukemias in less than one year (Bedigian et al., 1984). The endogenous retrovirus inserts into the genome and thereby alters gene expression; usually the retroviral enhancer deregulates and enhances expression of nearby genes, and occasionally retroviral mutagenesis results in loss of gene expression. The BXH-2 and related insertional mutagenesis systems have identified numerous loci that may contribute to hematopoietic neoplasms in mice. These loci are listed at the website of the Mouse Retroviral Tagged Cancer Gene Database. A number of these loci have been implicated in human leukemias. The leukemias that arise in BXH-2 mice have a diversity of insertion sites and a diversity of morphologic phenotypes. They therefore represent a reservoir of varied AMLs that could be exploited for preclinical therapeutic studies.

Additional translocation fusion proteins associated with human myeloid leukemias including AML, tAML, and myeloid blast phase of human CML have been introduced into mice by retroviral transduction. *NUP98/HOXA9* and *AML1/MDS1/EVI1* models are discussed below in the section on models of blast-phase CML.

Models of Human Chronic Myeloproliferative Diseases

Activated tyrosine kinases have been described in association with myeloid neoplasms in humans, and these abnormal proteins can generate neoplasms in mice. In this section, the BCR/ABL fusion protein is discussed in the context of modeling human CML, whereas alternative human activated tyrosine kinase fusions are discussed below in the context of modeling human MD/MPDs. The myeloid disorders typically generated in mice by these activated tyrosine kinases share many common features, but depending on peripheral blood findings, dissemination within the host, and aggressiveness, the various models may be classified as MPD or as MPD-like myeloid leukemia.

Models of Chronic Myelogenous Leukemia. Human CML is a relatively common neoplasm characterized by expanded myelopoiesis with, initially, retention of nearly normal differentiation. The peripheral blood of patients with CML typically demonstrates a leukocytosis of neutrophilic granulocytes, including all stages of differentiation, with mature neutrophils present in greater numbers than bands, metamyelocytes, and myelocytes. Small numbers of promyelocytes and blasts are also present. Other features of the peripheral blood include increased basophils and often increased platelets. The bone marrow shows a marked expansion of myeloid elements. Chronic-phase CML progresses to a blast phase that exhibits features of AML or precursor B-cell ALL. As mentioned above, the defining feature of CML is the presence of a *BCR/ABL* chimeric gene that fuses the multidomain BCR protein with ABL and thereby causes constitutive ABL tyrosine kinase activity.

Models of CML, Chronic Phase. Retroviral transduction of bone marrow and transgenic approaches have been utilized to generate models of BCR/ABL diseases in mice (reviewed in Wong and Witte, 2001). BCR/ABL expression leads to both lymphoid and myeloid neoplasms. Lymphoid neoplasms that develop in retroviral and transgenic models have been valuable in understanding BCR/ABL-positive ALL, including lymphoid blast phase of CML (discussed in Chapter 14). BCR/ABL transgenic mice can develop myeloid neoplasms and have been used for pathogenic studies, but the greatest amount of work on BCR/ABL myeloid diseases has been in retroviral bone marrow transduction models (system reviewed in Van Etten, 2001).

The spectrum of disease that results from bone marrow transduction with BCR/ABL with subsequent reconstitution of recipients can be influenced by strain background, infection conditions, and the particular retrovirus used. Recipients can develop not only disease resembling CML but also B- or T-cell lymphomas, histiocytic sarcomas, erythroid leukemias, or mast cell neoplasms. The CML-like disease parallels human CML in exhibiting splenomegaly and hepatomegaly accompanied by elevated neutrophilic leukocytes in the peripheral blood. These neutrophilic cells show essentially normal differentiation. The rapidly fatal nature of this disease in many of the retroviral transduction models defines the disease as an MPD-like leukemia.

The BCR/ABL retroviral models have been used extensively in studies of pathogenesis. In contrast to standard transgenic models, the retroviral models permit rapid assessment of a variety of constructs. These studies have shown that the oligomerization domain of BCR, the Grb2 binding site of BCR, and the SH2 and SH3 domains of ABL play important roles in leukemogenesis by the chimeric fusion (Gross et al., 1999; Million and Van Etten, 2000; Roumiantsev et al., 2001; Zhang et al., 2001a,b). Furthermore, because extensive cross-breeding of mice is not required, the retroviral system has also facilitated examining BCR/ABL-induced leukemogenesis in mice lacking a number of proteins, including both interleukin 3 (IL-3) and granulocyte-macrophage colony-stimulating factor (GM-CSF) (Li et al., 2001) as well as both STAT5A and STAT5B (Sexl et al., 2000). These types of experiments allow critical mediators of leukemogenesis to be delineated.

Models of CML, Myeloid Blast Phase. Blast phase can occur spontaneously in mice that have BCR/ABL-induced MPD-like leukemias and MPDs, particularly upon serial transplantation. Loss of p53 can facilitate blastic transformation (Skorski et al., 1996). In addition, studies of the combination of BCR/ABL with other translocation fusion genes have been particularly illuminating.

NUP98/HOXA9 and AML1/MDS1/EVI1 fusion proteins are observed in a number of myeloid neoplasms, including myeloid blast phase of CML and therapy-related MDS/AML. NUP98/HOXA9 is the result of a t(7;11) aberration. It is a fusion of the NUP98 protein, which regulates transport between the nucleus and cytoplasm, with HOXA9, a homeobox transcription factor. The t(7;11) is only one of a number of recurrent chromosomal aberrations that involve nuclear pore proteins and/or homeobox factors. Retroviral transduction of bone marrow with NUP98/HOXA9 leads to a MPD that transforms to AML in the presence of another homeobox protein, MEIS1 (Kroon et al., 2001). A recurrent translocation of chromosomes 3 and 21 results in fusion of the *AML1* gene with the MDS1/EVI1 transcription factor locus. Retroviral transduction of bone marrow with AML1/MDS1/EVI1 initially has little effect, but mice go on to develop an acute myelomonocytic leukemia (Cuenco et al., 2000). The combination of BCR/ABL with either NUP98/HOXA9 (Dash et al., 2002; Mayotte et al., 2002) or AML1/MDS1/EVI1 (Cuenco and Ren, 2001) results in

AML with short latency. These experiments indicate that the combination of an activated tyrosine kinase with an abnormal transcription factor can be rapidly leukemogenic.

Additional BCR/ABL Models of Nonlymphoid Hematopoietic Neoplasms. Expression of BCR/ABL in mice has also provided models of a few neoplasms uncommon in humans, including histiocytic sarcomas and mast cell sarcomas. Histiocytic sarcomas are common in recipients of BCR/ABL transduced bone marrow but also are a frequent spontaneous tumor in aging mice. In humans, systemic mastocytosis (of which mast cell sarcoma is one type) is closely associated with activation of the KIT tyrosine kinase. It is therefore perhaps not surprising that BCR/ABL can result in mast cell transformation.

Models of Other Chronic Myeloproliferative Diseases. Chronic idiopathic myelofibrosis (CIMF) is a disease characterized by expansion of myelopoiesis and megakaryopoiesis, bone marrow fibrosis, and extramedullary hematopoiesis. Although unbalanced chromosomal rearrangements are associated with human CIMF, the particular genetic events driving this disease in humans are unknown. Mouse models of CIMF that recapitulate progressive bone marrow fibrosis have been described. Retroviral overexpression of thrombopoietin (TPO) with resulting increased megakaryopoiesis induced a CIMF-like phenotype (Villeval et al., 1997; Yan et al., 1995). Mice with decreased function of the GATA-1 transcription factor had impaired thrombopoiesis and also developed myelofibrosis (Vannucchi et al., 2002). Future work should shed light on the value of these models for improving understanding of and therapy for human CIMF.

Models of Human Myelodysplastic/Myeloproliferative Disorders

Models of Juvenile Myelomonocytic Leukemia. Human JMML is a myeloid neoplasm of children characterized by increased numbers of monocytes and granulocytes, in which myeloid progenitors exhibit hypersensitivity to GM-CSF. The disease is associated with a number of genetic lesions, including RAS activation, monosomy for chromosome 7, and inactivation of *NF1*. The NF1 protein serves as a RAS GTPase-activating protein: it can convert RAS from an active to an inactive state. Human children with neurofibromatosis type 1 are susceptible to developing JMML, and mice in which the *Nf1* gene is inactivated have served as a model for this disorder.

Homozygous deletion of the *Nf1* gene is lethal during development (Brannan et al., 1994; Jacks et al., 1994). *Nf1* $^{+/-}$ mice can develop a MPD late in life, but a more robust model of JMML was created by reconstituting lethally irradiated recipient animals with fetal liver hematopoietic cells from *Nf1* $^{-/-}$ embryos (Largaespada et al., 1996). These mice develop a MPD characterized by expanded myelopoiesis in the bone marrow and spleen, eventually extending to other tissues including the liver. The disorder is well differentiated and includes numerous neutrophilic cells as well as a population of abnormal monocytes, but the peripheral blood platelet count is generally normal. The disease is slowly progressive, and animals with this disorder can live with little outward signs of illness for more than a year. The disease does occasionally progress spontaneously to AML, a process that is accelerated by mutagens (Largaespada et al., 1996; Mahgoub et al., 1999b).

Like human JMML, the myeloid progenitors of mice lacking *Nf1* exhibit hypersensitivity to GM-CSF. The animal model has been used to demonstrate that in vivo the MPD phenotype is critically dependent upon endogenous GM-CSF (Birnbaum et al., 2000). Other studies have utilized this JMML model to (i) show that normal hematopoietic elements can limit the ability of *Nf1* loss to induce MPD (Zhang et al., 2001c) and (ii) assess the efficacy of therapeutics that target the RAS signaling pathway (Mahgoub et al., 1999a).

Models of Other Human Myelodysplastic/Myeloproliferative Disorders. A fusion of the gene encoding the ETS family transcription factor TEL (*ETV6*) with the platelet-derived growth factor receptor beta polypeptide gene (*PDGFRB*) was identified in a few cases of human CMML, often with eosinophilia. Subsequently, fusions of *TEL* or *PDGFRB* with other partners have been described in a spectrum of human malignancies, including CMML, aCML, and AML. The *TEL-PDGFBR* fusion caused a MPD when expressed as a retrovirus or a transgene in myeloid cells of mice (Ritchie et al., 1999; Tomasson et al., 2000). Eosinophilia was not described in these models. *TEL-ABL, TEL-JAK2*, and *TEL-TRKC* fusions also induced well-differentiated myeloid diseases in mice (Liu et al., 2000; Million et al., 2002; Schwaller et al., 1998). The rapidly fatal nature of these neoplasms suggests that they should be classified as MPD-like myeloid leukemias.

Areas in Which Models Are Needed

There has been much success in modeling human myeloid neoplasms for which associated genetic events have been delineated. Chimeric proteins that arise as a result of recurrent balanced chromosomal aberrations have been introduced into the hematopoietic systems of mice. Selected intragenic mutations have also been utilized to generate murine myeloid neoplasms. These studies have generated a large number of models of AML as well as MPDs and MD/MPDs. Without question, the technologies

Table 13.5. Selected Mouse Models of Human Myeloid Malignancies

Genetic Change	Associated Human Myeloid Neoplasms	Mouse Models	Diagnoses (Mouse, Nonlymphoid)
PML/RARA	Acute promyelocytic leukemia	MRP8 *PML/RARA* transgenic	AML without maturation
		Cathepsin G *PML/RARA* transgenic	AML with maturation
		PML/RARA bone marrow transduction	AML without maturation, AML with maturation
PLZF/RARA, *RARA/PLZF*	Acute promyelocytic leukemia	Cathepsin G *PLZF/RARA* transgenic	MPD-like myeloid leukemia
		Cathepsin G *PLZF/RARA* transgenic × cathepsin G *RARA/PLZF* transgenic	AML with maturation
AML1/ETO	AML with t(8;21)	ENU mutagenesis of conditional *Aml1/ETO* knockin	Granulocytic sarcoma
CBFB/MYH11	AML with inv(16)	ENU mutagenesis of *Cbfb/MYH11* knockin (chimeras or conditional knockin)	Acute myelomonocytic leukemia
MLL fusions	AML with 11q23 abnormalities	*Mll-AF9* knockin, *Mll-AF9* conditional knockin, *Mll-Af9* conditional chromosomal translocation	Myeloid leukemias
		Bone marrow transduction with MLL fusions (including *MLL/ENL, MLL/CBP, MLL/ELL, MLL/AF10, MLL/AFX, MLL/FKHRL1, MLL/GAS7*)	Nonlymphoid leukemias including AML without maturation, acute myelomonocytic leukemia, acute monocytic leukemia, and biphenotypic leukemia
Various	AML and MDS, therapy related	Ionizing radiation	Myeloid leukemias
Various	Various	Recombinant inbred mice with endogenous retrovirus (e.g., BXH-2)	Myeloid leukemias
BCR/ABL	Chronic myelogenous leukemia	*BCR/ABL* bone marrow transduction	MPD-like leukemia, histiocytic sarcoma, mast cell sarcoma
		BCR/ABL bone marrow transduction combined with *NUP98/HOXA9* or *AML1/MDS1/EVI1* bone marrow transduction	AMLs
NF1 inactivation	Juvenile myelomonocytic leukemia	*Nf1* knockout	Myeloproliferative disease
TEL/PDGFBR	Chronic myelomonocytic leukemia	*TEL/PDGFBR* bone marrow transduction	Myeloproliferative disease, MPD-like leukemia
MYC amplification	AML	*MYC* repressible transgenic	Myeloid leukemias

(*continued overleaf*)

Table 13.5. (*continued*)

Genetic Change	Associated Human Myeloid Neoplasms	Mouse Models	Diagnoses (Mouse, Nonlymphoid)
RAS activation	AML, MD/MPD, MDS	Eμ-activated *NRAS* transgenic	Histiocytic sarcoma
		Activated *NRAS* bone marrow transduction	Myeloproliferative disease, myeloid leukemia
Deregulated expression of *BCL-2*	*BCL-2* is expressed in many AMLs	MRP8-*BCL2* transgenic × *lpr*/*lpr* mutant mice	Myeloid leukemias
		Eμ-BCL2 transgenic treated with ionizing radiation	Biphenotypic leukemia
Deregulated expression of *HOXA9*	*HOXA9* expression is associated with AML	*HOXA9* bone marrow transduction	Myeloid leukemias
		HOXA9 and *Meis1* bone marrow transduction	AMLs
NUP98/HOXA, *NUP98/HOXD*	AML, CML in blast phase	*NUP98/HOXA9* bone marrow transduction	MPD, AMLs
		NUP98/HOXA9 and *Meis1* bone marrow transduction	AMLs
		NUP98/HOXD13 and *Meis1* bone marrow transduction	AMLs
E2A/PBX1	None (E2A/PBX1 is associated with precursor B-cell acute lymphoblastic leukemia)	*E2A/PBX1* bone marrow transduction	Myeloid leukemias
Unknown	Acute erythroid leukemia (pure erythroid)	Friend virus complex	Myeloproliferative disease, erythroid leukemia

Source: For most models see text. Additional sources include Felsher and Bishop, 1999; Fredrickson and Harris, 2000; Gibbons et al., 1999; Harris et al., 1988; Kamps and Baltimore, 1993; Kroon et al., 1998, 2001; MacKenzie et al., 1999; Ney and D'Andrea, 2000; Pineault et al., 2003; Traver et al., 1998.

utilized thus far, including transgenic mice, retroviral transduction of mouse bone marrow, knockouts, knockins, and conditional knockins will continue to be used to generate many additional models based upon genetic changes already identified in humans.

The areas in which mouse models have lagged correspond to the areas in which they are perhaps most needed, that is, the largely incurable myelodysplastic syndromes and acute myeloid leukemias with complex karyotypes. Although a few models have been described that have some features of MDS, such as increased blasts and peripheral blood cytopenias (Iwasaki-Arai et al., 2002; Kogan et al., 2001), these models are not based upon the apparently complex genetic changes that underpin these diseases in humans. One obstacle to generating mouse models of MDS and poor-prognosis AML has been the incomplete understanding of their pathogenesis: The genes driving chromosomal gains and losses in human myeloid neoplasms have not been delineated. The identification of myeloid tumor suppressor genes putatively localized on human chromosomes (including 5q, 7q, 12p, and 20q) may yet occur from the study of human materials. It is however possible that mouse models of these diseases developed through the use of chromosome engineering will precede the identification of the relevant tumor suppressor genes (Shannon et al., 2001).

CONCLUSIONS

A large number of models of myeloid malignancies have been generated in mice (examples discussed above plus additional selected examples are summarized in Table 13.5). Specific models have been utilized to elucidate the mechanisms through which particular genetic changes contribute to myeloid neoplasia. These models have also been used for preclinical studies of molecularly targeted therapies. Further studies utilizing these models to understand pathogenesis and improve therapy will inevitably lead to additional insights.

More broadly, what have these models illuminated in regard to human neoplasia? First, the myeloid models have definitively demonstrated that recurrent genetic changes associated with particular neoplastic phenotypes in humans are pathogenically responsible for these illnesses. Second, the myeloid models have served to illustrate both the power and impotency of single genetic events: Whereas expression of an activated tyrosine kinase such as BCR/ABL can profoundly disturb hematopoiesis, expression of such fusion proteins as PML/RARα has little effect unless it is in the context of other changes. Third, the myeloid models have served to dramatize the diversity of genetic changes that can contribute to neoplastic transformation and have shown that this range of underlying genetic changes is responsible for pathologic diversity of human cancer. Mouse models of myeloid malignancy will continue to help us understand how genetic changes combine to generate cancers and how we may reverse these lethal processes.

ACKNOWLEDGMENTS

The author thanks Daphne Haas-Kogan, Philip Chan, and Endi Wang for their assistance in preparing this chapter. The author is grateful to Lucio Castilla, Michael Cleary, James Downing, Torgny Fredrickson, David Largaespada, Catherine Lavau, P. Paul Liu, Warren Pear, Kevin Shannon, and Richard Van Etten for providing materials from mouse models of myeloid neoplasms and for useful discussions.

REFERENCES

Akashi K, Traver D, Miyamoto T, Weissman IL (2000): A clonogenic common myeloid progenitor that gives rise to all myeloid lineages. *Nature* 404:193–197.

Ariyama Y, Fukuda Y, Okuno Y, Seto M, Date K, Abe T, Nakamura Y, Inazawa J (1998): Amplification on double-minute chromosomes and partial-tandem duplication of the MLL gene in leukemic cells of a patient with acute myelogenous leukemia. *Genes Chromosomes Cancer* 23:267–272.

Bedigian HG, Johnson DA, Jenkins NA, Copeland NG, Evans R (1984): Spontaneous and induced leukemias of myeloid origin in recombinant inbred BXH mice. *J Virol* 51:586–594.

Birnbaum RA, O'Marcaigh A, Wardak Z, Zhang YY, Dranoff G, Jacks T, Clapp DW, Shannon KM (2000): Nf1 and Gmcsf interact in myeloid leukemogenesis. *Mol Cell* 5:189–195.

Brannan CI, Perkins AS, Vogel KS, Ratner N, Nordlund ML, Reid SW, Buchberg AM, Jenkins NA, Parada LF, Copeland NG (1994): Targeted disruption of the neurofibromatosis type-1 gene leads to developmental abnormalities in heart and various neural crest-derived tissues. *Genes Dev* 8:1019–1029.

Brown D, Kogan S, Lagasse E, Weissman I, Alcalay M, Pelicci PG, Atwater S, Bishop JM (1997): A PMLRAR alpha transgene initiates murine acute promyelocytic leukemia. *Proc Natl Acad Sci USA* 94:2551–2556.

Cardiff RD (2001): Panel discussion on validation: Pathology summary report. Paper presented at Mouse Models of Human Cancers Consortium, Steering Committee Meeting, Orlando, FL.

Castilla LH, Garrett L, Adya N, Orlic D, Dutra A, Anderson S, Owens J, Eckhaus M, Bodine D, Liu PP (1999): The fusion gene Cbfb-MYH11 blocks myeloid differentiation and predisposes mice to acute myelomonocytic leukaemia. *Nat Genet* 23:144–146.

Castilla LH, Wijmenga C, Wang Q, Stacy T, Speck NA, Eckhaus M, Marin-Padilla M, Collins FS, Wynshaw-Boris A, Liu PP (1996): Failure of embryonic hematopoiesis and lethal hemorrhages in mouse embryos heterozygous for a knocked-in leukemia gene CBFB-MYH11. *Cell* 87:687–696.

Cheng GX, Zhu XH, Men XQ, Wang L, Huang QH, Jin XL, Xiong SM, Zhu J, Guo WM, Chen JQ, Xu SF, So E, Chan LC, Waxman S, Zelent A, Chen GQ, Dong S, Liu JX, Chen SJ (1999): Distinct leukemia phenotypes in transgenic mice and different corepressor interactions generated by promyelocytic leukemia variant fusion genes PLZF-RARalpha and NPM-RARalpha. *Proc Natl Acad Sci USA* 96:6318–6323.

Cline MJ (1994): The molecular basis of leukemia. *N Engl J Med* 330:328–336.

Collins EC, Pannell R, Simpson EM, Forster A, Rabbitts TH (2000): Interchromosomal recombination of Mll and Af9 genes mediated by cre-loxP in mouse development. *EMBO Rep* 1:127–132.

Corral J, Lavenir I, Impey H, Warren AJ, Forster A, Larson TA, Bell S, McKenzie AN, King G, Rabbitts TH (1996): An Mll-AF9 fusion gene made by homologous recombination causes acute leukemia in chimeric mice: A method to create fusion oncogenes. *Cell* 85:853–861.

Cuenco GM, Nucifora G, Ren R (2000): Human AML1/MDS1/EVI1 fusion protein induces an acute myelogenous leukemia (AML) in mice: A model for human AML. *Proc Natl Acad Sci USA* 97:1760–1765.

Cuenco GM, Ren R (2001): Cooperation of BCR-ABL and AML1/MDS1/EVI1 in blocking myeloid differentiation and rapid induction of an acute myelogenous leukemia. *Oncogene* 20:8236–8248.

Dash AB, Williams IR, Kutok JL, Tomasson MH, Anastasiadou E, Lindahl K, Li S, Van Etten RA, Borrow J, Housman D, Druker B, Gilliland DG (2002): A murine model of CML blast crisis induced by cooperation between BCR/ABL and NUP98/HOXA9. *Proc Natl Acad Sci USA* 99:7622–7627.

de Guzman CG, Warren AJ, Zhang Z, Gartland L, Erickson P, Drabkin H, Hiebert SW, Klug CA (2002): Hematopoietic stem cell expansion and distinct myeloid developmental abnormalities in a murine model of the AML1-ETO translocation. *Mol Cell Biol* 22:5506–5517.

DiMartino JF, Ayton PM, Chen EH, Naftzger CC, Young BD, Cleary ML (2002): The AF10 leucine zipper is required for leukemic transformation of myeloid progenitors by MLL-AF10. *Blood* 99:3780–3785.

DiMartino JF, Miller T, Ayton PM, Landewe T, Hess JL, Cleary ML, Shilatifard A (2000): A carboxy-terminal domain of ELL is required and sufficient for immortalization of myeloid progenitors by MLL-ELL. *Blood* 96:3887–3893.

Dobson CL, Warren AJ, Pannell R, Forster A, Lavenir I, Corral J, Smith AJ, Rabbitts TH (1999): The mll-AF9 gene fusion in mice controls myeloproliferation and specifies acute myeloid leukaemogenesis. *Embo J* 18:3564–3574.

Dobson CL, Warren AJ, Pannell R, Forster A, Rabbitts TH (2000): Tumorigenesis in mice with a fusion of the leukaemia oncogene Mll and the bacterial lacZ gene. *Embo J* 19:843–851.

Dong F, Brynes RK, Tidow N, Welte K, Lowenberg B, Touw IP (1995): Mutations in the gene for the granulocyte colony-stimulating-factor receptor in patients with acute myeloid leukemia preceded by severe congenital neutropenia. *N Engl J Med* 333:487–493.

Druker BJ, Talpaz M, Resta DJ, Peng B, Buchdunger E, Ford JM, Lydon NB, Kantarjian H, Capdeville R, Ohno-Jones S, Sawyers CL (2001): Efficacy and safety of a specific inhibitor of the BCR-ABL tyrosine kinase in chronic myeloid leukemia. *N Engl J Med* 344:1031–1037.

Felsher DW, Bishop JM (1999): Reversible tumorigenesis by MYC in hematopoietic lineages. *Mol Cell* 4:199–207.

Fredrickson TN, Harris AW (2000): *Atlas of Mouse Hematopathology*. Harwood Academic, Amsterdam.

Gibbons DL, MacDonald D, McCarthy KP, Cleary HJ, Plumb M, Wright EG, Greaves MF (1999): An E mu-BCL-2 transgene facilitates leukaemogenesis by ionizing radiation. *Oncogene* 18:3870–3877.

Grimwade D, Walker H, Oliver F, Wheatley K, Harrison C, Harrison G, Rees J, Hann I, Stevens R, Burnett A, Goldstone A (1998): The importance of diagnostic cytogenetics on outcome in AML: Analysis of 1,612 patients entered into the MRC AML 10 trial. The Medical Research Council Adult and Children's Leukaemia Working Parties. *Blood* 92:2322–2333.

Grisolano JL, Wesselschmidt RL, Pelicci PG, Ley TJ (1997): Altered myeloid development and acute leukemia in transgenic mice expressing PML-RAR alpha under control of cathepsin G regulatory sequences. *Blood* 89:376–387.

Gross AW, Zhang X, Ren R (1999): Bcr-Abl with an SH3 deletion retains the ability to induce a myeloproliferative disease in mice, yet c-Abl activated by an SH3 deletion induces only lymphoid malignancy. *Mol Cell Biol* 19:6918–6928.

Guillemin MC, Raffoux E, Vitoux D, Kogan S, Soilihi H, Lallemand-Breitenbach V, Zhu J, Janin A, Daniel MT, Gourmel B, Degos L, Dombret H, Lanotte M, De The H (2002): in vivo activation of cAMP signaling induces growth arrest and differentiation in acute promyelocytic leukemia. *J Exp Med* 196:1373–1380.

Harris AW, Langdon WY, Alexander WS, Hariharan IK, Rosenbaum H, Vaux D, Webb E, Bernard O, Crawford M, Abud H, Adams JM, Cory S (1988): Transgenic mouse models for hematopoietic tumorigenesis. *Curr Top Microbiol Immunol* 141:82–93.

Hayata I, Seki M, Yoshida K, Hirashima K, Sado T, Yamagiwa J, Ishihara T (1983): Chromosomal aberrations observed in 52 mouse myeloid leukemias. *Cancer Res* 43:367–373.

He LZ, Bhaumik M, Tribioli C, Rego EM, Ivins S, Zelent A, Pandolfi PP (2000): Two critical hits for promyelocytic leukemia. *Mol Cell* 6:1131–1141.

He LZ, Guidez F, Tribioli C, Peruzzi D, Ruthardt M, Zelent A, Pandolfi PP (1998): Distinct interactions of PML-RARalpha and PLZF-RARalpha with co-repressors determine differential responses to RA in APL. *Nat Genet* 18:126–135.

He LZ, Tolentino T, Grayson P, Zhong S, Warrell RP, Jr., Rifkind RA, Marks PA, Richon VM, Pandolfi PP (2001): Histone deacetylase inhibitors induce remission in transgenic models of therapy-resistant acute promyelocytic leukemia. *J Clin Invest* 108:1321–1330.

He LZ, Tribioli C, Rivi R, Peruzzi D, Pelicci PG, Soares V, Cattoretti G, Pandolfi PP (1997): Acute leukemia with promyelocytic features in PML/RARalpha transgenic mice. *Proc Natl Acad Sci USA* 94:5302–5307.

Higuchi M, O'Brien D, Kumaravelu P, Lenny N, Yeoh EJ, Downing JR (2002): Expression of a conditional AML1-ETO oncogene bypasses embryonic lethality and establishes a murine model of human t(8;21) acute myeloid leukemia. *Cancer Cell* 1:63–74.

Iwasaki-Arai J, Zhang P, Huettner CS, Fenyus M, Lekstrom-Himes J, Tenen DG (2002): C/EBPalpha deficiency in hematopoiesis induces accumulation of non-malignant myeloblasts mimicking acute myelogenous leukemia. *Blood* 100:61a.

Jacks T, Shih TS, Schmitt EM, Bronson RT, Bernards A, Weinberg RA (1994): Tumour predisposition in mice heterozygous for a targeted mutation in Nf1. *Nat Genet* 7:353–361.

Jaffe ES, Harris NL, Stein H, Vardiman JW (2001): *World Health Organization Classification of Tumours. Pathology and Genetics of Tumours of Haematopoietic and Lymphoid Tissues*. IARC Press, Lyon.

Kamps MP, Baltimore D (1993): E2A-Pbx1, the t(1;19) translocation protein of human pre-B-cell acute lymphocytic leukemia, causes acute myeloid leukemia in mice. *Mol Cell Biol* 13:351–357.

Kelly LM, Gilliland DG (2002): Genetics of myeloid leukemias. *Annu Rev Genom Hum Genet* 3:179–198.

Kelly LM, Kutok JL, Williams IR, Boulton CL, Amaral SM, Curley DP, Ley TJ, Gilliland DG (2002): PML/RARalpha and FLT3-ITD induce an APL-like disease in a mouse model. *Proc Natl Acad Sci USA* 99:8283–8288.

Kogan SC, Brown DE, Shultz DB, Truong BTH, Lallemand-Breitenbach V, Guillemin MC, Lagasse E, Weissman IL, Bishop JM (2001): BCL-2 cooperates with promyelocytic leukemia retinoic acid receptor alpha chimeric protein (PMLRAR alpha) to block neutrophil differentiation and initiate acute leukemia. *J Exp Med* 193:531–543.

Kogan SC, Hong SH, Shultz DB, Privalsky ML, Bishop JM (2000): Leukemia initiated by PMLRARalpha: The PML domain plays a critical role while retinoic acid-mediated transactivation is dispensable. *Blood* 95:1541–1550.

Kogan SC, Ward JM, Anver MR, Berman JJ, Brayton C, Cardiff RD, Carter JS, de Coronado S, Downing JR, Fredrickson TN, Haines DC, Harris AW, Harris NL, Hiai H, Jaffe ES, MacLennan IC, Pandolfi PP, Pattengale PK, Perkins AS, Simpson RM, Tuttle MS, Wong JF, Morse HC, 3rd (2002): Bethesda proposals for classification of nonlymphoid hematopoietic neoplasms in mice. *Blood* 100:238–245.

Kondo M, Weissman IL, Akashi K (1997): Identification of clonogenic common lymphoid progenitors in mouse bone marrow. *Cell* 91:661–672.

Kroon E, Krosl J, Thorsteinsdottir U, Baban S, Buchberg AM, Sauvageau G (1998): Hoxa9 transforms primary bone marrow cells through specific collaboration with Meis1a but not Pbx1b. *Embo J* 17:3714–3725.

Kroon E, Thorsteinsdottir U, Mayotte N, Nakamura T, Sauvageau G (2001): NUP98-HOXA9 expression in hemopoietic stem cells induces chronic and acute myeloid leukemias in mice. *Embo J* 20:350–361.

Lallemand-Breitenbach V, Guillemin MC, Janin A, Daniel MT, Degos L, Kogan SC, Bishop JM, de The H (1999): Retinoic acid and arsenic synergize to eradicate leukemic cells in a mouse model of acute promyelocytic leukemia. *J Exp Med* 189:1043–1052.

Landrette S, Perrat P, Hensen K, de Ven V, Castilla L (2002): Oncogenes PLAG1 and PLAG2 synergize with CBFB-MYH11 in AML. *Blood* 100:85a.

Largaespada DA, Brannan CI, Jenkins NA, Copeland NG (1996): Nf1 deficiency causes Ras-mediated granulocyte/macrophage colony stimulating factor hypersensitivity and chronic myeloid leukaemia. *Nat Genet* 12:137–143.

Lavau C, Du C, Thirman M, Zeleznik-Le N (2000a): Chromatin-related properties of CBP fused to MLL generate a myelodysplastic-like syndrome that evolves into myeloid leukemia. *Embo J* 19:4655–4664.

Lavau C, Luo RT, Du C, Thirman MJ (2000b): Retrovirus-mediated gene transfer of MLL-ELL transforms primary myeloid progenitors and causes acute myeloid leukemias in mice. *Proc Natl Acad Sci USA* 97:10,984–10,989.

Lavau C, Szilvassy SJ, Slany R, Cleary ML (1997): Immortalization and leukemic transformation of a myelomonocytic precursor by retrovirally transduced HRX-ENL. *Embo J* 16:4226–4237.

Le Beau MM, Larson RA (2000). Cytogenetics and neoplasia. In Hoffman R, Benz EJ, Shattil SJ, Furie B, Cohen HJ, Silberstein LE, McGlave P (Eds), *Hematology: Basic Principles and Practice*, 3rd ed. Livingstone, New York. pp 848–869.

Li S, Gillessen S, Tomasson MH, Dranoff G, Gilliland DG, Van Etten RA (2001): Interleukin 3 and granulocyte-macrophage colony-stimulating factor are not required for induction of chronic myeloid leukemia-like myeloproliferative disease in mice by BCR/ABL. *Blood* 97:1442–1450.

Liu Q, Schwaller J, Kutok J, Cain D, Aster JC, Williams IR, Gilliland DG (2000): Signal transduction and transforming properties of the TEL-TRKC fusions associated with t(12;15)(p13;q25) in congenital fibrosarcoma and acute myelogenous leukemia. *Embo J* 19:1827–1838.

Look AT (1997): Oncogenic transcription factors in the human acute leukemias. *Science* 278:1059–1064.

Luo RT, Lavau C, Du C, Simone F, Polak PE, Kawamata S, Thirman MJ (2001): The elongation domain of ELL is dispensable but its ELL-associated factor 1 interaction domain is essential for MLL-ELL-induced leukemogenesis. *Mol Cell Biol* 21:5678–5687.

MacKenzie KL, Dolnikov A, Millington M, Shounan Y, Symonds G (1999): Mutant N-ras induces myeloproliferative disorders and apoptosis in bone marrow repopulated mice. *Blood* 93:2043–2056.

Mahgoub N, Taylor BR, Gratiot M, Kohl NE, Gibbs JB, Jacks T, Shannon KM (1999a): In vitro and in vivo effects of a farnesyltransferase inhibitor on Nf1-deficient hematopoietic cells. *Blood* 94:2469–2476.

Mahgoub N, Taylor BR, Le Beau MM, Gratiot M, Carlson KM, Atwater SK, Jacks T, Shannon KM (1999b): Myeloid malignancies induced by alkylating agents in Nf1 mice. *Blood* 93:3617–3623.

Major IR, Mole RH (1978): Myeloid leukaemia in x-ray irradiated CBA mice. *Nature* 272:455–456.

Marshall CJ, Thrasher AJ (2001): The embryonic origins of human haematopoiesis. *Br J Haematol* 112:838–850.

Mayotte N, Roy DC, Yao J, Kroon E, Sauvageau G (2002): Oncogenic interaction between BCR-ABL and NUP98-HOXA9 demonstrated by the use of an in vitro purging culture system. *Blood* 100:4177–4184.

Million RP, Aster J, Gilliland DG, Van Etten RA (2002): The Tel-Abl (ETV6-Abl) tyrosine kinase, product of complex (9;12) translocations in human leukemia, induces distinct myeloproliferative disease in mice. *Blood* 99:4568–4577.

Million RP, Van Etten RA (2000): The Grb2 binding site is required for the induction of chronic myeloid leukemia-like disease in mice by the Bcr/Abl tyrosine kinase. *Blood* 96:664–670.

Minucci S, Monestiroli S, Giavara S, Ronzoni S, Marchesi F, Insinga A, Diverio D, Gasparini P, Capillo M, Colombo E, Matteucci C, Contegno F, Lo-Coco F, Scanziani E, Gobbi A, Pelicci PG (2002): PML-RAR induces promyelocytic leukemias with high efficiency following retroviral gene transfer into purified murine hematopoietic progenitors. *Blood* 100:2989–2995.

Morrison SJ, Weissman IL (1994): The long-term repopulating subset of hematopoietic stem cells is deterministic and isolatable by phenotype. *Immunity* 1:661–673.

Morse HC, 3rd, Anver MR, Fredrickson TN, Haines DC, Harris AW, Harris NL, Jaffe ES, Kogan SC, MacLennan IC, Pattengale PK, Ward JM (2002): Bethesda proposals for classification of lymphoid neoplasms in mice. *Blood* 100:246–258.

Mueller BU, Pabst T, Osato M, Asou N, Johansen LM, Minden MD, Behre G, Hiddemann W, Ito Y, Tenen DG (2002):

Heterozygous PU.1 mutations are associated with acute myeloid leukemia. *Blood* 100:998–1007.

Muller AM, Medvinsky A, Strouboulis J, Grosveld F, Dzierzak E (1994): Development of hematopoietic stem cell activity in the mouse embryo. *Immunity* 1:291–301.

Ney PA, D'Andrea AD (2000): Friend erythroleukemia revisited. *Blood* 96:3675–3680.

North T, Gu TL, Stacy T, Wang Q, Howard L, Binder M, Marin-Padilla M, Speck NA (1999): Cbfa2 is required for the formation of intra-aortic hematopoietic clusters. *Development* 126:2563–2575.

Nowell P (1960): A minute chromosome in human chronic granulocytic leukemia. *Science* 132:1497–1500.

Pabst T, Mueller BU, Zhang P, Radomska HS, Narravula S, Schnittger S, Behre G, Hiddemann W, Tenen DG (2001): Dominant-negative mutations of CEBPA, encoding CCAAT/enhancer binding protein-alpha (C/EBP alpha), in acute myeloid leukemia. *Nat Genet* 27:263–270.

Pineault N, Buske C, Feuring-Buske M, Abramovich C, Rosten P, Hogge DE, Aplan PD, Humphries RK (2003): Induction of acute myeloid leukemia in mice by the human leukemia-specific fusion gene NUP98-HOXD13 in concert with Meis1. *Blood* 101:4529–4538.

Pollock JL, Westervelt P, Kurichety AK, Pelicci PG, Grisolano JL, Ley TJ (1999): A bcr-3 isoform of RARalpha-PML potentiates the development of PML-RARalpha-driven acute promyelocytic leukemia. *Proc Natl Acad Sci USA* 96:15,103–15,108.

Rego EM, He LZ, Warrell RP, Jr., Wang ZG, Pandolfi PP (2000): Retinoic acid (RA) and As2O3 treatment in transgenic models of acute promyelocytic leukemia (APL) unravel the distinct nature of the leukemogenic process induced by the PML-RARalpha and PLZF-RARalpha oncoproteins. *Proc Natl Acad Sci USA* 97:10,173–10,178.

Rego EM, Wang ZG, Peruzzi D, He LZ, Cordon-Cardo C, Pandolfi PP (2001): Role of promyelocytic leukemia (PML) protein in tumor suppression. *J Exp Med* 193:521–529.

Rhoades KL, Hetherington CJ, Harakawa N, Yergeau DA, Zhou L, Liu LQ, Little MT, Tenen DG, Zhang DE (2000): Analysis of the role of AML1-ETO in leukemogenesis, using an inducible transgenic mouse model. *Blood* 96:2108–2115.

Ritchie KA, Aprikyan AA, Bowen-Pope DF, Norby-Slycord CJ, Conyers S, Bartelmez S, Sitnicka EH, Hickstein DD (1999): The Tel-PDGFRbeta fusion gene produces a chronic myeloproliferative syndrome in transgenic mice. *Leukemia* 13: 1790–1803.

Roumiantsev S, de Aos IE, Varticovski L, Ilaria RL, Van Etten RA (2001): The Src homology 2 domain of Bcr/Abl is required for efficient induction of chronic myeloid leukemia-like disease in mice but not for lymphoid leukemogenesis or activation of phosphatidylinositol 3-kinase. *Blood* 97:4–13.

Russell ES, Bernstein SE (1966). Blood and blood formation. In Green EL (Ed), *Biology of the Laboratory Mouse*, 2nd ed. McGraw-Hill, New York, pp. 351–372.

Schwaller J, Frantsve J, Aster J, Williams IR, Tomasson MH, Ross TS, Peeters P, Van Rompaey L, Van Etten RA, Ilaria R, Jr., Marynen P, Gilliland DG (1998): Transformation of hematopoietic cell lines to growth-factor independence and induction of a fatal myelo- and lymphoproliferative disease in mice by retrovirally transduced TEL/JAK2 fusion genes. *Embo J* 17:5321–5333.

Schwieger M, Lohler J, Friel J, Scheller M, Horak I, Stocking C (2002): AML1-ETO inhibits maturation of multiple lymphohematopoietic lineages and induces myeloblast transformation in synergy with ICSBP deficiency. *J Exp Med* 196:1227–1240.

Sexl V, Piekorz R, Moriggl R, Rohrer J, Brown MP, Bunting KD, Rothammer K, Roussel MF, Ihle JN (2000): Stat5a/b contribute to interleukin 7-induced B-cell precursor expansion, but abl- and bcr/abl-induced transformation are independent of Stat5. *Blood* 96:2277–2283.

Shannon KM, Le Beau MM, Largaespada DA, Killeen N (2001): Modeling myeloid leukemia tumor suppressor gene inactivation in the mouse. *Semin Cancer Biol* 11:191–2000.

Silver A, Moody J, Dunford R, Clark D, Ganz S, Bulman R, Bouffler S, Finnon P, Meijne E, Huiskamp R, Cox R (1999): Molecular mapping of chromosome 2 deletions in murine radiation-induced AML localizes a putative tumor suppressor gene to a 1.0 cM region homologous to human chromosome segment 11p11-12. *Genes Chromosomes Cancer* 24:95–104.

Skorski T, Nieborowska-Skorska M, Wlodarski P, Perrotti D, Martinez R, Wasik MA, Calabretta B (1996): Blastic transformation of p53-deficient bone marrow cells by p210bcr/abl tyrosinekinase. *Proc Natl Acad Sci USA* 93:13,137–13,142.

So CW, Cleary ML (2002): MLL-AFX requires the transcriptional effector domains of AFX to transform myeloid progenitors and transdominantly interfere with forkhead protein function. *Mol Cell Biol* 22:6542–6552.

So CW, Cleary ML (2003): Common mechanism for oncogenic activation of MLL by forkhead family proteins. *Blood* 101:633–639.

So CW, Karsunky H, Passegue E, Cozzio A, Weissman IL, Cleary ML (2003): MLL-GAS7 transforms multipotent hematopoietic progenitors and induces mixed lineage leukemias in mice. *Cancer Cell* 3:161–171.

Sohal J, Phan VT, Chan PV, Davis EM, Patel B, Kelly LM, Abrams TJ, O'Farrell AM, Gilliland DG, Le Beau MM, Kogan SC (2003): A model of APL with FLT3 mutation is responsive to retinoic acid and a receptor tyrosine kinase inhibitor, SU11657. *Blood* 101:3188–3197.

Tanaka K, Takechi M, Nishimura S, Oguma N, Kamada N (1993): Amplification of C-Myc oncogene and point mutation of N-Ras oncogene point mutation in acute myelocytic leukemias with double minute chromosomes. *Leukemia* 7:469–471.

Tomasson MH, Sternberg DW, Williams IR, Carroll M, Cain D, Aster JC, Ilaria RL, Jr., Van Etten RA, Gilliland DG (2000): Fatal myeloproliferation, induced in mice by TEL/PDGFbetaR expression, depends on PDGFbetaR tyrosines 579/581. *J Clin Invest* 105:423–432.

Traver D, Akashi K, Weissman IL, Lagasse E (1998): Mice defective in two apoptosis pathways in the myeloid lineage develop acute myeloblastic leukemia. *Immunity* 9:47–57.

Valent P, Horny HP, Li CY, Longley BJ, Metcalfe DD, Parwaresch RM, Bennett JM (2001). Mastocytosis. In Jaffe ES, Harris NL, Stein H, Vardiman JW (Eds), *World Health Organization Classification of Tumours. Pathology and Genetics of Tumours of Haematopoietic and Lymphoid Tissues*. IARC Press, Lyon, pp 293–302.

Van Etten RA (2001): Retroviral transduction models of Ph+ leukemia: Advantages and limitations for modeling human hematological malignancies in mice. *Blood Cells Mol Dis* 27:201–205.

Vannucchi AM, Bianchi L, Cellai C, Paoletti F, Rana RA, Lorenzini R, Migliaccio G, Migliaccio AR (2002): Development of myelofibrosis in mice genetically impaired for GATA-1 expression (GATA-1(low) mice). *Blood* 100:1123–1132.

Villeval JL, Cohen-Solal K, Tulliez M, Giraudier S, Guichard J, Burstein SA, Cramer EM, Vainchenker W, Wendling F (1997): High thrombopoietin production by hematopoietic cells induces a fatal myeloproliferative syndrome in mice. *Blood* 90:4369–4383.

Wechsler J, Greene M, McDevitt MA, Anastasi J, Karp JE, Le Beau MM, Crispino JD (2002): Acquired mutations in GATA1 in the megakaryoblastic leukemia of Down syndrome. *Nat Genet* 32:148–152.

Wong S, Witte ON (2001): Modeling Philadelphia chromosome positive leukemias. *Oncogene* 20:5644–5659.

Yan XQ, Lacey D, Fletcher F, Hartley C, McElroy P, Sun Y, Xia M, Mu S, Saris C, Hill D (1995): Chronic exposure to retroviral vector encoded MGDF (mpl-ligand) induces lineage-specific growth and differentiation of megakaryocytes in mice. *Blood* 86:4025–4033.

Yergeau DA, Hetherington CJ, Wang Q, Zhang P, Sharpe AH, Binder M, Marin-Padilla M, Tenen DG, Speck NA, Zhang DE (1997): Embryonic lethality and impairment of haematopoiesis in mice heterozygous for an AML1-ETO fusion gene. *Nat Genet* 15:303–306.

Yuan Y, Zhou L, Miyamoto T, Iwasaki H, Harakawa N, Hetherington CJ, Burel SA, Lagasse E, Weissman IL, Akashi K, Zhang DE (2001): AML1-ETO expression is directly involved in the development of acute myeloid leukemia in the presence of additional mutations. *Proc Natl Acad Sci USA* 98:10,398–10,403.

Zhang XW, Wong R, Hao SX, Pear WS, Ren RB (2001a): The SH2 domain of Bcr-Abl is not required to induce a murine myeloproliferative disease; however, SH2 signaling influences disease latency and phenotype. *Blood* 97:277–287.

Zhang XW, Subrahmanyam R, Wong R, Gross AW, Ren RB (2001b): The NH2-terminal coiled-coil domain and tyrosine 177 play important roles in induction of a myeloproliferative disease in mice by Bcr-Abl. *Mol Cell Biol* 21: 840–853.

Zhang Y, Taylor BR, Shannon K, Clapp DW (2001c): Quantitative effects of Nf1 inactivation on in vivo hematopoiesis. *J Clin Invest* 108:709–715.

WEB RESOURCES

http://emice.nci.nih.gov/mouse_models/organ_models/hema_models

http://genome2.ncifcrf.gov/RTCGD/

14

LYMPHOID MALIGNANCIES

MICHAEL A. TEITELL

Department of Pathology, David Geffen School of Medicine at University of California—Los Angeles, Los Angeles, California

PIER PAOLO PANDOLFI

Molecular Biology Program and Department of Pathology, Memorial Sloan Kettering Cancer Center, New York, New York

Malignant transformation may occur at any point during lymphoid cell maturation, resulting in stage- and lineage-specific cancers of the immune system. Many in-bred and engineered mouse strains develop specific lymphoid cancers, providing potentially useful models of human malignancy with insight for pathogenetic mechanisms. Newer techniques, targeting dysregulated gene expression to specific developmental stages portend newer, more accurate mouse models in the future. These models will likely have an increasing role in evaluating targeted therapeutic strategies for efficacy.

INTRODUCTION

Mammals rely on lymphocytes and dendritic cells for protection from pathologic processes such as infection and cancer. Unfortunately, transformation may occur at any point during lymphoid cell maturation, resulting in stage-specific malignancies. Tumor susceptibility seems to depend on specific genetic or epigenetic errors, host background, and immune status. Many in-bred and engineered mouse strains develop lymphoid cancer, providing potentially useful models for human malignancy. Desirable models tend to recapitulate the origin, progression, and outcome of comparable human diseases. In this chapter we briefly review lymphoid cell development as a framework for discussions on tumor analysis and classification schemes, followed by examples of genetically engineered models of lymphoid cancer in mice. Many excellent reviews on lymphoid cell development in mice and humans are available (see captions for Figs. 14.1 and 14.2).

LYMPHOID CELL DEVELOPMENT IN MICE

Postfetal lymphoid cells originate from renewable hematopoietic stem cells (HSCs) in the bone marrow (BM [1–4]). According to current models of hematopoiesis, HSC-derived common lymphoid progenitors (CLPs) develop into B, T, natural killer (NK), and NK/T cells while common myeloid progenitors (CMPs; see Chapter 13) yield erythrocytes, granulocytes, megakaryocytes, and monocytes [5, 6]. CLPs and CMPs, or a dendritic cell (DC) precursor, develop into lymphoid- and myeloid-dendritic cells, respectively [7–10].

B-Cell Development

Stages in B-cell maturation are marked by specific patterns of gene expression, protein expression and immunoglobulin (Ig) gene rearrangements (Fig. 14.1 [11–13]). These stages are controlled by key transcription factors that are expressed in response to signals from the environment [14–18]. In the BM, Ikaros and titered levels of PU.1 expression enable HSC-to-CLP differentiation [19]. B-lineage development is then guided by early B-cell factor (EBF) and E2A gene splice variants E12 and E47

Mouse Models of Human Cancer, edited by Eric C. Holland
ISBN 0-471-44460-X

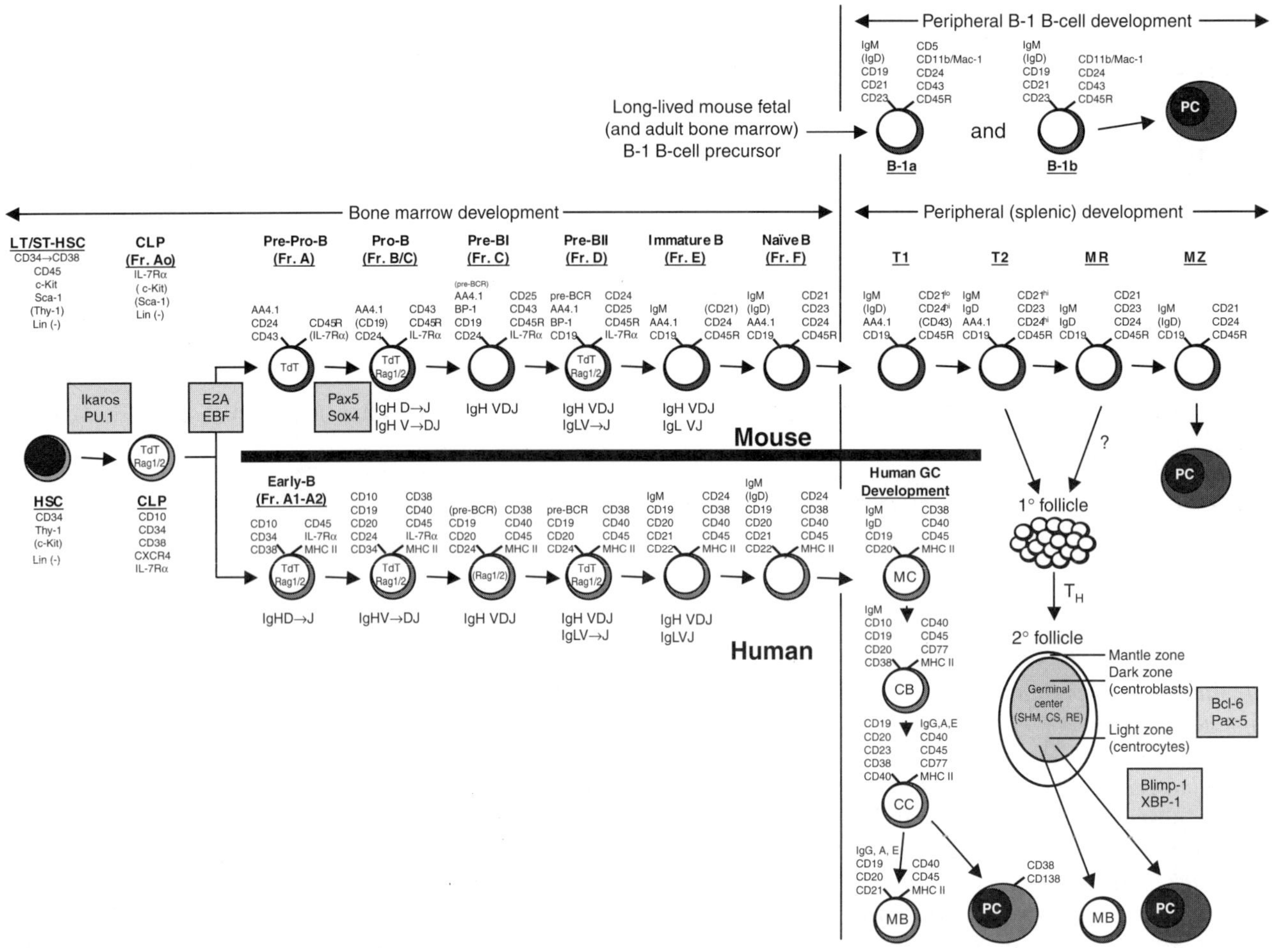

Figure 14.1. Mouse and human (below thick line) bone marrow and peripheral B-cell development. Useful cell surface, cytoplasmic, and nuclear stage–specific proteins are shown, along with essential mouse transcription factors (boxed). (Adapted from Refs. 2, 3, 12–14, 43, and 381–384.) Malignant transformation at distinct developmental stages yields characteristic patterns of protein expression. Mouse developmental stages are from Hardy and Melchers [12, 382]. LT/ST-HSC, long-term/short-term hematopoietic stem cell; CLP, common lymphoid precursor; T1, transitional 1; T2, transitional 2; MR, mature, recirculating B-cell; MZ, marginal zone B-cell; T_H, T-cell help; SHM, somatic hypermutation; CS, class switch recombination; RE, receptor editing; MC, naive mantle cell; CB, centroblast; CC, centrocyte; MB, memory B-cell; PC, plasma cell.

during the CLP-to-pre-pro-B-cell transition [20–22]. Pre-pro-B cells retain myeloid potential, which is abolished by subsequent expression of the B-lineage commitment factor Pax5 [23, 24]. Following successful immunoglobulin heavy-chain (IgH) gene rearrangements, pro-B cells transition to pre-B cells expressing the surface pre-B-cell receptor (pre-BCR [14, 25]). With subsequent Ig kappa (Igκ) or lambda (Igλ) light-chain (IgL) gene rearrangements and expression, pre-B cells develop into immature/naive B cells that coexpress surface IgM and IgD receptors.

IgM+/IgD+ coexpressing B cells enter the circulation and home to the splenic white pulp as transitional (T1) B cells. T1 B cells develop into T2 and then mature naive B cells under the influence of local splenic factors and antigen-dependent selection [26–30]. Mature naive B cells recirculate and populate lymphoid tissues such as lymph nodes, intestinal Peyer patches, and the spleen. Intraorgan localization of these immigrant B cells depends on specific molecular signals from the local microenvironment [31–37]. In the spleen, mature naive B cells reside in clusters called primary follicles within the white pulp. Additional seeded B cells stop recirculating and populate a thin marginal zone (MZB) at the outer edges of the white pulp [38]. MZBs develop directly into plasma cells and mainly participate

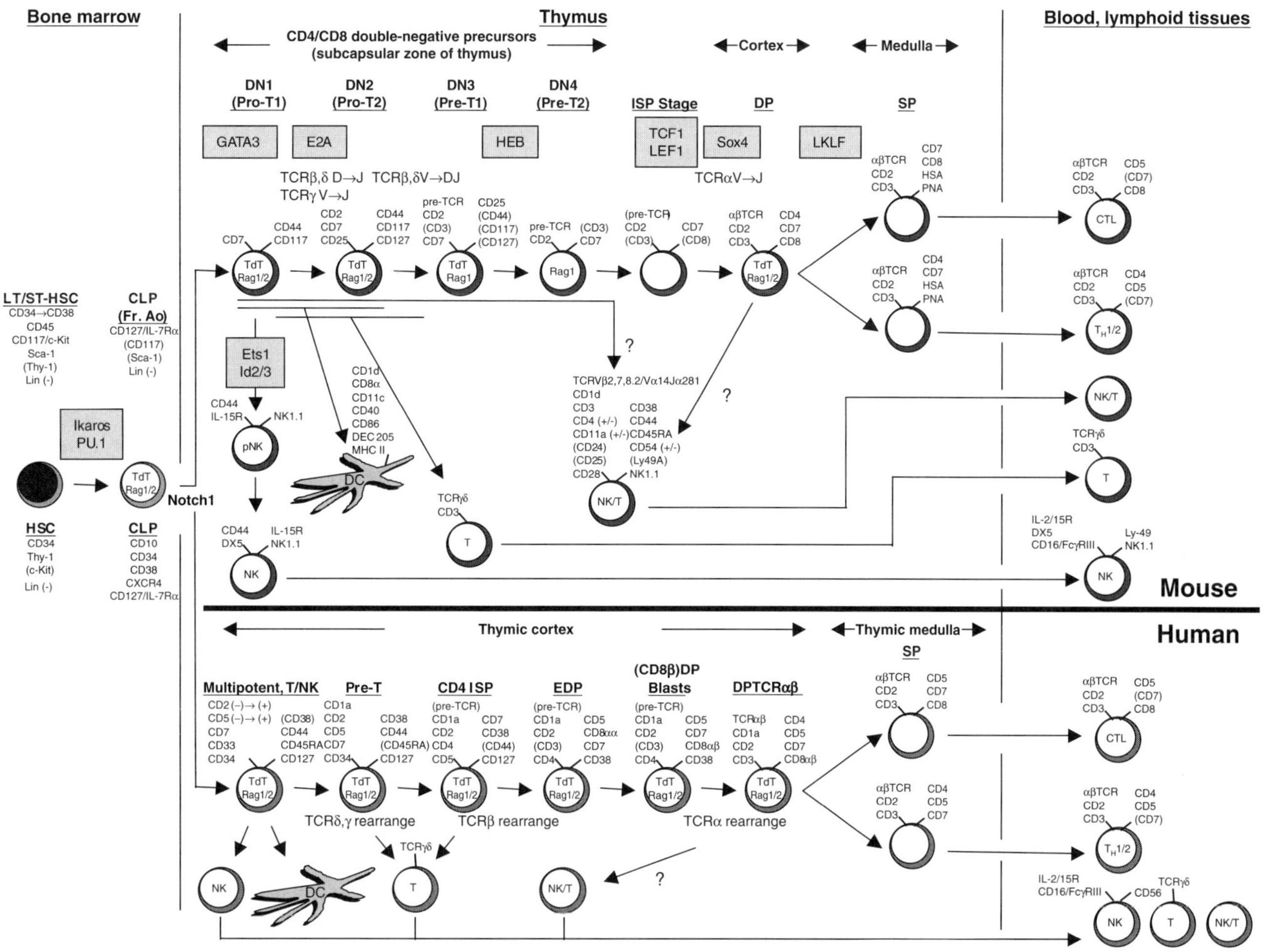

Figure 14.2. T, NK, NK/T, and DC development in mouse (above thick line) and human thymus. Precursor cells for some myeloid cell types may also be contained within the DN1 (Pro-T1) population [53, 54]. (Mouse T-lineage development is adapted from Refs. 20, 61, 62, 69, and 384–386. Mouse NK-lineage development is adapted from Refs. 20, 69, and 84–87. Mouse NK/T-lineage development is adapted from Refs. 99, 100, 102, and 103. Human T-lineage development is adapted from Refs. 72–76.) Essential mouse transcription factors are boxed. Malignant transformation at distinct developmental stages yields characteristic patterns of protein expression.

in front-line T-independent antigenic responses, while primary follicular B cells form secondary follicles under stimulation of T-dependent antigens. Secondary follicles exhibit a mantle zone of nonresponders and a germinal center (GC) of responding B cells undergoing expansion coupled to antibody refinement and selection [39–42]. Large, cycling centroblasts occupying the GC dark zone progress to small, noncycling centrocytes of the GC light zone. Surviving centrocytes yield high-affinity, class-switched, and somatically mutated Ig-secreting plasma cells and memory B cells. These GC emigrants either take up residence in interfollicular regions of the spleen or recirculate to populate additional hematopoietic organs such as the BM. As in the BM, transcription factors (Bcl-6, Pax-5, Blimp-1, XBP-1, IRF-4, NF-Atc, octamer binding proteins, and others) regulated by environmental signals control peripheral B-cell development [43].

A subpopulation of B cells called B-1 B cells are found mainly in the peritoneal and pleural body cavities and, like MZBs, form an innate natural antibody defense [14, 33, 44–47]. Although they share many features with traditional (B-2-type) MZBs, B-1 cells likely arise from a separate developmental pathway (Fig. 14.1 [48–50]). In humans, B-1a B cells were considered a precursor of B-chronic lymphocytic leukemia (B-CLL), mainly due to CD5 expression on malignant cells, although recent data suggests a memory B-cell origin [47, 51, 52].

T, NK, NK/T, and Dendritic Cell Development

T-cell development initiates with thymic seeding by BM-derived precursors (Fig. 14.2 [53, 54]). Signaling through the Notch-1 pathway specifies T-lineage fate and blocks B-lineage infidelity [55–59]. Early double-negative (DN1; CD4−/CD8−) GATA3-dependent progenitor cells retain B, T, NK, NK/T, and DC developmental potential [60]. By the DN2 stage, B- and NK-cell potential is lost [20, 61–63]. During the transition to DN3 cells, T-cell receptor (TCR) β, γ, and δ genes begin rearranging and DC potential is lost [20]. DN3 cells combine rearranged TCRβ, invariant pre-Tα, and specific CD3 complex components into a surface pre-TCR for early (so-called β) selection [64, 65]. Transition to DN4 cells is coupled to a pre-TCR-mediated down regulation in E47 expression [3, 20]. Cycling CD8, immature single-positive (ISP) cells transition to TCRα-rearranging, double-positive (DP; CD4+/CD8+)/αβTCR+ cortical thymocytes [66–69]. DP T cells undergo positive and negative selection, or death by neglect, to form naive CD4 or CD8 single-positive (SP) cells in the thymic medulla [70, 71]. αβTCR+, CD4, and CD8 SP T cells emigrate the thymus, recirculate, and take up residence in lymph nodes, Peyer patches, spleen, BM, and other sites. Human T cell development also begins in the thymus and largely resembles mouse T cell maturation (Fig. 14.2 [72–76]).

T-lineage commitment parallels CD25 expression, TCRβ, γ, and δ gene rearrangements and the DN1-to-DN2 transition. At this point, developing T cells become either TCRγδ or TCRαβ lineage restricted. The mechanism for this choice remains controversial [77–80]. The γδ T cells have limited TCR V-region gene usage and pass through the thymus in waves to seed mainly epithelial barriers of the gut, skin, and reproductive tract [81]. Human γδ T cells develop in roughly the same way and emigrate the thymus to seed peripheral epithelial barriers [82].

NK cells develop from early mouse precursor cells and from CD34+ thymic, BM, and umbilical cord blood precursors in humans (Fig. 14.2 [75, 83–87]). They retain germline Ig and TCR gene configurations and are interleukin-15 (IL-15) dependent [88–93]. The IL-15R complex and Id2, Id3, and Ets family transcription factors control early NK-cell development [20, 69, 84, 86, 94–96]. Up to 15% of human peripheral blood lymphocytes are CD56+/CD16+/CD3− NK cells while mouse NK cells lack a CD56 homologue. Both mouse and human NK cells control viral infections and malignancy through a balance of inhibitory and activating receptor stimulation [83, 97, 98]. NK cells are widely dispersed, produce cytokines [IFN-γ, tumor necrosic factor β (TNF-β) IL-10, IL-13 and others] and have cytolytic activity against target cells that are low or deficient in major histocompatibility complex (MHC).

NK/T cells share features and immune regulatory functions that overlap both T and NK cells [99–102]. Their Ets transcription factor–dependent origin from αβTCR+, DP T cells, or pre-TCR gene rearrangement precursors is unresolved [96, 103]. Human and mouse NK/T cells are usually DN or CD4 SP, have restricted TCR V gene usage, express NK1.1 (NKR-P1) and CD122 surface proteins, and respond to CD1d MHC elements loaded with glycolipids by secreting IL-4 or IFN-γ [100].

The origin(s) of antigen-presenting DC are unresolved [7]. Some studies indicate that DC arise separately from CLP and CMP [8, 104–112]. The wide tissue distribution and unique patterns of surface protein expression also suggest multiple independent DC origins [9, 113–116]. However, a common DC precursor capable of generating all DC subtypes has been recently shown [10]. Lymphoid DCs function in the thymus and in the periarteriolar lymphoid sheaths (PALSs) of the spleen and lymph nodes, where they present antigens to the innate and adaptive immune system [117–119]. Human DCs develop in similar stages and are also split into myeloid and lymphoid subtypes, although some controversy surrounds the existence of human lymphoid DCs [120–123]. DC neoplasms are rare in humans and a mouse model has not been described thus far.

CLASSIFICATION AND DIAGNOSIS OF LYMPHOID MALIGNANCIES

Despite strong similarities, differences exist in the development of lymphoid cells between humans and mice. For example, IL-7 and its receptor are essential for mouse but not for human early B-cell development [124, 125]. The location of predominant involvement by lymphoid cancers is also different between species. For example, B-cell lymphomas typically develop in the mouse spleen and then disseminate while lymph nodes are the most frequent source for B-cell lymphomas in humans. Not surprisingly, cancers from similar cell types or due to similar genetic insults result in common and distinct features between species. Still, accurate murine models of human lymphoid cancer remain prized for dissecting pathogenetic mechanisms or preclinical testing of tumor targets for efficacy. Schemes for human and mouse lymphoid malignancies that reproducibly classify tumors within and between species are therefore essential.

In 2001 the World Health Organization (WHO) adopted a classification algorithm for human lymphoid tumors [126, 127]. The WHO updates the Revised European-American Classification of Lymphoid Neoplasms (REAL [128]), which is the latest iteration in a series of lymphoid tumor classification schemes [129–131]. Like its 1994 predecessor, the WHO classification divides tumors into lymphoid, histiocytic/dendritic, myeloid, and mast cell categories.

Tumors in each category are further stratified by morphologic, immunophenotypic, genetic, and syndromic features. A cell of origin is postulated for each listed entity. Within the lymphoid category B-cell, T- and/or NK-cell, and Hodgkin lymphoma tumor types are recognized. Criteria for each WHO diagnosis are presented elsewhere [126, 127]. Table 14.1 lists the major lymphoid cell neoplasms of the WHO classification.

In 2002 a subcommittee of the Mouse Models of Human Cancers Consortium (MMHCC) published the "Bethesda classification" of mouse hematopoietic neoplasms based upon a combination of histologic, phenotypic, and molecular features [132, 133]. The Bethesda proposal is built on prior classification schemes, stratifies malignancies according to cell origin, and parallels related human cancers whenever possible [134–136]. The major recognized subgroups include lymphoid and nonlymphoid hematologic malignancies. Criteria for each Bethesda diagnosis are given elsewhere [132]. Table 14.1 lists the major lymphoid malignancies in this classification and "best-fit" comparisons to categories in the WHO classification.

A comprehensive approach is required for accurate lymphoid tumor diagnoses (see Chapter 2). Premortem, mice are observed for behavior changes along with periodic blood cell counts, blood smear, flow cytometric analysis, and, when indicated, clonality determination. At euthanasia, the body is weighed and inspected for cutaneous lesions and subcutaneous lumps. Blood studies include a cell count with differential, blood and BM smears, and potential flow cytometry. The skin is removed to expose superficial lymph node groups and then body cavities opened to collect extravasated fluids that may represent tumor ascites. The spleen is weighed and inspected. Lymphoid tumors infiltrating the white pulp may exhibit a white/gray nodular appearance while those involving the red pulp retain a red/pink smooth consistency. The thymus is also weighed and inspected. Deep lymph node groups are isolated if suspicious. The lungs, liver, kidneys, and brain are removed and examined for tumor infiltration. Thin sliced samples from each organ and the femurs following decalcification are placed in an appropriate fixative (10% neutral buffered formalin, Bouin's solution) and thin sectioned on a microtome, and slides are stained with hematoxylin and eosin for microscopic evaluation.

Additional studies may be required to determine the presence, type, and spread of malignancy. To perform these procedures, blood and tissues must be properly preserved (see Chapter 2). For immunohistochemistry, tissues can be fixed in 10% neutral buffered formalin or fresh frozen in cold-polymerization OCT (optimal cutting temperature) compound. Blood samples and cell suspensions from organs can be examined with cytospin cell-buttons. Cell suspensions can be evaluated by flow cytometry. DNA, RNA, and protein content are accessible by lysing cells in specific preservative solutions, such as phenol-saturated guanidinium hydrochloride for RNA analysis.

Determination of T- and B-cell tumor clonality by immunohistochemistry and flow cytometry is standard in humans using antibodies to CD4, CD8, Igκ, and Igλ proteins. Here, the normal ratios of CD4-to-CD8 and Igκ-to-Igλ expression in mature lymphocytes are close and changes from these ratios will appear in tumors. However, the ratio of Igκ to Igλ is roughly 10:1 in mice, and most B-cell tumors remain strongly Igκ positive, obscuring clonal evolution.

Of particular importance are determinations of IgH, TCRβ, and γ gene rearrangements. IgH VDJ rearrangements mark B-lineage cells, while TCR V(D)J rearrangements mark T-lineage cells. T-cell lymphomas can sometimes have IgH D→J, but not V→DJ, rearrangements and should not be interpreted as mixed T/B-cell tumors. Southern blot hybridization with specific IgJ_H or TCRβ probes demonstrate multiple, typically faint rearranged bands in lymphoid hyperplasia or early oligoclonal selection versus distinct, typically intense monoclonal bands in intermediate and advanced stages of leukemia/lymphoma. Polymerase chain reaction (PCR) analysis with primers spanning IgH D→J and TCRβ or γ D→J junctions is also useful for determining oligo- or monoclonal lymphocyte expansions and malignancies. For human B-cell lesions, evidence of somatic hypermutation (SHM) in the Ig locus and in selected non-Ig genes indicates a tumor origin from GC B cells or beyond, although SHM in the absence of GC has been documented [137–139]. While the rate of SHM in mice is about 10-fold less than in human B cells, such determinations can be informative [140].

Cytogenetic studies of human tumors use metaphase-frozen chromosome spreads to determine ploidy, translocations, and deletions. These studies are difficult with mouse chromosomes, which are acrocentric and very similar in overall size. A partial solution is spectral karyotyping (SKY), which combines fluorescence in situ hybridization (FISH) with spectral image analysis [141–144]. Mouse SKY employs 21 distinctly colored chromosome "paints" that color each chromosome uniquely (19 autosomes plus X and Y), facilitating detection of large interchromosomal rearrangements, such as translocations. For smaller lesions, such as gene amplifications or deletions, array-based comparative genomic hybridization (array CGH) provides a potential detection method [145–148].

A powerful method for diagnosing, classifying, and predicting outcomes of human lymphoid malignancies is DNA microarray analysis [149–151]. This technique accurately replicates the traditional classification of acute lymphoblastic versus acute myeloid leukemias [152]. Expression profiling and gene cluster analysis also subclassifies diffuse large B-cell lymphoma (DLBCL) into aggressive,

Table 14.1. Classification of Lymphocyte Neoplasms in Humans and Mice

Type	2001 WHO Classification (Human)	2002 Bethesda Classification (Mice)
	B-Cell Neoplasms	
Precursor B cells	Precursor B lymphoblastic leukemia/lymphoma	Precursor B-cell lymphoblastic leukemia/lymphoma (pre-B-LBL)
Mature B cells	Chronic lymphocytic leukemia/small lymphocytic lymphoma	Small B-cell leukemia/lymphoma (SBL)
	B-cell prolymphocytic leukemia	
	Lymphoplasmacytic lymphoma/Waldenstrom macroglobulinemia	
	Mantle cell lymphoma	
	Hairy cell leukemia	
	Splenic marginal zone lymphoma	Splenic marginal zone lymphoma (SMZL)
	Follicular lymphoma	Follicular B-cell lymphoma (FBL)
	Burkitt lymphoma	Classic Burkitt lymphoma (BL)
	Morphologic variants	Morphologic variants
	Burkittlike lymphoma	Burkittlike lymphoma (BLL)
	Plasmacytoid differentiation	
	Diffuse large B-cell lymphoma	Diffuse large B-cell lymphoma (DLBCL)
	Morphologic variants	Morphologic variants
	Centroblastic	Centroblastic (CB)
	Immunoblastic	Immunoblastic (IB)
	Histiocyte/T-cell rich	Histiocyte associated (HA)
	Subtypes	Subtypes
	Mediastinal (thymic)	Primary mediastinal (thymic) (PM)
	Intravascular	
	Primary effusion	
	Extranodal MZL-MALT type	
	Extranodal MZL	
	—	B natural killer cell lymphoma (BNKL)
	Plasma cell neoplasms	Plasma cell neoplasms
	—	Plasmacytoma (PCT)
	Extraosseous plasmacytoma	Extraosseous plasmacytoma (PCT-E)
	—	Anaplastic plasmacytoma (PCT-A)
	Solitary plasmacytoma of bone	
	Primary amyloidosis	
	Heavy-chain diseases	
	MGUS	
	Plasma cell myeloma	
Hodgkin lymphoma	Nodular lymphocyte predominant	
	Classical type	
	Morphologic variants	
	Nodular sclerosis	
	Lymphocyte rich	
	Mixed cellularity	
	Lymphocyte depleted	
	T/NK-Cell Neoplasms	
Precursor T/NK cells	Precursor T lymphoblastic leukemia/lymphoma	Precursor T-cell lymphoblastic lymphoma/leukemia (pre-T-LBL)
	Blastic NK-cell lymphoma	

(*continued overleaf*)

Table 14.1. (*continued*)

Type	2001 WHO Classification (Human)	2002 Bethesda Classification (Mice)
Mature T/NK cells	—	Small T-cell lymphoma (STL)
	T-cell prolymphocytic leukemia	
	T-cell large granular lymphocytic leukemia	
	—	T natural killer cell lymphoma (TNKL)
	Aggressive NK-cell leukemia	
	Adult T-cell leukemia/lymphoma	
	Extranodal NK/T-cell lymphoma, nasal	
	Enteropathy-type T-cell lymphoma	
	Hepatosplenic T-cell lymphoma	
	Subcutaneous panniculitis-like T-cell lymphoma	
	Mycosis fungoides	
	Sezary syndrome	
	Primary cutaneous anaplastic large-cell lymphoma	
	Peripheral T-cell lymphoma, unspecified	
	Angioimmunoblastic T-cell lymphoma	
	Anaplastic large-cell lymphoma	
		Large-cell anaplastic lymphoma (TLCA)

poor outcome (activated DLBCL) versus prolonged survival (GC-derived DLBCL) subtypes [153, 154]. An independent subclassification of DLBCL showed that specific pathway alterations, in addition to cell-of-origin grouping, provide reliable diagnosis and outcome predictions [155]. Expression profiling may also precipitate mergers of seemingly disparate diseases into unified entities. For example, B-CLL exhibits two types of behaviors that segregate with the presence or absence of Ig locus SHM [156, 157]. Tumors lacking SHM predict an aggressive, rapidly fatal course while those with SHM result in less aggressive disease. This correlation suggests that B-CLL arises from two types of mature B cells that differ in having or lacking passage through the GC, as indicated by the SHM status [157–163]. However, studies now show that both B-CLL subtypes have highly similar gene expression profiles and likely originate from memory B cells [51, 52].

Microarray analysis of over 90 mouse lymphoid tumors and cell lines yields molecular groupings consistent with groupings using traditional classification criteria [164]. A "lymphochip" containing target DNA sequences implicated in lymphoid malignancies is useful for screening human samples and is in development for mice [165]. Future gene expression profile comparisons between mouse and human lymphoid malignancies will help determine the extent to which a mouse tumor resembles a comparable human lymphoid malignancy [166].

MOUSE MODELS OF LYMPHOID CELL NEOPLASMS

Many methods exist for discovering candidate tumorigenic genes in mice, including ethylnitrosourea and gene trap mutagenesis, cre-lox-mediated inversions and deletions, and proviral insertion mutagenesis [167–175]. Genes identified by these and other approaches and known tumor-promoting genes in humans form the group of genes that are studied in mouse models of cancer. Somatic cell gene transfer and reconstitution and transgenic and gene "knockin" (KI) and "knockout" (KO) are the technologies most frequently employed (see Chapters 4,5). Some models are made to mimic a human lymphoid cancer while others develop unexpected lymphoid tumors. Each model, intended or not, has value for dissecting lymphomagenic mechanisms. However, no standardized evaluation scheme for mouse tumors exists, unlike the situation in human surgical pathology. This may cause uncertainty in the classification of certain modeled malignancies.

PREMALIGNANT LYMPHADENOPATHIES AND LYMPHOID CELL EXPANSIONS

Increased numbers of lymphoid cells are caused by homeostatic disorders that fail to properly regulate cell proliferation and/or death. Increases in proliferation may be so great that they mask concomitant increases in cell death, as may be seen in c-*MYC* overexpression [176, 177]. However, too many cells do not constitute a lymphoid cancer. Additional fundamental cell changes must occur, including independence from growth signals, insensitivity to antigrowth signals, blockade of apoptosis, and immortalization [178]. During immortalization, human cells overcome both replicative senescence and cellular crisis by disabling the *p53* and *Rb* tumor suppressor pathways and activating telomerase [179, 180]. In contrast, many mouse cells are immortalized by repression of the $p19^{ARF}$-*p53* signaling pathway alone [181]. Clearly, human and mouse cells have fundamental differences on the road to cancer.

FAS AND FAS LIGAND

The *lpr* (lymphoproliferation) or *gld* (generalized lymphoproliferative disease) mice have mutant *Fas* (*CD95/Apo1*) or *FasL* (*CD95L*) genes and signaling defects [182, 183]. The *gld* mice develop polyclonal expansions of DN T cells and altered B-cell subsets with generalized lymphadenopathy and a lupuslike autoimmune disease due to decreased cell death [184]. A similar pathology occurs in humans with *Fas/FasL* mutations in the autoimmune lymphoproliferative syndrome (ALPS), also known as Canale–Smith syndrome [185].

In general, mouse models with a single gene alteration experience a lag to cancer formation until companion mutations required for transformation occur. Strain-specific modifying factors or crosses between complementary tumor inducers, such as *Pim-1* and c-*Myc*, may decrease this lag time significantly [186]. Also, strain differences strongly influence the tumor type that ultimately occurs [187]. For example, a c-*Myc* transgene in C3H/HeJ mice causes T-cell lymphomas while the same transgene integrated at the same genomic locus causes B-cell lymphomas in C57BL/6 mice [188].

MODELS OF B-CELL CANCER

Mouse genetic models of B-cell cancer may be caused by inappropriately regulated signaling molecules (*Bcr-Abl, Blk, Pim1, Fas*, and *FasL*), transcription factors (c-*Myc, TEL/AML1, Aiolos, HOX11*), tumor suppressors (*p53*, $p16^{INK4a}/p19^{ARF}$, *Bcl-2*), cytokines (*IL-6, IL-7*), and adapter molecules (*TCL1*) (Fig. 14.3). These examples form one or more B-cell tumor type as defined in the MMHCC classification and discussed in sections under their tumor-type headings.

Precursor B-Cell Lymphoblastic Lymphoma/Leukemia (pre-B-LBL)

Pre-B-LBL form sheets of medium-sized blast cells with distinct histologic features [132]. These tumors are high grade and have numerous mitoses and apoptotic cells. Pre-B-LBL arise from BM precursors and have a sIg−/B220+/CD19+/TdT+ immunophenotype. They contain IgH but not IgL rearrangements and may present with splenomegaly and/or lymphadenopathy. Mature B-LBL show the same general features as pre-B-LBL but lack TdT expression and are sIg+ with IgH and IgL rearrangements. The human counterpart malignancies are pro-B and pre-B acute lymphoblastic leukemia (B-ALL) or B lymphoblastic lymphoma (B-LBL).

Philadelphia Chromosome (Bcr-Abl). The Philadelphia chromosome (Ph) originates from a reciprocal translocation between chromosomes 9 and 22 that results in a short der(22) chromosome [189, 190]. A hallmark of chronic myelogenous leukemia (CML; see Chapter 13), the Ph also occurs in 20% of adults and 5% of children with B-ALL [191]. A characteristic t(9;22)(q34;q11) causes a head-to-tail fusion of the breakpoint cluster region (*BCR*) gene on chromosome 22 with the cellular homologue of the Abelson (c-*ABL*) viral oncogene on chromosome 9, placing fused *BCR-ABL* oncogenes under control of the ubiquitously expressed *BCR* promoter [192]. *BCR-ABL* oncogenes encode functional BCR-ABL fusion oncoproteins with constitutive tyrosine kinase activity [193–196]. These fusion oncoproteins occur in two main isoforms of distinct molecular weights, p190 and p210, based mainly upon differences in the translocation breakpoints in the *BCR* gene [197–199]. The p210 isoform is expressed in >90% of CML cases while p190 and p210 are each expressed in about 50% of adult Ph-positive B-ALL cases [195, 200, 201]. Roughly 90% of Ph-positive B-ALL cases in children express p190, compared to 10% that express p210 [190].

Infection of 5-fluorouracil- (5-FU-) treated mouse BM with a $BCR\text{-}ABL^{p210}$-expressing retrovirus followed by reconstitution into lethally irradiated hosts yields clonal B-ALL in DBA/2 strain recipients and T-ALL in C57BL/6 mice (Table 14.2 [202]). Separately, infection of 5-FU-treated BALB/c BM with a $BCR\text{-}ABL^{p210}$-expressing retrovirus results in CML, B-ALL, T-lymphoma, and possibly macrophage-derived malignancies in BALB/c recipient mice [203]. In a third study, 5-FU-treated BALB/c BM infected with a $BCR\text{-}ABL^{p210}$-expressing retrovirus transferred into BALB/c mice results in CML and B-ALL [204].

Metallothionein 1 (mt)-promoter-$BCR\text{-}ABL^{p190}$ transgenic mice develop either CML in blast crisis (see Chapter 13) or B-ALL (Table 14.3 [205, 206]). The mt-promoter-$BCR\text{-}ABL^{p210}$ transgenic mice develop B- and T-ALL at equal frequencies with a slower onset than $BCR\text{-}ABL^{p190}$ mice [207]. Independently, studies with mt-promoter-$BCR\text{-}ABL^{p210}$ transgenic mice demonstrate T-ALL [208]. Founder tec-promoter $BCR\text{-}ABL^{p210}$ transgenic mice develop T-ALL (and possibly B-ALL) while transgenic progeny develop myeloproliferative disorders resembling CML [209]. IgV_H-promoter–Eμ or retroviral long terminal repeat (LTR)–controlled transgenic mice expressing a synthetic BCR-ABL gene created by fusing *bcr* with v-*abl* sequences develop either B-ALL or thymic-based T lymphomas, most of which would likely classify as T-ALL [210]. A tetracycline-inducible system with $BCR\text{-}ABL^{p210}$ expressed when

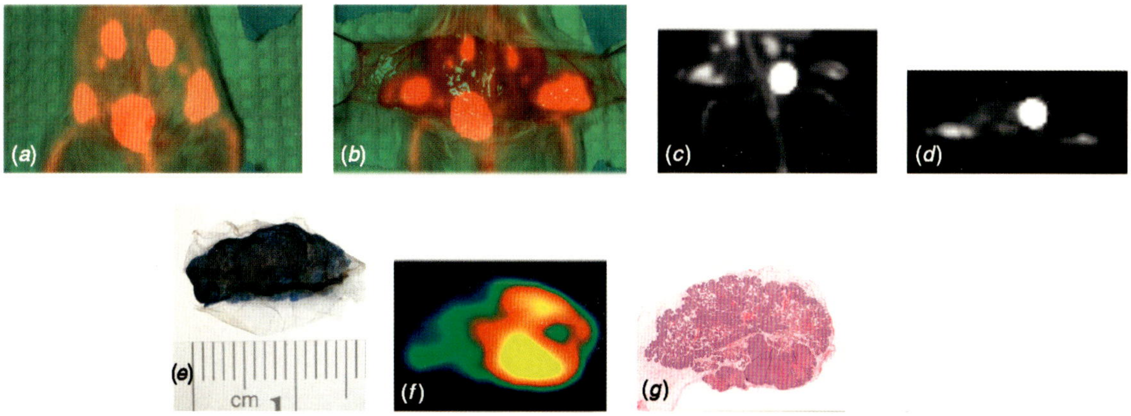

Plate 2.1. Ex vivo imaging. Validation of in vivo images may require detailed dissections to verify that the image detected in the animal represents a specific pathologic lesion. This figure illustrates how the in vivo image is removed and reimaged to prove that it represents a tumor mass. (*a*) Mammary tumors transplanted into the mammary fat pad are detected in vivo using micro–positron emission tomography (DET). (*b*) To determine whether the images correspond to the mammary mass, an incision was made and the skin that contains the mammary fat pad was spread to expose the tumors. The photographic and PET images are superimposed to demonstrate that the microPET images stay with the skin and the tumors. (*c*) Another animal went through the same procedure. The skin was spread. The microPET images of the tumor-bearing mouse show (*c*) the supine and (*d*) the coronal views. One of the tumors was (*e*) removed and placed on a slide and (*f*) reimaged to show that the in vivo image was the same as the ex vivo image. (*g*) The microscopic slide is used by the pathologist to verify the diagnosis. (Images courtesy of Simon Cherry, Craig Abbey, and Alexander Borowsky, University of California—Davis.)

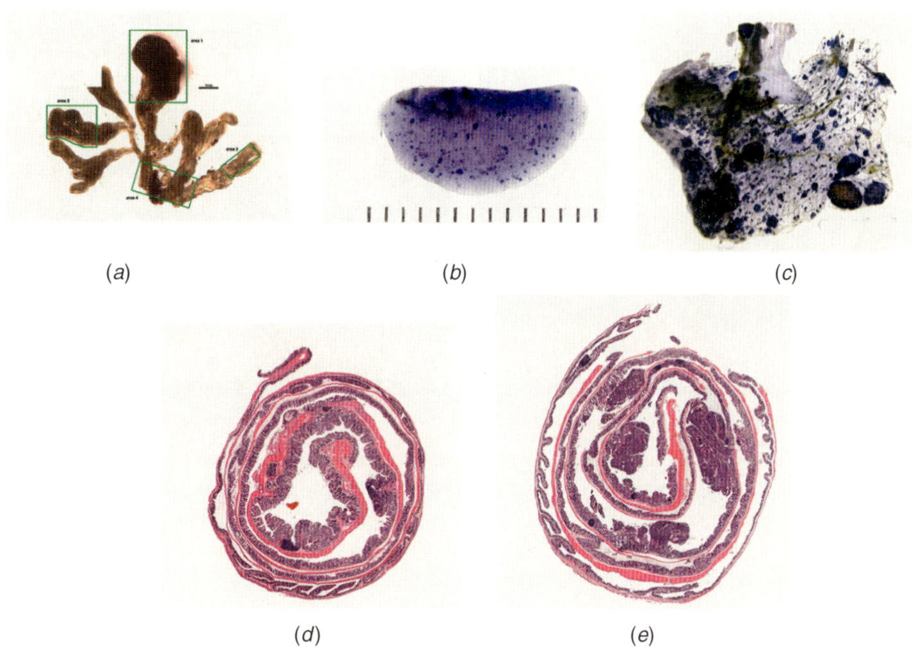

Plate 2.2. Special types of preparations. Special preparation techniques: whole-mount preparations of (*a*) prostate, (*b*) lung, and (*c*) mammary gland that allow the investigator to visualize number and distribution of tumors in the entire organ sampled. (*a*) Fresh, unstained prostate viewed via transmitted backlight. This image of a transgenic adenocarcinoma mouse prostate (TRAMP) was provided by A. A. Donjacour, University of California—San Francisco. (*b*) Fixed, hematoxylin-stained and cleared preparation of a mouse lung from a PyV-mT mouse. (Preparation and image by Erik T. McGoldrick, University of California—Davis.) The lung has numerous small dark spots that represent multiple metastatic masses. (*c*) Mammary gland whole mount from an *erbB* transgenic mouse. (Preparation and image by Erik T. McGoldrick, University of California—Davis). The normal architecture is disrupted by multiple small masses. (*d*, *e*) Histology obtained with Swiss roll preparations of the mouse gastrointestinal tract. Part (*d*) illustrates a normal mouse bowel and (*e*) illustrates multiple polyps in mice provided by Robert Coffee and Kay Washington, Vanderbilt Medical School.

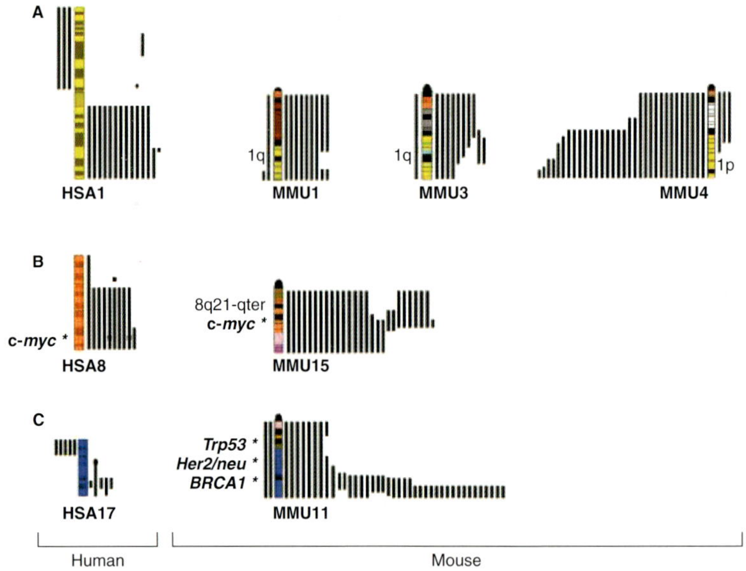

(*a*)

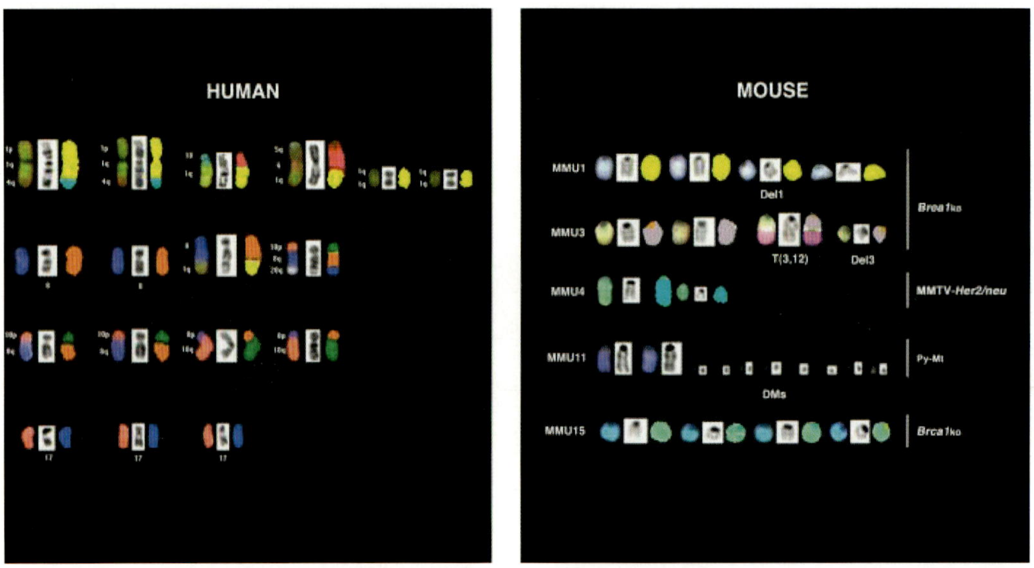

(*b*)

Plate 7.1. (*a*) **Summary of CGH and SKY analyses of human breast carcinomas and mouse mammary gland adenocarcinomas.** (A) In human breast cancers, chromosome arm 1p is lost, while copy number increases on chromosome arm 1q are common. The orthologous regions in the mouse genome are subject to similar copy number variations, although with different relative frequencies. These summary graphs are based on the analysis of 20 human breast carcinomas and 70 mouse tumors. (B) The comparative analysis of human chromosome 8 reveals similar results. The frequent gain of chromosome 8q and the c-myc oncogene (which maps to band 8q24) is recapitulated in murine tumor models. Note that the minimally overlapping region of genomic amplification on mouse chromosome 15 includes the locus of the c-myc oncogene. (C) The gain of the long arm of chromosome 17, which is commonly observed in human breast cancers, correlates well with genomic amplification of distal mouse chromosome 11. A divergence, however, exists. While copy number decreases that map to chromosome arm 17p are rather common in breast cancers, loss of orthologous regions in the mouse genome is an exception. The inactivation of the tumor suppressor gene TP53, which is a recurrent aberration in breast cancer, seems to be less important in mouse mammary gland tumorigenesis, or occurs by mechanisms, other than chromosome loss. (*b*) **Examples of involvement of human chromosomes 1, 8 and 17, and mouse chromosomes 1, 3, 4, 11, and 15 in rearrangements in BRCA1 deficient human breast cancer and murine tumor models that emerge in mice conditionally KO mutants for BRCA1.** In the human cell line, chromosome 1 is involved in numerous chromosomal translocations, which result in copy number increases of the long arm, yet loss of chromosome 1p. Equally, multiple segments of chromosome 8q are dispersed in this highly rearranged karyotype. Chromosome 17 appears to be normal. Orthologous chromosome in BRCA1 deficient murine tumors: Note that copy number losses of mouse chromosome 4 and gains of mouse chromosome 15 are similar to copy number changes in human BRCA1 related breast cancers.

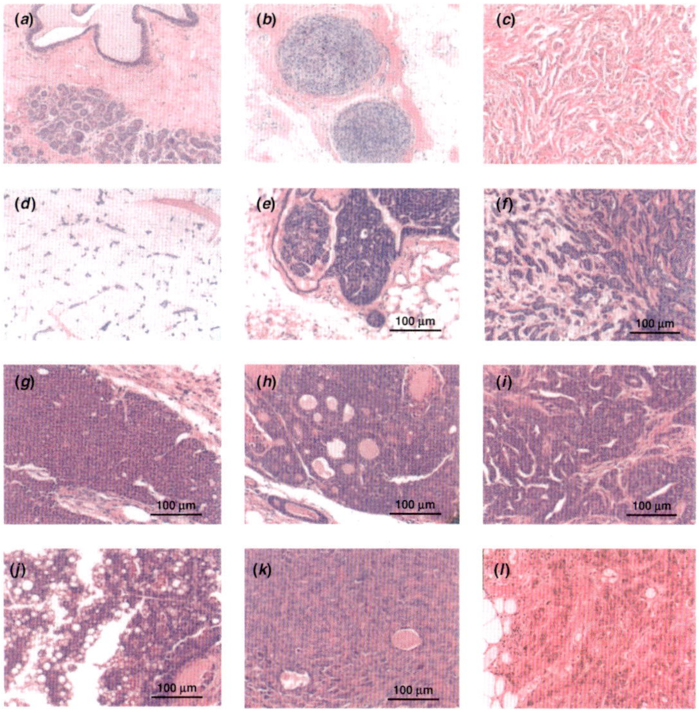

Plate 7.3. Histologic comparisons of human and mouse mammary tumors: (*a*) normal human non-lactating mammary gland; (*b*) human DCIS (20X); (*c*) human poorly differentiated infiltrating breast carcinoma. Nuclei are highly pleomorphic with tumor cells surrounded by dense connective tissue; (*d*) normal mouse adult virgin FVB/N mammary gland (20X); (*e*) mammary intraepithelial neoplasia in C3(1)/TAg mouse resembling human DCIS.; (*f*) invasive adenocarcinoma in C3(1)/TAg mouse resembling human tumor in C; (*g*) MMTV-her2/neu tumor with typical undifferentiated, solid, nodular appearance. Tumor is highly vascularized with uniform nuclei and red/pink cytoplasm; (*h*) MMTV-myc tumor with cystic and necrotic features. Large blue cells with prominent nucleoli are typical of myc-induced tumors; (*i*) MMTV-PyMT tumor with well differentiated and poorly differentiated characteristics. Papillary and invasive features with significant fibrosis are visible; (*j*) MMTV-TGFα adenosquamous carcinoma with secretory and papillary characteristics; (*k*) poorly differentiated adenocarcinoma from p53−/− transplant model; (*l*) tumor from BRCA1−/− mouse model. Images A-K and diagnoses courtesy of Robert Cardiff, UC Davis (Mammary Cancer in Humans and Mice; a tutorial for comparative pathology CD ROM) and image L courtesy of Chuxia Deng, NIDDK, NIH.

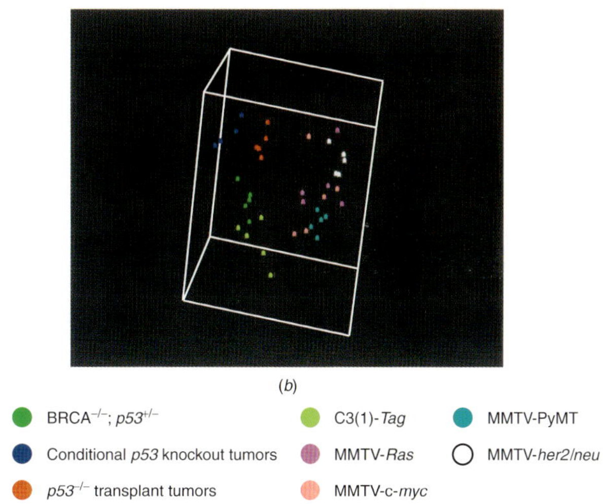

Plate 7.4b. Graphical distinctions between gene expression profiles of mouse models of human breast cancer represented in three dimensions by multidimensional scaling.

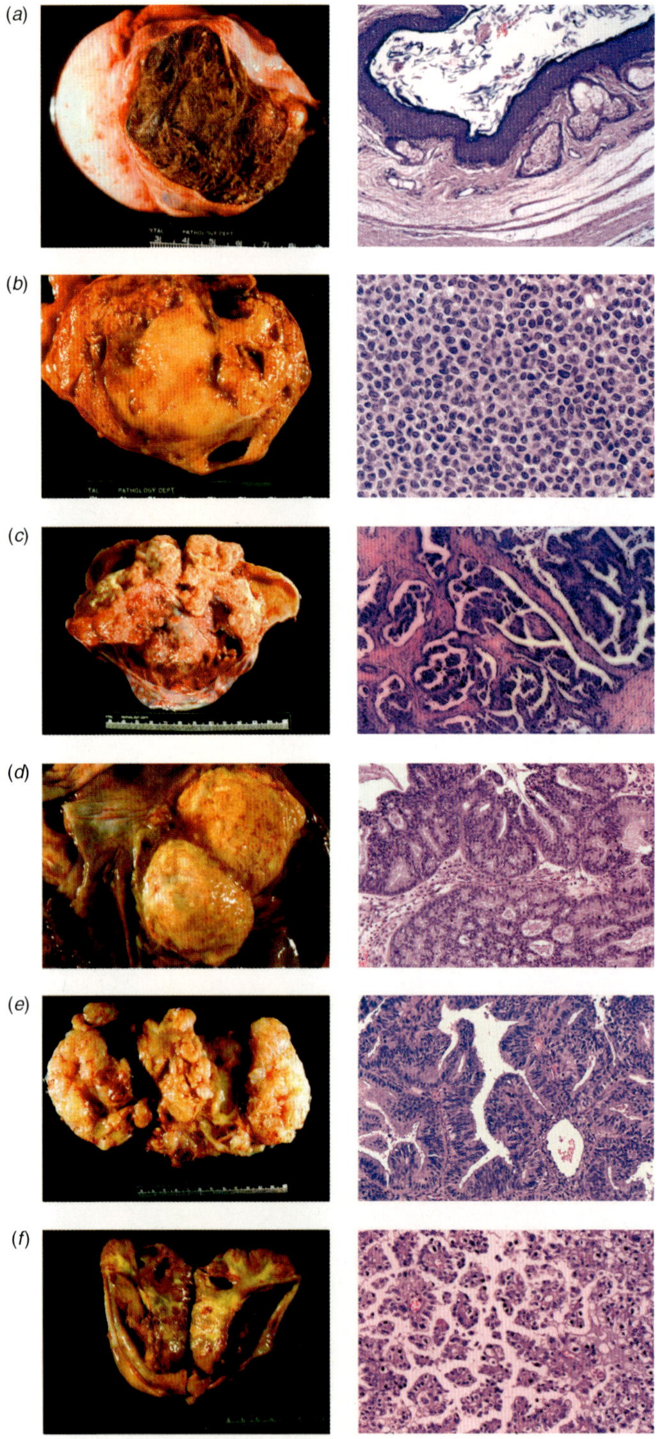

Plate 10.3. Histologic types of ovarian cancer. (*a*) Mature cystic teratoma, the most common subtype in the category of germ cell tumors. Hair is grossly seen. (*b*) Granulosa cell tumor (subtype of sex cord–stromal cell tumor). Four main human epithelial ovarian cancer subtypes: (*c*) serous, (*d*) mucinous, (*e*) endometrioid, and (*f*) clear-cell carcinoma. (Courtesy of E. Oliva.)

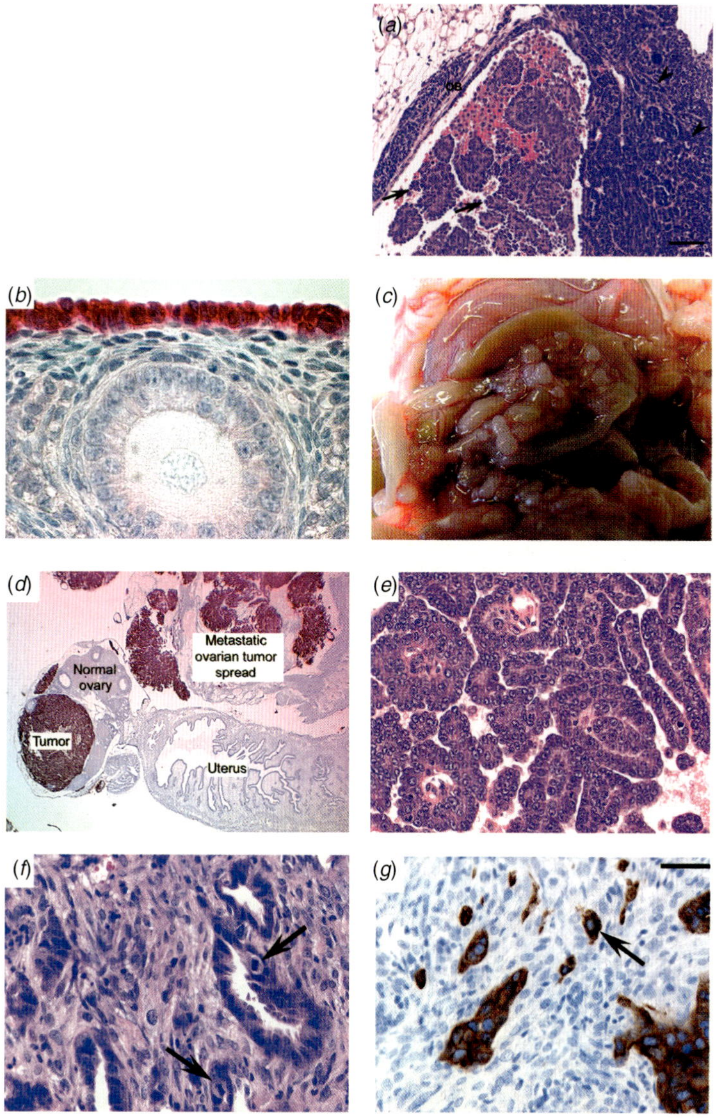

Plate 10.6. Mouse models of epithelial ovarian cancer. (*a*) The ovarian carcinoma in the mouse chimeric for expression of the simian virus 40 T antigen (*SV40 TAg*) under control of the Müllerian inhibitory substance type II receptor (*MISIIR*) promoter. Neoplastic cells form tubular (arrowheads) and papillary (arrows) structures in the ovary and in the intrabursal space, respectively, and invade the ovarian bursa (OB). (*b–e*) An RCAS-TVA mouse model for ovarian epithelial cancer. (*b*) Ovary section from a keratin 5-TVA transgenic mouse stained with antibody against the avian retroviral TVA receptor. Ectopic expression of the TVA receptor renders the cells susceptible to infection with avian RCAS viruses, which are used as vehicles for gene transfer. Cell-specific expression of the TVA receptor restricts the infection to the cells of the ovarian surface epithelium, which is the presumptive precursor tissue for ovarian carcinoma. (*c*) Intraperitoneal carcinomatosis in a mouse model for epithelial ovarian cancer resembles human ovarian cancer spread. (*d*) Metastatic ovarian tumor spread. An influenza hemaglutinin (HA)-tagged *Akt* oncogene is detected with an HA-probe in the primary ovarian tumor and in ovarian metastases. (*e*) H&E staining of an ovarian tumor induced in the RCAS-TVA mouse model. Ovarian tumors induced in the mouse model histologically resemble human ovarian papillary serous carcinoma. (*f,g*) Serous adenocarcinoma induced by selective AdCre-LoxP mediated inactivation of *p53* and *Rb1* in the ovarian surface epithelium of $p53^{loxP/loxP}Rb^{loxP/loxP}$ mice. (*f*) H&E staining representing mitotic (arrows) carcinoma cells that form glandular structures in dense fibrous tissue. (*g*) Immunohistochemical detection of keratin 8 in invasive neoplastic cells (arrow). (Courtesy of A. Yu. Nikitin.)

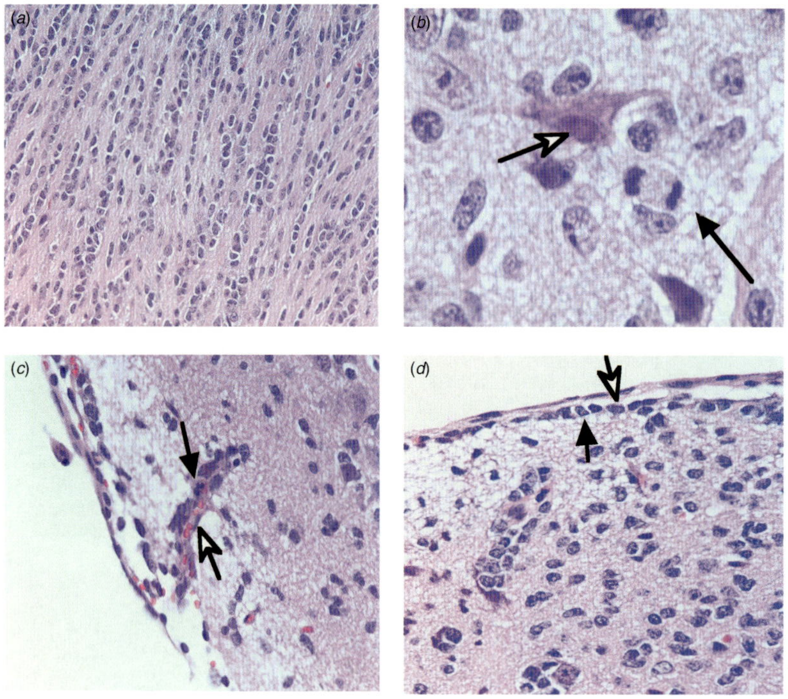

Plate 12.1. Glioma cells invading normal brain structures. Photomicrographs of mouse glioma models showing invasive features indistinguishable from the characteristic found in human gliomas. Glioma cells are seen migrating, along white matter tracts (*a*), glioma cells (closed arrow) surrounding neurons (open arrow) (*b*), glioma cells (closed arrow) lined up along a blood vessel (open arrow) (*c*), and glioma cells (closed arrow) accumulating in the subpial space (open arrow) (*d*).

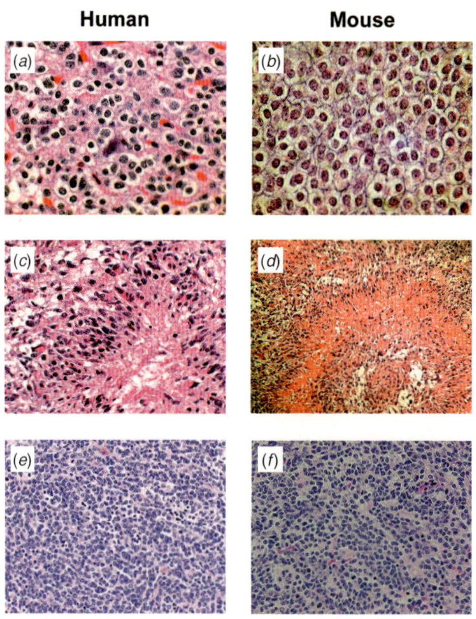

Plate 12.2. Comparative histology of (*a, c, e*) human and (*b, d, f*) mouse brain tumors. (*a*) Human oligodendroglioma with characteristic oval cells with nuclei surrounded by a white halo. (*b*) Mouse oligodendroglioma from *Pdgf* somatic cell gene transfer showing similar histologic features. (*c*) Human glioblastoma with characteristic pseudopalisading necrosis. (*d*) Pseudopalisading necrosis in a mouse glioblastoma induced by combined activation of *Ras* and *Akt* using somatic cell gene transfer. (*e*) Human medulloblastoma with characteristic small blue pleomorphic hypercellularity. (*f*) Similar histology in a mouse model of medulloblastoma induced by combined *Shh* and *Igf2* using somatic cell gene transfer. (*a*, *c*, *e*) Courtesy of Marc Rosenblum, MD, Department of Pathology, Memorial Sloan-Kettering Cancer Center, New York, NY. (*f*) Courtesy of Daniel W. Fults, MD, University of Utah School of Medicine, Salt Lake City, Utah.

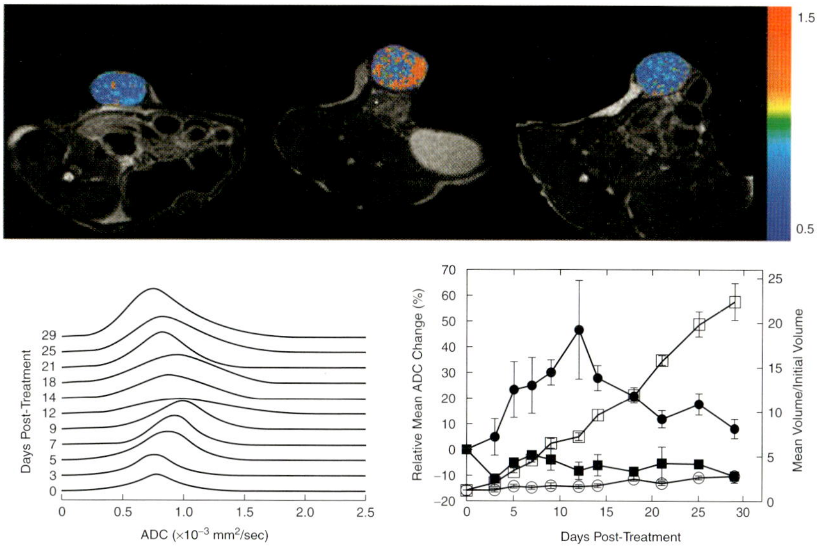

Plate 22.7. Diffusion-weighted MRI study of a subcutaneous PC-3 (prostate) tumor in a nude mouse treated with 100 mg/kg (i.p.) Camptosar. (Top panel) The three greyscale anatomic T2-weighted images are overlayed with a diffusion map (color) of the tumor, from left to right (days 3, 12, and 29 posttreatment). Note the increase in tumor diffusion values (red) on day 12 posttreatment. Tumor data are also displayed in a histogram format over time (bottom left) from the same animal as the images. The plot of "mean relative ADC change (%)" versus time is the change in mean ADC relative to day zero of three treated and three control animals. Solid symbols are ADC (circle = treated, square = control) and open symbols are volume (circle = treated, square = control).

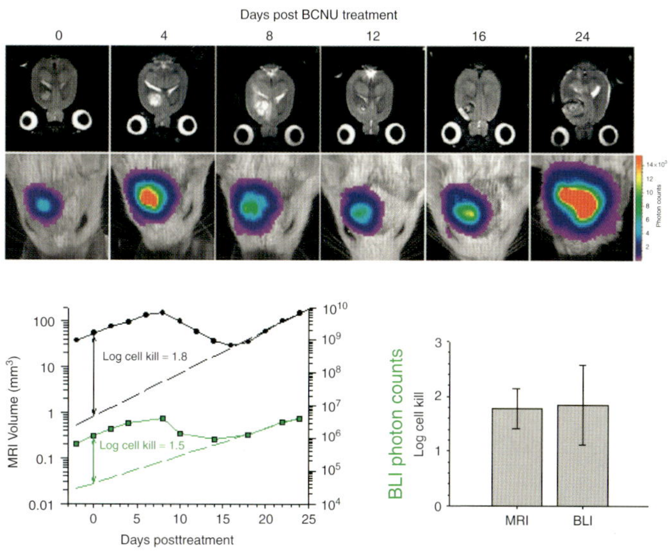

Plate 22.8. Temporal analysis of the response of a 9L^{Luc} tumor to BCNU chemotherapy. Tumor cells were implanted 16 days prior to treatment. Tumor volume was monitored with T2-weighted MRI (top panel, top row), and intratumoral luciferase activity was monitored with BLI (top panel, bottom row). The days post-BCNU therapy on which the images were obtained are indicated at the top. The scale to the right of the BL images describes the color map for the photon count. Quantitative analysis of tumor progression and response to BCNU chemotherapy is shown in the bottom left panel. Tumor volumes (●) and total tumor photon emission (■) obtained by T2-weighted MRI and BLI, respectively, are plotted versus days post-BCNU treatment. The dashed lines are the regression fits of the exponential tumor repopulation following therapy. The solid vertical lines denote the apparent tumor volume and photon production losses elicited by BCNU on the day of treatment from which log cell kill values were calculated. Comparison of log cell kill values determined from MRI and BLI measurements is shown in the bottom right panel. Log cell kill elicited by BCNU chemotherapy was calculated using MRI (1.78 ± 0.36) and BLI (1.84 ± 0.73). Data are represented as mean ± SEM for each animal ($n = 5$). There was no significantly significant difference between the log kills calculated using the MRI and BLI data ($p = 0.951$). (Used with permission from Rehemtulla et al., 2000.)

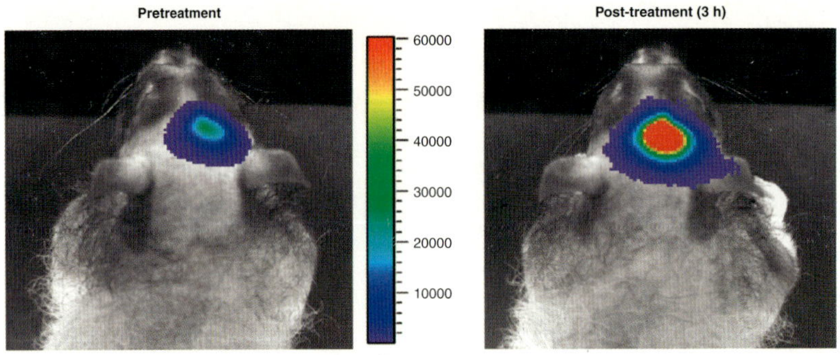

Plate 22.10. In vivo imaging of Caspase-3 activation. In vivo bioluminescence image of a rat with an intracerebral D54-Luc$_{ER}$ glioma obtained (*a*) before and (*b*) 3 h following intratumoral TRAIL treatment. Note the significant increase in photons collected from the same animal following TRAIL treatment.

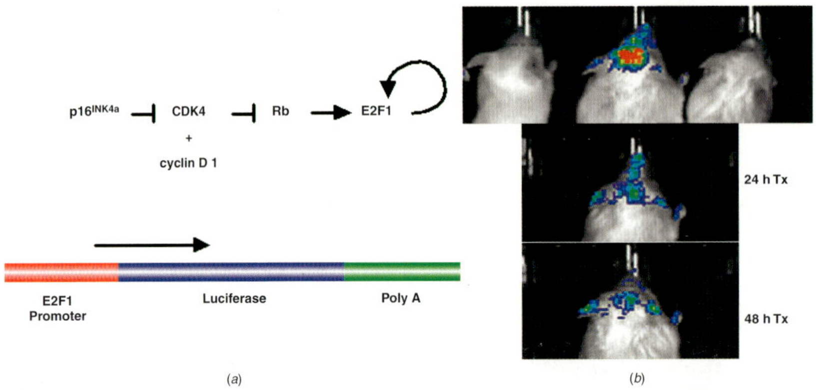

Plate 22.11. Strategy for imaging luciferase tumors using Rb. (*a*) The Eflux transgene construct driving luciferase from the E2F1 promoter that responds to Rb activity and cell proliferation. (*b*) Three Eflux transgenic mice imaged for bioluminescence, the center mouse harbors a PDGF-induced glioma. That mouse was subsequently treated with a small molecule inhibitor of the PDGF receptor kinase with resultant decrease in luciferase activity.

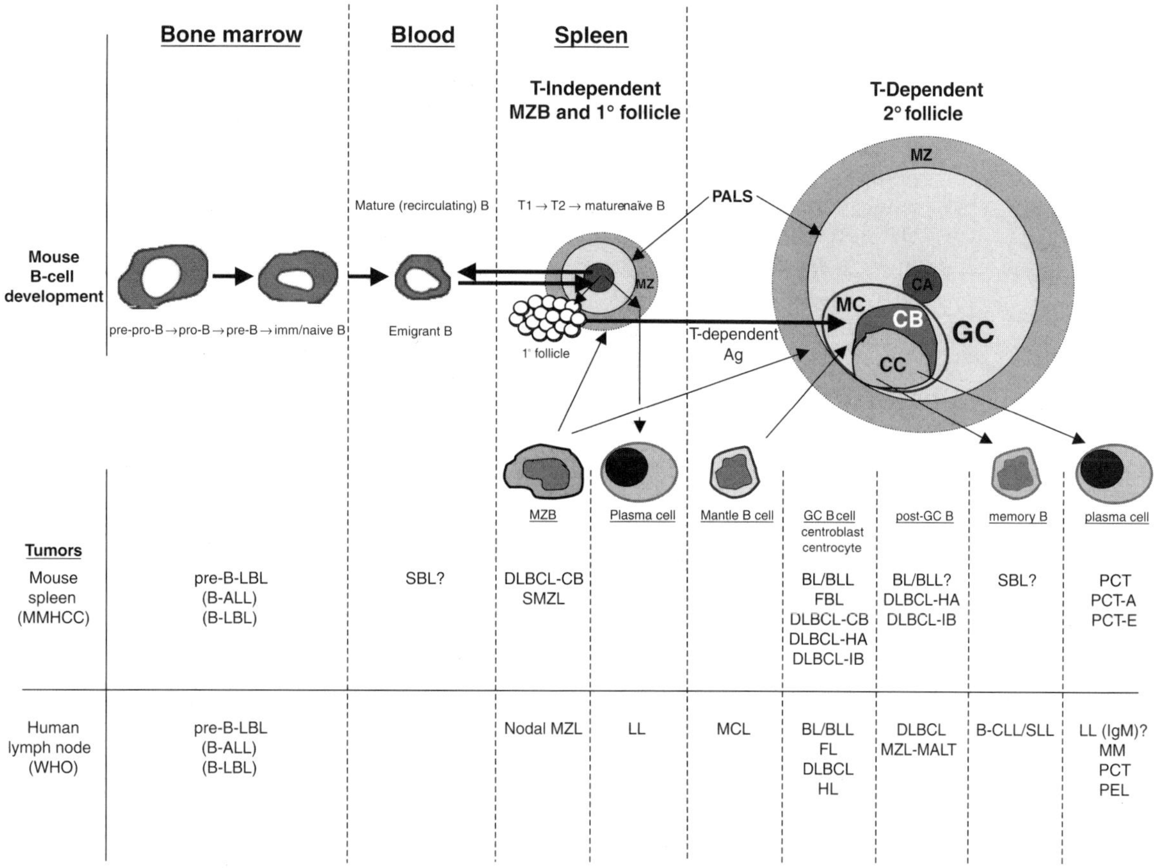

Figure 14.3. Origins of mouse and human B-cell malignancies. Normal mouse B-cell development in BM, blood, and spleen is shown. Below this is a list of tumors that arise from distinct stages of development as described in the MMHCC (Bethesda proposal) classification for mice [132] and the WHO classification for humans [126, 127]. B-cell tumors in the WHO classification that lack a presumed cell of origin are not included. The origin of B-CLL/SLL is from Refs. 47, 51, and 52. CA, central arteriole; CB, centroblasts; CC, centrocytes; FL, follicular lymphoma; GC, germinal center; HL, Hodgkin lymphoma; LL, lymphoblastic lymphoma; MC, mantle cell zone; MCL, mantle cell lymphoma; MM, multiple myeloma; MZ, marginal zone; PALS, periarteriolar lymphoid sheath; PEL, primary effusion lymphoma.

Table 14.2. Gene Transfer and Bone Marrow Reconstitution Models Developing Lymphoma/Leukemia

Cancer	Construct	Host Strain	MMHCC Classification	References
CML	MPZen(*bcr-abl*)	DBA/2	Pre-B-LBL	202
CML	MPZen(*bcr-abl*)	C57BL/6	Pre-T-LBL	202
CML	pGD210 (*bcr-abl*)	BALB/c	Pre-B-LBL, pre-T-LBL	203
CML	JW-RX (*bcr-abl*)	BALB/c	Pre-B-LBL	204
B-ALL	MING-TEL/AML1	C57BL/6	Pre-B-LBL, pre-T-LBL	223

tetracycline is removed from the drinking water (*tet* off; see Chapter 4) evaluates if continuous transgene expression is necessary for malignancy [211]. Maintenance of *BCR-ABL*^{p210} transgenic mice on tetracycline silences transgene expression. Removal of tetracycline causes the induction of B-ALL that is reversed by adding tetracycline to resuppress *BCR-ABL*^{p210} expression. Insertion of *BCR-ABL*^{p190} in-frame into exon 1 of the endogenous *bcr* locus by homologous recombination establishes an expression system that is controlled by the endogenous mouse *bcr* regulatory elements, mimicking half of the reciprocal translocation seen in Ph-positive

Table 14.3. Transgenic Mice Developing Lymphoma/Leukemia

Cancer	Transgene	MMHCC Classification	References
B-ALL, CML	mt-*BCR-ABL*^{p190}	Pre-B-LBL	205
CML	mt-*BCR-ABL*^{p210}	Pre-B-LBL, pre-T-LBL	207
CML	mt-*BCR-ABL*^{p210}	Pre-T-LBL	208
CML	tec-*BCR-ABL*^{p210}	Pre-T-LBL	209
CML	tTA-*BCR-ABL*^{p210}	Pre-B-LBL	211
nm	EμV_H(*bcr-v-abl*)	Pre-B-LBL, pre-T-LBL	210
nm	LTR(*bcr-v-abl*)	Pre-T-LBL	210
BL	Eμ-*myc*	Pre-B-LBL	216
nm	H-2K-*blk(Y495F)*	Pre-B-LBL, pre-T-LBL	215
nm	EμPμ-IL-7	(B, γδTCR lymphomas)	240
nm	E_α-IL-7	Pre-B-LBL	241
T-PLL/T-CLL	Eμ-*TCL1*	SBL	250
T-ALL	pGEM*HOX11*	SMZL	255
nm	Eμ-*Pim-1*	FBL, DLBCL, pre-T-LBL	217, 263, 264
T-PLL/T-CLL	PEμ-B29-*TCL1*	FBL, BLL, DLBCL	140
BL	Myc-Eλ-*c-MYC*	T1/T2 B lymphoma	275
BL	Ig/Myc (YAC)	T1/T2 B lymphoma	277, 278
PCT, MM	H2-L^d-*IL-6*	PCT-E, PCT-A	291
nm	SV40-Eμ-*v-abl*	PCT-E, PCT-A	292
FL	SV40-Eμ-*bcl-2*	PCT, pre-B-LBL	293
T-ALL	Lck(prox)-*Tal1*	Pre-T-LBL	332
T-ALL	Lck(prox)-*Tal1(R188/9G)*	Pre-T-LBL	333
T-ALL	mt-*Rbtn2*	Pre-T-LBL	326, 327
T-PLL/T-CLL	plck-*TCL1*	STL	368
T-PLL/T-CLL	CD2-*p13(MTCP1)*	STL	369
nm	H-2D^d-*IL-15*	TNKL	375
nm	pRc/CMV-HMGI-C/T	TNKL	376

Table 14.4. Knockin and Knockout Mice Developing Lymphoma/Leukemia

Cancer	Construct	Locus	MMHCC Classification	References
B-ALL	*BCR-ABL*^{p190}	*bcr* KI	Pre-B-LBL	212
B/T-ALL, NHL	*p16*^{INK4a}/*p19*^{ARF}	KO	Pre-B-LBL	235
nm	*Aiolos*	KO	Pre-B-LBL	230
Many types	*p53*	KO	SMZL, pre-T-LBL	260
Leukemias	*Msh2*	KO	Pre-T-LBL	347
Many types	*Atm*	KO	Pre-T-LBL	301–303
Many types	*Pten*	(+/−)	Pre-T-LBL	342, 343
nm	*Ku70*	KO	Pre-T-LBL	356, 357
Breast/ovarian	*Brca2*	KO	Pre-T-LBL	360
nm	*E2a*	KO	Pre-T-LBL	308
Many types	*p53*	KO	Pre-T-LBL	334, 335
B/T-ALL, CML	*Ikaros*	KO	Pre-T-LBL	311

cases (see Chapter 4 [212]). These *BCR-ABL*^{p190} KI mice develop pre-B-ALL (Table 14.4). In sum, mice containing *BCR-ABL*-infected BM or expressing either form of *BCR-ABL* (or *bcr*-v-*abl*) using transgenic, tetracycline-inducible or KI technologies develop mainly B- or T-ALL with variable penetrance and strain dependence [213]. The p210 isoform of BCR-ABL generally does not result in B-ALL in humans.

Blk. Blk is a Src family tyrosine kinase that associates with and signals from the pre-BCR and BCR during development [214]. Mice with a constitutively active Blk (Y495F) cDNA flanked by a H-2K promoter and Eμ develop pro-B- and pre-B-LBL, along with ISP-stage thymic pre-T-LBL (Table 14.3 [215]). B-lineage tumors occur in 45% of mice by six months. Activating lesions for the *Blk* homologue in human B- and T-cell tumors have not yet been described.

c-Myc. Pre-B-LBL arise in C57BL/6 × SJL mice with an integrated murine retroviral EμLTR-*myc* transgene (Table 14.3 [186, 216]). Most tumors lack IgL gene rearrangements and are sIg−. Crossing EμLTR-*myc* with Eμ-*pim-1* transgenic mice, which develop low-frequency pre-T-LBL, results in congenital pre-B-ALL/LBL [186, 217]. Eμ dysregulation of *c-Myc* does not appear to cause Burkitt lymphoma (BL), as in humans with IgH- or IgL-to-c-*MYC* gene rearrangements [218, 219].

TEL/AML1. A t(12;21)(p13;q22) occurs in 25% of pediatric and 3% of adult B-ALL cases, resulting in a *TEL/AML1* oncogene [220–222]. TEL/AML1 retains the amino portion of the TEL protein fused to the AML1 DNA binding domain. TEL/AML1 may cause B-ALL by direct repression of AML1 target genes or by TEL inhibition of other ETS family proteins via binding through its pointed domain [223–225]. Transduction of wild-type or $p16^{Ink4a}/p19^{Arf}$-null BM with a murine stem cell LTR-TEL/AML1 retrovirus was followed by adoptive transfer into lethally irradiated C57BL/6 mice [223]. A low frequency of pre-B-ALL and T-ALL from the transduced wild-type donor cells and a higher frequency of an undetermined leukemia type from the $p16^{Ink4a}/p19^{Arf}$-deficient cells is seen (Table 14.2 [223]).

Aiolos. Ikaros, Aiolos, Helios, Eos, Pegasus, and Daedalus are members of a kruppel-like zinc finger transcription factor family that binds DNA and regulates B- and T-cell development and function [226–229]. Targeted inactivation of *Aiolos* results in hyperproliferation and constitutive B-cell activation followed by lymphadenopathy and evolution into probable B-LBL in 20% of mice (Table 14.4 [230]). Alternatively spliced *Aiolos* isoforms lacking a full complement of DNA binding domains in normal and leukemic human B cells have been reported, but their role in B-cell transformation is not resolved [227, 231].

p16(Ink4a)/p19(Arf). The overlapping $p16^{INK4a}/p19\,(p14)^{ARF}$ genes encode tumor suppressor proteins that inhibit cancer formation [232]. Loss of $p16^{Ink4a}$, a cyclin-dependent kinase inhibitor, increases cell cycling by increased Rb phosphorylation while loss of mouse $p19^{Arf}$ (or human $p14^{ARF}$) increases Mdm2-mediated ubiquitination of p53, leading to decreased p53 activity [233, 234]. Mice deficient in $p16^{Ink4a}/p19^{Arf}$ develop mainly soft-tissue sarcomas or mitotically active B220+ B-cell lymphomas with prominent nucleoli that efface lymph nodes [235]. Although not further characterized, these features suggest a B-LBL phenotype (Table 14.4). Deletion of $p16^{INK4a}/p14^{ARF}$ occurs in 20% of B-ALL and 60% of T-ALL arising during childhood [236]. Also, DNA methylation of the $p16^{INK4a}$ (but not the $p14^{ARF}$) locus occurs in multiple types of non-Hodgkin lymphoma (NHL) and acute myeloid leukemia (AML) accompanied by loss of $p16^{INK4a}$ expression [237].

IL-7. IL-7 and the IL-7/γcR regulate B and T lymphopoiesis (Figs. 14.1 and 14.2 [238]). Four distinct transgenic strains aberrantly express IL-7 [239–242]. An IgH promoter–Eμ–IL-7 transgene causes thymic enlargement, B- and T-cell proliferations with adenopathy, skin-homing T-cell infiltrates, and B- and T-cell lymphomas in 100% of animals by 4.5 months (Table 14.3 [240]). Clonal B220+/sIgM+ B-cell and DN/Thy1+/γδTCR+ T-cell lymphomas are observed but not further characterized. Separately, a MHC-II E_α promoter–IL-7 transgene causes λ5 and EBF-expressing pre-pro-B/pro-B lymphoid tumors with germline Ig genes [241]. Clonality is difficult to establish and femurs are packed with lymphoid blast cells accompanying lymphadenopathy. The tumor incidence varies by strain, with almost 100% of BALB/c and 25% of C57BL/6 mice developing morbid disease. Isolated clones develop into macrophages in culture, consistent with a bipotential B-cell/macrophage stage in early BM and fetal lymphocyte development [243, 244]. Human lymphoid malignancies from IL-7 dysregulation have not been reported.

Small B-cell leukemia/lymphoma (SBL)

SBL occurs in old mice of different strains and appears as small round cells with scant cytoplasm, condensed chromatin, and a mature sIg+/B220+/CD19+ immunophenotype [132]. Mitoses and apoptotic cells are rare. IgH and IgL genes are rearranged and tumors show monoclonal banding patterns. CD5, a discriminating surface molecule for B-CLL/SLL and mantle cell lymphomas (MCLs) in humans, is less useful for classifying mouse B-cell tumors, including SBL, since 85% of all mouse B-cell tumors express CD5 [127, 164]. SBL usually originates from the spleen or lymph nodes and may spill over into the blood, forming a leukemic phase, which contrasts with human B-CLL/SLL, which usually begins with a leukemic phase.

TCL1. Dysregulated *TCL1* expression causes mature, chronic T-cell malignancies in humans [245, 246]. The tumorigenic mechanism likely depends on augmented AKT (protein kinase B) activity leading to increased cell survival and proliferation [247–249]. Mice with a IgV_H promoter–Eμ–*TCL1* transgene develop sIgM+/CD5+/Mac-1(CD11b)+ small-cell leukemias originating as peritoneal expansions [250]. B cells then spread to the spleen and BM before entering a leukemic phase with monoclonality at 13–18 months of age (Table 14.3). Spleens are infiltrated and marginal zones expanded by cells without surface antigen expression typical of MZB. The peritoneal origin and surface antigen expression pattern suggests possible transformation of a B-1a B-cell precursor [33, 251].

Splenic Marginal Zone Lymphoma (SMZL)

SMZL is common in nonmanipulated NFS.V+, AKXD R1, and NZB strains and demonstrates a sIgM+/B220+/CD19+ mature B-cell immunophenotype with clonal IgH and IgL rearrangements [164]. Splenomegaly is typical while involvement beyond regional lymph nodes is uncommon. SMZLs start as low-grade tumors of MZB and are made of medium-sized cells with abundant pale cytoplasm and few mitoses. A thin marginal zone initially expands, which may be followed by invasion of both red and white pulp and conversion to an aggressive lesion with centroblastic morphology and increased mitoses [252]. Human splenic marginal zone lesions are mainly from infiltration by tumors of nearby mantle and follicular cell origin. The major human counterpart lesion could be a nodal MZL [127]. Human MZL of mucosa-associated lymphoid tissue (MALT) type, which initiates in the GI tract from chronic *Helicobacter pylori* infection, is likely a post-GC B-cell malignancy [127]. In mice, polyclonal *H. pylori*–associated MALT accumulations have been reported [253].

***HOX11*.** Human T-ALL exhibit t(10;14)(q24;q11), which moves the *HOX11* homeodomain-containing transcription factor coding sequence near *TCR* α/δ regulatory regions resulting in ectopic expression of *HOX11* in thymocytes [254]. About 20% of CD-1 strain mice with an IgH promoter–Eμ–*HOX11* transgene develop splenic marginal zone hyperplasia, followed at 10–20 months by 85% of mice developing an IgM+/IgD+ SMZL (Table 14.3 [255, 256]). Transformation of MZB is suggested, although T2 and B1 B cells are IgM/IgD positive and locate in the marginal zone of mice. An association between *HOX11* expression and B-cell cancers in humans has not yet been reported.

***p53*.** The p53 tumor suppressor actively inhibits cell cycle progression until DNA damage has been repaired [257, 258]. *p53* is the most frequently mutated gene in humans and promotes multiple types of malignancy in KO mice [259]. Male C57BL/6 × 129 strain, *p53* KO mice develop B220+/CD5+/IgM+ SMZL and thymic pre-T-LBL by 12 weeks (Table 14.4 [187, 260]).

Follicular B-Cell Lymphoma (FBL)

FBL is the most common sIgM+/B220+/CD19+ mature B-cell lymphoma in many mouse strains. Irregularly shaped, diffuse white pulp expansions from enlarging follicles begin in the spleen, lymph nodes, or Peyer patches, resulting in organomegaly. FBL originates from GC B cells and contains large (centroblasts, immunoblasts), small (centrocytes), or a mixture of large and small tumor cells with scant cytoplasm and smooth or cleaved nuclei with or without prominent nucleoli, respectively. Tumor cells are clonal for IgH and IgL rearrangements and mitotic activity parallels the number of centroblasts present. FBLs are typically low-grade lesions resembling human follicular lymphomas. Unlike human follicular lymphomas, mouse FBL is not associated with *Bcl-2* gene rearrangement [132].

Pim1

Pim1 was originally identified as a candidate mouse T-cell lymphomagenic gene by proviral insertion [261]. It encodes a serine/threonine kinase of unknown function [262]. LTR-Eμ-*Pim-1* transgenic mice develop pre-T-LBL between 7 and 10 months (Table 14.3 [217, 263]). The rate of tumor formation is accelerated to 7–8 weeks by Moloney murine leukemia virus (MuLV) insertions that activate c-*myc* and N-*myc* proto-oncogenes. Animals not developing pre-T-LBL by 10 months develop FBL with mixed centroblast and centrocyte populations, a variable mitotic rate, and usual involvement of mesenteric lymph nodes [263, 264]. They also develop DLBCL, probably multiple subtypes [263]. Human *Pim1* shows point mutations caused by aberrant SHM in most DLBCL but not in normal GC B cells or follicular lymphomas [265]. A role for *Pim1* SHM in causing DLBCL has not been demonstrated yet.

***TCL1*.** Human *TCL1* is expressed at high levels in mantle zone B cells and reduced levels in GC B cells and is silenced in post-GC memory B and plasma cells [266]. B29 promoter–Eμ–*TCL1* transgenic mice develop multiple types of GC-based B-cell lymphoma, including Burkittlike lymphoma (BLL), FBL, and DLBCL [140]. One analyzed FBL is Bcl-6+/TdT− and demonstrates SHM of IgJ_H. Many human B-cell lymphomas, including FL, BL, and DLBCL express abundant *TCL1*, suggesting a role for *TCL1* overexpression in these tumors [266–270].

Burkitt and Burkittlike Lymphoma (BL, BLL)

BL is a sIgM+/B220+/CD19+, high-grade lesion that appears as sheets of lymphoblasts with abundant mitoses and a "starry sky" diffusely infiltrative pattern from ingestion of apoptotic debris by macrophages. Hematopoietic (spleen, lymph nodes, sometimes thymus) and nonhematopoietic organs are involved. A GC or post-GC B-cell origin is postulated. The main difference between BL and pre-B-LBL/B-LBL is the lack of TdT and presence of c-*Myc* activation by translocation or other means in BL. BLL, also called lymphoblastic lymphoma (LL) or DLBCL-LL, is phenotypically similar to BL with fewer mitotic and apoptotic cells. It is histologically identical to pre-B-LBL and pre-T-LBL and is distinguished by sIg expression [132]. BLL lacks functional c-*Myc* activation and often has a leukemic phase. Some inbred mouse strains develop BLL with age.

c-Myc. Human BL is associated with infection by the Epstein–Barr virus (EBV) and translocations of the c-*MYC* locus at 8q24.12 with 14q32 (IgH), 2p12 (Igκ), or 22q11.2 (Igλ) loci, causing aberrant c-*MYC* expression [219, 271–274]. BL does not occur spontaneously in mice. EμLTR-*myc* transgenic mice develop pre-B-LBL and not BL [186, 216]. c-Myc promoter–Eλ–c-*Myc* transgenic mice develop tumors that histologically resemble human BL but continue to express pre-B-cell antigens and lack SHM and Bcl-6 expression, both signatures of a GC-B cell experience (Table 14.3 [164, 275]). Crossing this λ-MYC mouse with a *Bcl-6–/–* mouse that lacks GC formation does not alter the kinetics of tumor development, confirming that this mouse is not a model of human BL [164, 276]. Inserting c-*MYC* into the IgH locus on a yeast artificial chromosome (YAC) vector followed by generation of transgenic mice results in IgM+/IgD+/CD43+ transitional stage B-cell lymphomas and not BL [277, 278]. The inability to model BL in mice by manipulating c-*Myc* may reflect a corequirement for EBV-related factors.

TCL1. Human BLL always demonstrates dysregulated c-*MYC* through *Ig*/c-*MYC* translocations [127]. B29 promoter–Eμ–*TCL1* transgenic mice develop IgM+/B220+/CD5+ BLL at a high rate between 4 and 12 months and demonstrate IgH SHM and Bcl-6 expression (Table 14.3 [140]).

Diffuse Large B-Cell Lymphoma (DLBCL)

DLBCL is an aggressive mature B-cell malignancy with centroblastic (CB), immunoblastic (IB), histiocyte-associated (HA), and primary mediastinal (PM, thymic) variants in mice [132]. Common to all subtypes is a sIgM+/B220+/CD19+ immunophenotype with clonal IgH and Ig rearrangements. CB, IB, and HA subtypes usually demonstrate splenomegaly or lymphadenopathy while PM shows mainly thymic enlargement. It is important to note that FBL was originally defined as having small-cell, large-cell, and mixed histologic phenotypes [136]. In the Bethesda classification, the original FBL, large-cell subtype is now considered a DLBCL [132].

DLBCL-CB originates from GC B cells or splenic MZB and must be distinguished from FBL or SMZL with increased centroblasts. Both IB and HA subtypes arise from GC or post-GC B cells while PM derives from thymic B cells. DLBCL-CB is comprised of at least 50% centroblasts that are medium sized with scant cytoplasm, round nuclei, prominent nucleoli, and numerous mitoses. This lesion overruns lymphoid structures, may originate from aggressive transformation of indolent FBL or SMZL, and is similar to human DLBCL, centroblastic variant. DLBCL-IB contains excess immunoblasts that are large with abundant cytoplasm, round nuclei with prominent nucleoli, and a high mitotic rate. A starry sky pattern may be seen with increased apoptotic cells. DLBCL-HA may involve the liver and contains many large histiocytes and foreign body giant cells intermixed with malignant centroblasts and immunoblasts. The pattern of growth is mainly nodular rather than diffuse, and this lesion resembles human histiocyte/T-cell-rich DLBCL [127]. It may be difficult to distinguish DLBCL-HA from histiocytic sarcoma, a tumor common to several mouse strains (see chapter 13 [279]). Lastly, DLBCL-PM demonstrates medium to large tumor cells with prominent nucleoli and abundant mitoses. Involvement begins in the thymic medulla with a diffuse, starry sky pattern of growth. There may not be a human counterpart lesion.

Pim-1, TCL1, and BM5def virus. DLBCL may arise de novo or by aggressive transformation of FBL and possibly SMZL. Ionizing radiation causes an increased frequency of pre-T-LBL, FBL, and DLBCL-CB in Eμ-*Pim-1* transgenic mice (Table 14.3 [164, 263]). DLBCL-HA occurs in older AKXD RI mice and in B29-Eμ-*TCL1* transgenic mice in a mixed C57BL/6 × C3H background [140, 280]. Mice coinfected with a replication-defective (BM5def) and a helper virus develop an immunodeficiency syndrome termed MAIDS [281, 282]. After a long latency, MAIDS mice develop DLBCL-PM [164].

Plasma Cell Neoplasms

Plasmacytoma (PCT) is a cytIg+/CD43+/CD138+ tumor of medium-sized Ig-secreting plasma cells [132]. PCT appears rarely in spleens and lymph nodes of some older inbred strains and forms at high frequency in mesenteric granulomas induced by pristane mineral oil injection of BALB/c, NZB, or F1 crossed mice [283–285].

Characteristic t(12;15) rearrangements of c-*Myc* and *IgH* loci promote pristane-induced PCT [286]. Activating c-*MYC* gene rearrangements also occur in 15% of plasma cell–derived human multiple myeloma (MM), although they may be secondary rather than initiating events [287–289]. Extraosseous PCT (PCT-E) is a mature plasma cell–derived variant that may involve the spleen, lymph nodes, or Peyer patches in nodular or diffuse patterns. Anaplastic PCT (PCT-A) is another variant that contains mixtures of less mature plasmablasts and immunoblasts in diffuse patterns involving the spleen or lymph nodes. These PCTs rarely have translocations involving c-*Myc* (C.-F. Qi and H. Morse, unpublished observations). Importantly, no agreed-upon model for an intramedullary BM-based plasma cell disease mimicking human MM has been reported in mice [290].

IL-6, v-abl, c-Myc, and Bcl-2. BALB/c mice with a H2-L^d-*IL-6* transgene develop mainly IgG-secreting PCT-E or rare PCT-A by 18 months of age (Table 14.3 [291]). These PCT-E variants almost always show activating c-*Myc/IgH* rearrangements and may coevolve with FBL or DLBCL. A simian virus 40 (SV40) promoter–Eμ–v-*abl* transgene causes clonal IgA or IgG secreting PCT-E with activating c-*Myc* translocations or PCT-A in 60% of C57BL/6 × SJL mice by one year of age [292]. Crossing these transgenic mice with SV40 promoter–Eμ–c-*Myc* mice, which normally develop pre-B-LBL, induces oligoclonal, IgM-secreting PCT of the peritoneum at an accelerated rate. Also, SV40 promoter–Eμ–*bcl-2* transgenic mice develop rare PCT with c-*Myc* rearrangements or early pro-B- or pre-B-LBL [293]. Human follicular lymphomas harbor a t(14;18)(q32;q21.3) that brings Eμ near the antiapoptotic *BCL-2* coding region, causing its overexpression and increased cell survival [294–298]. SV40 promoter–Eμ–*bcl-2* transgenics do not develop FBL, the mouse equivalent of human follicular lymphoma.

Fas or FasL. Early on, *lpr* or *gld* lymphocytes are not immortal or transformed, and they do not grow indefinitely or cause malignancy in adoptively transferred animals. At 6–15 months, up to 60% of BALB/c and 30% of C3H strain *gld* mice contain emerging clonal PCTs that are obscured among a robust DN T-cell expansion [299]. None of the PCTs have c-*Myc* genomic rearrangements, and transplantation of tumor cells into immune-deficient *scid* mice results in PCT.

MODELS OF T- AND NK-CELL MALIGNANCIES

Genetic models of T- and NK-cell cancer may be caused by inappropriately regulated signaling molecules (*Atm*), transcription factors (*E2A, Ikaros, Rbtn2, Tal1*), tumor suppressors (*p53, Pten*), DNA repair proteins (*Msh2, Ku70, Brca2*), cytokines (*IL-15*), and adapter molecules (*MTCP1, TCL1*) (Fig. 14.4). As with B-cell cancer models, these examples are discussed in sections under their MMHCC tumor-type headings.

Precursor T-Cell Lymphoblastic Leukemia/Lymphoma (pre-T-LBL)

The histologic appearance of pre-T-LBL is similar to that described for pre-B-LBL. The immunophenotype suggests transformation at distinct stages of intrathymic (or late BM) T-cell development, with immature, CD3+, CD4/CD8 DN, CD4/CD8 DP, CD4 or CD8 SP, TCR+, and cytTdT+ staining patterns observed [132]. The TCRβ chain gene is rearranged. The human counterpart malignancies are precursor T-cell lymphoblastic lymphoma (T-LBL) and T-cell acute lymphoblastic leukemia (T-ALL).

Atm. The *ATM* gene encodes a phosphatidylinositol 3-kinase (PI3K) family member protein that responds to DNA damage by phosphorylating key substrates involved in DNA repair and/or cell cycle regulation [300]. Targeted deletion of *Atm* results in rapid induction of pre-T-LBL before four months of age in several mouse strains (Table 14.4 [301–303]). *ATM* defects cause increased sensitivity to ionizing radiation or other agents causing DNA damage and appear to promote T-PLL, B-CLL, MCL, breast cancer, and additional malignancies in humans [304–307].

E2A. *E2A*-encoded proteins E12 and E47 control early B-cell development (Fig. 14.1). Targeted deletion of *E2A* in FVB/NJ mice results in blurring of thymic cortical–medullary junctions and increased numbers of T lymphoblasts that transform into mainly CD4/CD8 DP pre-T-LBL between three and nine months of age (Table 14.4 [308]). Most tumors display trisomy 15 and upregulated c-*Myc* proto-oncogene expression. The pattern of *E2A* expression is not altered in human T-ALL, although E47 interactions with aberrantly expressed TAL1 or TAL2 proteins may promote T-ALL by abolishing its normal transcription-regulating activity [309, 310].

Ikaros. Targeted removal of the *Ikaros* DNA-binding domain is embryonic lethal, but hemizygous mice develop thymic-based T-cell lymphoproliferations, followed by clonal DN, CD4 or CD8 SP, TCR+ pre-T-LBL in 100% of animals at three to six months of age (Table 14.4 [311]). Tumors uniformly demonstrate loss of heterozygosity (LOH) of the wild-type *Ikaros* allele. Increased levels of Ikaros isoforms lacking DNA binding domains are seen in B-ALL, T-ALL, and CML in blast

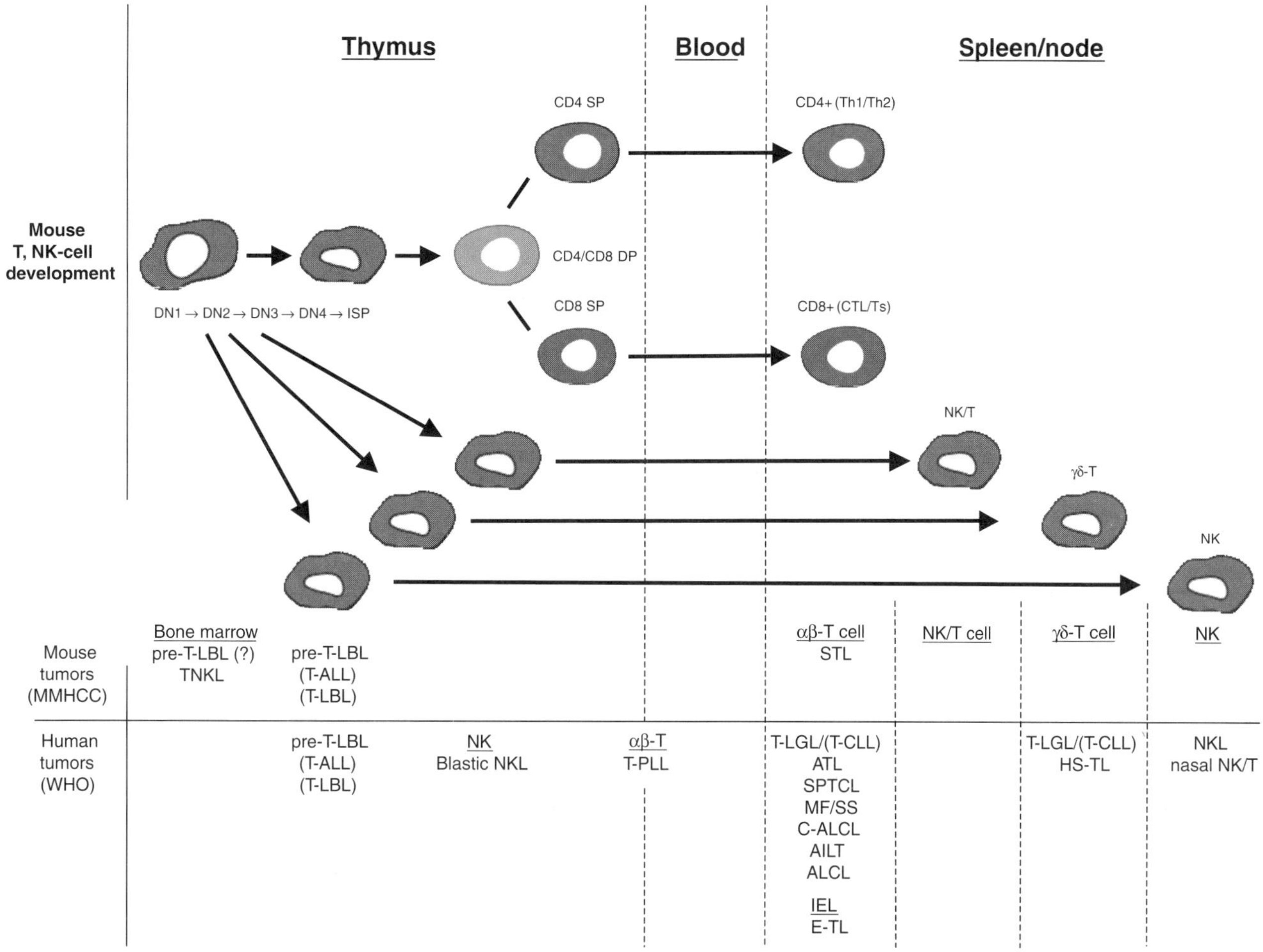

Figure 14.4. Origins of mouse and human T-cell malignancies. Normal mouse T-cell development in BM is shown. Below this is a list of tumors that arise from distinct stages of development as described in the MMHCC (Bethesda proposal) classification for mice [132] and the WHO classification for humans [126, 127]. T-cell tumors in the WHO classification that lack a presumed cell of origin are not included. AILT, angioimmunoblastic T-cell lymphoma; ALCL, anaplastic large-cell lymphoma; ATL, adult T-cell leukemia/lymphoma; C-ALCL, primary cutaneous anaplastic large-cell lymphoma; E-TL, enteropathy-type T-cell lymphoma; HS-TL, hepatosplenic T-cell lymphoma; IEL, intraepithelial lymphocyte; MF/SS, mycosis fungoides/Sezary syndrome; NKL; natural killer cell lymphoma; SPTCL, subcutaneous panniculitis-like T-cell lymphoma; T-LGL, T-cell large granular lymphocytic leukemia; T-PLL, T prolymphocytic leukemia.

crisis [312–319]. Overexpression of DNA-binding defective Helios isoforms, another Ikaros transcription factor family member, in T-ALL and HTLV-1-positive adult T-cell leukemia/lymphoma is also observed [320, 321]. The tumorigenic effect of dominant-negative Ikaros family isoforms in mice suggests a potential role for DNA-binding defective Ikaros isoforms in human lymphoid tumors.

Rbtn2**.** *Rbtn2* (*Lmo2*) encodes a cysteine-rich, tandem LIM domain-containing protein that interacts with a variety of nuclear factors, including Tal1/Sc1, during normal development [322–324]. In humans, t(11;14)(p13;q11) rearranges the *RBTN2* gene to the TCRδ locus, causing dysregulated expression in T-ALL [322, 325]. About 10% of C57BL/6 × SJL mice with a mt promoter–*Rbtn2* transgene develop CD4 or CD8 SP, CD4/CD8 DP, and CD4/CD8 DN pre-T-LBL between 1 and 1.5 years (Table 14.3 [326]). Also, 75% of CD2 promoter–*Rbtn2* transgenic mice develop a similar spectrum of αβTCR and γδTCR pre-T-LBL by 10 months of age [327].

Tal1**.** *TAL1* (*SCL*) encodes a helix–loop–helix DNA-binding transcription factor and is one of the most common mutation targets in childhood T-ALL [328–330].

t(1;14)(p32;q11), t(1;14)(p34;q11), and deletions aberrantly activate *TAL1* during thymocyte maturation, promoting transformation. Thymocyte-expressing CD2 promoter–*scl* transgenic mice fail to develop T-lineage malignancies unless crossed with N-*ras* transgenic or *p53*-null mice [331]. In contrast, proximal lck promoter–*Tal1* transgenics develop pre-T-LBL (Table 14.3 [332]). Transgenics lacking the *Tal1* DNA-binding domain still develop CD8 SP pre-T-LBL, suggesting the *Tal1* tumor-promoting effect is likely from interference with additional transacting factors, such as E2A proteins [310, 333].

p53. *p53*-null C57BL/6 × 129Sv mice develop lymphoblastic lymphomas originating in the thymus by three or six months, likely representing pre-T-LBL (Table 14.4 [187, 260, 334]). Separately, a *p53* KO model also develops CD4/CD8 DP pre-T-LBL as the main tumor type [335]. Crosses of *p53–/–* mice with a variety of KO or transgenic mouse models (*Rb+/–*, *Msh2–/–*, *Atm–/–*, *Rag1/2–/–*, *MMTV*-c-*Myc*, *CD2-myc*, *CK2*, and *CD2-scl*) result in oncogenic cooperation with increased rates and/or altered types of lymphoid malignancy [336]. Loss of *p53* occurs in all types of human hematopoietic malignancy at reduced rates compared with rates in solid-tissue tumors [337]. The predisposition for *p53*-null mice developing pre-T-LBL has an unknown etiology.

Pten. PTEN is a tumor suppressor with lipid and protein phosphatase activity that opposes the receptor tyrosine kinase/PI3K-induced activation of AKT [338–341]. Targeted removal of *Pten* is embryonic lethal while *Pten*-heterozygous CD1 strain mice survive and develop thymocyte expansions followed by pre-T-LBL at high frequency (Table 14.4 [342]). Analyzed tumors demonstrate LOH at the wild-type *Pten* locus. Inbred 129Sv and C57BL/6 *Pten*(+/–) strain mice do not develop T-cell leukemia/lymphoma, indicating strain-specific effects contribute to *Pten* tumor suppression [343]. Multiple human tumor types and patients with Cowden disease, Bannayan–Zonana syndrome, and Lhermitte–Duclos syndrome demonstrate diverse types of *PTEN*-inactivating mutations, although *PTEN* is rarely mutated or lost in human lymphoid malignancies [344–346].

Msh2. DNA mismatch repair (MMR) genes maintain fidelity of genomic replication in mammals. Targeted deletion of the *Msh2* MMR gene results in TdT-, Rbtn2-, and Tal1-positive pre-T-LBL that spans the CD4/CD8 DN, CD4 or CD8 SP, and CD4/CD8 DP developmental spectrum (Table 14.4 [347–349]). Human pre-T-LBLs are characterized by aberrant *HOX11, RBTN2*, and *TAL1* expression, suggesting a role for defective *MSH2* or additional MMR genes in certain human T lymphomas [350–352].

Ku70. Ku70 protein forms a heterodimeric complex with Ku80 that targets the DNA-PK serine/threonine kinase to sites of double-stranded DNA breaks and V(D)J recombination in B and T cells [353–355]. The majority of *Ku70–/–* mice develop CD4/CD8 DP, pre-T-LBL by eight to nine months of age [356]. Separately, *Ku70* KO mice have significant B-cell defects and develop CD4/CD8 DP, TCRβ-negative pre-T-LBL between two and seven months of age [357]. Preliminary studies indicate 6 of 17 human lymphomas have mutations within *Ku70* core domains responsible for heterodimerization and DNA binding, although a direct role in cancer has not been established [356].

Brca2. *Brca2* promotes homologous recombination but not nonhomologous end-joining pathways of DNA double-stranded break repair [358, 359]. Mice with targeted *Brca2* inactivation show embryonic lethality with incomplete penetrance. Those mice that do survive go on to develop mainly CD4 or CD8 SP pre-T-LBL by 14 weeks of age [360]. Humans carrying *BRCA2* mutations are at increased risk for male and female breast cancer and ovarian cancer, although it is unclear if there is an increased risk for leukemia/lymphoma development [361–364].

Small T-Cell Lymphoma (STL)

The histologic appearance of STL is similar to that described for SBL. STL occurs rarely in inbred strains and shows splenomegaly and lymphadenopathy but not thymic enlargement. The immunophenotype is that of a mature T cell with CD3+, TCR+, CD4, or CD8 SP and cytoTdT antibody staining. The presumed cell of origin is a recirculating small mature T lymphocyte. Human peripheral T-cell malignancies are usually CD4 or CD8 SP, TCR+ tumors that are subdivided into groups based on genetic, morphologic, and clinical features [127].

MTCP1 and TCL1. Human T-cell prolymphocytic leukemia (T-PLL) is usually a malignancy of mature CD4 SP postthymic T cells that may present with splenomegaly, lymphadenopathy, and a leukemic phase [127]. The histology of T-PLL does not exactly resemble that of T-cell chronic lymphocytic leukemia (T-CLL) in humans or STL in mice since the cells are a bit larger with more cytoplasm, coarse chromatin, round to irregular shaped nuclei, and visible nucleoli. Nevertheless, T-PLL demonstrates a prolonged polyclonal premalignant state followed by a clonal mature T-cell malignancy that may take many years to evolve in humans, similar to the prolonged time course in most STL mice. It is typically a malignancy of older individuals or children with the genomic instability syndrome ataxia telangiectasia and exhibits gene rearrangements between TCR regulatory elements and *TCL1* or

MTCP1 proto-oncogenes [365, 366]. *TCL1* and *MTCP1* are grouped in the same gene family by homology and encode small 13–14-kDa proteins that bind and increase AKT serine/threonine kinase activity [248, 367]. *TCL1* transgenic mice generated with the proximal lck promoter and *MTCP1* transgenic mice made with a CD2 promoter result in polyclonal mature T-cell expansions that evolve into monoclonal CD8 SP T-PLL at 15–20 months of age (Table 14.3 [368, 369]). Both transgenic models show a leukemic phase and diffuse splenic infiltrates along with involvement of nonhematologic tissues such as the liver. These *TCL1* and *MTCP1* transgenic mouse models demonstrate many features resembling human T-PLL and T-CLL and provide evidence for the oncogenicity of gene rearrangements involving these loci in mature T-cell malignancies of humans.

T Natural Killer Cell Lymphoma (TNKL)

TNKL originates in the BM and is composed of large CD3+/αβTCR+ tumor cells with variable CD8 and DX5 expression and prominent nucleoli. TNKL grows in a diffusely infiltrative pattern that may involve abdominal viscera, resulting in hepatosplenomegaly and lymphadenopathy [132].

IL-15. An association between chronic inflammation due to infection and the development of cancer has been suggested, such as recurrent *H. pylori* and gastric carcinoma, schistosomiasis, and transitional cell carcinoma of the bladder and hepatitis C virus and hepatocellular carcinoma [370]. A role for inflammation in lymphoid malignancies has also been shown. IL-15 is a proinflammatory and growth-stimulating cytokine that regulates the development, survival, and function of NK cells [88–93]. Regulation of IL-15 expression is mainly posttranscriptional, with multiple checkpoint sequences located in the 5′ untranslated, signal peptide and carboxy-terminal portions of the messenger RNA [371–374]. A transgenic mouse overexpressing secreted IL-15 in multiple tissue types was created by removal of these checkpoint sequences (Table 14.3 [375]). These mice develop early (six to eight weeks) peripheral lymphocytic expansions of DX5+/Ly49D+/CD3− NK and CD8 SP memory T cells followed by dense skin infiltrates, alopecia, and multiorgan lymphocytic infiltrates of clonal, malignant NK/T cells at 12–30 weeks. IL-15 transgenic mice have features in common with human large granular lymphocytic (LGL) leukemia, including extralymphoid tissue involvement, a chronic course that progresses to an aggressive expansion, and a predominance of T-cell subtypes. Furthermore, these mice also exhibit features of human NK/T-cell malignancy, with cutaneous involvement and similar histopathologic findings. Independently, transgenic mice expressing a truncated form of the HMGI-C protein controlled by a cytomegalovirus (CMV) promoter upregulates IL-2 and IL-15 proteins and their receptors with 35% of mice developing TNKL beginning at one year of age (Table 14.3 [376]). Unlike TNKL, aggressive NK/T tumors in humans probably originate from peripheral NK cells and mainly involve the sinonasal passages [127, 377–379].

CONCLUSIONS

Clearly, mice are not small people. Genes that cause human lymphoid cancers may or may not cause similar tumors in genetically modified mice. Some major categories of human lymphoid cancer have no defined counterpart in mice. For B-cell-derived cancers, these include hairy cell leukemia, lymphoplasmacytic lymphoma/Waldenstrom macroglobulinemia, MCL, extranodal marginal zone lymphoma-MALT type, MM, and Hodgkin lymphoma. For T-, NK-, and NK/T-cell-derived cancers, these include adult T-cell leukemia/lymphoma, mycosis fungoides/Sezary syndrome, and large-cell anaplastic lymphoma. There are no mouse models for DC cancers to date. Also, mice develop lymphoid cancers that may not have a human counterpart. One example is B natural killer cell lymphoma, which occurs in thymectomized (SL/Kh X AKR/Ms) F1 mice [380]. Even minor variations likely reflect differences in lymphoid cell development and transformation mechanisms between species. For example, *TCL1* dysregulation in both human and mouse T cells cause a similar disease (T-PLL/T-CLL vs. STL) with the human lesion almost always CD4 SP and the mouse form always CD8 SP.

Some tumor categories seem preferred and some lacking in mice. There are many genes that generate mouse models of BM-derived pre-B-LBL and pre-T-LBL compared with relatively few genes that create mature B- and T-cell mouse malignancies. Newer techniques, targeting dysregulated gene expression to peripheral lymphoid tissues, promise to somewhat balance this discrepancy. Some genes may also be tumorigenic in one species and not in the other, while some genes in one species may yield distinct tumor types in the other. For example, c-*MYC* promotes BL in humans while it drives pre-B-LBL and PCT in mice. Despite all of these imperfect comparisons, current mouse models and newer improved mouse models will continue to provide insight for transformation of human lymphoid cells into the foreseeable future.

ACKNOWLEDGMENTS

The Teitell laboratory is supported by National Cancer Institute/National Institutes of Health grants CA107300 and

CA90571 and CMISE, a NASA University Research, Engineering and Technology Institute, award number NCC2-1364. M. A. T. is a scholar of leukemia and lymphoma. We thank Jerrold M. Ward and Herbert C. Morse III for comments and insightful discussions.

REFERENCES

1. Morrison SJ, Uchida N, Weissman IL (1995): *Annu Rev Cell Dev Biol* 11, 35–71.
2. Busslinger M, Nutt SL, Rolink AG (2000): *Curr Opin Immunol* 12, 151–158.
3. Rothenberg EV (2000): *Curr Opin Genet Dev* 10, 370–379.
4. Kondo M, Scherer DC, King AG, Manz MG, Weissman IL (2001): *Curr Opin Genet Dev* 11, 520–526.
5. Kondo M, Weissman IL, Akashi K (1997): *Cell* 91, 661–672.
6. Akashi K, Traver D, Miyamoto T, Weissman IL (2000): *Nature* 404, 193–197.
7. Ardavin C, et al. (2001): *Trends Immunol* 22, 691–700.
8. Banchereau J, et al. (2000): *Annu Rev Immunol* 18, 767–811.
9. Manz MG, Traver D, Miyamoto T, Weissman IL, Akashi K (2001): *Blood* 97, 3333–3341.
10. del Hoyo GM, et al. (2002): *Nature* 415, 1043–1047.
11. Rolink A, et al. (1995): *Semin Immunol* 7, 155–167.
12. Hardy RR, Hayakawa K (2001): *Annu Rev Immunol* 19, 595–621.
13. LeBien TW (2000): *Blood* 96, 9–23.
14. Dorshkind K, Rawlings DJ (2003): in Hoffman R, Benz EJ, Shattil SJ, Silberstein LE, McGlave P (Eds), *Hematology: Basic Principles and Practice*. Churchill Livingstone, Philadelphia.
15. O'Riordan M, Grosschedl R (2000): *Immunol Rev* 175, 94–103.
16. Kee BL, Murre C (2001): *Curr Opin Immunol* 13, 180–185.
17. Reya T, Grosschedl R (1998): *Curr Opin Immunol* 10, 158–165.
18. Henderson A, Calame K (1998): *Annu Rev Immunol* 16, 163–200.
19. DeKoter RP, Singh H (2000): *Science* 288, 1439–1441.
20. Quong MW, Romanow WJ, Murre C (2002): *Annu Rev Immunol* 20, 301–322.
21. O'Riordan M, Grosschedl R (1999): *Immunity* 11, 21–31.
22. Kee BL, Murre C (1998): *J Exp Med* 188, 699–713.
23. Nutt SL, Heavey B, Rolink AG, Busslinger M (1999): *Nature* 401, 556–562.
24. Rolink AG, Nutt SL, Melchers F, Busslinger M (1999): *Nature* 401, 603–606.
25. Max EE (1999): In Paul WE (Ed), *Fundamental Immunology*, 4th ed. Lippincott-Raven, Philadelphia, pp 111–182.
26. Batten M, et al. (2000): *J Exp Med* 192, 1453–1466.
27. Kurosaki T (1999): *Annu Rev Immunol* 17, 555–592.
28. Loder F, et al. (1999): *J Exp Med* 190, 75–89.
29. Petro JB, et al. (2002): *J Biol Chem* 277, 48,009–48,019.
30. Su TT, Rawlings DJ (2002): *J Immunol* 168, 2101–2110.
31. Lu TT, Cyster JG (2002): *Science* 297, 409–412.
32. Lo CG, Lu TT, Cyster JG (2003): *J Exp Med* 197, 353–361.
33. Martin F, Kearney JF (2001): *Curr Opin Immunol* 13, 195–201.
34. Ngo VN, et al. (1999): *J Exp Med* 189, 403–412.
35. Hsu BL, Harless SM, Lindsley RC, Hilbert DM, Cancro MP (2002): *J Immunol* 168, 5993–5996.
36. Fu YX, Chaplin DD (1999): *Annu Rev Immunol* 17, 399–433.
37. Forster R, et al. (1996): *Cell* 87, 1037–1047.
38. Martin F, Kearney JF (2002): *Nat Rev Immunol* 2, 323–335.
39. Liu YJ, Johnson GD, Gordon J, MacLennan IC (1992): *Immunol Today* 13, 17–21.
40. Liu YJ, Zhang J, Lane PJ, Chan EY, MacLennan IC (1991): *Eur J Immunol* 21, 2951–2962.
41. Kroese FG, Wubbena AS, Seijen HG, Nieuwenhuis P (1987): *Eur J Immunol* 17, 1069–1072.
42. Jacob J, Kelsoe G (1992): *J Exp Med* 176, 679–687.
43. Calame KL, Lin KI, Tunyaplin C (2003): *Annu Rev Immunol* 21, 205–230.
44. Macpherson AJ, et al. (2000): *Science* 288, 2222–2226.
45. Kearney JF, et al. (1997): *Int Rev Immunol* 15, 207–241.
46. Kantor AB, Herzenberg LA (1993): *Annu Rev Immunol* 11, 501–538.
47. Berland R, Wortis HH (2002): *Annu Rev Immunol* 20, 253–300.
48. Hardy RR, Hayakawa K (1991): *Proc Natl Acad Sci USA* 88, 11550–11554.
49. Kantor AB, Stall AM, Adams S, Herzenberg LA (1992): *Proc Natl Acad Sci USA* 89, 3320–3324.
50. Solvason N, Lehuen A, Kearney JF (1991): *Int Immunol* 3, 543–550.
51. Rosenwald A, et al. (2001): *J Exp Med* 194, 1639–1647.
52. Klein U, et al. (2001): *J Exp Med* 194, 1625–1638.
53. Allman D, et al. (2003): *Nat Immunol* 4, 168–174.
54. Montecino-Rodriguez E, Dorshkind K (2003): *Nat Immunol* 4, 100–101.
55. Radtke F, et al. (1999): *Immunity* 10, 547–558.
56. Guidos CJ (2002): *Semin Immunol* 14, 395–404.
57. Ordentlich P, et al. (1998): *Mol Cell Biol* 18, 2230–2239.
58. Pear W, et al. (1999): *Cold Spring Harb Symp Quant Biol* 64, 33–38.
59. Pui JC, et al. (1999): *Immunity* 11, 299–308.
60. Ting CN, Olson MC, Barton KP, Leiden JM (1996): *Nature* 384, 474–478.

61. Benoist C, Mathis D (1999): In Paul WE (Ed), *Fundamental Immunology*, 4th ed. Lippincott-Raven, Philadelphia, pp. 367–409.
62. Engel I, Murre C (2001): *Nat Rev Immunol* 1, 193–199.
63. Engel I, Johns C, Bain G, Rivera RR, Murre C (2001): *J Exp Med* 194, 733–745.
64. Wang B, et al. (1999): *J Immunol* 162, 88–94.
65. Michie AM, Zuniga-Pflucker JC (2002): *Semin Immunol* 14, 311–323.
66. Schilham MW, Moerer P, Cumano A, Clevers HC (1997): *Eur J Immunol* 27, 1292–1295.
67. Okamura RM, et al. (1998): *Immunity* 8, 11–20.
68. Verbeek S, et al. (1995): *Nature* 374, 70–74.
69. Kuo CT, Leiden JM (1999): *Annu Rev Immunol* 17, 149–187.
70. Jameson SC, Bevan MJ (1998): *Curr Opin Immunol* 10, 214–219.
71. Robey E, Fowlkes BJ (1994): *Annu Rev Immunol* 12, 675–705.
72. Blom B, et al. (1999): *Blood* 93, 3033–3043.
73. Blom B, Res PC, Spits H (1998): *Crit Rev Immunol* 18, 371–388.
74. Carrasco YR, Navarro MN, de Yebenes VG, Ramiro AR, Toribio ML (2002): *Semin Immunol* 14, 325–334.
75. Spits H, et al. (1998): *Immunol Rev* 165, 75–86.
76. Spits H (2002): *Nat Rev Immunol* 2, 760–772.
77. MacDonald HR, Radtke F, Wilson A (2001): *Curr Opin Immunol* 13, 219–224.
78. Fehling HJ, Gilfillan S, Ceredig R (1999): *Adv Immunol* 71, 1–76.
79. Kang J, Raulet DH (1997): *Semin Immunol* 9, 171–179.
80. Robey E, Fowlkes BJ (1998): *Curr Opin Immunol* 10, 181–187.
81. Bluestone JA, Khattri R, Sciammas R, Sperling AI (1995): *Annu Rev Cell Dev Biol* 11, 307–353.
82. Plum J, et al. (2000): *Ann N Y Acad Sci* 917, 724–731.
83. Cooper MA, et al. (2001): *Blood* 97, 3146–3151.
84. Williams NS, et al. (1998): *Immunol Rev* 165, 47–61.
85. Raulet DH (1999): *Curr Opin Immunol* 11, 129–134.
86. Ikawa T, Fujimoto S, Kawamoto H, Katsura Y, Yokota Y (2001): *Proc Natl Acad Sci USA* 98, 5164–5169.
87. Carlyle JR, Zuniga-Pflucker JC (1998): *Immunol Rev* 165, 63–74.
88. Bamford RN, et al. (1994): *Proc Natl Acad Sci USA* 91, 4940–4944.
89. Carson WE, et al. (1994): *J Exp Med* 180, 1395–1403.
90. Mrozek E, Anderson P, Caligiuri MA (1996): *Blood* 87, 2632–2640.
91. Carson WE, et al. (1997): *J Clin Invest* 99, 937–943.
92. Kennedy MK, et al. (2000): *J Exp Med* 191, 771–780.
93. Waldmann TA, Tagaya Y (1999): *Annu Rev Immunol* 17, 19–49.
94. Liu CC, Perussia B, Young JD (2000): *Immunol Today* 21, 113–116.
95. Heemskerk MH, et al. (1997): *J Exp Med* 186, 1597–1602.
96. Lacorazza HD, et al. (2002): *Immunity* 17, 437–449.
97. Cerwenka A, Lanier LL (2001): *Nat Rev Immunol* 1, 41–49.
98. Kalberer CP, Siegler U, Wodnar-Filipowicz A (2003): *Blood* 13, 13.
99. MacDonald HR (2002): *Curr Opin Immunol* 14, 250–254.
100. Godfrey DI, Hammond KJ, Poulton LD, Smyth MJ, Baxter AG (2000): *Immunol Today* 21, 573–583.
101. Elewaut D, Kronenberg M (2000): *Semin Immunol* 12, 561–568.
102. Kronenberg M, Gapin L (2002): *Nat Rev Immunol* 2, 557–568.
103. Gapin L, Matsuda JL, Surh CD, Kronenberg M (2001): *Nat Immunol* 2, 971–978.
104. Scheicher C, Mehlig M, Zecher R, Reske K (1992): *J Immunol Methods* 154, 253–264.
105. Inaba K, et al. (1992): *J Exp Med* 176, 1693–1702.
106. Ardavin C, Wu L, Li CL, Shortman K (1993): *Nature* 362, 761–763.
107. Wu L, Li CL, Shortman K (1996): *J Exp Med* 184, 903–911.
108. Vremec D, et al. (1992): *J Exp Med* 176, 47–58.
109. Steinman RM, Pack M, Inaba K (1997): *Adv Exp Med Biol* 417, 1–6.
110. Steinman RM, Pack M, Inaba K (1997): *Immunol Rev* 156, 25–37.
111. Steinman RM, Inaba K (1999): *J Leukoc Biol* 66, 205–208.
112. Saunders D, et al. (1996): *J Exp Med* 184, 2185–2196.
113. Vremec D, Pooley J, Hochrein H, Wu L, Shortman K (2000): *J Immunol* 164, 2978–2986.
114. Shortman K, Caux C (1997): *Stem Cells* 15, 409–419.
115. Ardavin C, Wu L, Ferrero I, Shortman K (1993): *Immunol Lett* 38, 19–25.
116. Henri S, et al. (2001): *J Immunol* 167, 741–748.
117. Leenen PJ, et al. (1998): *J Immunol* 160, 2166–2173.
118. Pulendran B, et al. (1997): *J Immunol* 159, 2222–2231.
119. De Smedt T, et al. (1996): *J Exp Med* 184, 1413–1424.
120. Galy A, Travis M, Cen D, Chen B (1995): *Immunity* 3, 459–473.
121. Hart DN (1997): *Blood* 90, 3245–3287.
122. Bell D, Young JW, Banchereau J (1999): *Adv Immunol* 72, 255–324.
123. Res P, et al. (1996): *Blood* 87, 5196–5206.
124. Peschon JJ, et al. (1994): *J Exp Med* 180, 1955–1960.
125. Puel A, Ziegler SF, Buckley RH, Leonard WJ (1998): *Nat Genet* 20, 394–397.
126. Harris NL, et al. (2000): *Hematol J* 1, 53–66.
127. Jaffe ES, Harris NL, Stein H, Vardiman JW (2001): In Kleihues P, Sobin LH (Eds), *Pathology and Genetics of*

Tumours of Haematopoietic and Lymphoid Tissues, World Health Organization Classification of Tumours. IARC Press, Lyon.

128. Harris NL, et al. (1994): *Blood* 84, 1361–1392.
129. Lukes RJ, Collins RD (1974): *Cancer* 34 (Suppl), 1488–1503.
130. Stansfeld AG, et al. (1988): *Lancet* 1, 292–293.
131. Rosenberg SA, et al. (1982): *Cancer* 49, 2112–2135.
132. Morse HC 3rd, et al. (2002): *Blood* 100, 246–258.
133. Kogan SC, et al. (2002): *Blood* 100, 238–245.
134. Fredrickson TN, Hartley JW, Morse HC 3rd, Chattopadhyay SK, Lennert K (1995): *Curr Top Microbiol Immunol* 194, 109–116.
135. Fredrickson TN, et al. (1985): *Am J Pathol* 121, 349–360.
136. Pattengale PK, Taylor CR (1983): *Am J Pathol* 113, 237–265.
137. William J, Euler C, Christensen S, Shlomchik MJ (2002): *Science* 297, 2066–2070.
138. Weller S, et al. (2001): *Proc Natl Acad Sci USA* 98, 1166–1170.
139. Matsumoto M, et al. (1996): *Nature* 382, 462–466.
140. Hoyer KK, et al. (2002): *Proc Natl Acad Sci USA* 99, 14392–14397.
141. Schrock E, Padilla-Nash H (2000): *Semin Hematol* 37, 334–347.
142. Liyanage M, et al. (1996): *Nat Genet* 14, 312–315.
143. Bayani J, Squire JA (2002): *Methods Mol Biol* 204, 85–104.
144. Bayani JM, Squire JA (2002): *Cancer Invest* 20, 373–386.
145. Pinkel D, et al. (1998): *Nat Genet* 20, 207–211.
146. Pollack JR, et al. (1999): *Nat Genet* 23, 41–46.
147. Kashiwagi H, Uchida K (2000): *Hum Cell* 13, 135–141.
148. Lichter P, Joos S, Bentz M, Lampel S (2000): *Semin Hematol* 37, 348–357.
149. Lockhart DJ, et al. (1996): *Nat Biotechnol* 14, 1675–1680.
150. Iyer VR, et al. (1999): *Science* 283, 83–87.
151. Schena M, Shalon D, Davis RW, Brown PO (1995): *Science* 270, 467–470.
152. Golub TR, et al. (1999): *Science* 286, 531–537.
153. Alizadeh AA, Staudt LM (2000): *Curr Opin Immunol* 12, 219–225.
154. Alizadeh AA, et al. (2000): *Nature* 403, 503–511.
155. Shipp MA, et al. (2002): *Nat Med* 8, 68–74.
156. Damle RN, et al. (1999): *Blood* 94, 1840–1847.
157. Hamblin TJ, Davis Z, Gardiner A, Oscier DG, Stevenson FK (1999): *Blood* 94, 1848–1854.
158. Schroeder HW Jr, Dighiero G (1994): *Immunol Today* 15, 288–294.
159. Maloum K, et al. (1995): *Blood* 86, 3883–3890.
160. Cai J, Humphries C, Richardson A, Tucker PW (1992): *J Exp Med* 176, 1073–1081.
161. Fais F, et al. (1998): *J Clin Invest* 102, 1515–1525.
162. Oscier DG, Thompsett A, Zhu D, Stevenson FK (1997): *Blood* 89, 4153–4160.
163. Sakai A, et al. (2000): *Blood* 95, 1413–1419.
164. Morse HC III (2004): Classification and characteristics of mouse B-cell lineage lymphomas. In Honjo T, Alt FW, Neuberger M (Eds), *Molecular Biology of B Cells*. Elsevier, New York.
165. Alizadeh A, et al. (1999): *Cold Spring Harb Symp Quant Biol* 64, 71–78.
166. Staudt LM, Brown PO (2000): *Annu Rev Immunol* 18, 829–859.
167. Perkins AS (2002): *Funct Integr Genomics* 2, 81–91.
168. Justice MJ (2000): *Nat Rev Genet* 1, 109–115.
169. Schimenti J, Bucan M (1998): *Genome Res* 8, 698–710.
170. Muller U (1999): *Mech Dev* 82, 3–21.
171. Mills AA, Bradley A (2001): *Trends Genet* 17, 331–339.
172. Smith AJ, Xian J, Richardson M, Johnstone KA, Rabbitts PH (2002): *Oncogene* 21, 4521–4529.
173. Rabbitts TH, et al. (2001): *Blood Cells Mol Dis* 27, 249–259.
174. Collins EC, Pannell R, Simpson EM, Forster A, Rabbitts TH (2000): *EMBO Rep* 1, 127–132.
175. Smith AJ, et al. (1995): *Nat Genet* 9, 376–385.
176. Evan GI, et al. (1992): *Cell* 69, 119–128.
177. Askew DS, Ashmun RA, Simmons BC, Cleveland JL (1991): *Oncogene* 6, 1915–1922.
178. Hanahan D, Weinberg RA (2000): *Cell* 100, 57–70.
179. Hahn WC, Weinberg RA (2002): *N Engl J Med* 347, 1593–1603.
180. Hahn WC, Weinberg RA (2002): *Nat Rev Cancer* 2, 331–341.
181. Kamijo T, et al. (1998): *Proc Natl Acad Sci USA* 95, 8292–8297.
182. Matiba B, Mariani SM, Krammer PH (1997): *Semin Immunol* 9, 59–68.
183. Nagata S (1998): *J Hum Genet* 43, 2–8.
184. Davidson WF, Dumont FJ, Bedigian HG, Fowlkes BJ, Morse HC 3rd. (1986): *J Immunol* 136, 4075–4084.
185. Drappa J, Vaishnaw AK, Sullivan KE, Chu JL, Elkon KB (1996): *N Engl J Med* 335, 1643–1649.
186. Verbeek S, et al. (1991): *Mol Cell Biol* 11, 1176–1179.
187. Donehower LA, et al. (1995): *Mol Carcinog* 14, 16–22.
188. Yukawa K, et al. (1989): *J Exp Med* 170, 711–726.
189. Sawyers CL (1999): *N Engl J Med* 340, 1330–1340.
190. Laurent E, Talpaz M, Kantarjian H, Kurzrock R (2001): *Cancer Res* 61, 2343–2355.
191. Catovsky D, et al. (1979): *Am J Clin Pathol* 72, 736–745.
192. Heisterkamp N, Stam K, Groffen J, de Klein A, Grosveld G (1985): *Nature* 315, 758–761.
193. Konopka JB, Watanabe SM, Witte ON (1984): *Cell* 37, 1035–1042.
194. Kloetzer W, et al. (1985): *Virology* 140, 230–238.
195. Kurzrock R, et al. (1987): *Nature* 325, 631–635.

196. Lugo TG, Pendergast AM, Muller AJ, Witte ON (1990): *Science* 247, 1079–1082.
197. Griffiths SD, et al. (1992): *Oncogene* 7, 1391–1399.
198. Shtivelman E, Lifshitz B, Gale RP, Canaani E (1985): *Nature* 315, 550–554.
199. Mes-Masson AM, McLaughlin J, Daley GQ, Paskind M, Witte ON (1986): *Proc Natl Acad Sci USA* 83, 9768–9772.
200. Chan LC, et al. (1987): *Nature* 325, 635–637.
201. Hermans A, et al. (1987): *Cell* 51, 33–40.
202. Elefanty AG, Hariharan IK, Cory S (1990): *Embo J* 9, 1069–1078.
203. Daley GQ, Van Etten RA, Baltimore D (1990): *Science* 247, 824–830.
204. Kelliher MA, McLaughlin J, Witte ON, Rosenberg N (1990): *Proc Natl Acad Sci USA* 87, 6649–6653.
205. Heisterkamp N, et al. (1990): *Nature* 344, 251–253.
206. Voncken JW, et al. (1992): *Cancer Res* 52, 4534–4539.
207. Voncken JW, et al. (1995): *Blood* 86, 4603–4611.
208. Honda H, et al. (1995): *Blood* 85, 2853–2861.
209. Honda H, et al. (1998): *Blood* 91, 2067–2075.
210. Hariharan IK, et al. (1989): *Mol Cell Biol* 9, 2798–2805.
211. Huettner CS, Zhang P, Van Etten RA, Tenen DG (2000): *Nat Genet* 24, 57–60.
212. Castellanos A, et al. (1997): *Blood* 90, 2168–2174.
213. Wong S, Witte ON (2001): *Oncogene* 20, 5644–5659.
214. Lowell CA, Soriano P (1996): *Genes Dev* 10, 1845–1857.
215. Malek SN, Dordai DI, Reim J, Dintzis H, Desiderio S (1998): *Proc Natl Acad Sci USA* 95, 7351–7356.
216. Adams JM, et al. (1985): *Nature* 318, 533–538.
217. van Lohuizen M, et al. (1989): *Cell* 56, 673–682.
218. Wright DH (1997): *J Pathol* 182, 125–127.
219. Pelicci PG, Knowles DM, Magrath I, Dalla-Favera R (1986): *Proc Natl Acad Sci USA* 83, 2984–2988.
220. Romana SP, et al. (1995): *Blood* 86, 4263–4269.
221. Romana SP, et al. (1995): *Blood* 85, 3662–3670.
222. Golub TR, et al. (1995): *Proc Natl Acad Sci USA* 92, 4917–4921.
223. Bernardin F, et al. (2002): *Cancer Res* 62, 3904–3908.
224. Fears S, et al. (1997): *Proc Natl Acad Sci USA* 94, 1949–1954.
225. Hiebert SW, et al. (1996): *Mol Cell Biol* 16, 1349–1355.
226. Cortes M, Wong E, Koipally J, Georgopoulos K (1999): *Curr Opin Immunol* 11, 167–171.
227. Schmitt C, et al. (2002): *Apoptosis* 7, 277–284.
228. Westman BJ, Mackay JP, Gell D (2002): *Int J Biochem Cell Biol* 34, 1304–1307.
229. Georgopoulos K (2002): *Nat Rev Immunol* 2, 162–174.
230. Wang JH, et al. (1998): *Immunity* 9, 543–553.
231. Liippo J, et al. (2001): *Eur J Immunol* 31, 3469–3474.
232. Lowe SW, Sherr CJ (2003): *Curr Opin Genet Dev* 13, 77–83.
233. Sanchez-Beato M, Sanchez-Aguilera A, Piris MA (2003): *Blood* 101, 1220–1235.
234. Sharpless NE, DePinho RA (1999): *Curr Opin Genet Dev* 9, 22–30.
235. Serrano M, et al. (1996): *Cell* 85, 27–37.
236. Takeuchi S, et al. (1995): *Blood* 86, 755–760.
237. French SW, et al. (2002): *Clin Immunol* 103, 217–230.
238. Noguchi M, et al. (1993): *Science* 262, 1877–1880.
239. Samaridis J, et al. (1991): *Eur J Immunol* 21, 453–460.
240. Rich BE, Campos-Torres J, Tepper RI, Moreadith RW, Leder P (1993): *J Exp Med* 177, 305–316.
241. Fisher AG, Burdet C, Bunce C, Merkenschlager M, Ceredig R (1995): *Int Immunol* 7, 415–423.
242. Uehira M, et al. (1993): *Int Immunol* 5, 1619–1627.
243. Montecino-Rodriguez E, Leathers H, Dorshkind K (2001): *Nat Immunol* 2, 83–88.
244. Montecino-Rodriguez E, Dorshkind K (2002): *Semin Immunol* 14, 371–376.
245. Teitell MA (2002): In Creighton TE (Ed), *Wiley's Encyclopedia of Molecular Medicine*, Wiley, New York, pp 3106–3108.
246. Pekarsky Y, Hallas C, Croce CM (2001): *Oncogene* 20, 5638–5643.
247. French SW, et al. (2002): *Biochemistry* 41, 6376–6382.
248. Laine J, Kunstle G, Obata T, Sha M, Noguchi M (2000): *Mol Cell* 6, 395–407.
249. Pekarsky Y, et al. (2000): *Proc Natl Acad Sci USA* 97, 3028–3033.
250. Bichi R, et al. (2002): *Proc Natl Acad Sci USA* 99, 6955–6960.
251. Fagarasan S, Watanabe N, Honjo T (2000): *Immunol Rev* 176, 205–215.
252. Fredrickson TN, Lennert K, Chattopadhyay SK, Morse HC 3rd, Hartley JW (1999): *Am J Pathol* 154, 805–812.
253. Enno A, et al. (1995): *Am J Pathol* 147, 217–222.
254. Dube ID, Raimondi SC, Pi D, Kalousek DK (1986): *Blood* 67, 1181–1184.
255. Hough MR, et al. (1998): *Proc Natl Acad Sci USA* 95, 13853–13858.
256. Rosic-Kablar S, Chan K, Reis MD, Dube ID, Hough MR (2000): *Proc Natl Acad Sci USA* 97, 13300–13305.
257. Vogelstein B, Kinzler KW (1992): *Cell* 70, 523–526.
258. Levine AJ, Momand J, Finlay CA (1991): *Nature* 351, 453–456.
259. Parant JM, Lozano G (2003): *Hum Mutat* 21, 321–326.
260. Ward JM, et al. (1999): *Lab Invest* 79, 3–14.
261. Cuypers HT, et al. (1984): *Cell* 37, 141–150.
262. Saris CJ, Domen J, Berns A (1991): *Embo J* 10, 655–664.
263. Repacholi MH, et al. (1997): *Radiat Res* 147, 631–640.
264. Utteridge TD, Gebski V, Finnie JW, Vernon-Roberts B, Kuchel TR (2002): *Radiat Res* 158, 357–364.
265. Pasqualucci L, et al. (2001): *Nature* 412, 341–346.
266. Said JW, et al. (2001): *Lab Invest* 81, 555–564.
267. Nakayama I, et al. (2000): *Pathol Int* 50, 191–199.
268. Narducci MG, et al. (2000): *Cancer Res* 60, 2095–2100.

269. Takizawa J, et al. (1998): *Jpn J Cancer Res* 89, 712–718.
270. Roos J, et al. (2001): *Pathobiology* 69, 59–66.
271. Lenoir GM, Preud'homme JL, Bernheim A, Berger R (1982): *Nature* 298, 474–476.
272. Kirsch IR, Morton CC, Nakahara K, Leder P (1982): *Science* 216, 301–303.
273. Burkitt DP (1983): *Cancer* 51, 1777–1786.
274. Zech L, Haglund U, Nilsson K, Klein G (1976): *Int J Cancer* 17, 47–56.
275. Kovalchuk AL, et al. (2000): *J Exp Med* 192, 1183–1190.
276. Ye BH, et al. (1997): *Nat Genet* 16, 161–170.
277. Butzler C, Zou X, Popov AV, Bruggemann M (1997): *Oncogene* 14, 1383–1388.
278. Palomo C, Zou X, Nicholson IC, Butzler C, Bruggemann M (1999): *Cancer Res* 59, 5625–5628.
279. Ward JM, Mann PC, Morishima H, Frith CH (1999): In Maronpot RR (Ed), *Pathology of the Mouse: Reference and Atlas*. Cache River Press, Vienna, pp 343–360.
280. Morse HC 3rd, et al. (2001): *Leuk Res* 25, 719–733.
281. Chattopadhyay SK, Morse HC 3rd, Makino M, Ruscetti SK, Hartley JW (1989): *Proc Natl Acad Sci USA* 86, 3862–3866.
282. Aziz DC, Hanna Z, Jolicoeur P (1989): *Nature* 338, 505–508.
283. Morse HC 3rd, Riblet R, Asofsky R, Weigert M (1978): *J Immunol* 121, 1969–1972.
284. Potter M, Wiener F (1992): *Carcinogenesis* 13, 1681–1697.
285. Anderson PN, Potter M (1969): *Nature* 222, 994–995.
286. Shen-Ong GL, Keath EJ, Piccoli SP, Cole MD (1982): *Cell* 31, 443–452.
287. Avet-Loiseau H, et al. (2001): *Blood* 98, 3082–3086.
288. Bergsagel PL, Kuehl WM (2001): *Oncogene* 20, 5611–5622.
289. Shou Y, et al. (2000): *Proc Natl Acad Sci USA* 97, 228–233.
290. Radl J (1981): *Am J Pathol* 105, 91–93.
291. Kovalchuk AL, et al. (2002): *Proc Natl Acad Sci USA* 99, 1509–1514.
292. Rosenbaum H, et al. (1990): *Embo J* 9, 897–905.
293. Strasser A, Harris AW, Cory S (1993): *Oncogene* 8, 1–9.
294. Yunis JJ, et al. (1982): *N Engl J Med* 307, 1231–1236.
295. Tsujimoto Y, et al. (1987): *Proc Natl Acad Sci USA* 84, 1329–1331.
296. Pegoraro L, et al. (1984): *Proc Natl Acad Sci USA* 81, 7166–7170.
297. Ngan BY, Chen-Levy Z, Weiss LM, Warnke RA, Cleary ML (1988): *N Engl J Med* 318, 1638–1644.
298. Cleary ML, Smith SD, Sklar J (1986): *Cell* 47, 19–28.
299. Davidson WF, Giese T, Fredrickson TN (1998): *J Exp Med* 187, 1825–1838.
300. Shiloh Y (2003): *Nat Rev Cancer* 3, 155–168.
301. Elson A, et al. (1996): *Proc Natl Acad Sci USA* 93, 13084–13089.
302. Barlow C, et al. (1996): *Cell* 86, 159–171.
303. Xu Y, et al. (1996): *Genes Dev* 10, 2411–2422.
304. Schaffner C, Stilgenbauer S, Rappold GA, Dohner H, Lichter P (1999): *Blood* 94, 748–753.
305. Schaffner C, Idler I, Stilgenbauer S, Dohner H, Lichter P (2000): *Proc Natl Acad Sci USA* 97, 2773–2778.
306. Stilgenbauer S, et al. (2000): *Ann Oncol* 11, 127–130.
307. Vorechovsky I, et al. (1997): *Nat Genet* 17, 96–99.
308. Bain G, et al. (1997): *Mol Cell Biol* 17, 4782–1491.
309. Voronova AF, Lee F (1994): *Proc Natl Acad Sci USA* 91, 5952–5956.
310. Hsu HL, Wadman I, Baer R (1994): *Proc Natl Acad Sci USA* 91, 3181–3185.
311. Winandy S, Wu P, Georgopoulos K (1995): *Cell* 83, 289–299.
312. Sun L, et al. (1999): *J Clin Oncol* 17, 3753–3766.
313. Sun L, et al. (1999): *Clin Cancer Res* 5, 2112–2120.
314. Sun L, et al. (1999): *Proc Natl Acad Sci USA* 96, 680–685.
315. Nakayama H, et al. (1999): *Cancer Res* 59, 3931–3934.
316. Nakase K, et al. (2000): *Cancer Res* 60, 4062–4065.
317. Olivero S, et al. (2000): *Br J Haematol* 110, 826–830.
318. Takanashi M, et al. (2002): *Br J Haematol* 117, 525–530.
319. Tonnelle C, Calmels B, Maroc C, Gabert J, Chabannon C (2002): *Leuk Lymphoma* 43, 29–35.
320. Fujii K, et al. (2003): *Br J Haematol* 120, 986–989.
321. Nakase K, et al. (2002): *Exp Hematol* 30, 313–317.
322. Yamada Y, et al. (1998): *Proc Natl Acad Sci USA* 95, 3890–3895.
323. Mao S, Neale GA, Goorha RM (1997): *Oncogene* 14, 1531–1539.
324. Mao S, Neale GA, Goorha RM (1997): *J Biol Chem* 272, 5594–5599.
325. Boehm T, Foroni L, Kaneko Y, Perutz MF, Rabbitts TH (1991): *Proc Natl Acad Sci USA* 88, 4367–4371.
326. Neale GA, Rehg JE, Goorha RM (1995): *Blood* 86, 3060–3071.
327. Larson RC, et al. (1994): *Oncogene* 9, 3675–3681.
328. Aplan PD, et al. (1997): *Embo J* 16, 2408–2419.
329. Xia Y, et al. (1991): *Proc Natl Acad Sci USA* 88, 11416–11420.
330. Aplan PD, et al. (1992): *Blood* 79, 1327–1333.
331. Curtis DJ, Robb L, Strasser A, Begley CG (1997): *Oncogene* 15, 2975–2983.
332. Condorelli GL, et al. (1996): *Cancer Res* 56, 5113–5119.
333. O'Neil J, Billa M, Oikemus S, Kelliher M (2001): *Oncogene* 20, 3897–3905.
334. Donehower LA, et al. (1992): *Nature* 356, 215–221.
335. Jacks T, et al. (1994): *Curr Biol* 4, 1–7.
336. Lozano G, Liu G (1998): *Semin Cancer Biol* 8, 337–344.
337. Krug U, Ganser A, Koeffler HP (2002): *Oncogene* 21, 3475–3495.
338. Maehama T, Dixon JE (1998): *J Biol Chem* 273, 13375–13378.

339. Myers MP, et al. (1997): *Proc Natl Acad Sci USA* 94, 9052–9057.
340. Myers MP, Tonks NK (1997): *Am J Hum Genet* 61, 1234–1238.
341. Cantley LC, Neel BG (1999): *Proc Natl Acad Sci USA* 96, 4240–4245.
342. Suzuki A, et al. (1998): *Curr Biol* 8, 1169–1178.
343. Di Cristofano A, Pesce B, Cordon-Cardo C, Pandolfi PP (1998): *Nat Genet* 19, 348–355.
344. Marsh DJ, et al. (1998): *J Med Genet* 35, 881–885.
345. Liaw D, et al. (1997): *Nat Genet* 16, 64–67.
346. Marsh DJ, et al. (1997): *Nat Genet* 16, 333–334.
347. Lowsky R, et al. (1997): *Blood* 89, 2276–2282.
348. Reitmair AH, et al. (1995): *Nat Genet* 11, 64–70.
349. de Wind N, Dekker M, Berns A, Radman M, te Riele H (1995): *Cell* 82, 321–330.
350. Ricciardone MD, et al. (1999): *Cancer Res* 59, 290–293.
351. Cline MJ (1994): *N Engl J Med* 330, 328–336.
352. Whiteside D, et al. (2002): *Cancer Res* 62, 359–362.
353. Hartley KO, et al. (1995): *Cell* 82, 849–856.
354. Takata M, et al. (1998): *Embo J* 17, 5497–5508.
355. Walker JR, Corpina RA, Goldberg J (2001): *Nature* 412, 607–614.
356. Li GC, et al. (1998): *Mol Cell* 2, 1–8.
357. Gu Y, et al. (1997): *Immunity* 7, 653–665.
358. Yang H, et al. (2002): *Science* 297, 1837–1848.
359. Xia F, et al. (2001): *Proc Natl Acad Sci USA* 98, 8644–8649.
360. Friedman LS, et al. (1998): *Cancer Res* 58, 1338–1343.
361. Lens D, Matutes E, Catovsky D, Coignet LJ (2000): *Leukemia* 14, 427–430.
362. Foroni L, Panayiotidis P, Hoffbrand AV (1998): *Br J Haematol* 100, 800.
363. Garcia-Marco JA, et al. (1996): *Blood* 88, 1568–1575.
364. Wooster R, et al. (1995): *Nature* 378, 789–792.
365. Stern MH, et al. (1993): *Oncogene* 8, 2475–2483.
366. Virgilio L, et al. (1994): *Proc Natl Acad Sci USA* 91, 12530–12534.
367. Laine J, Kunstle G, Obata T, Noguchi M (2002): *J Biol Chem* 277, 3743–3751.
368. Virgilio L, et al. (1998): *Proc Natl Acad Sci USA* 95, 3885–3889.
369. Gritti C, et al. (1998): *Blood* 92, 368–373.
370. Cordon-Cardo C, Prives C (1999): *J Exp Med* 190, 1367–1370.
371. Grabstein KH, et al. (1994): *Science* 264, 965–968.
372. Bamford RN, Battiata AP, Burton JD, Sharma H, Waldmann TA (1996): *Proc Natl Acad Sci USA* 93, 2897–2902.
373. Onu A, Pohl T, Krause H, Bulfone-Paus S (1997): *J Immunol* 158, 255–262.
374. Tagaya Y, et al. (1997): *Proc Natl Acad Sci USA* 94, 14444–14449.
375. Fehniger TA, et al. (2001): *J Exp Med* 193, 219–231.
376. Baldassarre G, et al. (2001): *Proc Natl Acad Sci USA* 98, 7970–7975.
377. Proulx GM, et al. (2003): *Am J Clin Oncol* 26, 6–11.
378. Cuadra-Garcia I, et al. (1999): *Am J Surg Pathol* 23, 1356–1369.
379. Lin CW, et al. (2003): *Lab Invest* 83, 55–64.
380. Lu LM, Hiai H (1999): *Jpn J Cancer Res* 90, 1218–1223.
381. Weissman IL, Anderson DJ, Gage F (2001): *Annu Rev Cell Dev Biol* 17, 387–403.
382. Melchers F, Rolink A (1999): In Paul WE (Ed), *Fundamental Immunology*, 4th ed. Lippincott-Raven, Philadelphia, pp 183–224.
383. Knowles DM (2001): In Knowles DM (Ed), *Neoplastic Hematopathology*, 2nd ed. Lippincott Williams & Wilkins, Philadelphia, pp 93–227.
384. Schattner EJ, Casali P (2001): In Knowles DM (Ed), *Neoplastic Hematopathology*, 2nd ed. Lippincott Williams & Wilkins, Philadelphia, pp 43–92.
385. Fischer A, Malissen B (1998): *Science* 280, 237–243.
386. Ceredig R, Rolink T (2002): *Nat Rev Immunol* 2, 888–897.

PART III

GENERAL ISSUES IN CANCER BIOLOGY

15

GENETIC MODIFIERS

DAVID W. THREADGILL

Department of Genetics and Lineberger Comprehensive Cancer Center, University of North Carolina, Chapel Hill, North Carolina

KENT W. HUNTER

Laboratory of Population Genetics, National Cancer Institute, National Institutes of Health, Bethesda, Maryland

FEI ZOU

Department of Biostatistics, University of North Carolina, Chapel Hill, North Carolina

KENNETH F. MANLY

Department of Cellular and Molecular Biology, Roswell Park Cancer Institute, Buffalo, New York

Cancer is caused by a complex interaction between an individual's intrinsic genetic composition and extrinsic environmental exposures. Cancer modifiers, a special class of cancer genes that accounts for the majority of the intrinsic genetic differences in cancer susceptibility among individuals, contribute to a significant proportion of all human cancers. Modifiers are often normal allelic variants of genes that do not produce abnormal phenotypes alone, but rather "modify" the action of other cancer genes. Frequently, these genes are also called low-penetrance cancer genes since, individually, they typically have small effects on cancer phenotypes. However, cumulatively, they can have major influences on all stages of cancer development from initiation through response to therapy. Similar to the unique characteristics that distinguish cancer modifiers from high-penetrance cancer-causing genes, the methods used to identify these genes are distinct. Because of the low-penetrance nature of their actions, this class of cancer gene is particularly difficult to detect using human populations. As such, the mouse has become the predominant model to identify cancer modifiers. The unique characteristics, methods for detection and identification of cancer modifiers, and their use in human studies are described in this chapter.

INTRODUCTION

Most cancers arise because of mutations or epigenetic changes within a limited number of genes but progress through an accumulation and selection for additional mutations. The development and progression of most cancers is also strongly influenced by natural allelic differences between individuals. The action of these alleles, called cancer modifiers, can range from small, difficult-to-detect influences to profound alterations on virtual any characteristic of cancer, from initiation to therapy response. Although cancer modifiers typically have minor individual effects, cumulatively, they have a major influence on cancer susceptibility and progression. As such, these genes are also known as low-penetrance cancer genes, as distinguished from high-penetrance genes described in previous chapters that are associated with familial cancer syndromes like the *APC* gene with familial adenomatous polyposis (Groden et al., 1991; Kinzler et al., 1991; Nishisho et al., 1991) or *BRAC1* with familial breast cancer (Miki et al., 1994). Despite this distinction, many low-penetrance cancer modifiers are also likely to modulate the effects of the high-penetrance cancer genes. For example, knockouts of *Trp53* are highly prone to tumors in many cell types. However,

Mouse Models of Human Cancer, edited by Eric C. Holland
ISBN 0-471-44460-X

the spectrum of tumors that develop is highly dependent upon the genetic background, and thus cancer modifiers, carrying the mutation (Backlund et al., 2001; Donehower et al., 1992, 1995; Ochiai et al., 2003).

The goals of cancer modifier analysis are to establish the number and location of polymorphic genes affecting cancer development, to investigate the function of alternative modifier alleles, and to determine how alleles of different modifier genes interact with each other, with high-penetrance cancer genes, and with the environment. Collectively, the information about alleles, genes, and environments is referred to as the architecture of the cancer phenotype. A model for the architecture of a cancer phenotype is usually built sequentially by the successive characterization of individual or groups of modifiers.

Technically, modifier genes are variants that have little or no effect themselves but influence aberrant phenotypes caused by one or more major effect genes. However, in practice, cancer modifier genes can also be considered interchangeable with quantitative trait loci (QTLs) since cancer modifiers often have quantitative effects on cancer phenotypes and the approach to their detection, localization, and identification is similar to QTL analysis. Because of their broad effects and complex interactions, the identification of cancer modifier genes, like QTLs, has historically been difficult. However, combined with current genome technologies, the mouse offers a powerful tool to identify these important cancer-associated genes.

Finding the genetic basis of cancer modifiers is dominated by a "Catch-22." Similar to QTL analysis, the power with which a cancer modifier can be detected and the precision with which it can be mapped are profoundly affected by the number, strength of effect, and interactions of the modifiers to be detected. That is, the ability to characterize cancer modifiers is controlled by the unknown properties of the modifiers themselves. If the cancer phenotype of interest is dominated by a small number of strong, unlinked modifiers with few interactions, characterizing and ultimately cloning those modifiers may be comparatively easy. In contrast, if the phenotype is controlled by a large number of weak modifiers, by several closely linked modifiers, or by modifiers with complex interactions, characterizing them may be a significant challenge. Further complicating cancer modifier analysis is the fact that individual modifiers can have positive or negative effects depending on alleles at other genes (Tripodis et al., 2001) and on the cancer type (Czarnomska et al., 2003).

Thus, a major challenge of cancer research is to move beyond the high-penetrance cancer genes associated with rare hereditary cancer syndromes to the low-penetrance cancer modifiers that cumulatively account for the vast majority of cancers in the human population. Although complementary mutagenesis approaches have been proposed to induce new mutations conferring cancer modifier phenotypes with potentially stronger effects than natural alleles (Schindewolf et al., 2000; Shima et al., 2003), this chapter concentrates on the approaches that have been successfully applied to characterize and clone natural alleles that act as cancer modifiers and on their use in human cancer studies.

PARTITIONING CANCER PHENOTYPES

Cancer is not a single disease but rather an amalgamation of diseases that all share the common trait of aberrant growth control. Similarly, the phenotypic characteristics contributing to any particular cancer or cancer model are numerous, often with multiple genes contributing to each particular characteristic. As such, studies that precisely define and partition the cancer phenotype under study have a greater chance to successfully identify modifier genes. These characteristics are often distinguishable and can include gross measurements like tumor incidence, cancer growth rate, or frequency of metastasis to fine cellular details of individual cancers such as quantification of vascularization, morphologic differentiation of the cancer cells, and the relative proportion of stroma or inflammation, among others. Ultimately, the decision of how to partition cancer phenotypes will depend upon the type and unique characteristics of the cancer under study.

A carcinogen-induced mouse model of lung cancer is an example of the interrelatedness between cancer phenotypes. Using this model, eight cancer modifiers affecting three-dimensional tumor shape have been localized (Tripodis and Demant, 2003). Interestingly, the nuclear morphology of these lung cancers was previously associated with polymorphisms at the *Kras2* gene (Tripodis and Demant, 2001); the putative *Kras2* modifier was originally mapped as a susceptibility modifier for pulmonary adenomas (Gariboldi et al., 1993). Similar to this example, of the 12 modifiers detected using a carcinogen-induced mouse model of skin cancer, only a subset of the modifiers were responsible for survival time of tumor-bearing mice (Nagase et al., 1999).

Intermediate molecular phenotypes leading to cancer development and progression can also be partitioned in a similar manner. Intermediate phenotypes are often referred to as endo-phenotypes and can range from the transcriptional differences across cancers, either for single genes or an entire transcriptional profile, to the expression or absence of specific molecular markers at the protein or other macromolecular level to changes in physiologic measures or serum markers. Furthermore, these characteristics can be autonomous to the cancer cells or nonautonomous through interactions with non–cancer cells or macromolecules. The importance of understanding the biologic basis of cancer modifiers is related to the primary goal of identifying genes modifying particular cancer phenotypes. The success of this endeavor and, in particular, the effort required to fine-map and identify modifier genes are inversely correlated to the level of biologic knowledge known about the particular cancer phenotype under study (Fig. 15.1).

Cancer modifiers do not typically display Mendelian patterns of inheritance, and because of their low penetrance, properly partitioning cancer phenotypes increases the probability of successful modifier identification in at least two ways. First, by precisely defining the cancer characteristics under study, more accurate mapping data can be produced providing a narrower genomic interval from which to develop lists of gene candidates, as described in the section on detection and localization. Since particular characteristics of cancers can be masked or otherwise altered through cellular or molecular-level epistasis, it is important to know what is actually being mapped. For example, gross tumor size can be caused by expansion in the number of cancer cells, accumulation of intra-tumor–stromal cells, or cystic inclusions. Different genes may, and probably do, control each of these characteristics. Mapping gross tumor size without properly defining what is causing variation in size may give a false impression of genetic heterogeneity when in fact the data are noisy because of inaccurate identification of the cancer phenotype. An example is the mapping of metastasis genes using the FVB/N-$TgN(MMTVPyVT)^{634Mul}$ mouse model of breast cancer (Lifsted et al., 1998). Modifiers for a variety of metastatic phenotypes could be analyzed, including presence or absence of macroscopic lesions and the size of individual secondary tumors. However, genes controlling the size of metastatic lesions can only be mapped in mice containing metastasis. In mice that do not have metastasis, it is unknown whether the genome has the metastasis modifiers or not since modifiers controlling whether metastasis occurs will probably be different than the modifiers controlling the size of the metastatic lesions.

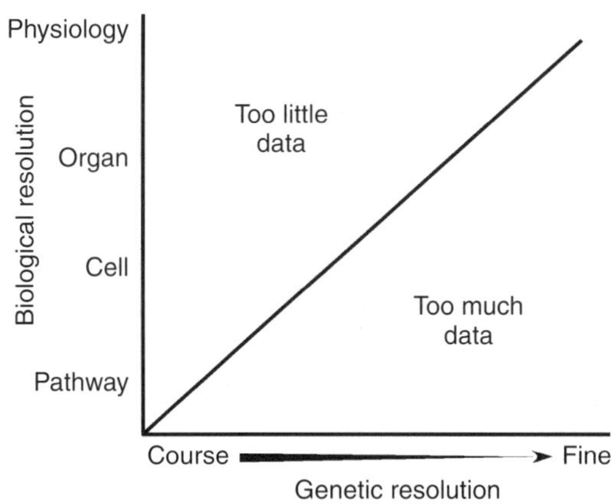

Figure 15.1. Graph depicting the relationship between biologic knowledge of cancer modifiers and genetic resolution required to select potential candidates from those residing in a defined genomic interval. Ideally, the information content for the two sources of data will be near the diagonal to maximize the balance between data collected and identification of strong candidates. If inadequate biologic or genetic data are collected, placing the total information content above the diagonal, there will be insufficient information to identify appropriate lists of candidate genes.

Proper partitioning of cancer phenotypes contributes to improving the odds of successfully identifying cancer modifier genes for a second reason. Knowing what characteristic is being mapped enhances the informed selection of potential candidates within genomic intervals known to harbor cancer modifier genes. The selected candidates must be further tested through validation as described in the section on fine mapping and localization; the fewer the candidates, the less work involved in the final step of cancer modifier analysis.

Illustrative of this effect, the *Apmt1* and *Apmt2* cancer modifiers in the FVB/N-$TgN(MMTVPyVT)^{634Mul}$ model are known to interact epistatically (Le Voyer et al., 2000) and to influence latency to tumor detection (Cozma et al., 2002). Therefore, it was hypothesized that the modifiers altered a cell growth phenotype. Using a bioinformatics approach, *Cdc25a* and *Myc*, two extensively characterized cell cycle genes, were found to be strong candidates for *Apmt1* and *Apmt2*, respectively, dramatically reducing the effort required to identify candidates for validation testing. In a second example, tumors arise earlier in hybrids between I/LnJ and FVB/N-$TgN(MMTVPyVT)^{634Mul}$ than on the parental FVB/N background (Le Voyer et al., 2001). However, the hybrids accumulate less tumor mass. While genetic mapping of modifiers contributing to this difference detected three loci, detailed analysis of the cancer phenotype revealed that the growth reduction correlated to a reduction in the tumors' ability to recruit microvessels, providing strong biologic data to assist in the eventual identification of candidate modifiers.

MOUSE STRAINS AND CROSSES

The first step in beginning a modifier study is to identify the strains of mice that differ in the genetic control of the cancer phenotype under study. Since cancer modifiers can influence any stage of tumor development, from initiation to therapy response, the selection of appropriate strains must be carefully weighted to ensure that they differ sufficiently in their genetic control of the cancer phenotype. A commonly used approach for identifying appropriate strains is a strain survey, a phenotypic analysis of a variety of different inbred strains using an assay to measure the cancer phenotype of interest. If the phenotype is dependent upon an engineered alteration, the transgenic or knockout cancer models can be outcrossed to a panel of other inbred strains to produce F1 hybrids or segregating N2 or F2 progeny before measuring the subsequent effects on the cancer phenotype as was done for the FVB/N-$TgN(MMTVPyVT)^{634Mul}$ mouse model of breast cancer metastasis (Lifsted et al., 1998). An important fact to keep in mind is that even though significant phenotypic variation across strains is not observed, strains can still have differences in their genetic control of the particular phenotype; the existence of genetic heterogeneity, a situation when individuals have similar phenotypes but different combinations of alleles, can only be determined by analyzing a pilot cross between strains and observing the phenotypic variation of the resulting progeny.

Most commonly used strains are available commercially, with The Jackson Laboratory (http://www.jax.org) having by far the largest collection of inbred strains. Although any set of strains can be chosen, ideally the selection of strains for a survey should encompass strains already selected for genetic diversity as part of the Mouse Phenome Project (MPP; http://aretha.jax.org/pub-cgi/phenome/mpdcgi?rtn=docs/home; Paigen and Eppig, 2000), a communitywide effort to produce baseline phenotypes for 40 genetically diverse inbred strains, including susceptibilities to various types of cancers. Since a large quantity of phenotypic data is being accumulated in the MPP database, meta-analysis with many other types of traits can be performed with the cancer phenotypes.

Although the choice of inbred strains for a cancer modifier project will be primarily driven by those that demonstrate phenotypic variation, other factors should also considered. If there is the option, C57BL/6J, 129S1/SvImJ, DBA/2J, or A/J mice should be considered since several genetic resources have been developed using these strains that greatly simplify fine mapping and candidate gene validation, as described later. Foremost is that these strains have been used as templates for the public and private mouse genome sequencing efforts (http://www.ncbi.nlm.nih.gov; http://www.celera.com; Waterston et al., 2002), creating significant sequence variation data that may enable more rapid narrowing of candidate intervals through haplotype analysis or identification of potential candidate genes. Embryonic stem cells and bacterial artificial chromosome (BAC) libraries are also commercially available for several of these strains (http://www.specialtymedia.com; http://bacpac.chori.org), supporting the construction of "knockin" or transgenic mouse models as described under Fine Mapping and Validation to directly test the effect of a particular polymorphism on the phenotype of interest. Other genetic resources that have been or are being built using these strains as founders include recombinant inbred and chromosome substitution (CS; also known as consomic) strains described below. Ultimately, the choice of the appropriate strains needs to be made based upon the cancer phenotype of interest. However, given a choice of strains, those with the greatest supporting resources greatly enhance the prospects for quickly identifying and validating cancer modifiers.

Additional cancer-specific resources are also available to help guide in strain selection. The organ-site reports of the Mouse Models of Human Cancer Consortium (MMHCC; http://emice.nci.nih.gov/) provide valuable information on current cancer models and strains that have been used to detect cancer phenotypes. The MMHCC has also compiled a database of cancer models that have been characterized on various genetic backgrounds (http://cancermodels.nci.nih.gov/mmhcc/index.jsp). Likewise, the Mouse Tumor Biology Database (http://tumor.informatics.jax.org/FMPro?-db=TumorInstance&-format=mtdp.html&-view; Bult et al., 2000) provides historical information on strain susceptibilities to specific types of cancers and genetic and histologic information for a wide variety of spontaneous and induced mouse tumors.

The design of the mapping cross to genetically detect and localize the cancer modifiers also depends on a variety of factors. When designing the experiments, it is essential to ensure that the experimental design provides enough statistical power to achieve the recommended level of genomewide significance, as described under Detection and Localization. This will depend on the number of modifiers segregating in a cross and the fraction of the total phenotypic variance for which each modifier accounts. As with any genetic mapping experiment, consultation with a statistician is strongly recommended during the design of the experiments. The cross design is often of personal preference, but specific characteristics of the modifiers may limit which cross can be chosen (Table 15.1; Fig. 15.2).

Backcrosses and Intercrosses

The most commonly used cross designs for cancer modifier analysis are backcrosses and intercrosses. These are robust designs for generating low-resolution localization of loci that modulate specific phenotypes. The choice between backcross and intercross depends on several factors (Darvasi, 1998). Intercrosses, generally requiring fewer animals than backcrosses since each intercross animal has twice the number of recombination events as a backcross animal, are frequently used to approximate the

Table 15.1. Comparison of Mapping Methods Frequently Used for Cancer Modifier Analysis

Mapping Method	Advantages	Disadvantages
Backcross	Any strain of mouse Unlimited numbers Low marker density needed for initial detection Can test dominance and interactions involving dominance	Each genome assayed once Only one homozygous class Full genotyping required Lower number of recombinations limit resolution for strong modifiers
Intercross	Any strain of mouse Unlimited numbers Low marker density needed for initial detection Can test dominance and interactions involving dominance All genotype combinations Twice the number of recombinations per mouse as backcrosses	Each genome assayed once Full genotyping required Lower number of recombinations limit resolution for strong modifiers
Interspecific backcross	High degree of marker polymorphisms Greater haplotype diversity Low marker density needed for initial detection Can test dominance and interactions involving dominance	Each genome assayed once Cannot perform intercrosses Only one homozygous class Full genotyping required Lower number of recombinations limit resolution for strong modifiers
Recombinant inbred	Unlimited number of identical mice Optimal for gene–environment interaction studies No genotyping required	Limited founder combinations Limited number of strains available Higher marker density required because of increased number of recombinants Cannot detect dominance effects
Recombinant congenic	Unlimited number of identical mice Donor genome partitioned, making is easier to identify epistatic interactions No genotyping required for initial detection Can test for dominance and interactions involving dominance	Limited founder combinations Limited number of strains available Only partial genome coverage Need to perform backcross for localization with partial genotyping

number of modifier loci present and to obtain estimates of their additive or dominance effects. Conversely, backcrosses, because there are only two genotypic classes at each locus, compared with three for intercrosses, are more efficient in detecting simple epistatic interactions between unlinked modifiers. For modifiers with additive effects, an intercross is somewhat more powerful than a backcross, but for dominant modifiers, a backcross can be twice as powerful as an intercross (Darvasi, 1998). One of the first cancer modifiers to be identified in mice was localized using backcrosses (Dietrich et al., 1993). The *Mom1* locus, which has a semidominant suppressor effect on tumor multiplicity in the Apc^{Min} mouse model of intestinal tumorigenesis (Gould et al., 1996), was localized in crosses between C57BL/6J and AKR/J.

A major strength of both the backcross and intercross methods, unlike many of the other strategies, is that any two genetically distinct strains of mice can be used to generate a mapping cross. Thus, the genetic components of any cancer phenotype that differs between two of the multitude of inbred mouse strains can be subjected to genetic dissection. This capability includes common (Mock et al., 1993) and ferally derived inbred strains (Nagase et al., 2001) as well as transgenic (Le Voyer et al., 2000) and knockout cancer models (Bolivar et al., 2001).

Interspecific Backcrosses

A specialized type of backcross that has been used successfully for cancer modifier studies are crosses between different species or subspecies of mice. The most common are between *Mus spretus* and any inbred strain of *Mus musculus* origin. Although these species diverged approximately 3 million years ago, they are still able to interbreed to produce fertile F1 female

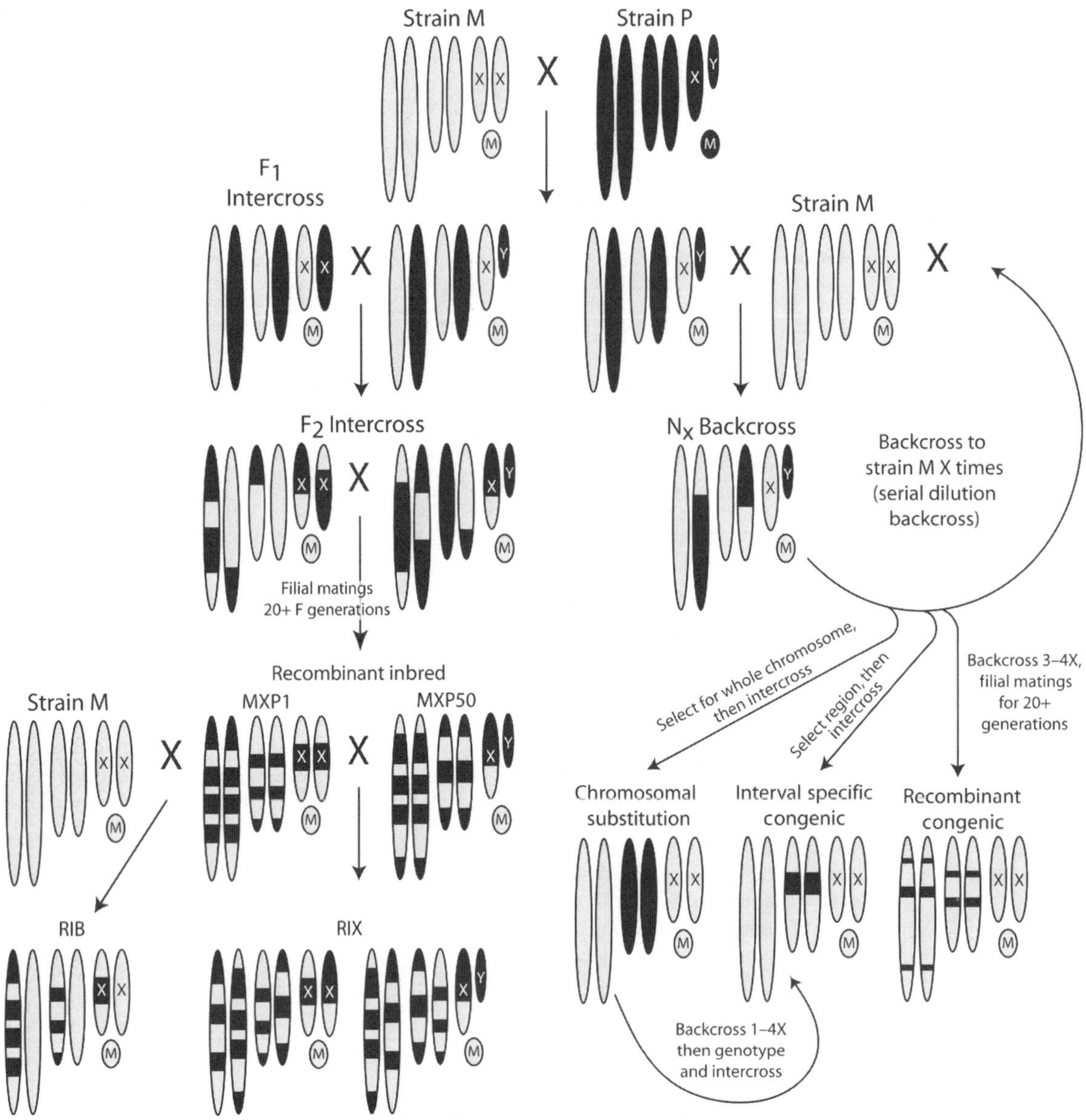

Figure 15.2. Schematic representation of different genetic mapping crosses and resources available for cancer modifier mapping. For illustration, two homologous pairs of autosomes, the sex chromosomes (X and Y) and mitochondrial genome (M) are shown for two inbred mouse strains, M (maternal) and P (paternal), whose DNA is represented in white and black, respectively. The color in subsequent crosses represents DNA from the respective founder strain. The origin of various types of crosses and derivative strains is shown with their consequences on mixing DNA from the two parental strains. Adapted from Hunter and Williams (2002).

progeny. Backcrosses can be carried out to either parent to map modifiers, although intercrosses are not possible because F1 males are sterile. Because of the large degree of genetic divergence between these species, all crosses between *M. spretus* and common laboratory strains are highly informative with numerous polymorphic differences that facilitate modifier mapping. This divergence also means that the genetic differences being surveyed in interspecific crosses are much greater than in intraspecific crosses between any two common laboratory strains that are much more closely related and therefore share alleles across larger proportions of their genomes. This has become particularly evident with the development of first-generation haplotype maps for several mouse strains (Wade et al., 2002; Wiltshire et al., 2003). Consequently, *M. spretus* typically shows a large number of major differences in cancer phenotypes when compared to *M. musculus*–derived strains (Guenet and

Bonhomme, 2003). Using *M. musculus* × *M. spretus* interspecific backcrosses, 12 cancer susceptibility loci influencing skin tumor development and progression in a carcinogen-induced model have been detected and localized (Nagase et al., 1995, 1999). Although not as optimal as the recombinant congenic strains described later, interspecific backcrosses can also be used to detect epistatic interactions (Nagase et al., 2001).

Recombinant Inbred Panels

Recombinant inbred (RI) strains offer compelling advantages for mapping cancer modifiers, particularly those that have low heritability or low penetrance or are otherwise associated with highly variable cancer phenotypes. Since a panel of RI lines originates from the inbreeding of independent F2 progeny, generated from two parental inbred strains, each RI genome is replicated in the form of an isogenic line and has roughly the same utility for mapping modifiers as 4–10 intercross progeny (Bailey, 1981); the number varies as a function of the recombination load carried by the RI lines. Mapping with RI panels is performed using an association between the differences in phenotypic trait values for each RI line and the strain distribution patterns (SDPs) across the panel for each marker; an SDP is the distribution of the two parental genotypes at each marker across the entire RI panel. This approach successfully localized three modifiers for a carcinogen-induced mouse model of lung cancer in the AXB and BXA panels of RI lines (Malkinson et al., 1985); AXB and BXA strains are reciprocal RI lines derived from the A/J and C57BL/6J inbred strains (Marshall et al., 1992).

A major advantage of RI panels is that measurement of replicate individuals can reduce the variance associated with experimental and uncontrolled error to very low levels (Belknap, 1998); this effectively elevates heritability and greatly improves prospects for mapping low-penetrance cancer modifiers. The RI design also permits testing gene effects under a spectrum of environmental perturbations to expose gene–environment interactions. The use of RI strains allows the average effect of a particular environment on the phenotype to be more accurately determined for a specific genotype, requiring fewer animals than conventional crosses. In contrast, gene–environment interactions are more difficult to study using other mapping populations since each animal represents a unique recombinant genome. Another advantage of RI strains is that phenotypes generated by different groups using a variety of methods can be compared when collected on the same RI strains, including genotypes for the strains. As a result, loci that segregate in RI sets can usually be localized without additional genotyping (Williams et al., 2001).

Although RI strains can significantly reduce the effort required to obtain linkage information, there are a number of limitations. First, RI sets exist for relatively few inbred strains. As a result, only the subset of polymorphic loci that differ among these strains can be mapped using this strategy. Fortunately, these strains, including C57BL/6J, A/J, BALB/cJ, and DBA/2J, are among the most widely used and characterized inbred mouse strains (Taylor, 1996). Second, despite the increased power associated with the ability to assay identical genotypes repeatedly, the limited number of lines in each RI panel precludes mapping cancer modifiers that do not account for a substantial fraction of the genetic variance (Belknap et al., 1996; Pataer et al., 1997). Mapping modifiers that account for less variance requires a two-tiered approach, identifying putative loci using RI analysis followed by a second experiment to confirm the modifier assignment in an independent cross. Third, like moderately sized backcross and intercross strategies, the precision with which modifiers can be mapped is generally poor, with candidate regions often spanning 10–20 centimorgans (cM).

RIX and RIB Derivatives

A straightforward extension to increase the power of RI mapping is through the generation of recombinant inbred intercrosses (RIXs; Threadgill et al., 2002). This method relies on a potentially large set of F1 hybrids generated by breeding all or a subset of pairwise combinations of RI strains within an RI panel; unlike the parental RI lines, RIX panels are similar to replicable F2 progeny with all three genotypic classes at each locus. RIX mice share all of the advantages of RI mapping (Bailey, 1981; Belknap et al., 1996) and have the added benefit of providing a large sample of unique but predictable genotypes. Two additional advantages are hybrid vigor of the F1 progeny used to assess phenotypes and the ability to make each F1 by reciprocal crosses, a feature that can be used to investigate parental effects. The RIX method can also be used to confirm a modifier and refine its position through virtual congenics by generating a subset of RIX that have known genotypes in the relevant interval hypothesized to contain the modifier.

Recombinant inbred strains can also be backcrossed to either a parental or another inbred strain to generate recombinant inbred backcrosses (RIBs). Unlike RI lines, the RIB crosses share segregation ratios similar to those of standard backcross, with a 1 : 1 ratio of homozygous to heterozygous genotypes. Similar to RIXs, each litter of RIB mice has a predictable genotype defined by the particular combination of parental genomes. RIB progeny can be used to solve one obvious limitation of RI lines, their lack of heterozygous genotypes. The RIB cross is particularly useful in that modifiers of dominant and semidominant mutations, engineered alleles, and transgenes can be easily mapped in a single generation cross without genotyping. This approach was used to map a modifier for breast cancer metastatic progression efficiency by crossing the FVB/N-$TgN(MMTV PyVT)^{634 Mul}$ mouse model to the AKXD RI panel (Hunter et al., 2001).

Recombinant Congenic and Chromosome Substitution Strains

Identifying the complex and often nonlinear interactions among many polymorphic genes and their products is difficult. Standard backcross, intercross, and recombinant inbred analyses often lack the power required to dissect multiple unlinked genes, particularly if they interact epistatically. To overcome this limitation, a variation of recombinant inbred panels, the recombinant congenic (RC) line, was developed (Groot et al., 1996). Like RI panels, RC strains are composites of the genomes of two inbred progenitor parent strains. Unlike the RI strains, which are generated by serial brother–sister matings after an initial outcross, two rounds of backcrossing to one of the progenitor strains precede the brother–sister matings in RC strains. The resulting RC strains contain, on average, a random 12.5% of the donor strain genome on the background of the recipient strain. Because of the reduced complexity of RC genomes, gene interactions and epistasis can be efficiently investigated (Fijneman et al., 1996, 1998a,b; Tripodis et al., 2001). The added power is due to a reduction in the amount of donor genome present in RC strains relative to other genetic resources, increasing the probability of separating interacting genes into individual strains where their effects can be studied in isolation. The OcB RC strains, developed by transferring O20/A donor genome onto a C57BL10/A recipient strain, were used to detect loci-modulating susceptibility to a carcinogen-induced lung cancer model (Tripodis et al., 2001). Despite assaying only half the genome, these studies detected and localized over 30 lung cancer modifiers. More striking was the fact that only a single locus had independent activity; all others showed epistatic interactions ranging from simple two-locus interactions to one locus having interactions with five other modifiers. Similarly, using the CcS RC strains, produced from the donor STS/A and recipient BALB/cHeA strains, over 10 colon cancer modifiers have been mapped (Moen et al., 1992, 1996; van Wezel et al., 1999), several of which show strong gene interaction effects (van Wezel et al., 1996). The AcB and BcA strains are a third, more extensive RC set containing reciprocal donor and recipient crosses using the widely available A/J and C57BL/6J strains (Fortin et al., 2001); these have not yet been used for cancer modifier studies.

Unlike the random distribution of donor genome present in RC strains, chromosome substitution (CS) strains have a defined donor chromosome on a recipient background. The percent of donor genome is determined by the size of the chromosome, ranging from 7.7% for chromosome 1 to 3.2% for chromosome 19. Both RC and CS resources can be used for modifier analysis by identifying recombinant lines that differ phenotypically from the recipient line. Unlike mapping using an entire RI panel, RC and CS strains are used in standard intercrosses or backcrosses with the recipient strain to identify the portion of the donor genome or chromosome, respectively, contributing to the cancer phenotype. Any phenotype in CS strains that differs from the recipient parental strains is obligatorily localized to the donor chromosome. For example, modifiers of testicular teratoma incidence on MOLF/Ei chromosome 19 have been detected using a 129.MOLF-Chr 19 CS strain (Matin et al., 1999).

If a full panel of 22 CS strains is constructed, chromosomes 1–19, X, Y, and the mitochondrial genome (also known as a conplastic strain), the chromosomal assignment of a modifier can be rapidly determined for any phenotype that differs between the parental strains. Like RI panels, RC strains, and congenic strains, CS strains have defined and fixed genotypes. Likewise, CS strains have increased power to detect modifiers because the variance due to environmental or experimental fluctuation can be averaged out by repeated analysis of each genotype. CS strains are also excellent starting points for high-resolution mapping, as described under Fine Mapping and Validation, since they can be used to develop series of overlapping interval-specific congenic animals. A disadvantage of this approach is that the initial modifier assignment is to a whole chromosome rather than to a chromosomal region. In addition, without additional crosses, CS strains cannot distinguish between single and multiple modifiers on a single donor chromosome. Although the use of reciprocal CS strains can identify the presence of interactions between host and donor loci, no linkage information on the host modifiers would be obtained. Furthermore, CS strains, by isolating individual donor chromosomes, preclude the possibility of detecting epistatic interactions between unlinked donor loci without additional breeding studies.

Genome Scans

All mapping approaches not using isogenic lines like RI panels require the progeny to be genotyped in order to statistically associate the cancer phenotype with particular allelic differences across the genome. The strategy of choice for performing genome scans has historically been microsatellite based. Over 6000 mouse microsatellite markers have been developed (http://www-genome.wi.mit.edu/cgi-bin/mouse/index#genetic; Dietrich et al., 1996), enabling the identification of polymorphic markers between any strains. All of the Massachusetts Institute of Technology (MIT)–developed microsatellites have been tested in 12 inbred strains. A subset of more than 400 markers has also been screened in greater than 50 strains (http://www.cidr.jhmi.edu/mouse/index.html). Thus, most, if not all, of the markers needed to perform an initial genome scan between any commonly used strains should already be available.

Primary genome scans can be performed in several ways. To reduce time and cost within individual laboratories, initial genome scans are frequently performed on the two extremes of the phenotypic distribution curve, often the top and bottom 10% to 15% of the animals. This subset of animals provides the bulk of the statistical power for modifier detection, enabling preliminary localization. Genomic regions that demonstrate the potential presence of modifier genes can then be typed on the entire set of animals, with an increased density of markers to confirm and improve linkage. Although increasing the number of markers typed in a putative modifier region increases the likelihood that existing modifiers will be detected, generally in the preliminary genome scan it is not necessary to exceed a 20-cM resolution; higher marker densities are unlikely to provide any greater statistical power because of the limited number of recombination sites present in single generation crosses (Darvasi et al., 1993; Dupuis and Siegmund, 1999). Publicly accessible genotyping services, supported by the National Institutes of Health (NIH), are also available through the Center for Inherited Disease Research (http://www.cidr.jhmi.edu) and the Marshfield Clinic (http://research.marshfieldclinic.org/genetics/Genotyping_Service/mgsver2.htm).

An alternative to microsatellite markers that is gaining in popularity is single-nucleotide polymorphism (SNP) markers. These have the advantage of being scored without electrophoresis. Representative gel-less technologies include MALDI-TOF and Taqman platforms (Ranade et al., 2001; Ross et al., 1998). Alternatively, fluorescent label techniques can be used to pool and analyze genotypes on DNA sequencers (Inazuka et al., 1997). At present, however, there are a number of disadvantages that limit these approaches. While a large number of SNPs have become available through public and private efforts (Wade et al., 2002; Wiltshire et al., 2003), the number of validated SNPs is low and these have been screened using a limited number of inbred strains. In addition, to screen SNPs in a high-throughput manner requires relatively sophisticated equipment that may not be available to many laboratories. Yet, even for small laboratories, SNPs can provide additional markers for regions in which there may not be a useful microsatellite marker and should therefore be considered either when attempting to fill in regions lacking sufficient polymorphic markers or during candidate gene searches. There are a number of web-accessible databases for mouse SNPs, including the subscription-only Celera database (www.celera.com), the Genetic Annotation Initiative of the Cancer Genome Anatomy Project (CGAP; http://cgap.nci.nih.gov), and the Whitehead Genome Center and Roche SNP collections, accessible through the MPP database (http://aretha.jax.org/pub-cgi/phenome/mpdcgi?rtn=studies/details&id=96&dsonly=1). Many of the putative SNPs in the Celera and CGAP databases have not been validated and should be tested prior to use in large-scale genotyping efforts.

DETECTION AND LOCALIZATION

Following production of the mapping cross and collection of genotypic and phenotypic data, there are two primary stages of cancer modifier mapping: detection and localization. In practice, the two stages overlap and together are often associated with the genome scan, but they are conceptually different and have distinct statistical requirements. Importantly, however, details of detection and localization need to be considered up front to ensure that the experimental phase of the modifier analysis is adequately powered to allow both detection and localization of existing modifier genes.

Modifier Detection

The detection of a cancer modifier depends on a statistical test. The power of this statistical test, defined as the probability of detecting a modifier given that one exists, is dependent on many variables, including the number of mice genotyped and phenotyped, the number and strength of the modifiers, potential interactions between modifiers, the type of cross, the magnitude of the environmental and genetic components of the phenotypic trait value variance, the method of analysis, and the false-positive rate tolerated by the analysis. The number of mice required for detecting a modifier is, roughly speaking, proportional to the variance of the phenotypic trait value contributed by nongenetic factors and inversely proportional to the square of the strength of the modifier. Consequently, weak modifiers, which account for a few percent of the trait variance, may require several hundred mice for even a statistical power of 50% (Darvasi, 1998).

If cancer modifiers are detected in small, low-power experiments, the modifiers detected will be those that are fortuitously overestimated in that cross. For example, simulation studies suggest that modifiers detected in experiments with a power of 10% will, on average, be overestimated by fourfold (Beavis, 1998). If several modifiers contribute equally to a trait, two low-power experiments may even detect distinct subsets of the actual modifiers, as each experiment will detect a different overestimated minority of the total. Some important factors that influence modifier detection are discussed in more detail below.

Heritability

The fraction of the total phenotypic trait value variability that can be attributed to genetic control is referred to as heritability, which is a summary measure for all loci contributing to a particular phenotype but does not imply any particular genetic model. Thus, heritability is proportional to the role that genetic differences contribute to phenotypic variation. Even highly heritable cancers like human breast cancer have heritabilities around 25%, meaning that the total genetic differences between individuals accounts for less than 25% of the differences in their breast cancer susceptibility (Czene et al., 2002). Cancer phenotypes with lower heritability have weaker genetic effects and, consequently, require larger populations to distinguish genetic effects in the context of large nongenetic phenotypic variance. Although mice within an inbred strain are genetically identical, they also exhibit variation in phenotypes due to nongenetic factors like developmental variation, often referred to as stochastic variation, or environmental differences.

Generally, cancer modifiers are easily detected and characterized if they account for at least 10% of the total phenotypic trait variance in an intercross. If the difference in average trait values between two inbred lines is twice the standard deviation of the trait in each line and the difference was due to two equal, additive modifiers, each would account for 12.5% of the variance in an intercross. Such a trait, therefore, would be a reasonable candidate for an attempt to identify modifiers. However, such a large difference between the parental lines is not absolutely necessary since strong modifiers of opposite effect could also control the trait. Starting with a cancer phenotype with a smaller parental difference makes being lucky more important.

Modifier Strength

The statistical power for detecting cancer modifiers is determined by modifier strength. The strength of a cancer modifier is measured with one of two quantities, either the allele substitution effect or the fraction (percent) of total phenotypic variance explained. The allele substitution effect is the expected change in trait value if the allele on one chromosome is changed to the alternative allele. This can be divided into additive and dominance components. The additive effect is half the difference in the trait value between individuals homozygous for one modifier allele and those homozygous for the other. The dominance allele substitution effect is the amount by which the trait value of a heterozygote differs from the expected additive effect. In other words, it is the difference between the trait value for a heterozygote and the average of the trait values for the two homozygotes. Although the strength of modifiers affecting a cancer phenotype is unknown when mapping experiments are planned, they are often estimated so that the minimum modifier strength that provides adequate power for modifier detection can be calculated.

Number of Mice for Detection

Although the number of mice needed to detect a cancer modifier is dependent upon the characteristics of the modifier, this can be easily estimated for a hypothetical modifier with an additive effect (Fig. 15.3). For example, Figure 15.3*a* shows the number of mice required for detection of a single additive or dominant modifier in a backcross or intercross population at a statistical power of 50%. This figure is derived from two sources, simulation-based relationships (Darvasi and Soller, 1997) and theoretical derivation (Lynch and Walsh, 1998) using genomewide significance levels (Lander and Kruglyak, 1995). The power is almost entirely a function of the strength of the modifier, measured as the fraction of the total phenotypic variance explained by the hypothetical modifier and not of the type of cross. Thus, detecting modifiers that control 10% of the phenotypic variance requires about 150 mice; detecting those that control 5% requires 300 mice.

The numbers of mice shown in Figure 15.3*a* should be considered a bare minimum. The statistical power for this number of mice is 50%, which means that half of the existing modifiers at the threshold strength will not achieve significance. Furthermore, both methods assume a high density of fully informative markers; using a lower marker density or less informative markers would reduce the power.

To be more realistic, Figure 15.3*b* shows the number of mice required to detect a modifier at a statistical power of 80%. This figure represents a more prudent experimental design and is based on the theoretical method used in Figure 15.3*a*. Detecting modifiers that control 10% of the variance requires about 200–250 mice; detecting those that control 5% requires 400–500 mice.

Significance of Modifier Detection

A common method for determining the significance of any potential modifier localization is the permutation test (Churchill and Doerge, 1994; Doerge and Churchill,

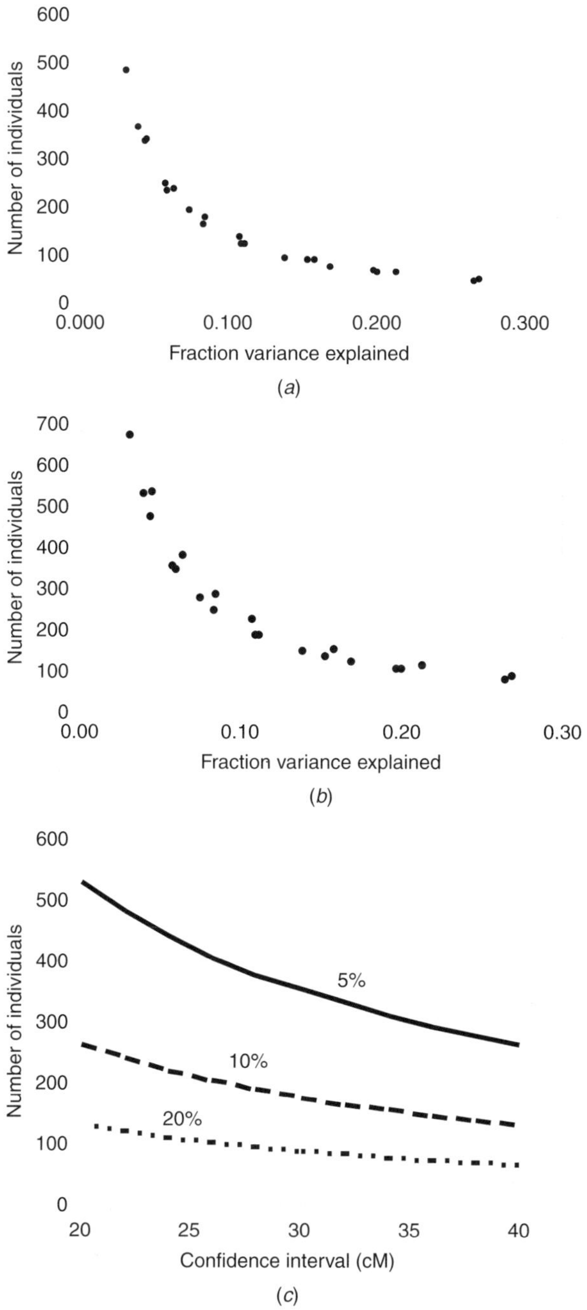

Figure 15.3. Theoretically determined number of progeny in experimental crosses needed for different stages of cancer modifier mapping. (*a*) Number of progeny needed to detect 50% of modifiers with a given strength as a function of the trait variance explained by the modifier. The figure includes both theoretical (Lynch and Walsh, 1998) and simulation-based (Darvasi and Soller, 1997) requirements for backcrosses or intercrosses with dominant or codominant alleles. (*b*) Number of progeny needed to detect 80% of modifiers of a given strength as a function of the trait variance explained by the modifier. (*c*) Expected size of a 95% confidence interval for a modifier location as a function of the number of progeny in the mapping cross. Based on a simple function fitted to the results of simulation experiments (Darvasi and Soller, 1997).

1996). The permutation test is an empirical test designed to determine the significance of an association in any given data set. The data are randomly permuted; that is, the trait values are randomly reassigned to other individuals to destroy any relationship between trait values and the genotypes of the marker loci used for the analysis. Using the permuted data sets, the maximum-likelihood score between the original trait value and the randomized genotypes is then determined by repeatedly mapping independently permuted data sets hundreds or thousands of times; this results in a distribution of likelihood values where there were no actual modifiers linked to any of the marker loci. Mapping values from the original, unpermuted data that exceed the 37th, 95th, and 99.9th percentiles of the maximum-likelihood scores determined by the frequency of association between specific markers and the trait values using the permuted data sets would be considered suggestive, significant, and highly significant according to widely adopted genomewide thresholds (Lander and Kruglyak, 1995).

Modifier Localization

The second stage in modifier mapping is estimating how many modifiers significantly affect a trait and the approximate location of those modifiers, usually by interval mapping methods (Lander and Botstein, 1989). Interval mapping consists of testing the effect of hypothetical modifiers at each marker locus and at regular intervals between the marker loci. The expected effect of a hypothetical modifier at each location is estimated from the genotypes at marker loci flanking each interval and assumes that intermarker recombinations are randomly distributed. Unlike single-marker analysis, interval mapping requires that the relative order of all markers is known. Furthermore, interval mapping can separate modifier location from modifier effect. A likelihood ratio test is performed at each locus across the entire genome. Scores for the likelihood ratio statistic (LRS; Haley and Knott, 1992), related to the logarithm of the odds (LOD) that is commonly used in human genetics, are the measures of the significance of associations. The genetic location with the highest LRS that is above the genomewide threshold is a putative modifier position while the value of the LRS at that position provides a measure of the statistical significance of the association between marker genotype and phenotype. Roughly speaking, the LRS can be considered as an approximation of a chi square χ^2 statistic.

The confidence interval for a modifier location determined by interval mapping is inversely proportional to the square of the strength of the modifier and the number of mice (Darvasi, 1998; Darvasi et al., 1993). Said another way, the stronger the modifier effect and the greater the number of mice used, the narrower the confidence interval will be. The spacing of the marker loci only marginally affects the size of the confidence interval, and except for large populations or strong modifiers, there is little to be gained by having marker loci spaced more closely together than 20 cM. Even with a strong modifier and large sample population, the confidence interval for a single-generation cross is unlikely to be less than 10 cM. At the other extreme, a weak modifier may have a confidence interval that covers an entire chromosome.

Number of Mice for Localization

The precision with which a cancer modifier is localized is limited by either of two factors. First, the number of mice in the mapping population for single-generation crosses limits the minimum size of the confidence interval for modifiers with weak to moderate strength. Figure 15.3*c* shows the number of mice needed for a 95% confidence interval. The results are a function of the strength of the modifier, measured as the fraction of total phenotypic trait variance explained (Darvasi and Soller, 1997). Modifiers that explain at least 20% of the trait variance can be localized to a 25-cM interval with approximately 100 mice. At the other end of the scale, a modifier that explains only 5% of the trait variance requires over 250 progeny to localize within 40 cM.

For confidence intervals below 20 cM, the precision of location is also limited by the amount of recombination in the cross. Simulations suggest a lower limit of about 11 cM for confidence intervals in a single-generation cross of practical size, regardless of the strength of the modifier (Darvasi et al., 1993).

Methods for Supporting Localization

In a seminal paper on interval mapping, a one-LOD support interval was proposed as an estimate of the confidence interval for the map position of a detected modifier (Lander and Botstein, 1989). A one-LOD support interval is set by identifying the maximum LOD and then subtracting 1. The interval covered on either side of the maximum LOD by the new LOD minus one value is considered the one-LOD support interval. Although this method has been widely used, others have found that this method yields confidence intervals that are too small, especially for weak modifiers that are often outside of this conservative estimate (Darvasi and Soller, 1997; Mangin et al., 1994; Ooijen, 1992). A two-LOD support interval should be used for a less stringent approximation if no other estimate is available (Broman, 2001; Dupuis and Siegmund, 1999).

Different solutions using both theoretical and empirical methods have been proposed to solve this problem. For example, theoretical approximations that apply to any genetic design have been proposed but can only be evaluated by simulation (Mangin et al., 1994). For backcross and intercross designs, a simple empirical formula was developed based on simulation experiments that applies to modifiers of moderate effect (Darvasi and Soller, 1997). This formula is $530/Nv$, where N is the number of individuals and v is the proportion of the trait variance explained by the modifier. However, the best solution is an empirical estimation by bootstrap (Visscher et al., 1996; Walling et al., 1998) or Bayesian methods that estimate a posterior probability as a function of location (Satagopan et al., 1996; Sen and Churchill, 2001; Sillanpää and Arjas, 1998). The recently adapted bootstrap method for evaluating confidence intervals is based upon a resampling procedure. Subsets of predetermined size from the total data set are randomly sampled followed by reanalysis of the sampled data. This is repeatedly performed, and the frequency that a particular genomic interval is detected, as defined by the markers used, provides a measure for the confidence interval containing the putative modifier.

Modifier Combinations and Interactions

The previous sections primarily dealt with detecting cancer modifiers that have a measurable additive effect, independent of any other modifiers or cofactors. However, as work with lung and colon cancer modifiers has shown (Tripodis et al., 2001; van Wezel et al., 1996), the vast majority of cancer modifiers will probably have independent as well as interactions with other modifiers. Some modifiers may even have effects that are only manifested when in combination with other genes such as the MOLF/Ei Chr 19 effect on testicular teratomas previously described (Matin et al., 1999). The detection and localization of these types of modifiers can be considerably more complicated, introducing additional computational challenges and requiring alternative approaches for analysis.

The first step to identify combinations of modifiers is to separate the analysis into a series of independent analyses. After a primary modifier has been detected and localized, the genomic interval containing the modifier can be held constant and the data reanalyzed using the same statistical approaches to identify other modifiers that may have been masked or not as easily detected because of the primary modifier. The first modifier is held constant using regression analysis; by fitting a linear regression curve, the fraction of the phenotypic trait values that are independent of the primary modifier and attributable to other modifiers can be determined and used for additional analyses. This approach, called composite interval mapping (CIM), uses a marker near the first modifier to partly control for that modifier's effects in subsequent calculations. Thus CIM can be used to detect modifiers, in the background of a primary modifier locus, that have additive effects independently of the primary modifier; although the total activity of these additional modifiers does not have to be independent, a measurable fraction does.

After sets of modifiers with independent activity have been detected and localized, additional analyses can be performed to develop multigene models that contribute nonlinearly to the cancer phenotype under study. This is performed by using the results of CIM with multiple interval mapping (MIM; Kao et al., 1999). Since the approaches for MIM are usually computationally intensive, techniques using computational approximations with defined error probabilities have been developed (Sen and Churchill, 2001).

Alternatively, rather than approaching modifier analysis individually, pairwise combinations of markers, also called two-dimensional genome scans, can be used to detect and identify genomic regions that act in concert (http://www.jax.org/staff/churchill/labsite/; Carlborg and Andersson, 2002; Sen and Churchill, 2002). These approaches have the potential to detect modifiers that have no activity alone but, when present with other modifiers, alter the phenotype. More complex modeling is usually not supported by current approaches because of a lack of power using the sizes of crosses and mapping populations commonly used.

FINE MAPPING AND VALIDATION

While initial phases of modifier mapping frequently leave large candidate intervals, often on the order of 10–30 cM, the goal of fine mapping is to narrow the candidate intervals to less than 5 cM and preferably to 1–2 cM. In mouse, this corresponds to approximately $1-4 \times 10^6$ bp and may still contain tens to hundreds of genes depending on gene density (Avner et al., 2001; Hudson et al., 2001; Waterston et al., 2002). Despite the coarse localizations obtained by initial mapping experiments, they are nonetheless highly informative with respect to providing insight into the genetic architecture of the cancer phenotype that is susceptible to modifier gene influence. However, these coarse localizations cannot distinguish whether there is a single strong modifier or multiple modifiers that are tightly linked and cumulatively having an effect on the phenotype. A classical example is the *Mom1* locus that is a major modifier of tumor multiplicity in the *Apc*^{Min} mouse model of intestinal cancer (Dietrich et al., 1993; Moser et al., 1992); *Mom1* was originally reported to be a single suppressor locus and *Pla2g2a* (secretory type II phospholipase A2) was identified as a strong candidate (MacPhee et al., 1995). The *Pla2g2a* gene was validated as contributing to *Mom1* through analysis of transgenic mice carrying allele-specific forms of *Pla2g2a* (Cormier et al., 1997). However, *Pla2g2a* did not account for the entire *Mom1* effect, and with further fine mapping, enabled by additional recombination sites within the *Mom1* interval, *Mom1* was revealed to be a complex locus whose cumulative effect was due to at least two tightly linked genes (Cormier et al., 2000). Fortuitously, both genes contributing to the *Mom1* effect were in the same polarity; that is, the two suppressor alleles were linked on a single haplotype. Thus, although an estimation of the number and strength of modifier genes can be attained through an initial analysis, fine mapping is required to ultimately identify the candidate genes contributing to the phenotype. As stated earlier, the better defined the phenotype is, the easier it will be to identify appropriate candidates from all possibilities within a critical genomic interval. Two primary approaches, congenics and haplotype analysis, have been successfully used to fine-map cancer modifiers.

Congenics

One of the most widely used strategies to verify the localization of cancer modifiers and to refine their positions is the generation of interval-specific congenic lines (Graff et al., 1966); this is the transfer through breeding of a defined genomic interval containing the modifier from a donor strain to a second, recipient strain. The donor and recipient strains are those used in the initial genetic mapping of the modifier loci. Occasionally, the two alleles are reciprocally exchanged between the two strains. If the modifier really exists, has been mapped to the correct interval, and has an independent effect, then phenotypes of the reciprocal congenic lines should each deviate from the recipient strains by a predictable amount. This also gives a more precise estimation of the additive effect of the modifier alleles. However, exceptions do occur. One happens when the initial localization was due to two or more linked genes each contributing to the phenotypic variance. Separation of the loci by breeding, for example, during the generation of a high-resolution congenic interval, can result in multiple, linked, but sometimes nonoverlapping, intervals contributing unequally to phenotypic differences with the recipient strain (Morel et al., 2001). A second exception occurs when the phenotype of interest is the result of the interaction between two unlinked modifiers or other genes in the genome of the donor for the congenic lines. However, under these circumstances, the absence of phenotype in the reciprocal congenic lines does not necessarily mean that the modifiers do not exist. In this instance, compound congenic mice with more than one donor chromosome would be required to confirm and fine-map the modifier loci.

The strength of the congenic method is that it will usually convert a polygenic trait to a monogenic (pseudo-Mendelian) trait. In other words, differences in the mean phenotypes of the congenic and the background, recipient strain should be generated by a single or multiple, tightly linked polymorphic loci in the introgressed interval. When the congenic lines have the expected phenotype, it is not difficult to narrow down the size of the interval progressively through additional breeding and selection for intra-congenic interval recombinations, producing subcongenics that concurrently fine-map the modifier, as was done for *Mom1*, a major modifier of intestinal tumor multiplicity in the Apc^{Min} mouse model (Gould et al., 1996).

The disadvantages of this strategy are that it requires a substantial genotyping effort and several generations of backcrossing, ranging from 4–7 generations when using marker-assisted selection for speed congenics (Markel et al., 1997) to 10 or more generations for standard congenics. Furthermore, each modifier needs its own resource generation. The subsequent generation of subcongenic mice requires large numbers of mice with new recombination sites to identify informative recombination sites that reduce the size of the congenic interval. As a result, while this strategy has a high probability of fine mapping the modifier loci, it does require significant time and resources to develop the appropriate genetic mapping reagents.

An alternative is to use premade congenic lines. Knockouts are generally made in 129-derived ES cells, and the subsequent animals backcrossed to C57BL/6J. In spite of backcross generations, a large interval of donor 129 genome, flanking the targeted gene, is often retained in the recipient C57BL/6J background, forming a ready-made congenic for that region. Flaherty and colleagues demonstrated that it was possible to use these congenics to rapidly identify specific intervals containing modifier genes for behavior traits (Bolivar et al., 2001). This strategy has the advantage of skipping the initial congenic derivation; however, it is limited to those traits that differ between 129 and C57BL/6J.

A variation on these themes is the CS strains (Nadeau et al., 2000). Like the interval-specific congenic mice, CS strains carry a genomic segment from a donor strain, in this instance a whole chromosome, in the background of a second inbred recipient strain. As described earlier under Mouse Strains and Crosses, CS strains can be used to fine-map modifier loci in addition to their initial detection. Starting with a CS strain, the initial congenic introgression step, typically taking 5–10 generations, can be skipped and the CS strains used directly to identify recombinants that partition the donor chromosome into smaller intervals. This is a powerful approach if the appropriate strains are represented in the CS strain set. Currently, there are three sets of CS strains, all with C57BL/6J as the background strain; B.A with donor A/J chromosomes (Nadeau et al., 2000), B.129 with donor 129S1/SvImJ chromosomes (under development; J. Nadeau, personal communication), and B.PWD with donor PWD chromosomes (J. Forejt, personal communication). Additionally, two subconsomic (congenic) series exist where each donor chromosome has been prepartitioned into approximately thirds of entire chromosomes, ranging from 8 to 58 cM. Again, C57BL/6J is the background for both congenic series; B.D with donor DBA/2J segments and B.CAST with donor CAST/Ei segments (Iakoubova et al., 2001).

The congenic-based strategy has a high probability of successfully identifying cancer modifier genes. The development of specialized congenic or CS strains permits the systematic and methodical delineation of sequentially smaller chromosomal intervals that contain the modifier gene(s) of interest. In addition, since they are inbred, any of the genotype combinations can be assayed ad infinitum, permitting detection of relatively subtle phenotypic variations.

Haplotype Analysis

A complementary approach that has also been successfully adapted to cancer modifier analysis is the use of outbred or panels of diverse inbred mice that have distinct haplotypes; haplotypes are linked polymorphisms that tend to be inherited together. Recent haplotype analysis of the mouse genome suggests that different commonly used mouse strains carry a limited number of haplotypes for any given locus, ranging from two to five (Wade et al., 2002; Wiltshire et al., 2003). The principle of this approach is to identify pedigrees with different haplotypes for the genomic region containing the putative modifier locus. Each pedigree with a different haplotype can be crossed to a reference inbred strain and the progeny tested to identify pedigrees with and without the effect of the putative modifier locus. Using the haplotype structure, an association study can be performed between a specific haplotype and the modifier effect, thus delimiting the effective size of the regions containing the modifier gene.

This approach has been successfully used to identify a skin cancer susceptibility modifier, *Stk6* (Ewart-Toland et al., 2003). By crossing common inbred strains with outbred *M. spretus* containing different haplotypes within the confidence interval, the interval containing the cancer modifier was reduced from approximately 20 to 1 cM. Subsequent verification, as described below, provided substantial evidence for the identity of the candidate gene. Similar to the CS strain approach, this method is also dependent upon having appropriate outbred stocks or related inbred strains with rearranged haplotypes segregating the modifiers.

Although restricted to cancer modifiers shared among strains, multiple crosses between resistant and susceptible strains can be combined to increase the resolution of modifier localization. This approach is particularly powerful when performed in conjunction with haplotype analysis (Hitzeman et al., 2000). For example, combining several moderately sized crosses between resistant and susceptible strains for breast cancer metastasis using the FVB/N-*TgN(MMTV-PyVT)*^{634Mul} mouse model narrowed an interval containing a cancer modifier for metastasis efficiency to 10 cM (Park et al., 2002). When the haplotypes within the 10-cM interval were compared across four strains used in the original crosses, the region of the interval shared between the two resistant and two susceptible strains but differing between the two sets of strains was used to narrow the candidate gene list from over 400 genes to only 23 (Park et al., 2003).

Other Approaches

Numerous alternatives, including advanced intercrosses (AI; Darvasi and Soller, 1995) and heterogeneous stocks (HS; Talbot et al., 1999), are also applicable for cancer modifier studies. Although not previously widely used for cancer modifier analysis, both AI and HS approaches have the capability for merging initial detection and localization with fine mapping.

Advanced intercrosses are similar to the classical intercrosses used to produce F2 progeny but are continued on for many more generations, often to F10 or higher. The effect is a continued accumulation of recombination sites, generating smaller and smaller intervals containing the putative modifiers. However, there are empirical limits to the number of recombinations that can be accumulated with additional generations; the recovery of new recombinations starts to be offset with skewing of allele frequencies and homozygosity through generations of intercrossing unless very large breeding populations are maintained. In practice, the approach can increase mapping resolution five- to eightfold over a standard intercross, but it also requires significantly greater genotyping density to detect the smaller recombination intervals. Recently this approach was applied to an (AI) at F11 between the lung cancer susceptible A/J and resistant C57BL/6J strains to narrow the interval containing the *Pas1* modifier of pulmonary adenoma susceptibility (Wang et al., 2003). Using a carcinogen-induced mouse model of lung cancer, analysis of AI lines refined the critical interval to less than 1 cM, which contained *Kras2*, a gene previously proposed as a candidate for *Pas1* (Lin et al., 1998).

Heterogeneous stocks are similar to AI in that they have been interbred for many generations, some greater than G60 (Mott et al., 2000). Consequently, HS can theoretically increase mapping resolution 30-fold over conventional intercrosses. A major difference between HS and all other cross designs previously described is that HS strains are usually formed from four to eight founder strains that are randomly bred in a manner to minimize inbreeding. Because of this random breeding design, each generation is designated with a "G" rather than the intercross designation "F." Unlike AI, analysis of HS lines requires a two-step approach for analysis because of the extensive number of recombinations potentially carried by HS mice. The first level of analysis is typically performed with mice from a limited number of pedigreed crosses with sparse genotyping to identify shared haplotypes. After this initial localization, very dense marker genotyping is required in candidate intervals to identify the recombination sites used to fine-map modifiers. A variation using outcrosses of HS mice to inbred strains was recently described that has the power to reduce confidence intervals to within 3 cM (Mott and Flint, 2002).

To accelerate the fine mapping process, a number of laboratories have explored and experimentally validated alternative methods for candidate gene fine mapping and identification from relatively large genomic intervals. For example, modifiers acting as oncogenes are typically overexpressed or amplified in tumors while those acting as tumor suppressors typically undergo loss of heterozygosity (LOH). Since cancers often amplify genomic intervals containing specific oncogenes and develop LOH by deletion of chromosomal DNA containing tumor suppressor genes, characterization of these genomic rearrangements in multiple cancers can be used to identify small overlapping intervals in regions where modifiers have been genetically mapped. However, this association is primarily restricted to cancer-autonomous modifiers.

DNA amplification and LOH can be detected by comparative genomic hybridization (CGH), originally performed by competitively hybridizing mitotic chromosome spreads with DNA from normal and cancer cells labeled with different fluorescent tags (Kallioniemi et al., 1992). The intensity of the two fluorescent tags that hybridized can be used to identify genomic regions within the tumor DNA that are overrepresented (amplified) or underrepresented (LOH). More recently, microarrays containing defined BACs or cDNAs have been used to perform array CGH, permitting much finer localization of DNA rearrangements (Pinkel et al., 1998; Pollack et al., 1999).

With the development of new databases containing functional information on gene products, genomic rearrangements in different tumor types, and the expansion of published research, novel bioinformatics-based approaches are also being tested to narrow candidate gene lists down for validation. For example, two modifiers that alter tumor latency via cell cycle regulation were previously localized using the FVB/N-*TgN(MMTV-PyVT)*^{634Mul} mouse model of breast cancer (Le Voyer et al., 2000). By querying abstracts in

PubMed (http://www.ncbi.nlm.nih.gov) for the coappearance of all pairwise combinations of genes residing within the two genomic intervals containing the modifiers, a single pair of genes, *Myc* and *Cdc25a*, were identified as strong candidates (Cozma et al., 2002).

Modifier Validation

After a genomic interval has been sufficiently narrowed to reduce the number of candidate genes that may account for the modifier effect, the individual candidates need to be evaluated and the correct gene validated. The criteria for including or excluding candidates for validation testing are often imprecise and dependent on previous experience. However, general guidelines can maximize the probability of correct selection. For instance, of the genes within the critical interval, those with nucleotide polymorphisms between strains that cause amino acid changes would be strong candidates for functional testing. Likewise, genes with altered expression or RNA splicing in the cancer, or another non–cancer cell type known to be biologically important for the cancer phenotype under study, would also identify high-quality candidates. Differential expression is being pursued to identify candidates for the many *Sluc* carcinogen-induced lung cancer modifiers by microarray analysis (Gariboldi et al., 2003).

After reducing the number of potential candidates, experimental validation needs to be performed to identify the correct cancer modifier gene. A genetically engineered knockin containing an allele substitution between strains used in the original mapping of the modifier to recapitulate the expected modifier effect is considered the gold standard for proof (Complex Trait Consortium (CTC), 2003; Glazier et al., 2002). However, this is not always practical since many nucleotide differences may distinguish the two allelic forms of the modifiers. As such, alternative strategies are also accepted.

Similar to QTL gene validation, the accepted proof for the correct identification of a cancer modifier is somewhat controversial. Guidelines have been proposed (Glazier et al., 2002) and modified and endorsed by a community of scientists active in the CTC (http://www.complextrait.org) (CTC, 2003). The basic consensus is that the actual nucleotide difference causing functional differences between allelic forms of cancer modifiers will not be easily identified, particularly if they are in regulatory elements or are due to the combined effects of several nucleotide changes. Therefore, a "preponderance of evidence" should be accepted as proof of correct identification rather than the more stringent requirements for Mendelian traits that are usually caused by easily identified, single nucleotide changes.

Accepted experimental validation includes altering the predicted modifier effect with a transgene from the alternative strain, as was done to validate *Pla2g2a* as a component of the *Mom1* locus that modifies tumor multiplicity in the Apc^{Min} model of intestinal cancer (Cormier et al., 1997). Similarly, creating a knockout would support, but not prove, the correct candidate. Either of these approaches needs additional functional data to validate the correct identification since they can both have outcomes consistent with correct identification but not actually be the correct gene. This would occur if additional genes that are involved in the cancer phenotype reside in the critical interval but are not functionally polymorphic between the strains.

Combined with known functions of the candidate genes, significant circumstantial evidence can also make a persuading case. For example, the identification of *Cdkn2a* that codes for p16 INK4a and p19 ARF as the gene responsible for the *Pctr1* modifier of plasmacytoma susceptibility was determined by functional analysis (Zhang et al., 1998). Sequence analysis of *Cdkn2a* in resistant and susceptible strains identified changes in an amino acid residue in the susceptible strains that was conserved in mice, rats, and humans. In vitro expression of the *Cdkn2a* allele carrying the amino acid change was inefficient in the normal inhibition of cyclin D2 activity using kinase assays. Interestingly, the susceptible strains also show aberrant expression of *Cdkn2a*, suggesting that the effect attributed to *Pctr1* may be a combination of reduced activity and abnormal expression (Zhang et al., 2003).

Two recent successes have also been based upon a preponderance of the evidence: the identification of *Stk6* as the gene responsible for the *Skts13* modifier detected in the carcinogen-induced skin cancer mouse model (Ewart-Toland et al., 2003) and *Ptprj* as the gene responsible for the *Scc1* modifier of carcinogen-induced colon cancer susceptibility (Ruivenkamp et al., 2002). Although no amino acid changes were found in *Stk6*, different alleles of this gene showed a small but statistically significant difference in expression level that was independent of carcinogen treatment. *Stk6* was validated as being responsible for *Skts13* by showing a suggestive association between the human homologue, *STK15*, and cancer cases in case–control studies and amplification or overexpression in several types of cancers. In vitro studies further supported *Stk6*. Using variants of *STK15*, it was shown that the cancer-associated allele has stronger transforming properties and the protein variants show differential interaction with a ubiquitin-conjugating enzyme that is important in DNA damage response.

Likewise, the link between *Ptprj* and *Scc1* has not been formally proven, but sufficient evidence exists to declare this cancer modifier identified. Only the *Ptprj* gene within the *Scc1* critical interval was not an olfactory receptor–like gene, and no obvious regulatory elements for other genes were identified. Sequence analysis revealed several amino acid changes and the human homologue is deleted in 50% of colorectal carcinomas as well as in lung and breast carcinomas. Subsequent analysis revealed that loss of human *PTPRJ* occurs early in colorectal cancer development, supporting its role as a colon cancer susceptibility modifier (Ruivenkamp et al., 2003). Thus, despite the lack for formal proof, many cancer modifiers are being cloned and validated.

The technique of BAC-mediated recombineering will be a powerful tool to greatly enhance validation studies and formally proving the identity of cancer modifiers. BAC recombineering is based upon homologous recombination in bacterial cells and is performed by using short homology arms to alter a BAC clone in bacterial cells specially engineered for this purpose (Copeland et al., 2001; Testa et al., 2003). This technology is now well established, with many reagents available to quickly make any desired alteration in BAC clones from a variety of strains before reintroducing into embryonic stem (ES) cells or to make transgenic mice from specific genetic backgrounds.

SOFTWARE RESOURCES

Irrespective of the experimental approach taken to map and localize cancer modifier genes, a significant statistical analysis effort will be required. Most single-marker analyses can be directly implemented using commercially or publicly available statistics software packages. However, for the vast majority of modifier studies where there are multiple genes with complex interactions, it is inefficient and difficult to use simple statistical programs; the statistical methods for modifier mapping are usually complicated and require specialized software packages. Although many specialized computer programs using different genetic and statistical models for various experimental crosses have been developed, most are based on single-marker analysis using standard interval mapping (Lander and Botstein, 1989), regression interval mapping (Haley and Knott, 1992), and composite interval mapping (Jansen, 1993; Zeng, 1993, 1994). All packages can usually handle commonly used mapping populations, such as backcrosses or intercrosses, and have implemented most of the methods listed above. For the same data set, all programs provide similar results. The differences among the software packages are data format, computer platforms, user interface, graphic output, and population types. Four public domain software packages that are widely used by biologists are described (Table 15.2). Details about other modifier mapping methods and supporting software can be found at http://linkage.rockefeller.edu/soft and http://mapmanager.org/qtsoftware.html.

Table 15.2. Comparison of Software Packages Commonly Used for Cancer Modifier Mapping

Software	URL	Functions	Populations	Platforms	Interface
Mapmaker/QTL	http://hpcio.cit.nih.gov/lserver/MAPMAKER_QTL.html	IM, multiple QTL	F2, BC, RI	Unix/Mac/Windows	Command line
QTL Cartographer	http://statgen.ncsu.edu/qtlcart/index.php	Single-marker IM, CIM, MIM, multiple-trait IM, permutation, bootstrap	F2, BC, RI	Unix/Windows	Graphic (Windows)
Map Manager QT/QTX	http://mapmanager.org/mmQTX.html	Single-marker IM, CIM, epistasis, permutation, bootstrap,	F2, BC, RI, RIX, AI	Mac/Windows	Graphic
R/qtl	http://www.biostat.jhsph.edu/~kbroman/qtl/	Single-marker IM, CIM, epistasis, nonparametric, permutation	F2, BC, four-way cross	Unix/Mac/Windows	Command line

Abbreviations: IM, interval mapping; CIM, composite interval mapping; MIM, multiple interval mapping; BC, backcross; RI, recombinant inbred; RIX, recombinant inbred intercross; AI, advanced intercross.

Mapmaker/QTL

One of the first available mapping programs was Mapmaker/QTL (Lincoln et al., 1992b). It is interactive, runs on most platforms, and uses standard interval mapping approaches. However, it uses a command-driven user interface and provides no graphic user interface, although output graphs can be saved as postscript files. A companion program, Mapmaker/EXP, is used to create genetic maps from marker recombination frequencies (Lincoln et al., 1992a).

QTL Cartographer

Although originally developed for implementing composite interval mapping, QTL Cartographer has been extended to multiple QTL mapping with epistasis and multiple-trait analysis (Basten et al., 1994). Similar to Mapmaker/QTL, the original software was command driven and not user friendly. However, a new Windows-compatible interface for QTL Cartographer, WinQtlCart, has become available providing a graphic–user interface. The graphing tool for presenting mapping results offers many output options.

Map Manager QT/QTX

Probably the most widely used software package for experimental mouse crosses is Map Manager QT/QTX (Manly and Olson, 1999; Manly et al., 2001). This program allows detection and localization of modifiers through single-locus association, interval mapping, and composite interval mapping. Furthermore, it has capabilities to search for interacting loci. One of the unique features of Map Manager QT/QTX is its ability to support advanced backcrosses and intercrosses as well as RI and RIX designs. Map Manager QT/QTX is the most user friendly with a well-developed graphic–user interface. A new web-accessible version, WebQTL (http://www.webqtl.org), also provides curated genotypes and an extensive database for correlations with previously measured phenotypes and mRNA transcript levels from several tissues for commonly used RI lines (Chesler et al., 2003; Manly et al., 2001).

R/qtl

One drawback of widely used software packages is that they are fairly restrictive in the implementation of specific analysis methods because they concentrate on particular computations instead of model building. Consequently, they usually cannot handle data with complex experimental designs. In practice, these programs are used by first generating data that are adjusted for the effects of nongenetic factors through regression analysis and then the adjusted values are used as input data for further modifier mapping. This two-stage analysis may not be effective in many situations where the underlying genetic model is complex. Directly incorporating both genetic and nongenetic factors through improved modeling, as performed by the new mapping software R/qtl (Broman et al., 2003), is generally more powerful. R/qtl is implemented as an add-on package for the freely available and widely used statistical language/software R that has powerful graphical capabilities (Ihaka and Gentleman, 1996). R/qtl can perform single-modifier and two-dimensional, two-modifier genome scans with the possible inclusion of covariates. R/qtl is the only software package that offers a choice of analysis methods: the maximum-likelihood method used by Mapmaker/QTL and QTL Cartographer, the regression method used by Map Manager QT/QTX, and a new multiple-imputation method (Sen and Churchill, 2001). Additionally, R/qtl is also the only program that can incorporate nonparametric approaches for data analysis. Since most modifier mapping algorithms assume that the underlying data have a normal distribution, parametric statistical analyses are used. However, for some cancer phenotypes like tumor multiplicity, the trait values typically do not have a normal distribution, being skewed toward one end of the spectrum. Analysis of this type of cancer phenotype requires nonparametric statistics (Kruglyak and Lander, 1995).

All modifier mapping software packages require a known linkage map containing the markers used for analysis. Mapmaker/QTL, MapManager QT, and R/qtl have their own companion packages to construct linkage maps using the provided genotyping data and to check for genotyping errors before modifier analysis. QTL Cartographer can directly incorporate the linkage results from Mapmaker/EXP into the analysis.

The software packages described above, except for Mapmaker/QTL, calculate empirical thresholds using a permutation procedure as described earlier under Detection and Localization (Churchill and Doerge, 1994) or other methods using approximations (Piepho, 2001). They also estimate the confidence intervals for modifier loci by bootstrap analysis, as described in the same section (Visscher et al., 1996).

APPLICATIONS TO HUMAN CANCER

Individual differences in cancer susceptibility, progression, and response to therapy exist for many types of human cancers. Epidemiologic and family-based studies have shown that these phenotypic differences are due to complex genetic and nongenetic interactions (reviewed in Mucci et al., 2001). Unfortunately, the genes that contribute to variation in cancer phenotypes, and particularly those involved with complex gene and environmental interactions, are virtually unknown. Despite this dearth of knowledge, there is optimism that with new genomic technologies and genome sequences that have been recently developed, combined with human–mouse comparative genomics, major advances in the dissection cancer susceptibility will be made in the coming years (Korstanje and Paigen, 2002).

As described in previous chapters, the high-penetrance cancer genes, contributing to familial forms of cancer development and progression, are highly similar between humans and mice. Likewise, there is every reason to predict that cancer modifiers will be equally conserved between the two species. Early successes in a wide range of biologic systems and diseases support this assertion (Korstanje and Paigen, 2002). Based upon these early observations and the historical difficulty in identifying cancer modifiers in humans, the mouse will continue to be the preeminent model to identify candidate cancer modifier genes that can be tested in appropriate human populations.

A seminal analysis of breast cancer susceptibility exemplifies the importance and difficulties of characterizing cancer modifiers in humans (Balmain et al., 2003). This study investigated the potential effects of breast cancer modifiers in human populations and, by analyzing different genetic models, predicted that, collectively, modifiers will account for a substantial proportion of cancer incidence but may individually have very little effect. A major difficulty encountered was that the human studies could not distinguish rare variant/common disease models from common variant/common disease models. However, the situation in mice is very different since most commonly used mouse crosses are between two inbred strains and are, by definition, obligate common variants/common disease models. Thus the phenotypic variation observed in mouse crosses is due to different combinations of common alleles, many of which have only weak or modest effects.

Consequently, the mouse offers significant advantages over human studies. For one, experiments using mice have better control of environmental variables. Likewise, the defined genetics of mouse strains provides more powerful statistics to detect genetic linkages with cancer phenotypes, primarily due to the reduced genetic heterogeneity that is present in randomly breeding human populations, causing extensive mixing of genetic variation.

Thus, one of the primary objectives of cancer modifier analysis in the mouse is to provide candidates that can be used in human population studies, particularly to identify individuals with heightened risk or as prognostic indicators of cancer recurrence. Other important contributions to human cancer studies are testing specific gene–environmental interactions detected in epidemiologic studies and the elucidation of complex combinatorial interactions that create individuals with extreme phenotypes caused by unique allelic combinations. Furthermore, even though individual cancer modifiers may only confer a modest increase in cancer risk, these genes will also be an important new class of potential therapeutic targets for cancer prevention or treatment.

Already, a large number of cancer modifiers have been localized in the mouse genome, most before the availability of the new genome tools and sequences (Fig. 15.4). From these early studies, it is clear that cancer-associated genes are not randomly positioned across the mammalian genome. Rather, there are clusters of cancer modifiers. As such, a major goal of future studies will be to determine whether there are different classes of modifiers, with some being cancer-type specific and others, currently mapped to clusters in the mouse genome, having pleiotrophic effects on a broad range of cancer types.

Despite the strong effects of particular genes on mouse cancer development, not all of them will be equally responsible for similar cancers in humans. For example, the *Mom1* modifier of cancer multiplicity in the Apc^{Min} mouse model of intestinal cancer does not appear to have polymorphisms associated with intestinal cancer in humans (Kahlenberg et al., 1998). However, expression differences of *Pla2g2a*, one of the components of the *Mom1* locus, are associated with prolonged survival and less frequent metastasis in gastric adenocarcinoma (Leung et al., 2002). Thus, even though the link with specific types of cancer is not conserved, the utility of cancer modifiers identified in mice is not diminished.

Future endeavors will be to develop systems approaches to understanding variation in cancer phenotypes. These holistic models should be much more powerful for accurately describing how genetic and environmental variation is translated into altered cancer susceptibility, progression, and response to therapy. Undoubtedly, the mouse will continue to be the major model organism to develop these approaches that can then be translated into a human clinic setting.

GLOSSARY

Additive Effect The component of the total phenotypic trait value variance that can be attributed to a single allele or gene independently of any other alleles or genes.

Advanced Intercross A line, produced by continued random intercrossing for many generations (F2, F3, F4, etc.), that accumulates increased numbers of recombination sites.

AI *See* Advanced Intercross.

Allele An alternative form of a gene.

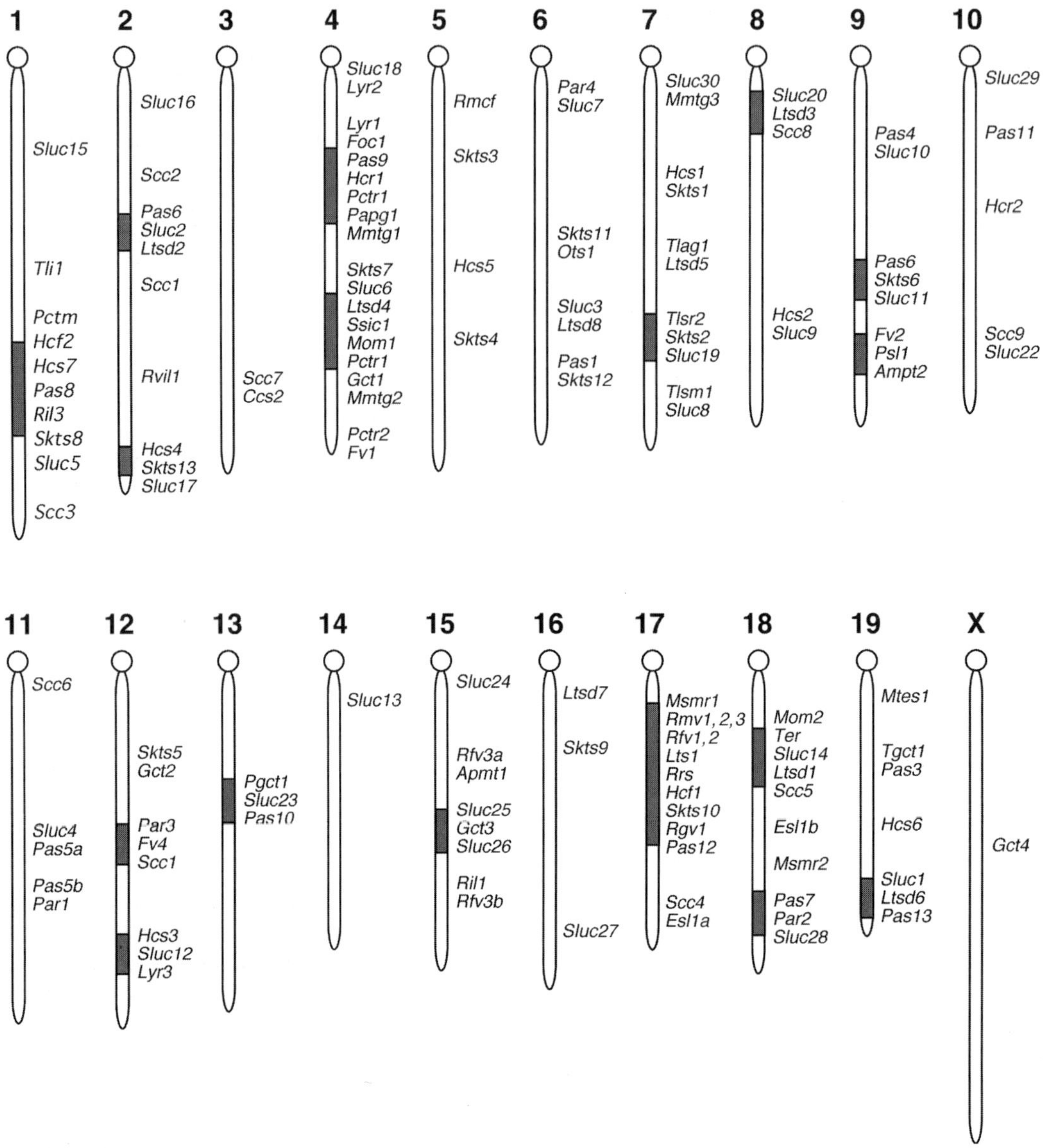

Figure 15.4. Map representing approximate locations of cancer modifiers that have been mapped in the mouse genome. Grey areas of chromosomes show clustering of cancer modifiers. Modifier gene designations: *Apmt*, mammary tumor latency (Cozma et al., 2002); *Esl*, B-cell lymphoma susceptibility (Hiai, 1996); *Foc*, B-cell lymphoma susceptibility (Hiai, 1996); *Fv*, Friend leukemia virus susceptibility (Miyazawa et al., 1992); *Gct*, germ cell tumor susceptibility (Matin et al., 1999); *Hcr*, hepatocarcinogen resistance (Manenti et al., 1994); *Hcs*, hepatocarcinogen susceptibility (Manenti et al., 1994); *Lts*, lung tumor susceptibility (Fijneman et al., 1995); *Ltsd*, lung tumor shape (Tripodis and Demant, 2003); *Lyr*, lymphoma resistance (Okumoto et al., 1990); *Mmtg*, mammary tumor growth (Le Voyer et al., 2001); *Mom*, modifier of multiple intestinal neoplasia (Dietrich et al., 1993; Silverman et al., 2002); *Msmr*, pre-B lymphoma resistance (Pataer et al., 1996); *Mtes*, mammary tumor metastatic progression (Hunter et al., 2001); *Ots*, susceptibility to ovarian teratomas (Lee et al., 1997); *Papg*, pulmonary adenoma progression (Zhang et al., 2002); *Par*, pulmonary adenoma resistance (Obata et al., 1996); *Pas*, pulmonary adenoma susceptibility (Malkinson et al., 1985; Wang et al., 2003); *Pctm*, susceptibility to plasmacytoma (Mock et al., 1993); *Pctr*, plasmacytoma resistance (Mock et al., 1997); *Pgct*, germ cell tumor susceptibility (Muller et al., 2000); *Psl*, susceptibility to skin cancer promotion (Angel et al., 1997); *Rfv*, recovery from Friend leukemia virus (Miyazawa et al., 1992); *Rgv*, susceptibility to Gross leukemia virus (Lilly, 1966); *Ril*, susceptibility to leukemia (Meruelo et al., 1981); *Rmcf*, lymphoma susceptibility (Gilbert et al., 1993); *Rmv*, resistance to Moloney leukemia virus (Debre et al., 1979); *Rrs*, susceptibility to Rous sarcoma virus tumors (Whitmore et al., 1978); *Rvil*, susceptibility to leukemia (Meruelo et al., 1983); *Scc*, susceptibility to colorectal cancer (van Wezel et al., 1999); *Sluc*, susceptibility to lung cancer (Tripodis et al., 2001); *Skts*, skin tumor susceptibility (Nagase et al., 2001); *Ssic*, susceptibility to small intestinal cancer (Fijneman and Demant, 1995); *Ter*, testicular teratoma susceptibility (Asada et al., 1994); *Tgct*, testicular germ cell tumor susceptibility (Matin et al., 1999); *Tlag*, attenuation of T lymphoma (Angel et al., 1989); *Tli*, resistance to T lymphoma (Wielowieyski et al., 1999); *Tlsm*, T-cell lymphoma susceptibility (Hiai, 1996); *Tlsr*, suppression of thymic lymphoma (Santos et al., 1996). Adapted from Balmain and Nagase (1998).

Allele Substitution Effect The effect on a phenotypic trait value of exchanging one allele for another allele.

Association Study A study design to statistically evaluate the relationship between the existence of a particular allele with a phenotype.

BAC *See* Bacterial Artificial Chromosome.

Backcross A cross between an F1 hybrid and one of the parental strains.

Bacterial Artificial Chromosome A cloning vector that replicates as a single copy in bacteria and is used to clone and propagate large genomic DNA inserts up to 300,000 bp.

Bayesian Inference An approach that uses the observed data information to produce a revised posterior probability for some parameter of interest and then the new posterior probability is used as the basis for statistical inference.

Bootstrap Sampling A technique for simulating new data sets, created by resampling from a larger original data set. This procedure can be used to evaluate the confidence interval containing a modifier. It is performed by sampling random subsets of pre-determined size from the total data set followed by reanalysis of the sampled subset. This is repeatedly performed, and the frequency that a particular genomic region is detected, as defined by the markers used, provides a measure for the confidence interval containing the putative modifier.

Cancer Genome Anatomy Project A National Cancer Institute–sponsored program to characterize human and mouse cancers for alterations in gene expression, DNA content, and other abnormalities. The project can be accessed at http://cgap.nci.nih.gov.

Centimorgan A measure of genetic distance defined as 1% recombination.

CGAP *See* Cancer Genome Anatomy Project.

CGH *See* Comparative Genomic Hybridization.

Chromosome Substitution An inbred strain containing one entire chromosome from a donor strain on the genetic background of another recipient strain.

CIM *See* Composite Interval Mapping.

cM *See* Centimorgan.

Co-Isogenic Strains that differ only by a single mutation, polymorphism, or genetically engineered alteration. These only exist for de novo mutations occurring within an inbred strain or a genetically engineered alteration maintained on the original genetic background.

Comparative Genomic Hybridization A technique used to identify gains and losses of DNA in cancers from specific genomic regions. It is performed by labeling DNA from cancer and normal cells with two different fluorescent labels and then competitively hybridizing to either metaphase chromosome spreads or microarrays containing cDNAs or BACs. The latter is also known as array CGH. Gain or loss of DNA in cancer cells can be detected by measuring the relative intensities of the two fluorescent labels.

Composite Interval Mapping A variation of interval mapping where the variation at one locus is held constant by regression analysis while the data are reanalyzed. This permits the detection of loci influencing phenotypes that are independent of the first, often primary, locus.

Confidence Interval The size of the genetic interval around a modifier localization that has a predefined, often 95%, probability of actually containing the modifier locus.

Congenic A strain containing a segment of DNA from a donor strain on a recipient genetic background that approximates a co-isogenic strain. These are produced by 10 generations of backcrossing or by marker-assisted selection as described for Speed Congenics.

Conplastic A strain with mitochondria from a donor strain on the genetic background of a recipient strain.

Consomic *See* Chromosome Substitution.

Cross Mating between individuals for the purpose of generating genetic recombinations to map loci controlling phenotypes.

CS *See* Chromosome Substitution.

Dominant An allele whose effect on a phenotype when present in one (heterozygous) or two (homozygous) copies is indistinguishable.

Endo-Phenotype A phenotype that occurs in the biologic flow of information between the primary genetic sequence and the terminal phenotype under study. These are usually measurements of RNA, protein, or other macromolecules.

Epigenetic Effects of alleles or genes on particular phenotypes that are not due to differences in the primary DNA sequence and include changes such as DNA methylation or histone acetylation.

Epistasis The masking, or other nonadditive effect, of one allele or gene on another.

Extrinsic Factors influencing a phenotype that originate external of the organism, such as diet or environmental exposures.

F1 Hybrid progeny resulting from an outcross between two genetically distinct individuals.

F2 Progeny resulting from the intercross of two F1 hybrid individuals.

G A designation for generation number that is used in mixed background matings.

Gene A segment of DNA that defines the unit of heredity.

Genome The entire DNA content of a cell.

Genome Scan A procedure by which many individual offspring from a cross containing genetic recombinations between parental strains are genotyped to identify the parental origin of each region of their genome.

Genotype The set of alleles at one or more loci carried by an organism.

Haplotype A set of alleles or polymorphic variants that are physically linked on a DNA molecule and tend to be inherited together.

Hemizygous Having a single copy of a gene.

Heritability The proportion of the total trait value variance for any phenotype that can be attributed to genetic differences between individuals.

Heterogeneity Identical or similar phenotypes caused by different genetic mechanisms.

Heterogeneous Stock A breeding population of heterozygous individuals with defined parental inputs, typically from multiple strains, that is bred to maintain the allelic diversity present in the parental strains.

Heterozygous Having two different alleles at a particular gene, polymorphism, or mutation.

Homozygous Having two identical alleles at a particular gene, polymorphism, or mutation.

HS *See* Heterogeneous Stock.

Imputation Generation of nonexistent marker data that are compatible with the actual marker genotype data used in a mapping experiment.

Inbred A strain that is homozygous for all loci. These are produced by 20 generations of brother–sister matings.

Incross A cross between genetically identical individuals to maintain inbred lines.

Intercross A cross between two F1 hybrids. The resulting offspring are F2 individuals.

Interspecific Cross A cross between individuals of different but closely related species. Although the male offspring are usually sterile, the female are fertile, allowing backcrosses to either parental strain.

Inheritance The mechanism by which genetic differences are transmitted from parent to offspring.

Interval Mapping A procedure for genetically mapping loci controlling particular phenotypes that allows the localization of putative modifiers to be between markers used in the study. The analysis assumes that intermarker recombinations are randomly distributed.

Intrinsic Factors influencing a phenotype that originate from genetic differences between individuals.

Introgress The process of transferring a specific segment of DNA from a donor genetic background to a recipient background through genetic crosses.

Isogenic Individuals that have identical DNA sequences.

Knockin Genetic engineering of a specific gene or locus by homologous recombination in embryonic stem cells resulting in the incorporation of a different allele or gene into the target locus.

Knockout Genetic engineering to disrupt the function or activity of a specific gene by homologous recombination in embryonic stem cells.

Likelihood Ratio Statistic The ratio of two probabilities of the same event under the alternative and null hypotheses.

Linkage The shared inheritance of genes due to being physically linked on the same stretch of DNA. Since this is a genetic term, genes can be physically linked on the same piece of DNA, but if the recombination between them is greater than 50%, then they are not linked. Compare with Synteny.

Locus A genetic location harboring DNA sequence variants that have a detectable effect on a particular phenotype. Plural is loci.

LOD *See* Logarithm of the Odds.

Logarithm of the Odds A logarithmic statistical score for measuring the probability of linkage, or the odds versus nonlinkage, between loci or the association between an allele and a phenotype.

LOH *See* Loss of Heterozygosity.

Loss of Heterozygosity Loss of one allele from a cell leaving the cell hemizygous, containing only one copy of the alternative allele, or homozygous for the alternative allele.

LRS *See* Likelihood Ratio Statistic.

Mapping The process of identifying the genomic region containing a locus that influences a particular phenotype. Coarse mapping generally refers to a localization that is larger than 5–10 cM while fine mapping refers to less than 5 cM.

Marker A polymorphism used to follow the genetic segregation at a specific location of the genome.

Maximum-Likelihood Score A criterion for estimating parameters from observed data that maximize the likelihood function for the parameters. In modifier mapping, the modifier location is the locus corresponding to the highest maximum likelihood.

Mendelian A trait that segregates in a manner consistent with being influenced by a single locus with a high genotype–phenotype correlation.

Meta-Analysis A statistical method to combine multiple, disparate data sets to identify previously unknown similarities or associations between data.

Microsatellite A type of polymorphic marker that is due to differences in the number of simple repeats, such as a repeated dinucleotide.

MIM *See* Multiple Interval Mapping.

MMHCC *See* Mouse Models of Human Cancer Consortium.

Modifier A genetic locus that influences the manifestation of a particular phenotype. In the classical sense, modifiers have no discernable effect on the phenotype alone but alter the phenotype caused by variation at another locus. In practice, modifiers can be considered interchangeable with QTLs. A weak modifier will have only a small effect on the phenotype while a strong modifier will have a large effect.

Monogenic A phenotypic trait controlled by a single gene.

Mouse Models of Human Cancer Consortium A consortium of laboratories sponsored by the National Cancer Institute with the goal of developing better mouse models of human cancer and using these models for preclinical studies. Details available at http://emice.nci.nih.gov.

Mouse Phenome Project An international project, based at The Jackson Laboratory, to characterize baseline phenotypes for a reference set of inbred mouse strains. Details are at http://aretha.jax.org/pub-cgi/phenome/mpdcgi?rtn=docs/home.

MPP *See* Mouse Phenome Project.

Multiple Interval Mapping A procedure that uses the outputs from composite interval mapping to identify combinations of alleles that are associated with a particular phenotype.

Mutation A change in the primary DNA sequence that frequently has adverse consequences.

N2 The progeny resulting from the backcross of an F1 hybrid to one of the parental strains.

Normal Distribution A continuous distribution of phenotypic trait values that form a bell-shaped curve and is symmetric with values more concentrated in the middle than in the tails. The width of the curve is determined by the standard deviation of the values.

Outcross A cross between genetically distinct individuals resulting in F1 progeny.

Parametric Statistical methods that assume a defined probability distribution such as a normal distribution of the data. The results are only valid if the data are distributed according to prediction. Nonparametric statistics do not make any assumptions about the distribution of the data.

Penetrance The proportion of individuals with a specific genotype that manifest a particular phenotypic trait.

Permutation Test The process of generating alternative arrangements of given data that is generally used to estimate the null distribution of test statistics. For modifier analysis, this typically entails randomly associating phenotypes with possible genotypes repeatedly with reanalysis of the data each time.

Phenotype The physical or other measurable traits of an individual.

Pleiotrophic A condition when a gene has an effect on many different phenotypes.

Polygenic A phenotypic trait controlled by more than one gene.

Polymorphism Genetic differences within an interbreeding population.

Posterior Probability The probability of an association after making use of the information in the data.

Power The probability of correctly rejecting a null hypothesis when the alternative is true. In modifier analysis, this refers to the frequencies of correctly detecting a modifier when it exists.

QTL *See* Quantitative trait locus.

Quantitative Trait Locus A locus that has a quantitative effect on a particular phenotypic trait value.

RC *See* Recombinant Congenic.

Recessive An allele that has no effect on a particular phenotype unless present in two copies (homozygous).

Recombinant Congenic An inbred line that has approximately 12.5% of a donor genome randomly distributed on a recipient genetic background. These are produced by inbreeding progeny after backcrossing two generations to the recipient strain.

Recombinant Inbred An inbred line that has an approximately equal mixture of alleles derived from two parental inbred lines. These are produced by inbreeding F2 progeny.

Recombinant Inbred Backcross The N2 individuals produced by backcrossing RI lines to a parental inbred strain used to produce the RI or another inbred strain.

Recombinant Inbred Intercross The F1 hybrids produced by intercrossing two different parental RI strains.

Regression Analysis A procedure to find and assess a linear relationship between a response variable, often the values of a trait of interest and one or more predictor variables. Single-marker analysis is a regression analysis used to assess the relationship between a quantitative trait value and a genetic marker.

RI *See* Recombinant Inbred.

RIB *See* Recombinant Inbred Backcross.

RIX *See* Recombinant Inbred Intercross.

SDP *See* Strain Distribution Pattern.

Semidominant Alleles that have different effects, often intermediate, on a particular phenotype when present in one (heterozygous) versus two (homozygous) copies.

Significance The degree to which the result could be explained by chance and is often signified with a *P*-value.

Single-Nucleotide Polymorphism A polymorphic difference between individuals at a single base pair.

SNP *See* Single-Nucleotide Polymorphism.

Speed Congenics The procedure of using genotypes to identify individuals with reduced amounts of donor genome during the derivation of congenics with the goal of reducing the number of generations required to produce congenic lines. Also called *marker-assisted selection.*

Strain Distribution Pattern The distribution of parental genotypes for a specific marker across a panel of recombinant inbred strains.

Strength A relative measure of the significance of an association.

Susceptibility The predisposition of an individual to develop or manifest a particular phenotype.

Synteny The physical association of genes on the same chromosome. This does not imply linkage which is a genetic association.

Trait Value The quantitative measurement value of a phenotype.

Variance A measure used to characterize the dispersion of trait values in a given population and is calculated as the average of the squares of the deviations of the trait values from their mean.

Variation Differences in phenotypic values between individuals within a population.

ACKNOWLEDGMENTS

The authors would like to thank Joe Nadeau and Jiri Forejt for providing information on ongoing resource development, Eric Holland for helpful suggestions on this chapter, and Allan Balmain, Hiroki Nagase, and Jian-Hua Mao for providing a preliminary figure showing mapped mouse modifiers. The preparation of this chapter was supported by National Institutes of Health grants CA-079869 and CA-105417 to DWT.

REFERENCES

Angel JM, Beltran L, Minda K, Rupp T, DiGiovanni J (1997): Association of a murine chromosome 9 locus (Psl1) with susceptibility to mouse skin tumor promotion by 12-*O*-tetradecanoylphorbol-13-acetate. *Mol Carcinog* 20:162–167.

Angel JM, Morizot DC, Richie ER (1989): Genetics of *N*-methyl-*N*-nitrosourea induction of thymic lymphomas in AKR/J mice: Assignment of a susceptibility gene to mouse chromosome 7. *J Natl Cancer Inst* 81:1652–1655.

Asada Y, Varnum DS, Frankel WN, Nadeau JH (1994): A mutation in the Ter gene causing increased susceptibility to testicular teratomas maps to mouse chromosome 18. *Nat Genet* 6:363–368.

Avner P, Bruls T, Poras I, Eley L, Gas S, Ruiz P, Wiles MV, Sousa-Nunes R, Kettleborough R, Rana A, Morissette J, Bentley L, Goldsworthy M, Haynes A, Herbert E, Southam L, Lehrach H, Weissenbach J, Manenti G, Rodriguez-Tome P, Beddington R, Dunwoodie S, Cox RD (2001): A radiation hybrid transcript map of the mouse genome. *Nat Genet* 29:194–200.

Backlund MG, Trasti SL, Backlund DC, Cressman VL, Godfrey V, Koller Bh (2001): Impact of ionizing radiation and genetic background on mammary tumorigenesis in p53-deficient mice. *Cancer Res* 61:6577–6582.

Bailey D (1981). Recombinant inbred strains and bilineal congenic strains. In Fox J (Ed), *The Mouse in Biomedical Research.* Academic, New York, pp 223–239.

Balmain A, Gray J, Ponder B (2003): The genetics and genomics of cancer. *Nat Genet* 33:238–244.

Balmain A, Nagase H (1998): Cancer resistance genes in mice: Models for the study of tumour modifiers. *Trends Genet* 14:139–144.

Basten CJ, Weir BS, Zeng Z-B (1994). Zmap—a QTL cartographer. In Smith C, Gavora JS, Benkel B, Chesnais J, Fairfull W, Gibson JP, Kennedy BW, Burnside EB (Eds), *Proceedings of the 5th World Congress on Genetics Applied to Livestock Production: Computing Strategies and Software.* Organizing Committee, 5th World Congress on Genetics Applied to Livestock Production, Guelph, Ontario, Canada.

Beavis W (1998). QTL analyses: Power, precision, and accuracy. In Paterson AH (Ed), *Molecular Dissection of Complex Traits.* CRC, Boca Raton, FL, pp 145–162.

Belknap JK (1998): Effect of within-strain sample size on QTL detection and mapping using recombinant inbred mouse strains. *Behav Genet* 28:29–38.

Belknap JK, Mitchell SR, O'Toole LA, Helms ML, Crabbe JC (1996): Type I and type II error rates for quantitative trait loci (QTL) mapping studies using recombinant inbred mouse strains. *Behav Genet* 26:149–160.

Bolivar VJ, Cook MN, Flaherty L (2001): Mapping of quantitative trait loci with knockout/congenic strains. *Genome Res* 11:1549–1552.

Broman KW (2001): Review of statistical methods for QTL mapping in experimental crosses. *Lab Anim (NY)* 30:44–52.

Broman KW, Wu H, Sen S, Churchill GA (2003): R/qtl: An extensible QTL mapping environment. *Bioinformatics.*

Bult CJ, Krupke DM, Sundberg JP, Eppig JT (2000): Mouse tumor biology database (MTB): Enhancements and current status. *Nucleic Acids Res* 28:112–114.

Carlborg O, Andersson L (2002): Use of randomization testing to detect multiple epistatic QTLs. *Genet Res* 79:175–184.

Chesler E, Lu L, Shou S, Qu Y, Wang J, Hsu HC, Mountz JD, Manly KF, Threadgill DW, Williams RW (2003): Massively parallel analysis of mouse brain gene transcription regulation and transcription networks.

Churchill GA, Doerge RW (1994): Empirical threshold values for quantitative trait mapping. *Genetics* 138:963–971.

Complex Trait Consortium (CTC) (2003): The nature and identification of quantitative trait loci: a community's view. *Nat Genet Rev* 4:911–916.

Copeland NG, Jenkins NA, Court DL (2001): Recombineering: A powerful new tool for mouse functional genomics. *Nat Rev Genet* 2:769–779.

Cormier RT, Bilger A, Lillich AJ, Halberg RB, Hong KH, Gould KA, Borenstein N, Lander ES, Dove WF (2000): The Mom1AKR intestinal tumor resistance region consists of Pla2g2a and a locus distal to D4Mit64. *Oncogene* 19:3182–3192.

Cormier RT, Hong KH, Halberg RB, Hawkins TL, Richardson P, Mulherkar R, Dove WF, Lander ES (1997): Secretory phospholipase Pla2g2a confers resistance to intestinal tumorigenesis. *Nat Genet* 17:88–91.

Cozma D, Lukes L, Rouse J, Qiu TH, Liu ET, Hunter KW (2002): A bioinformatics-based strategy identifies c-Myc and Cdc25A as candidates for the Apmt mammary tumor latency modifiers. *Genome Res* 12:969–975.

Czarnomska A, Krysiak E, Piskorowska J, Sitarz M, Pysniak K, Pilcik T, Demant P (2003): Opposite effects of modifiers of the ApcMin mutation in intestine and mammary gland. *Cancer Res* 63:4533–4537.

Czene K, Lichtenstein P, Hemminki K (2002): Environmental and heritable causes of cancer among 9.6 million individuals in the Swedish Family-Cancer Database. *Int J Cancer* 99:260–266.

Darvasi A (1998): Experimental strategies for the genetic dissection of complex traits in animal models. *Nat Genet* 18:19–24.

Darvasi A, Soller M (1995): Advanced intercross lines, an experimental population for fine genetic mapping. *Genetics* 141:1199–1207.

Darvasi A, Soller M (1997): A simple method to calculate resolving power and confidence interval of QTL map location. *Behav Genet* 27:125–132.

Darvasi A, Weinreb A, Minke V, Weller JI, Soller M (1993): Detecting marker-QTL linkage and estimating QTL gene effect and map location using a saturated genetic map. *Genetics* 134:943–951.

Debre P, Gisselbrecht S, Pozo F, Levy JP (1979): Genetic control of sensitivity to Moloney leukemia virus in mice. II. Mapping of three resistant genes within the H-2 complex. *J Immunol* 123:1806–1812.

Dietrich WF, Lander ES, Smith JS, Moser AR, Gould KA, Luongo C, Borenstein N, Dove W (1993): Genetic identification of *Mom-1*, a major modifier locus affecting *Min*-induced intestinal neoplasia in the mouse. *Cell* 75:631–639.

Dietrich WF, Miller J, Steen R, Merchant MA, Damron-Boles D, Husain Z, Dredge R, Daly MJ, Ingalls KA, O-Connor TJ, Evans CA, DeAngelis MM, Levinson DM, Kruglyak L, Goodman N, Copeland NG, Jenkins NA, Hawkins TL, Stein L, Page DC, Lander ES (1996): A comprehensive genetic map of the mouse genome. *Nature* 380:149–151.

Doerge RW, Churchill GA (1996): Permutation tests for multiple loci affecting a quantitative character. *Genetics* 142:285–294.

Donehower LA, Harvey M, Slagle BL, McArthur MJ, Montgomery CAJ, Butel JS, Bradley A (1992): Mice deficient for p53 are developmentally normal but susceptible to spontaneous tumors. *Nature* 356:215–221.

Donehower LA, Harvey M, Vogel H, McArthur MJ, Montgomery CAJ, Park SH, Thompson T, Ford RJ, Bradley A (1995): Effects of genetic background on tumorigenesis in p53-deficient mice. *Mol Carcinog* 14:16–22.

Dupuis J, Siegmund D (1999): Statistical methods for mapping quantitative trait loci from a dense set of markers. *Genetics* 151:373–386.

Ewart-Toland A, Briassouli P, De Koning JP, Mao JH, Yuan J, Chan F, MacCarthy-Morrogh L, Ponder BA, Nagase H, Burn J, Ball S, Almeida M, Linardopoulos S, Balmain A (2003): Identification of Stk6/STK15 as a candidate low-penetrance tumor-susceptibility gene in mouse and human. *Nat Genet* 34:403–412.

Fijneman RJ, Demant P (1995): A gene for susceptibility to small intestinal cancer, Ssic1, maps to the distal part of mouse chromosome 4. *Cancer Res* 55:3179–3182.

Fijneman RJ, de Vries SS, Jansen RC, Demant P (1996): Complex interactions of new quantitative trait loci, Sluc1, Sluc2, Sluc3, and Sluc4, that influence the susceptibility to lung cancer in the mouse. *Nat Genet* 14:465–467.

Fijneman RJ, Jansen RC, van der Valk MA, Demant P (1998a): High frequency of interactions between lung cancer susceptibility genes in the mouse: Mapping of Sluc5 to Sluc14. *Cancer Res* 58:4794–4798.

Fijneman RJ, Oomen LC, Snoek M, Demant P (1995): A susceptibility gene for alveolar lung tumors in the mouse maps between Hsp70.3 and G7 within the H2 complex. *Immunogenetics* 41:106–109.

Fijneman RJ, van der Valk MA, Demant P (1998b): Genetics of quantitative and qualitative aspects of lung tumorigenesis in the mouse: Multiple interacting susceptibility to lung cancer (Sluc) genes with large effects. *Exp Lung Res* 24:419–436.

Fortin A, Diez E, Rochefort D, Laroche L, Malo D, Rouleau GA, Gros P, Skamene E (2001): Recombinant congenic strains derived from A/J and C57BL/6J: A tool for genetic dissection of complex traits. *Genomics* 74:21–35.

Gariboldi M, Manenti G, Canzian F, Falvella FS, Radice MT, Pierotti MA, Della Porta G, Binelli G, Dragani TA (1993): A major susceptibility locus to murine lung carcinogenesis maps on chromosome 6. *Nat Genet* 3:132–136.

Gariboldi M, Spinola M, Milani S, Pignatiello C, Kadota K, Bono H, Hayashizaki Y, Dragani TA, Okazaki Y (2003): Gene expression profile of normal lungs predicts genetic predisposition to lung cancer in mice. *Carcinogen* 24:1819–1826.

Gilbert DJ, Neumann PE, Taylor BA, Jenkins NA, Copeland NG (1993): Susceptibility of AKXD recombinant inbred mouse strains to lymphomas. *J Virol* 67:2083–2090.

Glazier AM, Nadeau JH, Aitman TJ (2002): Finding genes that underlie complex traits. *Science* 298:2345–2349.

Gould KA, Dietrich WF, Borenstein N, Lander ES, Dove WF (1996): Mom1 is a semi-dominant modifier of intestinal adenoma size and multiplicity in Min/+ mice. *Genetics* 144:1769–1776.

Graff RJ, Hildemann WH, Snell GD (1966): Histocompatibility genes of mice. VI. Allografts in mice congenic at various non-H-2 histocompatibility loci. *Transplantation* 4:425–437.

Groden J, Thliveris A, Samowitz W, Carlson M, Gelbert L, Albertson H, Joslyn G, Stevens J, Spirio L, Robertson M, Sargeant L, Krapcho K, Wolff E, Burt R, Hughes JP, Warrington J, McPherson J, Wasmuth J, LePaslier D, Abderrahim H, Cohen D, Leppert M, White R (1991): Identification and characterization of the familial adenomatous polyposis coli gene. *Cell* 66:589–600.

Groot P, Moen C, Hart A, Snoek M, Demant P (1996). Recombinant congenic strains—genetic composition. In Brown S (Ed), *Genetic Variations and Strains of the Laboratory Mouse*. Oxford University Press, New York, pp 1660–1670.

Guenet JL, Bonhomme F (2003): Wild mice: An ever-increasing contribution to a popular mammalian model. *Trends Genet* 19:24–31.

Haley CS, Knott SA (1992): A simple regression method for mapping quantitative trait loci in line crosses using flanking markers. *Heredity* 69:315–324.

Hiai H (1996): Genetic predisposition to lymphomas in mice. *Pathol Int* 46:707–718.

Hitzeman R, Demarest K, Koyner J, Cipp L, Patel N, Rasmussen E, McCaughran J (2000): Effect of genetic cross on the detection of quantitative trait loci and a novel approach to mapping QTLs. *Pharmacol Biochem Behav* 67:767–772.

Hudson TJ, Church DM, Greenwaway S, Nguyen H, Cook A, Steen RG, Van Etten WJ, Castle AB, Strivens MA, Trickett P, Heuston C, Davison C, Southwell A, Hardisty R, Varela-Carver A, Haynes AR, Rodriguez-Tome P, Doi H, Ko MS, Pontius J, Schriml L, Wagner L, Maglott D, Brown SD, Lander ES, Schuler G, Denny P (2001): A radiation hybrid map of mouse genes. *Nat Genet* 29:201–205.

Hunter KW, Broman KW, Voyer TL, Lukes L, Cozma D, Debies MT, Rouse J, Welch DR (2001): Predisposition to efficient mammary tumor metastatic progression is linked to the breast cancer metastasis suppressor gene Brms1. *Cancer Res* 61:8866–8872.

Hunter KW, Williams RW (2002): Complexities of cancer research: Mouse genetic models. *Ilar J* 43:80–88.

Iakoubova OA, Olsson CL, Dains KM, Ross DA, Andalibi A, Lau K, Choi J, Kalcheva I, Cunanan M, Louie J, Nimon V, Machrus M, Bentley LG, Beauheim C, Silvey S, Cavalcoli J, Lusis AJ, West DB (2001): Genome-tagged mice (GTM): Two sets of genome-wide congenic strains. *Genomics* 74:89–104.

Ihaka R, Gentleman R (1996): R: A language for data analysis and graphics. *J Comput Graphical Statist* 5:299–314.

Inazuka M, Wenz HM, Sakabe M, Tahira T, Hayashi K (1997): A streamlined mutation detection system: Multicolor post-PCR fluorescence labeling and single-strand conformational polymorphism analysis by capillary electrophoresis. *Genome Res* 7:1094–1103.

Jansen RC (1993): Interval mapping of multiple quantitative trait loci. *Genetics* 135:205–211.

Kahlenberg MS, Stoler DL, Petrelli NJ, Rodriguez-Bigas M, Weber TK, Anderson GR (1998): Absence of frequent involvement of modifier of Min (APC) in sporadic colorectal cancer. *Ann Surg Oncol* 5:181–185.

Kallioniemi A, Kallioniemi OP, Sudar D, Rutovitz D, Gray JW, Waldman F, Pinkel D (1992): Comparative genomic hybridization for molecular cytogenetic analysis of solid tumors. *Science* 258:818–821.

Kao CH, Zheng ZB, Teasdale RD (1999): Multiple interval mapping for quantitative trait loci. *Genetics* 152:1203–1216.

Kinzler KW, Nilbert MC, Su LK, Vogelstein B, Bryan TM, Levy DB, Smith KJ, Preisinger AC, Hedge P, McKechnie D, et al. (1991): Identification of FAP locus genes from chromosome 5q21. *Science* 253:661–665.

Korstanje R, Paigen B (2002): From QTL to gene: The harvest begins. *Nat Genet* 31:235–236.

Kruglyak L, Lander ES (1995): A nonparametirc approach for mapping quantitative trait loci. *Genetics* 139:1421–1428.

Lander E, Kruglyak L (1995): Genetic dissection of complex traits: Guidelines for interpreting and reporting linkage results. *Nat Genet* 11:241–247.

Lander ES, Botstein D (1989): Mapping Mendelian factors underlying quantitative traits using RFLP linkage maps. *Genetics* 121:185–199.

Lee GH, Bugni JM, Obata M, Nishimori H, Ogawa K, Drinkwater NR (1997): Genetic dissection of susceptibility to murine ovarian teratomas that originate from parthenogenetic oocytes. *Cancer Res* 57:590–593.

Leung SY, Chen X, Chu KM, Yuen ST, Mathy J, Ji J, Chan AS, Li R, Law S, Troyanskaya OG, Tu IP, Wong J, So S, Botstein D, Brown PO (2002): Phospholipase A2 group IIA expression in gastric adenocarcinoma is associated with prolonged survival and less frequent metastasis. *Proc Natl Acad Sci USA* 99:16,203–16,208.

Le Voyer T, Lu Z, Babb J, Lifsted T, Williams M, Hunter K (2000): An epistatic interaction controls the latency of a transgene-induced mammary tumor. *Mamm Genome* 11:883–889.

Le Voyer T, Rouse J, Lu Z, Lifsted T, Williams M, Hunter KW (2001): Three loci modify growth of a transgene-induced mammary tumor: Suppression of proliferation associated with decreased microvessel density. *Genomics* 74:253–261.

Lifsted T, Le Voyer T, Williams M, Muller W, Klein-Szanto A, Buetow KH, Hunter KW (1998): Identification of inbred mouse strains harboring genetic modifiers of mammary tumor age of onset and metastatic progression. *Int J Cancer* 77:640–644.

Lilly F (1966): The inheritance of susceptibility to the Gross leukemia virus in mice. *Genetics* 53:529–539.

Lin L, Festing MF, Devereux TR, Crist KA, Christiansen SC, Wang Y, Yang A, Svenson K, Paigen B, Malkinson AM, You M (1998): Additional evidence that the K-ras protooncogene is a candidate for the major mouse pulmonary adenoma susceptibility (Pas-1) gene. *Exp Lung Res* 24:481–497.

Lincoln SE, Daly MJ, Lander ES (1992a): *Constructing Genetic Maps with MAPMAKER/EXP 3.0*, Technical report, Whitehead Institute Technical Report, Cambridge, MA.

Lincoln SE, Daly MJ, Lander ES (1992b): *Mapping Genes Controlling Quantitative Traits with MAPMAKER/QTL*, Technical report, Whitehead Institute Technical Report, Cambridge, MA.

Lynch M, Walsh B (1998): *Genetics and Analysis of Quantitative Traits*. Sinauer Associates, Sunderland, MA.

MacPhee M, Chepenik KP, Liddell RA, Nelson KK, Siracusa LD, Buchberg AM (1995): The secretory phospholipase A2 gene is a candidate for the *Mom1* locus, a major modifier of Apc^{Min}-induced intestinal neoplasia. *Cell* 81:957–966.

Malkinson AM, Nesbitt MN, Skamene E (1985): Susceptibility to urethan-induced pulmonary adenomas between A/J and C57BL/6J mice: Use of AXB and BXA recombinant inbred lines indicating a three-locus genetic model. *J Natl Cancer Inst* 75:971–974.

Manenti G, Binelli G, Gariboldi M, Canzian F, De Gregorio L, Falvella FS, Dragani TA, Pierotti MA (1994): Multiple loci affect genetic predisposition to hepatocarcinogenesis in mice. *Genomics* 23:118–124.

Mangin B, Goffinet B, Rebai A (1994): Constructing confidence intervals for QTL location. *Genetics* 138:1301–1308.

Manly KF, Cudmore Jr RH, Meer JM (2001): Map Manager QTX, cross-platform software for genetic mapping. *Mamm Genome* 12:930–932.

Manly KF, Olson JM (1999): Overview of QTL mapping software and introduction to Map Manager QT. *Mamm Genome* 10:327–334.

Markel P, Shu P, Ebeling C, Carlson GA, Nagle DL, Smutko JS, Moore KJ (1997): Theoretical and empirical issues for marker-assisted breeding of congenic mouse strains. *Nat Genet* 17:280–284.

Marshall JD, Mu JL, Cheah YC, Nesbitt MN, Frankel WN, Paigen B (1992): The AXB and BXA set of recombinant inbred mouse strains. *Mamm Genome* 3:669–680.

Matin A, Collin GB, Asada Y, Varnum D, Nadeau JH (1999): Susceptibility to testicular germ-cell tumors in a 129.MOLF-Chr 19 chromosome substitution strain. *Nat Genet* 23:237–240.

Meruelo D, Offer M, Flieger N (1981): Genetics of susceptibility for radiation-induced leukemia. Mapping of genes involved to chromosomes 1, 2, and 4, and implications for a viral etiology in the disease. *J Exp Med* 154:1201–1211.

Meruelo D, Offer M, Rossomando A (1983): Induction of leukemia by both fractional x-irradiation and radiation leukemia virus involves loci in the chromosome 2 segment H-30-A. *Proc Natl Acad Aci USA* 80:462–466.

Miki Y, Swensen J, Shattuck-Eidens D, Futreal PA, Harshman K, Tavtigian S, Liu Q, Cochran C, Bennett LM, Ding W, Bell R, Rosenthal J, Hussey C, Tran T, McClure M, Frye C, Hattier T, Phelps R, Haugen-Strano A, Katcher H, Yakumo K, Gholami Z, Shaffer D, Stone S, Bayer S, Wray C, Bogden R, Dayanath P, Ward J, Tonin P, Narod S, Bristow PK, Norris FH, Helvering L, Morrison P, Rosteck P, Lai M, Barrett JC, Lewis C, Neuhausen S, Cannon-Albright L, Goldgar D, Weisman R, Kamb A, Skolnick MH (1994): A strong candidate for the breast and ovarian cancer susceptibility gene BRCA1. *Science* 266:66–71.

Miyazawa M, Nishio J, Wehrly K, Jay G, Melvold RW, Chesebro B (1992): Detailed mapping of the Rfv-1 gene that influences spontaneous recovery from Friend retrovirus-induced leukaemia. *Eur J Immunogenet* 19:159–164.

Mock BA, Hartley J, Le Tissier P, Wax JS, Potter M (1997): The plasmacytoma resistance gene, Pctr2, delays the onset of tumorigenesis and resides in the telomeric region of chromosome 4. *Blood* 90:4092–4098.

Mock BA, Krall MM, Dosik JK (1993): Genetic mapping of tumor susceptibility genes involved in mouse plasmacytomagenesis. *Proc Natl Acad Aci USA* 90:9499–9503.

Moen CJA, Groot PC, Hart AAM, Snoek M, Demant P (1996): Fine mapping of colon tumor susceptibility (*Scc*) genes in the mouse, different from the genes known to be somatically mutated in colon cancer. *Proc Natl Acad Sci USA* 93:1082–1086.

Moen CJA, Snoek M, Hart AAM, Demant P (1992): *Scc-1*, a novel colon cancer susceptibility gene in the mouse: Linkage to *CD44* (*Ly-24, Pgp-1*) on chromosome 2. *Oncogene* 7:563–566.

Morel L, Blenman KR, Croker BP, Wakeland EK (2001): The major murine systemic lupus erythematosus susceptibility locus, Sle1, is a cluster of functionally related genes. *Proc Natl Acad Sci USA* 98:1787–1792.

Moser AR, Dove WF, Roth KA, Gordon JI (1992): The *Min* (multiple intestinal neoplasia) mutation: Its effect on gut epithelial cell differentiation and interaction with a modifier system. *J Cell Biol* 116:1517–1526.

Mott R, Flint J (2002): Simultaneous detection and fine mapping of quantitative trait loci in mice using heterogeneous stocks. *Genetics* 160:1609–1618.

Mott R, Talbot CJ, Turri MG, Collins AC, Flint J (2000): A method for fine mapping quantitative trait loci in outbred animal stocks. *Proc Natl Acad Aci USA* 97:12,649–12,654.

Mucci LA, Wedren S, Tamimi RM, Trichopoulos D, Adami HO (2001): The role of gene-environment interaction in the aetiology of human cancer: Examples from cancers of the large bowel, lung and breast. *J Intern Med* 249:477–493.

Muller AJ, Teresky AK, Levine AJ (2000): A male germ cell tumor-susceptibility-determining locus, pgct1, identified on murine chromosome 13. *Proc Natl Acad Aci USA* 97:8421–8426.

Nadeau JH, Singer JB, Matin A, Lander ES (2000): Analysing complex genetic traits with chromosome substitution strains. *Nat Genet* 24:221–225.

Nagase H, Bryson S, Cordell H, Kemp CJ, Fee F, Balmain A (1995): Distinct genetic loci control development of benign and malignant skin tumours in mice. *Nat Genet* 10:424–429.

Nagase H, Mao JH, Balmain A (1999): A subset of skin tumor modifier loci determines survival time of tumor-bearing mice. *Proc Natl Acad Sci USA* 96:15,032–15,037.

Nagase H, Mao JH, de Koning JP, Minami T, Balmain A (2001): Epistatic interactions between skin tumor modifier loci in interspecific (spretus/musculus) backcross mice. *Cancer Res* 61:1305–1308.

Nishisho I, Nakamura Y, Miyoshi Y, Miki Y, Ando H, Horii A, Koyama K, Utsunomiya J, Baba S, Hedge P (1991): Mutations of chromosome 5q21 genes in FAP and colorectal cancer patients. *Science* 253:665–669.

Obata M, Nishimori H, Ogawa K, Lee GH (1996): Identification of the Par2 (pulmonary adenoma resistance) locus on mouse chromosome 18, a major genetic determinant for lung carcinogen resistance in BALB/cByJ mice. *Oncogene* 13:1599–1604.

Ochiai Y, Tamura Y, Saito Y, Matsuki A, Wakabayashi Y, Aizawa Y, Niwa O, Kominami R (2003): Mapping of genetic modifiers of thymic lymphoma development in p53-knockout mice. *Oncogene* 22:1098–1102.

Okumoto M, Nishikawa R, Imai S, Hilgers J (1990): Genetic analysis of resistance to radiation lymphomagenesis with recombinant inbred strains of mice. *Cancer Res* 50:3848–3850.

Ooijen JWV (1992): Accuracy of mapping quantitative trait loci in autogenous species. *Theor Appl Genet* 84:803–811.

Paigen K, Eppig JT (2000): A mouse phenome project. *Mamm Genome* 11:715–717.

Park YG, Clifford R, Buetow KH, Hunter KW (2003): Multiple cross and inbred strain haplotype mapping of complex-trait candidate genes. *Genome Res* 13:118–121.

Park YG, Lukes L, Yang H, Debies MT, Samant RS, Welch DR, Lee M, Hunter KW (2002): Comparative sequence analysis in eight inbred strains of the metastasis modifier QTL candidate gene Brms1. *Mamm Genome* 13:289–292.

Pataer A, Kamoto T, Lu LM, Yamada Y, Hiai H (1996): Two dominant host resistance genes to pre-B lymphoma in wild-derived inbred mouse strain MSM/Ms. *Cancer Res* 56:3716–3720.

Pataer A, Nishimura M, Kamoto T, Ichioka K, Sato M, Hiai H (1997): Genetic resistance to urethan-induced pulmonary adenomas in SMXA recombinant inbred mouse strains. *Cancer Res* 57:2904–2908.

Piepho H-P (2001): A quick method for computing approximate thresholds for quantitative trait loci detection. *Genetics* 157:425–432.

Pinkel D, Segraves R, Sudar D, Clark S, Poole I, Kowbel D, Collins C, Kuo W-L, Chen C, Zhai Y, Dairkee SH, Ljung B-M, Grey JW, Albertson DG (1998): High resolution analysis of DNA copy number variation using comparative genomic hybridization to microarrays. *Nat Genet* 20:207–211.

Pollack JR, Perou CM, Alizadeh AA, Eisen MB, Pergamenschikov A, Williams CF, Jeffrey SS, Botstein D, Brown PO (1999): Genome-wide analysis of DNA copy-number changes using cDNA microarrays. *Nat Genet* 23:41–46.

Ranade K, Chang MS, Ting CT, Pei D, Hsiao CF, Olivier M, Pesich R, Hebert J, Chen YD, Dzau VJ, Curb D, Olshen R, Risch N, Cox DR, Botstein D (2001): High-throughput genotyping with single nucleotide polymorphisms. *Genome Res* 11:1262–1268.

Ross P, Hall L, Smirnov I, Haff L (1998): High level multiplex genotyping by MALDI-TOF mass spectrometry. *Nat Biotechnol* 16:1347–1351.

Ruivenkamp C, Hermsen M, Postma C, Klous A, Baak J, Meijer G, Demant P (2003): LOH of PTPRJ occurs early in colorectal cancer and is associated with chromosomal loss of 18q12-21. *Oncogene* 22:3472–3474.

Ruivenkamp CA, van Wezel T, Zanon C, Stassen AP, Vlcek C, Csikos T, Klous AM, Tripodis N, Perrakis A, Boerrigter L, Groot PC, Lindeman J, Mooi WJ, Meijjer GA, Scholten G, Dauwerse H, Paces V, van Zandwijk N, van Ommen GJ, Demant P (2002): Ptprj is a candidate for the mouse colon-cancer susceptibility locus Scc1 and is frequently deleted in human cancers. *Nat Genet* 31:295–300.

Santos J, Perez de Castro I, Herranz M, Pellicer A, Fernandez-Piqueras J (1996): Allelic losses on chromosome 4 suggest the existence of a candidate tumor suppressor gene region of about 0.6 cM in gamma-radiation-induced mouse primary thymic lymphomas. *Oncogene* 12:669–676.

Satagopan JM, Yandell BS, Newton MA, Osborn TC (1996): A Bayesian approach to detect quantitative trait loci using Markov chain Monte Carlo. *Genetics* 144:805–816.

Schindewolf C, Lobenwein K, Trinczek K, Gomolka M, Soewarto D, Fella C, Pargent W, Singh N, Jung T, Hrabe de Angelis M (2000): Comet assay as a tool to screen for mouse models with inherited radiation sensitivity. *Mamm Genome* 11:552–554.

Sen S, Churchill GA (2001): A statistical framework for quantitative trait mapping. *Genetics* 159:371–387.

Shima N, Hartford SA, Duffy T, Wilson LA, Schimenti KJ, Schimenti JC (2003): Phenotype-based identification of mouse chromosome instability mutants. *Genetics* 163:1031–1040.

Sillanpää MM, Arjas E (1998): Bayesian mapping of multiple quantitative trait loci from incomplete inbred line cross data. *Genetics* 148:1373–1388.

Silverman KA, Koratkar R, Siracusa LD, Buchberg AM (2002): Identification of the modifier of Min 2 (Mom2) locus, a new mutation that influences Apc-induced intestinal neoplasia. *Genome Res* 12:88–97.

Talbot CJ, Nicod A, Cherny SS, Fulker DW, Collins AC, Flint J (1999): High-resolution mapping of quantitative trait loci in outbred mice. *Nat Genet* 21:305–308.

Taylor B (1996). Recombinant inbred strains. In Brown S (Ed), *Genetic Variations and Strains of the Laboratory Mouse*. Oxford University Press, New York, pp 1597–1659.

Testa G, Zhang Y, Vintersten K, Benes V, Pijnappel WW, Chambers I, Smith AJ, Smith AG, Stewart AF (2003): Engineering the mouse genome with bacterial artificial chromosomes to create multipurpose alleles. *Nat Biotechnol* 21:443–447.

Threadgill DW, Hunter KW, Williams RW (2002): Genetic dissection of complex and quantitative traits: from fantasy to reality via a community effort. *Mamm Genome* 13:175–8.

Tripodis N, Demant P (2001): Genetic linkage of nuclear morphology of mouse lung tumors to the Kras-2 locus. *Exp Lung Res* 27:185–196.

Tripodis N, Demant P (2003): Genetic analysis of three-dimensional shape of mouse lung tumors reveals eight lung tumor shape-determining (Ltsd) loci that are associated with tumor heterogeneity and symmetry. *Cancer Res* 63:125–131.

Tripodis N, Hart AA, Fijneman RJ, Demant P (2001): Complexity of lung cancer modifiers: Mapping of thirty genes and twenty-five interactions in half of the mouse genome. *J Natl Cancer Inst* 93:1484–1491.

van Wezel T, Ruivenkamp CA, Stassen AP, Moen CJ, Demant P (1999): Four new colon cancer susceptibility loci, Scc6 to Scc9 in the mouse. *Cancer Res* 59:4216–4218.

van Wezel T, Stassen AP, Moen CJ, Hart AA, van der Valk MA, Demant P (1996): Gene interaction and single gene effects in colon tumour susceptibility in mice. *Nat Genet* 14:468–470.

Visscher PM, Thompson R, Haley CS (1996): Confidence intervals in QTL mapping by bootstrapping. *Genetics* 143:1013–1020.

Wade CM, Kulbokas EJ, 3rd, Kirby AW, Zody MC, Mullikin JC, Lander ES, Lindblad-Toh K, Daly MJ (2002): The mosaic structure of variation in the laboratory mouse genome. *Nature* 420:574–578.

Walling GA, Visscher PM, Haley CS (1998): A comparison of bootstrap methods to construct confidence intervals in QTL mapping. *Genet Res* 71:171–180.

Wang M, Lemon WJ, Liu G, Wang Y, Iraqi FA, Malkinson AM, You M (2003): Fine mapping and identification of candidate pulmonary adenoma susceptibility 1 genes using advanced intercross lines. *Cancer Res* 63:3317–3324.

Waterston RH, Lindblad-Toh K, Birney E, Rogers J, Abril JF, Agarwal P, Agarwala R, Ainscough R, Alexandersson M, An P, Antonarakis SE, Attwood J, Baertsch R, Bailey J, Barlow K, Beck S, Berry E, Birren B, Bloom T, Bork P, Botcherby M, Bray N, Brent MR, Brown DG, Brown SD, Bult C, Burton J, Butler J, Campbell RD, Carninci P, Cawley S, Chiaromonte F, Chinwalla AT, Church DM, Clamp M, Clee C, Collins FS, Cook LL, Copley RR, Coulson A, Couronne O, Cuff J, Curwen V, Cutts T, Daly M, David R, Davies J, Delehaunty KD, Deri J, Dermitzakis ET, Dewey C, Dickens NJ, Diekhans M, Dodge S, Dubchak I, Dunn DM, Eddy SR, Elnitski L, Emes RD, Eswara P, Eyras E, Felsenfeld A, Fewell GA, Flicek P, Foley K, Frankel WN, Fulton LA, Fulton RS, Furey TS, Gage D, Gibbs RA, Glusman G, Gnerre S, Goldman N, Goodstadt L, Grafham D, Graves TA, Green ED, Gregory S, Guigo R, Guyer M, Hardison RC, Haussler D, Hayashizaki Y, Hillier LW, Hinrichs A, Hlavina W, Holzer T, Hsu F, Hua A, Hubbard T, Hunt A, Jackson I, Jaffe DB, Johnson LS, Jones M, Jones TA, Joy A, Kamal M, Karlsson EK, Karolchik D, Kasprzyk A, Kawai J, Keibler E, Kells C, Kent WJ, Kirby A, Kolbe DL, Korf I, Kucherlapati RS, Kulbokas EJ, Kulp D, Landers T, Leger JP, Leonard S, Letunic I, Levine R, Li J, Li M, Lloyd C, Lucas S, Ma B, Maglott DR, Mardis ER, Matthews L, Mauceli E, Mayer JH, McCarthy M, McCombie WR, McLaren S, McLay K, McPherson JD, Meldrim J, Meredith B, Mesirov JP, Miller W, Miner TL, Mongin E, Montgomery KT, Morgan M, Mott R, Mullikin JC, Muzny DM, Nash WE, Nelson JO, Nhan MN, Nicol R, Ning Z, Nusbaum C, O'Connor MJ, Okazaki Y, Oliver K, Overton-Larty E, Pachter L, Parra G, Pepin KH, Peterson J, Pevzner P, Plumb R, Pohl CS, Poliakov A, Ponce TC, Ponting CP, Potter S, Quail M, Reymond A, Roe BA, Roskin KM, Rubin EM, Rust AG, Santos R, Sapojnikov V, Schultz B, Schultz J, Schwartz MS, Schwartz S, Scott C, Seaman S, Searle S, Sharpe T, Sheridan A, Shownkeen R, Sims S, Singer JB, Slater G, Smit A, Smith DR, Spencer B, Stabenau A, Stange-Thomann N, Sugnet C, Suyama M, Tesler G, Thompson J, Torrents D, Trevaskis E, Tromp J, Ucla C, Ureta-Vidal A, Vinson JP, Von Niederhausern AC, Wade CM, Wall M, Weber RJ, Weiss RB, Wendl MC, West AP, Wetterstrand K, Wheeler R, Whelan S, Wierzbowski J, Willey D, Williams S, Wilson RK, Winter E, Worley KC, Wyman D, Yang S, Yang SP, Zdobnov EM, Zody MC, Lander ES (2002): Initial sequencing and comparative analysis of the mouse genome. *Nature* 420:520–562.

Whitmore AC, Babcock GF, Haughton G (1978): Genetic control of susceptibility of mice to Rous sarcoma virus tumorigenesis. II. Segregation analysis of strain A.SW-associated resistance to primary tumor induction. *J Immunol* 121:213–220.

Wielowieyski A, Brennan LA, Jongstra J (1999): Tli1, a resistant locus for carcinogen-induced T-lymphoma. *Mamm Genome* 10:623–627.

Williams RW, Gu J, Qi S, Lu L (2001): The genetic structure of recombinant inbred mice: High-resolution consensus maps for complex trait analysis. *Genome Biol. 2 RESEARCH0046.*

Wiltshire T, Pletcher MT, Batalov S, Barnes SW, Tarantino LM, Cooke MP, Wu H, Smylie K, Santrosyan A, Copeland NG, Jenkins NA, Kalush F, Mural RJ, Glynne RJ, Kay SA, Adams MD, Fletcher CF (2003): Genome-wide single-nucleotide polymorphism analysis defines haplotype patterns in mouse. *Proc Natl Acad Sci USA* 100:3380–3385.

Zeng Z-B (1993): Theoretical basis for separation of multiple linked gene effects in mapping quantitative trait loci. *Proc Natl Acad Sci USA* 90:10,972–10,976.

Zeng Z-B (1994): Precision mapping of quantitative trait loci. *Genetics* 136:1457–1468.

Zhang S, Qian X, Redman C, Bliskovski V, Ramsay ES, Lowy DR, Mock BA (2003): p16 INK4a gene promoter variation and differential binding of a repressor, the ras-responsive zinc-finger transcription factor, RREB. *Oncogene* 22:2285–2295.

Zhang S, Ramsay ES, Mock BA (1998): Cdkn2a, the cyclin-dependnet kinase inhibitor encoding p16INK4a and p19ARF, is a candidate for the plasmacytoma susceptibility locus, Pctr1. *Proc Natl Acad Aci USA* 95:2429–2434.

Zhang Z, Wang Y, Herzog CR, Liu G, Lee HW, DePinho RA, You M (2002): A strong candidate gene for the Papg1 locus on mouse chromosome 4 affecting lung tumor progression. *Oncogene* 21:5960–5966.

16

MODELING BLOOD VESSEL FORMATION IN THE MOUSE

ROBERT BENEZRA

Department of Cancer Biology and Genetics, Memorial Sloan-Kettering Cancer Center, New York, New York

SHAHIN RAFII

Department of Hematology-Oncology and Genetic Medicine, Cornell University Medical College, New York, New York

DAVID LYDEN

Department of Pediatrics and the Children's Blood Foundation Laboratories, Cornell University Medical College, New York, New York

Much of our current understanding of the development and maintenance of blood vessels in mammals comes from the analysis of mice. We summarize our current understanding of the developmental, cell, and molecular biology of blood vessel formation both in the mouse embryo and postnatally. The reader is provided with an understanding of how this information may impact on our ability to control blood vessel formation for the management of human cancers.

INTRODUCTION

Much of our current understanding of the development and maintenance of blood vessels in mammals comes from the analysis of mice. Observations during normal murine development coupled with the analysis of effects of genetic manipulation on murine blood vessel formation are beginning to provide a clear picture of the signaling events that orchestrate the establishment of the complex blood vessel network essential for life. In this chapter we will summarize our current understanding of the developmental, cell, and molecular biology of blood vessel formation both in the mouse embryo and postnatally. We will also attempt to provide the reader with an understanding of how this information may impact on our ability to control blood vessel formation for the management of human cancers.

BLOOD VESSEL FORMATION DURING DEVELOPMENT

The development of blood vessels in the embryo can be grouped into two distinct phases: vasculogenesis and angiogenesis. During vasculogenesis, endothelial cells derived from mesoderm precursors migrate to the sites of blood vessel formation and differentiate into cells capable of forming lumenized vessels. This process generates the heart primordia, most of the large vessels of the animal, as well as a primary capillary plexus which surrounds the embryo and is incorporated into extraembryonic structures such as the yolk sac. In the second phase of blood vessel formation (angiogenesis), the endothelial cells found in the primary capillary network proliferate, migrate, and remodel to form both large and small vessels (see Fig. 16.1). While vasculogenesis was originally thought to take place exclusively in the embryo, recent advances on the identification of the origin of the endothelial cells that populate adult tissue during tumor formation suggest that recruitment of endothelial cells from the bone marrow

Mouse Models of Human Cancer, edited by Eric C. Holland
ISBN 0-471-44460-X

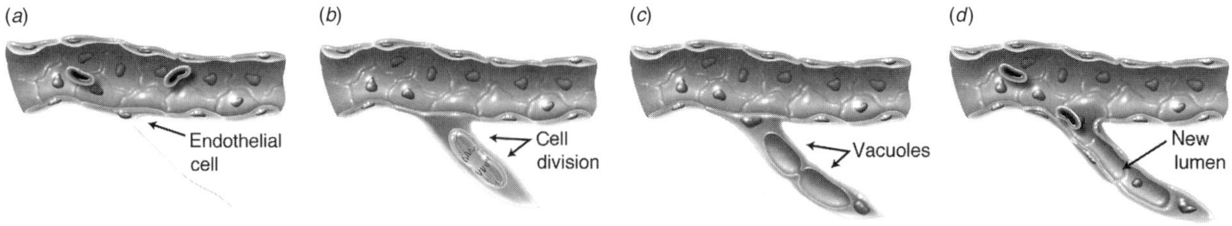

Figure 16.1. Process of angiogenesis. The branching and sprouting of new blood vessels from a preexisting network take place when an endothelial cell branches off from a small capillary. (*a*) First the endothelial cell sends off a pseudopodial process which invades the surrounding connective tissue. (*b*) The endothelial cell divides and then (*c*) the vacuoles form in adjacent cells. (*d*) The vacuoles fuse to form the luminal extension. (Adapted from B. Alberts, D. Bray, J. Lewis, M. Raff, K. Roberts, J.D. Watson, 1994. *Molecular Biology of the Cell*, Garland Publishing, New York.).

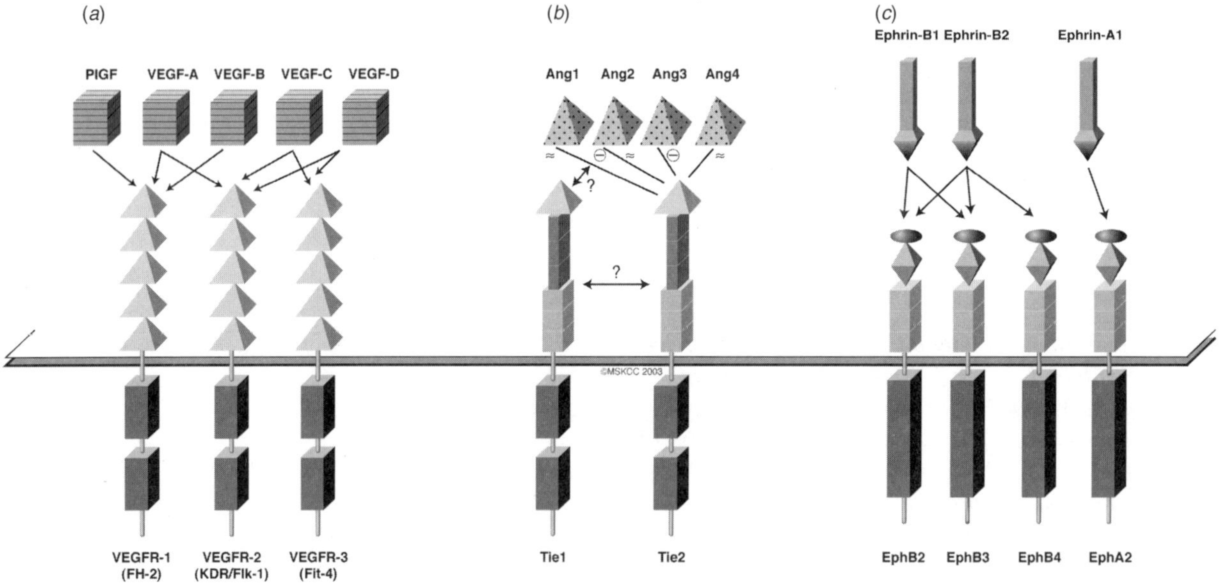

Figure 16.2. Receptor tyrosine kinases and their ligands. The three family of ligands and their receptors involved in blood vessel formation and maturation are depicted schematically. Note that the plus and minus signs below the angiopoietins refer to positive or negative stimuli for the Tie2 receptor. Only the ephrin ligands involved in blood vessel development are depicted.

may take place in a process resembling vasculogenesis (see discussion below).

Receptor Tyrosine Kinases and Their Ligands

Three tyrosine receptor kinase families and their ligands are known to control various aspects of vasculogenesis and angiogenesis in mammals. They are vascular endothelial growth factors (VEGF A–D) and its receptors VEGFR-1 and 2, the angiopoietins (Ang 1–4) and their receptors Tie1,2 receptors, and the ephrins (EphA1, EphB1, and EphB2) and their receptors (EphB2–4, EphA2) (see Fig. 16.2). Much of what we know about these receptor tyrosine kinases and their ligands are derived from mouse knockout studies and the results are summarized here. VEGF, the first identified growth factor specific for endothelial cells, was initially characterized for its ability to induce vascular permeability (Senger et al., 1990, 1983) and endothelial cell proliferation (Connolly et al., 1989) and is sometimes referred to as VEGF-A. As shown in Figure 16.1, VEGF family members have overlapping abilities to interact with the VEGF receptors on the surface of cells. VEGFR-2 [also called Flk1 or KDR (Quinn et al., 1993; Terman et al., 1991, 1992)] seems to mediate the growth and permeability activities of VEGF, whereas VEGFR-1 (de Vries et al., 1992) may act to "trap" VEGF or inhibit VEGFR-2 signaling, thereby acting as a negative regulator of the signaling pathway.

Mice engineered to lack VEGFR-2 do not develop a vasculature and have dramatically reduced numbers of endothelial cells consistent with its positive role in blood vessel development (Shalaby et al., 1995). On the other hand, mice deficient for VEGFR-1 have an excess of endothelial cells which form disorganized vessels (Fong et al., 1995). Interestingly, mice which only express a truncated form of VEGFR-1 are essentially normal, an observation which supports the decoy hypothesis or the idea that a negative regulatory role is independent of kinase signaling (Hiratsuka et al., 1998). Finally, VEGFR-3 and its ligand VEGF-C appear to be involved primarily with lymphatic development based on genetic manipulation studies in the mouse (for review see Olofsson et al., 1999).

The ligands for the VEGF receptors have also been shown to be critical for embryonic as well as postnatal angiogenesis. Loss of both copies of VEGF leads to a phenotype which resembles the VEGFR-2 knockout with a near-complete absence of a vasculature (Carmeliet et al., 1996; Ferrara et al., 1996). Even loss of one copy of the *VEGF* gene leads to severe vasculature abnormalities (Carmeliet et al., 1996; Ferrara et al., 1996), demonstrating the need for tight control over VEGF levels for normal vascular development. Conditional inactivation of *VEGF* using Cre-lox-mediated excision leads to death when performed several weeks after birth (Gerber et al., 1999a) but is much less traumatic when performed in adult animals. Affected structures in adult mice that have lost VEGF include the bone growth plates and ovarian corpus lutei, tissues which continue to undergo active remodeling in the adult (Ferrara et al., 1998; Gerber et al., 1999b). Whether other organs in humans require vascular remodeling is a critically important question to address as VEGF and VEGFR inhibitors move into clinical trial in the treatment of vascular diseases such as diabetic retinopathies and cancer. A relative of VEGF called PlGF (placental growth factor) seems to interact specifically with VEGFR-1 and is thought to be involved primarily with vascular remodeling and perhaps tumor angiogenesis (Carmeliet, 2000; Carmeliet et al., 2001; Persico et al., 1999).

The Tie receptors (1 and 2) and their ligands the angiopoietins (Ang1–4) comprise another signaling component required for normal vascular development and contain positively and negatively acting elements as in the VEGF pathway. Tie2 is the more well studied of the receptors and can engage both Ang1 and Ang2, with Ang1 providing a positive signaling stimulus and Ang2 a negative one (for review see Yancopoulos et al., 2000). Embryos lacking either Tie2 or Ang1 develop a normal appearing primary vascular network but fail to undergo normal remodeling in certain tissue types (Dumont et al., 1994; Sato et al., 1995; Suri et al., 1996). The most pronounced defect take place in the heart, in which the endothelial cells fail to associate properly with the underlying myocardial cells, and in large vessels, where the endothelial cells fail to associate properly with support cells. Overexpression of Ang1 in transgenic mice leads to an increase in vessel size as opposed to vessel number (Thurston et al., 1999), observed in VEGF overexpression experiments. Importantly these vessels are resistant to leak, again in stark contrast to VEGF overexpression experiments where the induced vessels are hemorrhagic. This tightening of vessels might be of therapeutic benefit in diseases in which vessel leakage or damage is a major problem, such as diabetic retinopathy or inflammation.

On the other hand, Ang2 seems to act as an antagonist of the Ang1/Tie2 signaling pathway, at least in certain cellular contexts. Overexpression of Ang in mouse endothelial cells results in embryonic lethality with defects resembling those of the Ang1 or Tie2 knockouts (Maisonpierre et al., 1997). Indeed, expression of Ang2 in the endothelium of vessels undergoing active remodeling or regression, such as those of the ovary or in certain tumors, is consistent with Ang2 being an antagonist of the Ang1/Tie2 signaling mechanism. Mice lacking Ang2 have defects in retinal vascular remodeling as well as disorganized lymphatic vessels (Gale et al., 2002). Interestingly, Ang1 can substitute for Ang2 to rescue the lymphatic defect, suggesting a possible positive role in this setting. It has been proposed that Ang2 destabilizes vessels in the absence of other factors but that a positive stimulus from VEGF may lead to a more "plastic" state in which vessels are more likely to undergo active remodeling, such as in the ovary or in the leading edge of certain tumor types.

The ephrin (Eph) receptor tyrosine kinases represent the third class of receptors important for blood vessel development and their ligands are referred to also as ephrins (see Figure 16.1). The ephrin ligands are unusual in that they need to be membrane bound in order to activate their respective receptors (Davis et al., 1994). These receptor–ligand pairs were originally thought to play a role primarily in the nervous system, but recent data demonstrate that the EphB4 receptor and the EphB2 ligand are important also for vascular development. Loss of either EphB2 or EphB4 results in a defect in angiogenic remodeling reminiscent of that observed in the Tie2 or Ang1 knockout animals (Adams et al., 1999; Wang et al., 1998). Expression analysis shows that the two are present in reciprocal distributions during vascular development with EphB2 expressed in primordial arterial vessels and EphB4 expressed in primordial venous vessels. These observations have suggested the possibility that these molecules are essential for arterial–venous identity and that they made be important for fusing the two vessel types together during development. The reexpression of EphB2 in tumors and during female reproductive angiogenesis (Shin et al., 2001) suggests that arterial angiogenesis, and not just venous angiogenesis, may be an import part of the adult angiogenic program, contrary to current dogma in the field.

Integrins

Integrins are cell surface receptors which are responsible for the attachment of cells to components of the extracellular matrix such a fibronectin, vitronectin, or collagen. The receptors are composed of an α and β subunit of which there are 18 members of the α family and 8 members of the β family and 24 known dimeric complexes. Some of the integrins on the surface of endothelial cells have been shown to be important for angiogenesis using mouse knockout technology. At the center of the analysis (and as we shall see current controversy) has been the α_v integrin subfamily. The α_v can partner with β_3, β_5, β_6 and β_8 as well as β_1 which itself partners with many other α subunits. Beta 1 knockout animals are embryonic lethal and indeed do have angiogenesis defects (Fassler and Meyer, 1995; Stephens et al., 1995). The α_v knockout animals do display vascular abnormalities in the forebrain (Bader et al., 1998), but this defect is likely due to the loss of $\alpha_v\beta_8$ (Zhu et al., 2002) (not detected on endothelial cells) but probably results from defective interactions between neurons and endothelial cells (McCarty et al., 2002). The problem arises in the analysis of the $\beta_{3,5}$ knockouts when compared to drug inhibition studies. Inhibition of $\alpha_v\beta_{3,5}$ by drug intervention was motivated by the appearance of these integrins on the surface of endothelial cell infiltrating tumors (Brooks et al., 1994). A monoclonal antibody targeting this integrin (called LM 609) and small molecules based on the RGD peptide motif recognized by these integrins were shown to be effective in inhibiting tumor angiogenesis (Brooks et al., 1994; Hammes et al., 1996). Therefore, the prediction from the mouse models in which β_3 and/or β_5 was disrupted would be an inhibition of tumor angiogenesis. To the contrary, the β_3 and β_5 double-knockout mice, which are viable and fertile, show enhanced tumor growth when challenged with tumor xenografts, suggesting that these integrins may in fact be antiangiogenic (Reynolds et al., 2002).

How can these two disparate observations be resolved? The "classical" model of integrin engagement maintains that monoclonal antibodies or RGD engagement of the integrins should activate the integrins and provide a survival or antiapoptotic signal through the activation of the phosphoinositol 3-kinase (PI3K)/Akt signaling cascade (for review see, Hynes, 2002). To accommodate the difference between the genetic analysis and the inhibitor studies, the "unligation" model has been proposed, which maintains that in the presence of antibodies or RGD molecules the integrins behave as if they are in an unliganded form, which leads to activation of the caspase pathway and apoptosis (Cheresh and Stupack, 2002). Two alternate possibilities also exist. It is possible that monoclonal antibodies or RGD peptides or mimetics displace positive stimuli such as vitronectin from the α_v integrins and allow for negative stimuli now to bind such as tumstatin (Maeshima et al., 2002). Finally, there is some evidence to suggest that RGD peptides themselves can directly activate the caspase apoptotic pathway independent of integrin engagement (Buckley et al., 1999). Further analysis using the mouse models described above will be required to distinguish between these possibilities and allow us to determine the most effective way to exploit the integrins for the design of antiangiogenic compounds.

Transcription Factors Controlling Blood Vessel Formation

A transcriptional response to lowered oxygen tension is required for the stimulation of new blood vessel growth in animals. This transcriptional response is mediated by the hypoxia-inducible transcription factor (HIF-1), which consists of a heterodimer between the aryl hydrocarbon receptor nuclear translocator (ARNT) and the hypoxia response factor HIF-1α (for review see Safran and Kaelin, 2003). The pathway is regulated by stabilization of the HIF-1α subunit when oxygen tension is low, leading in turn to the transcriptional activation of genes required for new blood vessel formation, such as VEGF (Mazure et al., 1996). Cell cycle withdrawal in response to low oxygen tension appears to be mediated primarily by the induction of the cyclin-dependent kinase inhibitors p21 and p27, leading to the hypophosphorylation of the Rb protein (Goda et al., 2003). Low oxygen levels prevent the activation of a prolyl hydoxylase which modifies a proline residue and targets the protein for degradation via the protein ubiquitination pathway (Ivan et al., 2001).

Genetic disruption of the HIF-1α subunit in mice has provided insights into the requirement for the oxygen-sensing pathway in mammals. Disruption of both copies of HIF-1α in embryonic stem cells leads to a dramatic reduction of solid tumor growth that is associated with a decrease in VEGF expression (Ryan et al., 1998). HIF-1α-null embryos exhibit clear morphologic differences by E8.0 and by E8.5 show reduced number of somites, abnormal neural fold formation, and increased hypoxia (Iyer et al., 1998; Ryan et al., 1998). Prior to embryonic death at E11.0 the HIF-1α embryos also display multiple cardiovascular defects, including abnormally dilated vascular structures in the cephalic region and an overall marked reduction in vascularity compared to wild-type control animals. These findings are remarkably similar to those observed in the Vegf−/− embryos and are consistent with the observation that Hif1a−/− embryonic stem (ES) cells and embryoid bodies have reduced Vegf expression. The overwhelming degree of cell death observed in the Hif1a−/− embryos most probably results from hypoxic stress due to the lack of adequate perfusion. In addition to preventing the formation of a sufficient vasculature, the loss of Hif1a also probably reduces the cell's ability to deal with hypoxic stress, compounding the loss of cell viability and ultimately the death of the embryo.

Another family of transcription factors shown to be critical for embryonic and postnatal angiogenesis are the Id proteins. These proteins act as dominant-negative antagonists of the helix–loop–helix transcription factors by virtue of the fact that Id proteins contain the helix–loop–helix dimerization motif but are missing the DNA-binding domain present on positively acting members of the family and thereby associate to form heteroligomeric complexes incapable of binding DNA (for review see Benezra et al., 2001). The main targets of the Id proteins are the ubiquitously expressed E proteins, which themselves heterodimerize with tissue-specific bHLH proteins (such as MyoD, NeuroD, etc.) to form high-affinity DNA-binding complexes. Thus by preventing the E proteins from binding DNA as homodimers (in B cells) or associating with other family members in other lineages, the Id proteins can inhibit the activity of bHLH transcription factors in a diverse array of cell types using the same biochemical mechanism. While there are four members of the Id family all of which are capable of binding and sequestering E proteins, only Id1 and Id3 have been demonstrated to be important in angiogenesis by genetic manipulations in mice (Lyden et al., 1999b). Mice with homozygous deletions of both Id1 and Id3 die at E13.5 with 100% penetrance, and the embryos were observed to have a characteristic hemorrhage in the ganglionic eminences of the forebrain. This phenotype is remarkably similar to the defect observed in avb3 integrin knockout animals (Bader et al., 1998), but the biologic significance of this is not yet clear since loss of av integrins does not change Id1 or Id3 expression levels in the blood vessels of the central nervous system (CNS) or vice versa. The confinement to the CNS of the angiogenic defect in the Id knockout animals might be explained by compensation by Id2 which is expressed in the vasculature outside the CNS exclusively, and indeed Id1–3 triple-knockout embryos have been observed to have more extensive hemorrhages throughout the embryo (R. Benezra, unpublished observation). How the loss of Id impacts on the activity or expression of other factors is first beginning to be explored.

But how does loss of Id affect tumor angiogenesis? Animals missing three out of four copies of the Id1,3 gene pair (i.e., Id1+/–, Id3–/– or Id1–/–, Id3+/–) are viable and fertile and when challenged with tumor xenografts show a dramatic inhibition of tumor angiogenesis (equivalent for the two genotypes) (Lyden et al., 1999a, 2001). This animal model has now been used to ask two very important follow-up questions: What is the origin of the endothelial cells that infiltrate tumors and do all tumors respond equally to antiangiogenic stress. Since the Id1–/–, Id3+/– animals fail to vascularize tumor xenografts, they could be used to test whether wild-type bone marrow (BM) is sufficient to reestablish the vasculature and allow tumor growth. After lethal irradiation, Id1–/–, Id3+/– animals were supplanted with wild-type BM derived from a ROSA 26 mouse so that the cells would be expressing β-galactosidase and therefore stain blue when treated with X-gal. These "chimeric" mice were then challenged with tumor xenografts. Indeed, these animals were now able to support the growth of the tumors and the blood vessels infiltrating these tumors were positive for β-galactosidase staining, indicating they were donor derived. This experiment demonstrates quite clearly that BM-derived precursor cells are sufficient for the formation of a tumor vasculature. To establish their necessity, the reverse experiment was performed in which BM from Id1–/–, Id3+/– animals was used to reconstitute a lethally irradiated wild-type mouse and these animals showed a dramatic delay in the onset of tumorigenesis. Thus the Id knockout model was used to determine that BM-derived precursor cells are necessary and sufficient for the vascularization and growth of tumors at least in the early phases of disease progression.

Importantly, the Id knockout animals, while displaying vascular defects in most tumor types examined, show a variable resistance to the growth of the tumor depending on tumor type (Lyden et al., 1999a). Lymphoma cells, for example, cannot grow beyond even a very limited size in the face of the Id vascular defect but Lewis lung carcinoma (LLC) cells grow rather extensively. LLC cells accomplish this by expanding a viable rim of cells around a centrally hypoxic and necrotic core. This type of "escape" mechanism has been observed in other settings (Yu et al., 2002) and suggests that not all tumor types are likely to respond equivalently to antiangiogenic stresses. Recently, dramatic synergy between loss of Id and a drug that can inhibit the growth of the viable rim of cells was reported in a transgenic model of breast cancer and may be a useful model for the extrapolation of predicted outcome in human clinical trials (de Candia et al., 2003).

POSTNATAL BLOOD VESSEL PROCESSES

Bone Marrow Contribution to Neoangiogenesis

Bone marrow–derived stem cells have been shown to play a critical role in the regulation of various postnatal processes, including wound healing, organ regeneration, and tumor growth as alluded to above. Tissue-specific stem cells reside in specific niches within the BM microenvironment where they are maintained in an undifferentiated and quiescent state (Cheng et al., 2000).

These niches are critical for regulating self-renewal and cell fate decisions as well as providing a dispensable source of stem cells for tissue vascularization and organogenesis (Krause et al., 2001)

During embryonic development, hematopoietic stem cells (HSCs) migrate from the yolk sac to the fetal liver and spleen and then settle within the BM. In common with all other hematopoietic cells, HSCs are motile, hence able to leave the marrow microenvironment in a multistage process called mobilization. Physiologic stress induces the release of angiogenic factors with chemotactic potential that promotes stem cell motility, thereby facilitating their entry into a permissive niche where they proliferate, differentiate, and are mobilized to the circulation (Gill et al., 2001; Hattori et al., 2001b; Peled et al., 1999; Takahashi et al., 1999). Mobilized stem cells either are incorporated into a specific organ or, as in the case of HSCs, expand in extramedullary sites and recirculate back to the marrow to replenish the stem cell pool (Lemischka et al., 1986).

Bone marrow–derived circulating endothelial progenitor cells (CEPs) have also been shown to be recruited from the marrow and to functionally contribute to neoangiogenesis during wound healing (Asahara et al., 1999a, b, 1997; Crisa et al., 1999), vascularization postmyocardial ischemia (Kocher et al., 2001), limb ischemia (Iwaguro et al., 2002; Kalka et al., 2000), endothelialization of vascular grafts (Bhattacharya et al., 2000; Shi et al., 1998), atherosclerosis (Sata et al., 2001), and retinal neovascularization (Grant et al., 2002; Otani et al., 2002). Growth of certain tumors is also dependent on the corecruitment of BM-derived CEPs as well as subsets of hematopoietic cells to the tumor vasculature. Release of angiogenic factors by tumor cells is essential to induce timely mobilization of CEPs and hematopoietic cells stimulating tumor angiogenesis and growth (Asahara et al., 1999b; Lyden et al., 2001; Reyes et al., 2002). Inhibition of the mobilization of the CEPs and hematopoietic cells results in retardation of the tumor growth, underscoring the significance of BM-derived cells in supporting tumor angiogenesis.

Collectively, these data suggest that selective expression of organ-specific chemokines promotes mobilization of BM-derived pluripotent cells, which is essential for tissue vascularization and organ regeneration. Recently several studies have partially elucidated the mechanism and the molecular mediators that orchestrate the mobilization of stem cells from BM to the respective organs. Release of angiogenic factors and hematopoietic cytokines seems to play a major role in the mobilization of endothelial progenitors and subsets of hematopoietic cells (Hattori et al., 2001a, b; Heissig et al., 2002). Shear stress may also activate molecular pathways that increase the response to chemokines and enhance mobilization of HSCs to the circulation (Grabovsky et al., 2000; Peled et al., 1999). These studies have laid the foundation for designing murine models where the mechanism for the mobilization of organ-specific stem cells can formally be studied. Identification of the molecular pathways that are essential for mobilization as well as proliferation and differentiation of vascular stem cells will open up new avenues to accelerate neoangiogenesis after vascular injury or ischemia.

Chemokine-induced activation of metalloproteinases is necessary for mobilization of stem and progenitor cells. The majority of quiescent stem cells reside in close cellular contact with stromal elements and the complex network of extracellular matrix, where they receive survival signal. Physiologic stress results in the release of stem cell active cytokines, such as Kit ligand, that promote rapid cycling of the stem cells, setting the stage for mobilization to the peripheral circulation. Remodeling of the extracellular matrix may be an essential step in the mobilization and recruitment of hematopoietic stem cells. We have shown that inhibition of matrix metalloproteases (MMPs), specifically MMP-9, influences the mobilization and release of mature hematopoietic cells, including platelets (Lane et al., 2000). Similarly, administration of a MMP inhibitor blocked chemokine-induced mobilization of 82±3% of cells with stem cell potential. Among the known MMPs, we have shown that gelatinase B (MMP-9) is produced by hematopoietic stem cells in response to chemokine activation and is required for migration through reconstituted extracellular matrix in vitro. To define the physiologic role of MMP-9 in vivo, we examined chemokine-induced mobilization in MMP-9-deficient mice. Mobilization of hematopoietic progenitors was partially blocked, whereas chemokine-driven mobilization of stem cells was significantly impaired, in MMP-9-deficient mice. MMP-9-deficient mice also displayed delayed hematopoietic reconstitution following treatment of the BM with the cell cycle–specific suppressive agent 5-fluorouracil (5-FU) (Rafii et al., 2002). In fact, 70% of mice treated with 5-FU succumbed to complications of prolonged marrow suppression.

Further investigation into a mechanism by which lack of MMP-9 resulted in a failure of hematopoietic reconstitution demonstrated that activation of MMP-9 was essential for the release of soluble Kit ligand (sKitL). Increase in the bioavailability of sKitL results in enhanced cycling of quiescent stem and progenitor cells, therefore physically translocating to a vascular niche that is conducive to stem cell proliferation, differentiation, and mobilization to the peripheral circulation. We have also demonstrated that angiogenic factors released as a result of vascular injury also activate MMP-9, resulting in the rapid increase in sKitL, thereby promoting recruitment of circulating endothelial progenitors (CEPs) and HSCs from marrow to the peripheral circulation (Heissig et al., 2002). These data suggest that angiogenic factor-mediated

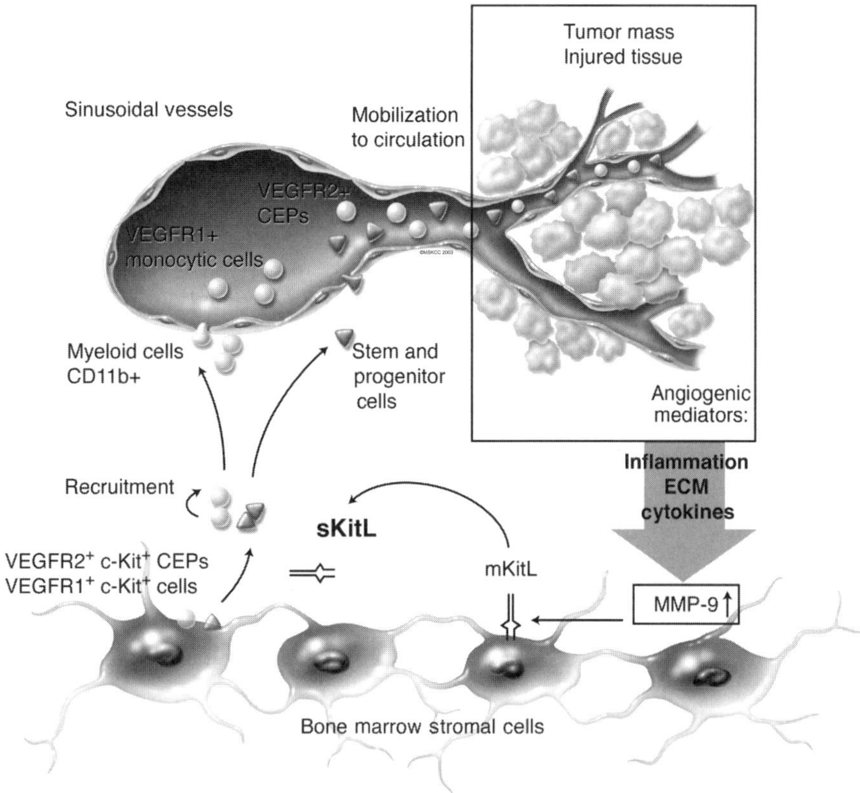

Figure 16.3. Summary of the events in postnatal angiogenesis. The essential events in the mobilization of VEGF-R1+ and VEGF-R2+ cells during postnatal angiogenesis are depicted schematically. MMP-9 provides an essential trigger by releasing soluble Kit ligand (sKitL), which facilitates the mobilization of the stem and progenitor cells into the periphery. These events are thought to be essential both for tumor neovascularization and in wound repair.

MMP-9 activation is the decisive checkpoint regulating mobilization of HSCs and CEPs (see Fig. 16.3).

Wound Healing

Wound repair is a dynamic and complex process consisting of hemostasis, inflammatory and cellular infiltration, and tissue regrowth and remodeling. Angiogenesis in response to tissue injury is highly regulated by signals from both serum and extracellular matrix. Initially, a complex series of timely events initiate the formation of a fibrin clot with activation of fibrin degradation products and extracellular matrix remodeling proteins, matrix metalloproteinases (MMPs), tissue inhibitors of metalloproteinases (TIMP), urokinase-type plasminogen activator (uPA), and (PAI-1), as well as the release of potent angiogenic cytokines such as VEGF-A, VEGF-C, fibroblast growth factor (FGF), platelet derived growth factor (PDGF), angiopoietins, and TGFβ. Cellular infiltrates consisting of inflammatory cells such as neutrophils, macrophages, and fibroblasts as well as proliferating endothelial cells and smooth-muscle cells establish a cell–matrix interaction (Coll 1, Coll 2, vitronectin, and integrins 3 and 5)

The migration of endothelial cells depends on the ECM components in both granulation tissue and endothelial basement membrane, establishing an interaction between the vascular cells and the wound provisional matrix and enabling the invasion of new capillaries into the established clot. The ECM provides scaffold support, signal roles, and a reservoir for growth factors with the upregulation of various collagens and laminins (8 and 10) that are highly expressed in blood vessels of skin wounds (Zhang et al., 2002). In the case of full-thickness cutaneous wounds, integrin $\alpha_v\beta_3$ is the vascular cell receptor expressed specifically on capillary sprouts for the wound-associated adhesive proteins (Clark et al., 1996).

The role of the thrombospondin (TSP) extracellular matrix proteins in tissue repair has remained controversial. Targeted overexpression of TSP-1 in the skin was associated with inhibition of fibroblast migration, diminished wound angiogenesis, and reduced granulation tissue following full-thickness skin wounds (Streit et al., 2000). Excisional wounds in mice with disruption of TSP-2 have abnormal collagen fibrils, and vascular density resulted in acceleration of wound healing, suggesting TSP-2 plays a crucial part in modulating fibroblast–matrix interactions in early wounds and regulating the extent of angiogenesis in late wounds (Kyriakides et al., 1999).

VEGF, through its cognate receptors VEGFR1 and VEGFR2, directly mediates wound vessel angiogenesis and optimizes healing in damaged tissues. For mouse models with impaired wound healing, VEGF is the major factor abnormally regulated during healing of full-thickness wounds in diabetic mice (Kirchner et al., 2003). Following brain injury, VEGF-A mRNA levels remain elevated for at least six days and play a significant role in blood–brain breakdown (BBB) and angiogenesis (Nag et al., 2002). VEGF-B, which is constitutively expressed in steady-state endothelium, increases later in proliferating endothelial cells and astrocytes at the lesion site. VEGF-B likely has a role in maintaining the BBB and stabilizes the repair processes during angiogenesis postinjury. Neuropilin-1, a coreceptor for VEGF, was abundantly expressed on neovessels in healing wounds, and mice treated with anti-neuropilin-1 antibodies exhibited a decrease in vascularity within these wounds, providing evidence that neuropilin may have a critical role in wound angiogenesis (Nag et al., 2002).

Along with VEGF, the angiopoietin factors are central to a normal and impaired wound-healing processes. For tissue regeneration during cutaneous wound repair in normal and diabetic mice, Ang1 and Tie2 are constitutively expressed with normal healing and absent in diabetes-impaired healing while Ang2 and Tie1 are strongly induced during repair with a prolonged expression in the diabetic mice (Kampfer et al., 2001). In an in vitro three-dimensional model of angiogenesis, hypoxia significantly increased platelet-derived growth factor (PDGF) and fibroblast growth factor (FGF) in sprout formation of mouse aortas (Humar et al., 2002). Platelets gather in the fibrin clot and release several growth factors, such as PDGF and TGFβ. VEGF binds to the fibrin clot and induces the proliferation of endothelial cells, smooth-muscle cells, and peripheral blood monocytes that produce interleukins IL-6 and IL-8 (Tezono et al., 2001). Neutrophils and fibroblasts debride the wound site of local contamination and are then engulfed by macrophages and induce apoptosis, preventing scarring. Macrophages secrete vimentin, which accelerates the homing and maturation of blood-derived monocytes into tissue macrophages and triggers bacterial killing and the generation of oxidative metabolites. The apoptotic fibroblasts release TSP1, which further stimulates the recruitment of macrophages (Moodley et al., 2003). Fibroblast assays cultured from diabetic mice exhibited reduction in migration and impairment in VEGF production and response to hypoxia, in contrast to normal fibroblasts during tissue repair (Lerman et al., 2003).

Subsets of hematopoietic cells such as hematopoietic progenitor cells (HPCs) and HSCs collaborate with BM-derived VEGFR2+ endothelial progenitor cells (EPCs) to support neoangiogenesis (Coussens et al., 2000; Rafii et al., 2002). Transplantation of a single green fluorescent protein (GFP)-expressing Sca1+cKit+lineage negative (Lin−) HSC into lethally irradiated mice reconstituted self-renewing HSCs with hemangioblast function (Grant et al., 2002). Following retinal ischemia by laser occlusion, donor-derived GFP+ hematopoietic and endothelial-derived cells were found in retinal neovessels. Mice with photocoagulated retinas were injected with Lin− HSC cells intravitreally and the Lin− cells were localized to the revascularized retinal vessels, confirming the functional incorporation of BM-derived progenitor cells into mature blood vessels. Transplanted monocytic cells are also capable of releasing angiogenic factors, such as VEGF, PDGF, and MMP-9, during the revascularization process that may facilitate the incorporation of EPCs.

Reproductive Angiogenesis

The endometrium is a dynamic tissue that undergoes cyclic degradation and renewal, which is heavily dependent on angiogenic pathways essential for endometrial repair. During menstruation, local factors such as hypoxia and ovarian steroids upregulate VEGF that can induce endometrial matrix metalloproteinases (MMPs) and tissue inhibitors of metalloproteinase (TIMPs) (Lamoreaux et al., 1998). MMP2 and MMP9 activities are thought to initiate the early phases of capillary formation, MMP1 promotes endothelial migration through the interstitial space, while TIMP1–4 and thrombospondin type 1 inhibit MMP2 and 9 and prevent neovascularization. Other local activated cells such as lymphomyeloid and mast cells influence angiogenesis by synthesis of cytokines and proteases such as MMPs, tumor necrosis factor α (TNFα), IL-1, histamine, heparin, tryptase, and chymase (Salamonsen et al., 2000).

Hallmarks of the implantation process requires increased vascular permeability and neovascularization of the blastocyst–uteran junction. During the attachment period of implantation, VEGF and its receptor Flk-1 mediate early phases of angiogenesis (Matsumoto et al., 2002).VEGF in conjunction with angiopoietins and Tie2 directs blood vessel formation during decidualization following implantation. Cyclooxygenase (COX-2), through its interaction with VEGF signaling, stabilizes vascular permeability and angiogenesis throughout implantation and decidualization.

Throughout gestation, blood vessel development varies considerably between the first trimester with robust proliferation compared to the last trimester with formation of terminal capillaries and a quiescent vascular bed. In the early placenta, VEGF is the predominant growth factor and correlates with the adherens junctional molecules, vascular endothelial cadherin and α- and β-catenin, while in the last trimester placenta, angiopoietin-1, and angiopoietin-2 are abundantly expressed and associate with tight junctional molecules occluding, claudin-1, and claudin-2 (Leach et al., 2002). Although placental growth factor (PlGF) is activated during placental development, gene inactivation demonstrates that the loss of PLGF does not affect reproductive angiogenesis, presumably by compensation of VEGF (De Falco et al., 2002).

Defining Markers and Targets for Tumor Angiogenesis

Since the pioneering work of Folkman and his collaborators which demonstrated that tumors cannot grow beyond a limited size in the absence of the recruitment of new blood vessels, a flurry of scientific activity has taken place to find agents that would specifically inhibit tumor vasculature since these agents should be useful anticancer drugs. This type of mad dash has often been seen in the annals of cancer therapy in the past (recall the promise of interferon and antisense strategies) and is often met with the so-called glitter-to-ashes syndrome: The public is lead to believe that antiangiogenic strategies will cure all cancers in a few years (i.e., glitter), a few clinical failures take place so the promise of the idea is gone (ashes), and many on the bandwagon give up and move onto the next fad. But as we shall see, now is the time scientists can do the real work to uncover the complexities of tumor angiogenesis, devise strategies to specifically inhibit it, and determine which tumors are likely to respond alone or in combination with other therapies.

If one wants to inhibit the formation of tumor vasculature, a reasonable way to go about it would be to define markers that are specifically expressed in tumor endothelium and then see if interfering with the activity of those markers can block tumor angiogenesis. Two such markers were described above (i.e., avb3 and the Id proteins) and work is progressing to exploit these markers for therapeutic intervention. A more comprehensive listing of tumor endothelial-specific markers was undertaken recently using a procedure called serial analysis of gene expression (or SAGE) (St Croix et al., 2000). In this method, RNA is isolated from purified endothelial cells present in tumor blood vessels and then after converting the RNA to double stranded DNA a small snippet of the sequence at the 3′ end is sequenced for the entire population of molecules. By comparing such libraries of sequences and comparing the frequency of occurrence of each of the different molecules (represented by the snippets), one can get a measure of relative expression levels of each of the molecules in different tissue types. Using serial analysis of gene expression (SAGE), Vogelstein and his colleagues (St. Croix et al., 2000) demonstrated that tumor endothelial cells express a set of 46 genes specifically expressed in the endothelial cells of tumors but not expressed in normal endothelium (TEMs, or tumor endothelial markers). Importantly, most of these genes were of unknown function and also expressed on endothelial cells in the granulation tissue of wounds, suggesting a common gene activation program in tumor angiogenesis and neovascularization after injury. The identification of these genes, their natural ligands (if any), and inhibition by small molecules or antibodies may lead to a new generation of compounds used in the management of human cancers.

In some cases effective inhibitors of tumor angiogenesis in certain animal models have been identified, and their utility in treating cancer is being evaluated using more and more sophisticated animal models. For example, angiostatin and endostatin were first shown to have efficacy in subcutaneous models in which human tumor cells are inoculated under the skin of immunocompromised severe combined immunodeficiency (SCID) mice and effects of drug therapies evaluated (Boehm et al., 1997). Angiostatin and endostatin both showed efficacy in inhibiting the blood flow to and growth of these types of tumors. When these agents were tested, however, in a more physiologically relevant tumor model in which T antigen overexpression is driven by the rat insulin promoter/enhancer in the pancreas (i.e., RIP-Tag), the effects were much less pronounced (Bergers et al., 1999). Tumors were modestly inhibited depending on the agent and time of administration, but complete eradication was not observed. Understanding why this trial was not completely successful will ultimately help us devise better treatment strategies. Other mouse models of human cancers in which oncogenes are overexpressed or tumor suppressor genes are deleted will also be of great utility. It has also been shown recently that colon carcinoma cells that are mutated for p53 and therefore resistant to the effects of hypoxic shock have an increased resistance to antiangiogenic stress (Yu et al., 2002), and such cells may therefore require another modality of treatment in order to inhibit their growth more effectively.

As described above, VEGF and its receptors also present attractive targets for antiangiogenic intervention, and indeed monoclonal antibodies and small-molecule inhibitors of these molecules are currently being tested for the ability to slow or reverse the growth of a variety of human cancers. As with most treatments in oncology, combinations of drugs will likely prove most effective, and targeting the tumor cell itself and the blood supply that feeds these cells is likely to be an important strategy in the management of human cancers.

ACKNOWLEDGMENTS

This work was supported by grants from the National Institutes of Health and Breast Cancer Research Foundation (R. B.), the National Cancer Institute and Children's Blood Foundation (D. L.), and the National Heart Lung and Blood Institute and American Cancer Society (S.R.).

REFERENCES

Adams RH, Wilkinson GA, Weiss C, Diella F, Gale NW, Deutsch U, Risau W, Klein R (1999): Roles of ephrinB ligands and EphB receptors in cardiovascular development: Demarcation of arterial/venous domains, vascular morphogenesis, and sprouting angiogenesis. *Genes Dev* 13:295–306.

Asahara T, Masuda H, Takahashi T, Kalka C, Pastore C, Silver M, Kearne M, Magner M, Isner JM (1999a): Bone marrow origin of endothelial progenitor cells responsible for postnatal vasculogenesis in physiological and pathological neovascularization. *Circ Res* 85:221–228.

Asahara T, Murohara T, Sullivan A, Silver M, van der Zee R, Li T, Witzenbichler B, Schatteman G, Isner JM (1997): Isolation of putative progenitor endothelial cells for angiogenesis. *Science* 275:964–967.

Asahara T, Takahashi T, Masuda H, Kalka C, Chen D, Iwaguro H, Inai Y, Silver M, Isner JM (1999b): VEGF contributes to postnatal neovascularization by mobilizing bone marrow-derived endothelial progenitor cells. *EMBO J* 18:3964–3972.

Bader BL, Rayburn H, Crowley D, Hynes RO (1998): Extensive vasculogenesis, angiogenesis, and organogenesis precede lethality in mice lacking all alpha v integrins. *Cell* 95:507–519.

Benezra R, Rafii S, Lyden D (2001): The Id proteins and angiogenesis. *Oncogene* 20:8334–8341.

Bergers G, Javaherian K, Lo KM, Folkman J, Hanahan D (1999): Effects of angiogenesis inhibitors on multistage carcinogenesis in mice. *Science* 284:808–812.

Bhattacharya V, McSweeney PA, Shi Q, Bruno B, Ishida A, Nash R, Storb RF, Sauvage LR, Hammond WP, Wu MH (2000): Enhanced endothelialization and microvessel formation in polyester grafts seeded with CD34(+) bone marrow cells. *Blood* 95:581–585.

Boehm T, Folkman J, Browder T, O'Reilly MS (1997): Antiangiogenic therapy of experimental cancer does not induce acquired drug resistance. *Nature* 390:404–407.

Brooks PC, Clark RA, Cheresh DA (1994): Requirement of vascular integrin alpha v beta 3 for angiogenesis. *Science* 264:569–571.

Buckley CD, Pilling D, Henriquez NV, Parsonage G, Threlfall K, Scheel-Toellner D, Simmons DL, Akbar AN, Lord JM, Salmon M (1999): RGD peptides induce apoptosis by direct caspase-3 activation. *Nature* 397:534–539.

Carmeliet P (2000): Mechanisms of angiogenesis and arteriogenesis. *Nat Med* 6:389–395.

Carmeliet P, Ferreira V, Breier G, Pollefeyt S, Kieckens L, Gertsenstein M, Fahrig M, Vandenhoeck A, Harpal K, Eberhardt C, Declercq C, Pawling J, Moons L, Collen D, Risau W, Nagy A (1996): Abnormal blood vessel development and lethality in embryos lacking a single VEGF allele. *Nature* 380:435–439.

Carmeliet P, Moons L, Luttun A, Vincenti V, Compernolle V, De Mol M, Wu Y, Bono F, Devy L, Beck H, Scholz D, Acker T, DiPalma T, Dewerchin M, Noel A, Stalmans I, Barra A, Blacher S, Vandendriessche T, Ponten A, Eriksson U, Plate KH, Foidart JM, Schaper W, Charnock-Jones DS, Hicklin DJ, Herbert JM, Collen D, Persico MG (2001): Synergism between vascular endothelial growth factor and placental growth factor contributes to angiogenesis and plasma extravasation in pathological conditions. *Nat Med* 7:575–583.

Cheng T, Rodrigues N, Shen H, Yang Y, Dombkowski D, Sykes M, Scadden DT (2000): Hematopoietic stem cell quiescence maintained by p21cip1/waf1. *Science* 287:1804–1808.

Cheresh DA, Stupack DG (2002): Integrin-mediated death: An explanation of the integrin-knockout phenotype? *Nat Med* 8:193–194.

Clark DE, Smith SK, Sharkey AM, Charnock-Jones DS (1996): Localization of VEGF and expression of its receptors flt and KDR in human placenta throughout pregnancy. *Hum Reprod* 11:1090–1098.

Connolly DT, Heuvelman DM, Nelson R, Olander JV, Eppley BL, Delfino JJ, Siegel NR, Leimgruber RM, Feder J (1989): Tumor vascular permeability factor stimulates endothelial cell growth and angiogenesis. *J Clin Invest* 84:1470–1478.

Coussens LM, Tinkle CL, Hanahan D, Werb Z (2000): MMP-9 supplied by bone marrow-derived cells contributes to skin carcinogenesis. *Cell* 103:481–490.

Crisa L, Cirulli V, Smith KA, Ellisman MH, Torbett BE, Salomon DR (1999): Human cord blood progenitors sustain thymic T-cell development and a novel form of angiogenesis. *Blood* 94:3928–3940.

Davis S, Gale NW, Aldrich TH, Maisonpierre PC, Lhotak V, Pawson T, Goldfarb M, Yancopoulos GD (1994): Ligands for EPH-related receptor tyrosine kinases that require membrane attachment or clustering for activity. *Science* 266:816–819.

de Candia P, Solit DB, Giri D, Brogi E, Siegel PM, Olshen AB, Muller WJ, Rosen N, Benezra R (2003): Angiogenesis impairment in Id-deficient mice cooperates with an Hsp90 inhibitor to completely suppress HER2/neu-dependent breast tumors *Proc Natl Acad Sci USA*, 100:12337–12342.

De Falco S, Gigante B, Persico MG (2002): Structure and function of placental growth factor. *Trends Cardiovasc Med* 12:241–246.

de Vries C, Escobedo JA, Ueno H, Houck K, Ferrara N, Williams LT (1992): The fms-like tyrosine kinase, a receptor for vascular endothelial growth factor. *Science* 255:989–991.

Dumont DJ, Gradwohl G, Fong GH, Puri MC, Gertsenstein M, Auerbach A, Breitman ML (1994): Dominant-negative and targeted null mutations in the endothelial receptor tyrosine kinase, tek, reveal a critical role in vasculogenesis of the embryo. *Genes Dev* 8:1897–1909.

Fassler R, Meyer M (1995): Consequences of lack of beta 1 integrin gene expression in mice. *Genes Dev* 9:1896–1908.

Ferrara N, Carver-Moore K, Chen H, Dowd M, Lu L, O'Shea KS, Powell-Braxton L, Hillan KJ, Moore MW (1996): Heterozygous embryonic lethality induced by targeted inactivation of the VEGF gene. *Nature* 380:439–442.

Ferrara N, Chen H, Davis-Smyth T, Gerber HP, Nguyen TN, Peers D, Chisholm V, Hillan KJ, Schwall RH (1998): Vascular endothelial growth factor is essential for corpus luteum angiogenesis. *Nat Med* 4:336–340.

Fong GH, Rossant J, Gertsenstein M, Breitman ML (1995): Role of the Flt-1 receptor tyrosine kinase in regulating the assembly of vascular endothelium. *Nature* 376:66–70.

Gale NW, Thurston G, Hackett SF, Renard R, Wang Q, McClain J, Martin C, Witte C, Witte MH, Jackson D, Suri C, Campochiaro PA, Wiegand SJ, Yancopoulos GD (2002): Angiopoietin-2 is required for postnatal angiogenesis and lymphatic patterning, and only the latter role is rescued by angiopoietin-1. *Dev Cell* 3:411–423.

Gerber HP, Hillan KJ, Ryan AM, Kowalski J, Keller GA, Rangell L, Wright BD, Radtke F, Aguet M, Ferrara N (1999a): VEGF is required for growth and survival in neonatal mice. *Development* 126:1149–1159.

Gerber HP, Vu TH, Ryan AM, Kowalski J, Werb Z, Ferrara N (1999b): VEGF couples hypertrophic cartilage remodeling, ossification and angiogenesis during endochondral bone formation. *Nat Med* 5:623–628.

Gill M, Dias S, Hattori K, Rivera ML, Hicklin D, Witte L, Girardi L, Yurt R, Himel H, Rafii S (2001): Vascular trauma induces rapid but transient mobilization of VEGFR2(+) AC133(+) endothelial precursor cells. *Circ Res* 88:167–174.

Goda N, Ryan HE, Khadivi B, McNulty W, Rickert RC, Johnson RS (2003): Hypoxia-inducible factor 1 alpha is essential for cell cycle arrest during hypoxia. *Mol Cell Biol* 23:359–369.

Grabovsky V, Feigelson S, Chen C, Bleijs DA, Peled A, Cinamon G, Baleux F, Arenzana-Seisdedos F, Lapidot T, van Kooyk Y, Lobb RR, Alon R (2000): Subsecond induction of alpha4 integrin clustering by immobilized chemokines stimulates leukocyte tethering and rolling on endothelial vascular cell adhesion molecule 1 under flow conditions. *J Exp Med* 192:495–506.

Grant MB, May WS, Caballero S, Brown GA, Guthrie SM, Mames RN, Byrne BJ, Vaught T, Spoerri PE, Peck AB, Scott EW (2002): Adult hematopoietic stem cells provide functional hemangioblast activity during retinal neovascularization. *Nat Med* 8:607–612.

Hammes HP, Brownlee M, Jonczyk A, Sutter A, Preissner KT (1996): Subcutaneous injection of a cyclic peptide antagonist of vitronectin receptor-type integrins inhibits retinal neovascularization. *Nat Med* 2:529–533.

Hattori K, Dias S, Heissig B, Hackett NR, Lyden D, Tateno M, Hicklin DJ, Zhu Z, Witte L, Crystal RG, Moore MA, Rafii S (2001a): Vascular endothelial growth factor and angiopoietin-1 stimulate postnatal hematopoiesis by recruitment of vasculogenic and hematopoietic stem cells. *J Exp Med* 193:1005–1014.

Hattori K, Heissig B, Tashiro K, Honjo T, Tateno M, Shieh JH, Hackett NR, Quitoriano MS, Crystal RG, Rafii S, Moore MA (2001b): Plasma elevation of stromal cell-derived factor-1 induces mobilization of mature and immature hematopoietic progenitor and stem cells. *Blood* 97:3354–3360.

Heissig B, Hattori K, Dias S, Friedrich M, Ferris B, Hackett NR, Crystal RG, Besmer P, Lyden D, Moore MA, Werb Z, Rafii S (2002): Recruitment of stem and progenitor cells from the bone marrow niche requires MMP-9 mediated release of kit-ligand. *Cell* 109:625–637.

Hiratsuka S, Minowa O, Kuno J, Noda T, Shibuya M (1998): Flt-1 lacking the tyrosine kinase domain is sufficient for normal development and angiogenesis in mice. *Proc Natl Acad Sci USA* 95:9349–9354.

Humar R, Kiefer FN, Berns H, Resink TJ, Battegay EJ (2002): Hypoxia enhances vascular cell proliferation and angiogenesis in vitro via rapamycin (mTOR)-dependent signaling. *FASEB J* 16:771–780.

Hynes RO (2002): Integrins: Bidirectional, allosteric signaling machines. *Cell* 110:673–687.

Ishibashi M, Ang SL, Shiota K, Nakanishi S, Kageyama R, Guillemot F (1995): Targeted disruption of mammalian hairy and enhancer of split homolog-1 (HES-1) leads to up-regulation of neural helix-loop-helix factors, premature neurogenesis, and severe neural tube defects. *Genes Dev* 9:3136–3148.

Ivan M, Kondo K, Yang H, Kim W, Valiando J, Ohh M, Salic A, Asara JM, Lane WS, Kaelin WG Jr (2001): HIFalpha targeted for VHL-mediated destruction by proline hydroxylation: Implications for O_2 sensing. *Science* 292:464–468.

Iwaguro H, Yamaguchi J, Kalka C, Murasawa S, Masuda H, Hayashi S, Silver M, Li T, Isner JM, Asahara T (2002): Endothelial progenitor cell vascular endothelial growth factor gene transfer for vascular regeneration. *Circulation* 105:732–738.

Iyer NV, Kotch LE, Agani F, Leung SW, Laughner E, Wenger RH, Gassmann M, Gearhart JD, Lawler AM, Yu AY, Semenza GL (1998): Cellular and developmental control of O_2 homeostasis by hypoxia-inducible factor 1 alpha. *Genes Dev* 12:149–162.

Kalka C, Masuda H, Takahashi T, Gordon R, Tepper O, Gravereaux E, Pieczek A, Iwaguro H, Hayashi SI, Isner JM, Asahara T (2000): Vascular endothelial growth factor(165) gene transfer augments circulating endothelial progenitor cells in human subjects. *Circ Res* 86:1198–1202.

Kampfer H, Pfeilschifter J, Frank S (2001): Expressional regulation of angiopoietin-1 and -2 and the tie-1 and -2 receptor tyrosine kinases during cutaneous wound healing: A comparative study of normal and impaired repair. *Lab Invest* 81:361–373.

Kirchner LM, Meerbaum SO, Gruber BS, Knoll AK, Bulgrin J, Taylor RA, Schmidt SP (2003): Effects of vascular endothelial growth factor on wound closure rates in the genetically diabetic mouse model. *Wound Repair Regen* 11:127–131.

Kocher AA, Schuster MD, Szabolcs MJ, Takuma S, Burkhoff D, Wang J, Homma S, Edwards NM, Itescu S (2001): Neovascularization of ischemic myocardium by human bone-marrow-derived angioblasts prevents cardiomyocyte apoptosis, reduces remodeling and improves cardiac function. *Nat Med* 7:430–436.

Krause DS, Theise ND, Collector MI, Henegariu O, Hwang S, Gardner R, Neutzel S, Sharkis SJ (2001): Multi-organ, multi-lineage engraftment by a single bone marrow-derived stem cell. *Cell* 105:369–377.

Kyriakides TR, Tam JW, Bornstein P (1999): Accelerated wound healing in mice with a disruption of the thrombospondin 2 gene. *J Invest Dermatol* 113:782–787.

Lamoreaux WJ, Fitzgerald ME, Reiner A, Hasty KA, Charles ST (1998): Vascular endothelial growth factor increases release of gelatinase A and decreases release of tissue inhibitor of metalloproteinases by microvascular endothelial cells in vitro. *Microvasc Res* 55:29–42.

Lane WJ, Dias S, Hattori K, Heissig B, Choy M, Rabbany SY, Wood J, Moore MA, Rafii S (2000): Stromal-derived factor 1-induced megakaryocyte migration and platelet production is dependent on matrix metalloproteinases. *Blood* 96:4152–4159.

Leach L, Babawale MO, Anderson M, Lammiman M (2002): Vasculogenesis, angiogenesis and the molecular organisation of endothelial junctions in the early human placenta. *J Vasc Res* 39:246–259.

Lemischka IR, Raulet DH, Mulligan RC (1986): Developmental potential and dynamic behavior of hematopoietic stem cells. *Cell* 45:917–927.

Lerman OZ, Galiano RD, Armour M, Levine JP, Gurtner GC (2003): Cellular dysfunction in the diabetic fibroblast: Impairment in migration, vascular endothelial growth factor production, and response to hypoxia. *Am J Pathol* 162:303–312.

Lyden D, Hattori K, Dias S, Costa C, Blaikie P, Butros L, Chadburn A, Heissig B, Marks W, Witte L, Wu Y, Hicklin D, Zhu Z, Hackett NR, Crystal RG, Moore MA, Hajjar KA, Manova K, Benezra R, Rafii S (2001): Impaired recruitment of bone marrow-derived endothelial and hematopoietic precursor cells blocks tumor angiogenesis and growth. *Nat Med* 7:1194–1201.

Lyden D, Young AZ, Zagzag D, Yan W, Gerald W, O'Reilly R, Bader BL, Hynes RO, Zhuang Y, Manova K, Benezra R (1999a): Id1 and Id3 are required for neurogenesis, angiogenesis and vascularization of tumour xenografts. *Nature* 401:670–677.

Lyden D, Young AZ, Zagzag D, Yan W, Gerald W, O'Reilly R, Bader BL, Hynes RO, Zhuang Y, Manova K, Benezra R (1999b): Id1 and Id3 are required for neurogenesis, angiogenesis and vascularization of tumour xenografts [see comments]. *Nature* 401:670–677.

Maeshima Y, Sudhakar A, Lively JC, Ueki K, Kharbanda S, Kahn CR, Sonenberg N, Hynes RO, Kalluri R (2002): Tumstatin, an endothelial cell-specific inhibitor of protein synthesis. *Science* 295:140–143.

Maisonpierre PC, Suri C, Jones PF, Bartunkova S, Wiegand SJ, Radziejewski C, Compton D, McClain J, Aldrich TH, Papadopoulos N, Daly TJ, Davis S, Sato TN, Yancopoulos GD (1997): Angiopoietin-2, a natural antagonist for Tie2 that disrupts in vivo angiogenesis. *Science* 277:55–60.

Matsumoto H, Ma WG, Daikoku T, Zhao X, Paria BC, Das SK, Trzaskos JM, Dey SK (2002): Cyclooxygenase-2 differentially directs uterine angiogenesis during implantation in mice. *J Biol Chem* 277:29260–29267.

Mazure NM, Chen EY, Yeh P, Laderoute KR, Giaccia AJ (1996): Oncogenic transformation and hypoxia synergistically act to modulate vascular endothelial growth factor expression. *Cancer Res* 56:3436–3440.

McCarty JH, Monahan-Earley RA, Brown LF, Keller M, Gerhardt H, Rubin K, Shani M, Dvorak HF, Wolburg H, Bader BL, Dvorak AM, Hynes RO (2002): Defective associations between blood vessels and brain parenchyma lead to cerebral hemorrhage in mice lacking alphav integrins. *Mol Cell Biol* 22:7667–7677.

Moodley Y, Rigby P, Bundell C, Bunt S, Hayashi H, Misso N, McAnulty R, Laurent G, Scaffidi A, Thompson P, Knight D (2003): Macrophage recognition and phagocytosis of apoptotic fibroblasts is critically dependent on fibroblast-derived thrombospondin 1 and CD36. *Am J Pathol* 162:771–779.

Nag S, Eskandarian MR, Davis J, Eubanks JH (2002): Differential expression of vascular endothelial growth factor-A (VEGF-A) and VEGF-B after brain injury. *J Neuropathol Exp Neurol* 61:778–788.

Olofsson B, Jeltsch M, Eriksson U, Alitalo K (1999): Current biology of VEGF-B and VEGF-C. *Curr Opin Biotechnol* 10:528–535.

Otani A, Kinder K, Ewalt K, Otero FJ, Schimmel P, Friedlander M (2002): Bone marrow-derived stem cells target retinal astrocytes and can promote or inhibit retinal angiogenesis. *Nat Med* 8:1004–1010.

Peled A, Petit I, Kollet O, Magid M, Ponomaryov T, Byk T, Nagler A, Ben-Hur H, Many A, Shultz L, Lider O, Alon RZ, Lapidot T (1999): Dependence of human stem cell engraftment and repopulation of NOD/SCID mice on CXCR4. *Science* 283:845–851.

Persico MG, Vincenti V, DiPalma T (1999): Structure, expression and receptor-binding properties of placenta growth factor (PlGF). *Curr Top Microbiol Immunol* 237:31–40.

Quinn TP, Peters KG, De Vries C, Ferrara N, Williams LT (1993): Fetal liver kinase 1 is a receptor for vascular endothelial growth factor and is selectively expressed in vascular endothelium. *Proc Natl Acad Sci USA* 90:7533–7537.

Rafii S, Lyden D, Benezra R, Hattori K, Heissig B (2002): Vascular and haematopoietic stem cells: Novel targets for anti-angiogenesis therapy? *Nat Rev Cancer* 2:826–835.

Reyes M, Dudek A, Jahagirdar B, Koodie L, Marker PH, Verfaillie CM (2002): Origin of endothelial progenitors in human postnatal bone marrow. *J Clin Invest* 109:337–346.

Reynolds LE, Wyder L, Lively JC, Taverna D, Robinson SD, Huang X, Sheppard D, Hynes RO, Hodivala-Dilke KM (2002): Enhanced pathological angiogenesis in mice lacking beta3 integrin or beta3 and beta5 integrins. *Nat Med* 8:27–34.

Ryan HE, Lo J, Johnson RS (1998): HIF-1 alpha is required for solid tumor formation and embryonic vascularization. *EMBO J* 17:3005–3015.

Safran M, Kaelin WG Jr (2003): HIF hydroxylation and the mammalian oxygen-sensing pathway. *J Clin Invest* 111:779–783.

Salamonsen LA, Zhang J, Hampton A, Lathbury L (2000): Regulation of matrix metalloproteinases in human endometrium. *Hum Reprod* 15 (Suppl 3):112–119.

Sata M, Luo Z, Walsh K (2001): Fas ligand overexpression on allograft endothelium inhibits inflammatory cell infiltration and transplant-associated intimal hyperplasia. *J Immunol* 166:6964–6971.

Sato TN, Tozawa Y, Deutsch U, Wolburg-Buchholz K, Fujiwara Y, Gendron-Maguire M, Gridley T, Wolburg H,

Risau W, Qin Y (1995): Distinct roles of the receptor tyrosine kinases Tie-1 and Tie-2 in blood vessel formation. *Nature* 376:70–74.

Senger DR, Connolly DT, Van de Water L, Feder J, Dvorak HF (1990): Purification and NH2-terminal amino acid sequence of guinea pig tumor-secreted vascular permeability factor. *Cancer Res* 50:1774–1778.

Senger DR, Galli SJ, Dvorak AM, Perruzzi CA, Harvey VS, Dvorak HF (1983): Tumor cells secrete a vascular permeability factor that promotes accumulation of ascites fluid. *Science* 219:983–985.

Shalaby F, Rossant J, Yamaguchi TP, Gertsenstein M, Wu XF, Breitman ML, Schuh AC (1995): Failure of blood-island formation and vasculogenesis in Flk-1-deficient mice. *Nature* 376:62–66.

Shi Q, Rafii S, Wu MH, Wijelath ES, Yu C, Ishida A, Fujita Y, Kothari S, Mohle R, Sauvage LR, Moore MA, Storb RF, Hammond WP (1998): Evidence for circulating bone marrow-derived endothelial cells. *Blood* 92:362–367.

Shin D, Garcia-Cardena G, Hayashi S, Gerety S, Asahara T, Stavrakis G, Isner J, Folkman J, Gimbrone MA Jr, Anderson DJ (2001): Expression of ephrinB2 identifies a stable genetic difference between arterial and venous vascular smooth muscle as well as endothelial cells, and marks subsets of microvessels at sites of adult neovascularization. *Dev Biol* 230:139–150.

St Croix B, Rago C, Velculescu V, Traverso G, Romans KE, Montgomery E, Lal A, Riggins GJ, Lengauer C, Vogelstein B, Kinzler KW (2000): Genes expressed in human tumor endothelium. *Science* 289:1197–1202.

Stephens LE, Sutherland AE, Klimanskaya IV, Andrieux A, Meneses J, Pedersen RA, Damsky CH (1995): Deletion of beta 1 integrins in mice results in inner cell mass failure and peri-implantation lethality. *Genes Dev* 9:1883–1895.

Streit M, Velasco P, Riccardi L, Spencer L, Brown LF, Janes L, Lange-Asschenfeldt B, Yano K, Hawighorst T, Iruela-Arispe L, Detmar M (2000): Thrombospondin-1 suppresses wound healing and granulation tissue formation in the skin of transgenic mice. *EMBO J* 19:3272–3282.

Suri C, Jones PF, Patan S, Bartunkova S, Maisonpierre PC, Davis S, Sato TN, Yancopoulos GD (1996): Requisite role of angiopoietin-1, a ligand for the TIE2 receptor, during embryonic angiogenesis. *Cell* 87:1171–1180.

Takahashi T, Kalka C, Masuda H, Chen D, Silver M, Kearney M, Magner M, Isner JM, Asahara T (1999): Ischemia- and cytokine-induced mobilization of bone marrow-derived endothelial progenitor cells for neovascularization. *Nature Med* 5:434–438.

Terman BI, Carrion ME, Kovacs E, Rasmussen BA, Eddy RL, Shows TB (1991): Identification of a new endothelial cell growth factor receptor tyrosine kinase. *Oncogene* 6:1677–1683.

Terman BI, Dougher-Vermazen M, Carrion ME, Dimitrov D, Armellino DC, Gospodarowicz D, Bohlen P (1992): Identification of the KDR tyrosine kinase as a receptor for vascular endothelial cell growth factor. *Biochem Biophys Res Commun* 187:1579–1586.

Tezono K, Sarker KP, Kikuchi H, Nasu M, Kitajima I, Maruyama I (2001): Bioactivity of the vascular endothelial growth factor trapped in fibrin clots: Production of IL-6 and IL-8 in monocytes by fibrin clots. *Haemostasis* 31:71–79.

Thurston G, Suri C, Smith K, McClain J, Sato TN, Yancopoulos GD, McDonald DM (1999): Leakage-resistant blood vessels in mice transgenically overexpressing angiopoietin-1. *Science* 286:2511–2514.

Wang HU, Chen ZF, Anderson DJ (1998): Molecular distinction and angiogenic interaction between embryonic arteries and veins revealed by ephrin-B2 and its receptor Eph-B4. *Cell* 93:741–753.

Yancopoulos GD, Davis S, Gale NW, Rudge JS, Wiegand SJ, Holash J (2000): Vascular-specific growth factors and blood vessel formation. *Nature* 407:242–248.

Yu JL, Rak JW, Coomber BL, Hicklin DJ, Kerbel RS (2002): Effect of p53 status on tumor response to antiangiogenic therapy. *Science* 295:1526–1528.

Zhang H, Li C, Baciu PC (2002): Expression of integrins and MMPs during alkaline-burn-induced corneal angiogenesis. *Invest Ophthalmol Vis Sci* 43:955–962.

Zhu J, Motejlek K, Wang D, Zang K, Schmidt A, Reichardt LF (2002): Beta8 integrins are required for vascular morphogenesis in mouse embryos. *Development* 129:2891–2903.

17

METASTASIS

Dawn S. Chandler and Guillermina Lozano
Department of Molecular Genetics, University of Texas M. D. Anderson Cancer Center, Houston, Texas

Metastasis is a compilation of many steps, a process mediated by numerous genes involved in cell adhesion, motility, degradation of the extracellular matrix, and angiogenesis. The most commonly used models to study metastasis alter expression of specific genes in human tumor cell lines and measure migration and colonization in a nude mouse. These studies are limiting because several steps in the metastatic pathway cannot be analyzed in these models. The use of genetically engineered mice has recently been exploited to study metastasis. For example, clear and direct roles for E-cadherin, L-selectin, and the hepatocyte growth factor/scatter factor (HGF/SF) in metastasis have been identified with these models. However, few models have been developed to study metastasis, and a flurry of activity in this important area is needed. In particular, genes directly involved in the process of metastasis need to be identified and examined as possible therapeutic targets.

INTRODUCTION

The ability of a single cell to change genetically, adapt to these changes, and finally survive is the underlying theme for evolution and has allowed a world of single-celled organisms to evolve into a diverse collection of uni- and multicellular beings. The same classes of events underlying the genetic diversity of the earth's organisms is also responsible for the great changes that take place when a normal cell becomes tumorigenic and eventually metastatic (Hanahan and Weinberg, 2000). Here, we will briefly review the known steps of the evolution of a locally tumorigenic cell into a metastatic cancer cell and the current animal models now available to study this transformation. It will become apparent that, although understanding the mechanisms of metastasis is paramount, relatively little is known about the molecular events that underlie this deadly transformation. Astonishingly, there are very few good animal models that exist to study the metastatic process. The study of metastasis in genetically modified mice has lagged considerably behind similar studies of primary tumor formation in mouse models. It is our hope that this review will pique the interest of scientists currently using genetically modified mouse models to study cancer and will further inspire them to generate new models to explore the complicated questions pertaining to cancer metastasis and the molecular mechanisms that drive the evolution of cells to the metastatic phenotype. New animal models will not only further our understanding of the progression of cancer but will also allow the eventual study of therapeutic possibilities.

A MYRIAD OF STEPS TO METASTASIS

Metastasis encompasses a series of steps by which a tumor cell leaves the local environment, migrates to a new location, and grows (Fig. 17.1). It involves a myriad of steps. First, the cell must detach from the primary tumor and invade the surrounding tissue. This process involves changes in or recognition of cell adhesion molecules. The detached cell must then migrate to a blood vessel and penetrate that vessel (intravasation). This step involves changes in motility that allow the cell to migrate to the vessel and alterations in cell shape that allow the

Mouse Models of Human Cancer, edited by Eric C. Holland
ISBN 0-471-44460-X

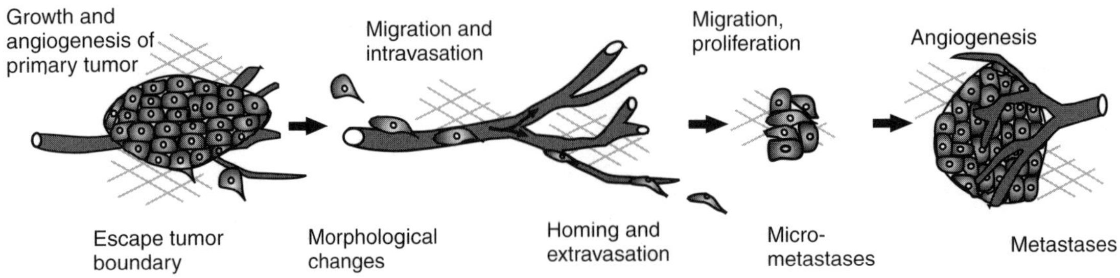

Figure 17.1. Schematic of the steps required for metastasis.

cancer cell to squeeze into the closed vessel system. Upon entering the circulatory (or lymphatic) system, the cell has to survive the onslaught of macrophages that kill invaders and the hydroforce of the liquid flowing through the vessels. Having evaded the immune system and survived the violent force of blood flow, the cell must next identify a home. Potential receptors on the cell surface may aid in this process. Once the cell recognizes or identifies a target tissue, the cell must leave the circulation in a process known as extravasation. Finally the cell must survive and grow in its new environment. This adaptation to the new environment is the least efficient step of the metastasis pathway and represents the most important obstacle to metastatic dissemination. The appropriate growth factors must be present to sustain growth. Given the number of steps required for a cell to metastasize, it seems unlikely that deregulation of a single gene results in metastasis. On the other hand, disruption of a single gene in the pathway, inhibiting any one of the steps noted above, may inhibit the process of metastasis, thus providing a therapeutic window.

ORTHOTOPIC VERSUS GENETICALLY MODIFIED MOUSE MODELS

Much of what is known about metastasis to date has been derived from quantitative analyses of human metastatic tumors, in vitro invasion assays, and transplantation models in which cells with possible metastatic potential are injected into immunocompromised mice. Several recent reviews interpret the vast amount of information that has thus far shaped and guided the field of metastasis (Fidler, 2001; Herzig, and Christofori, 2002; Price, 1994; Webb and Vande Woude, 2000). Although the information gathered from these assays has been pivotal to our current understanding of the metastatic process, their weaknesses cannot be denied. Quantitation of genes in human metastatic tumors directs us to a narrowed list of possible candidate genes by correlation but does not give any clue as to causation. Similarly, in vitro assays tell us what genes are responsible for migration in vitro but ignore many factors that affect metastasis within the more complex context of an animal. Finally, transplantation models do take into account the whole animal environment but are somewhat artificial in that (1) orthotopic injection models account for only the final stages of metastasis by being injected directly into the blood stream (skipping the migration through intravasation steps) and (2) both the direct and orthotopic transplantation models are most often performed using human cells injected into nude or severe combined immunodeficiency (SCID) mice and thus ignore completely the immune response's interplay with metastasis.

Many of the weaknesses of orthotopic models can be overcome by utilizing genetically modified mouse models to study metastasis. To date, only a limited number of these models have been generated and utilized to further our understanding of the individual steps leading to metastasis. The current mouse metastatic models have been built on the ideas gained from the existing orthotopic models and range from transgenics to knockouts to more complex genetic manipulations. The problem with transgenic models is that one may understand the effect of a specific gene in the process of metastasis but not necessarily the specific step it affects. Thus, the intersection of these types of models should lead to a more thorough understanding of this complicated process.

A number of studies aimed at discerning the function of proteins suspected to be involved in the metastatic process have utilized variations of Hanahan's Rip1Tag2 transgenic mice (Hanahan, 1985). This model, which will be described here, is the cornerstone for many of the genetic manipulations that have furthered our understanding of metastasis. Rip1Tag2 transgenic mice express the large-T antigen of simian virus 40 (SV40) under the control of the rat insulin promoter and elicit tissue-specific expression of the transgene in the endocrine pancreas. T antigen inactivates two important tumor suppressor pathways by interacting with and inhibiting the functions of p53 and Rb. These transgenic mice succumb to a well-characterized progression of multistep tumorigenesis occurring specifically in the pancreatic islets of Langerhans of transgenic animals. The

Table 17.1. Summary of Mouse Models Using Rip1Tag2-Sensitized Background

Gene	Genetic Manipulation	Findings	References
E-cadherin	A dominant negative truncated form and a wild-type form of *E-cadherin* were ectopically expressed in the Rip1Tag2 background	*E-cadherin* plays a role in maintaining intracellular adhesion and thus counteracts the invasive phenotype. The loss of *E-cadherin*, however, causes the transition from adenomas to carcinomas and contributes significantly to metastatic potential.	Perl et al., 1998
N-CAM	*N-CAM* deletion and ectopic expression of isoform 120 were tested in the Rip1Tag2 mice	Loss of *N-CAM* expression is essential for tumor metastasis. Forced expression of *N-CAM120* transgene prevents ß-cell metastasis in the Rip1Tag2:N-CAM−/− mice.	Perl et al., 1999
VEGF-A165	*VEGF-A165* ectopically expressed in the Rip1Tag2 mice	Overexpression of *VEGF-A165* increases the rate of transition from adenoma to carcinoma as well as the rate of onset of tumor growth and mortality primarily as a result of increased levels of tumor vascularization. Expression of VEGF-A is not sufficient to induce metastasis.	Gannon et al., 2002
VEGF-C	*VEGF-A165* ectopically expressed in the Rip1Tag2 mice	Overexpression of *VEGF-C* induces lymphangiogenesis and mediates dissemination of the tumor and metastasis to the lymph node.	Mandriota et al., 2001
L-selectin	*L-selectin* and *LacZ* expressed ectopically in the Rip1Tag2 mice	*L-selectin* facilitates homing of tumor cells to the lymph nodes and therefore facilitates metastasis. Since mice expressing *LacZ* and *L-selectin* in the absence of the T antigen displayed no homing phenotype, primary tumor formation is necessary for the distribution of the cells.	Qian et al., 2001
IGF-1 Receptor	*IGF-1* receptor ectopically expressed in the Rip1Tag2 mice	Overexpression of *Igf-1R* results in invasive carcinoma and metastasis correlated with down-regulated E-cadherin.	Lopez and Hanahan, 2002
MMP2	*MMP2* deletion was tested in the Rip1Tag2 mice	Loss of *MMP2* does not alter the number of tumors or the metastatic potential of the tumors in the Rip1Tag2 mice	Bergers et al., 2000
MMP9	*MMP9* deletion was tested in the Rip1Tag2 mice	Loss of *MMP9* in conjunction with TAg expression results in decreased angiogenesis at the primary tumor site and in fewer and smaller tumors as compared to the Rip1Tag2 mice	Bergers et al., 2000

tumorigenic phenotype in these mice has three phases: (1) the development of disordered islets, (2) hyperplasia, and (3) the formation of β-cell solid tumors. Though a stepwise progression of tumorigenesis is characteristic of these mice, the progression does not proceed fully to metastatic disease as solid tumors in this model are not well vascularized and metastasis is not observed. These mice therefore present a "sensitized" background from which additional genetic manipulations can be monitored for their ability to promote metastasis (summarized in Table 17.1).

DETACHMENT AND MOTILITY

In the first step of the progression toward metastasis, changes at the cell surface are required for the tumor cell to detach from the primary tumor (Fig. 17.1). The expression of numerous cell adhesion proteins is altered. Mouse models containing deletions in these genes are listed in Table 17.2. These proteins include cell–cell adhesion molecules (CAMs) and cadherins, both of which mediate interactions between cells, and integrins, which link cells to substrates in the extracellular matrix (Cavallaro and Christofori, 2001;

Table 17.2. Available Null Alleles of Genes Implicated in Adhesion and Motility

Gene	References
MMP2 (gelatinase A)	Itoh et al., 1997
MMP3 (stromelysin-1)	Mudgett et al., 1998
MMP7 (matrilysin)	Wilson et al., 1997
MMP9 (gelatinase B)	Vu et al., 1998
MMP11 (stromelysin-3)	Masson et al., 1998
MMP12 (metalloelastase)	Shipley et al., 1996
MMP14 (MT1-MMP)	Holmbeck et al., 1999
TIMP1	Nothnick, 2000
TIMP3	Leco et al., 2001
ICAM1	Sligh et al., 1993
N-CAM	Cremer et al., 1994; Tomasiewicz et al., 1993
L-selectin	Arbones et al., 1994
IGF-1R	Holzenberger et al., 2003a
α_v *integrin*	Yang et al., 1995
B_1 *integrin*	Fassler and Meyer, 1995; Stephens et al., 1995
B_3 *integrin*	Hodivala-Dilke et al., 1999
B_5 *integrin*	Huang et al., 2000
B_6 *integrin*	Huang et al., 1996
B_7 *integrin*	Wagner et al., 1996

Hood and Cheresh, 2002). Disintegration of these links must occur for the cell to become motile. In tumorigenesis, one of the CAMs, N-CAM, has received considerable attention because embryonic isoforms encoded by alternative splicing of a single gene are reexpressed in tumors. Isoform switching occurs between a highly adhesive splicing isoform and a nonadhesive one. In the Rip1Tag2 pancreatic tumor model, loss of one or two copies of *N-CAM* yielded increased tumor burden and metastasis to mesenteric lymph nodes, testes, liver, adrenal gland, and kidney in approximately 50% of mice (Perl et al., 1999). To directly test the role of different N-CAM isoforms, transgenic mice expressing the adhesive form of N-CAM (N-CAM120) were crossed into the Rip1Tag2 and *N-CAM*$^{-/-}$ background. The production of N-CAM120 prevented formation of metastases (Perl et al., 1999). This in vivo system illustrates the direct role of N-CAM in metastatic pancreatic carcinomas.

The role of cadherins in tumorigenesis has also been studied extensively (Cavallaro and Christofori, 2001). Again, different cadherins have opposing function in metastasis. Some, such as E-cadherin, inhibit invasion and metastasis, while others, such as N-cadherin, can boost metastasis (Nieman et al., 1999). In humans, E-cadherin function is lost in the majority of epithelial cancers (Christofori and Semb, 1999). Similarly, in Rip1Tag2 mice, E-cadherin expression is absent in pancreatic carcinomas. Overexpression of E-cadherin in islets using the rat insulin promoter causes a decrease in the incidence of carcinomas (Perl et al., 1998). More interestingly, when Rip1Tag2 mice are mated to mice expressing a dominant-negative form of E-cadherin, one quarter of the mice develop metastases to the pancreatic lymph node. These data support the original observation that increased E-cadherin levels in epithelial cells in an orthotopic model led to decreased invasion (Vleminckx et al., 1991). Thus, in two different examples, changes in cell adhesion by direct alteration of specific genes altered metastatic potential. These data raise the possibility that any perturbation of the ability of cells to stick together will affect their ability to metastasize.

Extracellular proteases that degrade the extracellular matrix may also be important in the process of detachment and migration. The two classes of proteases are matrix metalloproteinases (MMPs), commonly upregulated in metastases, and tissue inhibitors of metalloproteinases (TIMPs), which are down modulated in metastases (Chambers and Matrisian, 1997; McCawley and Matrisian, 2001). Table 17.2 lists many of the genes deleted in mice used to study the function of these molecules, and earlier transgenic models are reviewed by Khokha et al. (1995). MMPs, of which there are more than 20, are secreted or transmembrane proteins that require zinc to degrade the extracellular matrix (Brinckerhoff and Matrisian, 2002). Different MMPs have different substrate specificities and degrade such diverse substrates as collagens, proteoglycans, glycoproteins, gelatin, elastin, and laminin. Four different TIMPs can complex with active MMPs to inhibit MMP function and thus keep them in check. Changes in the levels of MMPs and TIMPs markedly affect metastasis in orthotopic models (Coussens et al., 2002).

Mice lacking *MMP2* and *MMP9* have been generated (Itoh et al., 1997, 1998; Vu et al., 1998). Loss of either *MMP9* or *MMP2* yields viable mice with few abnormalities. *MMP9*$^{-/-}$ mice develop fewer angiogenic islets in the developing bone. In the RipTag2 model of spontaneous pancreatic carcinogenesis, deletion of *MMP9* results in decreased angiogenesis and a smaller number and size of tumors (Bergers et al., 2000). In this same model, loss of *MMP2* does not alter the number of tumors, but they are smaller in size (Bergers et al., 2000). Similarly, loss of *MMP9* in a squamous cell carcinoma mouse model resulted in mice with a decreased tumor incidence, but the tumors were more malignant than those in control mice expressing *MMP9* (Coussens et al., 2000).

In both studies, a role of *MMP9* and *MMP2* in angiogenesis and proliferation of tumors is obvious. Neither study assigned a direct role in metastasis for the MMPs. However, these studies were performed in squamous cell carcinomas and pancreatic adenocarcinomas and so do not exclude a possible role for MMPs in metastasis of other cancer types. In fact, *MMP9* seems to be required in lung metastasis (Hiratsuka et al., 2002).

Another class of proteins with roles in cell adhesion are the integrins (Eliceiri and Cheresh, 1999; van der Flier and Sonnenberg, 2001). Integrins are cell surface receptors that mediate interactions between cells and between cells and the extracellular matrix. The α and β subunits combine to form numerous integrins with cell-type specificity. The role of integrins in angiogenesis has been controversial, although recent evidence using mice with deletions in various integrin subunits points to a role as inhibitors of angiogenesis (Hynes, 2002). Mice lacking the β_3 or both β_3 and β_5 subunits showed enhanced tumor growth and angiogenesis of the injected tumor cells (Reynolds et al., 2002). Any analysis of tumor growth in vivo awaits crossing these mice with other tumor-prone mice.

The process of cell motility, an important aspect of metastasis, has not been studied in genetically engineered mice. However, the rho family of proteins has been implicated in motility and cell adhesion by virtue of their ability to reorganize the actin cytoskeleton (Narumiya et al., 1997; Sahai and Marshall, 2002). One of the effectors of Rho, ROCK1, was recently identified in an in vitro screen for genes affecting cell migration (Suyama et al., 2003). Using a chemotaxis assay, ribozymes that inhibited cell movement were cloned. Two ribozymes were specific to ROCK1. Unfortunately, no genetically engineered mice exist with which to study the role of the rho family of proteins in metastasis. Thc potential for lethality complicates the engineering of such models, although, as has been seen for the integrins, deletions of some have no deleterious effects on development. Development of these mice will be useful in assaying the role of ROCK in metastasis in vivo.

ANGIOGENESIS AND INTRAVASATION

Disrupting cell–cell and cell–matrix attachments in the initial steps of metastasis is critical and involves many changes. The subsequent step is also very important: entering the lymphatic or circulatory system via intravasation. Genes that regulate blood vessel formation (angiogenesis) are integral in this process because an increase in the number of blood vessels means a cell has to move a shorter distance to reach a vessel and therefore has increased access to the circulation. Vascular endothelial growth factor (VEGF) is a potent angiogenic factor that stimulates endothelial cells to proliferate, migrate, and differentiate. It is a common marker of angiogenesis. VEGF-A is the first of five family members that has been extensively studied. In the Rip1Tag2 model of pancreatic tumorigenesis, overexpression of VEGF-A in transgenic mice accelerates the onset of angiogenesis and the proliferation of the tumors but does not cause tumor metastasis (Gannon et al., 2002). This VEGF-A study suggests that inducing blood vessel formation does not contribute to metastasis but does contribute to primary tumor growth.

VEGF-C, another member of this family, is a lymphangiogenic factor. Transgenic mice expressing VEGF-C in the pancreas develop metastases to the pancreatic lymph nodes when crossed with the Rip1Tag2 mice, indicating an important and direct role for VEGF-C in metastasis (Mandriota et al., 2001). VEGF-C overexpression in human cancer cell lines yielded increased lymphatic metastases when the cell lines were injected orthotopically in mice, sealing its importance in both experimental systems (Padera et al., 2002).

HOMING AND EXTRAVASATION

Metastatic cells in the blood or lymphatic system reach many organs but usually identify a target organ in a process called homing. Identification of a target organ is independent of vascular anatomy, rate of blood flow, and number of tumor cells in the circulation (Chambers et al., 2002). Homing is followed by extravasation, the steps by which a cell leaves the circulation and begins to proliferate in its new environment. These processes contribute to the organ-specific pattern of spread that is especially evident in the cases of prostate and breast cancer (Debies and Welch, 2001). The fact that only certain types of tumor cells can grow in certain congenial host organ sites gives rise to the “seed-and-soil hypothesis” of metastasis (Fidler, 2001).

L-selectin is a molecule expressed on the surface of lymphocytes and is known to be responsible for their homing to lymph nodes (Siegelman et al., 1989). The Rip1Tag2 mouse model was utilized to test the importance of L-selectin on homing and metastasis of ß-cell solid tumors (Qian et al., 2001). When the LacZ marker was coexpressed in the pancreas of Rip1Tag2 mice, LacZ-positive cells could be identified in the bloodstream but not in the lymph nodes, and no metastasis was observed. However, when L-selectin was also expressed in the pancreas, LacZ-positive cells were identified in the bloodstream and lymph nodes and 20% of the mice had lymph node metastases. First, these experiments show that the inability of the tumors in the Rip1Tag2 mice to metastasize is not due to the lack of tumor shedding into the bloodstream. In addition, Rip1Tag2 ß-tumor cells become metastatically competent because of their homing capabilities upon coexpression of the L-selectin transgene. Thus, entry of tumor cells into the bloodstream is necessary but insufficient for metastasis to occur.

Once a tumor cell has settled in a distant organ site as a micrometastasis, further colonization and growth of the tumor requires vasculature, and thus angiogenesis again plays an important role, this time in the final stages of metastasis. Compared with other steps of metastasis, colonization is sparsely described and has not been studied in a mouse model.

METASTASIS SUPPRESSORS

Genes that must be deleted or down regulated for metastases to occur have been classified as metastasis suppressor genes (Debies and Welch, 2001; Steeg, 2003). Metastasis suppressor genes, by definition, do not affect the rate or size of primary tumor growth. Eleven metastasis suppressor genes have been identified, of which eight have been validated by demonstrations of loss of metastatic capability following their transfection into metastatically competent tumor cell lines. Four metastasis suppressors, nonmetastatic 23 (*nm23*), breast cancer metastasis suppressor 1 (*BRMS1*), *KiSS1*, and *MKK1*, were identified using microcell-mediated fusion experiments in which transfer of a portion of a human chromosome into metastatic cell lines suppressed the metastatic phenotype. Validation experiments performed using metastatic melanoma (*nm23, BRMS1*, and *KiSS1*), breast cancer cell lines (*nm23* and *BRMS1*), and prostate and ovarian cancer cell lines (*MKK1*) subsequently proved that all were able to suppress metastasis without affecting tumorigenicity. An additional metastasis suppressor, *Kai1* (CD82), was identified by virtue of its differential expression in metastatic prostate cancers and has been shown to act functionally as a metastasis suppressor in prostate and breast cancers. Three of the most recently identified metastasis suppressors, *RHOGD12, CRSP3*, and *VDUP1*, were identified using microarray technology. *RHOGD12* was identified from T24 bladder carcinoma cell lines with low metastatic potential and is known to be an inhibitor of Rho, which contributes to cell motility, and Rac, which contributes to cell adhesion. Overexpression of *RHOGD12* was shown to inhibit metastatic invasion in organotypic bladder invasion assays. Both *CRSP3* and *VDUP1* were identified by virtue of their up regulation in chromosome-6-suppressed melanoma cells and were able to suppress melanoma metastasis without affecting primary tumor formation in subsequent functional assays.

None of these metastasis suppressors have been tested functionally in genetically modified mice. One rate-limiting step for the production of mice with deletions in metastasis suppressors is the importance of many of these genes in cell signaling and communication: Deletions thus may cause lethal developmental phenotypes. Tissue-specific deregulation of these genes is one way to circumvent this problem. Alternatively, these genes may be expressed as tissue-specific transgenes in mouse models that already exhibit metastatic phenotype for the tumor type in question and assayed for the ability to suppress metastasis.

The TRAMP prostate cancer model, which exhibits a high frequency of metastasis (Table 17.3 and discussed below), may serve as an avenue to test the functionality of prostate cancer metastasis suppressors (such as *Kai1* and *MKK1*) in the context of the whole animal. Similarly, HGF/SF transgenic mice which have malignant melanoma (Table 17.3 and discussed below) may serve as a good model system to test the metastasis-suppressing phenotype of *nm23, BRMS1, KiSS1, CRSP3*, and *VDUP1*, all of which were shown to be suppressors of melanoma metastasis in in vitro and orthotopic studies. As no mouse model has been generated for metastatic breast cancer, these types of validations of metastasis-suppressing activity in the context of the whole animal are not yet an option for genes affecting the metastatic potential of breast cancer. This again underscores a gap in mouse models of metastasis and points to a need for further investigation.

CELL SIGNALING AND METASTASIS

Alterations in cell signaling of tumor cells can trigger changes in the metastatic potential. Since changes in cell adhesion, motility, and shape are all thought to secondarily affect cell signaling, this is perhaps not surprising. There are three mouse models that exemplify the relationship between cell signaling and the metastatic phenotype.

The first model introduces a cytokine, HGF/SF and its receptor, the tyrosine kinase c-*met*. HGF/SF and c-*met* function together as regulators of such cellular activities as the stimulation of angiogenesis, degradation of extracellular matrix, and production of cell adhesion molecules (Matsumoto and Nakamura, 1996; Rosen et al., 1997). These cellular activities make them potential regulators of the metastatic phenotype. Furthermore, the role of c-*met* in cancer is undeniable, as it is able to transform NIH3T3 cells (fibroblasts) in vitro and has been shown to be overexpressed in a number of human and mouse tumors (Jeffers et al., 1997; Lee et al., 2000; Rong et al., 1994; Schmidt et al., 1999). The ectopic expression of HGF/SF in transgenic mice using the metallothionein promoter and the consequent autocrine activation of c-*met* results in the formation of cutaneous malignant melanomas, carcinomas, and sarcomas (Otsuka et al., 1998; Yu and Merlino, 2002). Approximately 20% of the primary melanomas were metastatic to the lymph nodes and visceral organs. Although other mouse models of melanoma do exist, this model is the first to recapitulate the highly metastatic potential of this deadly cancer as observed in human melanoma patients. For this reason, this mouse model provides a useful tool for further studies to identify genes that contribute to the metastatic

Table 17.3. Available Mouse Models of Metastasis

Model Name	Genetic Manipulation	Findings	Sites of Metastasis	Prospectives	References
TRAMP (transgenic adenocarcinoma mouse prostate model)	Simian virus 40 (SV40) large-tumor T antigen driven by the rat probasin prostate-specific promoter	Expression of TAg in the prostate causes progressive prostatic neoplasia with distant metastasis as early as 12 weeks with 100% of mice harboring metastatic disease by 28 weeks; tumor progression is correlated to loss of E-cadherin.	*Common*: periaortic lymph nodes and lungs *Less common*: kidney, adrenal gland, and bone	Model for prostate metastasis with a pattern of metastatic spread that closely resembles that of human prostate cancer	Gingrich et al., 1996
T7-PKCε (protein kinase c)	T7 epitope-tagged PKCε driven by the K14 skin-specific promoter	Following an initiation–promotion protocol, mice-suppressed papilloma formation but more often developed carcinoma that metastasized within 3 weeks of tumor appearance.	Regional lymph nodes	Model for metastatic squamous cell carcinoma	Jansen et al., 2001
Merlin/NF2 (neurofibromatosis 2; heterozygotes)	Targeted deletion of *NF2* by homologous recombination	Mice develop osteosarcomas, fibrosarcomas, and hepatocellular carcinomas that frequently metastasize, though with a long latency period.	*Common*: lung and liver *Less common*: kidney and spleen	Although originally designed to model NF2, more promising as metastatic model	McClatchey et al., 1998
p53R172HΔg	Targeted missense mutation in *p53* by homologous recombination	Varied tumor spectrum and increased incidence of metastasis for osteosarcomas (69%) and carcinomas (38%) in $p53R172H\Delta g/+$ compared to $p53^{+/-}$.	Lung and liver	Model for metastasis that may allow for further dissection of p53 function in tumorigenesis	Liu et al., 2000

(*continued overleaf*)

Table 17.3. ***(continued)***

Model Name	Genetic Manipulation	Findings	Sites of Metastasis	Prospectives	References
HGF/SF (hepatocyte growth factor/scatter factor)	*HGF/SF* expressed under the control of the metallothionein 1 (MT-1) promoter	Mice were predisposed to carcinomas and sarcomas with a drastic increase in the induction of malignant metastatic melanomas as compared to WT mice; transplants of two melanomas from the HGF/SF mice into nude mice maintained metastatic potential.	Liver, spleen, skin, and lymph nodes in *HGF/SF* mice; in transplant assays—liver, spleen, skin, lung, lymph nodes, and ovaries	Model for malignant metastatic melanoma	Otsuka et al., 1998
Smad3	Targeted deletion of *Smad3* by homologous recombination	Mice become moribund with metastatic colorectal carcinomas at 4–6 months of age.	Regional and distant lymph nodes	Model for malignant colorectal cancer and for assaying role of transforming growth factor β (TGFβ) signaling in metastasis	Zhu et al., 1998

phenotype and to eventually test therapeutic regimens that target these genes or other components of these pathways.

Signals at the cell surface relay critical messages to the cell. The ectopic expression of another cell surface receptor, the insulinlike growth factor 1 receptor (IGF-1R), also yields metastasis in the Rip1Tag2 model of pancreatic cancer. IGF-1R, the receptor tyrosine kinase for IGF-I, is variably upregulated in early steps of tumorigenesis and focally overexpressed at the invasive margins of carcinomas. The ectopic expression of *IGF-1R* was studied in the Rip1Tag2 mouse model to determine its role in the invasive and metastatic phenotype. Seventy-seven percent of double transgenic mice analyzed died of metastatic ß-cell tumors that expressed dramatically decreased levels of E-cadherin. Since these mice are more than 20 times as likely to develop metastases than the dominant-negative E-cadherin Rip1Tag2 mice, it can be concluded that *IGF-1R* signaling regulates additional factors for the metastatic phenotype aside from down regulation of E-cadherin.

TGFß signaling has also been implicated in cancer and metastasis. TGFß is a cytokine that binds to specific cell surface receptors to signal growth inhibition of many cells (Derynck, 1994; Liu et al., 1997; Massague, 1996). Loss of TGFß signaling results in the reversal of its growth inhibitory properties and has been associated with human cancers (Eppert et al., 1996; Markowitz et al., 1995; Polyak, 1996; Thiagalingam et al., 1996; Wang et al., 1995). Genetically modified mice containing a deletion of individual components of the TGFß pathway, including TGFß ligands and receptors and intracellular transducers of the TGFß signal such as Smad2 and Smad4, do not develop cancer (Dickson et al., 1995; Kaartinen et al., 1995; Kulkarni et al., 1993; Oshima et al., 1996; Sanford et al., 1997; Shull et al., 1992; Sirard et al., 1998; Waldrip et al., 1998; Yang et al., 1998). However, mice harboring a deletion in Smad3, another intracellular TGFß signal transducer, do develop metastatic colorectal cancer, indicating that the inactivation of the TGFß pathway leads to the formation of tumors and eventually to metastasis (Zhu et al., 1998). Furthermore, inhibition of TGF signaling by expression of a dominant-negative mutant of the TGFß type II receptor results in decreased breast cancer metastasis to bone in an orthotopic model system (Yin et al., 1999). Taken together, these results emphasize the importance of further investigation of the TGFß signaling pathway in metastasis.

OTHER IN VIVO MODELS OF METASTASIS

Other genetically modified mouse models have been developed that exhibit metastasis and thus provide models to study the progression of tumors to the metastatic phenotype. Two of these models, p53R172HΔG and TRAMP, will be discussed here. Although the initiating event that leads to tumorigenesis is known for both of the models, p53 mutation for the first and expression of the SV40 T antigen for the second, the underlying mechanism that supports the metastatic phenotype remains a mystery.

The *p53* model gives rise to spontaneous sarcomas and carcinomas that metastasize (Liu et al., 2000). In attempts to examine the importance of p53 missense mutations in tumorigenesis, mice that express one of the hot-spot mutations, an arg-to-his mutation at amino acid 175 (corresponding to amino acid 172 in the mouse), were generated. Due to an additional regulatory mutation, these mice, *p53R172HΔg*, inherit a hypomorphic allele that expresses low levels of mutant p53. The *p53R172HΔg* heterozygous mice spontaneously develop tumors that metastasize at high frequency. The three main tumor types in these mice are lymphomas, sarcomas, and carcinomas, which develop at 54%, 29%, and 16% frequencies, respectively. Approximately 70% of the osteosarcomas produce gross metastasis to the lung and liver. Carcinomas were more common in *p53R172HΔg* heterozygous mice than in $p53^{+/-}$ mice and 40% of these metastasized also to the lung and liver. Importantly, not all tumors metastasized as fibrosarcomas, which were also common in these mice, never metastasized. These data suggest that the determinants of metastasis may be present in the tumor to begin with and may not be the same for all tissue types. This will be an important in vivo model with which to probe the mechanisms of metastasis.

The TRAMP model mimics human prostate cancer, both in the stepwise progression characteristic of prostate cancer and in its ability to form metastasis with a pattern of spread that closely resembles that seen in the human condition (Gingrich et al., 1996). These mice have been utilized in a number of subsequent studies of the progression of cancer and possible therapeutic targets. For instance, the TRAMP mice were used to test the effect of castration (androgen deprivation) on tumor progression and metastasis. It was determined that, although overall tumor burden in the castrated mice was decreased, there was no decrease or delay in the formation of metastasis (Gingrich et al., 1996). Additionally, markers of prostate tumor progression such as an increase in IGF-1 levels (Kaplan et al., 1999), a decrease in C-CAM levels (Pu et al., 1999), and an increase in expression of Huntington-interacting protein 1 (HIP1) (Rao et al., 2002) have been identified in the TRAMP mice. These factors represent putative prognostic factors and therapeutic targets. Furthermore, cells derived from the tumors in these mice have been utilized to determine the effects of various proteins on metastasis. Maspin, a serine protease inhibitor, was transfected into TRAMP tumor cell lines and shown to decrease metastatic potential in in vitro metastasis assays (Abraham et al., 2003).

The TRAMP mouse model has been additionally exploited as a preclinical model for testing various therapies. The effect of green tea polyphenols (GTPs) on prostate tumor progression and metastasis was tested: GTPs delayed tumor incidence and burden, decreased the appearance of cancer progression markers, and almost completely inhibited distant site metastasis in the TRAMP mouse model (Gupta et al., 2001). Additionally, E-7869 (R-flurbiprofen) drug administration and addition of a dominant-negative TGFβ type II receptor have been tested in TRAMP mice. Both reduced the incidence of metastasis (Shah et al., 2002; Wechter et al., 2000). The promising outcome relating to treatment of prostate cancer in mice with the drug E-7869 has led to two concurrent phase 1–2a clinical trials in humans for its use in prostate and colon cancer (Keegan and Loughman, 2001). Thus the TRAMP mouse has served as an important tool for examining prostate cancer at the molecular level as well as for testing therapeutic regimens in preclinical trials. It serves as a paradigm for the study of these types of metastatic models.

The few existing mouse models of metastasis (summarized in Table 17.3) have begun to be exploited to determine the underlying molecular mechanisms and possible clinical interventions that can eventually lead to the prevention of metastatic disease. These models are limiting, and the generation of new models with more in-depth analysis will lead to a better understanding of the molecular changes that promote metastasis and which steps may be potential targets for therapeutic intervention.

CONCLUSIONS

In summary, we have outlined the numerous steps that are involved in metastasis by first defining the process and then discussing the dearth of genetically engineered mouse models that have probed metastasis in vivo. One theme that has emerged is the unexpected or unpredictable nature of the loss-of-function mutations in what are thought to be critical proteins in cell and matrix adhesion. Considering the importance of these molecules in development, it is remarkable that both *MMP2* and *MMP9* deletions result in viable mice. The issue of redundancy and the prospect of generating mice that contain deletions in two or more of these types of genes will therefore need to be addressed to determine more clearly their roles in metastasis. Furthermore, the role of many of the gene products with suspected roles in metastasis have been studied in the context of pancreatic cancer using the RipTag2 mice. Perhaps the success achieved by using this model is in part due to the preexisting tumor phenotype, which initiates a cascade of events that are necessary but insufficient for metastasis to occur. The introduction of additional alterations into the Rip1Tag2 background then provides the final steps sufficient to confer the metastatic phenotype. Models like the p53R172HΔg model with metastatic capabilities will be priceless to study the function of the metastasis suppressors in vivo. No comprehensive study exists for metastasis in other tumor models. Additionally, no gene expression profiles of metastasis from mouse tumors have been performed. These studies are essential in light of recent controversial observations in human tumors that the molecular signature of some metastatic lesions most closely resembles the primary tumor (Fidler and Kripke, 2003; Ramaswamy et al., 2003a, b). New metastatic mouse models therefore provide exciting opportunities to delineate individual steps, even early ones, in the progression of primary tumors to metastasis.

Considering the number of steps the cell must complete to metastasize, it is amazing that metastasis occurs at all. Yet metastasis is the cause of death in 90% of cancer patients (Sporn, 1996). Plainly, additional mouse models will be invaluable to the understanding of this complicated process and will facilitate research with direct clinical implications for the treatment of metastatic disease.

ACKNOWLEDGMENT

This work was supported by National Institutes of Health grant number CA82577 and The Kadoorie Charitable Foundation.

REFERENCES

Abraham S, Zhang W, Greenberg N, Zhang M (2003): Maspin functions as tumor suppressor by increasing cell adhesion to extracellular matrix in prostate tumor cells. *J Urol* 169(3):1157–1161.

Arbones ML, Ord DC, Ley K, Ratech H, Maynard-Curry C, Otten G, Capon DJ, Tedder TF (1994): Lymphocyte homing and leukocyte rolling and migration are impaired in L-selectin-deficient mice. *Immunity* 1(4):247–260.

Bergers G, Brekken R, McMahon G, Vu TH, Itoh T, Tamaki K, Tanzawa K, Thorpe P, Itohara S, Werb Z, Hanahan D (2000): Matrix metalloproteinase-9 triggers the angiogenic switch during carcinogenesis. *Nat Cell Biol* 2(10):737–744.

Brinckerhoff CE, Matrisian LM (2002): Matrix metalloproteinases: A tail of a frog that became a prince. *Nat Rev Mol Cell Biol* 3(3):207–214.

Cavallaro U, Christofori G (2001): Cell adhesion in tumor invasion and metastasis: Loss of the glue is not enough. *Biochim Biophys Acta* 1552(1):39–45.

Chambers AF, Groom AC, MacDonald IC (2002): Dissemination and growth of cancer cells in metastatic sites. *Nat Rev Cancer* 2(8):563–572.

Chambers AF, Matrisian LM (1997): Changing views of the role of matrix metalloproteinases in metastasis. *J Natl Cancer Inst* 89(17):1260–1270.

Christofori G, Semb H (1999): The role of the cell-adhesion molecule E-cadherin as a tumour-suppressor gene. *Trends Biochem Sci* 24(2):73–76.

Coussens LM, Fingleton B, Matrisian LM (2002): Matrix metalloproteinase inhibitors and cancer: Trials and tribulations. *Science* 295(5564):2387–2392.

Coussens LM, Tinkle CL, Hanahan D, Werb Z (2000): MMP-9 supplied by bone marrow-derived cells contributes to skin carcinogenesis. *Cell* 103(3):481–490.

Cremer H, Lange R, Christoph A, Plomann M, Vopper G, Roes J, Brown R, Baldwin S, Kraemer P, Scheff S, Barthels D, Rajewsky K, Willie W (1994): Inactivation of the N-CAM gene in mice results in size reduction of the olfactory bulb and deficits in spatial learning. *Nature* 367(6462):455–459.

Debies MT, Welch DR (2001): Genetic basis of human breast cancer metastasis. *J Mammary Gland Biol Neoplasia* 6(4):441–451.

Derynck R (1994): TGF-beta-receptor-mediated signaling. *Trends Biochem Sci* 19(12):548–553.

Dickson MC, Martin JS, Cousins FM, Kulkarni AB, Karlsson S, Akhurst RJ (1995): Defective haematopoiesis and vasculogenesis in transforming growth factor-beta 1 knock out mice. *Development* 121(6):1845–1854.

Eliceiri BP, Cheresh DA (1999): The role of alpha integrins during angiogenesis: Insights into potential mechanisms of action and clinical development. *J Clin Invest* 103(9):1227–1230.

Eppert K, Scherer SW, Ozcelik H, Pirone R, Hoodless P, Kim H, Tsui LC, Bapat B, Gallinger S, Andrulis IL, Thomsen GH, Wrana JL, Attisano L (1996): MADR2 maps to 18q21 and encodes a TGFbeta-regulated MAD-related protein that is functionally mutated in colorectal carcinoma. *Cell* 86(4):543–552.

Fassler R, Meyer M (1995): Consequences of lack of beta 1 integrin gene expression in mice. *Genes Dev* 9(15):1896–1908.

Fidler IJ (2001): Seed and soil revisited: Contribution of the organ microenvironment to cancer metastasis. *Surg Oncol Clin N Am* 10(2):257–269, vii–viiii.

Fidler IJ, Kripke ML (2003): Genomic analysis of primary tumors does not address the prevalence of metastatic cells in the population. *Nat Genet* 34(1):23.

Gannon G, Mandriota SJ, Cui L, Baetens D, Pepper MS, Christofori G (2002): Overexpression of vascular endothelial growth factor-A165 enhances tumor angiogenesis but not metastasis during beta-cell carcinogenesis. *Cancer Res* 62(2):603–608.

Gingrich JR, Barrios RJ, Morton RA, Boyce BF, DeMayo FJ, Finegold MJ, Angelopoulou R, Rosen JM, Greenberg NM (1996): Metastatic prostate cancer in a transgenic mouse. *Cancer Res* 56(18):4096–4102.

Gupta S, Hastak K, Ahmad N, Lewin JS, Mukhtar H (2001): Inhibition of prostate carcinogenesis in TRAMP mice by oral infusion of green tea polyphenols. *Proc Natl Acad Sci USA* 98(18):10350–10355.

Hanahan D (1985): Heritable formation of pancreatic beta-cell tumours in transgenic mice expressing recombinant insulin/simian virus 40 oncogenes. *Nature* 315(6015):115–122.

Hanahan D, Weinberg RA (2000): The hallmarks of cancer. *Cell* 100(1):57–70.

Herzig M, Christofori G (2002): Recent advances in cancer research: Mouse models of tumorigenesis. *Biochim Biophys Acta* 1602(2):97–113.

Hiratsuka S, Nakamura K, Iwai S, Murakami M, Itoh T, Kijima H, Shipley JM, Senior RM, Shibuya M (2002): MMP9 induction by vascular endothelial growth factor receptor-1 is involved in lung-specific metastasis. *Cancer Cell* 2(4):289–300.

Hodivala-Dilke KM, McHugh KP, Tsakiris DA, Rayburn H, Crowley D, Ullman-Cullere M, Ross FP, Coller BS, Teitelbaum S, Hynes RO (1999): Beta3-integrin-deficient mice are a model for Glanzmann thrombasthenia showing placental defects and reduced survival. *J Clin Invest* 103(2):229–238.

Holmbeck K, Bianco P, Caterina J, Yamada S, Kromer M, Kuznetsov SA, Mankani M, Robey PG, Poole AR, Pidoux I, Ward JM, Birkedal-Hansen H (1999): MT1-MMP-deficient mice develop dwarfism, osteopenia, arthritis, and connective tissue disease due to inadequate collagen turnover. *Cell* 99(1):81–92.

Holzenberger M, Dupont J, Ducos B, Leneuve P, Geloen A, Even PC, Cervera P, Le Bouc Y (2003a): IGF-1 receptor regulates lifespan and resistance to oxidative stress in mice. *Nature* 421(6919):182–187.

Hood JD, Cheresh DA (2002): Role of integrins in cell invasion and migration. *Nat Rev Cancer* 2(2):91–100.

Huang X, Griffiths M, Wu J, Farese Jr RV, Sheppard D (2000): Normal development, wound healing, and adenovirus susceptibility in beta5-deficient mice. *Mol Cell Biol* 20(3):755–759.

Huang XZ, Wu JF, Cass D, Erle DJ, Corry D, Young SG, Farese Jr RV, Sheppard D (1996): Inactivation of the integrin beta 6 subunit gene reveals a role of epithelial integrins in regulating inflammation in the lung and skin. *J Cell Biol* 133(4):921–928.

Hynes RO (2002): A reevaluation of integrins as regulators of angiogenesis. *Nat Med* 8(9):918–921.

Itoh T, Ikeda T, Gomi H, Nakao S, Suzuki T, Itohara S (1997): Unaltered secretion of beta-amyloid precursor protein in gelatinase A (matrix metalloproteinase 2)-deficient mice. *J Biol Chem* 272(36):22389–22392.

Itoh T, Tanioka M, Yoshida H, Yoshioka T, Nishimoto H, Itohara S (1998): Reduced angiogenesis and tumor progression in gelatinase A-deficient mice. *Cancer Res* 58(5):1048–1051.

Jansen AP, Verwiebe EG, Dreckschmidt NE, Wheeler DL, Oberley TD, Verma AK (2001): Protein kinase C-epsilon transgenic mice: a unique model for metastatic squamous cell carcinoma. *Cancer Res* 61(3):808–812.

Jeffers M, Schmidt L, Nakaigawa N, Webb CP, Weirich G, Kishida T, Zbar B, Vande Woude GF (1997): Activating mutations for the met tyrosine kinase receptor in human cancer. *Proc Natl Acad Sci USA* 94(21):11445–11450.

Kaartinen V, Voncken JW, Shuler C, Warburton D, Bu D, Heisterkamp N, Groffen J (1995): Abnormal lung development and cleft palate in mice lacking TGF-beta 3 indicates defects of epithelial-mesenchymal interaction. *Nat Genet* 11(4):415–421.

Kaplan PJ, Mohan S, Cohen P, Foster BA, Greenberg NM (1999): The insulin-like growth factor axis and prostate cancer: Lessons from the transgenic adenocarcinoma of mouse prostate (TRAMP) model. *Cancer Res* 59(9):2203–2209.

Keegan P, Loughman BE (2001): Early clinical trials of chemopreventive and biologic agents: Designs, populations, and endpoints. *Urology* 57(4 Suppl 1):216–219.

Khokha R, Martin DC, Fata JE (1995): Utilization of transgenic mice in the study of matrix degrading proteinases and their inhibitors. *Cancer Metastasis Rev* 14(2):97–111.

Kulkarni AB, Huh CG, Becker D, Geiser A, Lyght M, Flanders KC, Roberts AB, Sporn MB, Ward JM, Karlsson S (1993): Transforming growth factor beta 1 null mutation in mice causes excessive inflammatory response and early death. *Proc Natl Acad Sci USA* 90(2):770–774.

Leco KJ, Waterhouse P, Sanchez OH, Gowing KL, Poole AR, Wakeham A, Mak TW, Khokha R (2001): Spontaneous air space enlargement in the lungs of mice lacking tissue inhibitor of metalloproteinases-3 (TIMP-3). *J Clin Invest* 108(6):817–829.

Lee JH, Han SU, Cho H, Jennings B, Gerrard B, Dean M, Schmidt L, Zbar B, Vande Woude GF (2000): A novel germ line juxtamembrane Met mutation in human gastric cancer. *Oncogene* 19(43):4947–4953.

Liu G, McDonnell TJ, Montes de Oca Luna R, Kapoor M, Mims B, El-Naggar AK, Lozano G (2000): High metastatic potential in mice inheriting a targeted p53 missense mutation. *Proc Natl Acad Sci USA* 97(8):4174–4179.

Liu X, Sun Y, Constantinescu SN, Karam E, Weinberg RA, Lodish HF (1997): Transforming growth factor beta-induced phosphorylation of Smad3 is required for growth inhibition and transcriptional induction in epithelial cells. *Proc Natl Acad Sci USA* 94(20):10669–10674.

Lopez T, Hanahan D (2002): Elevated levels of IGF-1 receptor convey invasive and metastatic capability in a mouse model of pancreatic islet tumorigenesis. *Cancer Cell* 1(4):339–353.

Mandriota SJ, Jussila L, Jeltsch M, Compagni A, Baetens D, Prevo R, Banerji S, Huarte J, Montesano R, Jackson DG, Orci L, Alitalo K, Christofori G, Pepper MS (2001): Vascular endothelial growth factor-C-mediated lymphangiogenesis promotes tumour metastasis. *EMBO J* 20(4):672–682.

Markowitz S, Wang J, Myeroff L, Parsons R, Sun L, Lutterbaugh J, Fan RS, Zborowska E, Kinzler KW, Vogelstein B, Brattain M, Willson JKV (1995): Inactivation of the type II TGF-beta receptor in colon cancer cells with microsatellite instability. *Science* 268(5215):1336–1338.

Massague J (1996): TGFbeta signaling: Receptors, transducers, and Mad proteins. *Cell* 85(7):947–950.

Masson R, Lefebvre O, Noel A, Fahime ME, Chenard MP, Wendling C, Kebers F, LeMeur M, Dierich A, Foidart JM, Basset P, Rio MC (1998): In vivo evidence that the stromelysin-3 metalloproteinase contributes in a paracrine manner to epithelial cell malignancy. *J Cell Biol* 140(6):1535–1541.

Matsumoto K, Nakamura T (1996): Emerging multipotent aspects of hepatocyte growth factor. *J Biochem (Tokyo)* 119(4):591–600.

McCawley LJ, Matrisian LM (2001): Matrix metalloproteinases: They're not just for matrix anymore! *Curr Opin Cell Biol* 13(5):534–540.

McClatchey AI, Saotome I, Mercer K, Crowley D, Gusella JF, Bronson RT, Jacks T (1998): Mice heterozygous for a mutation at the Nf2 tumor suppressor locus develop a range of highly metastatic tumors. *Genes Dev* 12(8):1121–1133.

Mudgett JS, Hutchinson NI, Chartrain NA, Forsyth AJ, McDonnell J, Singer II, Bayne EK, Flanagan J, Kawka D, Shen CF, Stevens K, Chen H, Trumbauer M, Visco DM (1998): Susceptibility of stromelysin 1-deficient mice to collagen-induced arthritis and cartilage destruction. *Arthritis Rheum* 41(1):110–121.

Narumiya S, Ishizaki T, Watanabe N (1997): Rho effectors and reorganization of actin cytoskeleton. *FEBS Lett* 410(1):68–72.

Nieman MT, Prudoff RS, Johnson KR, Wheelock MJ (1999): N-cadherin promotes motility in human breast cancer cells regardless of their E-cadherin expression. *J Cell Biol* 147(3):631–644.

Nothnick WB (2000): Disruption of the tissue inhibitor of metalloproteinase-1 gene results in altered reproductive cyclicity and uterine morphology in reproductive-age female mice. *Biol Reprod* 63(3):905–912.

Oshima M, Oshima H, Taketo MM (1996): TGF-beta receptor type II deficiency results in defects of yolk sac hematopoiesis and vasculogenesis. *Dev Biol* 179(1):297–302.

Otsuka T, Takayama H, Sharp R, Celli G, LaRochelle WJ, Bottaro DP, Ellmore N, Vieira W, Owens JW, Anver M, Merlino G (1998): c-Met autocrine activation induces development of malignant melanoma and acquisition of the metastatic phenotype. *Cancer Res* 58(22):5157–5167.

Padera TP, Kadambi A, di Tomaso E, Carreira CM, Brown EB, Boucher Y, Choi NC, Mathisen D, Wain J, Mark EJ, Munn LL, Jain RK (2002): Lymphatic metastasis in the absence of functional intratumor lymphatics. *Science* 296(5574): 1883–1886.

Perl AK, Dahl U, Wilgenbus P, Cremer H, Semb H, Christofori G (1999): Reduced expression of neural cell adhesion molecule induces metastatic dissemination of pancreatic beta tumor cells. *Nat Med* 5(3):286–291.

Perl AK, Wilgenbus P, Dahl U, Semb H, Christofori G (1998): A causal role for E-cadherin in the transition from adenoma to carcinoma. *Nature* 392(6672):190–193.

Polyak K (1996): Negative regulation of cell growth by TGF beta. *Biochim Biophys Acta* 1242(3):185–199.

Price JE (1994): Analyzing the metastatic phenotype. *J Cell Biochem* 56(1):16–22.

Pu YS, Luo W, Lu HH, Greenberg NM, Lin SH, Gingrich JR (1999): Differential expression of C-CAM cell adhesion molecule in prostate carcinogenesis in a transgenic mouse model. *J Urol* 162(3 Pt 1):892–896.

Qian F, Hanahan D, Weissman IL (2001): L-selectin can facilitate metastasis to lymph nodes in a transgenic mouse model of carcinogenesis. *Proc Natl Acad Sci USA* 98(7):3976–3981.

Ramaswamy S, Ross KN, Lander ES, Golub TR (2003a): A molecular signature of metastasis in primary solid tumors. *Nat Genet* 33(1):49–54.

Ramaswamy S, Ross KN, Lander ES, Golub TR (2003b): Reply to "Genomic analysis of primary tumors does not address the prevalence of metastatic cells in the population" and "Genetic background is an important determinant of metastatic potential." *Nat Genet* 34(1):25.

Rao DS, Hyun TS, Kumar PD, Mizukami IF, Rubin MA, Lucas PC, Sanda MG, Ross TS (2002): Huntington-interacting protein 1 is overexpressed in prostate and colon cancer and is critical for cellular survival. *J Clin Invest* 110(3):351–360.

Reynolds LE, Wyder L, Lively JC, Taverna D, Robinson SD, Huang X, Sheppard D, Hynes RO, Hodivala-Dilke KM (2002): Enhanced pathological angiogenesis in mice lacking beta3 integrin or beta3 and beta5 integrins. *Nat Med* 8(1):27–34.

Rong S, Segal S, Anver M, Resau JH, Vande Woude GF (1994): Invasiveness and metastasis of NIH 3T3 cells induced by Met-hepatocyte growth factor/scatter factor autocrine stimulation. *Proc Natl Acad Sci USA* 91(11):4731–4735.

Rosen EM, Lamszus K, Laterra J, Polverini PJ, Rubin JS, Goldberg ID (1997): HGF/SF in angiogenesis. *Ciba Found Symp* 212:215–226; discussion 227–229.

Sahai E, Marshall CJ (2002): RHO-GTPases and cancer. *Nat Rev Cancer* 2(2):133–142.

Sanford LP, Ormsby I, Gittenberger-de Groot AC, Sariola H, Friedman R, Boivin GP, Cardell EL, Doetschman T (1997): TGFbeta2 knockout mice have multiple developmental defects that are non-overlapping with other TGFbeta knockout phenotypes. *Development* 124(13):2659–2670.

Schmidt L, Junker K, Nakaigawa N, Kinjerski T, Weirich G, Miller M, Lubensky I, Neumann HP, Brauch H, Decker J, Vocke C, Brown JA, Jenkins R, Richard S, Bergerheim U, Gerrard B, Dean M, Linehan WM, Zbar B (1999): Novel mutations of the MET proto-oncogene in papillary renal carcinomas. *Oncogene* 18(14):2343–2350.

Shah AH, Tabayoyong WB, Kundu SD, Kim SJ, Van Parijs L, Liu VC, Kwon E, Greenberg NM, Lee C (2002): Suppression of tumor metastasis by blockade of transforming growth factor beta signaling in bone marrow cells through a retroviral-mediated gene therapy in mice. *Cancer Res* 62(24):7135–7138.

Shipley JM, Wesselschmidt RL, Kobayashi DK, Ley TJ, Shapiro SD (1996): Metalloelastase is required for macrophage-mediated proteolysis and matrix invasion in mice. *Proc Natl Acad Sci USA* 93(9):3942–3946.

Shull MM, Ormsby I, Kier AB, Pawlowski S, Diebold RJ, Yin M, Allen R, Sidman C, Proetzel G, Calvin D, Annunziata N, Doetschman T (1992): Targeted disruption of the mouse transforming growth factor-beta 1 gene results in multifocal inflammatory disease. *Nature* 359(6397):693–699.

Siegelman MH, van de Rijn M, Weissman IL (1989): Mouse lymph node homing receptor cDNA clone encodes a glycoprotein revealing tandem interaction domains. *Science* 243(4895):1165–1172.

Sirard C, de la Pompa JL, Elia A, Itie A, Mirtsos C, Cheung A, Hahn S, Wakeham A, Schwartz L, Kern SE, Rossant J, Mak TW (1998): The tumor suppressor gene Smad4/Dpc4 is required for gastrulation and later for anterior development of the mouse embryo. *Genes Dev* 12(1):107–119.

Sligh JE Jr, Ballantyne CM, Rich SS, Hawkins HK, Smith CW, Bradley A, Beaudet AL (1993): Inflammatory and immune responses are impaired in mice deficient in intercellular adhesion molecule 1. *Proc Natl Acad Sci USA* 90(18):8529–8533.

Sporn MB (1996): The war on cancer. *Lancet* 347(9012): 1377–1381.

Steeg PS (2003): Metastasis suppressors alter the signal transduction of cancer cells. *Nat Rev Cancer* 3(1):55–63.

Stephens LE, Sutherland AE, Klimanskaya IV, Andrieux A, Meneses J, Pedersen RA, Damsky CH (1995): Deletion of beta 1 integrins in mice results in inner cell mass failure and peri-implantation lethality. *Genes Dev* 9(15):1883–1895.

Suyama E, Kawasaki H, Kasaoka T, Taira K (2003): Identification of genes responsible for cell migration by a library of randomized ribozymes. *Cancer Res* 63(1):119–124.

Thiagalingam S, Lengauer C, Leach FS, Schutte M, Hahn SA, Overhauser J, Willson JK, Markowitz S, Hamilton SR, Kern SE, Kinzler KW, Vogelstein B (1996): Evaluation of candidate tumour suppressor genes on chromosome 18 in colorectal cancers. *Nat Genet* 13(3):343–346.

Tomasiewicz H, Ono K, Yee D, Thompson C, Goridis C, Rutishauser U, Magnuson T (1993): Genetic deletion of a neural cell adhesion molecule variant (N-CAM-180) produces distinct defects in the central nervous system. *Neuron* 11(6):1163–1174.

van der Flier A, Sonnenberg A (2001): Function and interactions of integrins. *Cell Tissue Res* 305(3):285–298.

Vleminckx K, Vakaet Jr L, Mareel M, Fiers W, van Roy F (1991): Genetic manipulation of E-cadherin expression by epithelial tumor cells reveals an invasion suppressor role. *Cell* 66(1):107–119.

Vu TH, Shipley JM, Bergers G, Berger JE, Helms JA, Hanahan D, Shapiro SD, Senior RM, Werb Z (1998): MMP-9/gelatinase B is a key regulator of growth plate angiogenesis and apoptosis of hypertrophic chondrocytes. *Cell* 93(3):411–422.

Wagner N, Lohler J, Kunkel EJ, Ley K, Leung E, Krissansen G, Rajewsky K, Muller W (1996): Critical role for beta7 integrins in formation of the gut-associated lymphoid tissue. *Nature* 382(6589):366–370.

Waldrip WR, Bikoff EK, Hoodless PA, Wrana JL, Robertson EJ (1998): Smad2 signaling in extraembryonic tissues determines anterior-posterior polarity of the early mouse embryo. *Cell* 92(6):797–808.

Wang J, Sun L, Myeroff L, Wang X, Gentry LE, Yang J, Liang J, Zborowska E, Markowitz S, Willson JK, Brattain MG (1995): Demonstration that mutation of the type II transforming growth factor beta receptor inactivates its tumor suppressor activity in replication error-positive colon carcinoma cells. *J Biol Chem* 270(37):22044–22049.

Webb CP, Vande Woude GF (2000): Genes that regulate metastasis and angiogenesis. *J Neurooncol* 50(1–2):71–87.

Wechter WJ, Leipold DD, Murray Jr ED, Quiggle D, McCracken JD, Barrios RS, Greenberg NM (2000): E-7869 (R-flurbiprofen) inhibits progression of prostate cancer in the TRAMP mouse. *Cancer Res* 60(8):2203–2208.

Wilson CL, Heppner KJ, Labosky PA, Hogan BL, Matrisian LM (1997): Intestinal tumorigenesis is suppressed in mice lacking the metalloproteinase matrilysin. *Proc Natl Acad Sci USA* 94(4):1402–1407.

Yang JT, Rayburn H, Hynes RO (1995): Cell adhesion events mediated by alpha 4 integrins are essential in placental and cardiac development. *Development* 121(2):549–560.

Yang X, Li C, Xu X, Deng C (1998): The tumor suppressor SMAD4/DPC4 is essential for epiblast proliferation and mesoderm induction in mice. *Proc Natl Acad Sci USA* 95(7):3667–3672.

Yin JJ, Selander K, Chirgwin JM, Dallas M, Grubbs BG, Wieser R, Massague J, Mundy GR, Guise TA (1999): TGF-beta signaling blockade inhibits PTHrP secretion by breast cancer cells and bone metastases development. *J Clin Invest* 103(2):197–206.

Yu Y, Merlino G (2002): Constitutive c-Met signaling through a nonautocrine mechanism promotes metastasis in a transgenic transplantation model. *Cancer Res* 62(10):2951–2956.

Zhu J, Motejlek K, Wang D, Zang K, Schmidt A, Reichardt LF (2002): Beta8 integrins are required for vascular morphogenesis in mouse embryos. *Development* 129(12):2891–2903.

Zhu Y, Richardson JA, Parada LF, Graff JM (1998): Smad3 mutant mice develop metastatic colorectal cancer. *Cell* 94(6):703–714.

18

IMMUNOLOGIC STUDY OF TUMORS IN MOUSE MODELS

TAHA MERGHOUB AND ALAN N. HOUGHTON
Immunology Program, Memorial Sloan-Kettering Cancer Center, New York, New York

Mouse tumor models have played a central role in the history of cancer immunology and more broadly in immunology. Studies of transplantable tumors opened the field of transplantation immunology. Serologic analyses of antibodies against tumors led to the discovery of major histocompatibility complex (MHC) antigens that determine transplantation immunity as well as other alloantigens and lymphocyte subset markers, including the well-known CD4+ and CD8+ subsets of T cells. In the present era, genetically manipulated mice are crucial tools to (1) understand the role of various components and cells of the immune system in promoting, accelerating, or rejecting tumors; (2) investigate immunity against cancers at different stages of tumor progression; and (3) develop and test new prevention and therapeutic strategies against cancer.

INTRODUCTION

The immune system evolved to protect the host from microbial infections based on the recognition of self versus nonself. However, the recognition of cancer cells by the immune system poses a problem given the fact that normal cells also express all genes expressed by cancer cells. For this reason, the challenge in tumor immunology is to use the remarkable filter of the immune system to detect altered molecules that are involved in the pathogenesis of cancers. In this scenario, the immune system discerns altered self and spares normal tissues or self. One of the surprising findings is that the immune system recognizes unaltered tissue-specific molecules even more frequently on cancer cells. Both altered and tissue-specific antigen targets could be used for the prevention, therapy, and diagnosis of cancer. Therefore, the major scope of the field of tumor immunology is to understand how the immune system can recognize and reject tumor cells.

Most milestones in tumor immunology have been established in mouse models. In fact, the experimental foundation for cancer immunology comes from studies of immune rejection of chemically and ultraviolet- (UV) induced tumors in syngeneic (genetically identical) mice. Following the development of inbred mouse strains in the 1940s–1950s, tumor transplantation studies were able to show that mice could be immunized against transplants of chemically induced tumors that arose in the same mouse strain. These experiments established that immunity to cancer was real and that immunity in transplantation experiments was not an artifact of transplantation immunity due to genetic origins of the tumor and recipient mice.

Over the past 20 years, immunology has made important contributions to new therapies to cancer, including some of the more successful recent therapies. Examples include monoclonal antibodies, interleukins, colony-stimulating factors, interferons (IFNs), and attenuated mycobacteria [bacille Calmette–Guérin (BCG) for bladder cancer]. Beyond cancer therapy, studies in tumor immunology led to the discovery of cellular immunity, MHC molecules, and several cytokines as well as to a better understanding of immunity against infectious organisms.

Mouse Models of Human Cancer, edited by Eric C. Holland
ISBN 0-471-44460-X

Discovery of MHC Antigens

Mice have been used in cancer research for over a century. Initially, mice were used for same-species tumor transplantation experiments and drug treatment studies. Tumor transplantation studies paved the way for the discovery of the MHC and the development of inbred mouse strains. The first systematic observations were carried out ~100 years ago, at the beginning of the last century, by Little and Tyzzer (1916), who analyzed the fate of tumors transplanted between mice. They frequently observed tumor rejection and demonstrated that several dominant genes influenced the outcome of allogenic tumor grafts (Little and Tyzzer, 1916). Subsequently, Bover (1927) found that skin transplants between identical twins (isotransplantation) were not rejected, which further indicated that genetic differences were involved in determining transplant outcome. The search for genetic factors that influence tumor transplant outcome continued.

In 1921, inbred strains of mice predisposed to tumors were identified and shared among cancer researchers. In 1929, with the founding of the Jackson Laboratory in Bar Harbor, Maine, more strains of mice were created and bred. This institution is now the largest supplier of mice in the United States. A major event occurred ~60 years ago, when Gorer (1936, 1937) serologically identified an MHC antigen in the mouse called "antigen II." The MHC locus was then characterized by Snell, working at Jackson Laboratory, using the sera against antigen II in transplantation studies in congenic mouse lines. In 1948, he proposed to designate the genes in the MHC histocompatibility locus as H genes (Snell, 1948). The designation antigen II was blended with the H locus to create the term H-2, which is used to designate the mouse MHC locus even today.

STUDIES WITH TRANSPLANTABLE TUMOR MODELS

Evidence of tumor-specific rejection of transplanted cancers was mainly established in studies of inbred mice that have been immunized against tumors induced with methylcholanthrene (MCA), a chemical carcinogen. Several groups showed indisputable evidence of tumor-specific rejection of transplanted cancers (Baldwin, 1955; Foley, 1953; Globerson and Feldman, 1964; Klein et al., 1960; Old et al., 1962; Prehn and Main, 1957). One of the most compelling pieces of evidence was brought forward by Prehn and Main (1957) in experiments showing that mice immunized with MCA-induced sarcomas derived from syngeneic mice were able to reject syngeneic tumors but failed to reject syngeneic skin grafts of mouse tumor origin. The concept of tumor-specific transplantation immunity was further consolidated by demonstrating that mice treated surgically to resect an established tumor became resistant to a second challenge with the same tumor (Klein et al., 1960) (Fig. 18.1). Interestingly, resistance was relatively specific to the immunizing tumor, but weak cross-resistance was observed against independently derived tumors of the same histologic type.

Subsequent studies showed that cancers induced with strong mutagens were typically very immunogenic. However, spontaneously arising tumors were usually weakly immunogenic. It has been argued that spontaneous tumors more accurately reflect most tumors that develop in humans, so the question of relative immunogenicity for spontaneous tumors is important. In general, spontaneous tumors can be recognized by the immune system and rejected if the immune system is manipulated, but spontaneous tumors by themselves are typically ineffective at priming or triggering initial immune responses (Uyttenhove et al., 1980). Antigens that confer protection against UV- or chemically induced transplanted tumors are referred to variously as unique antigens, tumor-specific antigens, tumor-specific transplantation antigens, or tumor-specific rejection antigens. The term "specific" refers to antigens that are detected in individual tumors and are not shared by other tumors or normal tissues. Immunity against a unique antigen expressed in individual UV- or chemically derived tumors will only protect against that specific tumor and not against independently derived, UV- or chemically induced tumors of the same histologic type, even if induced with the same chemical carcinogen in the same mouse. Nevertheless, cross-protection between independently induced tumors is observed with UV- or chemically induced tumors, albeit at a lower level. These observations suggest that UV- or chemically induced tumors express strong unique tumor rejection antigens and much weaker shared antigens that are not as effective in inducing tumor immunity. Spontaneous tumors are generally characterized by shared antigens, without detectable strong unique antigens. Tumor antigens that are not restricted to individual tumors are generally shared with normal tissues. In general, unique antigens are much more effective at priming immunity than shared antigens. The molecular basis of unique and shared antigens in cancer is discussed below. These studies showed that the method of tumor induction influences greatly the immunogenicity of the tumor. Spontaneous tumors tend to be less immunogenic than induced tumors (generated through exposure to carcinogen or DNA tumor viruses) in transplantation studies. These tumors may resemble more closely human tumors and thus can serve as better models to understand tumor immunity.

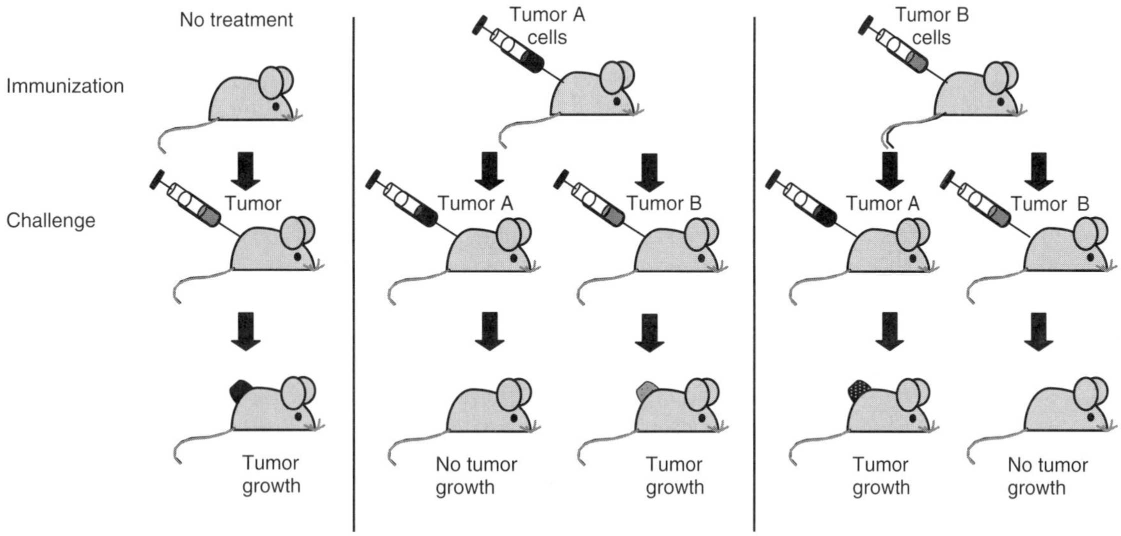

Figure 18.1. Experiments demonstrating existence of transplantation antigens on MCA-induced tumors. In the first step, mice are immunized by inoculation with cells from tumor A or B ("immunization" step). Subsequently, tumors are removed. In the second step, "challenge," the immunized mice are challenged by injection of tumor A or B cells, as specified. Study of tumor progression in these mice shows that only the mice that had been immunized and challenged with the same type of tumor cells rejected their tumors; that is, mice immunized with tumor A reject tumor A but not tumor B, and mice immunized with tumor B reject tumor B but not tumor A. These results suggest that mice can be immunized against tumors/antigen to which they have been previously exposed.

TUMOR ANTIGENS

There have been numerous classifications of tumor antigens, although no consensus has been reached. We will present our favorite classification that includes most tumor antigens described so far.

Unique Antigens

Genes that contain mutations can encode unique antigens. Mutations can lead to increased visibility of an epitope to the immune system through enhanced binding to MHC molecules or T-cell receptors or by changes in protein stability or trafficking. In the case of mutated forms of cyclin-dependent kinase 4 (CDK4) (target of the cell cycle inhibitory kinase p16/INK4a) and β-catenin, the immune system can recognize the products of point mutations that are implicated in the pathogenesis of malignancy. In these cases, point mutations altered amino acid residues leading to enhanced binding of peptides to MHC molecules or T-cell receptors. The immune system can also see products from alternative transcripts, including those from cryptic start sites and alternative reading frames, as well as pseudogenes and antisense DNA strands (Moreau-Aubry et al., 2000; Van Den Eynde et al., 1999).

Shared Antigens

Shared antigens represent the largest class of tumor antigens that have been discovered to date. These antigens are expressed by multiple independent cancers and by selective normal tissues. There are two major subcategories of shared antigens.

Germ Cell Antigens These proteins are expressed by spermatozoa but are normally silenced in somatic cells. Given that germline cells do not express classical MHC molecules, these antigens have been considered as operationally tumor specific during cancer progression. Although germ cells may not be capable of presenting antigen directly to T cells because of deficient MHC expression, if germ cells are taken up by phagocytic antigen-presenting cells (APCs), such as macrophages or dendritic cells (DCs), germ cell antigens could be presented normally to the immune system through cross-presentation. During cancer progression, these genes can be reexpressed. The prototype of this family is the MAGE-1 (melanoma-associated antigen 1) protein, which was the first human tumor antigen to be molecularly identified as being recognized by CD8+ T cells (van der Bruggen et al., 1991). The list of shared germ cell antigens has expanded to products of approximately a dozen genes. Despite their restricted expression, germ cell antigens seem to be poorly immunogenic in mouse models.

Differentiation Antigens These antigens are the most widely recognized and best studied tumor antigens. They are expressed on the cancer cells and their normal counterpart. For example, in melanoma, proteins required in melanin synthesis are differentiation antigens that are relevant to tumor immunity. Their expression is restricted to melanoma cells and normal melanocytes. T-cell tolerance to these antigens during development does not always occur, perhaps because these antigens are not expressed in the thymus. Therefore, reactive T cells are not deleted, although it is likely that T cells with high affinity are deleted. T cells with lower affinity to differentiation antigens remain in the immune repertoire, with the capacity to become responsive upon appropriate priming and activation.

Differentiation antigens have been studied in patients with breast, colon, and pancreatic cancer, but immunity to differentiation antigens has been best characterized in melanoma (Barnd et al., 1989; Disis et al., 1994; Lloyd, 1991). Melanoma and normal melanocytes share the melanosomal differentiation antigens, which are proteins involved in the synthesis of melanin and correlate with various stages of melanocytic differentiation. The two prototypes of these antigens are the carbohydrate antigens, particularly gangliosides, and the melanocyte-specific antigens, particularly the melanosome membrane proteins. In patients with melanoma and in mouse melanoma models, the melanosome proteins are recognized by autoantibodies, CD4+, and CD8+ T cells.

Viral Antigens

Viral antigens are implicated in the pathogenesis of some cancers. The Epstein–Barr virus (EBV) is associated with several tumors, including lymphoid and epithelial nasopharyngeal malignancies. Human papillomavirus (HPV) is involved in the etiology of both benign respiratory papillomas and squamous cell carcinomas. Human T-cell lymphotropic virus type 1 (HTLV-1) is implicated in adult T-cell leukemia/lymphoma. Hepatitis B infection is associated with an increased risk of hepatocellular carcinoma development. Cells infected by viruses provide strong antigens that can be recognized by T cells, but these viruses have evolved mechanisms to escape the immune system.

IMMUNE SURVEILLANCE

The description of tumor antigens that could be recognized by the immune system led Macfarlane Burnet and Lewis Thomas to independently propose the theory of cancer immunosurveillance. The concept of immune surveillance states that the immune system is able to survey for incipient tumors and destroy them, thus protecting immunocompetent hosts against tumor development. Burnet (1964, 1970) proposed that lymphocytes acted by recognizing and eliminating continuously arising nascent transformed cells. Subsequently, several groups tried to test this hypothesis by inducing tumors in immunocompromised animals (Burstein and Law, 1971; Grant and Miller, 1965; Nishizuka et al., 1965; Sanford et al., 1973; Stutman, 1975; Trainin et al., 1967). The identification of mice with inherited immunologic deficiencies leading to absence of thymus (*nu/nu* nude mice) provided immunologists with a tool to test this hypothesis. Stutman found no difference in tumor formation between athymic *nu/nu* mice and syngeneic wild-type mice, with both chemically induced and spontaneous tumors (Stutman, 1974). Other groups have corroborated this observation (Outzen et al., 1975; Rygaard and Povlsen, 1974), and therefore, experiments led to the abandonment of the immunosurveillance theory by most immunologists.

However, nude mice do not completely lack functional T cells (Ikehara et al., 1984) and they do have natural killer (NK) cells, macrophages, and natural antibodies that can mediate tumor rejection. Follow-up experiments with mice specifically deficient in particular molecules on the immune system revived the theory of immune surveillance. IFN-γ and the cytotoxic pore-forming molecule perforin are required to prevent tumor formation in mice, and mice deficient in either of these molecules are more susceptible to induction of cancers. These observations have led to renewed interest in the immune surveillance hypothesis. The definitive evidence supporting the existence of immunosurveillance came from the use of gene-targeted mice that lack the *recombinase-activating gene 1* (*RAG-1*) or *RAG-2*. These genes are required for development of adaptive immune system with functional T and B cells, respectively. RAG-1 and RAG-2 are enzymes involved in the repair of double-stranded DNA breaks, which are exclusively expressed in the lymphoid compartment. Mice deficient for one of these genes fail to rearrange lymphocyte antigen receptors and thus lack T and B cells (Mombaerts et al., 1992; Shinkai et al., 1992). Using the RAG-deficient model system, Schreiber and colleagues demonstrated that mice deficient in adaptive immunity were highly sensitive to MCA-induced and spontaneous tumors (Shankaran et al., 2001). Studies utilizing inbred mouse strains with mutations critical for other components of the immune system show that both innate and adaptive immunity are important in cancer immunosurveillance. T cells, NKT cells, IFNs, and perforin are just some of the immune components that can play a role in immune surveillance. These studies suggest that the immune system may have also evolved to destroy incipient tumors. Most importantly, experiments with mice deficient in specific components of the immune system are valuable in implicating immunity

Table 18.1. Models of Immune-Deficient Mice

Immune Element or Function	Genotype	Phenotype[a]	References
Antigen presentation	*beta-2 microglobulin*$^{-/-}$	Deficient in MHC class I protein expression and Ag presentation	Koller et al., 1990
	MHC II$^{-/-}$	Deficient in MHC class II protein expression and Ag presentation	Hermans et al., 1999; Kontgen et al., 1993; Liljedahl et al., 1998
	TAP-1$^{-/-}$	Deficient in peptide transport and Ag presentation	Dorfman et al., 1997; Grandea et al., 2000
Lymphocytes and myeloid cells	*Prkdc*^{scid}	Deficient in T and B cells	Engel et al., 1997
	RAG1$^{-/-}$	Deficient in T, B, and NKT cells	Mombaerts et al., 1992
	RAG2$^{-/-}$	Deficient in T, B, and NKT cells	Shankaran et al., 2001; Shinkai et al., 1992
	CD4$^{-/-}$	Deficient in helper T cells	Rahemtulla et al., 1991
	CD8$^{-/-}$	Deficient in cytotoxic T cells	Fung-Leung et al., 1991
	$\alpha\beta^{-/-}$	Deficient in functional αβ T cells	Girardi et al., 2001
	$\gamma\delta^{-/-}$	Deficient in functional γδ T cells	Girardi et al., 2001
	*J*α*281*$^{-/-}$	Deficient in NKT cells	Cui et al., 1997
	*Ig*μ *chain*$^{-/-}$	Deficient in B cells	Chow and Bennet, 1989; Kitamura et al., 1991
	CD 19$^{-/-}$	Deficient in B cells	Rickert et al., 1995
	TCR transgenic	T-cell repertoire recognizes one single main peptide	Pircher et al., 1989; Rahemtulla et al., 1991
	*FcR*γ$^{-/-}$	Deficient in activity of Fc receptor	Takai et al., 1994
	*Fc*γ*RII*$^{-/-}$	Deficient in inhibition of Fc receptor	Takai et al., 1996
Cytokines and chemokines	*IFN-*γ$^{-/-}$	Deficient in IFN-γ expression	Kaplan et al., 1998; Shankaran et al., 2001
	*IFN-*γ*R*$^{-/-}$	Deficient in IFN-γ receptor expression	Kaplan et al., 1998; Shankaran et al., 2001
	IL-12$^{-/-}$	Deficient in IL-12 expression	Smyth et al., 2000a
	Stat-1$^{-/-}$	Deficient in STAT-1 expression	Kaplan et al., 1998; Shankaran et al., 2001
	MIP-1 α$^{-/-}$	Deficient in MIP-1α expression	Cook et al., 1995
	CCR7$^{-/-}$	Deficient in CCR7 expression	Forster et al., 1999
Effector molecules	*Perforin*$^{-/-}$	Deficient in cytolytic T cells and NK cells	Smyth et al., 2000b; Street et al., 2001; van den Broek and Hengartner, 2000
	TRAIL$^{-/-}$	Deficient in effector system mediated by T cells and macrophages	Sedger et al., 2002

[a] Ag = antigen; *CCR7* = CC receptor 7; MIP-1 = macrophage inflammatory protein-1.

in tumor growth and in dissecting immune mechanisms (Table 18.1).

APPROACHES TO STUDY MECHANISMS OF TUMOR REJECTION

In experimental studies, mice can be immunized by administering inactivated (e.g., irradiated) tumor cells or transduced or transfected tumor cells secreting inflammatory cytokines or by removal of a primary tumor. Once a mouse has developed immunity, this can be transferred from an immune mouse to a naive host through either T cells or sera (antibodies). Many different components of the immune system (nonspecific and specific, humoral and cellular) can participate in immunity against tumors.

Given that cancer represents many diseases, it is not surprising that each tumor model may be relevant for one type of cancer and have little relevance to other tumors. One possibility is that different sets of components of the immune system have different roles for each tumor type.

Recent studies using genetically deficient transgenic mice have established that both the adaptive and innate immune systems play roles in tumor immunity.

Innate Immunity

The innate immune system is broadly defined as immunity mediated by soluble and cellular components that are present during development, in contrast to the incredibly polymorphic adaptive immunity (T and B cells) that is a response and a result of exposure to pathogens. The innate immune system relies on pattern recognition receptors that are hardwired into macrophages, DCs, neutrophils, and NK cells. Although innate immunity is often ignored as an important component of antitumor immune response, lessons learned from immunity to infectious microorganisms indicate that early host defense determines the nature of downstream adaptive immune responses.

NK Cells. NK cells represent a subtype of lymphocytes that can recognize and kill tumor cells without prior sensitization and without the requirement for MHC restriction. NK cells are thought to have a role in the elimination of nascent tumors. Two main strategies are employed by NK cells to recognize tumor targets. These cells express both inhibitory and activation receptors that can compete for regulating NK cells. Inhibitory receptors are induced to signal by binding to "self" class I MHC molecules, and inhibition is relieved if a cell does not express self-MHC. Many tumor cells downregulate their class I MHC molecules, thus releasing the NK cell from the inhibition provided by class I MHC-specific inhibitory receptors ("missing self-recognition"). More recently, it has become clear that stimulatory receptors expressed by NK cells, NKT cells, T cells, and macrophages (called NKG2D activation receptors) recognize ligands [MHC class I chain related (MIC), H6O, retinoic acid early inducible (Rae1), and UL16 binding proteins (ULBP)] that are upregulated on tumor cells and virally infected cells, but are not expressed by normal cells. These ligands may be specifically upregulated during malignant transformation and probably other stress conditions. Expression of these ligands on tumor cells leads to the rejection of tumors in vivo. Importantly, mice that previously rejected ligand-positive tumor cells develop T-cell immunity to parental (ligand-negative) tumor cells. The recognition of stress-induced self-ligands as a strategy to recognize abnormal self sets a precedent for a new immune recognition strategy of the innate immune system.

The increase in spontaneous tumor incidence with aging in mice (endogenous NK activity decreases with age) is consistent with a role of NK cells in tumor protection. It is worth noting that nude mice, which lack T cells but not NK cells, are able to resist some tumors in tumor challenge experiments through NK cells.

Macrophages. Macrophages that have been activated can mediate cytolysis of tumor cells. The antitumor effect of macrophages can involve cell contact, secretion of cytotoxic molecules, or phagocytosis. Macrophages in varying states of activation differ in their ability to perform antibody-independent macrophage-mediated tumor cytotoxicity and antibody-dependent cellular cytotoxicity. Tumor-infiltrating macrophages secrete cytokines, including Fas ligand, tumor necrosis factor α (TNF-α), and TNF-related apoptosis-inducing ligand (TRAIL), in addition to nitric oxide and other reactive species.

Adaptive Immunity

B Cells and Antibodies. Many studies using tumor challenge in mice have shown the development of antibodies reactive to tumor cells or tumor antigens (Hara et al., 1995; Katsumata et al., 1995; Zhu et al., 2003). These antibodies can prevent tumor growth and protect against tumor challenge. Antibodies, used either as whole serum or as monoclonal antibodies, can be passively transferred for tumor protection and even rejection of established tumors. Some of the most successful new therapies for cancer in humans use monoclonal antibodies. In mice, antibodies have been shown to mediate tumor rejection in leukemia, lymphoma, and melanoma models. An example is the B16 melanoma cell line, a spontaneous, poorly immunogenic tumor (Hara et al., 1995). B16 grows aggressively in syngeneic C57BL/6 mice as intradermal, subcutaneous, or metastatic lung tumor. Similar to human melanomas, B16 expresses several immunogenic melanosomal differentiation antigens, such as tyrosinase-related protein 1 (product of the *brown* locus), also known as the gp75 antigen. Although gp75 is mainly intracellular, a fraction is present on the cell surface (Takechi et al., 1996). Initial studies of antibody-mediated immunity to gp75 involved the use of *passive immunity*, where monoclonal antibodies against gp75 were passively transferred to tumor-bearing mice or to tumor-free mice that were subsequently challenged with tumor cells. *Active immunization*, where the animals are vaccinated against gp75, uses different methods to elicit the production of antibodies, including gp75 xenogeneic cells, xenogeneic human gp75 protein, and xenogeneic human gp75 cDNA. In both passive and active immunity, antibodies against gp75 can induce B16 tumor rejection. In the case of active immunization, antibody responses were sufficient for protection against challenge with B16, which was demonstrated by the fact that immunity could be passively transferred by serum. Mice deficient in T cells (CD8+) retained full immunity.

Use of genetically deficient mice allows a better understanding of the mechanisms responsible for antibody-mediated tumor rejection. Antibody molecules contain two regions. The Fab domain contains the variable region that forms the antigen binding site, and the Fc constant domain recruits either complement or Fc receptor-positive cells as immune effector cells. Immune effector cells mediate clearance of pathogens, autoimmunity, and destruction of tumor cells. These cells express Fc receptors that bind to the Fc domains of antibodies. The binding of Fc receptors to antibodies can trigger signals that activate and recruit immune and inflammatory cells or send inhibitory signals that downregulate immunity, depending on the receptor that binds to the antibody. FcγRIII is an activation receptor that activates macrophages and NK and mast cells, while FcγRII is a receptor expressed on macrophages that inhibits FcγRIII signaling.

Clynes et al. (1998) investigated the mechanism of antibody-mediated B16 tumor rejection using FcR-deficient mice. FcγRI- and FcγRIII-deficient mice had impaired tumor protection, suggesting that macrophages and/or NK cells engage antibodies against gp75 through FcγRI and or FcγRIII to kill gp75-positive tumor cells. In contrast, gp75 antibodies mediate even more effective tumor protection in FcγRII-deficient mice (Clynes et al., 2000). These observations support the notion that monocytes/macrophages are the critical effector cells involved in antibody-mediated tumor immunity and that the Fc domain can modulate both positive and negative signaling.

T Lymphocytes. More than 40 years ago, Prehn and Main (1957) were among the first to describe the specificity of immunity to chemically induced tumors in syngeneic mice. It is now clear that T cell–mediated immunity in mice is of critical importance for the rejection of virally, UV-, and chemically induced tumors. Mice immunized with irradiated MCA-induced sarcoma cells are fully protected against a subsequent challenge with that same tumor, but not with independently derived tumors. A large body of work using antibody depletion of T cells and genetically deficient knockout mice has revealed that both CD8+ and CD4+ T lymphocytes are crucial for tumor rejection in these models. The importance of the various T-cell subsets has been a subject of discussion. Many studies have implicated a central role for cytotoxic CD8+ T cells, but less is known about the role of CD4+ T cells. However, a number of experimental models have shown that CD4+ T cells are required for tumor immunity, and in several cases they appear to be sufficient for tumor rejection in the absence of CD8+ T cells. In some cases, tumor inhibition depends on the direct lytic activity of cytotoxic CD8+ T lymphocytes (Boon et al., 1994; Hanson et al., 2000; Sahin et al., 1997; Ward et al., 1990), whereas CD4+ T cells activated through the indirect presentation of antigen by APCs are required to support a CD8+ T-cell response (Greenberg et al., 1981; Pericle et al., 1994). Tumor eradication also depends on noncytotoxic CD4+ (Forni and Giovarelli, 1984) and CD8+ (Pericle et al., 1994) lymphocytes releasing proinflammatory cytokines that recruit distinct effector cells at the tumor site. This series of events gives rise to a complex delayed-type hypersensitivity reaction in the tumor site that involves distinct inflammatory cell types, including macrophages, neutrophils, and eosinophils (Forni et al., 1988).

Mice Deficient in Components of the Immune System

Natural mouse mutants, such as nude *(nu/nu)* and *scid* mice, have taught valuable lessons about the development of the immune system, but the precise mutations in these animals have been only recently described. In the past decade, advances in genetic manipulation, including introduction of constitutive and inducible transgenes and inherited and conditional gene deficiencies (knockouts), have produced mouse strains that have advanced our understanding of the physiology of the immune system. Analysis of mice lacking certain components of the immune system has clarified important roles for cell types, secreted molecules, intracellular signaling pathways, and transcription complexes and has shed light on regulatory mechanisms that govern normal immune response, autoimmunity, and tumor immunity. We will describe a few examples of these mouse models and their contributions to the field of tumor immunology.

Perforin-Deficient Mice. Perforin is a component of the cytolytic granules of cytolytic T cells and NK cells that is important in mediating lymphocyte-dependent killing of target cells, including tumors. Perforin enables the entry of caspase-activating granzymes that play a key role in target cell DNA fragmentation, although these serine esterases, granzymes, are not critical for tumor cell death mediated by effector cells (Trapani et al., 1998).

Perforin-deficient mice on a C57BL/6 background were more prone to MCA-induced tumor formation than their wild-type counterparts (van den Broek and Hengartner, 2000). In this model, perforin has been demonstrated to protect the host against tumor initiation, tumor growth, and tumor metastasis (Smyth et al., 2000b; van den Broek and Hengartner, 2000). Moreover, perforin-deficient mice have been crossed with p53-deficient mice and a higher incidence of tumors was observed in the perforin–p53 double-deficient mice, suggesting a role for perforin in tumor protection in this system as well (Smyth et al., 2000b).

IFN-γ-Deficient Mice. IFN-γ is a pleiotropic cytokine that plays a central role in both innate and adaptive arms of the host immune defense (Boehm et al., 1997; Farrar and Schreiber, 1993). IFN-γ is secreted by NK cells and thymus-derived T cells. Under certain conditions of activation, IFN-γ acts in various ways on host and tumor cells to favor tumor regression (Seder et al., 1994; Trinchieri, 1995).

The study of mice lacking IFN-γ showed that IFN-γ contributes to the control of tumor initiation, growth, and metastasis (Street et al., 2002). Comparison of mice deficient in either IFN-γ or perforin demonstrates that, although both are critical for tumor suppression of lymphomagenesis, the level of protection afforded by IFN-γ was strain specific. Lymphomas arising in IFN-γ-deficient mice were very nonimmunogenic compared with those derived from perforin-deficient mice, suggesting a comparatively weaker immunoselective pressure by IFN-γ (Street et al., 2002).

RAG-Deficient Mice. Mice deficient for recombinase activity (encoded by the *RAG* genes) fail to rearrange lymphocyte antigen receptors and thus lack NKT, T, and B cells (Shinkai et al., 1992). This mouse constitutes a model strain with an engineered genetic alteration leading to absence of adaptive immune responses. The first strong evidence supporting the existence of immunosurveillance came from experiments performed in these mice. Other mouse models with alterations in one or more immune components that are essential for immune surveillance have been studied (Table 1). These molecules are implicated in antigen processing and presentation, lymphoid development, and chemokine and cytokine production.

Issues in Use of Deficient Mice in Tumor Studies. Although mice deficient in different components of the immune system constitute excellent reagents for studying tumor immunity, they have to be used with caution, given that they present some disadvantages. Mice constitutively deficient for one or more components of the immune system may compensate for their defects by recruitment of alternative mechanisms during development that could lead to a milder phenotype. Moreover, the inactivation of one gene does not always lead to the complete absence of function of the targeted immune component. In these cases, data have to be cautiously interpreted.

The strain and the background of the mouse need to be taken into consideration. Knockout or transgenic mice often have mixed genetic backgrounds due to the technical steps involved in the generation of these models. In that case, the investigator has to perform a minimum of 11–12 backcrosses, sometimes 20 or more, against a pure genetic background, where the donor strain carrying the mutation is crossed with a host strain with defined genetic background (i.e., an inbred strain). This process is very time consuming, given that backcrosses may take two to three years. Recently, it became possible to reduce this time by half. The speed congenic process uses molecular markers at each generation of the backcross to select mice for the next cross. However, this process is still time consuming and expensive.

APPROACHES TO STUDY IMMUNE ESCAPE

Tumor cells can evade the host immune system through many mechanisms. One of the main reasons tumors are not recognized is immune tolerance to self. This can occur via (1) deletion of T and B cells that recognize self; (2) anergy, where B or T cells recognizing the tumor are paralyzed upon antigen recognition because costimulatory molecules are not available; and (3) suppression of T cells that recognize the tumor by T regulatory cells. Tumor cells can be quite effective at inducing tolerance because of the close relationship of identity of molecules expressed by tumor cells and normal cells. Most tumor cells are not effective antigen-presenting cells. In fact, immunity to cancer is largely elicited through cross-priming, where antigens from tumor cells are taken up by professional APCs, specifically DCs, which present antigens to T cells for initial priming and activation of the adaptive immune response (Fig. 18.2). As tumors progress, defective antigen presentation by DCs can be observed.

Moreover, the immune system may be overwhelmed by the rapid growth and spread of the tumors. Even at very early stages of tumor development, when tumor cells are immunogenic, multiple escape mechanisms can quickly emerge to allow cancer cells to avoid recognition by the host immune system. Other components of the immune and inflammatory systems can be downregulated or upregulated during tumor progression (Table 18.2).

SAMPLES OF TRANSGENIC AND GENETICALLY DEFICIENT ANIMAL MODELS USED IN TUMOR IMMUNOLOGY

In the past decade, advances in genetic engineering and mouse technology allowed the generation of multiple mouse models of cancer that reflect different biologic properties and disease stages of the human counterpart. For example, models frequently used are $Trp53^{-/-}$ mice and RAS-transgenic mice, which have been described elsewhere in the book. Transgenic mice in which the expression of the simian virus 40 (SV40) large-T antigen (TAg) is targeted to a particular tissue represent another model that has been frequently utilized. The transforming

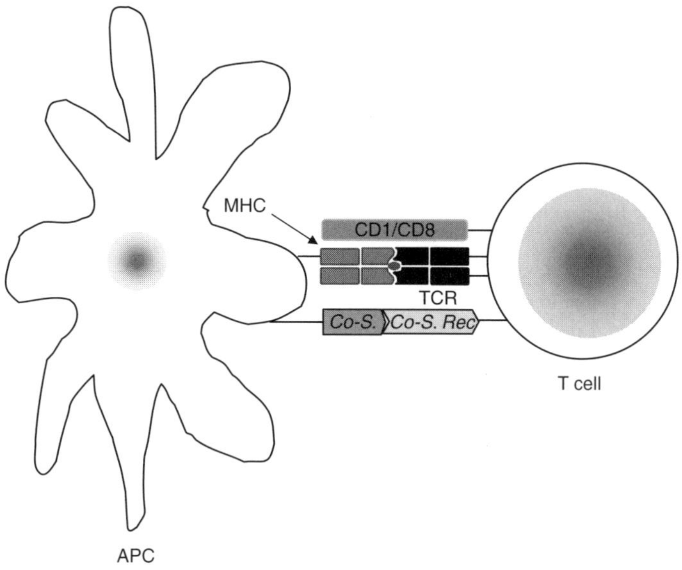

Figure 18.2. Interactions occurring between antigen-presenting cells (APCs) and T cells during antigen presentation. Picture is a simplified scheme that depicts the minimal number of molecules involved. MHC class I or II presents a peptide epitope to the T-cell receptor (TCR, black). Additional molecules, including CD4 or CD8, participate in binding to MHC molecules. TCR is not sufficient to activate T cells. Costimulatory molecules on the APC (Co-S) bind to the costimulatory receptors (Co-S. Rec) on the T cell to provide additional signals required for T-cell activation.

activity of TAg depends on its binding to cellular proteins involved in the control of the cell cycle (p53, pRb, p300) and its ability to interfere with these pathways. Mice expressing TAg under the control of a prostate-specific Probasin promoter, called TRAMP mice, develop prostate disease that resembles human prostate cancer in terms of histopathology and lethality (Greenberg et al., 1995). An additional group of mice has been generated that expresses TAg under the control of the rat prostate steroid binding protein (PSBP) promoter (Green et al., 2000). This promoter targets the expression of TAg to the epithelium of both the mammary and prostate glands, resulting in models of mammary and prostate cancers that histologically resemble the human diseases. In addition, in another mouse model, TAg expressed under control of the rat insulin promoter (RIP) induced pancreatic β-cell tumors producing insulin, causing progressive hypoglycemia (Hanahan, 1985). It should be noted that the TAg protein itself can be recognized by T cells of some mouse strains.

T-cell recognition is dictated by the T-cell receptor (TCR). The TCR is a heterodimer constituted of two chains, α and β (the majority of T cells are α:β T cells). These membrane-bound proteins, which are covalently linked by a disulfide bond, are the products of gene rearrangement that occurred during T-cell differentiation (Davis, 1990). Transgenic mice expressing specific TCRs against defined antigens have been generated (Ohashi et al., 1991, 1993; Pircher et al., 1989). Taking advantage of these models, induction and maintenance of cytotoxic lymphocyte (CTL) activity specific for a primary endogenous tumor expressing a transgene recognized by a specific TCR have been investigated in vivo. The lymphocytic choriomeningitis virus (LCMV) glycoprotein (GP) was expressed under the control of the RIP promoter as an endogenous tumor antigen in mice bearing the pancreatic β-cell tumors (induced by TAg). LCMV infection of these transgenic mice and their controls induced an antitumor CTL response that efficiently reduced the tumor mass in double-transgenic GP × TAg mice (Speiser et al., 1997). Subsequently, crossing these double-transgenic mice with transgenic mice expressing a TCR specific for the LCMV GP (Pircher et al., 1989), it was possible to study whether tumor growth promotes cross-presentation and tolerance of tumor-specific T cells (Nguyen et al., 2002). A large number of studies have reported failure of tumors to generate a tumor-specific T-cell response (ignorance) or a suppression of an active immune response (tolerance or deletion) (Engelhard et al., 2002; Mizoguchi et al., 1992).

Tolerance may be induced by cross-presentation, which is akin to cross-priming, but involves antigen presentation in later stages of T-cell differentiation, following priming. Although the factors influencing cross-presentation have not been well defined, cross-presentation of self-antigens can mediate T-cell tolerance in transgenic mice (Adler et al., 1998; Kurts et al., 1997). In the case of high tumor

Table 18.2. Mechanisms of Tumor Escape

Mechanism	Example	Effect	References
Loss or decrease in MHC I expression	Loss of β_2 microglobulin	Decreased Ag presentation leads to lack of recognition by CD8+ T cells	Koller et al., 1990; Milling et al., 2002; Zijlstra et al., 1990
Loss of Ag expression	Tumor loses expression of a cell surface Ag recognized by host immune system	Antibodies or T cells against the tumor no longer effective	Jager et al., 1996
Lack of Ag processing	Loss or reduction of molecules involved in proteosomal processing and peptide transport (LMP2 or TAP1)	Decrease or loss of recognition of tumor cells by CD4+ or CD8+ T cells	Alimonti et al., 2000; Hayashi and Faustman, 2002
Regulatory T cells	Deficiency of regulatory T cells (FOXP3/Scurfin mice) or blockade of regulatory T cells with anti-CTLA-4 antibody	Deficient regulatory T cells lead to enhanced tumor immunity	Fontenot et al., 2003; Khattri et al., 2003; Waterhouse et al., 1995
Regulatory molecules	Suppression of T cells mediated by • Interleukin-10 (IL-10) • Transforming growth factor β(TGF-β) • Prostaglandins I(2) and E(2)	Blockade of inhibitory ligands or their receptors enhances tumor immunity	Bettelli et al., 1998; Gorelik and Flavell, 2000; Shinomiya et al., 2001; Wynn et al., 1997
Macrophage/granulocyte	Elimination of granulocytes in vivo by a specific antigranulocyte antibody	Depletion of granulocytes enhances tumor immunity; mechanism not understood	Pekarek et al., 1995
Tumor stroma	Failure of host effectors to reach the tumor	Stroma provides a barrier against immune system	Singh et al., 1992; Spiotto et al., 2002
Expression of killer domain molecules by tumors	Expression of Fas L on tumor cells	Fas L on tumors induces apoptosis of infiltrating T cells	Hahne et al., 1996; O'Connell et al., 1996; Strand et al., 1996

Ag = antigen; L = ligand; LMP2 = low molecular weight peptide 2; TAP1 = transporter associated with antigen processing 1; FOXP3 = Forkhead box P3; CTLA-4 = cytotoxic T-lymphocyte antigen 4.

burden, where there are large numbers of tumor cells and spontaneous tumor cell death, cross-presentation may lead to tolerance of tumor-specific T cells. The triple-transgenic mice described above were used to investigate the effect of tumor growth on cross-presentation of tumor antigens and tolerance of tumor-specific T cells. This study demonstrated that advanced tumor burden enhanced cross-presentation of tumor antigens to high-avidity tumor T cells, inducing T-cell proliferation and limited effector function in vivo. Moreover, tumor growth did not induce tolerance (Nguyen et al., 2002).

A number of laboratories have generated transgenic mice expressing the rat *Her-2/neu* oncogene in mammary epithelial cells. These mice have slight variations in tumor onset and frequency, but they all develop spontaneous mammary tumors that overexpress *Her-2/neu* (Bouchard et al., 1989; Guy et al., 1996; Muller et al., 1988). This oncogene was initially isolated from chemically induced rat neuroblastoma and was shown to encode a 185-kD growth factor receptor that has close homology to the EGF receptor (Shih et al., 1981). Ligand binding to *Her-2/neu* receptor results in homo- or heterodimerization and activation of intrinsic kinase activity followed by phosphorylation of specific tyrosine residues in the receptor cytoplasmic tail (Schlessinger, 2000). The human homologue of *Her-2/neu* (c-*ERB-2*) was shown to be amplified and expressed in several human primary breast cancers. Amplified levels provide a poor prognostic sign for the evolution of the disease.

Antibodies directed against oncogenic tyrosine kinase receptors directly impair the proliferative activity of target cells. Among these receptors, the protein product of the *Her-2/neu* oncogene has been studied extensively. Antibodies may induce a Her-2/neu functional block (Klapper et al., 1997), downregulate its expression on the cell membrane, and impede its ability to form homo- or heterodimers that spontaneously transduce proliferative signals to the cells (Drebin et al., 1985; Katsumata et al., 1995; Klapper et al., 1997). In the mouse system, antibodies hamper the three-dimensional growth of *Her-2/neu* mouse mammary carcinoma cells (Nanni et al., 2000) and play an important role in delaying the onset of mammary carcinomas in *Her-2/neu* transgenic mice (Dakappagari et al., 2000; Katsumata et al., 1995; Nanni et al., 2001; Rovero et al., 2000). Although in a few systems antibodies appear to play an important role in the rejection of $p185^{neu}$ tumors (Drebin et al., 1988; Reilly et al., 2001), in others, T-cell reactivity appears to be sufficient for tumor rejection (Chen et al., 1998; Foy et al., 2001; Mukai et al., 2002; Pilon et al., 2001).

These two models illustrate the fact that the immune system can use both humoral and cellular immunity to eliminate neoplastic cells. Also, the mechanisms of tumor rejection are different, depending on the etiology of the tumor. Therefore, the mouse model needs to be chosen and designed in accordance.

LIMITATIONS OF THE KNOCKOUT AND TRANSGENIC MOUSE MODELS IN TUMOR IMMUNOTHERAPY

Transgenic mice that express TAg constitute tumor models that arise de novo; however, the fact that they express a potentially immunogenic viral antigen might be misleading since the immune protection in these mice might be enhanced due to the presence of class I epitopes that are not present in bona fide, less contrived cancers.

Most transgenic mice that have been generated to date were engineered to express an oncogene in a particular tissue. The majority of these transgenic models express a gene that is from a different species (generally, human). For example, in the case of *Her-2/neu*, most transgenic models express the rat *Her-2/neu* gene under the control of the MMTV (murine mammary tumor virus) promoter, which limits expression of *Her-2* to mammary tissue. This does not recapitulate the expression of human *Her-2* (c-*ERB-2*), which is more broadly expressed in a constitutive, not just inducible, manner. Thus, tolerance to *Her-2* may be much more readily induced in humans compared with these contrived mouse transgene systems expressing a xenogeneic oncogene. A transgenic *Her-2/neu* mouse line has been described that expresses a constitutively active mutated rat *Her-2/neu* gene. This gives a better tumor incidence but is divergent from human disease, in which the *Her-2/neu* gene is amplified, not mutated.

Another limitation of these mouse models is technical. In mice challenged with tumors after immunization, the time frame of the experiments is relatively short (typically 1–4 weeks for melanoma). In the case of spontaneous tumor models, the time frame for each experiment is very long (6–24 months); most of these mice do not develop cancer before 6 months of age, and the time frame over which tumors appear is large (generally, 2–3 months).

Table 18.3. Agents Used for Cancer Immune Therapy

Immune modulators
BCG (Bladder cancer)
IFN-α (melanoma, kidney, leukemia)
IL-2 (melanoma, kidney)
Monoclonal antibodies
Anti-CD20 (lymphoma)
Anti-HER2 (breast cancer)
Vaccines
Hepatitis vaccine
BCG?

CANCER IMMUNOTHERAPY

Immunotherapy has been used as adjunct to traditional cancer treatments. Both active and passive means of stimulating the nonspecific and specific immune systems have been applied, in some cases with significant success, and immunotherapies are now part of standard cancer care (Table 18.3).

Passive Immunotherapy

Passive immunotherapy involves passive transfer of immune components to the host with cancer. Passive immunotherapy uses monoclonal antibodies or adaptively transferred lymphocytes. This type of therapy does not require the immune system to take an "active" role and can be effective even if the immune system is weakened. Monoclonal antibody therapy is particularly promising, and several monoclonal antibodies are approved for the clinical treatment of cancer. One of the limitations of this therapy, however, is the development of immunity against the monoclonal antibody itself, limiting the amount and duration of the treatment. Strategies to humanize monoclonal antibodies have been developed to avoid human anti-mouse response.

Active Immunotherapy

Active immunotherapy is the basis of vaccination. A major obstacle for cancer vaccines is that most potential antigens on tumor cells are also present on normal cells.

Nonspecific Immune Modulation

A number of immunotherapies use cytokines, bacteria, and other nonspecific agents to stimulate the immune system. Examples include IL-2, which stimulates T cells and NK cells; type I IFNs, which activate NK cells and certain subsets of DCs; and IL-12, which activates NK and T cells. IL-12 antitumor activity was first observed in mouse models, and it is presently being evaluated in clinical trials. The antitumor activity of IL-2 has also been first noticed in transplantable tumor models, and thus IL-2 is presently approved for treatment of renal cell carcinoma and melanoma.

Another strategy that is being tested experimentally in mice and humans uses blockade of CTLA-4 with antibodies or recombinant CTLA-4-Ig constructs to enhance T-cell responses. CTLA-4 blockade to treat cancer was first demonstrated in mouse models and is presently being evaluated in clinical trials.

The final example is BCG, an attenuated form of *Tuberculosis mycobacterium* that is used for vaccines against tuberculosis in underdeveloped countries. BCG is also a potent activator of APCs, especially DCs, and was shown to effectively enhance immunity to tumors in mice more than 50 years ago (Old et al., 1959). BCG is an effective nonspecific immune therapy for early invasive bladder cancer in humans.

ACKNOWLEDGMENTS

The authors are grateful for support from Swim Across America, the Lymphoma Foundation, TJ Martell Foundation, and the National Cancer Institute.

REFERENCES

Adler AJ, Marsh DW, Yochum GS, Guzzo JL, Nigam A, Nelson WG, Pardoll DM (1998): CD4+ T cell tolerance to parenchymal self-antigens requires presentation by bone marrow-derived antigen-presenting cells, *J Exp Med* 187(10): 1555–1564.

Alimonti J, Zhang QJ, Gabathuler R, Reid G, Chen SS, Jefferies WA (2000): TAP expression provides a general method for improving the recognition of malignant cells in vivo, *Nat Biotechnol* 18(5):515–520.

Baldwin RW (1955): Immunity to methylcholanthrene-induced tumors in inbred rats following atrophy and regression of the implanted tumors. *Br J Cancer* 9:652–657.

Barnd DL, Lan MS, Metzgar RS, Finn OJ (1989): Specific, major histocompatibility complex-unrestricted recognition of tumor-associated mucins by human cytotoxic T cells, *Proc Natl Acad Sci USA* 86(18):7159–7163.

Bettelli E, Das MP, Howard ED, Weiner HL, Sobel RA, Kuchroo VK (1998): IL-10 is critical in the regulation of autoimmune encephalomyelitis as demonstrated by studies of IL-10- and IL-4-deficient and transgenic mice, *J Immunol* 161(7):3299–3306.

Boehm U, Klamp T, Groot M, Howard JC (1997): Cellular responses to interferon-gamma, *Annu Rev Immunol* 15:749–795.

Boon T, Cerottini JC, Van den Eynde B, van der Bruggen P, Van Pel A (1994): Tumor antigens recognized by T lymphocytes. *Annu Rev Immunol* 12:337–365.

Bouchard L, Lamarre L, Tremblay PJ, Jolicoeur P (1989): Stochastic appearance of mammary tumors in transgenic mice carrying the MMTV/c-neu oncogene. *Cell* 57(6):931–96.

Bover KH (1927): Homoisotransplantation van Epidermis bei eineigen Zwillingen. *Beitr Klin Chir* 141:442–447.

Burnet FM (1964): Immunological factors in the process of carcinogenesis. *Br Med Bull* 20:154–158.

Burnet FM (1970): The concept of immunological surveillance. *Prog Exp Tumor Res* 13:1–27.

Burstein NA, Law LW (1971): Neonatal thymectomy and non-viral mammary tumours in mice, *Nature* 231(5303):450–452.

Chen Y, Hu D, Eling DJ, Robbins J, Kipps TJ (1998): DNA vaccines encoding full-length or truncated Neu induce protective immunity against Neu-expressing mammary tumors. *Cancer Res* 58(9):1965–1971.

Chow DA, Bennet RD (1989): Low natural antibody and low in vivo tumor resistance, in xid-bearing B-cell deficient mice. *J Immunol* 142(10):3702–3706.

Clynes R, Takechi Y, Moroi Y, Houghton A, Ravetch JV (1998): Fc receptors are required in passive and active immunity to melanoma, *Proc Natl Acad Sci USA* 95(2):652–656.

Clynes RA, Towers TL, Presta LG, Ravetch JV (2000): Inhibitory Fc receptors modulate in vivo cytoxicity against tumor targets, *Nat Med* 6(4):443–446.

Cook DN, Beck MA, Coffman TM, Kirby SL, Sheridan JF, Pragnell IB, Smithies O (1995): Requirement of MIP-1 alpha for an inflammatory response to viral infection, *Science* 269(5230):1583–1585.

Cui J, Shin T, Kawano T, Sato H, Kondo E, Toura I, Kaneko Y, Koseki H, Kanno M, Taniguchi M (1997): Requirement for Valpha14 NKT cells in IL-12-mediated rejection of tumors. *Science* 278(5343):1623–1626.

Dakappagari NK, Douglas DB, Triozzi PL, Stevens VC, Kaumaya PT (2000): Prevention of mammary tumors with a chimeric HER-2 B-cell epitope peptide vaccine. *Cancer Res* 60(14):3782–3789.

Davis MM (1990): T cell receptor gene diversity and selection. *Annu Rev Biochem* 59:475–496.

Disis ML, Calenoff E, McLaughlin G, Murphy AE, Chen W, Groner B, Jeschke M, Lydon N, McGlynn E, Livingston RB, Moe R, Cheever MA (1994): Existent T-cell and antibody immunity to HER-2/neu protein in patients with breast cancer. *Cancer Res* 54(1):16–20.

Dorfman JR, Zerrahn J, Coles MC, Raulet DH (1997): The basis for self-tolerance of natural killer cells in beta2-microglobulin- and TAP-1- mice. *J Immunol* 159(11):5219–5225.

Drebin JA, Link VC, Greene MI (1988): Monoclonal antibodies specific for the neu oncogene product directly mediate anti-tumor effects in vivo. *Oncogene* 2(4):387–394.

Drebin JA, Link VC, Stern DF, Weinberg RA, Greene MI (1985): Down-modulation of an oncogene protein product and reversion of the transformed phenotype by monoclonal antibodies. *Cell* 41(3):697–706.

Engel AM, Svane IM, Rygaard J, Werdelin O (1997): MCA sarcomas induced in scid mice are more immunogenic than

MCA sarcomas induced in congenic, immunocompetent mice. *Scand J Immunol* 45(5):463–470.

Engelhard VH, Bullock TN, Colella TA, Sheasley SL, Mullins DW (2002): Antigens derived from melanocyte differentiation proteins: Self-tolerance, autoimmunity, and use for cancer immunotherapy. *Immunol Rev* 188(1):136–146.

Farrar MA, Schreiber RD (1993): The molecular cell biology of interferon-gamma and its receptor. *Annu Rev Immunol* 11:571–611.

Foley EJ (1953): Antigenic properties of methylcholanthrene-induced tumors in mice of the strain of origin. *Cancer Res* 13:835–837.

Fontenot JD, Gavin MA, Rudensky AY (2003): Foxp3 programs the development and function of CD4+CD25+ regulatory T cells. *Nat Immunol* 4(4):330–336.

Forni G, Fujiwara H, Martino F, Hamaoka T, Jemma C, Caretto P, Giovarelli M (1988): Helper strategy in tumor immunology: Expansion of helper lymphocytes and utilization of helper lymphokines for experimental and clinical immunotherapy. *Cancer Metastasis Rev* 7(4):289–309.

Forni G, Giovarelli M (1984): In vitro reeducated T helper cells from sarcoma-bearing mice inhibit sarcoma growth in vivo. *J Immunol* 132(1):527–533.

Forster R, Schubel A, Breitfeld D, Kremmer E, Renner-Muller I, Wolf E, Lipp M (1999): CCR7 coordinates the primary immune response by establishing functional microenvironments in secondary lymphoid organs. *Cell* 99(1):23–33.

Foy TM, Bannink J, Sutherland RA, McNeill PD, Moulton GG, Smith J, Cheever MA, Grabstein K (2001): Vaccination with Her-2/neu DNA or protein subunits protects against growth of a Her-2/neu-expressing murine tumor. *Vaccine* 19(17–19): 2598–2606.

Fung-Leung WP, Schilham MW, Rahemtulla A, Kundig TM, Vollenweider M, Potter J, van Ewijk W, Mak TW (1991): CD8 is needed for development of cytotoxic T cells but not helper T cells. *Cell* 65(3):443–449.

Girardi M, Oppenheim DE, Steele CR, Lewis JM, Glusac E, Filler R, Hobby P, Sutton B, Tigelaar RE, Hayday AC (2001): Regulation of cutaneous malignancy by gammadelta T cells. *Science* 294(5542):605–609.

Globerson A, Feldman M (1964): Antigenic specificity of benzo(a)pyrene-induced sarcomas. *J Natl Cancer Inst* 32: 1229–1243.

Gorelik L, Flavell RA (2000): Abrogation of TGFbeta signaling in T cells leads to spontaneous T cell differentiation and autoimmune disease. *Immunity* 12(2):171–181.

Gorer PA (1936): The detection of antigenic differences in mouse erythrocytes by the employment of immune sera. *Br J Exp Pathol* 17:42–50.

Gorer PA (1937): The genetic and antigenic basis for tumor transplantation. *J Pathol Bacteriol* 44:691–697.

Grandea AG 3rd, Golovina TN, Hamilton SE, Sriram V, Spies T, Brutkiewicz RR, Harty JT, Eisenlohr LC, Van Kaer L (2000): Impaired assembly yet normal trafficking of MHC class I molecules in Tapasin mutant mice. *Immunity* 13(2): 213–222.

Grant GA, Miller JF (1965): Effect of neonatal thymectomy on the induction of sarcomata in C57 BL mice. *Nature* 205(976):1124–1125.

Green JE, Shibata MA, Yoshidome K, Liu ML, Jorcyk C, Anver MR, Wigginton J, WiltroutR, Shibata E, Kaczmarczyk S, Wang W, Liu ZY, Calvo A, Couldrey C (2000): The C3(1)/SV40 T-antigen transgenic mouse model of mammary cancer: Ductal epithelial cell targeting with multistage progression to carcinoma, *Oncogene* 19(8):1020–1027.

Greenberg NM, DeMayo F, Finegold MJ, Medina D, Tilley WD, Aspinall JO, Cunha GR, Donjacour AA, Matusik RJ, Rosen JM (1995): Prostate cancer in a transgenic mouse. *Proc Natl Acad Sci USA* 92(8):3439–3443.

Greenberg PD, Cheever MA, Fefer A (1981): Eradication of disseminated murine leukemia by chemoimmunotherapy with cyclophosphamide and adoptively transferred immune syngeneic Lyt-1+2- lymphocytes. *J Exp Med* 154(3):952–963.

Guy CT, Cardiff RD, Muller WJ (1996): Activated neu induces rapid tumor progression. *J Biol Chem* 271(13):7673–7678.

Hahne M, Rimoldi D, Schroter M, Romero P, Schreier M, French LE, Schneider P, Bornand T, Fontana A, Lienard D, Cerottini J, Tschopp J (1996): Melanoma cell expression of Fas(Apo-1/CD95) ligand: Implications for tumor immune escape. *Science* 274(5291):1363–1366.

Hanahan D (1985): Heritable formation of pancreatic beta-cell tumours in transgenic mice expressing recombinant insulin/simian virus 40 oncogenes. *Nature* 315(6015): 115–122.

Hanson HL, Donermeyer DL, Ikeda H, White JM, Shankaran V, Old LJ, Shiku H, Schreiber RD, Allen PM (2000): Eradication of established tumors by CD8+ T cell adoptive immunotherapy. *Immunity* 13(2):265–276.

Hara I, Takechi Y, Houghton AN (1995): Implicating a role for immune recognition of self in tumor rejection: Passive immunization against the brown locus protein. *J Exp Med* 182(5):1609–1614.

Hayashi T, Faustman DL (2002): Development of spontaneous uterine tumors in low molecular mass polypeptide-2 knockout mice. *Cancer Res* 62(1):24–27.

Hermans IF, Ritchie DS, Daish A, Yang J, Kehry MR, Ronchese F (1999): Impaired ability of MHC class II−/− dendritic cells to provide tumor protection is rescued by CD40 ligation. *J Immunol* 163(1):77–81.

Ikehara S, Pahwa RN, Fernandes G, Hansen CT, Good RA (1984): Functional T cells in athymic nude mice. *Proc Natl Acad Sci USA* 81(3):886–888.

Jager E, Ringhoffer M, Karbach J, Arand M, Oesch F, Knuth A (1996): Inverse relationship of melanocyte differentiation antigen expression in melanoma tissues and CD8+ cytotoxic-T-cell responses: Evidence for immunoselection of antigen-loss variants in vivo. *Int J Cancer* 66(4):470–476.

Kaplan DH, Shankaran V, Dighe AS, Stockert E, Aguet M, Old LJ, Schreiber RD (1998): Demonstration of an interferon gamma-dependent tumor surveillance system in immunocompetent mice. *Proc Natl Acad Sci USA* 95(13):7556–7561.

Katsumata M, Okudaira T, Samanta A, Clark DP, Drebin JA, Jolicoeur P, Greene MI (1995): Prevention of breast tumour development in vivo by downregulation of the p185neu receptor. *Nat Med* 1(7):644–648.

Khattri R, Cox T, Yasayko SA, Ramsdell F (2003): An essential role for Scurfin in $CD4^+CD25^+$ T regulatory cells. *Nat Immunol* 4(4):337–342.

Kitamura D, Roes J, Kuhn R, Rajewsky K (1991): A B cell-deficient mouse by targeted disruption of the membrane exon of the immunoglobulin mu chain gene. *Nature* 350(6317):423–426.

Klapper LN, Vaisman N, Hurwitz E, Pinkas-Kramarski R, Yarden Y, Sela M (1997): A subclass of tumor-inhibitory monoclonal antibodies to ErbB-2/HER2 blocks crosstalk with growth factor receptors. *Oncogene* 14(17):2099–2109.

Klein G, Sjogren HO, Klein E, Hellstrom KE (1960): Demonstration of resistance against methylcholanthrene-induced sarcomas in the primary autochthonous host. *Cancer Res* 20:1561–1572.

Koller BH, Marrack P, Kappler JW, Smithies O (1990): Normal development of mice deficient in beta 2 M, MHC class I proteins, and CD8+ T cells. *Science* 248(4960):1227–1230.

Kontgen F, Suss G, Stewart C, Steinmetz M, Bluethmann H (1993): Targeted disruption of the MHC class II Aa gene in C57BL/6 mice. *Int Immunol* 5(8):957–964.

Kurts C, Kosaka H, Carbone FR, Miller JF, Heath WR (1997): Class I-restricted cross-presentation of exogenous self-antigens leads to deletion of autoreactive CD8(+) T cells. *J Exp Med* 186(2):239–245.

Liljedahl M, Winqvist O, Surh CD, Wong P, Ngo K, Teyton L, Peterson PA, Brunmark A, Rudensky AY, Fung-Leung WP, Karlsson L (1998): Altered antigen presentation in mice lacking H2-O. *Immunity* 8(2):233–243.

Little CC, Tyzzer EE (1916): Further experimental studies on the inheritance of susceptibility to a transplantable tumor, carcinoma (JWA) of the Japanese waltzing mouse. *J Med Res* 33:393–453.

Lloyd KO (1991): Humoral immune responses to tumor-associated carbohydrate antigens. *Semin Cancer Biol* 2(6): 421–431.

Milling SW, Silvers WK, Sai T, Mintz B (2002): Decline in MHC class I expression with increasing thickness of cutaneous melanomas in standard-strain transgenic mouse models. *Melanoma Res* 12(3):221–230.

Mizoguchi H, O'Shea JJ, Longo DL, Loeffler CM, McVicar DW, Ochoa AC (1992): Alterations in signal transduction molecules in T lymphocytes from tumor-bearing mice. *Science* 258(5089):1795–1798.

Mombaerts P, Iacomini J, Johnson RS, Herrup K, Tonegawa S, Papaioannou VE (1992): RAG-1-deficient mice have no mature B and T lymphocytes. *Cell* 68(5):869–877.

Moreau-Aubry A, Le Guiner S, Labarriere N, Gesnel MC, Jotereau F, Breathnach R (2000): A processed pseudogene codes for a new antigen recognized by a CD8(+) T cell clone on melanoma. *J Exp Med* 191(9):1617–1624.

Mukai K, Yasutomi Y, Watanabe M, Kenjo A, Aota T, Wang L, Nishikawa H, Ishihara M, Fujita T, Kuribayashi K, Shiku H (2002): HER2 peptide-specific CD8(+) T cells are proportionally detectable long after multiple DNA vaccinations. *Gene Ther* 9(13):879–888.

Muller WJ, Sinn E, Pattengale PK, Wallace R, Leder P (1988): Single-step induction of mammary adenocarcinoma in transgenic mice bearing the activated c-neu oncogene. *Cell* 54(1):105–115.

Nanni P, Nicoletti G, De Giovanni C, Landuzzi L, Di Carlo E, Cavallo F, Pupa SM, Rossi I, Colombo MP, Ricci C, Astolfi A, Musiani P, Forni G, Lollini PL (2001): Combined allogeneic tumor cell vaccination and systemic interleukin 12 prevents mammary carcinogenesis in HER-2/neu transgenic mice. *J Exp Med* 194(9):1195–1205.

Nanni P, Pupa SM, Nicoletti G, De Giovanni C, Landuzzi L, Rossi I, Astolfi A, Ricci C, De Vecchi R, Invernizzi AM, Di Carlo E, Musiani P, Forni G, Menard S, Lollini PL (2000): p185(neu) protein is required for tumor and anchorage-independent growth, not for cell proliferation of transgenic mammary carcinoma. *Int J Cancer* 87(2):186–194.

Nguyen LT, Elford AR, Murakami K, Garza KM, Schoenberger SP, Odermatt B, Speiser DE, Ohashi PS (2002): Tumor growth enhances cross-presentation leading to limited T cell activation without tolerance. *J Exp Med* 195(4): 423–435.

Nishizuka Y, Nakakuki K, Usui M (1965): Enhancing effect of thymectomy on hepatotumorigenesis in Swiss mice following neonatal injection of 20-methylcholanthrene. *Nature* 205:1236–1238.

O'Connell J, O'Sullivan GC, Collins JK, Shanahan F (1996): The Fas counterattack: Fas-mediated T cell killing by colon cancer cells expressing Fas ligand. *J Exp Med* 184(3):1075–1082.

Ohashi PS, Oehen S, Burki K, Pircher H, Ohashi CT, Odermatt B, Malissen B, Zinkernagel RM, Hengartner H (1991): Ablation of "tolerance" and induction of diabetes by virus infection in viral antigen transgenic mice. *Cell* 65(2):305–317.

Ohashi PS, Zinkernagel RM, Leuscher I, Hengartner H, Pircher H (1993): Enhanced positive selection of a transgenic TCR by a restriction element that does not permit negative selection. *Int Immunol* 5(2):131–138.

Old LJ, Clarke DA, Benacerraf B (1959): Effect of bacillus Calmette-Guerin (BCG) infection on transplanted tumors in the mouse. *Nature* 184:291–293.

Old LJ, Boyse EA, Clarke DA, Carswell EA (1962): Antigenic properties of chemically induced tumors. *Ann NY Acad Sci* 101:80–106.

Outzen HC, Custer RP, Eaton GJ, Prehn RT (1975): Spontaneous and induced tumor incidence in germfree "nude" mice. *J Reticuloendothel Soc* 17(1):1–9.

Pekarek LA, Starr BA, Toledano AY, Schreiber H (1995): Inhibition of tumor growth by elimination of granulocytes. *J Exp Med* 181(1):435–440.

Pericle F, Giovarelli M, Colombo MP, Ferrari G, Musiani P, Modesti A, Cavallo F, Di Pierro F, Novelli F, Forni G (1994): An efficient Th2-type memory follows CD8+ lymphocyte-driven and eosinophil-mediated rejection of a spontaneous mouse mammary adenocarcinoma engineered to release IL-4. *J Immunol* 153(12):5659–5673.

Pilon SA, Piechocki MP, Wei WZ (2001): Vaccination with cytoplasmic ErbB-2 DNA protects mice from mammary tumor growth without anti-ErbB-2 antibody. *J Immunol* 167(6):3201–3206.

Pircher H, Burki K, Lang R, Hengartner H, Zinkernagel RM (1989): Tolerance induction in double specific T-cell receptor transgenic mice varies with antigen. *Nature* 342(6249):559–561.

Prehn RT, Main JM (1957): Immunity to methylcholanthrene-induced sarcomas. *J Natl Cancer Inst* 18:769.

Rahemtulla A, Fung-Leung WP, Schilham MW, Kundig TM, Sambhara SR, Narendran A, Arabian A, Wakeham A, Paige CJ, Zinkernagel RM, Miller RG, Mak TW (1991): Normal development and function of CD8+ cells but markedly decreased helper cell activity in mice lacking CD4. *Nature* 353(6340):180–184.

Reilly RT, Machiels JP, Emens LA, Ercolini AM, Okoye FI, Lei RY, Weintraub D, Jaffee EM (2001): The collaboration of both humoral and cellular HER-2/neu-targeted immune responses is required for the complete eradication of HER-2/neu-expressing tumors. *Cancer Res* 61(3):880–883.

Rickert RC, Rajewsky K, Roes J (1995): Impairment of T-cell-dependent B-cell responses and B-1 cell development in CD19-deficient mice. *Nature* 376(6538):352–355.

Rovero S, Amici A, Carlo ED, Bei R, Nanni P, Quaglino E, Porcedda P, Boggio K, Smorlesi A, Lollini PL, Landuzzi L, Colombo MP, Giovarelli M, Musiani P, Forni G (2000): DNA vaccination against rat her-2/Neu p185 more effectively inhibits carcinogenesis than transplantable carcinomas in transgenic BALB/c mice. *J Immunol* 165(9):5133–5142.

Rygaard J, Povlsen CO (1974): The mouse mutant nude does not develop spontaneous tumours. An argument against immunological surveillance. *Acta Pathol Microbiol Scand [B] Microbiol Immunol* 82(1):99–106.

Sahin U, Tureci O, Pfreundschuh M (1997): Serological identification of human tumor antigens. *Curr Opin Immunol* 9(5):709–716.

Sanford BH, Kohn HI, Daly JJ, Soo SF (1973): Long-term spontaneous tumor incidence in neonatally thymectomized mice. *J Immunol* 110(5):1437–1439.

Schlessinger J (2000): Cell signaling by receptor tyrosine kinases. *Cell* 103(2):211–225.

Seder RA, Germain RN, Linsley PS, Paul WE (1994): CD28-mediated costimulation of interleukin 2 (IL-2) production plays a critical role in T cell priming for IL-4 and interferon gamma production. *J Exp Med* 179(1):299–304.

Sedger LM, Glaccum MB, Schuh JC, Kanaly ST, Williamson E, Kayagaki N, Yun T, Smolak P, Le T, Goodwin R, Gliniak B (2002): Characterization of the in vivo function of TNF-alpha-related apoptosis-inducing ligand, TRAIL/Apo2L, using TRAIL/Apo2L gene-deficient mice. *Eur J Immunol* 32(8):2246–2254.

Shankaran V, Ikeda H, Bruce AT, White JM, Swanson PE, Old LJ, Schreiber RD (2001): IFNgamma and lymphocytes prevent primary tumour development and shape tumour immunogenicity. *Nature* 410(6832):1107–1111.

Shih C, Padhy LC, Murray M, Weinberg RA (1981): Transforming genes of carcinomas and neuroblastomas introduced into mouse fibroblasts. *Nature* 290(5803):261–264.

Shinkai Y, Rathbun G, Lam KP, Oltz EM, Stewart V, Mendelsohn M, Charron J, Datta M, Young F, Stall AM, Alt FW (1992): RAG-2-deficient mice lack mature lymphocytes owing to inability to initiate V(D)J rearrangement. *Cell* 68(5):855–867.

Shinomiya S, Naraba H, Ueno A, Utsunomiya I, Maruyama T, Ohuchida S, Ushikubi F, Yuki K, Narumiya S, Sugimoto Y, Ichikawa A, Oh-ishi S (2001): Regulation of TNFalpha and interleukin-10 production by prostaglandins I(2) and E(2): Studies with prostaglandin receptor-deficient mice and prostaglandin E-receptor subtype-selective synthetic agonists. *Biochem Pharmacol* 61(9):1153–1160.

Singh S, Ross SR, Acena M, Rowley DA, Schreiber H (1992): Stroma is critical for preventing or permitting immunological destruction of antigenic cancer cells. *J Exp Med* 175(1):139–146.

Smyth MJ, Thia KY, Street SE, Cretney E, Trapani JA, Taniguchi M, Kawano T, Pelikan SB, Crowe NY, Godfrey DI (2000a): Differential tumor surveillance by natural killer (NK) and NKT cells. *J Exp Med* 191(4):661–668.

Smyth MJ, Thia KY, Street SE, MacGregor D, Godfrey DI, Trapani JA (2000b): Perforin-mediated cytotoxicity is critical for surveillance of spontaneous lymphoma. *J Exp Med* 192(5):755–760.

Snell GD (1948): Methods for the study of histocompatibility genes. *J Genet* 49:87–108.

Speiser DE, Miranda R, Zakarian A, Bachmann MF, McKall-Faienza K, Odermatt B, Hanahan D, Zinkernagel RM, Ohashi PS (1997): Self antigens expressed by solid tumors Do not efficiently stimulate naive or activated T cells: Implications for immunotherapy. *J Exp Med* 186(5):645–653.

Spiotto MT, Yu P, Rowley DA, Nishimura MI, Meredith SC, Gajewski TF, Fu YX, Schreiber H (2002): Increasing tumor antigen expression overcomes "ignorance" to solid tumors via crosspresentation by bone marrow-derived stromal cells. *Immunity* 17(6):737–747.

Strand S, Hofmann WJ, Hug H, Muller M, Otto G, Strand D, Mariani SM, Stremmel W, Krammer PH, Galle PR (1996): Lymphocyte apoptosis induced by CD95 (APO-1/Fas) ligand-expressing tumor cells—a mechanism of immune evasion? *Nat Med* 2(12):1361–1366.

Street SE, Cretney E, Smyth MJ (2001): Perforin and interferon-gamma activities independently control tumor initiation, growth, and metastasis. *Blood* 97(1):192–197.

Street SE, Trapani JA, MacGregor D, Smyth MJ (2002): Suppression of lymphoma and epithelial malignancies effected by interferon gamma. *J Exp Med* 196(1):129–134.

Stutman O (1974): Tumor development after 3-methylcholanthrene in immunologically deficient athymic-nude mice. *Science* 183(124):534–536.

Stutman O (1975): Immunodepression and malignancy. *Adv Cancer Res* 22:261–422.

Takai T, Li M, Sylvestre D, Clynes R, Ravetch JV (1994): FcR gamma chain deletion results in pleiotropic effector cell defects. *Cell* 76(3):519–529.

Takai T, Ono M, Hikida M, Ohmori H, Ravetch JV (1996): Augmented humoral and anaphylactic responses in Fc gamma RII-deficient mice. *Nature* 379(6563):346–349.

Takechi Y, Hara I, Naftzger C, Xu Y, Houghton AN (1996): A melanosomal membrane protein is a cell surface target for melanoma therapy. *Clin Cancer Res* 2(11):1837–1842.

Trainin N, Linker-Israeli M, Small M, Boiato-Chen L (1967): Enhancement of lung adenoma formation by neonatal thymectomy in mice treated with 7,12-dimethylbenz(a)anthracene or urethan. *Int J Cancer* 2(4):326–336.

Trapani JA, Jans P, Smyth MJ, Froelich CJ, Williams EA, Sutton VR, Jans DA (1998): Perforin-dependent nuclear entry of granzyme B precedes apoptosis, and is not a consequence of nuclear membrane dysfunction. *Cell Death Differ* 5(6):488–496.

Trinchieri G (1995): Natural killer cells wear different hats: Effector cells of innate resistance and regulatory cells of adaptive immunity and of hematopoiesis. *Semin Immunol* 7(2):83–88.

Uyttenhove C, Van Snick J, Boon T (1980): Immunogenic variants obtained by mutagenesis of mouse mastocytoma P815. I. Rejection by syngeneic mice. *J Exp Med* 152(5):1175–1183.

van den Broek MF, Hengartner H (2000): The role of perforin in infections and tumour surveillance. *Exp Physiol* 85(6):681–685.

Van Den Eynde BJ, Gaugler B, Probst-Kepper M, Michaux L, Devuyst O, Lorge F, Weynants P, Boon T (1999): A new antigen recognized by cytolytic T lymphocytes on a human kidney tumor results from reverse strand transcription. *J Exp Med* 190(12):1793–1800.

van der Bruggen P, Traversari C, Chomez P, Lurquin C, De Plaen E, Van den Eynde B, Knuth A, Boon T (1991). A gene encoding an antigen recognized by cytolytic T lymphocytes on a human melanoma. *Science* 254(5038):1643–1647.

Ward PL, Koeppen HK, Hurteau T, Rowley DA, Schreiber H (1990): Major histocompatibility complex class I and unique antigen expression by murine tumors that escaped from CD8+ T-cell-dependent surveillance. *Cancer Res* 50(13):3851–3858.

Waterhouse P, Penninger JM, Timms E, Wakeham A, Shahinian A, Lee KP, Thompson CB, Griesser H, Mak TW (1995): Lymphoproliferative disorders with early lethality in mice deficient in Ctla-4. *Science* 270(5238):985–988.

Wynn TA, Morawetz R, Scharton-Kersten T, Hieny S, Morse HC 3rd, Kuhn R, Muller W, Cheever AW, Sher A (1997): Analysis of granuloma formation in double cytokine-deficient mice reveals a central role for IL-10 in polarizing both T helper cell 1- and T helper cell 2-type cytokine responses in vivo. *J Immunol* 159(10):5014–5023.

Zhu Z, Hattori K, Zhang H, Jimenez X, Ludwig DL, Dias S, Kussie P, Koo H, Kim HJ, Lu D, Liu M, Tejada R, Friedrich M, Bohlen P, Witte L, Rafii S (2003): Inhibition of human leukemia in an animal model with human antibodies directed against vascular endothelial growth factor receptor 2. Correlation between antibody affinity and biological activity. *Leukemia* 17(3):604–611.

Zijlstra M, Bix M, Simister NE, Loring JM, Raulet DH, Jaenisch R (1990): Beta 2-microglobulin deficient mice lack CD4-8+ cytolytic T cells. *Nature* 344(6268):742–746.

PART IV

IMAGING TECHNOLOGIES

19

MICRO–COMPUTED TOMOGRAPHY OF MOUSE CANCER MODELS

Jamey P. Weichert

Department of Radiology, University of Wisconsin Medical School, Madison, Wisconsin

As mouse cancer models continue to more accurately depict the human disease, the ability to detect and monitor tumor development and regression has become increasingly important. Recent developments in microimaging techniques now offer researchers the ability to accurately detect and monitor tumors less than a millimeter in diameter in live mouse models. Introduced commercially in 2000, micro–computed tomography (CT) scanners are now capable of scanning live mice with spatial resolution exceeding 20 μm. Although X-ray computed tomography provides exquisite skeletal images due to the inherent high density of bone, this chapter is intended to provide an overview of microCT imaging of mouse cancer models, including soft-tissue tumors, which are particularly challenging. A further goal is to highlight the strengths and expose the weaknesses of this amazing new technology.

INTRODUCTION

Since its clinical introduction in the early 1970s, X-ray computed tomography (CT) has revolutionized our ability to noninvasively examine tissues and organs deep within the body. Despite strong challenges by other anatomic imaging modalities, including magnetic resonance imaging (MRI), CT has remained the clinical workhorse for diagnostic anatomic evaluation of abdominal organs and tissues. Although capable of achieving submillimeter spatial resolution in humans and large research animals, clinical CT scanners afford only limited imaging utility in small rodents, including rats and mice. Recently, however, the introduction of commercial ultrahigh-resolution laboratory microCT scanners has revolutionized our ability to noninvasively examine tissues and organs in living experimental animal models, including mice, at spatial resolutions of less than 20 μm, which in a three-dimensional sense makes it possible to image 1 μL volume.

Principles

Since their commercial introduction in 2000, laboratory microCT imaging systems have generally fallen into one of two categories, wherein the sample is either rotated within a fixed X-ray source and flat-panel detector system or alternatively where the X-ray source and detector system rotate around a stationary sample. The former, capable of attaining sub-10-μm spatial resolution, is optimized for use with excised or preserved tissues, whereas the latter is capable of imaging intact live rodents with resolution typically exceeding 20 μm. In either case, however, hundreds of X-ray projection views are acquired at equiangular positions around the object, thus affording a truly three-dimensional data set. Several excellent reviews have recently appeared that discuss in detail the physical differences as well as the advantages and limitations associated with each type of system (Holdsworth and Thornton, 2002; Paulus et al., 2000, 2001). Moreover, sophisticated image processing and visualization software programs now allow the user to simultaneously generate and display ultrahigh-resolution two-dimensional planer and three-dimensional volumetric images in mice, which

Mouse Models of Human Cancer, edited by Eric C. Holland
ISBN 0-471-44460-X

in some instances technologically rival or even exceed those obtained on clinical scanners, albeit at an order-of-magnitude better spatial resolution. Although the majority (80%) of microCT publications involve bone or skeletal studies, this report will focus on the imaging of intact live mouse models due to tremendous recent advances in the development of mouse models of human cancer.

MicroCT, unlike its clinical counterpart, acquires a true three-dimensional density map of the object being evaluated. On the CT scan, high-density tissues like bone absorb more X-rays than low-density tissues like lung. Therefore bone appears white relative to low-density tissues like lung. Fat appears as dark or black structures. In clinical CT, tissue density is represented by Houndsfield units (HU), which range from −1000 HU for air to +1000 HU for bone. It is the relative inherent difference in tissue density that affords the image its contrast range. It is generally thought that only a 3% to 5% difference in tissue density is necessary to allow sufficient in vivo tissue differentiation. Although it is easy to generate useful three-dimensional images of high-density tissues like bone, visualizing subtle changes in adjacent soft tissues is much more difficult. In clinical CT, soft-tissue contrast can be enhanced substantially with use of contrast-enhancing imaging agents. These agents are generally polyiodinated water-soluble agents that are injected intravenously and undergo nonspecific and extracellular biodistribution prior to rapid (1–2 min) renal clearance. Tissue density is transiently enhanced while the agent is passing through the tissue of interest. In some cases tumors can be visualized due to contrast agent entering the interstitial space in tumors via their *leaky* vasculature. In most cases the resulting contrast differential between tumor and surrounding tissue only lasts for a couple of minutes at best due to the fact that the agent freely diffuses back into the circulation prior to rapid renal elimination. The use of clinical contrast agents in microCT scanning is virtually precluded by the relatively long image acquisition times required by microCT scanners due to their cone beam architecture. Although the acquisition times are longer (5–20 min depending on desired resolution) in microCT than in clinical CT, the data set produced in microCT is truly three dimensional.

Rationale for Use

In a typical research scenario, laboratory mice are serially sacrificed and examined via dissection in order to quantitate the phenomena (phenotype, tumors, therapy response) of interest. In the development of new cancer models in mice, for example, groups of mice are euthanized at various times and subsequently examined by dissection in order to monitor the location, number, and size of existing tumors. In many cases it is necessary to sacrifice and dissect hundreds or even thousands of mice in order to ensure sufficient statistical significance. Imaging affords the potential to use many fewer mice and allows each animal to serve as its own control. Although it is certainly possible to acquire meaningful CT scans on rats using a clinical CT unit, attempts to image mice have failed. Using the smallest field of views on a clinical scanner, it is possible to detect only the largest of structures and then only at very low resolution.

Although commercial microCT scanners have only been available for about three years, several excellent review articles thoroughly describe the physical nature of the systems as well as many of their potential applications. One of these (Paulus et al., 2001) also describes other microimaging modalities, including microMRI, microPET (position emission tomography), microSPECT (single-photon-emission CT), and bioluminescence.

IN VIVO SMALL-ANIMAL MICROCT IMAGING

Anesthesia

Since microCT affords the ability to serially monitor an animal over long periods of time, it is desirable to minimize the stress and maintain the overall health of the animal. It is necessary, therefore, to anesthetize the animal prior to scanning in order to minimize image artifacts created by motion (e.g., respiratory or peristaltic) of the mouse. Although a variety of injectable or volatile anesthetics are appropriate for use with microCT scanning, inhalation anesthetics (isoflurane) are generally preferable, especially in serial scanning studies. The use of intraperitoneally administered barbiturates is convenient for single imaging studies; however, this method must be monitored in long-term repetitive imaging studies involving the lower abdomen due to the formation of large adhesions in the peritoneal cavity near the injection sites. In some cases these adhesions can be quite extensive and exceed 1 cm in diameter.

Radiation Exposure

Exposure of the mouse to ionizing radiation is unavoidable in microCT scanning. Although image reconstruction can be performed on a subset of the entire data acquisition set, microCT scanners typically acquire a large series (180–400) of X-ray projections through the entire subject. The potential impact of significant radiation exposure to the subject must be considered in any serial imaging study. Such exposure may not only affect the health of the mouse but could theoretically affect tumor growth and interfere with therapeutic response to other treatment regimens. Median organ exposure during a typical medium-resolution microCT scan with 20 min acquisition time may approach 0.4 Gy (Paulus et al., 2000).

Image Artifacts

Several artifacts found in clinical CT scanning are also present in microCT scanning. The most noticeable are streak artifacts, which originate from extremely dense structures, including bones, metal surgical clips, and highly concentrated contrast agents. In microCT scanning of mice, the most common source of high-density artifacts, however, are caused by bone meal in the intestinal tract. The diet of the mouse plays a very important role in the quality of the CT scan, especially during imaging of the lower abdomen and intestinal tract. Normal mouse chow pellets contain relatively high amounts of bone meal as filler. While in transit through the intestinal tract the bone meal can cause severe streak artifacts, especially on axial images. In order to circumvent this problem, mice can be put on a diet consisting of fruits and vegetables as well as unsalted raw sunflower nuts (no shells) at least two days prior to scanning. This simple "low-density" diet affords virtually artifact-free images. Several commercial liquid diets are currently being evaluated as potential substitutes for the "veggie" diet due to ease of handling and more accurate quantification of consumption.

Skeleton

MicroCT is perhaps the best suited of all the microimaging modalities for examining bone structure and density due to its extremely high image contrast relative to soft tissue (Genant et al., 2000; Issever et al., 2003; Laib et al., 2001). In fact, most early home-built microCT machines were designed for high-resolution scanning of small pieces of bone tissue. It is also possible to quantitate bone mineral density by comparison of measured bone density with a series of known hydroxyapatite reference standards. A majority of literature reports (80%) utilize microCT for skeletal or bone imaging; however, with the introduction of microCT-compatible contrast agents, this trend is expected to shift to tumor detection and monitoring.

Soft-Tissue Challenge

The greatest challenge of microCT scanning is soft-tissue imaging. In many cases, density gradients between adjacent tissues are so slight that it is often difficult to distinguish neighboring organs. Similarly, detection of a tumor located deep within an organ can also be difficult. Soft-tissue contrast can be influenced by three main factors: X-ray energy level, X-ray flux, and the use of contrast-enhancing agents. Regarding energy level, there is typically little difference in soft-tissue contrast in the 30–80 kVp range. The lower energy X-rays provide more soft-tissue contrast; however, longer scan times are necessary to gain an equivalent X-ray flux. If acquisition time is not crucial, then it is desirable to decrease the energy, increase the exposure time of each step, and take more projection images. Gaining soft-tissue contrast via these methods results in a higher cumulative radiation dose to the mouse, however, which can limit the frequency of scanning in serial imaging studies.

Organs and Tissues

MicroCT scanning of organs and soft tissues ranges from those that are easy to visualize due to low background tissue density, including lung and adipose tissue, to those that are all fairly similar, including liver, spleen, and kidneys. Other tissues, including adrenal glands and several peripheral lymph nodes, despite being very small, are easy to visualize because they are located within adipose tissue. The heart is also easy to monitor because it is located adjacent to the lungs. Quite unexpectedly, however, resolution in the heart and lungs is much better than expected despite the respiratory motion of these organs. It is speculated that this is due to tissue volume averaging over 5–20 min acquisition time. The prostate gland and associated seminal vesicles are also easy to locate because they are located directly behind the bladder. It is possible to image the presence of very small tumors in the liver and spleen utilizing hepatocyte-selective contrast agents (Weichert et al., 1995, 2000), and the spleen can also be opacified with macromolecular vascular contrast agents. Pancreatic imaging, however, has proven to be a challenge due to its ubiquitous nature in the mouse. Initial attempts to image mouse brain have been problematic due to the inherent low level of soft-tissue contrast differential within the brain.

Vasculature

In contrast to skeletal elements, blood vessels afford very little natural contrast for CT imaging. In clinical radiology, it is possible to accurately image both arteries and veins utilizing dynamic scanning protocols in conjunction with water-soluble contrast agents. Accordingly, rapid scanning is performed immediately during injection and for an additional 15 s in order to image the arterial vessels and then continued for up to a minute thereafter to capture the venous side. This procedure is possible due to the ability of modern multidetector clinical CT scanners to scan the entire abdomen of a human in only a few seconds. Such dynamic procedures are not possible on current microCT scanners due to relatively long data acquisition times on the order of several minutes.

In vitro vascular images have been constructed in mouse models with the use of latex casts wherein the latex has been doped with a high-density material (Bentley et al., 2002, Garcia-Sanz et al., 1998; Wan et al., 2000). In vivo vascular imaging is possible, however, by using long-acting, macromolecular blood pool contrast agents. In contrast to commercial water-soluble contrast agents that rapidly diffuse out of the vascular space, these new agents are macromolecular in nature and as such remain entirely within the vascular space for up to several hours after injection (Weichert et al., 1998, 2000). A pegylated chylomicron remnant–like macromolecule (described below) has been developed that works quite well within the acquisition time frame of microCT. Using this approach it is possible to produce exquisite three-dimensional images of normal vasculature as well as angiogenic vessels associated with tumor development. It is unlikely, however, that it will be possible to separate arterial from venous structures using current methods due to the elevated respiratory rate of mice relative to humans and the relatively long acquisition times of current microCT scanners. It is anticipated that microCT using long-acting vascular contrast agents will play a major role in the evaluation of antiangiogenic cancer therapy agents in mouse models.

Strengths and Weaknesses

The benefits of high-resolution microCT scanning are numerous. Its ability to provide ultrahigh-resolution anatomic images in living mice is unsurpassed. Moreover, as new longer acting tissue and cell-selective contrast agents are developed, soft-tissue and tumor imaging will be outstanding. With the aid of a hepatocyte-selective contrast agent, ITG, for example, it is already possible to accurately detect primary and metastatic liver tumors less than 300 μm in diameter in live mice using a first-generation microCT scanner (Fig. 19.1). It is likely that it will be possible to detect tumors less than 100 μm in diameter using newer second-generation units. The impact

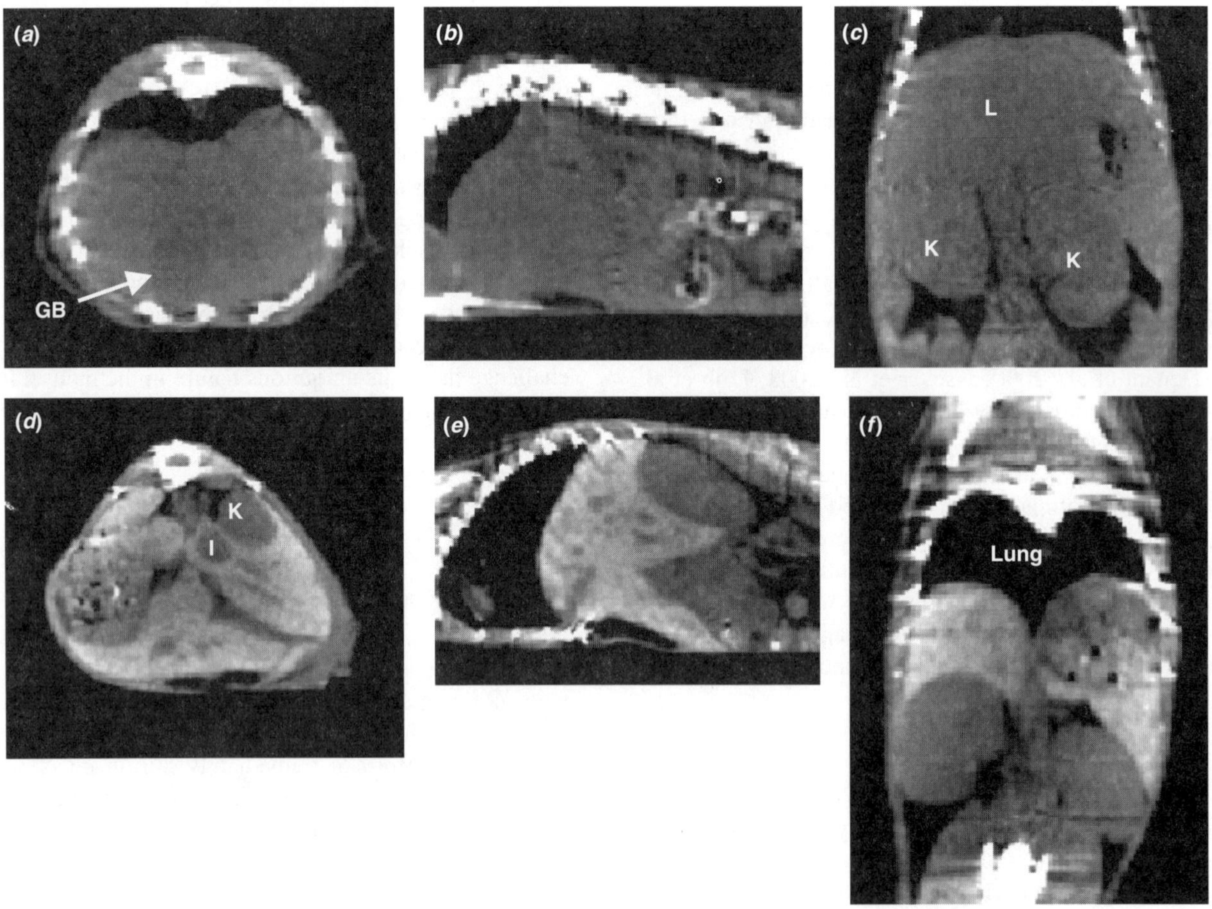

Figure 19.1. (*a–c*) Non-contrast-enhanced and (*d–f*) contrast-enhanced microCT scans of a normal anesthetized mouse abdomen following intravenous injection of ITG. Liver lobes are brighter (*d–f*) due to presence of ITG contrast agent. Axial (*a,d*), sagittal (*b,e*), and coronal (*c,f*) views are presented (GB = gall bladder, K = kidney, L = liver, I = inferior vena cava).

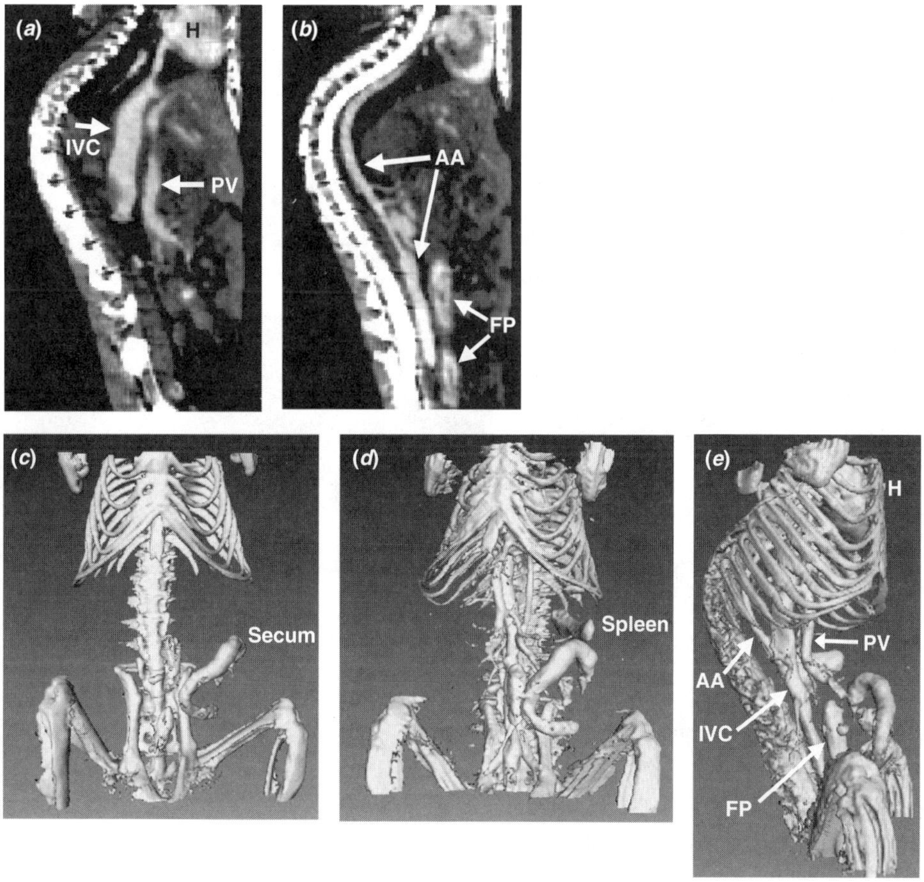

Figure 19.2. Vascular images of an anesthetized mouse abdomen following intravenous administration of BP20 vascular contrast agent. (*a,b*) Major vessels (IVC = inferior vena cava, AA = abdominal aorta, PV = portal vein) are enhanced in the two-dimensional sagittal images. Three-dimensional surface rendered images of the same mouse (*c*) prior to and (*d,e*) after injection of BP20 reveal major vessels as well as spleen. The chambers of the heart (H) are visible through the rib cage in image (*e*). The secum and two fecal pellets [FP in (*b*) and (*e*)] are also seen due to residual bone meal remaining in the intestinal tract from mouse chow.

of this would be significant given the volume of such a tumor would conservatively only represent about a few thousand tumor cells.

The potential limitations of microCT scanners are outlined in Table 19.1. Many of the early limitations have already been overcome by technological scanner improvements afforded by the manufacturers. As novel targeted contrast agents are developed, microCT will most likely offer the most utility in terms of spatial resolution and animal throughput of many of the new microimaging modalities.

Data archiving and image interpretation and analysis remain the most important issues facing microCT researchers. The file sizes are large (>800 MB for a single medium-resolution scan) and several thousand contiguous axial images can be generated in a mouse abdomen alone. Visual interpretation of the images is definitely the rate-limiting step in the entire process. When looking only at two-dimensional axial images, it is a common practice to only reconstruct every twentieth axial slice in order to save time and disk space. To minimize interpretation time, data sets are typically only reconstructed three dimensionally in order to focus attention on specific areas of interest. However, it is still possible to simultaneously view two-dimensional images anywhere within the three-dimensional image.

CONTRAST-ENHANCING AGENTS

Contrast-enhancing agents play a significant role in clinical CT scanning of soft tissues and tumors. Until several years ago, patients would receive both a baseline non-contrast-enhanced scan followed by a contrast-enhanced scan for comparison. Currently, however, especially in abdominal scanning, the majority of patients receive only

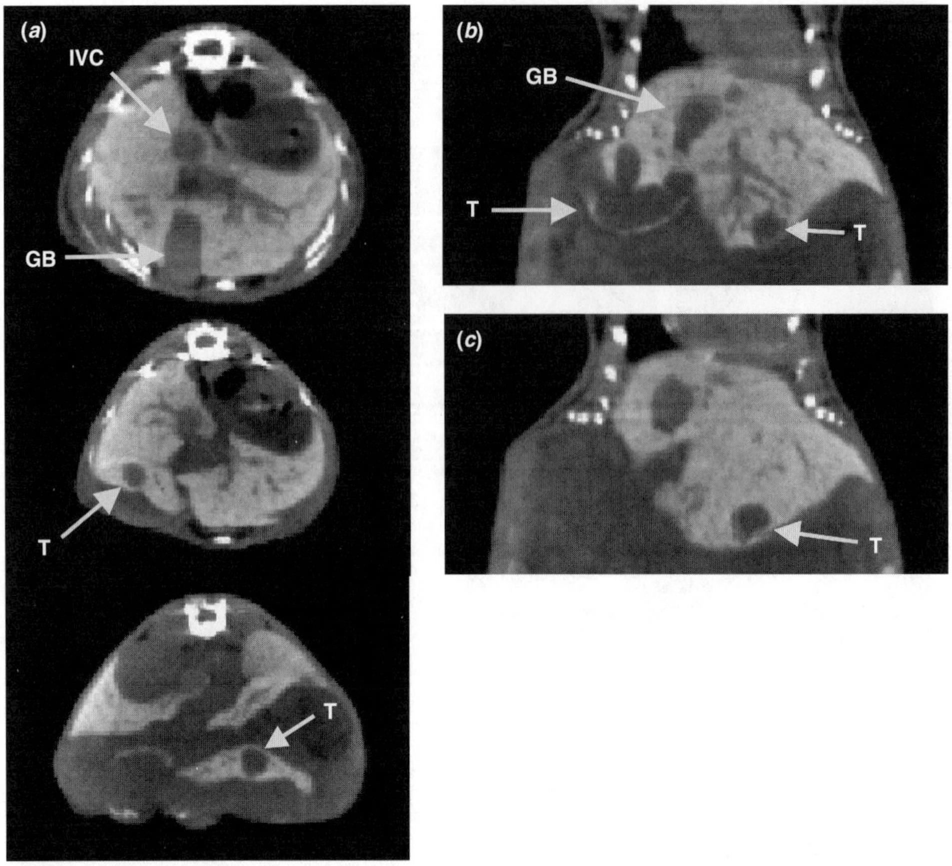

Figure 19.3. (*a*) Axial and (*b,c*) coronal microCT images obtained in a Balb/C mouse bearing 2 CT-26 (colon) tumors in the liver. The smaller of the tumors, which is about 2 mm in diameter, is easy to detect following opacification of healthy hepatocytes with ITG. Axial images in (*a*) represent three different anatomical slice levels through the liver.

a contrast-enhanced scan. The majority of commercial CT contrast agents are comprised of water-soluble iodinated agents in very high concentrations. Due to extremely rapid renal clearance of these iodinated agents, scanning occurs within the first few seconds to minutes of injection. These agents are free to diffuse into and out of interstitial spaces and therefore rapidly become diluted. It is not uncommon for a patient to receive 200 mL of contrast material injected intravenously via a power injector at a rate of 3–5 mL/s. These agents are nonspecific in that they are not designed to target any particular tissue of interest. Moreover, due to the relatively long data acquisition times associated with microCT scanners, the use of commercial contrast agents with research microCT scanners is severely limited. Unlike commercial agents, which undergo nearly complete renal elimination within a couple of minutes of injection in humans, agents suitable for use in microCT must reside in the tissue of interest for the duration of the acquisition. Although several biomimetic organ-selective CT contrast agents are being developed for use in humans, because these agents display long residence times (hours) in the tissues of interest, they are ideally suitable for use in microCT scanning.

Hepatocyte-Selective Agent

ITG is a polyiodinated triglyceride [glyceryl 2-oleoyl 1,3-bis(7-(3-amino-2, 4, 6-triiodophenyl) heptanoate)] formulated into a chylomicron remnant–like delivery vehicle that specifically targets normal hepatocytes via the Apo-E receptor pathway (Weichert et al., 2000). This agent has undergone extensive preclinical development, including acute and chronic toxicology studies. The agent is administered via tail vein injection and accumulates selectively in hepatocytes prior to undergoing metabolism and biliary excretion. It does not accumulate in tumors located within the liver. Unlike other rodents and dogs, wherein peak liver enhancement is achieved within 1 h, liver enhancement in mice occurs by 3–4 hours using typical injection volumes of around 300 μL for a 20-g mouse. The delay in liver uptake in mice is most likely due to lower levels of lipid metabolic enzymes in mice and rabbits.

Table 19.1. Potential Limitations of MicroCT Scanners for In Vivo Imaging[a]

Parameter	Effect	First-Generation Units	Second-Generation Units
Radiation dose	Cell death	0.4 Gy/390 step acquisition/40 kVp/800 μA	0.4 Gy/390 step acquisition/40 kVp/800 μA
Soft-tissue contrast	Tissue/tumor detection	Modest	better
Image acquisition time	Radiation dose/temporal resolution/limited commercial contrast agent use	10–25 min	5–15 min
Image reconstruction time	Throughput	20 min to hours	Parallel
Large data file size	Archiving	1 GB/med-resolution scan	1 GB/med-resolution scan
DICOM file compatibility	Image fusion/analysis	Only recently	Yes
Image interpretation time	Operator/radiologist time	1024 slices/mouse abdomen	≈2000 slices/mouse abdomen
Object size (detector size)	Rats or mice	3 × 3 in. "large" 1024^2 detector	6 × 6 in./4096^2 detector
Subvolume image acquisition	File size/acquisition time; dynamic studies possible	No	Not yet
Subvolume image reconstruction	File size and reconstruction time	Yes	Yes
Built-in anesthesia unit	Animal stress/motion	Yes	Yes
Source energy range	Tissue contrast/penetration	<50 kVp	<100 kVp
High throughput	Animals per hour	No	Yes
Self-shielding	Large box	No	Yes

[a]Based on our experience with Imtek units only.

Vascular Agent

The blood pool (BP) agent is a modified version of ITG in that PEG (polyethylene glycol) moieties are incorporated into the surface of the vehicle. This prevents association of Apo-E, which is responsible for hepatocyte receptor recognition. Invisible to the hepatocytes, the agent remains in the blood for several hours (Fig. 19.2) until nonspecific plasma hydrolases remove the PEG moieties, at which time the complex is transformed into the hepatocyte agent and is cleared via the hepatobiliary pathway. PEG-modified agents have been approved by the Food and Drug Administration (FDA). The agent is given intravenously via tail vein injection. Typical volumes are around 300 μL to a 20-g mouse. Both of these agents have been described in detail (Weichert et al., 2000).

Water-Soluble Contrast Agents

Although there are literally dozens of water-soluble commercial contrast agents available for clinical use today, all are similar in that they are all renally eliminated from the blood very rapidly following intravenous administration. All are nonspecific in their biodistribution and freely diffuse into and out of interstitial spaces. Moreover, renal elimination occurs extremely quickly wherein most of the agents are completely removed from the blood in a matter of just a few minutes in humans. Although investigators have reported obtaining microCT kidney images from mice following intravenous administration of water-soluble agents, the images have been less than optimal. Typically, only urine in the bladder is opacified by the end of image acquisition. In order to slow renal clearance of water-soluble contrast agents in mice, it is possible to inject clinical water-soluble agents (0.5 mL/iohexol-300) subcutaneously between the shoulder blades to obtain usable kidney images. The relatively slow absorption of the agent ensures a constant supply of contrast in the kidneys for the duration of the acquisition sequence. It is important to inject the agent anatomically high enough toward the neck so that it will not overlap with the kidney field of view. Timing is important and the images will most likely be limited to relatively low resolution due to the extreme dynamic nature of the agents themselves.

Commercial water-soluble agents do show some utility for imaging tumors of the intestinal tract and prostate, however, by administering them intraperitoneally (IP). In this case, the commercial agent (iohexol-300) is diluted 1 : 9 with sterile water prior to IP injection of about

Table 19.2. Mouse Organ MicroCT Imaging Sensitivity

Organ	Rating	Need Contrast Agent?	Comment
Adrenals	Excellent	No	Easy to see
Bladder	Fair	No	Easy to see urine due to density of minerals in water. Feed distilled water 2 days prior to see bladder wall itself.
Blood pool/vessels	Excellent	Yes	Long-acting BP needed
Bone	Outstanding	No	Trabecula easy to see
Brain	Poor	No	Low contrast
Heart–muscle	Good	No	Within lung field
Heart–blood pool	Excellent	Yes	Long-acting BP, motion not as bad as anticipated
Intestine–lumen	Good	Yes/No	Dilute barium oral or rectal for positive contrast/liquid or vegetable diet 2 days in advance for low contrast
Intestine–exterior	Good	Yes	Dilute water-soluble injected into peritoneal cavity
Kidney	Good	Yes	Diluted water-soluble subcutaneous between shoulder blades
Liver	Excellent	Yes	Need long-acting CM/sub-300 μm tumor detection
Lung	Excellent	No/yes	Natural low-density, tumors (1-mm)/vessels (with BP) easy to see
Lymph nodes	Peripheral are excellent	No	Easy to see, can inject CM into drainage field to enhance
Mammary glands	Good	No	Usually in fat pad so easily to detect sub-mm tumors
Muscle	Good	No	Low density
Ovary	Good	Yes	Use BP to follow artery to ovary
Pancreas	Poor	No	Ubiquitous nature of organ
Prostate/seminal vesicles	Good	No	Easy-to-see seminal vesicles, especially when hypertrophic, as they are usually within fat bed
Spleen	Excellent	Yes	Particulate CM, sub-300-μm tumor detection; can deliver tumor cells directly to liver by splenic injection
Skin	Poor	No	Do not see skin and hair but do see body wall
Skull	Excellent	No	Very high resolution possible of skull and bony structures in head
Testicles	Good	No	Fairly easy to see
Tumor	Depends on location	No/Yes	Can get tumor "blush" with long-acting BP contrast agents; easy to see in fat pads

Note: BP = long-acting blood pool contrast agent; CM = contrast media.

1 mL/25 g body weight. The agent surrounds organs and intestines located in the lower abdomen and renders them more visible relative to each other (Fig. 19.3). These agents are extremely safe and are absorbed and eventually eliminated from the peritoneal cavity.

The ability to image each mouse organ by microCT is summarized in Table 19.2. Each organ has particular characteristics which affect its density and therefore sensitivity to X-rays. Motion is typically a concern; surprisingly, however, it is not as big a problem as one might anticipate. This may be due, in part, to volume averaging over the acquisition period.

COMPARISON WITH OTHER MICROIMAGING MODALITIES

Relative to other microimaging modalities, microCT affords many advantages (Table 19.3). Foremost among

Table 19.3. Comparison of MicroCT with Other Microimaging Modalities

Parameter	MicroCT	MicroMRI	US	MicroPET	Bioluminescence
Anatomy	5	5	2	1	2
Function/biochemistry	2	2	3	5	4
Contrast agents	Y	Y	Y	Y	Y
Examination speed, min	5–20	10–30	Real time	30–60	20–40
Three-dimensional capability	5	5	2	3	Coming
Radiation dose	2	NA	NA	4	NA
Spatial resolution	<18 μm	50 μm	mm	2 mm	mm
Temporal resolution	2	3	5	3	2
Equipment costs, $	200,000–400,000	0.8–1.3 million	<200,000	500,000–750,00	<300,000

Note: Best = 5, worse = 1.

these is high spatial resolution. In terms of high-resolution modalities, microCT units are currently much less expensive than microPET and easily less than half the cost of a high-field, small-bore MRI unit. Although PET provides exquisite quantitative functional imaging information, it usually requires a team of dedicated PET chemists to provide the necessary short-lived imaging agents used in conjunction with this technique. High-field MRI units, on the other hand, afford high-resolution anatomic as well as dynamic imaging capabilities and no radiation dose concerns. In addition to being the most expensive of the microimaging modalities, microMRI requires cryogenic cooling as an added cost. Bioluminescence, while currently limited to two-dimensional output, affords high sensitivity, reasonable cost, and the ability to image gene function in vivo with the use of specific reporting groups. While each modality has its own advantages and limitations, they are all complimentary with each other and offer researchers powerful methods of noninvasively monitoring molecular events deep within living mice.

TYPICAL IMAGING PROTOCOL

Tumor-bearing mice are submitted for microCT scanning under an animal use protocol specific for this modality. Each animal is anesthetized and placed prone on an index card in order to minimize abdominal wall respiratory motion (as opposed to supine orientation wherein respiratory motion is unrestricted), and the paws are gently taped to the index card prior to the scanning procedure. The acquisition time typically lasts from 8 to 20 min depending on the degree of spatial resolution desired. The scanner has a spatial resolution exceeding 40 μm. If the imaging protocol calls for the use of a contrast-enhancing agent, it is administered just prior to scanning with the exception of our hepatocyte-selective agent, which requires 3–4 h for optimal liver opacification. Following scanning, each animal is placed on a covered heating pad preset to 35°C (low setting) during the recovery period. Recovery is monitored and recorded. Animals have been serially scanned up to two to three times per week for up to three weeks under these conditions to identify and monitor tumor development or regression with no apparent adverse effects. Because the microCT facility examines mice from several animal facilities on campus as well as mice from other institutions, animals are not returned to their original animal rooms following a scan but rather are housed in a quarantine room designated by the animal care unit. Mouse tail vein injections can be challenging and require a sufficient level of experience. Typically the tail of the anesthetized mouse (or restrained conscious mouse) is placed in warm water prior to injection with a 1-mL syringe fitted with either a 27- (for mice >20 g bw) or 30-gauge (for mice <20 g bw) needle. A magnifier light is also helpful for localizing the tail veins, which are located laterally on each side of the tail. Tail vein injection is more difficult in some mouse strains than in others. It usually takes about 20–30 min to inject and scan a single mouse.

FUTURE DIRECTIONS

As technology begins to bridge the gap between clinical whole-body imaging scanners and microscopic molecular imaging systems, our ability to visualize anatomy and functional events in living animal models will continue to improve. Moreover, hybrid systems capable of acquiring high-resolution anatomical CT images simultaneously with functional PET scans are now appearing in clinics. The high spatial resolution afforded by the CT scan allows precise localization of functional PET imaging agents. Microversions of hybrid scanners, including microCT/SPECT, are now becoming available.

Although the spatial resolution of current scanners is outstanding and in many cases exceeds that necessary for most mouse tumor monitoring studies, the temporal resolution is limited. It would be advantageous to be able to scan the same anatomic slice rapidly over a period of time that is compatible with the function event occurring in the tissue. Pharmacodynamic analyses of contrast agent uptake and clearance from organs or tumors, for example, are examples of processes that could be evaluated if a dynamic scanning mode were available. In the latest scanners, it is now possible to electronically "crop" the flat-panel detector by only activating a few vertical rows of detector elements at a time, thus decreasing acquisition time and disk space significantly. It is feasible that single slice images could be acquired in a matter of a few seconds.

With the development and integration of powerful image visualization and analysis programs such as Amira with microCT, it is now possible to perform virtual colonoscopy and, with the aid of a long-acting vascular contrast agent, virtual angiography in live mice in vessels less than 170 μm in diameter.

As the spatial resolution of scanners continues to improve and cell-selective molecular imaging agents are introduced into the research community, it may someday be feasible to image tumors consisting of only a few cells in live mice.

ACKNOWLEDGMENTS

This work was supported in part by the National Institutes of Health (NIH 1 P20 CA86278 Molecular Imaging Pre-Center Grant) and both the Comprehensive Cancer Center and Department of Radiology at the University of Wisconsin.

REFERENCES

Bentley MD, Ortiz MC, Ritman EL, Romero JC (2002): The use of microcomputed tomography to study microvasculature in small rodents. *Am J Physiol Regulatory Integrative Comp Physiol* 282:R1267–R1279.

Garcia-Sanz A, Rodriguez-Barbero A, Bentley MD, Ritman EL, Romero JC (1998): Three-dimensional microcomputed tomography of renal vasculature in rats. *Hypertension* 31:440–444.

Genant HK, Gordon C, Jiang Y, Link TM, Hans D, Majumdar S, Lang TF (2000): Advanced imaging of the macrostructure and microstructure of bone. *Horm Res* 54(Suppl 1):24–30.

Holdsworth DW, Thornton MM (2002): MicroCT in small animal and specimen imaging. *Trends Biotech* 20(8):S34–S39 (Suppl.).

Issever AS, Walsh A, Lu Y, Burghardt A, Lotz JC, Majumdar S (2003): Microcomputed tomography evaluation of trabecular bone structure on loaded mice tail vertebrae. *Spine* 28:123–128.

Laib A, Kumer JL, Majumdar S, Lane NE (2001): The temporal changes of trabecular architecture in ovariectomized rats assessed by microCT. *Osteoporosis Int* 12:936–941.

Paulus MJ, Gleason SS, Easterly ME (2001): A review of high-resolution x-ray computed tomography and other imaging modalities for small animal research. *Lab Animal* 30(3):1–10.

Paulus MJ, Gleason SS, Kennel SJ, Hunsicker PR, Johnson DK (2000): High-resolution x-ray computed tomography: An emerging tool for small animal cancer research. *Neoplasia* 2(1–2):62–70.

Wan SY, Kiraly AP, Ritman EL, Higgins WE (2000): Extraction of the hepatic vasculature in rats using 3-D micro-CT images. *IEEE Trans Med Imag* 19:964–971.

Weichert JP, Lee FT Jr, Chosy SG, Longino MA, Kuhlman JE, Heisey D, Leverson G (2000): Combined hepatocyte-selective and blood pool contrast agents for CT detection of experimental liver tumors. *Radiology* 216:865–871.

Weichert JP, Lee FT Jr, Longino MA, Chosy SG, Counsell RE (1998): Lipid-based blood pool CT imaging of the liver. *Acad Radiol* 5:S16–19.

Weichert JP, Longino MA, Bakan DA, Spigarelli MG, Chou T, Schwendner SW, Counsell RE (1995): Polyiodinated triglyceride analogs as potential computed tomography imaging agents for the liver. *J Med Chem* 38:636–646.

20

MAGNETIC RESONANCE IMAGING IN MOUSE CANCER MODELS

MANICKAM MURUGANANDHAM AND JASON A. KOUTCHER

Departments of Medical Physics, Radiology, and Medicine, Memorial Sloan-Kettering Cancer Center, New York, New York

Currently MRI is the imaging modality of choice for a wide variety of medical diagnostic questions and has an important role in the study of small-animal tumor physiology. We provide a brief overview of the physical principles of MRI and describe some of the areas wherein new developments in MRI affect preclinical oncology studies.

INTRODUCTION

The phenomenon of nuclear magnetic resonance (NMR) in bulk material was first demonstrated by Edward Purcell and Felix Bloch independently in 1946, for which they shared the 1952 Nobel Prize in Physics [1, 2]. Over the past 50 years, NMR has expanded from the exclusive domain of physics to include other fields. Particularly, clinical and biomedicine have benefited by advances in NMR, both as an imaging modality, magnetic resonance imaging (MRI), and as a method for obtaining physiologic and biochemical information, magnetic resonance spectroscopy (MRS).

The idea of using NMR and magnetic field gradients for imaging was published in 1973 [3]. The first image of a tumor in the body of a live mouse was published shortly thereafter [4, 5], followed by the first cross-sectional image of the human body in 1977 [6a, 6b]. Currently MRI is the imaging modality of choice for a wide variety of medical diagnostic questions and has an important role in the study of small-animal tumor physiology. Magnetic resonance spectroscopy provides unique information about in vivo biochemistry and physiology of tissues by permitting the detection of the presence of many different metabolites in vivo, although it will not be discussed in this chapter. The goal of this chapter is to provide a brief overview of the physical principles of MRI and to describe some of the areas wherein new developments in MRI affect preclinical oncology studies.

BRIEF REVIEW OF THE PHYSICS OF MAGNETIC RESONANCE IMAGING

The underlying principle of NMR is covered in depth in a number of excellent texts [7–11]. In brief, NMR is based on the interaction between an external magnetic field and an atomic nucleus that possesses spin, more precisely nuclear spin angular momentum. Qualitatively, in the presence of an externally applied magnetic field, nuclei precess at some frequency around the axis of the magnetic field. In spin-1/2 systems, such as ^{1}H, ^{13}C, ^{31}P, and ^{19}F, the direction of the nuclear spin can be thought of as lying approximately with (parallel) or approximately against (antiparallel) the main magnetic field. The spins aligned with the external main magnetic field (B_0) are in a slightly lower energy state than those that are antiparallel. Transitions between these two states are a resonant phenomenon, that is, the frequency of the energy used to excite the lower energy nuclei and induce a transition must be identical to the nuclear precession frequency. As described by the Larmor's relation:

$$\omega = \gamma B_0 \qquad (20.1)$$

Mouse Models of Human Cancer, edited by Eric C. Holland
ISBN 0-471-44460-X

the frequency of precession (ω) is linearly dependent upon the strength of the applied main magnetic field (B_0) as well as the magnetogyric ratio (γ) of the nuclide under study. After absorbing the energy, spins release this energy and are said to "relax" to their initial equilibrium state. The rate at which relaxation and loss of signal occurs is governed by two time constants, T_1 (spin–lattice relaxation) and T_2 (spin–spin relaxation). These "relaxation times" depend upon the chemical environment of the molecules and provide a mechanism of contrast between different tissues. Two important outcomes of the Larmor equation are that the resonant frequency at a given magnetic field strength is different for each nuclide (since each nuclide has a different value of γ) allowing one to "tune" in to different nuclei; that is, one can detect nuclei such as ^{13}C, ^{31}P, ^{19}F, and so on, with minimal interference from the signal from other nuclei. The other important point is that the signal strength (which is dependent on ω) increases with increasing magnetic field. This has been one of the main driving forces for higher field studies since one requires higher signal if one needs to obtain higher spatial resolution.

BASIS OF MRI

MRI mainly maps the distribution of magnetization of ^1H nuclei (*referred hereafter as "spins"*) from water and fat molecules, which are present most abundantly (in molar concentration) in tissues. Briefly, the imaging procedure involves the use of a specific combination of radiofrequency (RF) pulses and magnetic field gradients applied in a preprogrammed manner, known as RF pulse scheme or sequence. Information about the spin's spatial distribution is encoded by applying external linear magnetic field gradients in three perpendicular directions at specific times during and after RF pulses. The magnetic field gradients map spatial information into an NMR signal that can be used to form an image. Basically, the spin's precessional frequency history is computed from the phase and amplitude of the received NMR signal. A number of techniques exist for encoding the spatial location of a signal by selectively exciting part (or all) of a sample and changing the total accumulated phase and frequency of spins at different locations when they are subsequently detected. The data are transformed to yield the signal intensity of different voxels: an image. Further details of how images are formed using phase and frequency encoding are provided in other texts [7–11].

MECHANISMS OF IMAGE CONTRAST FORMATION IN MRI

As stated earlier, MRI is mainly based on visualizing the distribution of hydrogen nuclei from water and fat molecules in tissues. The image contrast—that is, the relative MRI signal intensity between tissues—is modulated by various intrinsic and extrinsic parameters. The major intrinsic contrast parameters are the density of hydrogen nuclei in tissues (spin density) and the tissue relaxation parameters T_1, T_2, and T_2^*. Extrinsic contrast mechanisms are based on the use of paramagnetic agents, which affect the inherent relaxation times of tissues.

Intrinsic Contrast Mechanisms

Among the major intrinsic contrast parameters, spin density—the number of nuclei that can be detected in a particular voxel—determines the maximum potential MRI signal intensity that can be achieved from that volume of tissue. The other parameters—T_1, T_2, and T_2^* relaxation time constants—as mentioned earlier, define how quickly the excited nuclei relax back to their initial (equilibrium) state. The longitudinal, or spin–lattice, relaxation time (T_1) refers to the amount of time it takes for the tissue magnetization to return to the equilibrium state in the longitudinal direction of the main magnetic field after excitation with a RF pulse of energy. The transverse, or spin–spin, relaxation time (T_2) denotes the transfer of excess energy (that is deposited in the tissue by the RF pulse) between tissues. This transfer of energy between interacting spins within tissues leads to loss of spin-phase coherency (spin dephasing) and thereby losses in the detected MR signal. However, processes other than these inherent spin–spin interactions such as inhomogeneities in the applied main magnetic field and intrinsic variations in magnetic susceptibilities of tissues also cause spins to dephase. Thus the signal loss (or decay) due to both these natural processes and magnetic field imperfections is characterized by the relaxation time constant T_2^* (known as apparent transverse relaxation). "T_1-weighted" or "T_2-weighted" images, reflecting respectively the T_1 and T_2 of the tissues, are obtained by specific schema of pulses that excite nuclei in the subjects under study. Tumors generally have longer T_1's and T_2's than normal tissue [12, 13], resulting in tumors usually having low signal on T_1-weighted images and high signal on T_2 weighted images (Fig. 20.1) than corresponding normal solid organs.

Tumor contrast (i.e., difference in signal intensity between normal and cancer) is also affected by field strength. As noted above, signal intensity increases with field strength. While image resolution is technically determined by magnetic gradient field strength, practically it is usually limited by lack of signal. This is best appreciated by realizing that if one chooses to double the in-plane resolution of an image (e.g., going from 1 to 0.5 mm resolution in both dimensions of the image plane), the size of the voxel (volume element) is decreased by a factor of 4, resulting in the signal only being 25% as strong

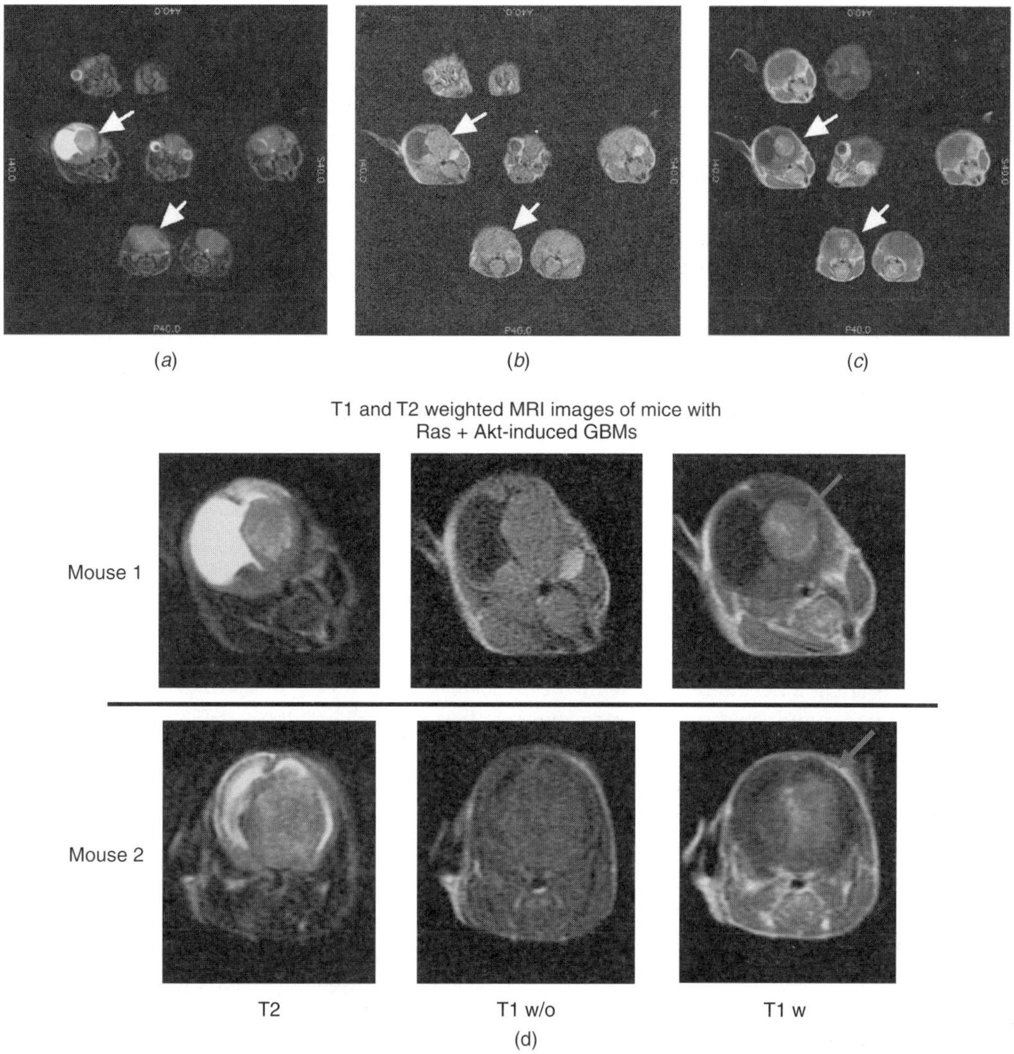

Figure 20.1. Images (arrow) obtained on Ntv-a mice that had been infected with the combination of RCAS-Akt and RCAS-Ras. Plates (*a*), (*b*), and (*c*) are T_2, T_1 (precontrast), and T_1 (postcontrast) from a single slice of multiple mice. Arrows indicate tumor. Areas of hyperintense signal in the T_2-weighted images correspond to areas of hydrocephalus, which are noted to not enhance with Gd-DTPA. Acquisition parameters included TR = 5500 ms, TE = 102 ms, NEX = 4, 512 × 512, 8 × 8 cm^2 (156 μm ×156 μm in-plane resolution), slice thickness 1.5 mm, gap 0.5 mm. (*d*) Expanded view of two tumor-bearing mice: T1w/o = T_1-weighted images without contrast; T1w = T_1-weighted image with contrast. Note that the tumor appears bright on T_2-weighted images but there is relatively poor contrast on T_1-weighted images without contrast at 1.5 T.

(i.e., loss of 75% of signal). Thus the need for a higher signal-to-noise ratio is evident if one wants to obtain higher spatial resolution. However, there are drawbacks to higher field strength. One concern is that at high field strengths (>1.0 T) one does not obtain good contrast on T_1-weighted images. At higher field strengths, the differences in T_1 between normal and pathologic tissues decrease and contrast is lost (Fig. 20.1). Fortunately, high contrast can be obtained from T_2-weighted images and other (diffusion-weighted) techniques.

Two important intrinsic contrast mechanisms, perfusion and diffusion, have been exploited in the realm of oncologic MRI to obtain images depending on tissue physiologic parameters such as blood flow (perfusion) or water molecular diffusion in tumors to reflect the tumor microenvironment. Movement or diffusion of bulk water between tissues in a random motion under an applied magnetic field gradient leads to spin dephasing and loss of phase coherence, causing a decrease in MRI signal. Destruction of cellular structures due to necrosis may lead to unrestricted motion of water molecules in the necrotic regions. These changes are expected to increase the mobility of water molecules in damaged tissue. Similarly, edematous tissue has different diffusing characteristics compared to organs or tumors. Thus, diffusion weighted (DW) MRI is highly sensitive to microscopic changes in tissue structure and physiology [14, 15]. Similarly, perfusion of blood in the microcirculation of the tissue being imaged also contributes to the spin dephasing and a decrease in the intensity of the MRI signal.

Extrinsic Contrast Mechanisms

Signal intensity in MRI images can also be modulated by externally administered paramagnetic contrast agents such as gadolinium diethylene triamine pentaacetic acid (Gd-DTPA). The gadolinium (Gd^{3+}) atom in Gd-DTPA is usually trapped in the protective cavity formed by $DTPA^{5-}$ to minimize the toxic effects of gadolinium. Even though not observed directly by MRI, the Gd^{3+} atoms, by virtue of possessing seven unpaired electrons in their outer orbital, exhibits strong paramagnetic interaction with the ^1H nuclei of neighboring water molecules and shorten the tissue water T_1 and T_2 relaxation times. Shortening of T_1 and T_2 provides very large signal changes, as one molecule of Gd^{3+} can affect very many molecules of water. Due to intrinsically short tissue T_2 values, the T_1 shortening effect predominates, which leads to increased signal intensity on T_1-weighted images (known as *positive enhancement*; Fig. 20.1) caused by leakage of Gd^{3+} from the vascular space into the extravascular tumor space. On the other hand, Gd-DTPA, when confined to blood capillaries, perturbs the local magnetic fields in their vicinity through magnetic susceptibility affect and shortens the T_2 and T_2^* relaxation times, leading to a drop in signal intensity on T_2- or T_2^*-weighted images (known generally as *negative enhancement*).

DYNAMIC CONTRAST-ENHANCED MRI (DCE-MRI)

In standard contrast-enhanced imaging, the subject is imaged prior to and after contrast (Gd-DTPA) administration. With the advent of faster pulse sequences and greater computer storage capability, it is possible to trace the uptake of the contrast agent in tissues using a dynamic imaging protocol, that is, to take a series of rapid images and detect changes in signal intensity that are due to delivery of Gd to the tissue. The kinetics of contrast agent in tissue is represented through the signal intensity–time curve (Fig. 20.2). The underlying physiology governing the shape of these curves is often analyzed using appropriate pharmacokinetic models. The images acquired in

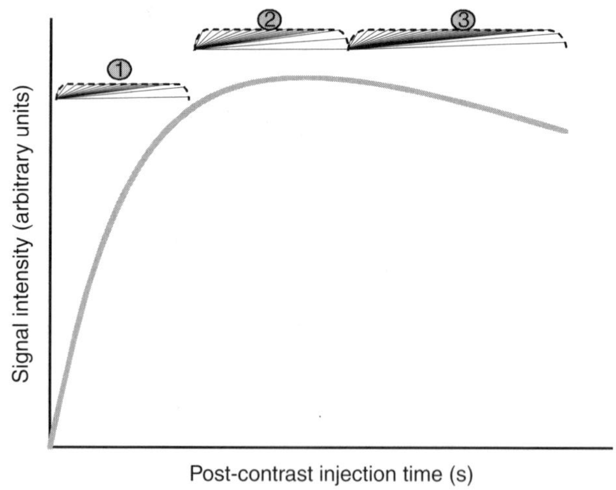

Figure 20.2. Schematics of typical Gd-DTPA contrast kinetics using T_1-weighted DCE-MRI protocol showing different phases of contrast enhancement: 1, initial steep incremental phase; 2, plateau phase; 3, wash-out or decline phase. Signal shown would represent a single voxel (volume element) and similar graphs would be generated for each voxel, leading to a parametric image of the rate uptake of contrast, which can be translated into various physiologic parameters.

DCE-MRI protocol using Gd-DTPA can be either T_1 or T_2^* weighted. In regions like the brain where the blood–brain barrier is intact, the injected Gd-DTPA remains within the vasculature and the T_2^*-weighted images obtained under such circumstances would provide information about vascular volume and flow. The T_1-weighted Gd-DTPA-based DCE-MRI method is widely used to study tumors outside the brain. Using specialized software tools and contrast kinetics analysis, the acquired data can be displayed as a parametric image, that is, a pixel-by-pixel image of blood volume, permeability, perfusion, diffusion, extravascular space, and so on.

ONCOLOGIC IMAGING OF SMALL ANIMALS: MULTIPLE MOUSE IMAGING

There has been increased use of MRI and other imaging modalities to study transgenic and knockout mice for phenotypic changes induced by genetic alterations. Determining phenotypic penetrance requires large numbers of mice for study. Noninvasive screening permits determination of the phenotypic penetrance of these genetic changes that induce tumor and allows these mice to also be used for other experiments and take maximum advantage of these valuable mice.

While previous studies have demonstrated the value of high-spatial-resolution MRI for quantifying tumor volume in a limited number of animals, the problem of screening large numbers of mice for anatomic abnormalities is potentially time and cost prohibitive. Screening experimental rodents can be described as attempting to detect small lesions caused by abnormal phenotypic expression at a time when they are not yet physiologically significant. Screening requires a high-throughput technique, possibly compromising on spatial resolution. It has been suggested that requirements for a screening study would be to be able to detect and quantify the volume of orthotopic tumors in mice with a goal of detecting tumors greater than ~2 mm. To achieve these requirements, an adequate signal-to-noise ratio is necessary for a spatial resolution greater than 300 μm in plane and a slice thickness of 1–2 mm. Preliminary high-spatial-resolution studies of multiple samples (primarily phantoms and postmortem animals) have been performed using complicated RF coil arrays [16–19], which have the potential advantage of greater signal-to-noise ratio and thus can be translated into higher spatial resolution. In these designs, MRI signals are collected from all the coils in the array and the data are combined from all the channels to construct an image, with the signal-to-noise ratio in each sample somewhat less but approaching that of a single coil in the array. An alternative approach to the problem has been to use clinical (1.5-T) systems with either standard clinical or specially designed coils that can study more than one animal at a time. A recent study [20] applied solenoidal resonators to maximize the signal-to-noise ratio and achieved higher spatial resolution (maximum resolution of 1500 μm × 156 μm × 156 μm, or 0.036 μL) than previous studies. The RF coil utilized had several attractive features, including feasibility of injecting contrast agents and imaging up to 13 mice simultaneously (Fig. 20.1).

An important component of small-animal imaging is the necessity of anesthesia. Attempting to obtain submillimeter resolution with an animal that is poorly anesthetized is obviously futile. A relatively safe method of anesthetizing mice is the use of inhalation anesthesias such as isofluorane. This technique has the advantages that the mice readily wake up and very low mortality. There are other short-acting drugs and combinations (e.g., ketamine and xylazine) which have also been used and have a good margin of safety, but the duration of anesthesia does not compare to that which can be obtained using inhalation techniques. Development of techniques to ensure prolonged (1-h) anesthesia for multiple mice will be necessary to exploit screening techniques described above.

MRI has been used extensively for studies of subcutaneous tissues providing significant information on tumor biology, physiology, and growth [21–24]. There has been increased use of MRI and other imaging modalities [optical imaging, computerized tomography (CT), radionuclide imaging] to study transgenic and knockout mice for phenotypic changes induced by genetic alterations. These imaging studies allow sequential studies of the same animals to determine the phenotypic effect of altering different genes and evaluating novel genetic therapies. The following discussion will focus on orthotopic tumors, including both xenografts and transgenic mice. We have broken up the text by anatomic regions to try to highlight the difficulties or special features associated with each organ.

Brain

Clinically, contrast-enhanced T_1-weighted images are often used for diagnosing brain tumors, and this technique is also used in small-animal studies of brain tumors. With the availability of higher field magnets for rodent studies, other sequences such as diffusion-weighted imaging have been applied very effectively for detection of tumor, quantitation of tumor volume, and monitoring response. MRI has been utilized for screening and detecting brain tumors in mice based upon their imaging characteristics and their pattern of enhancement.

It has been shown that many murine viral-induced tumors have imaging characteristics similar to patient tumors [25]. Imaging of these tumors reveals many similarities to those observed in humans with identical pathology. Specifically, high-grade murine gliomas have histologic characteristics of glioblastoma multiforme (GBM) with contrast enhancement after intravenous administration of Gd-DTPA, implying disruption of the blood–brain barrier in these tumors. In contrast, low-grade murine oligodendrogliomas do not undergo contrast enhancement, similar to human tumors. MRI can be used to identify mice with brain neoplasms as inclusion criteria in preclinical trials.

MRI has been used to measure tumor growth rates in various organs, including brain and liver tumors in rats [25–28]. A variety of techniques exist to estimate tumor volumes, but a simple method often utilized is to add tumor areas on different slices. This technique has been verified by measurements against phantoms. Using this technique, it has been shown that intracranial 9L tumor growth is exponential, with a doubling time of 66 h [26]. Treatment with 1,3-bis(2-chloroethyl)-1-nitrosourea (BCNU) was shown by MRI to induce 2 logs of cell kill. It has been suggested that [29] quantitative diffusion MRI in rat brain tumor models is more accurate at tumor localization than T_2-weighted images, although it remains difficult to detect a border between edema and tumor. The diffusion coefficient of untreated tumors has been shown not to be altered by size or growth rate, but changes in response to chemotherapy are observed before tumor regression occurs, although the diffusion coefficients subsequently return to pretreatment levels [30, 31]. The change in diffusion coefficient has been associated with an increase in extracellular space histologically, which might explain the increase.

Since MRI can estimate tumor volume quantitatively, it has also been used extensively to evaluate response to novel therapies such as viral and gene therapy. These types of novel treatments include the effects of ganciclovir treatment on C6 brain tumors that had been transfected with herpes simplex viral thymidine kinase (HSVtk), including studying the effect of timing of therapy on response [32], the effects of transfection with cytosine deaminase (CD) and subsequent treatment with 5-fluorocytosine (5FC) [33–35], and evaluation of response. Similar studies have been done with interleukin-12 (IL-12) treatment [36] and immunotherapy with *Corynebacterium parvum (C. parvum)* [37].

Changes in vascular morphology have also been studied in brain tumors using MRI. These parallel the clinical studies wherein measurements of cerebral blood volume and functional imaging have been applied clinically. Based on differences in the relaxation rates ($1/T_2$ and $1/T_2^*$) after injection of an intravascular iron contrast agent, it is possible to image differences in tumor angiogenesis between gray matter and a rat glioma [38]. Heterogeneity of tumor vasculature with the highest blood vessel density located in the peripheral areas of the tumor has also been demonstrated [38, 39].

Liver

The quantitative study of diffuse organ metastases represents a different problem than the detection of a single primary tumor such as is typically the problem in the brain and many other organs. These problems include the necessity of detecting multiple very small nodules, thereby requiring higher resolution than is required for detecting a single larger primary, and also quantitating tumor burden. Two early studies [27, 28] have demonstrated the quantitative value of MRI studies for detecting tumors in the liver using rats as the host, and these are appropriate paradigms for orthotopic studies of hepatic metastases in liver. Qin et al. studied the DHD/K12 rat colon tumor, which produces hepatic metastases when injected under the liver capsule or via the portal vein. Liver tumor volumes were estimated by adding the tumor volume in each two-dimensional slice and then compared with tumor volume measured by water displacement after the animals were sacrificed and tumors removed. They were able to detect tumors as small as 1 mm in length. Tumor volumes measured by MRI and by water volume displacement method were closely correlated. An alternative method of demonstrating the accuracy of MRI is by comparison with histology [28]. This method has also been used to demonstrate a very good correlation between MRI and histologic measurements of both tumor and liver.

Clinically, imaging the liver often requires either respiratory gating or having the patients hold their breath since respiratory motion causes liver motion. Fichtner et al. [40] addressed this issue in their study of adoptive immunotherapy in a mouse model of metastatic mouse lymphoma. To reduce motion artifact caused by respiration, the images were obtained with respiratory gating. Images with a 1-mm-thick slice and 100 μm in-plane resolution were obtained in 20 min with the use of a high-field (7.0-T) magnet. The authors were able to detect lesions that were $\geq$0.3 mm and were able to detect complete regressions to treatment without the use of contrast agents. Adoptive immunotherapy has also been studied in experimental colorectal liver metastases using a single metastatic nodule wherein it was [41] demonstrated that MRI was much more sensitive than CT or ultrasound.

A novel approach to contrast studies of the liver is to develop agents directed toward receptors on the liver cells. The feasibility of detecting liver cancer in rats by injection of particles that are bound by the asialoglycoprotein (ASG) receptors on hepatocytes [42, 43] has been demonstrated. Ultrasmall (12-nm) particles of iron oxide (USPIO) were targeted to these receptors by attaching arabinogalactan to their surface. The presence of iron causes local magnetic fields to be generated near the iron atoms, which causes a decrease in T_2^*. Both metastatic and hepatocellular carcinomas were found not to bind these particles since the ASG receptors are not present on malignant cells. The authors found a considerable benefit in terms of dose reduction of contrast agent by targeting the particles to hepatocytes instead of the mononuclear phagocytes, which are typically what take up larger iron particles and provide less specific contrast. An alternative method [44] is the use of a gadolinium complex bound to polylysine (polydiethylenetriamine pentaacetic acid polyneogalactosyl polylysine), which binds to the ASG receptor. Enhancement of tumor contrast in the liver was noted both by MRI and nuclear imaging wherein technetium-99 m was attached to the molecule. Since tumors do not have the ASG receptor, they do not bind the molecule and therefore do not enhance.

Lung

Imaging of the lung is complicated by respiratory and cardiac motion. Thus, to obtain high-quality images of the chest, it is necessary to "gate," or trigger, the acquisitions to be synchronous to the phases of respiratory and cardiac motion. While this technology is quite mature for patients, it is slowly becoming available for small animals such as mice and rats. The use of short TR interval imaging techniques decreases but does not eliminate these artifacts. Fortunately, there is inherent contrast between the normal air-filled lungs (low signal areas) and solid tumors (high signal area), which lead to good contrast to noise with almost any pulse sequence chosen. The need to use a short TR interval usually results in selection of T_1-weighted images (short TR and short TE).

Recent images of tumors in the lung are indicative of the potential to detect these tumors with modern MRI equipment. Fisher et al. [45] studied the role of K-Ras in development of lung adenocarcinomas. They studied transgenic mice wherein they could turn "on" or "off" K-Ras4b (G12D) by administering doxycycline to the mice. These effects were monitored by MRI using both cardiac and respiratory gating and demonstrated that with appropriate gating techniques high-quality images of tumors in the lung can be quickly obtained.

Bladder

Orthotopic animal models of bladder cancer are important for determining the activity of novel potential antineoplastic agents [46–48]. Compared to subcutaneous tumors, orthotopic bladder tumor models more closely mimic the behavior of human bladder cancer. Imaging of tumors in the bladder is complicated by the need to image small tumors and the poor tissue contrast that occurs when the bladder is collapsed. However, accurate noninvasive assessment of established tumors in the orthotopic model has been difficult, particularly for early-stage superficial bladder tumors.

To increase tumor contrast, previous reports relied upon putting fluid in the bladder which distends the bladder and provides high contrast relative to solid tumors. Initial studies [49] utilized Gd-DTPA instilled in the bladder of mice and demonstrated that MRI can be used for monitoring mouse bladder tumors treated with intravesically administered agents such as tumor necrosis factor α and bacillus Calmette-Geurin [50]. The in-plane tumor area in their studies arranged from 30.4 to 203.5 mm^2. The excellent contrast between tumor and the bladder when it is distended by fluid combined with high-quality RF coils allows use [51] of 1.5-T clinical magnets for detecting tumors less than 10 mm^2 (in plane) and assessing the growth kinetics of MBT-2 bladder cancer. Contrast was optimized by injecting 50 μL of Gd-DTPA plus 50 μL of air into the bladder and orthotopically implanted MBT-2 tumors were evaluated (10 and 14 days after tumor inoculation). Small superficial bladder tumors of 1.5 mm in diameter or less as well as advanced large tumors could readily be detected with an accuracy rate of 87.5% (Fig. 20.3).

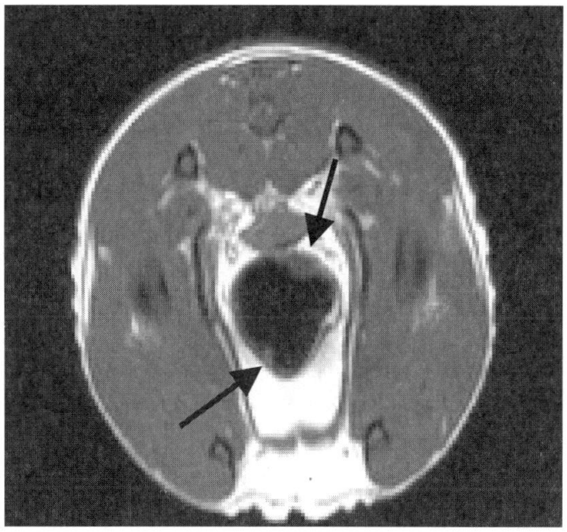

Figure 20.3. T_1-weighted image of orthotopically implanted MBT-2 bladder tumor. The bladder had been filled with 50 μL Gd-DTPA +50 μL air. Images resolution 156 μm × 156 μm in-plane with 1.5-mm-thick slice. Arrow indicates tumor.

Prostate

Imaging of prostate tumors in animal models has been done extensively, although most studies have focused on subcutaneous models. The latter have provided insights into (a) the effects of contrast agents, (b) tumor response to hormonal manipulation and chemotherapy, and (c) oxygenation. The prostate is very small, thus requiring very high spatial resolution in addition to high contrast. Its presence near the bladder also is problematic for positron emission tomography studies since the trace (often fluorodeoxyglucose) concentrates in the bladder make it difficult to discern whether the signal is from the prostate or adjacent tissue.

Since the prostate is small, successful studies of this organ requires high-resolution imaging in addition to high contrast. A prospective longitudinal study on cohort of transgenic mice at high risk for cancer development due to expression of simian virus 40 (SV40) T antigen [Transgenic Adenocarcinoma of the Mouse Prostate (TRAMP)] demonstrated that tumors were detectable by MRI before they were palpable [52]. With sequential MRI studies, this paradigm could be used for evaluating chemoprevention agents to inhibit development of prostate cancer [53].

High-resolution MRI has also been used to study the inhibition of carcinogenesis in the prostate using the TRAMP model [54]. The effect of oral infusion of green tea polyphenols on the development of prostate cancer was studied. The polyphenolic fraction that can be isolated from green tea was administered to TRAMP mice at a level equivalent to a human drinking six cups of tea per day and resulted in greater than 50% inhibition of tumor growth. MRI has also been used to monitor formation and progression. Song et al. [55] studied a transgenic mouse (CR2-TAg) which originated from neuroendocrine cells wherein SV40 large-T antigen expression is controlled

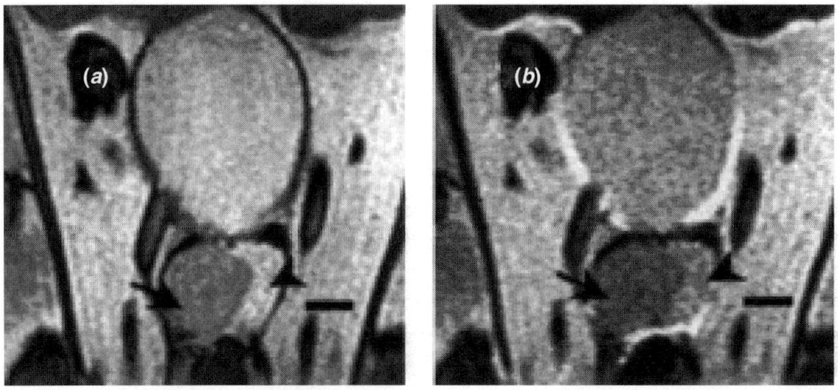

Figure 20.4. T_2- weighted ^{1}H MRI of prostate cancer in a CR2-TAg transgenic mouse. Both images are T_2 weighted, although (*b*) has greater T_2 weighting (TE = 57 ms) compared to (*a*) (TE = 30 ms). Position of the tumor is indicated by a *black arrow* and normal prostate by an *arrowhead*. Displayed field of view in all images is 15 × 15 mm. Scale bars = 1 mm. (Reproduced with permission from S. K. Song et al., *Cancer Research*, Vol. 62, pp. 1555–1558, March 2002.).

by the mouse cryptdin-2 (CR2) gene. Using both T_2-weighted and diffusion-weighted imaging, they detected small tumors and their progression. They showed that normal prostate had an apparent diffusion coefficient (ADC) twice as great as tumor, which provided very good contrast to noise (Fig. 20.4). A parallel study by Jennings et al. [56] in subcutaneous tumors also showed the efficacy of diffusion-weighted imaging to detect tumor response to chemotherapy. Thus diffusion-weighted imaging, particularly at higher magnetic field strengths, has been shown to be very useful for detecting prostate tumors.

An important concern is the detection of nodal metastases from prostate tumors. Weissleder et al. [57] demonstrated that using iron oxide (AMI-25) it was feasible to detect metastatic lymph nodes from a hepatoma model. Vassallo et al. [58] demonstrated that the use of an iron oxide contrast agent (AMI-227) could lead to selective detection of lymph nodes with metastatic prostate cancer, which could be discerned, from inflammatory lymph nodes. They studied rats injected with R3327-MAT-LyLu prostate cancer cells or Freund's adjuvant to generate ipsilateral lymph node metastases or lymphadenitis. Contrast between the node and adjacent fat increased for normal or inflammatory nodes but was unchanged in tumor-bearing nodes, indicating that this agent might be useful in detecting prostate cancer metastatic to lymph nodes.

MRI OF TUMOR FUNCTION

In addition to providing excellent high-spatial-resolution tumor morphologic images, advances in MRI pulse sequences, hardware, and computational technology combined with the increased understanding of MRI contrast (intrinsic and extrinsic) mechanisms, as discussed earlier, facilitate study of tumor functions in a noninvasive manner at high temporal and spatial resolution. Some of these applications are discussed in the following sections.

Applications of DW-MRI—Evaluation of Tumor Necrosis, Cellularity, and Response to Treatment

Diffusion-weighted MRI has been exploited for tumor assessment in both preclinical [59, 60] and clinical settings [61–63], including tumor characterization and prediction of tumor response. DW-MRI has been used to compare viable tumor with necrosis and edema based on higher ADCs of water molecules in the latter regions [64–68]. A correlation between diffusion measurements and cell density in human gliomas has also been demonstrated; regions with high cell density had lower diffusion constants (ADC) than regions with low cell density [69]. Similarly, an inverse correlation between tumor water ADC and signal intensity of choline-containing compounds (as determined by ^{1}H MRS) in human gliomas has been demonstrated which suggested a relationship between tumor ADC and cellularity [70]. These findings were further supported by a correlation between tumor cell density and ADC observed in melanoma xenografts [71].

In addition to tumor characterization, DW-MRI has been widely exploited to evaluate the response of tumors to various cytotoxic treatments in animal tumor models. Several studies showed an increase in ADCs upon treatment with anticancer drugs [30, 31, 72] and gene therapy [30, 31, 34, 72–76]. These elevated tumor ADCs after treatment occurred before changes in tumor volume were detected, suggesting that the changes in ADC could be an early indication of response. Radiation-induced fibrosarcoma (RIF-1) tumors treated with cyclophosphamide showed an increase in tumor water ADC prior to reduction in tumor volume [75], but this elevated ADC reversed upon tumor regrowth, suggesting diffusion measurements to be an early index of favorable cytotoxic treatment response. Similarly, in the orthotopic 9L glioma model 55% increase in tumor water ADC on day 6 after treatment with high-dose BCNU chemotherapy was noted; these changes also occurred before tumor growth arrest and regression [30]. These consistent observations of an increase in tumor water diffusion following different treatment strategies in various tumor models suggest that elevated tumor water ADC may be a universal response of tumors to successful therapeutic interventions.

Applications of DCE-MRI—Evaluation of Tumor Vascularity and Physiology

Dynamic contrast-enhanced MRI facilitates the assessment of tumor vascular function. The inherent ability of tumors to form new blood vessels (tumor angiogenesis) as they progress along their malignant growth leads to the most frequently observed tumor features, such as an increased blood volume (which is often histologically associated with elevated microvessel density counts) and hyperpermeability of blood vessels to blood-borne solutes, including MRI contrast agents. Thus, in general, malignant tumors differ from normal tissue in their vascular blood flow and volume characteristics, including vascular permeability, and an increased fractional volume of the extracellular extravascular space (EES) compartment [77, 78]. These underlying physiologic changes in the tumor vasculature can be assessed by probing the kinetics of externally applied MRI contrast agents. Typical kinetics of Gd-DTPA from a T_1-weighted imaging protocol is as shown in Figure 20.2 and can be characterized by three main phases: an initial steep *incremental phase* followed by a *plateau* and finally a *wash-out* or *decline phase*. These observed characteristics are analyzed using various pharmacokinetic models in order to relate it to the underlying physiology [79–85]. The parameters of particular interest are (i) the volume of extravascular extracellular space (EES) per unit volume of tissue, which determines the contrast distribution volume (v_e); (ii) the volume transfer constant between blood plasma and EES, known as the *transfer constant* (K_{trans}), which represents the vascular endothelial permeability product surface area; and (iii) the rate constant between EES and blood plasma—the *rate constant* (k_{ep}). Functional information of the tumor vasculature obtained from DCE-MRI facilitates the assessment of the tumor angiogenesis status, which in turn can be exploited for tumor characterization, metastasis screening, and therapy monitoring. For example, this technique was used to compare permeability and hypoxia in fast- and slow-growing tumors [86]. The fast-growing E102 tumor was found to exhibit a more homogenous distribution of the vascular permeability surface area across the tumor, whereas the slow-growing E106 tumor exhibited a more heterogeneous pattern. Three distinct perfused areas with differences in vascular permeability were identified in these tumors: well-perfused periphery with high vascular permeability, perfused capillaries with low vascular permeability located at the interior tumor regions, and perfused capillaries with high vascular permeability located adjacent to the necrotic portions of the tumors. These results were found to have good correlation with the histologic estimates of hypoxic fractions in these tumors. The E102 tumors having homogeneous permeability distribution had low hypoxic fraction, and the fast-growing E106 tumor with the heterogeneous vascular permeability pattern had high hypoxic fractions.

Vascular differences between metastatic and nonmetastatic breast and prostate tumor xenografts have also been studied using albumin-Gd-DTPA-based DCE-MRI [87]. Metastatic tumors formed vasculature with significantly higher vascular permeability or vascular volume. In addition, the permeability profiles matched VEGF (vascular endothelial growth factor) expression in these tumors. Within tumors, regions of high vascular volume exhibited low permeability, whereas regions of low vascular volume showed high permeability. Moreover, areas of cell death and necrosis were found to be typically associated with low or nondetectable vascular volume, although areas of low vascular volumes are not always associated with necrosis in histologic sections.

Dynamic contrast-enhanced MRI has been shown to be useful in monitoring the effects of various forms of tumor therapy, including fractionated radiation therapy on tumor vascular function [88] and estrogen therapy in breast tumor models. In the case of tumors treated with the antiestrogenic drug tamoxifen, the derived tumor permeability and leakage space (i.e., interstitial space or EES) maps before and after treatment show (Fig. 20.5) marked growth of intratumoral permeable capillaries around the necrotic regions, developed as a result of treatment. Immunostaining of vascular endothelial cells of these tumors, as shown in Figure 20.5, has further confirmed this localized angiogenic stimulation.

Targeting against tumor angiogenesis is a promising approach for cancer therapy [89–92]. Various antiangiogenic agents are being developed and successfully tested in preclinical settings. Minimally invasive optical studies on tumor vascularization and perfusion are limited to special model systems which allow easy and direct inspection of tumor vasculature, such as dorsal skin and cranial

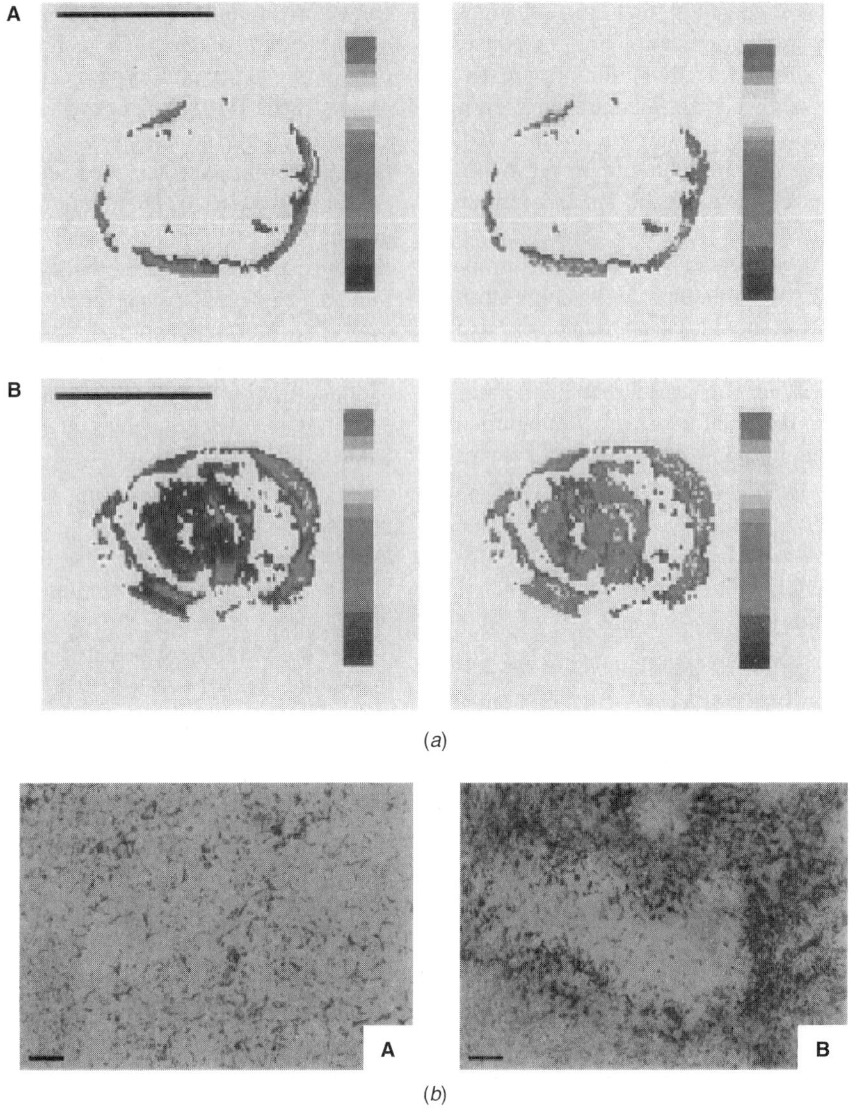

Figure 20.5. (*a*) Maps of permeability factor (K) and fraction of free leakage space (v_1) of MCF7 tumor before and during tamoxifen treatment. (A) Before tamoxifen treatment. *Left*: Permeability factor; the scale of K ranges from 0 to 0.19 min^{-1}. *Right*: Fraction of leakage space; the scale of v_1 ranges from 0 to 1. Scale bar = 1 cm. (B) Thirteen days of tamoxifen treatment. *Left*: Permeability factor; the scale of K ranges from 0 to 0.25 min^{-1}. *Right*: Fraction of leakage space; the scale of v_1 ranges from 0 to 1. Scale bar = 1. (Reproduced with permission from Furman-Haran E et al., *Proceedings of the National Academy of Science USA*, Vol. 93, pp. 6247–6251, June 1996.) (*b*). Microscopic view of endothelial cells in histologic sections of MCF7 tumors stained with *Griffonia simplicifolia* lectin. (A) Endothelial cells in a viable cellular area of a tumor grown in the presence of estrogen. (B) Endothelial cells localized at the boundary of necrosis in a tumor treated with tamoxifen for 2 weeks. Scale bar = 30 μm. (Reproduced with permission from Furman-Haran E et al., *Proceedings of the National Academy of Science USA* Vol. 93, pp. 6247–6251, June 1996).

chamber models. However, these techniques cannot be translated directly to clinical use. In this context, DCE-MRI has great potential to identify a suitable index of tumor vascular function for the evaluation of antiangiogenic therapy. Moreover, it can be easily translated to the clinic. The effect of anti-VEGF antibody therapy in U-87 glioblastoma multiforme has been assessed [93] by measuring tumor vessel permeability and vascular volume using DCE-MRI with Gd-DTPA contrast agent. Both the values of the transfer constant (K_{trans}) and the vascular volume (estimated through fractional plasma volume, fPV) derived from the contrast kinetics were significantly lower with anti-VEGF-treated tumors as compared with the saline-treated sham controls. Anti-VEGF antibody treatment in human breast adenocarcinoma (MDA-MB-435 cell line) assessed using albumin-Gd-DTPA-based DCE-MRI caused a 98% decline in vascular permeability 24 h after treatment, indicating that DCE-MRI with macromolecular agents is highly sensitive to interference with the angiogenesis process as soon as one day after a single treatment [94]. However, vascular volume did not change significantly in this study, implying that vascular volume is dependent on vessel morphology and is less sensitive than permeability to acute changes in angiogenesis. A similar finding of decrease in vascular permeability after the anti-VEGF antibody treatment has been reported in an experimental human ovarian cancer model [95]. Thus existing literature data suggest that the decrease in vascular permeability can serve as a reliable index of tumor response to antiangiogenic treatments.

CONCLUSION

Magnetic resonance imaging has become a valuable tool for clinical medicine. As technology has improved, this technique has been extended to preclinical studies. MRI is extremely important to maximize data that can be obtained from transgenic and knockout mice. The main disadvantage of MRI is that it is costly and somewhat time consuming. The latter problem is likely to be ameliorated by the use of higher magnetic field strengths and novel RF coil designs, which will allow multiple animal imaging to be done simultaneously. As in clinical medicine, MRI can be used to detect tumors, stratify mice into different treatment arms to investigate novel drugs, evaluate response, and probe tumor physiology.

ACKNOWLEDGMENTS

The authors acknowledge generous support from the National Institutes of Health R24CA8308, U.S. Army Breast Cancer Research Program (BC022020), and the Helen and Samuel Haber Foundation.

REFERENCES

1. Purcell EM, Torrey HC, Pound RV (1946): Resonance absorption by nuclear magnetic moments in a solid. *Phys. Rev* 69:37.
2. Bloch F, Hansen WW, Packard M (1949): Nuclear induction. *Phys Rev* 69:127.
3. Lauterbur PC (1973): Image formation by induced local interactions: Examples employing nuclear magnetic resonance. *Nature* 242:190–191.
4. Damadian R, Minkoff L, Goldsmith M, Stanford M, Koutcher JA (1976): Tumor imaging in a live animal by focusing NMR (FONAR). *Physiol Chem Phys* 8(1):61–65.
5. Damadian R, Minkoff L, Goldsmith M, Stanford M, Koutcher JA (1976): Field focusing nuclear magnetic resonance (FONAR): Visualization of a tumor in a live animal. *Science* 194(4272):1430–1432.
6. (a) Damadian R, Goldsmith M, Minkoff L (1977): NMR in cancer: XVI. FONAR image of the live human body. *Physiol Chem Phys* 9(1):97–100, 108. (b) Damadian R, Minkoff L, Goldsmith M, Koutcher JA (1978): Field-focusing nuclear nuclear magnetic resonance (FONAR). *Naturwissenschaften* 65(5):250–252.
7. Mansfield P, Morris PG (1982): *NMR Imaging in Biomedicine*, Academic, New York.
8. Callaghan PT (1991): *Principles of Nuclear Magnetic Resonance Microscopy*, Oxford University Press, Oxford.
9. Gadian DG (1996): *NMR and Its Applications to Living Systems*, 2nd ed. Oxford University Press, New York.
10. Hornark JP (1996–2004): *The Basics of MRI*. Online HTML textbook at http://www.cis.rit.edu/htbooks/mri/.
11. Haacke EM, Brown WR, Thompson RM, Venkatesan R (1999): *Magnetic Resonance Imaging: Physical Principles and Sequence Design*. Wiley, New York.
12. Damadian R (1971): Tumor detection by nuclear magnetic resonance. *Science* 171(976):1151–1153.
13. Damadian R, Zaner K, Hor D, DiMaio T, Minkoff L, Goldsmith M (1973): Nuclear magnetic resonance as a new tool in cancer research: Human tumors by NMR. *Ann N Y Acad Sci* 222:1048–1076.
14. Le Bihan D (1995): Molecular diffusion, tissue microdynamics and microstructure. *NMR Biomed* 8(7–8):375–386.
15. Herneth AM, Guccione S, Bednarski M (2003): Apparent diffusion coefficient: A quantitative parameter for in vivo tumor characterization. *Eur J Radiol* 45(3):208–213.
16. Bock NA, Konyer NB, Henkelman RM (2001): Multiple-mouse MRI. *Proc Intl Soc Mag Reson Med* 9:1108 (Abs).
17. Wright SM, McDougall M, Brown DG, Hazle J (2001): Arrays of birdcage coils for imaging multiple samples. *Proc Intl Soc Mag Reson Med* 9:18 (Abs).
18. Morris HD, Chesnick S (2001): Two methods for multi-sample imaging: applications to mouse phenotyping via MRI. *Proc Intl Soc Mag Reson Med* 9:19 (Abs).
19. Bock NA, Konyer NB, Henkelman RM (2001): Multiple-mouse MRI. *Proc Intl Soc Mag Reson Med* 9:1108 (Abs).
20. Xu S, Gade TPF, Matei C, Zakian K, Alfieri AA, Hu X, Holland EC, Soghomonian S, Tjuvajev J, Ballon D, Koutcher JA (2003): *In vivo* multiple-mouse imaging at 1.5T. *Magn Reson Med* 49(3):551–557.
21. Evelhoch JL, Sapareto SA, Nussbaum GH, Ackerman JJ (1986): Correlations between 31P NMR spectroscopy and 15O perfusion measurements in the RIF-1 murine tumor in vivo. *Radiat Res* 106(1):122–131.

22. Raghunand N, He X, van Sluis R, Mahoney B, Baggett B, Taylor CW, Paine-Murrieta G, Roe D, Bhujwalla ZM, Gillies RJ (1999): Enhancement of chemotherapy by manipulation of tumour pH. *Br J Cancer* 80(7):1005–1011.
23. Pilatus U, Shim H, Artemov D, Davis D, van Zijl PC, Glickson JD (1997): Intracellular volume and apparent diffusion constants of perfused cancer cell cultures, as measured by NMR. *Magn Reson Med* 37(6):825–832.
24. Robinson SP, McIntyre DJ, Checkley D, Tessier JJ, Howe FA, Griffiths JR, Ashton SE, Ryan AJ, Blakey DC, Waterton JC (2003): Tumour dose response to the antivascular agent ZD6126 assessed by magnetic resonance imaging. *Br J Cancer* 88(10):1592–1597.
25. Koutcher JA, Hu X, Xu S, Gade TPF, Leeds N, Zhou XJ, Zagzag D, Holland EC (2002): MR imaging of mouse models for gliomas show similarities to humans and can be used to identify mice for preclinical trials. *Neoplasia* 4:480–485.
26. Kim B, Chenevert TL, Ross BD (1995): Growth kinetics and treatment response of the intracerebral rat 9L brain tumor model: A quantitative in vivo study using magnetic resonance imaging. *Clin Cancer Res* 1(6):643–650.
27. Qin Y, Van Cauteren M, Osteaux M, Willems G (1992): Quantitative study of the growth of experimental hepatic tumors in rats by using magnetic resonance imaging. *Int J Cancer* 51(4):665–670.
28. Mahmood U, Devitt ML, Kocheril PG, Sutanto-Ward E, Ballon D, Sigurdson ER, Koutcher JA (1992): Quantitation of total metastatic tumor volume in the rat liver: Correlation of MR and histologic measurements. *J Magn Reson Imaging* 2(3):335–340.
29. Els T, Eis M, Hoehn-Berlage M, Hossmann KA (1995): Diffusion-weighted MR imaging of experimental brain tumors in rats. *MAGMA* 3(1):13–20.
30. Chenevert TL, McKeever PE, Ross BD (1997): Monitoring early response of experimental brain tumors to therapy using diffusion magnetic resonance imaging. *Clin Cancer Res* 3:1457–1466.
31. Chenevert TL, Stegman LD, Taylor JM, Robertson PL, Greenberg HS, Rehemtulla A, Ross BD (2000): Diffusion magnetic resonance imaging: An early surrogate marker of therapeutic efficacy in brain tumors. *J Natl Cancer Inst* 92:2029–2036.
32. Maron A, Gustin T, Le Roux A, Mottet I, Dedieu JF, Brion JP, Demeure R, Perricaudet M, Octave JN (1996): Gene therapy of rat C6 glioma using adenovirus-mediated transfer of the herpes simplex virus thymidine kinase gene: Long-term follow-up by magnetic resonance imaging. *Gene Ther* 3(4):315–322.
33. Bouali-Benazzouz R, Laine M, Vicat JM, Boisseau S, Remy C, Fouilhe N, Thomas F, Nissou MF, Benabid AL, Berger F (1999): Therapeutic efficacy of the thymidine kinase/ganciclovir system on large experimental gliomas: A nuclear magnetic resonance imaging study. *Gene Ther* 6(6):1030–1037.
34. Stegman LD, Rehemtulla A, Hamstra DA, Rice DJ, Jonas SJ, Stout KL, Chenevert TL, Ross BD (2000): Diffusion MRI detects early events in the response of a glioma model to the yeast cytosine deaminase gene therapy strategy. *Gene Ther* 7(12):1005–1010.
35. Adachi Y, Tamiya T, Ichikawa T, Terada K, Ono Y, Matsumoto K, Furuta T, Hamada H, Ohmoto T (2000): Experimental gene therapy for brain tumors using adenovirus-mediated transfer of cytosine deaminase gene and uracil phosphoribosyltransferase gene with 5-fluorocytosine. *Hum Gene Ther* 11(1):77–89.
36. Roy EJ, Gawlick U, Orr BA, Rund LA, Webb AG, Kranz DM (2000): IL-12 treatment of endogenously arising murine brain tumors. *J Immunol* 165(12):7293–7299.
37. Blankenberg F, Conley FK, Sayre J, Enzmann D (1991): MR imaging in an experimental model of brain tumor immunotherapy. *Am J Neuroradiol* 12(3):543–548.
38. Dennie J, Mandeville JB, Boxerman JL, Packard S, Rosen BR, Weisskoff RM (1998): NMR imaging of changes in vascular morphology due to tumor angiogenesis. *Magn Reson Med* 40:793–799.
39. Le Duc G, Peoc'h M, Remy C, Charpy O, Muller RN, Le Bas JF, Decorps M (1999): Use of T2 weighted susceptibility contrast MRI for mapping the blood volume in the glioma bearing rat brain. *Magn Reson Med* 42:754–761.
40. Fichtner KP, Schirrmacher V, Griesbach A, Hull WE (1997): In vivo 1H-NMR microimaging with respiratory triggering for monitoring adoptive immunotherapy of metastatic mouse lymphoma. *Magn Reson Med* 38(3): 440–455.
41. Lechaux D, Gervais A, Dazord L, Eliat PA, Franconi F, Lemaire L, Rioux-Leclercq N, Catros-Quemener V (2002): Adoptive immunotherapy monitored by micro-MRI in experimental colorectal liver metastasis. *Anticancer Res* 22(1A):151–158.
42. Reimer P, Weissleder R, Lee AS, Wittenberg J, Brady TJ (1990): Receptor imaging: Application to MR imaging of liver cancer. *Radiology* 177(3):729–734.
43. Reimer P, Weissleder R, Brady TJ, Yeager AE, Baldwin BH, Tennant BC, Wittenberg J (1991): Experimental hepatocellular carcinoma: MR receptor imaging. *Radiology* 180(3):641–645.
44. Vera DR, Buonocore MH, Wisner ER, Katzberg RW, Stadalnik RC (1995): A molecular receptor-binding contrast agent for magnetic resonance imaging of the liver. *Acad Radiol* 2(6):497–506.
45. Fisher GH, Wellen SL, Klimstra D, Lenczowski JM, Tichelaar JW, Lizak MJ, Whitsett JA, Koretsky A, Varmus HE (2001): Induction and apoptotic regression of lung adenocarcinomas by regulation of a K-Ras transgene in the presence and absence of tumor suppressor genes. *Genes Dev* 15(24):3249–3262.
46. Cozzi PJ, Malhotra S, McAuliffe P, Kooby DA, Federoff HJ, Huryk B, Johnson P, Scardino PT, Heston WD, Fong Y (2001): Intravesical oncolytic viral therapy using attenuated, replication-competent herpes simplex viruses G207 and Nv1020 is effective in the treatment of bladder cancer in an orthotopic syngeneic model. *FASEB J* 15:1306.

47. Horiguchi Y, Larchian WA, Kaplinsky R, Fair WR, Heston WD (2000): Intravesical liposome-mediated interleukin-2 gene therapy in orthotopic murine bladder cancer model. *Gene Ther* 7:844.
48. Larchian WA, Horiguchi Y, Nair SK, Fair WR, Heston WD, Gilboa E (2000): Effectiveness of combined interleukin 2 and B7.1 vaccination strategy is dependent on the sequence and order: A liposome-mediated gene therapy treatment for bladder cancer. *Clin Cancer Res* 6:2913.
49. Chin J, Kadhim S, Garcia B, Kim YS, Karlik S (1991): Magnetic resonance imaging for detecting and treatment monitoring of orthotopic murine bladder tumor implants. *J Urol* 145:1297.
50. Kadhim SA, Chin JL, Batislam E, Karlik SJ, Garcia B, Skamene E (1997): Genetically regulated response to intravesical bacillus Calmette Guerin immunotherapy of orthotopic murine bladder tumor. *J Urol* 158:646.
51. Kikuchi E, Xu S, Ohori M, Matei C, Lupu M, Menendez S, Koutcher JA, Bochner BH (2003): Detection and quantitative analysis of early stage orthotopic murine bladder tumor using in vivo magnetic resonance imaging. *J Urol* 170(4 Pt 1):1375–1378.
52. Hsu CX, Ross BD, Chrisp CE, Derrow SZ, Charles LG, Pienta KJ, Greenberg NM, Zeng Z, Sanda MG (1998): Longitudinal cohort analysis of lethal prostate cancer progression in transgenic mice. *J Urol* 160(4):1500–1505.
53. Eng MH, Charles LG, Ross BD, Chrisp CE, Pienta KJ, Greenberg NM, Hsu CX, Sanda MG (1999): Early castration reduces prostatic carcinogenesis in transgenic mice. *Urology* 54(6):1112–1119.
54. Gupta S, Hastak K, Ahmad N, Lewin JS, Mukhtar H (2001): Inhibition of prostate carcinogenesis in TRAMP mice by oral infusion of green tea polyphenols. *Proc Natl Acad Sci USA* 98(18):10350–10355.
55. Song SK, Qu Z, Garabedian EM, Gordon JI, Milbrandt J, Ackerman JJ (2002): Improved magnetic resonance imaging detection of prostate cancer in a transgenic mouse model. *Cancer Res* 62(5):1555–1558.
56. Jennings D, Hatton BN, Guo J, Galons JP, Trouard TP, Raghunand N, Marshall J, Gillies RJ (2002): Early response of prostate carcinoma xenografts to docetaxel chemotherapy monitored with diffusion MRI. *Neoplasia* 4(3):255–262.
57. Weissleder R, Elizondo G, Josephson L, Compton CC, Fretz CJ, Stark DD, Ferrucci JT (1989): Experimental lymph node metastases: Enhanced detection with MR lymphography. *Radiology* 171(3):835–839.
58. Vassallo P, Matei C, Heston WDW, McLachlin SJ, Koutcher JA, Castellino RA (1994): BMS 180549 enhanced lymphography: Utility for differentiating reactive from tumor bearing lymph nodes. *Radiology* 193:501–506.
59. Le Bihan D (1991): Molecular diffusion nuclear magnetic resonance imaging. *Magn Reson Q* 7(1):1–30.
60. Kauppinen RA (2002): Monitoring cytotoxic tumour treatment response by diffusion magnetic resonance imaging and proton spectroscopy. *NMR Biomed* 15(1):6–17.
61. Castillo M, Mukherji SK (2000): Diffusion-weighted imaging in the evaluation of intracranial lesions. *Semin Ultrasound CT MR* 21(6):405–416.
62. Le Bihan D, Douek P, Argyropoulou M, Turner R, Patronas N, Fulham M (1993): Diffusion and perfusion magnetic resonance imaging in brain tumor. *Top Magn Reson Imaging* 5(1):25–31.
63. Sinha S, Sinha U (2002): Functional magnetic resonance of human breast tumors: Diffusion and perfusion imaging. *Ann NY Acad Sci* 980:95–115.
64. Tsuruda JS, Chew WM, Moseley ME, Norman D (1991): Diffusion-weighted MR imaging of extraaxial tumors. *Magn Reson Med* 19(2):316–320.
65. Maeda M, Kawamura Y, Tamagawa Y, Matsuda T, Itoh S, Kimura H, Iwasaki T, Hayashi N, Yamamoto K, Ishii Y (1992): Intravoxel incoherent motion (IVIM) MRI in intracranial, extraaxial tumors and cysts. *J Comput Assist Tomogr* 16(4):514–8.
66. Brunberg JA, Chenevert TL, McKeever PE, Ross DA, Junck LR, Muraszko KM, Dauser R, Pipe JG, Betley AT (1995): In vivo MR determination of water diffusion coefficients and diffusion anisotropy: Correlation with structural alteration in gliomas of the cerebral hemispheres. *AJNR Am J Neuroradiol* 16(2):361–371.
67. Eis M, Els T, Hoehn-Berlage M (1995): High resolution quantitative relaxation and diffusion MRI of three different experimental brain tumors in rat. *Magn Reson Med* 34(6):835–844.
68. Maier CF, Paran Y, Bendel P, Rutt BK, Degani H (1997): Quantitative diffusion imaging in implanted human breast tumors. *Magn Reson Med* 37(4):576–581.
69. Sugahara T, Korogi Y, Kochi M, Ikushima I, Shigematu Y, Hirai T, Okuda T, Liang L, Ge Y, Komohara Y, Ushio Y, Takahashi M (1999): Usefulness of diffusion-weighted MRI with echo-planar technique in the evaluation of cellularity in gliomas. *J Magn Reson Imaging* 9(1):53–60.
70. Gupta RK, Sinha U, Cloughesy TF, Alger JR (1999): Inverse correlation between choline magnetic resonance spectroscopy signal intensity and the apparent diffusion coefficient in human glioma. *Magn Reson Med* 41(1):2–7.
71. Lyng H, Haraldseth O, Rofstad EK (2000): Measurement of cell density and necrotic fraction in human melanoma xenografts by diffusion weighted magnetic resonance imaging. *Magn Reson Med.* 43:828–836.
72. Galons JP, Altbach MI, Paine-Murrieta G, Taylor CW, Gilles RJ (1999): Early increases in breast tumor xenograft water mobility in response to paclitaxel therapy detected by non-invasive diffusion magnetic resonance imaging. *Neoplasia* 1:113–117.
73. Poptani H, Puumalainen A-M, Gröhn OHJ, Loimas S, Kainulainen R, Ylä-Herttuala S, Kauppinen RA (1998): Monitoring thymidine kinase and ganciclovir-induced changes in rat malignant glioma *in vivo* by nuclear magnetic resonance imaging. *Cancer Gene Ther.* 5:101–109.

74. Hakumaki JM, Poptani H, Puumalainen AM, Loimas S, Paljärvi LA, Ylä-Herttuala S, Kauppinen RA (1998): Quantitative 1H nuclear magnetic resonance diffusion spectroscopy of BT4C rat glioma during thymidine kinase-mediated gene therapy in vivo: Identification of apoptotic response. *Cancer Res* 58:3791–3799.

75. Zhao M, Pipe JG, Bonnett J, Evelhoch JL (1996): Early detection of treatment response by diffusion weighted 1H-NMR spectroscopy in a murine tumour *in vivo*. *Br J Cancer* 73:61–64.

76. Chinnaiyan AM, Prasad U, Shankar S, Hamstra DA, Shanaiah M, Chenevert TL, Ross BD, Rehemtulla A (2000): Combined effect of tumor necrosis factor-related apoptosis-inducing ligand and ionizing radiation in breast cancer therapy. *Proc Natl Acad Sci USA* 97(4):1754–1759.

77. Taylor JS, Tofts PS, Port R, Evelhoch JL, Knopp M, Reddick WE, Runge VM, Mayr N (1999): MR imaging of tumor microcirculation: Promise for the new millennium. *J Magn Reson Imaging* 10(6):903–907.

78. Griebel J, Mayr NA, de Vries A, Knopp MV, Gneiting T, Kremser C, Essig M, Hawighorst H, Lukas PH, Yuh WT (1997): Assessment of tumor microcirculation: A new role of dynamic contrast MR imaging. *J Magn Reson Imaging* 7(1):111–119.

79. Parker GJ, Tofts PS (1999): Pharmacokinetic analysis of neoplasms using contrast-enhanced dynamic magnetic resonance imaging. *Top Magn Reson Imaging* 10(2):130–142.

80. Tofts PS, Brix G, Buckley DL, Evelhoch JL, Henderson E, Knopp MV, Larsson HB, Lee TY, Mayr NA, Parker GJ, Port RE, Taylor J, Weisskoff RM (1999): Estimating kinetic parameters from dynamic contrast-enhanced T(1)-weighted MRI of a diffusable tracer: Standardized quantities and symbols. *J Magn Reson Imaging* 10(3):223–232.

81. Tofts PS, Kermode AG (1991): Measurement of the blood-brain barrier permeability and leakage space using dynamic MR imaging—1. Fundamental concepts. *Magn Reson Med* 17:357–367.

82. Larsson HB, Stubgaard M, Frederiksen JL, Jensen M, Henriksen O, Paulson OB (1990): Quantitation of blood-brain barrier defect by magnetic resonance imaging and gadolinium-DTPA in patients with multiple sclerosis and brain tumors. *Magn Reson Med* 16(1):117–131.

83. Brix G, Semmler W, Port R, Schad LR, Layer G, Lorenz WJ (1991): Pharmacokinetic parameters in CNS Gd-DTPA enhanced MR imaging. *J Comput Assist Tomogr* 15(4):621–628.

84. Hoffmann U, Brix G, Knopp MV, Hess T, Lorenz WJ (1995): Pharmacokinetic mapping of the breast: A new method for dynamic MR mammography. *Magn Reson Med* 33(4):506–514.

85. van der Sanden BP, Rozijn TH, Rijken PF, Peters HP, Heerschap A, van der Kogel AJ, Bovee WM (2000): Noninvasive assessment of the functional neovasculature in 9L-glioma growing in rat brain by dynamic 1H magnetic resonance imaging of gadolinium uptake. *J Cereb Blood Flow Metab* 20(5):861–870.

86. Verhoye M, van der Sanden BP, Rijken PF, Peters HP, Van der Kogel AJ, Pee G, Vanhoutte G, Heerschap A, Van der Linden A (2002): Assessment of the neovascular permeability in glioma xenografts by dynamic T(1) MRI with Gadomer-17. *Magn Reson Med* 47(2):305–313.

87. Bhujwalla ZM, Artemov D, Natarajan K, Ackerstaff E, Solaiyappan M (2001): Vascular differences detected by MRI for metastatic versus nonmetastatic breast and prostate cancer xenografts. *Neoplasia* 3(2):143–153.

88. Griebel J, Mayr NA, de Vries A, Knopp MV, Gneiting T, Kremser C, Essig M, Hawighorst H, Lukas PH, Yuh WT (1997): Assessment of tumor microcirculation: A new role of dynamic contrast MR imaging. *J Magn Reson Imaging* 7(1):111–119.

89. Folkman J (2002): Role of angiogenesis in tumor growth and metastasis. *Semin Oncol* 29(6 Suppl 16):15–18.

90. Ranieri G, Gasparini G (2001): Angiogenesis and angiogenesis inhibitors: A new potential anticancer therapeutic strategy. *Curr Drug Targets Immune Endocr Metabol Disord* 1(3):241–253.

91. Nesbit M (2000): Abrogation of tumor vasculature using gene therapy. *Cancer Metastasis Rev* 19(1–2):45–49.

92. Harris AL (1997): Antiangiogenesis for cancer therapy. *Lancet* 349 (Suppl 2):SII13–5.

93. Gossmann A, Helbich TH, Kuriyama N, Ostrowitzki S, Roberts TP, Shames DM, van Bruggen N, Wendland MF, Israel MA, Brasch RC (2002): Dynamic contrast-enhanced magnetic resonance imaging as a surrogate marker of tumor response to anti-angiogenic therapy in a xenograft model of glioblastoma multiforme. *J Magn Reson Imaging* 15(3):233–240.

94. Brasch R, Turetschek K (2000): MRI characterization of tumors and grading angiogenesis using macromolecular contrast media: status report. *Eur J Radiol* 34(3): 148–155.

95. Gossmann A, Helbich TH, Mesiano S, Shames DM, Wendland MF, Roberts TP, Ferrara N, Jaffe RB, Brasch RC (2000): Magnetic resonance imaging in an experimental model of human ovarian cancer demonstrating altered microvascular permeability after inhibition of vascular endothelial growth factor. *Am J Obstet Gynecol* 183(4): 956–963.

21

BIOLUMINESCENCE IMAGING OF MOUSE MODELS OF HUMAN CANCER

CHRISTOPHER H. CONTAG

Departments of Pediatrics, Radiology, and Microbiology and Immunology, Stanford School of Medicine, Stanford, California

The increasing sophistication of rodent models of human neoplastic disease has led to models that more closely approximate human disease where lesions can arise at various times and locations. The uncertainty of initiation and progression of disease and the complexity of these models have necessitated advancement and refinement of noninvasive imaging technologies that can reveal cellular and molecular changes in the context of the living animal. One of these technologies, in vivo *bioluminescent imaging, uses internal biologic sources of light, luciferases, as reporters of biologic change in these models. This optical imaging modality is based on the transmission of light through mammalian tissues, and the sensitivity of detecting photonic signals from within the body, as with other optical modalities, is governed by the absorbing and scattering properties of tissue.* In vivo *imaging of luciferase activity has been used to reveal patterns of transcription, levels of protein–protein interaction, extent of tumor burden, sites of metastatic disease, location of immune cell migration, and mechanisms of oncogenesis. High signal-to-noise ratios result in sensitive* in vivo *measurements, and the ability to label a variety of steps in oncogenesis enables unprecedented* in vivo *cellular and molecular dissection of disease processes.*

OVERVIEW

Clinical imaging modalities such as positron emission tomography (PET), magnetic resonance imaging (MRI), X-ray computed tomography (CT), and single-photon-emission computed tomography (SPECT) utilize the penetrating radiation of positrons, radio waves, X rays, and gamma rays. These have all been successfully applied to imaging of animal models of human biology and disease [1–7]. Alternatively, energies in the visible and near-infrared (NIR) regions of the electromagnetic spectrum offer a number of opportunities for visualizing biologic processes at high resolution and with great sensitivity in animal models. Most optical imaging approaches use external sources of light and rely on measurements of reflectance, fluorescence, absorption, or scatter as an indication of the presence, or nature, of selected biologic molecules or processes [8, 9]. These approaches comprise a set of powerful tools for biomedical research with significant potential for in vivo clinical analyses. Use of external sources of light requires that noise from the external illumination be filtered [8, 9] or otherwise separated by time or frequency [10–15] from the specific signals that are relevant to the target to be imaged and measured. To enhance the signal-to-noise ratio (SNR) in these approaches, optical contrast agents have been developed, and more recently activatable agents have been demonstrated as a means of further reducing the noise in NIR fluorescence imaging [8, 16]. In vivo bioluminescence imaging (BLI) offers an alternative optical imaging approach where the light originates internally [17, 18]. In BLI, luciferases are used as biologic indicator lights that can be imaged through living tissue using sensitive photon detectors; this modality as it applies to mouse models of human cancer is the subject of this review.

Since mammalian tissues do not emit significant levels of intrinsic bioluminescence and photon emission is

Mouse Models of Human Cancer, edited by Eric C. Holland
ISBN 0-471-44460-X

restricted to cells engineered to express the bioluminescent reporter gene, there is an inherently low background. This is demonstrated by the fact that virtually no signal can be detected from naive cells or tissues that have not been engineered to express luciferase. This exceptional SNR is what yields the exquisite sensitivity of BLI for measuring bioluminescent signals in living animals. Nonetheless, penetration of photons through mammalian tissues, whether from internal or external sources, is limited by both absorption and scatter [9, 10, 19]. Despite the internal origin of light in BLI, the same principles that govern light transmission through tissues from external sources govern the migration of photons from bioluminescent sources. Basic principles of photon migration through absorbing and scattering media, such as mammalian tissue, have been the subject of intense investigation [9, 10, 14, 19], and advances in this field serve as an important guide for the advancement of molecular imaging modalities that use light and for the development of reagents for in vivo optical imaging. The amount of light from a given source that passes through mammalian tissue is determined by the amount of scatter and absorption, which is determined by the tissue depth and type, the number of tissue boundaries that are encountered along the path of the photons, and the concentration of absorbing pigments in the tissue. The primary absorbing pigment in albino and hairless mice is hemoglobin, and in animals with pigmented skin and fur melanin also contributes to loss of signal due to absorbance [20].

The absorbance spectra of oxy- and deoxy-hemoglobin are known, and the absorbance by these molecules is most significant in the blue and green regions of the spectrum [9, 10, 19]. Above 600 nm (orange region) the loss of optical signal due hemoglobin absorbance is minimal and scatter is the primary barrier to photon migration. The lower absorbance of longer wavelengths of light has a corresponding reduction in tissue autofluorescence, which together improve SNR in modalities based on NIR light. For these reasons NIR fluorescence imaging has greater potential than fluorescence imaging, which relies on excitation and emission of shorter wavelengths. Unfortunately, all of the fluorescent proteins that have been described to date [21–24] are excited by and emit light in the region of the spectrum where hemoglobin absorbs the signal. This has limited the use of green fluorescent protein (GFP), its derivatives, and related proteins as in vivo reporters and the greatest sensitivity has been obtained at superficial tissue sites or after removal of overlying tissues [25–27]. The interaction of light with biologic materials comprises the field of biophotonics, and the discoveries in this area have a tremendous impact on the interpretation of optical signals that originate from within the body.

As biologic indicators in living cells and tissues, luciferases are intimately tied to the physiology of the cell. This is an advantage in many situations where markers of metabolism are needed. In addition, marking cells with genes that encode signal-producing reporter proteins is not affected by the constraint of dilution of the label due to cell division because the reporter genes can be integrated into the genome and then propagate proportionately with each cell division in the animal. All described luciferases are oxygenases, requiring oxygen and energy as well as a substrate. Although there is indication of greater diversity of bioluminescent biochemistries in nature [28–30], among the enzymes that are used as reporter genes in biomedical research there are three basic chemistries that luciferases use for light production. These can roughly be grouped into those that arose in terrestrial eukaryotic organisms (e.g., firefly luciferase, fLuc for the enzyme and *fluc* for the gene), those from marine eukaryotes (e.g., *Renilla* luciferase; rLuc), and those from luminescent bacteria (Lux). These three chemistries vary in the substrate that is used (luciferin, coelenterazine, and long-chain fatty aldehydes, respectively) and the required energy source [adenosine triphosphate (ATP), coelenterazine, and flavin mononucleotide ($FMNH_2$), respectively]. These chemistries have been the subject of many published reports and reviews and will not reviewed in detail here [29, 30].

When luciferase activity is assayed in cell lysates, the substrate, energy, and cofactors are present in excess in the assay buffer, and oxygen is also in excess, such that the signal intensity is dependent on the concentration of the enzyme. Thus, light production is an indication of the levels of transcription and translation when luciferases are used as reporter genes. If any one of the reaction components becomes limiting, bioluminescence is an indication of the concentration of the most limiting reactant. The use of fLuc as the basis of the most sensitive ATP assay depends on this fact. Luciferase-based ATP assays are performed in reaction conditions where ATP is eliminated from the assay buffer, and photon output is dependent on the ATP levels in the sample [31]. This is relevant to BLI in that in vivo measurements of luciferase activity will not be quantitative unless the cellular oxygen and ATP are in excess of the reaction requirements, and the substrate can be readily supplied to the cells expressing the enzyme. Through specific study [18] and nearly a decade of in vivo use, the biodistribution of luciferin after exogenous delivery has been found to distribute to all tissues in the body. Luciferase and its substrate, luciferin, have not been shown to be toxic to mammalian cells, and no functional differences have been reported, thus far, between cells expressing luciferase (Luc+) compared to parental cell lines. Firefly luciferase is only one of a range of bioluminescent reactions that have been described and characterized [28–30]. As these other reactions are developed for in vivo use, it is imperative that the reaction conditions be biocompatible, that at least part of the emission wavelength be in a range that penetrates tissues (red to NIR), and that the substrate distributes to tissues evenly so that the reaction is dependent on enzyme concentration, not substrate availability, for a duration that is long enough to be imaged.

The method of BLI is fairly straightforward and can be performed by investigators with a modest amount of training. For imaging, mice are generally anesthetized and an aqueous solution of luciferin [150 mg/kg intraperitoneally (i.p.)] is injected prior to imaging. A number of anesthetics have been used for BLI, and nose cones and gas manifolds have been installed into imaging systems for the convenience of using gas anesthetics rather than injectables. The delivery of luciferin is generally via i.p. injection, again for convenience; however, the substrate can be injected intravenously (i.v.), and this accelerates the biodistribution of the compound. Injection into the tail vein of mice can be difficult, and multiple injections may not be possible; retro-orbital i.v. injections can circumvent some of these limitations and can be considered. Since the kinetics of substrate biodistribution are slower after i.p. injection than after i.v., imaging should be performed at 20–30 min post-i.p. substrate injection. In each new animal model it is prudent to determine the time of peak light emission. In general, delivery to brain and skin is slow (30 min) and to cells in the peritoneal cavity is fast (1 min) [18]. Substrate can also be delivered orally; however, the kinetics of biodistribution for this route have not been well studied and have only been used for rapid screening of animals. Imaging takes place in a light-tight chamber, and the quality of these chambers can vary. A grayscale body surface reference image is taken under weak illumination; this enables anatomic localization of the signal. After switching off the light source, photons emitted from luciferase-expressing cells within the animal's body and transmitted through the tissue are quantified over a defined period of time ranging from 1 s to 5 min. A pseudocolor representation of bioluminescent signal intensity is then superimposed over the grayscale reference image for localization. In practice, the method is not complicated; however, the biodistribution of substrate, the principles of photon migration, the biology of the model, and the specifications of the detector that is used will influence BLI data. Knowledge of these parameters greatly improves interpretation of the data.

DETECTOR TECHNOLOGIES FOR LOW-LIGHT IMAGING

Sensitive photodetectors based on charge-coupled device (CCD) cameras are used in BLI, and there are several different architectures and designs of systems. These include intensified and cooled CCDs or variations thereof [32, 33]. The two basic approaches for low light imaging are to increase the signal or decrease the noise. Intensified systems increase the signal using photocathodes to convert captured photons to electrons that can be amplified, then converted back to photons using a phosphorscreen, and finally detected on the CCD. The CCD in these cameras can be a standard detector, and the image intensity is the costly component. The spectral sensitivity of these systems is limited by the materials in the photocathode and is more constrained than that of cooled detectors. Intensified systems are generally more sensitive in the blue region of the spectrum, and given that mammalian tissues filter out these wavelengths, intensified CCD systems can be less sensitive to in vivo signals that arise deep within the tissue [32]. The other approach is to reduce the noise, and CCD detectors can be cooled to 120°C to reduce thermal noise, thereby greatly improving the SNR while preserving the spectral sensitivity of the CCD. Below −120°C the CCD is less efficient and a point of diminishing returns is reached. By cooling the CCD, the spectral properties of the chip are preserved and systems based on cooled CCD detectors are sensitive to light across the entire visible spectrum and into NIR wavelengths (230–1100 nm). Cooled CCD detectors are typically deployed in systems used for in vivo imaging applications because of spectral sensitivity to the longer wavelengths.

VALIDATION OF BLI USING BIOCHEMICAL ASSAYS AND IMAGING

Since BLI is a relatively new modality, validation of this tool as an indicator of biologic function was necessary and has been performed by several groups [20, 34–39], and continued research with this tool will further define its parameters. The dependence of the luciferase reaction on cell physiology suggests that validation should be performed in any new model system that employs this method where cellular metabolism may vary from normal. There are several published reports where in vivo luciferase activity measurements correlated well with the tagged biologic process, and the sensitivity of detecting internal sources of bioluminescent light has also been addressed [20, 34–37, 39]. The sensitivity of detection for any expressed optical signal is dependent upon several parameters, including the level of expression within the cell, the depth of labeled cells within the animal's body (i.e., the distance that the photons must travel through mammalian tissue), the wavelength of emission, and the sensitivity of the detection system. Even in its early state of development, BLI has proven to be one of the most sensitive technologies described to date for macroscopic detection of cancer cells in small research animals.

In syngeneic BALB/c mouse models of leukemia and lymphoma, tumor cells could be detected with high sensitivity in internal organs like lung, liver, spleen, lymph nodes, and bone marrow [40]. In an A20 lymphoma model, 7000 cells, delivered i.v., produced a detectable signal above background immediately after injection when the cells were still systemically distributed. As the disease was allowed to progress, tumor infiltration of the spleen was observed earlier in the disease course and with higher sensitivity than could be detected by tissue removal and flow cytometry [40]. To address the question as to what extent light emission accurately reflects the tumor burden in the animal, 15 mice were selected from different stages of disease after injection of fLuc and GFP-expressing lymphoma cells (cell line designated BCLI). These animals displayed a range of signal intensities in BLI that likely related to a broad range of tumor burden. After BLI the animals were sacrificed, and the cells from livers and spleen were analyzed by flow cytometry for the percentages of GFP^+ lymphoma cells. A linear relationship between the tumor cell number, as determined by flow cytometry, and the BLI signal intensity was observed with a correlation $r = 0.989$. The smallest number of GFP^+ tumor cells that were detectable by cytometry in the spleen was 2.3×10^4. However, animals with undetectable signals in cytometry produced detectable signals in BLI. Therefore, the in vivo measurements were significantly more sensitive than the most sensitive cell-counting method [40]. Xenograft models have also been used to determine the reliability and sensitivity of measuring tumor burden using BLI. Injections of a known number of prostate cancer cells (PC3M) or cervical cancer cells (HeLa) into the peritoneal cavities of mice have demonstrated that BLI is sensitive to signals from 100 cells in the peritoneal cavity [35, 41]. The PC3M cells in these studies produced 80 photons per cell per second. The relationship between cell number and signal intensity was linear in both studies with an excellent correlation.

Measurements of the minimal detectable cell number are useful for a given model where the luciferase expression per cell is not variable. However, there are a number of published reports that use promoters to express luciferase that can be regulated by various endogenous and exogenous signals [42–48] or use luciferases that are expressed as inactive forms and are activated posttranslationally [49–52], and in these models it is necessary to use other measures of sensitivity. In a model of gene transfer to fetal tissues, the levels of luciferase protein were determined using quantitative luciferase assays on tissue lysates from animals that had detectable signals in the liver—as measured in vivo via BLI [39]. Three animals with different levels of reporter gene expression and a control, naive animal were selected from adult animals (eight months of age) that received in utero adeno-associated viral (AAV) vectors that contained a EF1a-luciferase reporter construct. In vivo and ex vivo luciferase assays, Western blots, and quantitative polymerase chain reactions (PCRs) were performed on these animals. The in vivo patterns of luciferase expression indicated that one of the injected animals had no detectable signal over the liver, one animal presented with weak signal from the liver, and one animal presented with a pattern of strong liver expression. In the animal with a weak liver signal 38 pg luciferase/g liver tissue was measured, no activity was found in animals without detectable in vivo signals, and about 400 pg/g liver tissue was measured in the animal with a strong liver signal. This level of expression resulted from about 1 AAV genome in 1×10^6 genome equivalents in the weak expresser and 1 AAV genome in 1000 genome equivalents in the strong expresser. By Western blot analyses a detectable signal was apparent only in the strongest expresser; the sensitivity of the Western blot detection was 400 pg luciferase in 1 g total liver protein (10 pg/mg total protein). These data indicated that BLI was as sensitive as the ex vivo luciferase assays and more sensitive than Western blot analyses [39].

Correlations between luciferase mRNA levels and in vivo measures of luciferase activity were determined in a transgenic mouse model of heme oxygenase (HO) regulation [44]. The transgene in this model was comprised of the HO-1 promoter and the *fluc* coding sequence. Expression of the transgene and the native HO-1 gene can be induced in this model by heavy-metal ions, and $CdCl_2$ was used in a dose–response study. Several different doses of $CdCl_2$ (5–20 μmol/kg) were used to treat mice, the levels of reporter gene expression were measured via BLI and then by Northern blot analyses, and real-time PCR (RT-PCR) assays were conducted for both *fluc* and HO-1 message levels in liver and kidney. BLI was used to assess levels of expression in dorsal and ventral views of each animal and then the tissues were removed and RNA extracted and analyzed. Dose-dependent increases in luciferase activity were observed in both liver and kidney by in vivo imaging. Northern blot analyses revealed detectable increases in *fluc* and HO-1 message only in the treatment group that received the highest dose of $CdCl_2$. Using RT-PCR with sensitivity to 100 copies of message, increases in both *fluc* and HO-1 message were apparent in all mice that received the inducer. In this study, detection by BLI and RT-PCR was comparable, and BLI was more sensitive than Northern blot hybridization. This demonstrates that in vivo imaging is not subject to the same sampling artifacts as ex vivo assays in that the entire organ can be interrogated by imaging, and tissue samples need to be selected for most biochemical assays. The principle of in vivo imaging, that tissue-sampling limitations can be obviated, is a significant strength of BLI for the study of small-animal models of human cancer.

Other imaging modalities have also been used to determine the reliability of BLI to measure in vivo tumor burden. In an orthotopic rat xenograft model of glioma, in vivo measurements of tumor burden were obtained via BLI and transverse T_2-weighted MRI [34]. Growth of the human glioma (9L) in the CNS, as measured by BLI, correlated well with the MRI data. Using BLI and MRI, therapeutic efficacy was measured in this glioma model. The MRI images revealed that the tumor continued to expand up to eight days after treatment and then was followed by regression and subsequent regrowth. The corresponding BLI images revealed parallel changes in photon emission over the same time period. A regression fit of the tumor exponential regrowth data was plotted using each modality. Mean log cell kill values were 1.78 ± 0.36 and 1.84 ± 0.73 for MRI and BLI measurements; these were statistically identical. These results indicated that BLI could be used as a rapid in vivo measure of tumor burden and response to treatment in otherwise difficult-to-study orthotopic models. The quantitative measurements of BLI and its rapid high-throughput nature suggest that its use will significantly accelerate drug development studies and reduce the costs of preclinical testing of experimental therapies. The sensitivity has further been validated using PET imaging of thymidine kinase (TK) expression [53]. In this comparison, BLI was found to be 1000 times more sensitive than the PET approach for expression in skeletal muscle and 10 times more sensitive for the study of liver expression.

Radiographic methods can be used to study osteolytic bone metastasis but may be limited to large lesions where there is significant damage. In a study by Wetterwald and co-workers, bone metastases were quantitatively evaluated in a xenograft human breast cancer model in immunodeficient mice using BLI [54]. Since bone micrometastatic lesions can elude radiographic detection, BLI was employed to monitor metastatic disease in this model. Using BLI, the presence and location of small metastatic lesions were detectable at early times after introduction of the tumor cells via intracardiac injection. In this study microscopic bone metastases with volumes as small as 0.5 mm^3, corresponding to approximately 2×10^4 cells, were detectable. Lesions of this size are not associated with osteolytic lesions and hence could not be detected by X-ray imaging. Application of BLI to these aspects of animal tumor models will greatly improve our ability to develop effective approaches for targeting small numbers of cells present in the early stages of disease progression.

Orthotopic models can be difficult to study using conventional methods due to the inaccessibility of the tumor to physical measurements. Relevant orthotopic models of leukemia and lymphoma and metastatic disease present the additional complication of being systemic in nature, and since BLI is a whole-body imaging approach, it can be used in these challenging models. However, the ex vivo study of the tumor cells in these models and histologic studies could be significantly enhanced by expressing both fLuc and a fluorescent protein in the tumor cells. For these reasons dual-function reporter genes have been developed that encode both a fluorescent marker [either GFP or yellow fluorescent protein (YFP)] for ex vivo cell sorting, and a bioluminescent marker for in vivo imaging. These dual-function reporters were used in the lymphoma model, described above, to study disease progression, including metastasis to the lung [40]. Expression of the dual-function marker facilitated labeling of the cell line. As this cell does not grow in culture, it was necessary to use the GFP expression to select the transduced cells; labeling was not otherwise possible since drug selection was not an option. Initial trafficking of the malignant cells through the lungs as well as organ-specific homing and orthotopic expansion over time was visualized and quantified. In this model, minimal residual disease following radiotherapy and bone marrow transplantation (BMT) was evaluated. The ability to monitor the entire disease course over time revealed tumor-specific escape strategies, as signals from the lymphoma cells could be detected in the vertebral column following high-dose irradiation and BMT (Fig. 21.1). For these complex models, BLI offers a real alternative to assays that require tissue excision for analysis.

ADVANCING SPONTANEOUS TUMOR MODELS THROUGH BLI

Orthotopic models are a significant improvement over subcutaneous models. However, to more effectively model human oncogenesis, mutant and transgenic (Tg) lines of mice that spontaneously develop malignancies have been developed [55–63]. These models have been designed to test key molecular mechanisms of oncogenesis; however, their evaluation can be compromised by not knowing the disease status prior to analysis. These powerful models offer opportunities to study the initial, or early, events in the disease processes. Evaluating these events in the context of intact organs and tissues would increase their predictive value. A conditional Tg mouse model for retinoblastoma-dependent sporadic cancer was established with *fluc* incorporated as an externally detectable marker [64]. BLI allowed monitoring development of pituitary tumors and the full disease course. Imaging in this model enabled the study of the onset of disease as well as disease progression and response to therapy. Coupling sensitive imaging modalities with the development of spontaneous tumor models, in which disease is dependent on defined genetic alterations, will facilitate the analysis of the critical steps in the initiation and progression

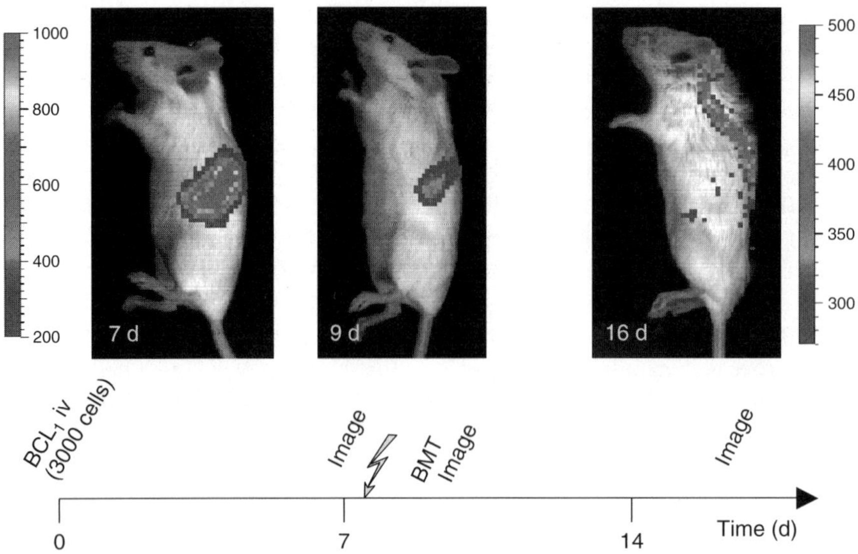

Figure 21.1. Monitoring the complete disease course and response to therapy in an orthotopic lymphoma model. Following radiation therapy and BMT, the signals from the primary tumor in the spleen were reduced in the syngeneic mouse lymphoma model (BCL1). Despite effective reduction of tumor burden in the spleen, detectable minimal residual disease in the spinal column remained. These extradural tumor cells continued to proliferate despite the therapy. Minimal residual disease is a therapeutic target that is being addressed by the development of immune cell therapies. Time line under images indicates time course of imaging and treatment.

of neoplastic disease and will accelerate the development of more effective anticancer therapies that target specific molecular events in oncogenesis.

IMAGING IMMUNE SURVEILLANCE MECHANISMS AND IMMUNE CELL THERAPIES

The therapy of cancer using surgery, chemotherapy, and radiation therapy, either alone or in combination, has resulted in significant treatment successes; however, drug and radiation resistance and relapse continue to be major causes of treatment failure. Novel strategies are therefore needed to improve treatment outcomes. Hijacking the immune mechanisms that the body uses to control malignancy is an appealing concept for the control and/or eradication of malignant cell growth, since there is compelling evidence that immune-based treatment strategies can control a variety of different malignancies under the appropriate conditions. One such approach to harness immune cells which is being used with increasing frequency has been to combine high-dose chemotherapy with hematopoietic cell transplantation (HCT) with either autologous or allogeneic sources of hematopoietic progenitor and stem cells.

BLI has been used to mark and track lymphocyte populations that have in vivo antitumor effects [40]. Culture conditions that result in the ex vivo expansion of a population of T cells which coexpress NK markers, termed cytokine-induced killer (CIK) cells, have been developed. These cells are readily expandable from both human and murine tissues. To monitor the migration of CIK cells in tumor-bearing animals the *gfp-fluc* fusion gene was introduced into these cells using retroviral transduction. In this instance, splenocytes are prepared and transduced with a *gfp-luc* retroviral vector. GFP^+ cells were selected and expanded under CIK culture conditions. Transduced CIK cells (5×10^6) were injected into animals which had A20 lymphoma cells injected subcutaneously 10 days prior to cell the immune cell transfer (Fig. 21.2). One hour following injection of cells, BLI analysis detected the CIK cells in the lungs; at 16 h the CIK cells began to migrate out of the lungs into the liver and spleen; and by 72 h postinjection CIK cells could be clearly seen infiltrating the tumor site, as can be seen in the circled section in the lateral view. Signals from the *gfp-fluc*-labeled cells were detectable for an additional 9 days at the tumor site with apparent regression of the tumor at this time. Untreated animals developed large tumor masses and died due to progressive disease within 12 weeks, whereas CIK-treated animals did not show any signs of tumor relapse over the 6-month observation period. These data demonstrate the ability to track effector populations to tumor sites and visualize the tempo and duration of immunotherapeutic response using BLI. Significant limitations to the study of

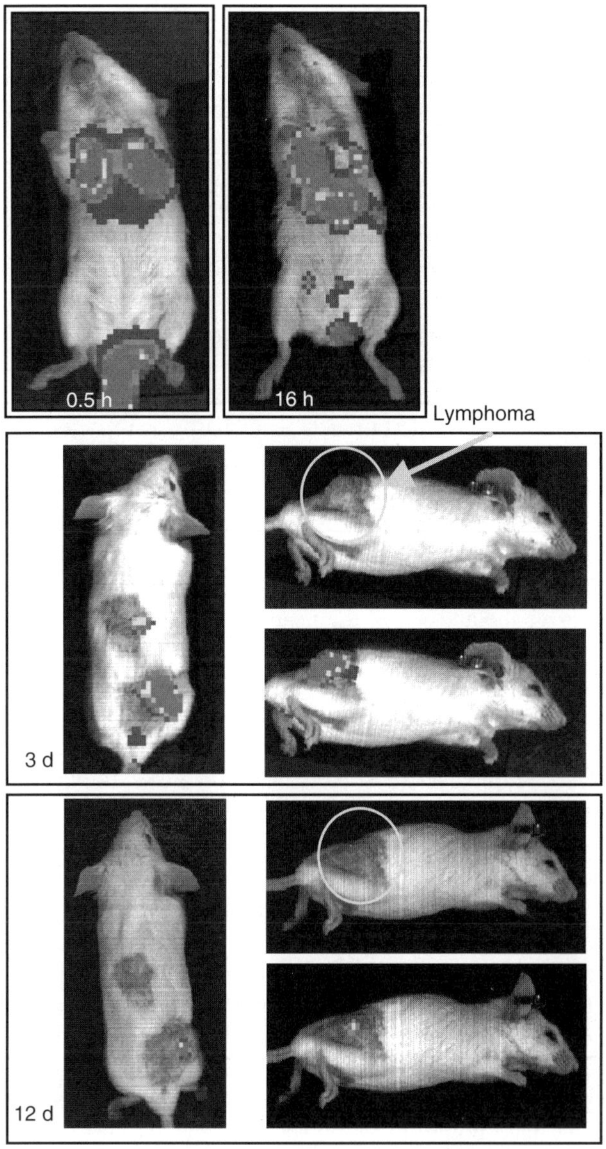

Figure 21.2. Trafficking of tumoricidal lymphocytes. CIK cells were labeled by retroviral transduction and transferred into Balb/c mice with subcutaneous tumors (A20 lymphoma). The tempo of migration from the lungs (0.5 h) to the liver and spleen (16 h) and then to the tumor site is apparent over the first three days (3 d). The signal from the lymphocytes was detectable up to 12 days (12 d) and after that the signals and the tumor were no longer detectable. This demonstrates both the migration patterns and tumoricidal activity of this T-cell population.

primary immune cells are the poor transduction efficiency of these cells and the loss of the reporter that can occur during the 3-week period that CIK cells are in culture for enrichment. An alternative to transduction is to generate a Tg donor animal in which all hematopoietic cells are labeled. This approach has recently been employed to study hematopoietic stem cells [65].

ENGRAFTMENT OF HEMATOPOIETIC STEM CELLS AND MULTILINEAGE RECONSTITUTION

Hematopoietic reconstitution from stem cells (HSC) rather than whole bone marrow may circumvent graft verses host disease (GVHD) while preserving and perhaps improving the graft verses lymphona (GVL) response. Therefore, it is essential that we understand the factors that control engraftment, proliferation, and reconstitution of these cells. Stem cell behavior and biology can best be evaluated in the context of the living body where the influences of tissues and organ systems remain intact. HSCs have been well characterized as self-renewing multipotent cells with potential for full hematopoietic reconstitution [66], and the cell markers that define this population of cells have been defined. Conventional assays of these cells have required removal of tissues from the body for analyses, and the sensitivity of these approaches has often limited the measurements to weeks or months after transplantation. Therefore, the early events that govern seeding of tissue sites, successful engraftment, and proliferation remain largely undefined. BLI is well suited to the study of some of these parameters and has been applied to the study of HSCs following transplantation of small numbers of highly purified, syngeneic mouse HSCs into lethally irradiated recipients [65]. Use of a Tg donor mouse obviated any potential heterogeneity in HSCs or their progeny due to variable sites of reporter gene integration or ex vivo cell culture, as would occur with labeling of HSCs via retroviral transduction, and enabled trafficking studies with these cells.

ADVANCES IN REPORTER GENE STRATEGIES AND IMAGING TECHNOLOGIES

Luciferases have been used as reporter genes and biosensors (e.g., for ATP levels) for decades, and there is a long and active history that we can draw on for development of novel tools for in vivo assays. The mechanism of light production and the constraints on light emission are understood to the point where the relative contribution of single-amino-acid residues to light emission has been determined [67–69]. This information has resulted in the development of tools for improving light production and splitting the protein for use in protein–protein interactions and has offered the possibility of multiplexing in vivo optical assays.

The luciferase from the sea pansy (*Renilla reniformis*), a blue-emitting luciferase that uses the substrate coelenterazine, has been used as a fusion protein with glial cell line–derived neurotropic factor (GDNF) and expressed in a glial cell line to monitor secretion in living animal models [70]. The GDNF-reporter fusion was measured optically in cerebrospinal fluid in one study, and although expression levels were not assayed in vivo, the initial report demonstrated that rLuc could be functionally fused to other genes and used to assay expression levels. Use of coelenterazine-utilizing enzymes in vivo presents some challenges, as indicated above, but does offer the possibility of multiparameter analyses using BLI since there should not be any crossover between substrate utilization by fLuc and rLuc [71]. Having the ability to monitor two, and possibly more, biologic processes in a single animal would greatly increase the utility of BLI. Distinguishing bioluminescent reporters spectrally is, at present, not possible since the current bioluminescent reporters emit light that is largely absorbed by mammalian tissues, and distinguishing the red components of the two emission spectra is not possible with the weak signals that are available from bioluminescence. Thus, using a luciferin-utilizing enzyme and coelenterazine-utilizing enzyme offers the opportunity to separate the expression of two reporters biochemically. The use of two different substrates by fLuc and rLuc suggests that these two reporters are, in one sense, an ideal pair for dual detection in vivo. The ability to image these two reporter genes in vivo has been demonstrated; however, the limitations of biodistribution and blue emission were not addressed [71], and considerable effort will be required for effective in vivo imaging of coelenterazine-based systems.

A number of reports have been published using gene fusions to create dual-function reporter genes to enhance the utility of a singe reporter. This strategy was first demonstrated by Day and co-workers to study expression patterns in fruit flies [72] This approach has been further advanced to the point where the light-producing enzyme excites the fluorescent protein. This has been called bioluminescence resonance energy transfer (BRET) as well as chemiluminescence resonance energy transfer (CRET) and luminescence resonance energy transfer (LRET) [73–77]—all referring to the process of energy transfer from an enzyme such as rLuc to a fluorescent protein such as GFP. This has been used to study protein–protein interaction in cultured cells [73–76] and more recently in animal models. The use of BRET to monitor protein–protein interaction in vivo may however be severely limited by the wavelengths of emission for these two reporters (in the blue-green region of the spectrum) where spectral resolution will not be possible. In another approach the components of the well-studied yeast two-hybrid system have been adapted for use in mammalian cells to activate transcription of fLuc [52]. As protein–protein interactions contribute to oncogenesis and disease progression, the incorporation of these tools into orthotopic and spontaneous tumor models would be a significant advance.

A protein-slicing mechanism has been adapted to luciferase to create a reporter gene strategy that would interrogate protein–protein interactions [50]. Protein splicing, first described in bacteria [78, 79], uses a nomenclature based on that originally developed for RNA splicing. The internal regions of proteins that are "spliced out" have been called inteins and the external, or spliced, proteins are referred to as exteins. As applied to luciferase, two complementary bacterial inteins were fused to two parts of the luciferase such that upon protein–protein interaction the inteins are excised out of the inactive precursor fusion proteins [80]. The luciferase exteins are immediately ligated into a contiguous luciferase polypeptide with enzymatic activity. The recovered activity in the original studies was only a fraction of wild-type activities, and little or no improvement of the efficiency has been reported. The poor recovery of activity may likely be a consequence of the amino acid residues left by the splicing event. There is no particular reason why the splicing sequences could not be placed at other sites in the protein where amino acid additions would not affect the enzymatic activity. This may improve the activity of the resulting enzyme.

SUMMARY AND CONCLUSIONS

BLI is a rapid, sensitive assay based on a versatile labeling process and accessible detection technologies. It offers the potential of applying many of the tools available to molecular biologists to the study of mouse models of human cancer where the contextual influences of living animals remain intact. The study of cell biology has benefited greatly from the development of responsive reporter genes, and as similar approaches are applied to animal models, we will similarly see tremendous advances in the study of mouse models. The data from these models will likely become more predictive of the human response as we incorporate these powerful strategies into relevant Tg mouse models.

ACKNOWLEDGMENTS

This work was supported, in part, by unrestricted gifts from the Mary L. Johnson and Hess Research Funds and by the following grants from the National Institutes of Health: R33CA88303, R24CA92862 and P20CA86312.

REFERENCES

1. Paulus MJ, Gleason SS, Kennel SJ, Hunsicker PR, Johnson DK (2000): High resolution X-ray computed tomography: An emerging tool for small animal cancer research. *Neoplasia* 2:62.
2. Gambhir SS, Barrio JR, Wu L, Iyer M, Namavari M, Satyamurthy N, Bauer E, Parrish C, MacLaren DC, Borghei AR,

Green LA, Sharfstein S, Berk AJ, Cherry SR, Phelps ME, Herschman HR (1998): Imaging of adenoviral-directed herpes simplex virus type 1 thymidine kinase reporter gene expression in mice with radiolabeled ganciclovir. *J Nucl Med* 39:2003.

3. Tjuvajev JG, Avril N, Oku T, Sasajima T, Miyagawa T, Joshi R, Safer M, Beattie B, DiResta G, Daghighian F, Augensen F, Koutcher J, Zweit J, Humm J, Larson SM, Finn R, Blasberg R (1998): Imaging herpes virus thymidine kinase gene transfer and expression by positron emission tomography. *Cancer Res* 58:4333.
4. Ross BD, Zhao YJ, Neal ER, Stegman LD, Ercolani M, Ben-Yoseph O, Chenevert TL (1998): Contributions of cell kill and posttreatment tumor growth rates to the repopulation of intracerebral 9 L tumors after chemotherapy: An MRI study. *Proc Natl Acad Sci USA* 95:7012.
5. Lewis JS, Achilefu S, Garbow JR, Laforest R, Welch MJ (2002): Small animal imaging. Current technology and perspectives for oncological imaging. *Eur J Cancer* 38:2173.
6. Strand SE, Ivanovic M, Erlandsson K, Franceschi D, Button T, Sjogren K, Weber DA (1994): Small animal imaging with pinhole single-photon emission computed tomography. *Cancer* 73:981.
7. Xu S, Gade TP, Matei C, Zakian K, Alfieri AA, Hu X, Holland EC, Soghomonian S, Tjuvajev J, Ballon D, Koutcher JA (2003): In vivo multiple-mouse imaging at 1.5 T. *Magn Reson Med* 49:551.
8. Ntziachristos V, Chance B (2001): Probing physiology and molecular function using optical imaging: Applications to breast cancer. *Breast Cancer Res* 3:41.
9. Chance B, Luo Q, Nioka S, Alsop DC, Detre JA (1997): Optical investigations of physiology: A study of intrinsic and extrinsic biomedical contrast. *Philos Trans R Soc Lond B Biol Sci* 352:707.
10. Benaron DA, Cheong WF, Stevenson DK (1997): Tissue optics. *Science* 276:2002.
11. Benaron DA, Stevenson DK (1993): Optical time-of-flight and absorbance imaging of biologic media. *Science* 259:1463.
12. So PT, Konig K, Berland K, Dong CY, French T, Buhler C, Ragan T, Gratton E (1998): New time-resolved techniques in two-photon microscopy. *Cell Mol Biol (Noisy-le-grand)* 44:771.
13. French T, So PT, Dong CY, Berland KM, Gratton E (1998): Fluorescence lifetime imaging techniques for microscopy. *Methods Cell Biol* 56:277.
14. Gratton E, Fantini S, Franceschini MA, Gratton G, Fabiani M (1997): Measurements of scattering and absorption changes in muscle and brain. *Philos Trans R Soc Lond B Biol Sci* 352:727.
15. Hawrysz DJ, Sevick-Muraca EM (2000): Developments toward diagnostic breast cancer imaging using near-infrared optical measurements and fluorescent contrast agents. *Neoplasia* 2:388.
16. Tung CH, Mahmood U, Bredow S, Weissleder R (2000): In vivo imaging of proteolytic enzyme activity using a novel molecular reporter. *Cancer Res* 60:4953.
17. Contag CH, Contag PR, Mullins JI, Spilman SD, Stevenson DK, Benaron DA (1995): Photonic detection of bacterial pathogens in living hosts. *Mol Microbiol* 18:593.
18. Contag CH, Spilman SD, Contag PR, Oshiro M, Eames B, Dennery P, Stevenson DK, Benaron DA (1997): Visualizing gene expression in living mammals using a bioluminescent reporter. *Photochem Photobiol* 66:523.
19. Tromberg BJ, Shah N, Lanning R, Cerussi A, Espinoza J, Pham T, Svaasand L, Butler J (2000): Non-invasive in vivo characterization of breast tumors using photon migration spectroscopy. *Neoplasia* 2:26.
20. Rocchetta HL, Boylan CJ, Foley JW, Iversen PW, LeTourneau DL, McMillian CL, Contag PR, Jenkins DE, Parr TR Jr (2001): Validation of a noninvasive, real-time imaging technology using bioluminescent *Escherichia coli* in the neutropenic mouse thigh model of infection. *Antimicrob Agents Chemother* 45:129.
21. Gurskaya NG, Fradkov AF, Terskikh A, Matz MV, Labas YA, Martynov VI, Yanushevich YG, Lukyanov KA, Lukyanov SA (2001): GFP-like chromoproteins as a source of far-red fluorescent proteins. *FEBS Lett* 507:16.
22. Yarbrough D, Wachter RM, Kallio K, Matz MV, Remington SJ (2001): Refined crystal structure of DsRed, a red fluorescent protein from coral, at 2.0-A resolution. *Proc Natl Acad Sci USA* 98:462.
23. Fradkov AF, Chen Y, Ding L, Barsova EV, Matz MV, Lukyanov SA (2000): Novel fluorescent protein from Discosoma coral and its mutants possesses a unique far-red fluorescence. *FEBS Lett* 479:127.
24. Matz MV, Fradkov AF, Labas YA, Savitsky AP, Zaraisky AG, Markelov ML, Lukyanov SA (1999): Fluorescent proteins from nonbioluminescent Anthozoa species. *Nat Biotechnol* 17:969.
25. Chishima T, Miyagi Y, Wang X, Yamaoka H, Shimada H, Moossa AR, Hoffman RM (1997): Cancer invasion and micrometastasis visualized in live tissue by green fluorescent protein expression. *Cancer Res* 57:2042.
26. Yang M, Jiang P, Sun FX, Hasegawa S, Baranov E, Chishima T, Shimada H, Moossa AR, Hoffman RM (1999): A fluorescent orthotopic bone metastasis model of human prostate cancer. *Cancer Res* 59:781.
27. Yang M, Baranov E, Jiang P, Sun FX, Li XM, Li L, Hasegawa S, Bouvet M, Al-Tuwaijri M, Chishima T, Shimada H, Moossa AR, Penman S, Hoffman RM (2000): Whole-body optical imaging of green fluorescent protein-expressing tumors and metastases. *Proc Natl Acad Sci USA* 97:1206.
28. Viviani VR, Hastings JW, Wilson T (2002): Two bioluminescent diptera: The North American *Orfelia fultoni* and the Australian *Arachnocampa flava*. Similar niche, different bioluminescence systems. *Photochem Photobiol* 75:22.
29. Wilson T, Hastings JW (1998): Bioluminescence. *Annu Rev Cell Dev Biol* 14:197.
30. Hastings JW (1996): Chemistries and colors of bioluminescent reactions: A review. *Gene* 173:5.

31. Khan HA (2003): Bioluminometric assay of ATP in mouse brain: Determinant factors for enhanced test sensitivity. *J Biosci* 28:379.

32. Rice BW, Cable MD, Nelson MB (2001): In vivo imaging of light-emitting probes. *J Biomed Opt* 6:432.

33. Oshiro M (1998): Cooled CCD versus intensified cameras for low-light video—applications and relative advantages. *Methods Cell Biol* 56:45.

34. Rehemtulla A, Stegman LD, Cardozo SJ, Gupta S, Hall DE, Contag CH, Ross BD (2000): Rapid and quantitative assessment of cancer treatment response using in vivo bioluminescence imaging. *Neoplasia* 2:491.

35. Sweeney TJ, Mailander V, Tucker AA, Olomu AB, Zhang W, Cao Y, Negrin RS, Contag CH (1999): Visualizing the kinetics of tumor-cell clearance in living animals. *Proc Natl Acad Sci USA* 96:12044.

36. Edinger M, Sweeney TJ, Tucker AA, Olomu AB, Negrin RS, Contag CH (1999): Noninvasive assessment of tumor cell proliferation in animal models. *Neoplasia* 1:303.

37. Contag CH, Stevenson DK (2001): In vivo patterns of heme oxygenase-1 transcription. *J Perinatol* 21(Suppl 1): S119.

38. Hardy J, Edinger M, Bachmann MH, Negrin RS, Fathman CG, Contag CH (2001): Bioluminescence imaging of lymphocyte trafficking in vivo. *Exp Hematol* 29:1353.

39. Lipshutz GS, Gruber CA, Cao Y, Hardy J, Contag CH, Gaensler KM (2001): In utero delivery of adeno-associated viral vectors: Intraperitoneal gene transfer produces long-term expression. *Mol Ther* 3:284.

40. Edinger M, Cao YA, Verneris MR, Bachmann MH, Contag CH, Negrin RS (2003): Revealing lymphoma growth and the efficacy of immune cell therapies using in vivo bioluminescence imaging. *Blood* 101:640.

41. Edinger M, Cao YA, Hornig YS, Jenkins DE, Verneris MR, Bachmann MH, Negrin RS, Contag CH (2002): Advancing animal models of neoplasia through in vivo bioluminescence imaging. *Eur J Cancer* 38:2128.

42. Zhang W, Feng JQ, Harris SE, Contag PR, Stevenson DK, Contag CH (2001): Rapid in vivo functional analysis of transgenes in mice using whole body imaging of luciferase expression. *Transgenic Res* 10:423.

43. Zhang J, Tan X, Contag CH, Lu Y, Guo D, Harris SE, Feng JQ (2002): Dissection of promoter control modules that direct Bmp4 expression in the epithelium-derived components of hair follicles. *Biochem Biophys Res Commun* 293:1412.

44. Zhang W, Contag PR, Hardy J, Zhao H, Vreman HJ, Hajdena-Dawson M, Wong RJ, Stevenson DK, Contag CH (2002): Selection of potential therapeutics based on in vivo spatiotemporal transcription patterns of heme oxygenase-1. *J Mol Med* 80:655.

45. Zhang W, Purchio A, Chen K, Burns SM, Contag CH, Contag PR (2003): In vivo activation of the human CYP3A4 promoter in mouse liver and regulation by pregnane X receptors. *Biochem Pharmacol* 65:1889.

46. Zhang N, Weber A, Li B, Lyons R, Contag PR, Purchio AF, West DB (2003): An inducible nitric oxide synthase-luciferase reporter system for in vivo testing of anti-inflammatory compounds in transgenic mice. *J Immunol* 170:6307.

47. Zhang W, Purchio AF, Chen K, Wu J, Lu L, Coffee R, Contag PR, West DB (2003): A transgenic mouse model with a luciferase reporter for studying in vivo transcriptional regulation of the human CYP3A4 gene. *Drug Metab Dispos* 31:1054.

48. Wu JC, Inubushi M, Sundaresan G, Schelbert HR, Gambhir SS (2002): Optical imaging of cardiac reporter gene expression in living rats. *Circulation* 105:1631.

49. Laxman B, Hall DE, Bhojani MS, Hamstra DA, Chenevert TL, Ross BD, Rehemtulla A (2002): Noninvasive real-time imaging of apoptosis. *Proc Natl Acad Sci USA* 99:16551.

50. Paulmurugan R, Gambhir SS (2003): Monitoring protein-protein interactions using split synthetic renilla luciferase protein-fragment-assisted complementation. *Anal Chem* 75:1584.

51. Paulmurugan R, Umezawa Y, Gambhir SS (2002): Noninvasive imaging of protein-protein interactions in living subjects by using reporter protein complementation and reconstitution strategies. *Proc Natl Acad Sci USA* 99:15608.

52. Ray P, Pimenta H, Paulmurugan R, Berger F, Phelps ME, Iyer M, Gambhir SS (2002): Noninvasive quantitative imaging of protein-protein interactions in living subjects. *Proc Natl Acad Sci USA* 99:3105.

53. Wu JC, Sundaresan G, Iyer M, Gambhir SS (2001): Noninvasive optical imaging of firefly luciferase reporter gene expression in skeletal muscles of living mice. *Mol Ther* 4:297.

54. Wetterwald A, van der Pluijm G, Que I, Sijmons B, Buijs J, Karperien M, Lowik CW, Gautschi E, Thalmann GN, Cecchini MG (2002): Optical imaging of cancer metastasis to bone marrow: A mouse model of minimal residual disease. *Am J Pathol* 160:1143.

55. Jacks T, Weinberg RA (2002): Taking the study of cancer cell survival to a new dimension. *Cell* 111:923.

56. Van Dyke T, Jacks T (2002): Cancer modeling in the modern era: Progress and challenges. *Cell* 108:135.

57. Tuveson DA, Jacks T (2002): Technologically advanced cancer modeling in mice. *Curr Opin Genet Dev* 12:105.

58. Reilly KM, Jacks T (2001): Genetically engineered mouse models of astrocytoma: GEMs in the rough? *Semin Cancer Biol* 11:177.

59. DePinho RA, Jacks T (2001): Introduction. The laboratory mouse in cancer research. *Semin Cancer Biol* 11:175.

60. Holland EC (2001): Brain tumor animal models: Importance and progress. *Curr Opin Oncol* 13:143.

61. Holland EC (2001): Gliomagenesis: Genetic alterations and mouse models. *Nat Rev Genet* 2:120.

62. Dai C, Holland EC (2001): Glioma models. *Biochim Biophys Acta* 1551:M19.

63. Holland EC (2000): A mouse model for glioma: Biology, pathology, and therapeutic opportunities. *Toxicol Pathol* 28:171.

64. Vooijs M, Jonkers J, Lyons S, Berns A (2002): Noninvasive imaging of spontaneous retinoblastoma pathway-dependent tumors in mice. *Cancer Res* 62:1862.

65. Cao Y-A, Wagers A, Beilhack A, Dusich J, Bachmann MH, Negrin RS, Weissman IL, Contag CH (2004): Shifting foci of hematopoiesis during reconstitution from single stem cells. *Proc Natl Acad Sci, USA* 101(1):221–226.

66. Spangrude GJ, Scollay R (1990): A simplified method for enrichment of mouse hematopoietic stem cells. *Exp Hematol* 18:920.

67. Sandalova TP, Ugarova NN (1999): Model of the active site of firefly luciferase. *Biochemistry (Mosc)* 64:962.

68. Conti E, Franks NP, Brick P (1996): Crystal structure of firefly luciferase throws light on a superfamily of adenylate-forming enzymes. *Structure* 4:287.

69. Branchini BR, Magyar RA, Murtiashaw MH, Anderson SM, Helgerson LC, Zimmer M (1999): Site-directed mutagenesis of firefly luciferase active site amino acids: A proposed model for bioluminescence color. *Biochemistry* 38:13223.

70. Liu H, Iacono RP, Szalay AA (2001): Detection of GDNF secretion in glial cell culture and from transformed cell implants in the brains of live animals. *Mol Genet Genomics* 266:614.

71. Bhaumik S, Gambhir SS (2002): Optical imaging of Renilla luciferase reporter gene expression in living mice. *Proc Natl Acad Sci USA* 99:377.

72. Day RN, Kawecki M, Berry D (1998): Dual-function reporter protein for analysis of gene expression in living cells. *Biotechniques* 25:848.

73. Boute N, Jockers R, Issad T (2002): The use of resonance energy transfer in high-throughput screening: BRET versus FRET. *Trends Pharmacol Sci* 23:351.

74. Issad T, Boute N, Pernet K (2002): A homogenous assay to monitor the activity of the insulin receptor using bioluminescence resonance energy transfer. *Biochem Pharmacol* 64:813.

75. Boute N, Pernet K, Issad T (2001): Monitoring the activation state of the insulin receptor using bioluminescence resonance energy transfer. *Mol Pharmacol* 60: 640.

76. Xu Y, Piston DW, Johnson CH (1999): A bioluminescence resonance energy transfer (BRET) system: Application to interacting circadian clock proteins. *Proc Natl Acad Sci USA* 96:151.

77. Wang Y, Wang G, O'Kane DJ, Szalay AA (2001): A study of protein-protein interactions in living cells using luminescence resonance energy transfer (LRET) from Renilla luciferase to Aequorea GFP. *Mol Gen Genet* 264: 578.

78. Clarke ND (1994): A proposed mechanism for the self-splicing of proteins. *Proc Natl Acad Sci USA* 91: 11084.

79. Colston MJ, Davis EO (1994): The ins and outs of protein splicing elements. *Mol Microbiol* 12:359.

80. Ozawa T, Kaihara A, Sato M, Tachihara K, Umezawa Y (2001): Split luciferase as an optical probe for detecting protein-protein interactions in mammalian cells based on protein splicing. *Anal Chem* 73:2516.

PART V

PRECLINICAL TRIALS

22

USE OF MAGNETIC RESONANCE IMAGING FOR EVALUATION OF TREATMENT RESPONSE

Brian D. Ross, Thomas L. Chenevert, Bradford A. Moffat, Alnawaz Rehemtulla, and Daniel E. Hall

Center for Molecular Imaging, University of Michigan, School of Medicine, Ann Arbor, Michigan

Patrick McConville and Jonathan Moody

Molecular Imaging Research Inc., Ann Arbor, Michigan

Rodent cancer models have become indispensable in the discovery and assessment of new anticancer agents. These models allow for easy quantitation of tumor growth and treatment response monitoring using caliper measurements of tumor volume. It is essential to use noninvasive imaging to facilitate the evaluation of the responsiveness of orthotopic tumors in preclinical therapeutic studies. We provide some examples and strategies for using imaging to quantitate the treatment response of orthotopically located tumors in rodent preclinical trials.

INTRODUCTION

Rodent cancer models have become indispensable in the discovery and assessment of new anticancer agents (Curt, 1994). Preclinical in vivo studies of experimental therapeutics utilize rapidly growing transplantable mouse or human tumor cell lines grown subcutaneously in syngeneic or immunodeficient rodents. These models allow for easy quantitation of tumor growth and treatment response monitoring using caliper measurements of tumor volume. A significant concern related to these traditional models is that the phenotypes of neoplastic cells may be modulated by interaction with surrounding normal tissue, and both treatment response and therapy-induced toxicities can vary at different organ sites (Dong et al., 1994; Fidler et al., 1994; Killion et al., 1998; Ross et al., 1989). Orthotopic tumor models may prove to be more appropriate models of human cancer because they may reproduce some of the organ-specific tumor interactions, such as, for example, blood flow, which may affect delivery of the therapeutic agent. However, quantitation of therapeutic-induced changes in tumor growth in orthotopic models is more difficult to obtain, traditionally requiring large numbers of animals to be sacrificed at multiple time points to overcome the interanimal variability. These issues underscore the need for using noninvasive imaging to facilitate the evaluation of the responsiveness of orthotopic tumors in preclinical therapeutic studies. This overview will provide some examples and strategies for using imaging to quantitate the treatment response of orthotopically located tumors in rodent preclinical trials.

USE OF MAGNETIC RESONANCE IMAGING (MRI) FOR EVALUATING TREATMENT RESPONSE

Lead compounds which are identified as having antitumor activity using in vitro screening assays are subsequently evaluated in vivo using rodent models (Crafts and Wilson, 1977; Peterson et al., 1994; Roosen and Rosenblum, 1995). As an example, animal survival, colony-forming efficiency (CFE) assays of cells disaggregated from solid tumors, and measurements of excised tumor weights have

Mouse Models of Human Cancer, edited by Eric C. Holland
ISBN 0-471-44460-X

been used to quantitate the efficacy of various treatments on the orthotopic 9L brain tumor (Barker et al., 1973, 1979; Gutin et al., 1984; Kimler, 1994; Kimler et al., 1992; Nomura et al., 1978; Rosenblum et al., 1975, 1976, 1977, 1980; Vats et al., 1982; Weizsaecker et al., 1981; Wheeler, 1981; Wheeler and Kaufman, 1980, 1981a,b; Wheeler and Wallen, 1980; Wheeler et al., 1979). Many of these studies have shown that 1,3-bis(2-chloroethyl)-1-nitrosourea (BCNU) is an effective treatment for the 9L brain tumor with a single bolus of BCNU achieving 3–4 log cell kill, producing a 97% increase in life span (Rosenblum et al., 1976).

More recently, MRI has been used to follow the growth and response to therapy of the orthotopic 9L brain tumor which has yielded new information. Figure 22.1*a* shows T_2-weighted coronal MRI images from an animal treated with a 2 × LD_{10} dose of BCNU. The first image (top left) was acquired within 2 h of BCNU administration, while the other three images were acquired at the times indicated by the arrows in Figure 22.1*b*. Each displayed image is from approximately the same coronal plane in which the tumor appears as a hyperintense mass in the right hemisphere. The tumor continued to expand after treatment (top right), then shrunk (bottom left) before subsequently regrowing (bottom right). Since the MRI data are multislice in nature (500-μm-thick slices) with 25 slices taken through the brain, the three-dimensional tumor volume can easily and accurately be determined by drawing a region of interest around the tumor in each slice on the computer screen. These data yield the tumor volume for each slice and the summation of all tumor slices provides an overall volumetric measure of the total tumor. Use of MRI in this fashion allows the tumor to be followed over time, providing an opportunity to quantitate the growth rates (tumor doubling time, T_d) and therapeutic response. Shown in Figure 22.1*b* is an example of a treated orthotopic 9L tumor in which the tumor volumes obtained from MRI measurements were plotted over time pre- and posttreatment. The individual volumetric data points obtained from MRI images are shown along with lines representing the least-squares fits. Shown in Figure 22.1*b* is a representative plot of intracerebral 9L tumor volumes from an individual rat treated with a 2 × LD_{10} dose of BCNU. Deviation from the pretreatment exponential growth pattern (solid lines)

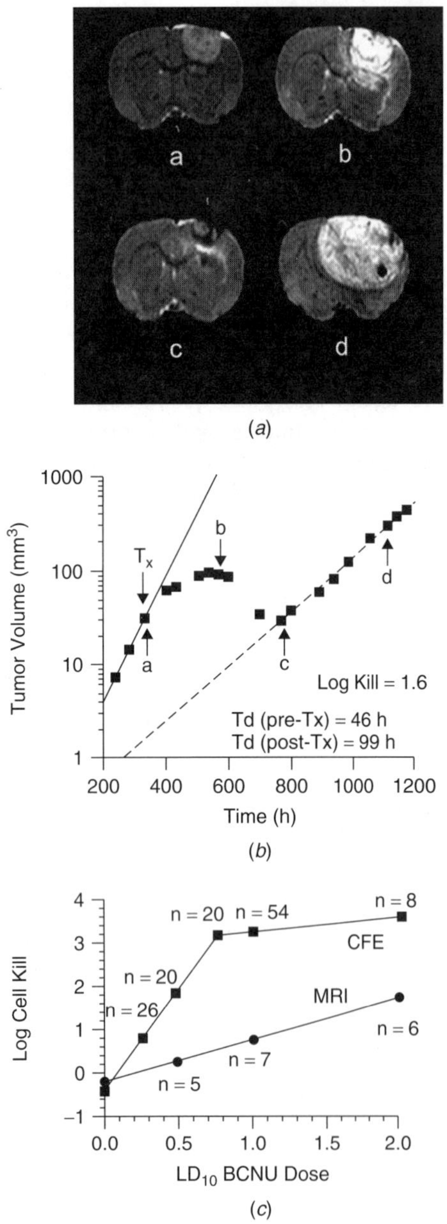

Figure 22.1. (*a*) A series of coronal T_2-weighted MRI images of a rat harboring a 9L tumor in the right hemisphere (a) within 2 h following BCNU treatment (2 × LD_{10}) and (b, c, d) at the time intervals depicted by the arrows in (*b*). Each displayed image is from approximately the same region of the brain. (Adapted from Ross et al., 1998.) (*b*) Tumor volumes obtained from serial MRI images of the rat shown in (*a*) are displayed versus time following 9L cell implantation. Individual tumor volume measurements are shown along with the line corresponding to the least-squares fit. Shown here is a representative plot of a BCNU-treated intracerebral 9L tumor which reveals the exponential growth characteristic of this tumor prior to treatment (Tx arrow, solid line) and posttreatment (dashed line) corresponding to the least-squares fit of exponential growth curve. This animal was treated with 2 × LD_{10} BCNU and the pre- and post-BCNU doubling times were 46 and 99 h, respectively, revealing a decrease in growth rate during posttreatment exponential tumor regrowth. (*c*) Calculated log cell kill values for BCNU doses of 0.5, 1, and 2 × LD_{10} are shown along with historical data obtained from the literature using colony-forming assays of cells taken from solid treated tumors. (Adapted from Ross et al., 1998).

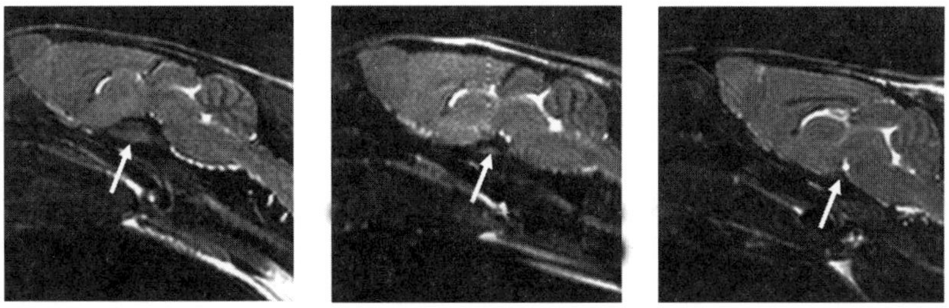

Figure 22.2. T2-weighted MRI of midsagittal slices through a transgenic mouse model for pituitary hyperplasia. MRI permits noninvasive quantitation of the pituitary glands in 9-month-old *Gsu−/−* mice at 0 and 21 days of thyroid hormone treatment (*a, b*) compared with aged matched wild-type littermates (*c*). (Adapted from Brinkmeier et al., 2001.)

(Kim et al., 1995) within two days of treatment reveals that tumor growth was inhibited at this dose. An increase in the delay until exponential tumor regrowth (dashed line) was observed; however, the exponential tumor regrowth occurred at a reduced rate compared with the pretreatment growth rate.

The use of MRI to quantitate in vivo log cell kill values for orthotopic tumor models using MRI-determined tumor volumes is both highly quantitative and elegant as it can be accomplished on a single animal which serves as its own control. Studies such as these have been used to evaluate the dose dependence of BCNU on the intracerebral 9L glioma, which revealed log kill values of 0.2 ± 0.1 ($\pm$ SE, $n = 5$), 0.8 ± 0.1 ($n = 7$), and 1.7 ± 0.2 ($n = 6$) for 0.5, 1, and 2 × LD_{10} doses of BCNU, respectively (Fig. 22.1*c*, circles). The data obtained using MRI were quantitative and the log kill values in each treatment group were found to be normally distributed (Kolmogorov–Smirnov normality test), and differences in the mean values were statistically significant [$p < 0.001$, one-way analysis of variance (ANOVA)] (Ross et al., 1998). The increases in repopulation time due to cell kill for the 0.5, 1, and 2 × LD_{10} doses of BCNU were $91.0 \pm 0.5\%$, $67.4 \pm 3.4\%$, and $46.7 \pm 14.7\%$, respectively, which revealed that animal survival was not solely dependent upon cell kill but was also a function of changes in tumor regrowth rates following therapy. These results have important implications for interpreting data obtained from orthotopic tumor studies which use animal survival as the only biologic readout of efficacy. If the contribution of slower tumor regrowth following treatment is not considered, the increased life span produced by an antineoplastic agent could be incorrectly interpreted as a larger cell kill than was actually achieved. Thus, basing a decision as to the effectiveness of a therapy using animal survival as the sole biologic endpoint could be misleading if imaging of the tumor was not accomplished to verify that changes in tumor growth kinetics had not occurred. Moreover, the MRI-determined cell kill values also revealed a dramatically lower kill than that obtained using traditional colony-forming assays from cells obtained from disaggregated solid tumors treated with BCNU (Rosenblum et al., 1977). The results from these studies reveal the value of noninvasive imaging for evaluating therapeutic agents in preclinical animal studies.

An additional example of using anatomical MRI to follow therapeutic response of an orthotopic tumor can be found in a recent paper using the αGsu−/− mouse model of pituitary hyperplasia (Brinkmeier et al., 2001). In this study, MRI was used to document the size of the pituitary gland in αGsu−/− mice prior to thyroid hormone treatment. As shown in Figure 22.2, MRI provided an opportunity to quantitate the size of the normal wild-type (+/+ control) pituitary in the mouse. Moreover, treatment of a hyperplastic pituitary with thyroid hormone resulted in a significant decrease in pituitary size by 21 days posttreatment. MRI provides a unique ability to determine the size of a lesion at the time of treatment and follow subsequent response dynamically over time in the same animal. The ability of imaging to remove the interanimal differences in orthotopic tumor size and growth rates is a powerful means to assess drug efficacy using in vivo cancer models.

The ability of quantitative anatomical MRI techniques to accurately quantify orthotopic tumor volume, growth rate, and cell kill in individual animals with a single noninvasive assay provides a new and more informative avenue for evaluating preclinical therapeutic studies. Differences in tumor growth rates and size at the time of initial treatment can be controlled, thereby reducing interanimal variability. MRI also requires fewer numbers of animals than CFE or tumor weight assays because groups of animals need not be sacrificed for each experimental time point. This may be especially useful for more costly genetically engineered tumor models with a less than 100% tumor formation rate. Finally, these studies demonstrate the important role that MRI can play in the preclinical drug development process by providing valuable insights into the therapeutic effectiveness of a specific experimental agent.

USE OF DIFFUSION-WEIGHTED MRI FOR QUANTITATING EARLY RESPONSE TO THERAPY

As described above, immediate applications of conventional MRI stems from the ability to noninvasively determine gross tumor morphology and volume and monitor changes with time and/or treatment. Tissue contrast in conventional MRI is obtained by exploiting differences in the various physiochemical properties of normal and tumor tissue. Differences in tissue properties such as alterations in water relaxation properties and vascular permeability to contrast agents are routinely used to diagnose the presence and extent of tumors. There is an important need for imaging methods which can provide accurate and timely quantitation of therapeutic response in the preclinical phase of drug development, which would dramatically facilitate drug dosage optimization. Currently, therapeutic response is assessed by monitoring changes in tumor volume by either calipers for subcutaneous tumors or anatomical imaging using conventional MRI. Earlier and more sensitive/predictive methods to determine treatment efficacy during the initial course of treatment would be extremely valuable. Since many therapeutic agents such as chemotherapy are frequently given in fractionated cycles, an early assessment of therapeutic response during the initial phase of administration would provide an opportunity to gauge the dosage effectiveness and provide feedback related to the dosage frequency, which would greatly facilitate planning of additional dosage protocols. In this regard, largely untapped potential resides in MRI methods known to be sensitive to tissue structure at the cellular level. Such information may be derived via quantitation of tissue properties that are reflective of dynamics in the microscopic environment.

During the course of successful therapeutic intervention, changes in cellular structure and physiology occur which precede macroscopic changes such as a reduction in the overall tumor volume. The diffusion of water in tissue is strongly affected by the viscosity of intra- and extracellular fluids, membrane permeability, active transport and flow, and directionality of cellular structures that enhance or impede its mobility. As cells are damaged and

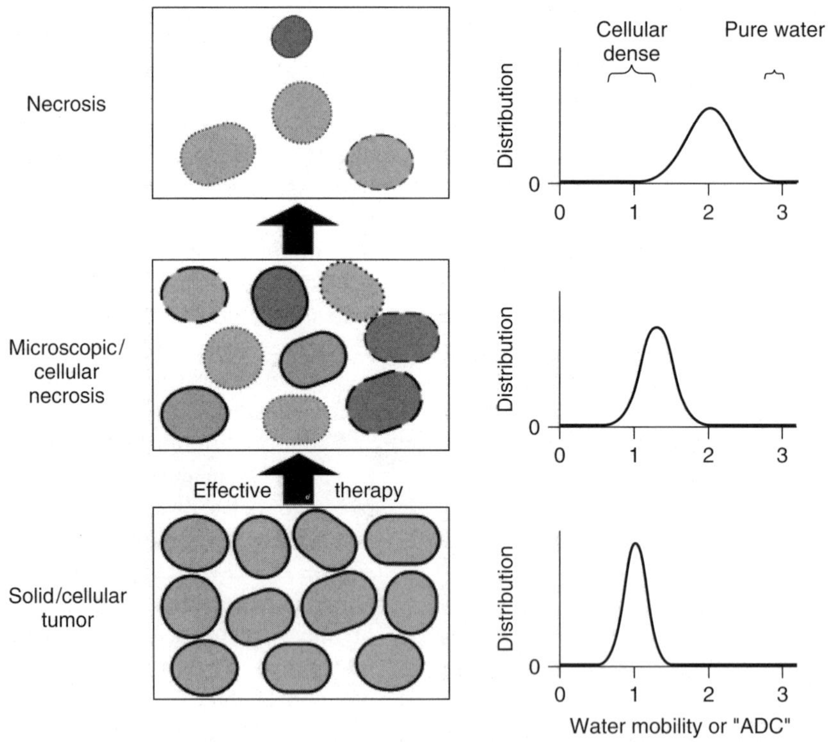

Figure 22.3. Schematic representation of the relationship between change in tissue cellularity and molecular water mobility measured as an ADC. At left is the evolution toward necrosis following an effective therapy. The increase in extracellular space and membrane permeability allow greater water mobility as illustrated by distributions of diffusion on the right. (Used with permission from Ross, BD et al., *Molecular Cancer Therapeutics*, 2003).

killed by therapeutic interventions, the integrity of cell membranes are compromised and the fractional volume of the interstitial space may increase due to apoptotic body formation and cell loss. Damage to tissue microvasculature may also lead to vasogenic edema, thereby reducing the viscosity of the interstitial fluid and increasing the volume of extracellular water. Thus, the water mobility within a tumor will increase over time following treatment and the magnitude of the change will be related to the effectiveness of the therapy, which will result in membrane damage with a subsequent reduction in cell density, as shown diagrammatically in Figure 22.3. The diffusibility of tissue water in vivo can be accurately and noninvasively quantified as an apparent diffusion coefficient (ADC) using diffusion-weighted MRI. Each voxel within the tumor is described by a single ADC value, and all of the voxels can be plotted in a histogram format (Fig. 22.3, right panel) to provide for a quantitative and visual readout of how the tumor diffusion values change over time following therapeutic intervention. In addition, the directionality of cellular structures can be probed by controlling the direction of the applied diffusion pulses; however, when monitoring tumors, this is avoided since tumors typically have a chaotic cytoarchitecture.

The ability to monitor the movement (diffusion) of water using MRI can be accomplished by making the MRI signal intensity dependent on water mobility by the application of additional pulsed magnetic field gradients incorporated within the normal MRI sequence (Le Bihan et al., 1986; Stejskal and Tanner, 1965; Taylor and Bushell, 1985). Individual nuclear spins of the tumor water molecules accumulate a phase shift in proportion to their spatial position in these additional gradient fields. After waiting a given evolution time for spins to diffuse, a second gradient pulse is applied to completely refocus stationary spins. The spin of the water molecules which have diffused during the evolution time do not experience an equivalent compensatory pulse and are therefore refocused incompletely. Thus, the paired gradient pulses attenuate signal in proportion to local tissue water mobility. Quantitative measurements of ADC values are obtained by measuring signal attenuation as a function of varying gradient strength and evolution time. The early clinical applications of this technique focused on its use for the detection of cerebral ischemia where it was demonstrated that early changes could be observed by diffusion-weighted MRI long before changes become detectable using conventional MRI methods (Moseley et al., 1990).

Recent work has demonstrated that diffusion MRI is highly sensitive to microscopic changes in tissue structure and physiology (Rowley et al., 1999). The goal of using diffusion MRI is to monitor early changes in tumor

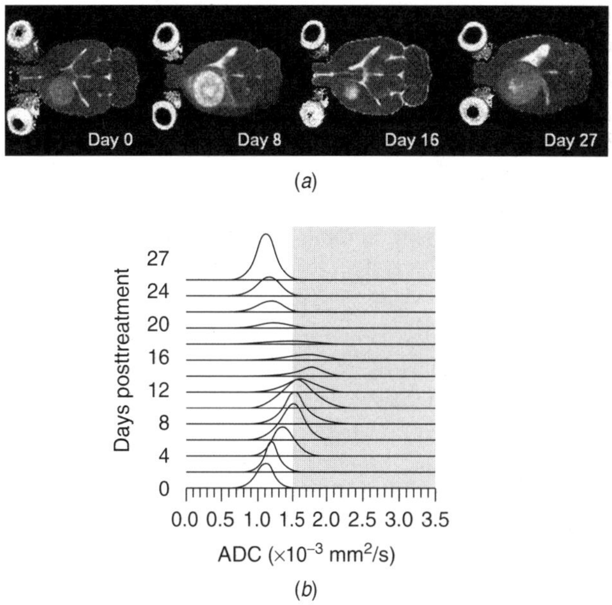

Figure 22.4. Illustration of a 9L rat glioma treated with a single dose of 13.3 mg/kg of BCNU chemotherapy. At this dose, this rat mimics a "responder" since tumor shrinks substantially. ADC maps on this animal (*a*) illustrate the degree of maximal tumor shrinkage on day 16. Despite continued tumor growth on day 7, the increased water mobility is visually apparent on the ADC maps, which suggests necrosis. This is also illustrated graphically in the diffusion histograms on the bottom. (Adapted from Chenevert et al., 2000.)

tissue following initiation of therapy where the magnitude of the changes will be related to the effectiveness of the treatment. Initial studies using diffusion MRI to detect changes in water diffusion in animal tumor models following high-dose chemotherapy were found to be very promising (Chenevert et al., 1997; Chinnaiyan et al., 2000; Galons et al., 1999; Hakumaki et al., 1998; Poptani et al., 1998; Ross et al., 1994, 1997; Stegman et al., 2000; Zhao et al., 1996). Changes in ADC values were observed to precede changes in tumor growth kinetics and tumor regression, indicating the potential for applying this approach to early predictive monitoring of cancer therapy. An example of the application of diffusion MRI for monitoring therapeutic-induced changes in tumor tissue is shown in Figure 22.4. In this figure, diffusion MRI was used to detect early changes in water diffusion values in a rat 9L brain tumor following treatment with BCNU (single dose of 26.6 mg/kg i.p.). ADC images are shown for a single slice acquired over several time points where the intensity of the signal is proportional to the diffusion value of the water molecules within that tissue region (voxel). Prior to chemotherapy, the signal within the tumor tissue is slightly enhanced relative to that of normal brain, as evidenced by the elevated signal intensity within the tumor mass. Following

BCNU administration, the diffusibility of water in the tumor increased significantly at eight days posttreatment (Fig. 22.4*a*, second image from left), although the tumor continued to expand, revealing that this technique is able to detect early events in the cell death process prior to regression of the tumor mass. The increase in water ADC values (signal intensity) was due to a significant increase in extracellular water within the tumor tissue as the density of cells decreased due to therapeutic-induced death. Removal of the cellular debris at later time points resulted in shrinkage of the primary mass followed by a return to previous diffusion values during subsequent regrowth of the tumor. Serial ADC histograms from this animal are shown in Figure 22.4*b* for this tumor and demonstrate a right shift in tumor water diffusion beginning at two days posttherapy, despite continued tumor growth reflected by an increase in the area under the histogram. As shown in Figure 22.3, this right shift toward a higher diffusion value in Figure 22.4 histograms reflects a significant cell kill effect, which eventually results in a decrease in overall tumor volume. Water diffusibility appeared to peak around day 8, after which the tumor began to regress. Eventually the small surviving fraction of tumor cells repopulate the tumor volume, resulting in a recurrent tumor with a water environment resembling that of the primary tumor prior to BCNU treatment. Such diffusion histograms permit segmentation of tumor pixels that exhibit "high-diffusion" properties, postulated to be subelements of the tumor responding to treatment. For example, the shaded region in the histograms illustrates the volume of tumor pixels above an arbitrarily selected ADC threshold of 1.5×10^{-3} mm^2/s, which may indicate the fractional volume of tumor that exhibits a strong therapeutic response.

Dose escalation experiments are common in preclinical drug development experiments; therefore the sensitivity of diffusion MRI for detection of relatively modest tumor "cell kill" value is an important issue to understand the relative sensitivity of this approach. To assess the dose dependence and sensitivity of treatment-induced diffusion changes, experiments were performed on intracranial 9L gliomas following administration of three different doses of BCNU. Figure 22.5 is a plot of changes in mean tumor ADC values relative to the mean baseline tumor ADC measured at the beginning of the experiment. Plotting the mean diffusion tumor values is another approach to quantitate and display the data over time. There is a clear dose dependence in both the magnitude and duration of tumor diffusion changes. Statistically significant differences between each treatment group and the control group were first apparent at day 4 posttreatment. Treatment of tumors with $0.5 \times LD_{10}$, $1 \times LD_{10}$, and $2 \times LD_{10}$ doses of BCNU resulted in log cell kill values of 0.2, 0.8, and 1.7, respectively (Ross et al., 1998). The observation of statistically significant changes in tumor ADC values at four days after administration of a $0.5 \times LD_{10}$ dose, producing a cell kill of 0.2 log, reveals that diffusion is very sensitive to small cell kill values, which could provide an important opportunity for detection of small therapeutic-induced tissue changes which may be evident at the initial lower doses of a dose escalation preclinical study.

The ability of diffusion MRI to retain spatial information related to water mobility also provides the possibility of assessing the regional/spatial heterogeneity of therapeutic response within a tumor. The heterogeneity of response may be accentuated in applications involving direct intratumoral administration of the therapeutic agent, as is done in certain therapeutic protocols involving, for example, cancer gene therapy. In this regard, diffusion MRI has been shown to evaluate the response of orthotopic 9L gliomas to the adenoviral delivered yeast cytosine deaminase (yCD) gene therapy paradigm. Flucytosine (5-fluorocytosine, 5FC) is the prodrug used in this gene therapy approach since 5FC itself does not elicit cytotoxicity, and its efficacy depends on the ability of the microbial enzyme CD to convert 5FC to the antimetabolite 5-fluorouracil (5FU). Expression of the CD gene in tumor cells followed by systemic administration of 5FC

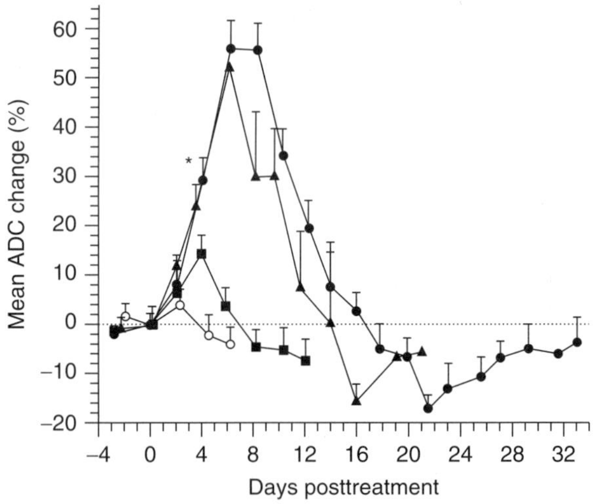

Figure 22.5. Dose dependence of BCNU-induced changes in tumor water diffusion. BCNU was given as a single i.p. bolus of 6.6 (■, $n = 5$), 13.3 (▲, $n = 6$), and 26.6 (●, $n = 6$) mg/kg body weight, corresponding to 0.5×, 1×, and 2× the LD_{10} dose, respectively. Control animals (○, $n = 5$) were sham treated with vehicle alone. Tumor ADC values are given as means ± SEM normalized to standard diffusion mapping ADC values. Differences between the treatment and control groups became statistically significant ($p < 0.019$) at posttreatment day 4 as determined by one-way ANOVA (*). (Adapted from Chenevert et al., 2000.)

will result in the generation of 5FU within the tumor. This localized production of 5FU chemotherapy avoids the systemic toxicity associated with intravenous 5FU therapy and has been developed to improve outcome by achieving higher localized intratumoral 5FU concentrations. As with all cancer gene therapy approaches, uniform delivery of the therapeutic transgene throughout the tumor mass and at high enough levels has been limiting the therapeutic effectiveness. The effects of the heterogeneous distribution of the yCD transgene expression on tumor cell viability was examined using multislice diffusion-weighted MRI. Shown in Figure 22.6 (upper left panel) are diffusion-weighted transverse MRI images of a rat that was treated with the adenoviral vector but received only phosphate-buffered saline (PBS) and no 5FC. In this image, the brightness of the tissue region (or voxels) is proportional to the diffusibility of water. The very bright regions within the normal brain correspond to cerebral spinal fluid located within the ventricular space. The tumor, located in the right hemisphere, has overall a slightly higher diffusion value than that of the normal brain parenchyma. In addition, the tumor image displays a relatively uniform pattern of diffusion values indicating a uniform distribution of cellular structures (e.g., intra- and extracellular space). All of the voxels from the entire tumor mass were added together and plotted in a histogram format (Fig. 22.6, upper right). While this format removes spatial information as related to anatomical position, it allows the display of all of the values obtained from the entire tumor in a single plot providing for visualization of the changes over time. Diffusion histograms plotted for this rat at days 21, 24, and 26 remained essentially unchanged in ADC units except that they increased in area due to increased tumor volume over this time period. In contrast, a rat treated identically except that 5FC was administered daily from day 18 through day 28 yielded an entirely different effect on diffusion MRI. The effects of 5FC treatment on tumor diffusion are easily seen in Figure 22.6 (lower left) as a heterogeneous distribution of bright voxels located throughout the tumor mass. The interpretation of these images is that the bright areas (high diffusion) represent regions of CD expression leading to conversion of 5FC to the cytotoxic product, 5FU, with subsequent cell death leading to regions of focal necrosis. Diffusion histogram analysis of this tumor over time is shown in Figure 22.6 (lower right).

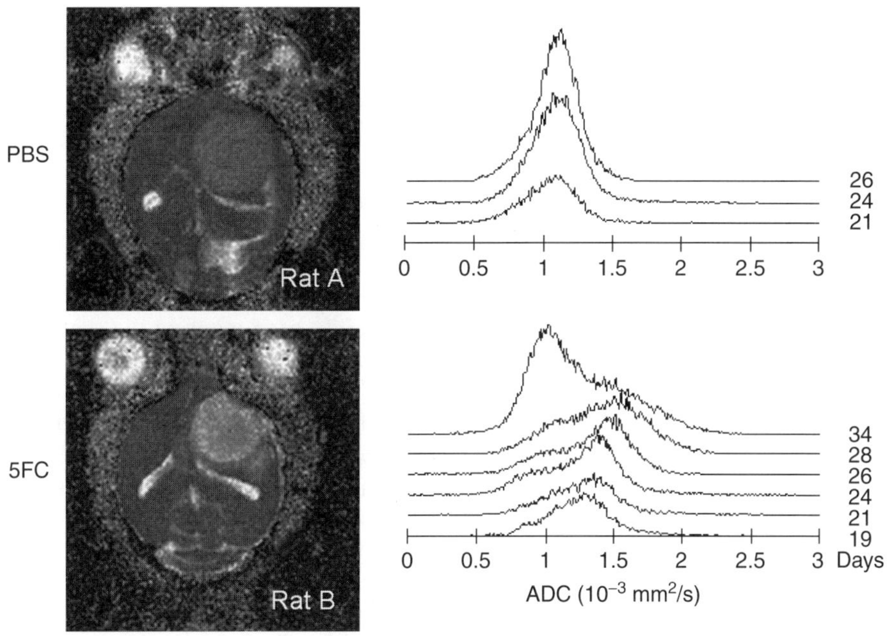

Figure 22.6. Representative ADC maps of PBS control (top left) and 5FC-treated tumors (bottom left) at 7 days postinitiation of 5FC administration. Note the areas of increased ADC (bright regions) within the 5FC-treated tumor as compared to the control animal. Serial histograms of tumor pixel ADC values from the control (top right) and the 5FC-treated animal (bottom right). The number of days postinitiation of 5FC treatment is shown to the right of each histogram. Administration of the adenoviral vector did not affect the average tumor diffusion values, as shown in the control. Treatment with 5FC produced a nonuniform increase in tumor diffusion values which is interpreted to be due to heterogeneous expression of the CD transgene within the tumor mass due to the limitations of direct intratumoral injection of the vector. (Used with permission from Rehemtulla et al., 2002.)

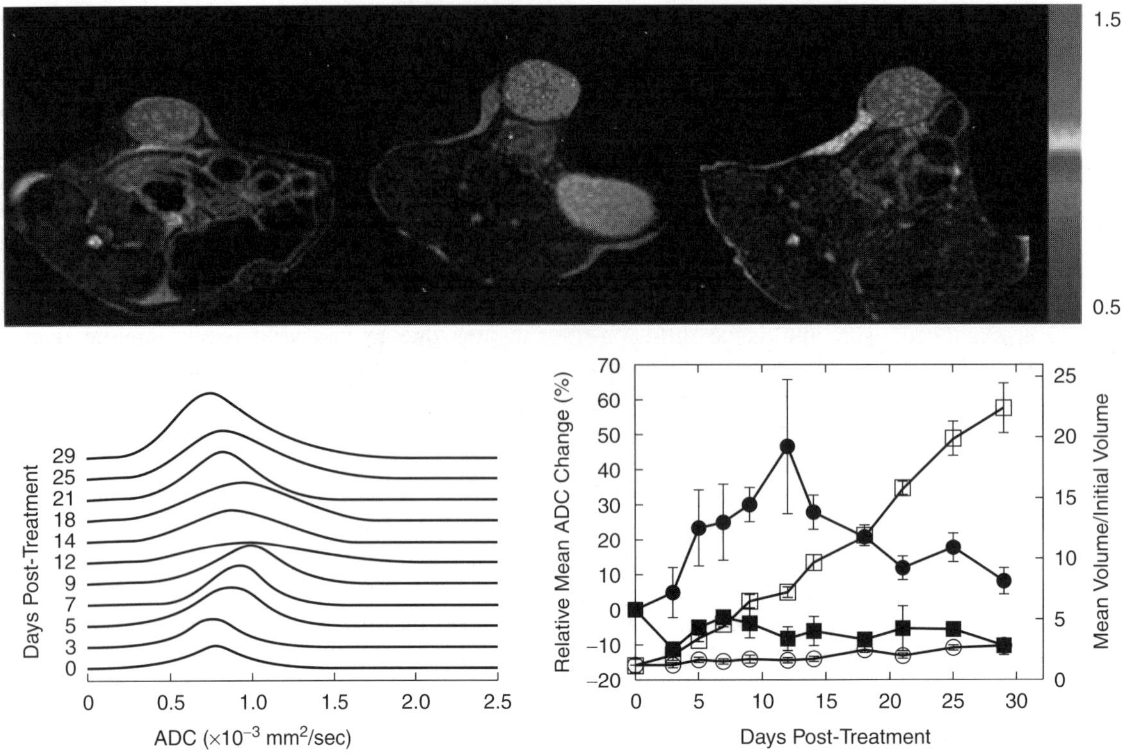

Figure 22.7. Diffusion-weighted MRI study of a subcutaneous PC-3 (prostate) tumor in a nude mouse treated with 100 mg/kg (i.p.) Camptosar. (Top panel) The three grayscale anatomic T2-weighted images are overlayed with a diffusion map (color) of the tumor, from left to right (days 3, 12, and 29 posttreatment). Note the increase in tumor diffusion values (red) on day 12 posttreatment. Tumor data are also displayed in a histogram format over time (bottom left) from the same animal as the images. The plot of "mean relative ADC change (%)" versus time is the change in mean ADC relative to day zero of three treated and three control animals. Solid symbols are ADC (circle = treated, square = control) and open symbols are volume (circle = treated, square = control). (See color insert)

In contrast to what was observed in the untreated animal (Fig. 22.6, upper right), the histogram for the 5FC-treated animal broadened with a fraction of the histogram area moving to the right (higher diffusion) to an approximate mean of 1.5 ADC units. The fraction of tumor tissue that exhibited an increase in diffusion is interpreted as the relative fraction of the tumor that is undergoing a significant therapeutic response. The region of the histogram in the treated animal that did not increase but remained at approximately 1–1.2 ADC units is most likely reflective of the cell fraction which was not exposed to sufficient 5FU concentrations. At day 28, administration of 5FC was terminated and the tumor subsequently began to regrow, as shown by the day 34 histogram (Fig. 22.6, lower right). This histogram shows that repopulation of the solid portion of the tumor is occurring as a loss of high-diffusion regions was observed. This in fact was consistent with the observed recurrent tumor using anatomical MRI of this animal. These data reveal that diffusion MRI has the capability of monitoring the spatial heterogeneity of therapeutic response within a tumor over time.

A final example of the use of diffusion MRI is provided from a study of a subcutaneous PC-3 (prostate) tumor grown in a nude mouse which was treated with 100 mg/kg (i.p.) Camptosar. As shown in Figure 22.7, the diffusion increased significantly following treatment, which also corresponded to a dramatic reduction (loss) of tumor growth versus the untreated (control) animal. These data, along with the above examples, clearly indicate that diffusion MRI can be used both in the evaluation of traditional in vivo tumor models and treatments and with novel, experimental treatment agents and transgenic models.

QUANTITATION OF TUMOR CELL KILL USING BIOLUMINESCENCE IMAGING (BLI)

Application of optical-based methods for in vivo tumor detection and evaluation of treatments is an active area

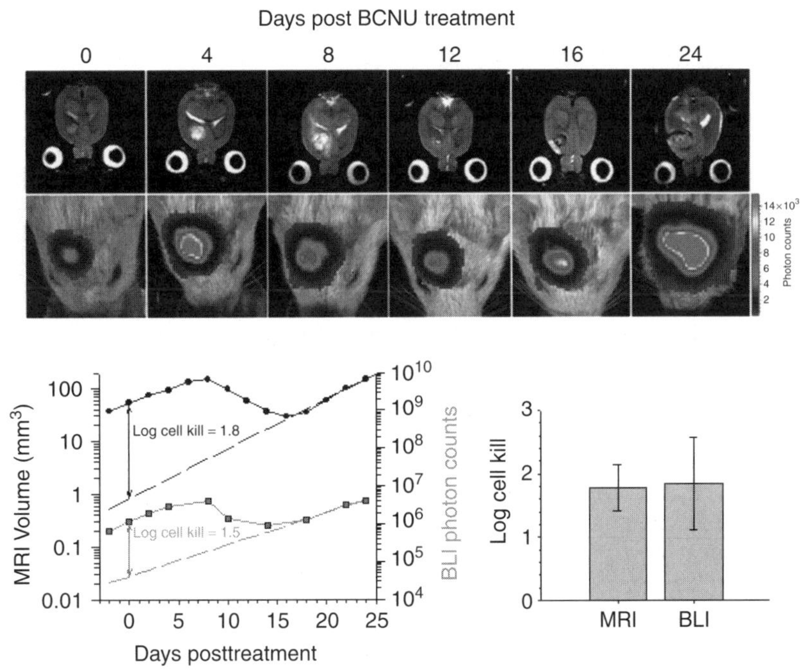

Figure 22.8. Temporal analysis of the response of a 9L^{Luc} tumor to BCNU chemotherapy. Tumor cells were implanted 16 days prior to treatment. Tumor volume was monitored with T2-weighted MRI (top panel, top row), and intratumoral luciferase activity was monitored with BLI (top panel, bottom row). The days post-BCNU therapy on which the images were obtained are indicated at the top. The scale to the right of the BL images describes the color map for the photon count. Quantitative analysis of tumor progression and response to BCNU chemotherapy is shown in the bottom left panel. Tumor volumes (●) and total tumor photon emission (■) obtained by T2-weighted MRI and BLI, respectively, are plotted versus days post-BCNU treatment. The dashed lines are the regression fits of the exponential tumor repopulation following therapy. The solid vertical lines denote the apparent tumor volume and photon production losses elicited by BCNU on the day of treatment from which log cell kill values were calculated. Comparison of log cell kill values determined from MRI and BLI measurements is shown in the bottom right panel. Log cell kill elicited by BCNU chemotherapy was calculated using MRI (1.78 ± 0.36) and BLI (1.84 ± 0.73). Data are represented as mean $\pm$ SEM for each animal ($n = 5$). There was no significantly significant difference between the log kills calculated using the MRI and BLI data ($p = 0.951$). (Used with permission from Rehemtulla et al., 2000). (See color insert)

of research. Fluorescence and bioluminescence optical approaches for cancer detection and monitoring are very promising. A recent study compared the use of MRI with BLI for quantitating therapeutic effects using the 9L brain tumor model. BLI necessitates detection of light emitted from expression of the bioluminescent enzyme firefly luciferase (Luc) from tumor cells (Contag et al., 2000). Implantation of tumor cells expressing Luc (9L^{luc}) into rats resulted in tumors which could be visualized by MRI as well as by BLI. Shown in Figure 22.8 (top panel) are a series of T2-weighted MRI images (top) with the corresponding optical images (bottom) acquired over time in an animal treated with BCNU. A quantitative plot of the MRI volumetric data and the BLI photon counts over time are also shown in Figure 22.8 (lower left). These data were used to calculate the log kill as described above. Comparison of log kill values using these two imaging techniques revealed similar results (Fig. 22.8, lower right) (Rehemtulla et al., 2000).

BLI provides a relatively rapid quantification of tumor size (tumor volume was found to be proportional to photon count) (Rehemtulla et al., 2000); however, it is currently limited to a single two-dimensional image with limited spatial resolution. MRI provides a three-dimensional tomographic image set which provides the investigator with additional information such as the spatial extent and heterogeneity of the tumor mass. The decision of which imaging modality to use will depend upon the type of data required for a particular study.

Additional applications of BLI for monitoring tumor response to therapy include monitoring of metastatic lesions following injection of PC-3M human prostate cells transfected with luciferase into the left ventricle (Rosol et al., 2003). This study showed that BLI could detect rapid localization of tumor cells to bones and other organs, such as the kidneys and lungs. Additional applications of BLI have included studies investigating the control of minimal residual disease by noninvasively quantitating tumor burden in a rodent model (Edinger et al., 2003). In this application, BLI was used to detect small numbers of cells to follow tumor growth within internal organs and monitor radiation and chemotherapy as well as immune-based strategies that employed the tumoricidal activity of ex vivo–expanded CD8(+) natural killer (NK)– T cells. These investigators were able to image the course of malignant disease, including engraftment, expansion, metastasis, response to therapy, and patterns of relapse. This study showed that complex cellular processes in bone marrow transplantation and antitumor immunotherapy could be revealed in real time in living animals using BLI.

IMAGING OF APOPTOSIS

As described above, the majority of imaging has been relegated to obtaining anatomical information based upon differences in physical parameters to generate image contrast. Significant strides have been made in developing approaches to use noninvasive imaging technologies to obtain information related to specific molecular events. A molecular imaging construct capable of detecting apoptosis has been recently developed to noninvasively image the activation of Caspase-3 in living animals (Laxman et al., 2002). This molecular reporter construct consisted of a hybrid recombinant reporter molecule (Fig. 22.9) wherein the activity of the reporter is silenced by fusion of Luc with the estrogen receptor (ER) regulatory domain. Significant silencing of the reporter's enzymatic activity was achieved using this hybrid molecule.

Inclusion of a protease cleavage site for caspase-3 (DEVD) between these two domains allowed for protease-mediated activation of the reporter molecule by separation of Luc from the silencing domain (ER) (Fig. 22.9). As most cells activate Caspase-3 during apoptosis, this engineered chimeric polypeptide was used as a novel molecular-based reporter for noninvasive detection of apoptosis using BLI. As shown in Figure 22.10, the expression of this molecule in a transplanted human glioma cell line in a nude rat revealed that Caspase-3 activation could be noninvasively imaged following intratumoral treatment with the apoptosis-inducing agent tumor necrosis factor-α (TNF-α)–related apoptosis-inducing ligand (TRAIL). The ability to image Caspase-3 activation noninvasively provides a unique tool for the evaluation of therapeutic efficacy of experimental therapeutic agents as well as for studies on the role of apoptosis in various

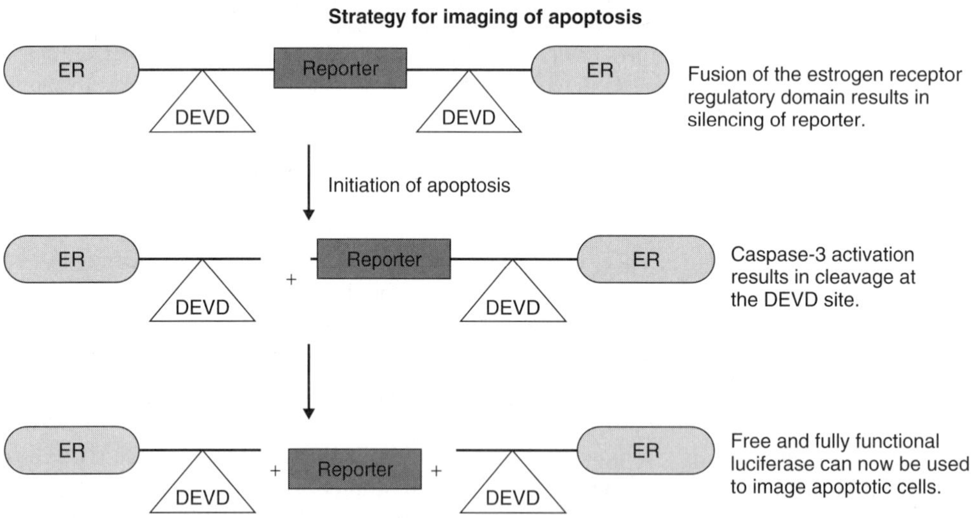

Figure 22.9. Strategy for imaging of apoptosis. Shown is the luciferase molecule with ER regulatory domains at the amino and carboxy termini. Inclusion of the Caspase-3 cleavage site between these domains provided for apoptosis-mediated activation of the reporter molecule following separation of the silencing domains (i.e., ER). Since most cells activate Caspase-3 during apoptosis, this reporter construct can be used for reporting (imaging) of apoptosis. The cleavage site of this molecular construct can be modified to "report" on the activation of a variety of proteases for detection of proteolytic activity in both cells and transgenic animals using noninvasive imaging methodologies. (Adapted from Laxman et al., 2002.)

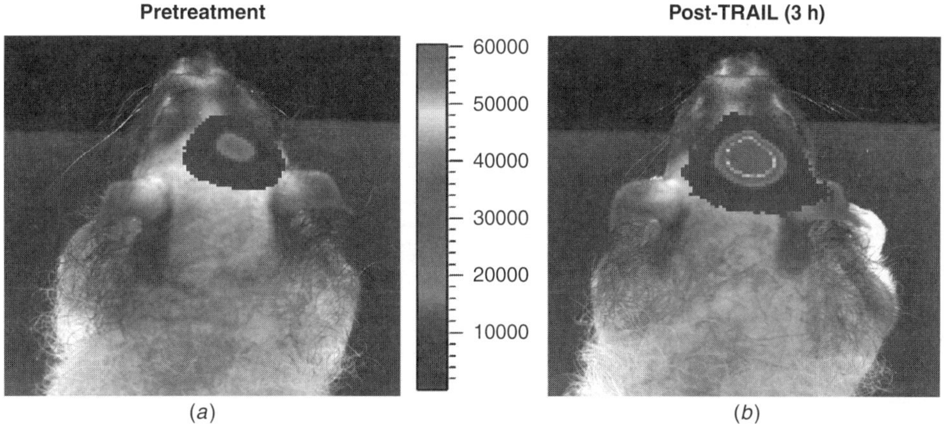

Figure 22.10. In vivo imaging of Caspase-3 activation. In vivo bioluminescence image of a rat with an intracerebral D54-Luc$_{ER}$ glioma obtained (*a*) before and (*b*) 3 h following intratumoral TRAIL treatment. Note the significant increase in photons collected from the same animal following TRAIL treatment. (See color insert)

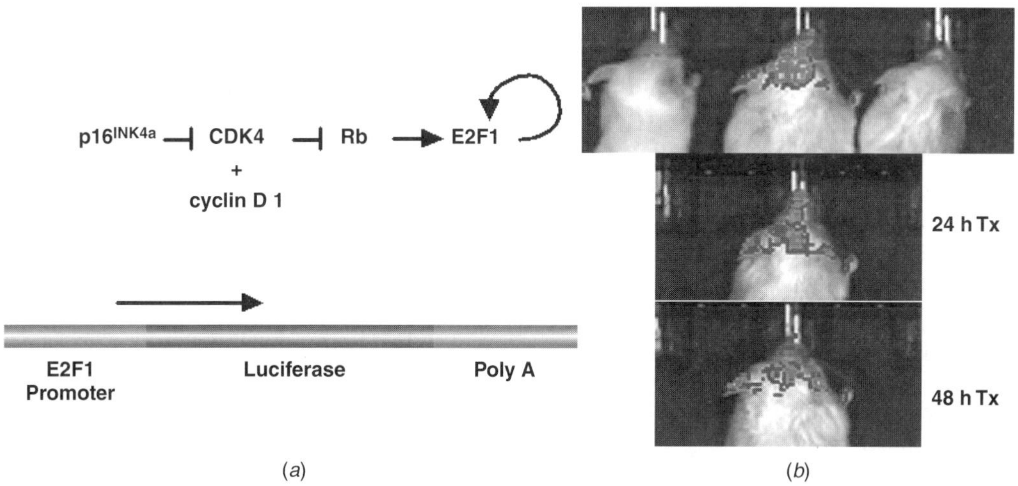

Figure 22.11. Strategy for imaging luciferase tumors using Rb. (*a*) The Eflux transgene construct driving luciferase from the E2F1 promoter that responds to Rb activity and cell proliferation. (*b*) Three Eflux transgenic mice imaged for bioluminescence, the center mouse harbors a PDGF-induced glioma. That mouse was subsequently treated with a small molecule inhibitor of the PDGF receptor kinase with resultant decrease in luciferase activity. (See color insert)

disease processes. In addition, the Caspase-3 cleavage site can be exchanged with a site for other proteases, including other caspases, which would provide for studying the action of other important proteases and therapeutic modulation of their activities in intact animals. Moreover, this reporter molecule also can be used to develop a transgenic animal wherein a specific proteolytic event could be noninvasively monitored in a tissue-specific manner. Thus, this molecular reporter approach could greatly facilitate the preclinical drug evaluation process for pro- or antiapoptotic agents using xenograft or transgenic models.

BIOLUMINESCENCE IMAGING OF RB PATHWAY ACTIVITY IN EFLUX TRANSGENIC MICE

Another use of BLI is for in vivo imaging of the activities of specific pathways. This can be achieved by the generation of reporter constructs that drive luciferase expression from promoters that are regulated by the activity of a pathway in question. One example of this approach is the construction of a luciferase reporter that utilizes the E2F1 promoter. The E2F1 promoter is responsive to the activity of the Rb pathway that is disregulated in most cancer types and roughly represents cell cycle progression. Pathologic disruption of the Rb pathway results in elevated expression from the E2F1 promoter, and in such cells or populations of cells in a tumor, luciferase expression and bioluminescence correlate with the degree of that pathway. This construct has been used to generate a transgenic mouse line referred to as Eflux. The Eflux line contains the Eflux construct in all cell types and should theoretically report on Rb activity in any cancer that arises in these mice (Fig. 22.11*a*). The Eflux line has been crossed to the Ntv-a mouse line that supports gene transfer to glial progenitors (see Chapter 5). These mice were infected with an RCAS vector that encodes platelet-derived growth factor (PDGF), resulting in the generation of an autocrine loop PDGFR activation and the generation of oligodendrogliomas. As illustrated in Figure 22.11*b*, bioluminescence imaging using Eflux is capable of detecting mice with gliomas. Furthermore, since the luciferase production is an ultimate readout of the oncogenic effects of PDGF in this case, BLI can be used as an assay for therapeutic effect in preclinical trials. The same mouse that was originally detected by BLI as having a PDGF-induced glioma was then treated with a small molecule inhibitor of PDGFR. As illustrated, the light emission from this mouse decreases over the 48 h of the treatment. Immunohistochemical staining for proliferating cell nuclear antigen, a readout of the proliferative activity of a tumor, indicates that posttreatment this tumor has reduced proliferating cells significantly more than leaving the gliomas untreated (data not shown). This approach of generating transgenic lines that respond with BLI to specific signaling pathways can obviously be extended to applications measuring other pathways in tumors noninvasively.

ACKNOWLEDGMENTS

This work was supported in part by the following grants from the National Institutes of Health and the National Cancer Institute: PO1CA85878, 1P50CA93990, and R24CA83099.

REFERENCES

Barker M, Deen DF, Baker DG (1979): BCNU and X-ray therapy of intracerebral 9L rat tumors. *Int J Radiat Oncol Biol Phys* 5:1581–1583.

Barker M, Hoshino T, Gurcay O, Wilson CB, Nielsen SL, Downie R, Eliason J (1973): Development of an animal brain tumor model and its response to therapy with 1,3-bis(2-chloroethyl)-1-nitrosourea. *Cancer Res* 33:976–986.

Brinkmeier ML, Stahl JH, Gordon DF, Ross BD, Sarapura VD, Dowding JM, Kendall SK, Lloyd RV, Ridgway EC, Camper SA (2001): Thyroid hormone-responsive pituitary hyperplasia independent of somatostatin receptor 2. *Mol Endocrinol* 15:2129–2136.

Chenevert TL, McKeever PE, Ross BD (1997): Monitoring early response of experimental brain tumors to therapy using diffusion magnetic resonance imaging. *Clin Cancer Res* 3:1457–1466.

Chenevert TL, Stegman LD, Taylor JM, Robertson PL, Greenberg HS, Rehemtulla A, Ross BD (2000): Diffusion magnetic resonance imaging: An early surrogate marker of therapeutic efficacy in brain tumors. *J Natl Cancer Inst* 92:2029–2036.

Chinnaiyan AM, Prasad U, Shankar S, Hamstra DA, Shanaiah M, Chenevert TL, Ross BD, Rehemtulla A (2000): Combined effect of tumor necrosis factor-related apoptosis-inducing ligand and ionizing radiation in breast cancer therapy. *Proc Natl Acad Sci USA* 97:1754–1759.

Contag CH, Jenkins D, Contag PR, Negrin RS (2000): Use of reporter genes for optical measurements of neoplastic disease in vivo. *Neoplasia* 2:41–52.

Crafts D, Wilson CB (1977): Animal models of brain tumors. *Natl Cancer Inst Monogr* 46:11–17.

Curt GA (1994): The use of animal models in cancer drug discovery and development. *Stem Cells* 12:23–29.

Dong Z, Radinsky R, Fan D, Tsan R, Bucana CD, Wilmanns C, Fidler IJ (1994): Organ-specific modulation of steady-state mdr gene expression and drug resistance in murine colon cancer cells. *J Natl Cancer Inst* 86:913–920.

Edinger M, Cao YA, Verneris MR, Bachmann MH, Contag CH, Negrin RS (2003): Revealing lymphoma growth and the efficacy of immune cell therapies using in vivo bioluminescence imaging. *Blood* 101:640–648.

Fidler IJ, Wilmanns C, Staroselsky A, Radinsky R, Dong Z, Fan D (1994): Modulation of tumor cell response to chemotherapy by the organ environment. *Cancer Metastasis Rev* 13:209–222.

Galons JP, Altbach MI, Paine-Murrieta GD, Taylor CW, Gillies RJ (1999): Early increases in breast tumor xenograft water mobility in response to paclitaxel therapy detected by non-invasive diffusion magnetic resonance imaging. *Neoplasia* 1:113–117.

Gutin PH, Bernstein M, Sano Y, Deen DF (1984): Combination therapy with 1,3-bis(2-chloroethyl)-1-nitrosourea and low dose rate radiation in the 9L rat brain tumor and spheroid models: Implications for brain tumor brachytherapy. *Neurosurgery* 15:781–786.

Hakumaki JM, Poptani H, Puumalainen AM, Loimas S, Paljarvi LA, Yla-Herttuala S, Kauppinen RA (1998): Quantitative 1H nuclear magnetic resonance diffusion spectroscopy of BT4C rat glioma during thymidine kinase-mediated gene therapy in vivo: Identification of apoptotic response. *Cancer Res* 58:3791–3799.

Killion JJ, Radinsky R, Fidler IJ (1998): Orthotopic models are necessary to predict therapy of transplantable tumors in mice. *Cancer Metastasis Rev* 17:279–284.

Kim B, Chenevert TL, Ross BD (1995): Growth kinetics and treatment response of the intracerebral rat 9L brain tumor model: A quantitative in vivo study using magnetic resonance imaging. *Clin Cancer Res* 1:643–650.

Kimler BF (1994): The 9L rat brain tumor model for pre-clinical investigation of radiation-chemotherapy interactions. *J Neurooncol* 20:103–109.

Kimler BF, Liu C, Evans RG, Morantz RA (1992): Intracerebral chemotherapy in the 9L rat brain tumor model. *J Neurooncol* 14:191–200.

Laxman B, Hall DE, Bhojani MS, Hamstra DA, Chenevert TL, Ross BD, Rehemtulla A (2002): Noninvasive real time imaging of apoptosis. *Proc Natl Acad Sci USA* 99:16551–16555.

Le Bihan D, Breton E, Lallemand D, Grenier P, Cabanis E, Laval-Jeantet M (1986): MR imaging of intravoxel incoherent motions: Application to diffusion and perfusion in neurologic disorders. *Radiology* 161:401–407.

Moseley ME, Cohen Y, Mintorovitch J, Chileuitt L, Shimizu H, Kucharczyk J, Wendland MF, Weinstein PR (1990): Early detection of regional cerebral ischemia in cats: Comparison of diffusion- and T2-weighted MRI and spectroscopy. *Magn Reson Med* 14:330–346.

Nomura K, Hoshino T, Knebel K, Deen DF, Barker M (1978): BCNU-induced perturbations in the cell cycle of 9L rat brain tumor cells. *Cancer Treat Rep* 62:747–754.

Peterson DL, Sheridan PJ, Brown WE, Jr (1994): Animal models for brain tumors: Historical perspectives and future directions. *J Neurosurg* 80:865–876.

Poptani H, Puumalainen AM, Grohn OH, Loimas S, Kainulainen R, Yla-Herttuala S, Kauppinen RA (1998): Monitoring thymidine kinase and ganciclovir-induced changes in rat malignant glioma in vivo by nuclear magnetic resonance imaging. *Cancer Gene Ther* 5:101–109.

Rehemtulla A, Stegman LD, Cardozo SJ, Gupta S, Hall DE, Contag CH, Ross BD (2000): Rapid and quantitative assessment of cancer treatment response using in vivo bioluminescence imaging. *Neoplasia* 2:491–495.

Rehemtulla A, Hall DE, Stegman LD, Prasad U, Chen G, Bhojani MS, Chenevert TL, Ross BD (2002): Molecular imaging of gene expression and efficacy following adenoviral-mediated brain tumor gene therapy. *Mol Imaging*, 1:43–55.

Roosen N, Rosenblum ML (1995): In Kay AH, Laws Jr ER (Ed), *Brain Tumors: An Encyclopedic Approach*. Churchill Livingstone; New York, pp. 387–404.

Rosenblum ML, Deen DF, Hoshino T, Dougherty DA, Williams ME, Wilson CB (1980): Comparison of clonogenic cell assays after in vivo and in vitro treatment of 9L gliosarcoma. *Br J Cancer Suppl* 41:307–308.

Rosenblum ML, Knebel KD, Vasquez DA, Wilson CB (1976): in vivo clonogenic tumor cell kinetics following 1,3-bis(2-chloroethyl)-1-nitrosourea brain tumor therapy. *Cancer Res* 36:3718–3725.

Rosenblum ML, Knebel KD, Vasquez DA, Wilson CB (1977): Brain-tumor therapy. Quantitative analysis using a model system. *J Neurosurg* 46:145–154.

Rosenblum ML, Wheeler KT, Wilson CB, Barker M, Knebel KD (1975): In vitro evaluation of in vivo brain tumor chemotherapy with 1,3-bis(2- chloroethyl)-1-nitrosourea. *Cancer Res* 35:1387–1391.

Rosol TJ, Tannehill-Gregg SH, LeRoy BE, Mandl S, Contag CH (2003): Animal models of bone metastasis. *Cancer* 97:748–757.

Ross BD, Ben-Yoseph O, Chenevert TL (1997): in vivo magnetic resonance imaging and spectroscopy: Application to brain tumors. *Magn Reson Spectro Imaging Neurochem* 8:145–178.

Ross BD, Chenevert TL, Kim B, Ben-Yoseph O (1994): Magnetic resonance imaging and spectroscopy: Application to experimental neuro-oncology. *Q Magn Reson Biol Med* 1:89–106.

Ross BD, Mitchell SL, Merkle H, Garwood M (1989): in vivo ^{31}P and ^{2}H NMR studies of rat brain tumor pH and blood flow during acute hyperglycemia: Differential effects between subcutaneous and intracerebral locations. *Magn Reson Med* 12:219–234.

Ross BD, Moffat BA, Lawrence TS, Mukherji SK, Gebarski SS, Quint DJ, Johnson TD, Junck L, Robertson PL, Muraszko KM, Dong Q, Meyer CR, Bland PH, McConville P, Geng H, Rehemtulla A, Chenevert TL (2003): Evaluation of cancer therapy using diffusion magnetic resonance imaging. *Mol Cancer Ther* 2:581–7.

Ross BD, Zhao YJ, Neal ER, Stegman LD, Ercolani M, Ben-Yoseph O, Chenevert TL (1998): Contributions of cell kill and posttreatment tumor growth rates to the repopulation of intracerebral 9L tumors after chemotherapy: an MRI study. *Proc Natl Acad Sci USA* 95:7012–7017.

Rowley HA, Grant PE, Roberts TP (1999): Diffusion MR imaging. Theory and applications. *Neuroimaging Clin N Am* 9:343–361.

Stegman LD, Rehemtulla A, Hamstra DA, Rice DJ, Jonas SJ, Stout KL, Chenevert TL, Ross BD (2000): Diffusion MRI detects early events in the response of a glioma model to the yeast cytosine deaminase gene therapy strategy. *Gene Ther* 7:1005–1010.

Stejskal EO, Tanner J (1965): Spin diffusion measurements: Spin-echoes in the presence of a time-dependent field gradient. *J Chem Phys* 42:288–292.

Taylor DG, Bushell MC (1985): The spatial mapping of translational diffusion coefficients by the NMR imaging technique. *Phys Med Biol* 30:345–349.

Vats TS, Kimler BF, Henderson SD, Morantz RA (1982): Combination chemotherapy for the 9L rat brain tumor model system. *Cancer Treat Rep* 66:575–579.

Weizsaecker M, Deen DF, Rosenblum ML, Hoshino T, Gutin PH, Barker M (1981): The 9L rat brain tumor: Description and application of an animal model. *J Neurol* 224: 183–192.

Wheeler KT (1981): Review of factors influencing the response of experimental brain tumors to therapy. *Cancer Treat Rep* 65:75–81.

Wheeler KT, Kaufman K (1980): Influence of fractionation schedules on the response of a rat brain tumor to therapy with BCNU and radiation. *Int J Radiat Oncol Biol Phys* 6: 845–849.

Wheeler KT, Kaufman K (1981a): Brain tumor therapy: Prospects for combining BCNU with conventional radiotherapy schedules. *Int J Radiat Oncol Biol Phys* 7: 1065–1068.

Wheeler KT, Kaufman K (1981b): Efficacy of continuous treatment with radiation in a rat brain-tumor model. *J Neurosurg* 55:52–54.

Wheeler KT Jr, Kaufman K, Feldstein M (1979): Response of a rat brain tumor to fractionated therapy with low doses of BCNU and irradiation. *Int J Radiat Oncol Biol Phys* 5:1553–1557.

Wheeler KT, Wallen CA (1980): Is cell survival a determinant of the in situ response of 9L tumours? *Br J Cancer Suppl* 41:299–303.

Zhao M, Pipe JG, Bonnett J, Evelhoch JL (1996): Early detection of treatment response by diffusion-weighted 1H-NMR spectroscopy in a murine tumour in vivo. *Br J Cancer* 73:61–64.

23

PHYSIOLOGICALLY BASED PHARMACOKINETIC AND PHARMACODYNAMIC MODELING

RAYMOND S. H. YANG AND JAMES E. DENNISON

Quantitative and Computational Toxicology Group, Center for Environmental Toxicology and Technology, Department of Environmental and Radiological Health Sciences, Colorado State University, Fort Collins, Colorado

MELVIN E. ANDERSEN

Department of Biomathematics and Physical Sciences, CIIT Centers for Health Research, Research Triangle Park, North Carolina

YING C. OU

Human Genome Sciences, Inc., Rockville, Maryland

KAI H. LIAO AND BRAD REISFELD

Quantitative and Computational Toxicology Group, Center for Environmental Toxicology and Technology, Department of Chemical Engineering, Colorado State University, Fort Collins, Colorado

We introduce the general concept and background knowledge of physiologically based pharmacokinetic/pharmacodynamic (PBPK/PD) modeling. Examples of application of PBPK/PD modeling are provided. We illustrate the utility of PBPK/PD modeling, particularly in the pharmaceutical drug development process. We project future development on "second-generation" PBPK/PD modeling and In silico toxicology. An emphasis is given to introducing the concepts of PBPK/PD modeling rather than the details of its techniques and processes.

INTRODUCTION

At first glance, the topic of physiologically based pharmacokinetic/pharmacodynamic (PBPK/PD) modeling appears to be outside the central theme of this book. However, with the advances of modern biology and computational technology, it is just a matter of time for any area of biomedical sciences to be integrated with computer science. Two aspects are particularly important for the application of PBPK/PD modeling in cancer research, thus underscoring the relevance for the inclusion of this chapter in the book: (1) incorporation of PBPK/PD modeling in any biomedical research can avoid unnecessary experiments, thereby conserving precious resources, and (2) computer simulations (i.e., in silico experimentation) using validated PBPK/PD models will minimize animal usage.

The intent of this chapter is to (1) introduce the general concept and background knowledge of PBPK/PD modeling; (2) provide some examples of application of PBPK/PD modeling; (3) illustrate the utility of PBPK/PD modeling, particularly in the pharmaceutical drug development process; and (4) project future development on "second-generation" PBPK/PD modeling and In silico toxicology. An emphasis was given to introducing the concepts of PBPK/PD modeling rather than the details of its techniques and processes. For more detailed background and conceptual information on PBPK modeling, the readers are referred to our earlier introductory discussion (Yang and Andersen, 1994) and the two "timeless papers" by Bischoff and Brown (1966) and Dedrick (1973). For more extensive information, particularly regarding modeling techniques, the reader

Mouse Models of Human Cancer, edited by Eric C. Holland
ISBN 0-471-44460-X

should consult the two future books from our laboratories (Reddy et al., 2004; Yang et al., 2005).

WHAT IS PBPK? WHAT ARE THE DIFFERENCES BETWEEN PBPK AND CLASSICAL PHARMACOKINETICS?

Physiologically based pharmacokinetics (PBPK), as the name implies, is a special branch of pharmacokinetics where physiology and anatomy of the animal or human body as well as the biochemistry of the chemical or chemicals of interest are incorporated into the conceptual model for computer simulation. *Classical pharmacokinetics* refer to those empirical noncompartmental or compartmental pharmacokinetic studies routinely practiced in the pharmaceutical industry (van de Waterbeemd and Gifford, 2003). As will be illustrated later, the compartments of a PBPK model have anatomic and physiologic significance. This is a major difference from empirical noncompartmental or compartmental pharmacokinetic modeling approaches. PBPK models can be used to describe concentration–time profiles in individual tissue/organ and in the plasma or blood. When the concentration of a certain target tissue, rather than the plasma concentration, is highly related to a compound's efficacy or toxicity, PBPK modeling will be a more useful tool than classical pharmacokinetic models for describing PBPK/PD relationships and thus make a better prediction of the time course of drug effects resulting from a certain dose regimen for the compound of interest. Furthermore, PBPK models in combination with absorption simulation and quantitative structure–activity relationship (QSAR) approaches will bring us closer to a full prediction of drug disposition for pharmaceutical new entities and help streamline the selection of lead drug candidates in the drug discovery process (van de Waterbeemd and Gifford, 2003). Lastly, unlike empirical noncompartmental and compartmental pharmacokinetics, PBPK modeling is a powerful tool for extrapolation, be it for interspecies, interroutes, interdoses, interlife stages, and so on.

The concept of PBPK had its embryonic development in the 1920s and 1940s; for a more detailed early history, readers are referred to two books which are in preparation (Reddy et al., 2005; Yang et al., 2005). PBPK modeling blossomed and flourished in the late 1960s and early 1970s in the chemotherapeutic area mainly due to the efforts of investigators with expertise in chemical engineering process design and control. Two notable pioneers in this development are Kenneth B. Bischoff,

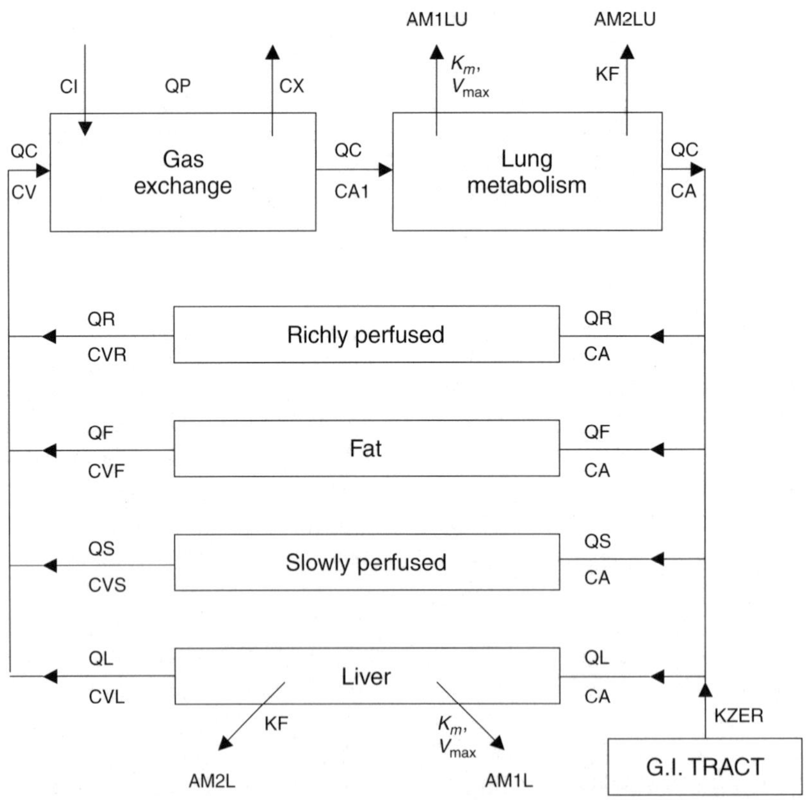

Figure 23.1. Graphical representation of a PBPK model for methylene chloride. CA and CV are concentrations in arterial and venous blood, with compartment abbreviations added in some cases. Q is flow rate and AM is amount metabolized. For further details, please see the original reference (Andersen et al., 1987).

then at the University of Texas, Austin, Texas, and Robert Dedrick of Biomedical Engineering and Instrumentation Branch, Division of Research Services, National Institutes of Health, Bethesda, Mariland. Two timeless publications by these investigators are, respectively, "Drug Distribution in Mammals" (Bischoff and Brown, 1966) and "Animal Scale-Up" (Dedrick, 1973); these articles are highly recommended to those who are interested in PBPK modeling. In the mid-1980s, two articles on PBPK modeling of styrene and methylene chloride (Andersen et al., 1987; Ramsey and Andersen, 1984) started yet another "revolution" in the toxicology and risk assessment arena. Today, there are more than 700 publications directly related to PBPK modeling on industrial chemicals, drugs, environmental pollutants, and simple and complex chemical mixtures (Reddy et al., 2005).

A PBPK model, graphically illustrated in Figure 23.1, reflects the incorporation of basic physiology and anatomy. The compartments actually correspond to anatomic entities such as liver, lung, and so on, and the blood circulation conforms to the basic mammalian physiology. In this specific model, a published example on methylene chloride, it is quite obvious that the exposure route of interest is inhalation because the lung and gas exchange compartments are prominently displayed with intake (CI) and exhalation (CX) vapor concentrations indicated. Oral and/or dermal exposures may be added easily to the gastrointestinal tract compartment or general venous circulation, respectively. Some tissues (e.g., richly or slowly perfused tissues in Fig. 23.1) are "lumped" together because there is insufficient evidence to conclude that each of these tissues is kinetically distinct enough for the specific chemical to warrant a separate compartment.

If one draws an analogy of the "scale-up" from a laboratory chemical engineering process to a chemical plant to the scale-up of a mouse to a human, one finds that both situations are governed by a great number of physical and chemical processes. In mammals, the physical processes (i.e., mass balance, thermodynamics, transport, and flow) often vary in a predictable way. However, chemical processes such as metabolic reactions may vary greatly and are less predictable among species. These physical and chemical processes interact in the body such that the pharmacokinetics of any given chemical between one species and another may be more (or less) predictable depending on the amount of background information available.

HOW DOES A PHYSIOLOGICALLY BASED PHARMACOKINETIC MODEL WORK?

A PBPK model applies fundamental physiologic, biochemical, and engineering principles to describe the distribution and disposition of chemicals in the body at any given time. The process and approach may be summarized in a flowchart (Fig. 23.2). Once the chemical of interest and the problems needing to be addressed are identified, a thorough literature evaluation is conducted.

The fundamentals of PBPK modeling are to identify the principal organs or tissues involved in the disposition of the chemical of interest and to correlate the chemical absorption, distribution, metabolism, and excretion within and among these organs and tissues in an integrated and biologically plausible manner. A scheme is usually formed where the normal physiology is followed in a graphical manner (i.e., a conceptual model as in Fig. 23.1). Within the boundary of the identified compartment (e.g., an organ or tissue or a group of organs or tissues), whatever "comes" in must be accounted for via whatever "goes out" or whatever is transformed into something else. This "mass balance" is expressed as a mathematical equation with appropriate parameters carrying biologic significance. A series of such equations representing all of the interlinked compartments are formulated to express a mathematical representation, or model, of the biologic system. This model can then be used for computer simulation to predict the time course behavior of any given parameter. Three sets of parameters are needed for PBPK model building: physiologic parameters (e.g., ventilation rates, cardiac output, organs as percent body weight), thermodynamic parameters (e.g., tissue partition coefficients, flow rates), and biochemical parameters (e.g., K_m and $V_{\max}$). Most, if not all, of the parameters for laboratory animals are available in relevant literature, such as the *Biological Data Book* and the special report by the International Life Sciences Institute (ILSI) on the compilation of physiologic parameters for PBPK models (Brown et al., 1997). When information gaps exist, the solution is either an empirical one via experimentation or through allometric extrapolation, usually based on a power function of the body weight (Lindstedt, 1987).

The U.S. Environmental Protection Agency (EPA) guidance document (2002, pp 37–38) includes a very nice discussion on "modeling the data." Although it is not necessarily for PBPK modeling, the discussion reflects some "dos" and "don'ts" on computer modeling. We quote some of the passages below:

> The selection of a mathematical model structure to fit the data being analyzed should be guided by the biology of the common mechanism of toxicity, the toxicokinetics of the chemicals, and the observed shapes of their dose-response curves and the experimental designs used to generate the data. If available, pharmacodynamic and pharmacokinetic data should be considered in order to account for tissue concentrations and to aid in defining dose-response relationships across different species, routes and time-frames of exposure . . . .

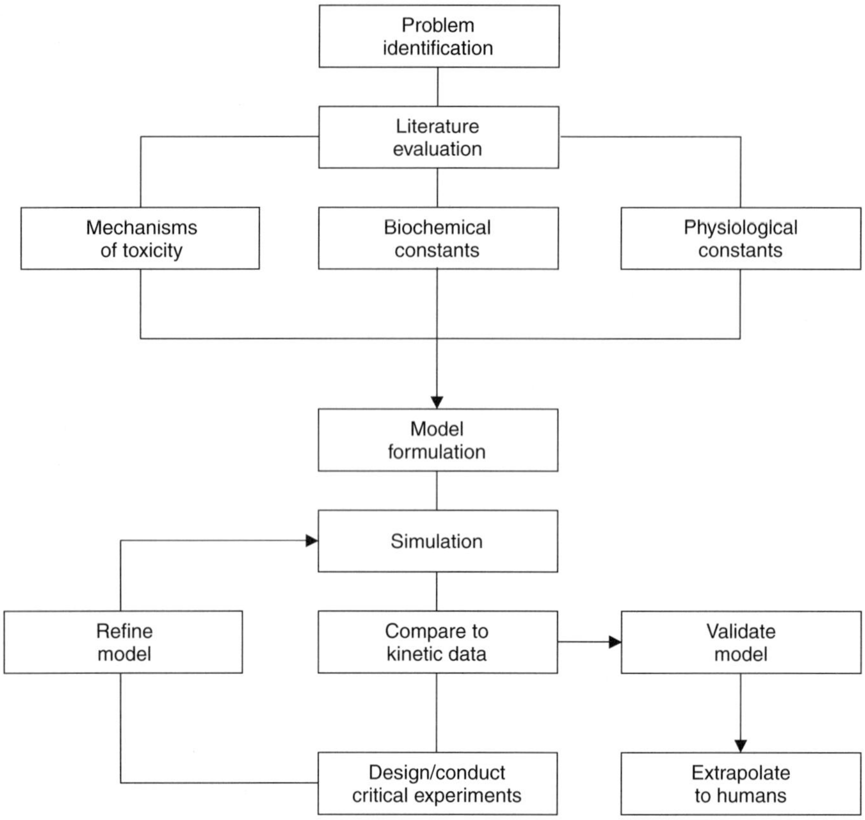

Figure 23.2. Flowchart illustrating processes involved in PBPK. (From M. E. Andersen, *Pharmacokinetics in Risk Assessment, Drinking Water and Health,* National Academy Press, Washington, D.C., 1987, pp. 8–23.)

... Although it is not possible to recommend the use of specific models, a few points that should be considered in modeling the data follow:

- Modeling of individual animal data is desirable; however, if this is not practical, then use of summary data such as means and standard deviations can be alternatives
- Care should be taken with modeling high-dose data (particularly extreme doses) because the model shape in the low-dose region can be influenced by high-dose data
- Log transformation of data should be justified because such a transformation may distort the dose-response curve
- Data variability should be described by appropriate statistical techniques and reflected in the potency estimate (e.g., by weighting the data in the fitting procedure)
- Confidence intervals or limits should be included in the analysis because they can be valuable for evaluating the influence of variability on the potency estimates
- An estimate for the uncertainty of the model used in the analysis should be included
- The statistical fitting method used must be clearly described.

For the most well-studied chemicals or drugs, it is likely that the biochemical constants such as K_m's and V_{max}'s are known and readily retrievable from the information data base. However, it must be made clear here that the K_m, V_{max}, and K_F (first-order rate constant) in a PBPK model (known as in vivo K_m, V_{max}, K_F for a given chemical) such as the ones given in Figure 23.1 are hybrid constants of all the saturable or linear metabolic pathways, respectively, for the chemical of interest in the organ and/or body. They are different from the in vitro K_m, V_{max}, or K_F of a given pure enzyme. While they are not directly interchangeable, the in vitro constants in the literature may be used to estimate in vivo constants for modeling purposes (Kedderis and Lipscomb, 2001; Lipscomb et al., 1998). Also, for most well-known chemicals, it is likely that enough is known about the mechanism of toxicity to be incorporated into the model for computer simulation. Physiologic constants such as organ volumes and blood flow rates are usually available in the literature for the common laboratory animals as well as humans. Therefore, at least in those instances of "well-known" chemicals, a model may be conceptually illustrated as in Figure 23.1 and mathematically represented by a number of mass balance differential equations. Computer simulations may be made for any number of desired time course endpoints such as the blood levels of the parent compound, liver level of a reactive metabolite, and similar information on different species, at lower or higher dose levels, and/or via a different route of exposure. The experimental pharmacokinetic data may then be compared with PBPK model simulation to see if they are superimposable upon each other. If this is indeed the case, the model is consistent with actual results. Validation of the PBPK model with data sets other than the working set (or training set) to develop the model is necessary. Once validated, the PBPK model is ready for extrapolation to other animal species, including humans. However, if the experimental data and PBPK model simulation are not consistent, the model might be deficient because critical scientific information might be missing or certain assumptions are incorrect. The investigator, with knowledge of the chemical and a general understanding of the physiology and biochemistry of the animal species, can design and conduct critical experiments for refining the model to reach consistency with experimentation (Fig. 23.2). This refinement process may be repeated again and again when necessary; such an iterative process is critically important for the development of a PBPK model. There is always the possibility that a good model may not be obtained at the time because of the limitation of our knowledge on the chemical. An emphasis must be made here that the investigator and the knowledge possessed are the single most important determinant of the outcome of the results; mathematical modeling and advanced computational capabilities are nothing more than good tools.

MYTHS ABOUT PHYSIOLOGICALLY BASED PHARMACOKINETICS; ARE THEY REAL?

Common belief is that PBPK modeling is highly resource intensive and very difficult to do, particularly involving interactive pharmacokinetics or pharmacodynamics. This warrants some special discussion: First, by the time chemicals such as drugs or pesticides reach the stage of commercialism, many types of studies have already been conducted, including pharmacokinetic studies. Thus, they are already resource intensive during the developmental stage to become successful chemicals in commerce. The important point is how may quantitative, time course data useful for PBPK model development be generated during the product development phase? The data required for PBPK modeling are really not too much different from those required for the classical pharmacokinetics in the present IND/NDA (investigational new drug application/new drug application) process. A slight new orientation to the existing battery of studies would generate adequate quantitative time course data for PBPK modeling. Thus, it is definitely not any more resource intensive from the existing requirements. Furthermore, if the incentive (e.g., regulatory guideline-driven scientific studies) is there, such quantitative, time course data would have been automatically generated during the product development phase. In fact, PBPK modeling, being a hypothesis testing tool in toxicology, may be utilized to conduct many different kinds of experiments on a computer (i.e., in silico toxicology).

A great deal of research effort has been devoted to the development of *simple*, *high-throughput*, and in vitro predictive tools in the pharmaceutical industry (van de Waterbeemd and Gifford, 2003). While these short-term, rapid assays certainly offer some utility, particularly in the early drug development process, an inherent danger of such tools is the "by-pass" of integrated mammalian physiology and architecture of the body. In that regard, development of in silico toxicology such as PBPK/PD modeling and other biologically based computer modeling has the advantage of integrating whole-body pharmacokinetics and pharmacodynamics with computational technology. The resulting predictive tools will minimize unnecessary experiments and improve the *attrition rate* of drug or chemical product development with much more scientific validity and confidence. In that sense, PBPK/PD modeling, once integrated as a part of the product development plan, will actually save expenses and resources as well as minimize unnecessary animal experiments.

Second, while PBPK modeling is by no means a very easy technology, it is not any more difficult than some of the classical pharmacokinetic and statistical modeling carried out in the routine product development process. Furthermore, excellent training opportunities (e.g., www.cvmbs.colostate.edu/enhealth/cett/) are available, and the development of software is such that more and more user-friendly tools are going to be available.

PHYSIOLOGICALLY BASED PHARMACODYNAMIC MODELING

Using plain English, pharmacokinetics can be considered as "What the body does to the chemicals," and pharmacodynamics can thus be considered as "What the chemicals do to the body." Physiologically based pharmacodynamic (PBPD) modeling is therefore computer simulation of pharmacologic or toxicologic effects of chemicals or drugs based on the biology of chemical/drug–receptor interactions. From the point at which a chemical or a drug enters into the body to the point of pharmacologic or toxicologic effect, it is a continuum of pharmacokinetics and pharmacodynamics. It is difficult to distinguish where pharmacokinetics end and pharmacodynamics start. When we consider the pharmacology or toxicology of a chemical or a drug, we must consider both pharmacokinetics and pharmacodynamics to have a full understanding. To consider either area alone is to understand only part of the picture.

PBPK modeling preceded PBPD modeling by many years because of the slower and later advances of mechanistic understanding of modes of action of chemicals and drugs. It is a natural course of evolution that PBPK blossomed and flourished first because it was developed based on the fundamentals of mammalian physiology, analytical chemistry, engineering principles of mass transport and mass balance, and desk-top computer hardware and software development. These were all well-developed areas in the earlier days. PBPD modeling is dependent upon the mechanistic basis of chemical/drug–receptor interactions, and the biology associated with it took time to mature. The recent trend is such that more and more PBPD modeling work is evident. PBPK modeling has matured to the point that research endeavors are centered around more complex systems (e.g., multichemical interactions, subcompartmentalization of organs) or special problem-driven models (e.g., dermal uptake of chemicals from showers, exposure dose reconstruction). As will be discussed later, the "delumping" of organ and tissue compartments and the linkage of PBPK models with other types of biologically based models may lead us to *second-generation PBPK/PD modeling*.

DATA REQUIREMENTS FOR PBPK OR PBPD MODELING

What are the specific data needed for building PBPK models? Obviously, well-conducted in vivo pharmacokinetic data are essential, and usually the more the data sets (e.g., different doses, routes, species), the better. In each study, time course blood and tissue concentration data are essential. These time course data should include at least the following tissues and organs: blood (or plasma if blood cell binding is not an issue), liver (organ of metabolism), kidney (representing well-perfused organs/tissues), muscle (representing slowly perfused organs/tissues), and target organ(s)/tissue(s). We also need other PBPK-modeling specific information, such as (1) physiologic constants, including body size, organ and tissue volumes, blood flow, and ventilation rates; (2) biochemical constants, including the chemical-specific metabolic rate constants such as $V_{\max}$ and K_m, partition coefficients for tissues; and (3) mechanistic factors such as target tissues, metabolic pathways, and receptor interactions. Enzyme kinetic data, particularly human data, of at least the key metabolic processes will be important for the PBPK model. In vitro determination of tissue partition coefficients and enzyme kinetic data is relatively straightforward and inexpensive. With modern genetic engineering technologies, many human enzymes are available commercially. Thus, heretofore unavailable human enzyme kinetic information for many of the environmentally important chemicals are within easy reach for many laboratories. These experiments should be performed.

For PBPD modeling, the data requirement is much more variable because of the many different types of chemical/drug–receptor interactions. Thus, it is much more of a compound-specific nature. However, as a rule of thumb, quality time course data on the key biologic processes to be modeled are essential.

PBPK OR PBPK/PD MODELS FOR CHEMICAL INTERACTIONS

Human exposure to chemicals is rarely, if ever, to single chemicals. In the area of clinical pharmacology, adverse drug interactions are undoubtedly serious concerns. For instance, Lazarou et al. (1998) estimated that there were over 2.2 million cases of serious adverse drug reactions (ADRs) in hospital patients in 1994 in the United States and among these cases 106,000 were fatal. During their hospital stay, the patients in the survey statistics were given an average of eight drugs. Comparing with other statistics of causes of death, these investigators (Lazarou et al., 1998) indicated that ADRs became the fourth to the sixth leading cause of death for that year in the United States. Thus, it is important to discuss the issues related to the development of interactive PBPK/PD modeling (see particularly the example given later on the work of Kanamitsu et al., 2000).

From the perspective of interactive PBPK/PD modeling, two aspects need to be addressed: pharmacokinetic interactions and pharmacodynamic interactions. Because PBPD modeling is a relatively recent effort and few such models are available for even single chemicals, we will concentrate all our discussion on interactive PBPK models. The most ideal and scientifically defensible data requirement for establishing an interactive PBPK model is that each component chemical in the mixture already have its respective established PBPK model and that there are many pharmacokinetic data sets in laboratory animals as well as in humans available for each of these component chemicals. The interactive PBPK model is then built on the basis of known pharmacokinetic interactions. For instance, the component chemicals may inhibit each others' biotransformation. The individual PBPK models may then be linked together at the liver compartment by introducing competitive inhibition (or other types of inhibition) terms in the mass balance differential equation.

In some cases, even more specific and stringent data requirements are needed; an example may be the ultralow-dose pharmacokinetic data in perinatal developmental stages of laboratory animals for extrapolation to human fetuses and babies using PBPK modeling. This area is still in its infancy and further development is anticipated.

APPLICATION OF PBPK/PD MODELING

Dosing Schedule of a Chemotherapeutic Agent

Methotrexate is a folate analogue and a well-known cancer chemotherapeutic agent. This is one of the chemotherapeutic agents which was studied extensively with PBPK modeling. The mechanism of toxicity is due to the binding of methotrexate to dihydrofolate reductase, a key enzyme in DNA synthesis, leading to cessation of DNA synthesis and cell death. In some organs (liver, kidney, intestine, and marrow) and many tumors, methotrexate undergoes metabolism to active polyglutamate derivatives (even more potent inhibitors for dihydrofolate reductase). Because these polyglutamates are retained in some tissues far longer than the parent compound, it has been suggested that this effect is of great importance for the antineoplastic property of methotrexate.

Table 23.1 shows the toxicity of methotrexate to mice under a variety of dosing schedules (Morrison et al., 1987). Obviously, toxicity does not directly correlate with total dose. Decreasing total dose by a factor of 117 (350/3) led to an increase, rather than a substantial decrease, in toxicity. In addition, the area under the plasma concentration–time curve (AUC), a frequently used pharmacokinetic parameter for bioavailability, did not correlate with toxicity. For instance, the AUC for a bolus dose of 350 mg/kg is about two orders of magnitude greater than that of the 96-h infusion at 0.8 μg/h, yet toxicity is higher with infusion.

The reason for the above phenomenon turned out to be intimately associated with the pharmacokinetics of methotrexate (and its polyglutamate metabolites) and the threshold and length of time that dihydrofolate reductase is inhibited (thus inhibition of DNA synthesis). At a bolus dose of 350 mg/kg, even though there is a short period of very high blood and tissue concentration, the inhibition of DNA synthesis did not persist long enough to cause lethality in at least some of the mice. At an infusion rate of 0.8 μg/h for 96 h, even though the blood and tissue levels were low, they were nevertheless high enough to cause sustained inhibition of DNA synthesis, which ultimately translated into higher lethality in the animals. These scientific discoveries eventually led to the revision of the methotrexate PBPK model by incorporating the inhibition of DNA synthesis into the model.

From the modeling perspective, the iterative process of making new scientific discoveries and then refining the PBPK model by incorporating such new information into the model is a wonderful illustration of what we discussed in relation to Figure 23.2. An even more significant illustration is the fact that a validated methotrexate PBPK model can be used to conduct all the experiments in Table 23.1 on a computer. With such complicated dosing schedules, it is apparent how the lives of hundreds of mice may be saved and how much time and resources may be diverted to other more efficient usages.

"Electronic Rats"

One of our earlier examples was the PBPK/PD modeling of a toxicologic interaction between kepone (also known as chlordecone) and carbon tetrachloride (CCl_4) based on mechanisms of interactive toxicity and the application of computer technology in acute toxicity studies. This was a collaboration among three research groups: Melvin E. Andersen, CIIT (presently CIIT Centers for Health Research); Harihara M. Mehendale, Northeast Louisiana

Table 23.1. Dosing Schedule Dependence of Methotrexate Toxicity in Mice

Dose (mg/kg)	Schedule	Total Dose (mg/kg)	Peak Plasma Concentration (*M*)	Effect
350	Single dose	350	10^{-3}	LD50
25	Twice daily	50	10^{-4}	LD50
3	Every 3 h, 5 times, rest 8 h, then every 3 h, 3 times	24	10^{-5}	>LD50[a]
0.5	Every 3 h, 20 times	10	10^{-6}	>LD50[a]
0.8 μg/h	Infusion 96 h	3	10^{-8}	>LD50[a]

[a] Higher toxicity than LD_{50}.

Source: Morrison et al., 1987.

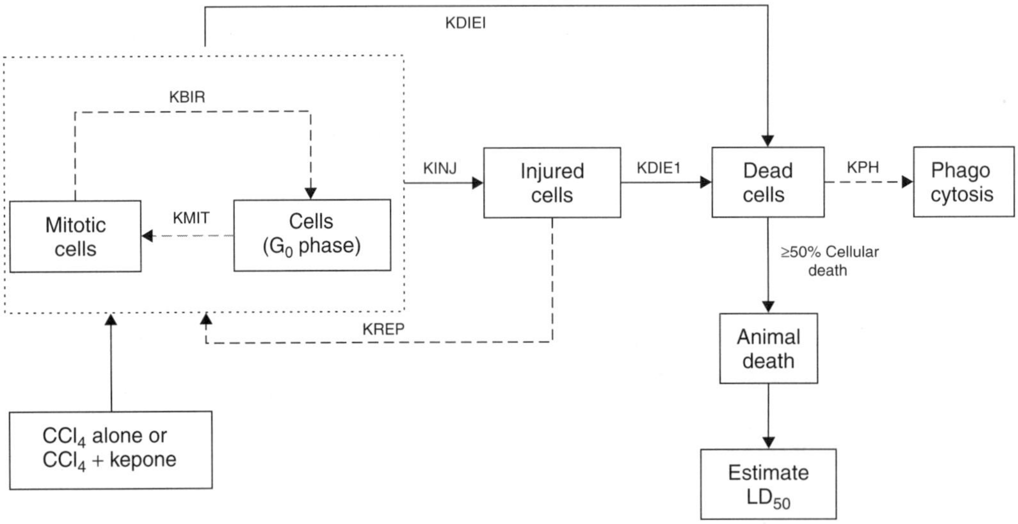

Figure 23.3. A conceptual PBPD model for CCl_4 and Kepone interaction. KMIT = rate constant for mitosis; KBIR = rate constant for cell birth; KINJ = rate constant for cell injury; KDIEI = rate constant for general cell death; KDIE1 = rate constant for cell death due to injury; KPH = rate constant for phagocytosis (El-masri et al., 1996).

University (presently University of Louisiana at Monroe); and our Quantitative and Computational Toxicology group at Colorado State University. The details of this study may be found elsewhere (El-masri et al., 1996). Briefly, CCl_4 is a well-known hepatotoxin. Following free-radical formation through the P450 enzyme system, the toxicity of CCl_4 can be an accumulation of lipids (steatosis, fatty liver) and degenerative processes leading to cell death (necrosis). Kepone is found in the environment as a result of photolytic oxidation of Mirex, a pesticide used for the control of fire ants, or as a pollutant from careless and irresponsible discharge. At relatively low levels (e.g., 10 ppm in the diet), even repeated dosing of kepone in the diet up to 15 days caused no apparent toxicity to the liver. The toxicologic interaction between kepone and CCl_4 was reported by Curtis et al. (1979). They illustrated that a 15-day dietary exposure of male rats to kepone at 10 ppm, an environmentally realistic level, markedly enhanced liver toxicity produced by an i.p. injection of a marginally toxic dose of CCl_4 (100 μL/kg). This toxicologic interaction is unique in that (1) unlike many other toxicologic interaction studies which were usually dealing with acute toxicity at very high doses, kepone in this instance was administered at a very low environmental level; (2) CCl_4 was also dosed at a low, marginally toxic level; and (3) the magnitude of toxicologic interaction, 67-fold, is very large. The mechanism of this toxicologic interaction was elucidated by Mehendale's group through a series of studies to be the impairment, by kepone, of the liver's regeneration process. These mechanistic studies were summarized in a number of publications (Mehendale, 1984, 1991, 1994).

As shown in Figure 23.3, a PBPD model was constructed by El-masri et al. (1996) based on the mechanism of toxicologic interaction between kepone and CCl_4. This PBPD model was verified by literature information, and it was capable of providing time course computer simulations of mitotic, injured, and pyknotic (dead) cells after treatment with CCl_4 alone or with kepone pretreatment. This PBPD model was further linked with Monte Carlo simulation to provide predictability of the acute lethality of CCl_4 alone and in combination with kepone. As shown in Table 23.2, the a priori predictions of lethality were in very good agreement with experimentally derived values except at very high CCl_4 levels. In this latter case, the underprediction of lethality was due to toxicity other than in the liver, that is, the neurotoxic effects of CCl_4 on the central nervous system. When this study was first presented at the International Congress of Toxicology Symposium in 1995, a reporter for *Food and Chemical News* wrote a section titled "Colorado Researchers Use Electronic Rats." Although it was somewhat amusing at the time, the term "*electronic rats*" nevertheless reflects our ultimate goal of in silico toxicology.

Clonal Growth in Relation to Carcinogenesis

The U.S. National Toxicology Program (NTP) and its predecessor, the National Cancer Institute's Carcinogenesis Bioassay Program, collectively, form the world's largest toxicology program (NTP, 2002). In its nearly 40 years of operation, fewer than 600 chemicals have been studied for carcinogenicity and other chronic toxicities

Table 23.2. PBPK/PD Modeling and Monte Carlo Simulation of Kepone/CCl_4 Toxicologic Interaction

Dose Given		Model Prediction[a]		Observed[b]	
Kepone (ppm)	CCl_4 (μL kg^{-1})	Dead Rats	% Dead	Dead Rats	% Dead
0	100	0	0.0	0	0.0
0	1000	1–2	13.2	1	11.1
0	3000	3	32.8	4	44.4
0	6000	4–5	47.8	8	88.8
10	10	0	0.0	0	0.0
10	50	4–5	47.5	4	44.4
10	100	8–9	84.0	8	88.8

[a]Mortalities in 48 h, given a hypothetical condition of $n = 9$; Monte Carlo simulation, $n = 1000$.
[b]Actual lethality studies ($n = 9$).
Source: Modified after Yang et al., *Toxicol. Lett.*, 82/82 : 497–504, 1995.

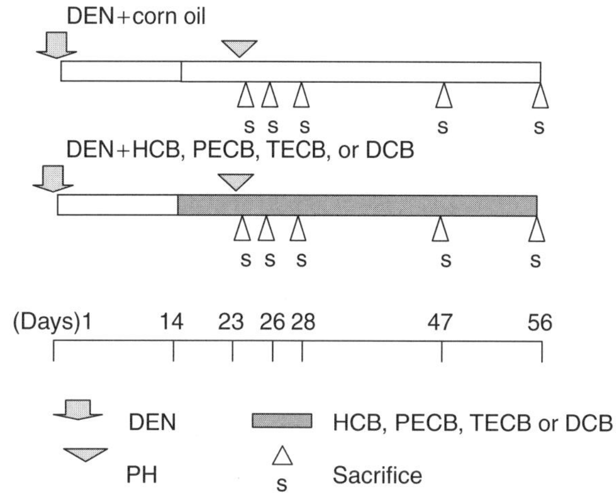

Figure 23.4. Experimental design for initiation–promotion study and estimated parameters for clonal growth model. The initiation agent, DEN, was administered i.p. (200 mg/kg) at week 0. HCB, PECB, TECB, or DCB was delivered by gavage starting week 2 at dose of 0.1 mmol/kg per day, 7 days/week. A two-thirds partial hepatectomy was performed on all animals on week 3. Liver tissues were collected 23, 26, 28, 47, and 56 days following initial DEN dosing. Concurrent controls without DEN initiation were also performed for all groups (Ou et al., 2003).

(NTP, 2002, 2003). These "gold standard" chronic toxicity/carcinogenicity studies are extremely expensive (i.e., up to several million dollars per chemical), require large number of animals (i.e., about 2000 animals per chemical), and are lengthy (i.e., 5–12 years per chemical). Thus, considering the approximately 80,000 chemicals in commerce (NTP, 2002), the number of compounds for which we have adequate toxicology information for risk assessment so far is miniscule. With the mode and rate of studying these chemicals as indicated above, it is unlikely that our society will ever have thorough toxicology information on the majority of the chemicals that we use now or may use in the future. Considering further the "real-world" issue of the health effects of chemical mixtures, it would be impossible to obtain adequate information on most of the chemicals or chemical mixtures that humans might be exposed to using the conventional approach (Yang, 1994, 1997). Thus, the PBPK/PD modeling approach described below represents a possible alternative to the expensive and time-consuming cancer bioassays.

We have used a modification of the medium-term bioassay of Ito and co-workers (1989a,b) to study the carcinogenicity of chlorobenzenes. The Ito assay involves the sequential administration of a potent initiator, diethylnitrosamine (DEN), followed by chemical treatment and mitogenic stimulation of hepatocyte growth via partial hepatectomy. As shown in Figure 23.4, this protocol allows the evaluation of carcinogenic potential within eight weeks by identification of glutathione *S*-transferase placental form (GST-P) positive preneoplastic foci as endpoint marker lesions. A large number of chemicals have been tested using this protocol. When compared with the two-year chronic bioassay, results from the Ito medium-term bioassay have correctly identified 97% of genotoxic hepatocarcinogens and 86% of the known nongenotoxic hepatococarcinogens (Ogiso et al., 1990). The principal modification of this protocol in our laboratory is the incorporation of pharmacokinetics and pharmacodynamics by conducting time course studies on tissue dosimetry, cell division rates, cell death rates, and GST-P foci formation.

To collect experimental data, briefly, as shown in Figure 23.4, male Fisher 344 rats, 8 weeks of age, were initiated with a single dose (200 mg/kg i.p.) of diethylnitrosamine. Two weeks later, animals were exposed to daily gavage consisting of 0.1 mmol/kg 1,4-dichlorobenzene (DCB), 1,2,4,5-tetrachlorobenzene (TECB), pentachlorobenzene (PECB), or hexachlorobenzene (HCB) in corn oil vehicle for 6 weeks. Partial hepatectomy was performed 3 weeks after initiation. Liver weight, 5-bromo-2′-deoxyuridine labeling index for analysis of cell division rate, and number and volume of GST-P positive foci were measured at 23, 26, 28, 47, or 56 days after DEN treatment.

We then used a clonal growth stochastic model (Ou et al., 2001, 2003; Thomas et al., 2000) to describe the dynamic growth of preneoplastic foci during the Ito medium-term bioassay. The clonal growth model is based on the two-stage model of carcinogenesis (Moolgavkar and Luebeck, 1990; Moolgavkar and Venzon, 2000), where the carcinogenesis process is described by the two critical rate-limiting steps: (a) from normal to initiated cells and (2) from initiated cells to malignant

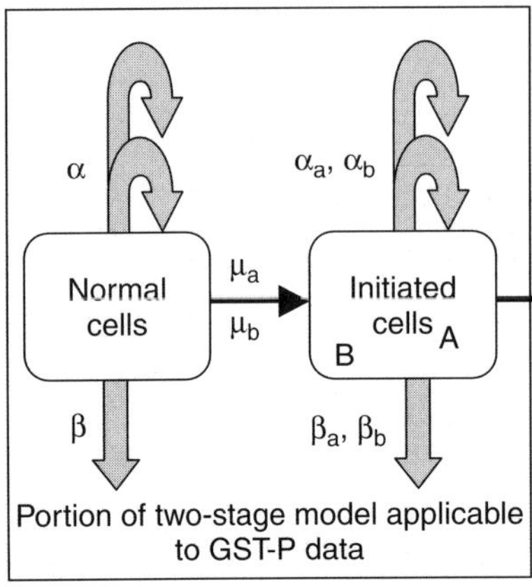

Figure 23.5. Simple two-stage model of carcinogenesis. The analysis presented here focuses on normal and initiated states (Ou et al., 2003).

states (Fig. 23.5). The model allows the incorporation of relevant biologic information such as the kinetics of tissue growth and differentiation and mutation rates. The clonal growth stochastic model adopts a discrete-time numerical approach, where the time axis is decomposed into a series of time intervals, where parameters are allowed to change between but not with segments. To represent the multiplicity of the cellular states and the time-varying nature of the numerous cell behavior variables, the numerical model resorts to a recursive simulation. The growth of normal liver is described deterministically, whereas other cellular events use stochastic simulation. This approach facilitates description of the complex biologic process with time-dependent values.

A combination of experimental toxicology and computer simulation described above was used to analyze clonal growth of GST-P enzyme-altered foci during liver carcinogenesis in an initiation–promotion regimen for DCB, TECB, PECB, and HCB. The clonal growth stochastic model (Ou et al., 2001, 2003; Thomas et al., 2000) incorporating the hypothesis of two initiated cell populations (referred to as A and B cells) (Fig. 23.5) was successfully used to describe the foci development data for four chlorobenzenes (Fig. 23.6 presented as an example for HCB and PECB). The B cells are initiated cells that display selective growth advantage under conditions that inhibit the growth of initiated A cells and normal hepatocytes. A sensitivity analysis of model parameters indicated that the net growth rate of B cells during the two-week regenerative period following partial hepatectomy is among the most sensitive parameters for determining the final foci volume. Furthermore, the estimated values of this model parameter among four chlorobenzenes appear to be positively correlated with the induction of CYP2B1/2, CYP1A2, c-*fos*, enlarged liver, and final foci volume, indicating that examining effects of chemicals on regenerative responses following partial hepatectomy may be a way to understand the carcinogenicity potential of chlorobenzene compounds. While TECB, PECB, and HCB all increased significant foci volume, only HCB had effects on normal hepatocyte proliferation. The use of a two-cell hypothesis for the description of DEN control data (with partial hepatectomy) also indicated the presence of multiple phenotypes of initiated clones following DEN treatment, with resistant phenotypes arising during early carcinogenesis.

As initiation–promotion protocols are widely used in the study of carcinogenesis, the clonal simulation of foci growth, in combination with PBPK modeling, will be a useful quantitative tool for examining the time course of a dose–concentration relationship at critical target tissues and concentration-dependent pharmacodynamic changes at cellular levels during carcinogenesis.

HOW ARE PBPK/PD MODELS VALIDATED? HOW CAN THEY BE USEFUL PREDICTIVE TOOLS?

To develop a PBPK/PD model, quantitative time course experimental data sets are essential for comparisons with computer simulations. As indicated earlier, the goal and the hope are that computer simulations are superimposable with experimental data. The data sets used for model building are working sets (or training sets). For model validation, it is very important that data sets other than the working sets are used to compare with the model simulation results. For instance, assuming a PBPK/PD model was developed using PK and PD data sets from intravenous dosing of two dose levels of drug A reported by laboratory X. The validation of this PBPK/PD model for drug A should be carried out using different data sets such as PK and PD following oral and dermal dosing of drug A reported, respectively, by laboratories Y and Z. Generally, the more different data sets used in the validation process, the more robust the PBPK/PD model.

Once a PBPK/PD model is validated, it may be used for predictive purposes. However, the present state-of-the-art is such that the predictive capability is limited to the very compound for which the PBPK/PD model is developed, or at the very least to close analogues with the same mechanism of action. For example, Kanamitsu et al. (2000), using a PBPK model, predicted in vivo drug–drug interaction between triazolam and erythromycin based on in vitro enzyme kinetic studies using human liver microsomes and recombinant human cytochrome P450 3A4 (CYP3A4). The

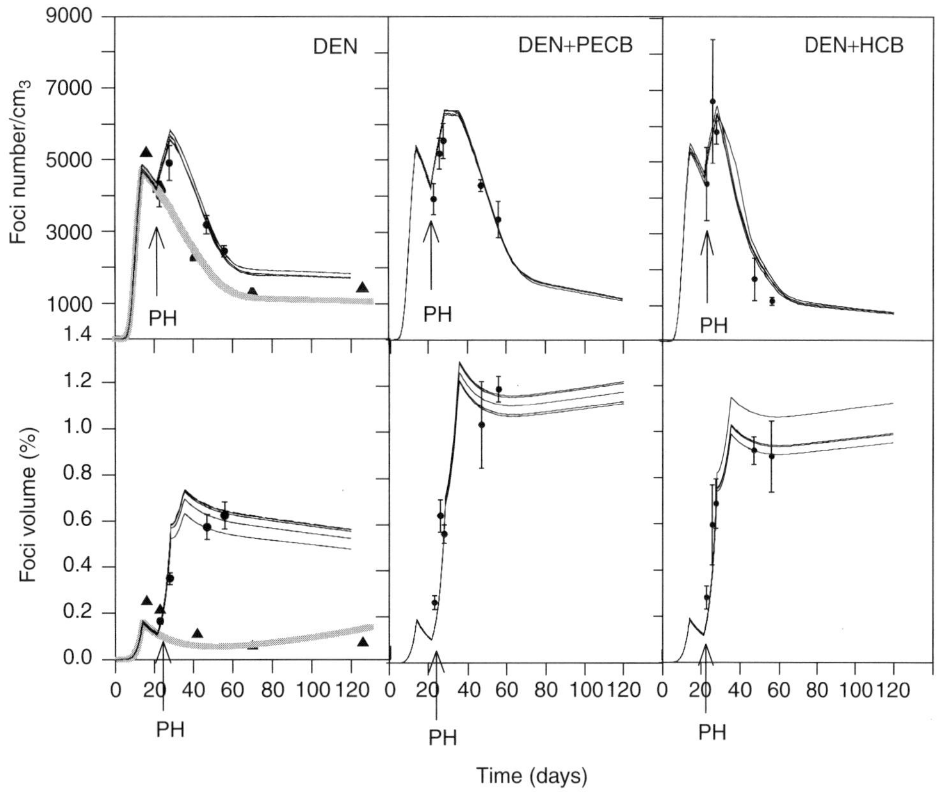

Figure 23.6. Comparison of the clonal model outputs with experimental measurements of foci growth for PECB and HCB congeners. Time-dependent changes in foci growth were measured in animals subjected to an initiation/promotion protocol using DEN as an initiator and PECB or HCB as a promoting agent. Using the standard stereologic methods, two-dimensional transection data of GST-P foci were converted to three-dimensional foci number (foci number/cm^3) and foci volume (volume percentage of foci) in the liver. To illustrate the stochastic nature of foci growth, the figures here show results from five runs of simulation. For comparison, experimental data of Jang et al. (1993) (triangle symbols) and simulation results for the DEN controls without partial hepatectomy (lower solid gray lines) are shown along with those of the DEN controls (Ou et al., 2003).

mechanistic basis for this drug–drug interaction which involved 15 fatalities in Japan in 1993 is a *mechanism-based inhibition* (or "suicide inhibition," meaning the enzyme metabolizes a chemical to a reactive metabolite which, in turn, irreversibly inactivates the enzyme) for macrolide antibiotics, such as erythromycin. Once the enzyme is permanently inactivated, its metabolism of the other coadministered drug is impaired, leading to a scenario of "drug overdose." With the present state-of-the-art, Kanamitsu et al. (2000) should be able to use their PBPK model to predict potential serious dose-limiting toxicities for combination therapies involving macrolide antibiotics and coadministered drugs metabolized by CYP3A4 provided certain basic PK parameters are known for the drugs involved as well as in vitro enzyme kinetic information using human CYP3A4. When more and more such mechanistic information is available for classes of chemicals, it is likely that reaction rules may be established for certain molecular attributes, and reaction network modeling (see the following section on second-generation PBPK/PD modeling) may be linked with PBPK/PD modeling such that more generalized prediction of toxicities might become a reality.

FUTURE PERSPECTIVES: IN SILICO TOXICOLOGY AND SECOND-GENERATION PBPK/PD MODELING

In essence, in silico toxicology means integrating computer modeling with focused, mechanistic animal experimentation such that experiments which are impractical (e.g., too large, too expensive) or impossible (e.g., human experiments with carcinogens) to perform are conducted on a computer. We believe that utilization of computer modeling is essential in the studies of toxicology of chemicals/drugs, chemical/drug mixtures, and their risk/safety assessment. The area of biology, in general, will be well served by the application of computer technology as an alternative research method to conserve resources and minimize the killing of laboratory animals. Looking into the future, we believe that the linkage of PBPK modeling and "reaction network modeling," described briefly in the following two subsections, has the potential of providing a computer simulation platform for complex biologic systems involving chemicals/drugs, chemical/drug mixtures, and/or multiple stressors. For a more detailed discussion, the readers are referred to three publications (Liao et al., 2002; Yang et al., 2004a,b) from our laboratory.

BUILDING A SECOND-GENERATION PBPK/PD MODEL

In classical pharmacokinetics and physiologically based pharmacokinetics, human or animal bodies were often described by a few compartments. By *second-generation PBPK/PD modeling*, we refer to integrating PBPK with reaction network modeling, thus including many more compartments (i.e., a delumping process). Our thoughts may best be explained in this way: If one draws a parallel between an oil refinery, where application of the reaction network modeling approach has been very successful, and a human body, the individual processing units in the oil refinery may be considered as equivalent to the vital organs of the human body. Even though the cell or organ may be much more complicated, the complex biochemical reaction networks in each organ may be modeled and linked much the same way as the modeling of the entire oil refinery through linkage of the individual processing units.

REACTION NETWORK MODELING

From the perspective of its original application in petroleum engineering, a *Reaction Network Model* is a tool for predicting the amounts of reactants, intermediates, and products as a function of time for a series of coupled chemical reactions (potentially numbering in the tens of thousands of reactions for some systems). It is usually a mathematical or symbolic formulation suitable for solution on the computer. A reaction network model builder is a tool for the computer generation of a reaction network model. The model builder can thus be used not only to solve the kinetic equations of interest but also to generate the reaction mechanisms, rate constants, and reaction equations themselves.

Essentially, the model builder works as follows:

1. The concentrations of the species to be reacted or metabolized are input to the model builder.
2. For each species in turn, the model builder performs a test against each of a set of "reaction rules" to determine whether or not the species is susceptible to a particular chemical reaction.
3. If it is not susceptible to any reactions, no further action is taken on this species.
4. If it is susceptible, a transformation of the species into one or more product species is performed based on the particular chemical reaction.
5. Each of these product species then undergoes the same susceptibility tests and a similar transformation sequence. This leads to a linking of all reactants with intermediates and, ultimately, with final products. This linking forms the structure of the chemical "reaction network."
6. After the reaction network is established, the rate constants for the reactions are retrieved or are computed.
7. The coupled differential equations governing the reaction kinetics for the network are then formulated by the model builder.
8. Finally, the kinetic equations, that is, the model equations, are solved numerically, leading to the concentrations of all species as a function of time.

More details on reaction network modeling, particularly the initial application to biomedical research, are available in a number of recent publications (Klein et al., 2002; Liao et al., 2002; Reisfeld and Yang, 2004; Yang et al., 2004a,b).

CONCLUSION

It is fitting to conclude with some recent testimonials from reputable scientists for the importance of the integration of different fields and the central role computer modeling will play in biology:

In an AAAS Plenary Lecture on February 13, 1998, Dr. Harold Varmus, then Director of the National Institutes of Health, emphasized, among others, two specific themes: "Discoveries in biology and medicine depend on progress in many fields of science" and "Methods that dramatically expand biological data also demand new modes of analysis and new ways to ask scientific questions." He said, "In short, biology is not only for biologists."

Craig Venter of the human genome fame stated, "If we hope to understand biology, instead of looking at one little protein at a time, which is not how biology works, we will need to understand the integration of thousands of proteins in a dynamically changing environment. A computer will be the biologist's number one tool." (Butler, 1999).

Tyson and colleagues (2001) indicated, "Many prominent molecular biologists have pointed out the pressing need for theoretical and computational tools to show the spatial and temporal organization implicit in the way the macromolecules are 'wired together' to create a living cell."

PBPK models have proven useful in uncovering determinants of disposition of carcinogens and other compounds in the body; PBPD cancer models have shown the role of mutation and cell proliferation in the time course of tumor development. Both of these types of models initially lacked considerable biologic detail due to limitations of our knowledge of fundamentals of cell and molecular biology. The revolution in genomic technologies in the past decade has revolutionized the database to support mechanistic modeling of chemical disposition and of biologic responses and these technologies now provide a basis for expansion of these models with increasing biologic detail. Reaction network models of cell constituents and genetic regulatory networks of cellular controls appear particularly attractive candidates for approaches to uncover the interactions and perturbations controlling neoplastic transformations and growth.

Another contemporary extension of PBPK/PD models is closely tied to current attempts to unravel the circuitry of living cells and to discover the manner in which these circuits lead to biologic function and health. Mathematical models of gene networks and their perturbation by disease or by chemical exposures are now being developed (Andersen and Barton, 1999). In some cases, simple prokaryotic cells with specific circuit elements (e.g., biologic oscillators, switches, amplifiers) have been produced and examined by laboratory experiments and by computation (Guet et al., 2002; Hasty et al., 2002; McMillen et al., 2002). These computational models evaluate the protein networks within cells, the genetic control of these networks, and the logic of cellular responses affected by these networks (Alm and Arkin, 2003; Davidson et al., 2002; Ferrell, 2002).

Gene network modeling in intact animals will inevitably draw on PBPK and reaction network models to track concentrations of endogenous and exogenous signaling compounds and on PBPD models to simulate consequences of the interactions of these compounds with signaling pathways within cells. Reverse engineering approaches attempt to uncover the circuitry of working cells (Csete and Doyle, 2002). Large-scale simulation modeling that has formed the major core of PBPK/PD models remains important in examining these genetic regulatory networks. However, these types of models are being augmented by Boolean approaches using on–off logic to increase the scope of genomic coverage (Bolouri and Davidson, 2002). Efforts in understanding genetic networks may be especially important in providing insights into neoplastic diseases in which cell signaling networks become impaired or deranged (Hahn and Weinberg, 2002; Hanahan and Weinberg, 2000).

ACKNOWLEDGMENTS

The concepts and work discussed in this presentation were partially contributed by many colleagues associated or collaborating with the Quantitative and Computational Toxicology group at Colorado State University; we are grateful to their contribution and intellectual stimulation. Any advances in science require funding support from many agencies. We thank NIEHS (Superfund Basic Research Program Project P42 ES05949; research grants RO1 ES09655 and RO3 ES10116 ZES1; training grant T32 ES 07321; and Career Development Award K25 ES11146), ATSDR (Cooperative Agreement U61/ATU881475), and U.S. Air Force (research grants F33615-91-C-0538; F49620-94-1-0304). Without the generous support of these agencies, the development of research described herein could have never been possible.

REFERENCES

Alm E, Arkin AP (2003): Biological networks. *Curr Opinion Struct Biol* 13:193–202.

Andersen ME, Barton HA (1999): Biological regulation of receptor-hormone complex concentrations in relation to dose response assessments for endocrine active compounds. *Toxicol Sci* 48:38–50.

Andersen ME, Clewell HJ, Gargas ML, Smith FA, Reitz RH (1987): Physiologically-based pharmacokinetics and the risk assessment for methylene chloride. *Toxicol Appl Pharmacol* 87:185–205.

Bischoff KB, Brown RG (1966): Drug distribution in mammals. *Chem Eng Prog Symp Series* 62:33–45.

Bolouri H, Davidson EH (2002): Modeling transcriptional regulatory networks, *BioEssays* 24:1118–1129.

Brown RP, Delp MD, Lindstedt SL, Rhomberg LR, Beliles RP (1997): Physiological parameter values for physiologically based pharmacokinetic models. *Toxicol Ind Health* 13: 407–484.

Butler D (1999): Computing 2010 from black holes to biology. *Nature* 402:C67–C70.

Csete ME, Doyle JC (2002): Reverse engineering of biological complexity. *Science* 295:1664–1669.

Curtis LR, Williams WL, Mehendale HM (1979): Potentiation of hepatotoxicity of carbon tetrachloride following preexposure to chlordecone (Kepone) in the male rat. *Toxicol Appl Pharmacol* 51:283–293.

Davidson EH, Rast JP, Oliveri P, Ransick A, Calestani C, Yuh CH, Minokawa T, Amore G, Hinman V, Arenas-Mena C, Otim O, Brown CT, Livi CB, Lee PY, Revilla R, Rust AG,

Pan Z, Schilstra MJ, Clarke PJ, Arnone MI, Rowen L, Cameron RA, McClay DR, Hood L, Bolouri H (2002): A genomic regulatory network for development. *Science* 295:1669–1678.

Dedrick RL (1973): Animal scale-up. *J Pharmacok Biopharm* 1:435–461.

El-Masri HA, Thomas RS, Sabados GR, Phillips JK, Constan AA, Benjamin SA, Andersen ME, Mehendale HM, Yang RSH (1996): Physiologically based pharmacokinetic/pharmacodynamic modeling of the toxicologic interaction between carbon tetrachloride and Kepone. *Arch Toxicol* 70:704–713.

Ferrell JE (2002): Self-perpetuating states in signal transduction: Positive feedback, double-negative feedback and bistability. *Curr Opin Chem Biol* 6:140–148.

Guet CC, Elowitz MB, Hsing W, Liebler S (2002): Combinatorial synthesis of genetic networks. *Science* 296:1466–1470.

Hahn WC, Weinberg RA (2002): Modelling the molecular circuitry of cancer. *Nature Rev Cancer* 2:331–341.

Hanahan D, Weinberg RA (2000): The hallmarks of cancer. *Cell* 100:57–70.

Hasty J, McMillen D, Collins JJ (2002): Engineered gene circuits. *Nature* 420:224–230.

Ito N, Imaida K, Hasegawa R, Tsuda H (1989a): Rapid bioassay methods for carcinogens and modifiers of hepatocarcinogenesis. *Crit Rev Toxicol* 19:385–415.

Ito N, Tatematsu M, Hasegawa R, Tsuda H (1989b): Medium-term bioassay system for detection of carcinogens and modifiers of hepatocarcinogenesis utilizing the GST-P positive liver cell focus as an endpoint marker. *Toxicol Pathol* 17:630–641.

Jang JJ, Henneman JR, Kurata Y, Uno H, Ward JM (1993): Alterations in populations of GST-p-immunoreactive single hepatocytes and hepatocellular foci after a single injection of N-nitrosodiethylamine with or without phenobarbital promotion in male F344/NCr rats. *Cancer Lett* 71:89–95.

Kanamitsu S, Ito K, Green CE, Tyson AA, Shimada M, Sugiyama Y (2000): Prediction of *in vivo* interaction between triazolam and erythromycin based on *in vitro* studies using human liver microsomes and recombinant human CYP3A4. *Pharmaceut Res* 17:419–426.

Kedderis GL, Lipscomb JC (2001): Application of *in vitro* biotransformation data and pharmacokinetic modeling to risk assessment. *Toxicol Ind Health* 17:315–321.

Klein MT, Hou G, Quann R, Wei W, Liao KH, Yang RSH, Campain JA, Mazurek M, Broadbelt LJ (2002): BioMOL: A computer-assisted biological modeling tool for complex chemical mixtures and biological processes at the molecular level. *Environ Health Perspect* 110(Suppl. 6):1025–1029.

Lazarou J, Pomeranz BH, Corey PN (1998): Incidence of adverse drug reactions in hospitalized patients. A meta-analysis of prospective studies. *JAMA* 279:1200–1205.

Liao KH, Dobrev I, Dennison Jr, JE, Andersen ME, Reisfeld B, Reardon KF, Campain JA, Wei W, Klein MT, Quann RJ, Yang RSH (2002): Application of biologically based computer modeling to simple or complex mixtures. *Environ Health Perspect* 110(Suppl. 6):957–963.

Lindstedt SL (1987): Allometry: Body size constraints in animal design, In *Pharmacokinetics in Risk Assessment. Drinking Water and Health*, Vol. 8, National Academy Press, Washington, DC, pp 65–79.

Lipscomb JC, Fisher JW, Confer PD, Byczkowski JZ (1998): *In vitro* to *in vivo* extrapolation for trichloroethylene metabolism in humans. *Toxicol Appl Pharmacol* 152:376–387.

McMillen D, Kopell N, Hasty J, Collins JJ (2002): Synchronizing genetic relaxation oscillators by intercell signaling. *Proc Nat Acad Sci* 99:679–684.

Mehendale HM (1984): Potentiation of halomethane hepatotoxicity: Chlordecone and carbon tetrachloride. *Fundam Appl Toxicol* 4:295–308.

Mehendale HM (1991): Role of hepatocellular regeneration and hepatolobular healing in the final outcome of liver injury. *Biochem Pharmacol* 42:1155–1162.

Mehendale HM (1994): Mechanism of the interactive amplification of halomethane hepatotoxicity and lethality by other chemicals. In Yang RSH (Ed), *Toxicology of Chemical Mixtures: Case Studies, Mechanisms, and Novel Approaches*, Academic, San Diego, pp 299–334.

Moolgavkar SH, Luebeck G (1990): Two-event model for carcinogenesis: Biological, mathematical, and statistical considerations. *Risk Anal* 10:323–341.

Moolgavkar SH, Venzon DJ (2000): Two-event model for carcinogenesis. *Math Biosci* 47:55–77.

Morrison PF, Dedrick RL, Lutz RJ (1987): Methotrexate: Pharmacokinetics and Assessment of Toxicity, In *Pharmacokinetics in Risk Assessment, Drinking Water and Health*, Vol. 8, National Academy Press, Washington, DC, pp 410–427.

National Toxicology Program (NTP) (2002): National Toxicology Program: Annual Plan for Fiscal Year 2001. National Toxicology Program, U.S. Department of Health and Human Services, Washington, DC.

National Toxicology Program (NTP) (2003): National Toxicology Program, Management Status Report, U.S. Department of Health and Human Services, Washington, DC, April 7.

Ogiso T, Tatematsu M, Tamano S, Hasegawa R, Ito N (1990): Correlation between medium-term liver bioassay system data and results of long-term testing in rats. *Carcinogenesis* 11:561–566.

Ou YC, Connolly RB, Thomas RS, Gustafson DL, Long ME, Dobrev I, Chubb L, Xu Y, Lapidot SA, Andersen ME, Yang RSH (2003): Stochastic simulations of liver foci development in a medium-term bioassay for four chlorobenzene congeners. *Toxicol Sci* 73:301–314.

Ou YC, Connolly RB, Thomas RS, Xu Y, Andersen ME, Chubb L, Pitot HC, Yang RSH (2001): A clonal growth model: Time-course simulations of liver foci growth following penta- or hexa-chlorobenzene treatment in a medium-term bioassay. *Cancer Res* 61:1879–1889.

Ramsey JC, Andersen ME (1984): A physiological model for the inhalation pharmacokinetics of inhaled styrene monomer in rats and humans. *Toxicol Appl Pharmacol* 73:159–175.

Reddy M, Yang RSH, Clewell III HJ, Andersen ME (2005): *Physiologically Based Pharmacokinetics: Science and Applications*. Wiley, Hoboken, NJ.

Reisfeld B, Yang RSH (2004): A reaction network model for CYP2E1-mediated metabolism of toxicant mixtures. *Environ Toxicol Pharmacol* (submitted for publication).

Thomas RS, Conolly RB, Gustafson DL, Long ME, Benjamin SA, Yang RSH (2000): A physiologically-based pharmacodynamic analysis of hepatic foci within a medium-term liver foci bioassay using pentachlorobenzene as a promoter and diethylnitrosamine as an initiator. *Toxicol Appl Pharmacol* 166:128–137.

Tyson JJ, Chen K, Novak B (2001): Network dynamics and cell physiology. *Nature Rev/Mol Cell Biol* 2:908–916.

U.S. Environmental Protection Agency (EPA) (2002): Guidance on Cumulative Risk Assessment of Pesticide Chemicals That Have a Common Mechanism of Toxicity. Office of Pesticide Programs, U.S. Environmental Protection Agency, Washington, DC 20460. (Online at www.epa.gov/oppfead1/trac/science/cumulative_guidance.pdf.)

van de Waterbeemd H, Gifford E (2003): ADMET *in silico* modeling: Towards prediction paradise. *Nature Rev* 2:192–204.

Yang RSH (1994): Introduction to the toxicology of chemical mixtures. In Yang RSH (Ed), *Toxicology of Chemical Mixtures: Case Studies, Mechanisms, and Novel Approaches*. Academic, San Diego, pp 1–10.

Yang RSH (1997): Toxicologic interactions of chemical mixtures. In Bond J (Ed), *Comprehensive Toxicology*. Vol. 1, *General Principles, Toxicokinetics, and Mechanisms of Toxicity*. Elsevier Science, Oxford, England, pp 189–203.

Yang RSH, Andersen ME (1994): Pharmacokinetics. In Hodgson E, Levi P (Eds), *Introduction to Biochemical Toxicology*, 2nd ed., Elsevier, New York, pp 49–73.

Yang RSH, Andersen ME, Clewell III HJ (2005): *Physiologically Based Pharmacokinetic/Pharmacodynamic Modeling: Principles, Methods, and Applications*. Academic, San Diego.

Yang RSH, El-Masri HA, Thomas RS, Dobrev I, Dennison Jr JE, Bae DS, Campain JA, Liao KH, Reisfeld B, Andersen ME, Mumtaz MM (2004a): Chemical mixture toxicology: From descriptive to mechanistic, and going on to *in silico* toxicology. *Environ Toxicol Pharmacol* (in press).

Yang RSH, Liao KH, Reisfeld B (2004b): The integration of computer modeling and experimental toxicology for the study of chemical mixtures and multiple stressors. *Arch Complex Environ Factors* In press.

24

TRIAL DESIGN AND BIOSTATISTICS

MICHAEL F. W. FESTING
Medical Research Council Toxicology Unit, University of Leicester, Leicester, United Kingdom

Better design and statistical analysis of experiments could save research resources and minimize ethical issues when they involve animals or humans. Each experiment should have clearly stated objectives, should avoid bias, have high power to detect treatment effects, and have a wide range of applicability but should not be so complex that mistakes are made in its execution. This can be achieved by choosing uniform material, such as healthy isogenic strains of rodents, and controlling all sources of extraneous variation using a formal experimental design. Wherever possible, sample size should be determined using power analysis calculations.

INTRODUCTION

Experiments in the life sciences are not always designed well or correctly analyzed. Altman (2002) observed that "research papers in medical journals often exhibit the use of inappropriate design and analysis, incorrect use of appropriate techniques, incorrect interpretation of results, selective reporting of results, selective citation of the literature, and drawing unjustified conclusions." He commented, "This is surely a scandal." Surveys of published papers commonly find that over half of them have obvious statistical errors, some of which may be so serious that they cast doubt on the validity of the conclusions (Altman, 1982; Festing, 1994; McCance, 1995). Where humans or animals are involved, such poor-quality work is surely unethical. As a specific example, a meta-analysis of 44 experiments on the effects of fluid resuscitation on animals found that only two said how animals had been allocated to treatments, none had sufficient power to detect reliably a halving in risk of death, there was substantial scope for bias, the results were highly heterogeneous due to the method of inducing the bleeding, and as a result the odds ratios were impossible to interpret. The authors questioned whether these experiments had any relevance to human medicine (Roberts et al., 2002).

The aim of this chapter is to review the principles of experimental design and the statistical analysis of experiments. However, this is not a textbook of statistics. The aim is only to give a broad overview of the main principles of experimental design so that research workers should be able to design and analyze simple experiments more effectively and consult a statistician more effectively with more complex situations. There are several textbooks on experimental design and related statistical analysis which can be consulted for more details, ranging from introductory texts (Festing et al., 2002; Roberts and Russo, 1999) to comprehensive monographs (Cox, 1958; Cox and Reid, 2000; Mead et al., 1993; Montgomery, 1997). Standard statistical texts tend to concentrate more on the statistical analysis rather than on the design of the experiments, but many are certainly worth consulting.

AIMS AND OBJECTIVES OF PRECLINICAL CANCER TRIALS AND PRINCIPLES OF STATISTICAL INFERENCE

Preclinical Studies

The aim of preclinical cancer research is to gain a fundamental understanding of the processes that lead to the development of cancer with a view to developing

Mouse Models of Human Cancer, edited by Eric C. Holland
ISBN 0-471-44460-X

therapies or strategies for prevention. It involves, for example, gaining an understanding of the action of carcinogens and identifying the genes associated with the initiation, promotion, and development of cancer. Such research involves a wide range of formal experiments from small one-off in vitro experiments through in vitro and in vivo screening experiments to a wide range of different types of animal experiments. Although the main principles of experimental design and statistical analysis are common to all these type of study, the details will vary depending on circumstances. For example, whereas it may be important to optimize design and sample size in an in vivo screen which may be repeated many times, it will be less so for a one-off in vitro experiment.

Study Objectives

A clear understanding of the aims of the proposed experiment is essential. These could be:

1. To test a hypothesis about the effects of some treatment, such as a potential drug.
2. To estimate some parameter such as a dose–response relationship or the mean of a character of interest for a group of subjects.

In both these cases descriptive statistics such as means, medians, and/or proportions as well as formal significance tests and confidence intervals will usually be needed.

3. To explore patterns of response to a treatment but without any clearly defined hypothesis. Many microarray studies are of this type. Such experiments will usually involve descriptive statistics, with less emphasis being placed on formal tests and they are often used to generate formal hypotheses for testing in subsequent experiments.
4. Pilot studies are sometimes used to test the logistics of a proposed experiment and to gain a rough idea of the likely outcome on which to base a definitive experiment. Pilot studies are often useful in showing that a particular approach is unlikely to be successful.

In the rest of this chapter it will be assumed that experiments are set up to test hypotheses, unless otherwise stated.

Understanding and Controlling Variation

When designing an experiment, it is vital to understand the various types of variation and the ways in which it can be controlled through selection of experimental material and in the design of the experiment. There are many sources of variation that can affect laboratory animals, see, for example, Howard (2002), and practical issues such as the availability of resources, personnel, and apparatus must not be neglected (Johnson and Besselsen, 2002).

EXPERIMENTAL UNIT

Every experiment involves a number of experimental subjects, or "units," such as cages of animals, individual animals, or petri dishes in the case of an in vitro study. In a crossover study an animal is given a treatment for a period of time followed by a different treatment, so the animal for a period of time is the experimental unit. Failure to identify the experimental unit correctly is a common error. For example, if a treatment is given in the diet or water to a cage of, say, 10 animals with another cage of 10 animals being kept as the controls, then it is the cage of animals, not the individual animals, that is the experimental unit. A statistical test comparing the means of some character based on individual animals in the two groups may show a statistically significant difference, but this could be because the social environment within the two cages differs (e.g., the animals may be fighting) and may be nothing to do with the treatment. So, where the treatment is given to a whole cage via the food or water, it is necessary to have several cages in each treatment group but with few animals per cage. However, if the animals within a cage can be individually treated, then the individual animal will be the experimental unit.

INTERINDIVIDUAL VARIATION

Inevitably, there will be individual differences between experimental units. Wells in a multiwell plate will differ depending on whether they are in the middle, on the edge, or in a corner of the plate. Animals of an outbred stock will be genetically and therefore phenotypically different. Even members of an inbred strain, though genetically identical, will have had different environments immediately following conception due to position in the uterus and postconception due to social hierarchy and interactions with microorganisms. Such interindividual variation creates "noise" which can obscure true treatment effects or be mistaken for the effect of a treatment, depending on chance. The more uniform the experimental subjects, the greater the *power* that an experiment will have to detect a true treatment effect or, alternatively, the smaller the number of subjects that will be needed to detect a treatment effect of a specified magnitude. Power and sample size are discussed below. Clearly it is most important to control those sources of variation, such as disease, genotype, or environment, which produce most noise. Experimental animals should be free of pathogenic microorganisms and of uniform age and body weight, and they should have been acclimatized to the environment for at least one and preferably two weeks, as it takes this long for the animal to adapt to a new environment.

MEASUREMENT ERROR

Measurement error will depend on the complexity of the characters being measured. For example, there will be little measurement error in weighing an animal, but it may be substantial in measuring the amount of a rare mRNA in the liver. In normal circumstance such error will be indistinguishable from interindividual variation, although the two can be separated if multiple measurements are made on each individual. In some circumstances this may be worthwhile. For example, optimizing the number of microscope fields to be counted for each individual when assessing histologic specimens could lead to useful cost savings. If measurement error is substantial, then it will be cheaper to count several fields for each individual, but if interindividual variation is large and measurement error is small, it may be cheaper to use more individuals (Cox, 1958). A detailed consideration is beyond the scope of this chapter, but where experiments are done repeatedly, these sorts of calculations may well be worthwhile.

USE OF ISOGENIC STRAINS

Scientists working in cancer research have a long tradition of using inbred strains of mice and rats. Such animals are usually phenotypically more uniform than outbred stocks, they stay genetically constant for long periods of time, there is substantial background information both on the phenotypic and genetic characteristics of each strain, and they are available internationally. There may be subline differences in the same strain from different suppliers due to new mutations, but such differences are smaller than the differences in nominally identical outbred stocks from different suppliers. Where it is important to detect genetic variation in response to a treatment, small numbers of animals of several strains can be used without increasing total sample size using a "factorial" experimental design (see below).

Inbred strains have one serious drawback: They often breed poorly. As a result, people developing genetically modified (GM) animals often cross strains to improve breeding performance through heterosis. The GM animals will then be on a genetically segregating background. However, unless the resulting colony is maintained in large numbers (e.g., 25 breeding pairs minimum) to avoid inbreeding, it will be subject to rapid genetic drift which may change the expression of the transgene over a period of a few generations. There may also be substantial phenotypic variation so that sample sizes need to be large when comparing groups. Geneticists therefore generally recommend that the transgene is backcrossed to an inbred strain. The use of genetic markers ("speed congenics") can speed up this process, which otherwise may take two to three years (Markel et al., 1997).

Toxicologists have traditionally used outbred stocks and justify this on the grounds that they are using the animals as models of humans, who are outbred. The problem is that if the outbred animals are truly phenotypically variable, then the experiments will lack statistical power and so may miss some toxic effects. Toxicologists do not use animals of widely differing body weights, for example, and so do not carry their arguments to their logical conclusion.

TEMPORAL AND SPATIAL EFFECTS

Measurements done in different laboratories, by different people, or at different times are often different. Table 24.1 shows glutathione-*S*-transferase (GST) activities in the liver of mice of four inbred strains treated with diallyl sulfide (present in garlic) using identical protocols but about two months apart. Note that in every case the measurements in the second are higher than those in the first replication. Such temporal differences are common, particularly with in vitro experiments or ones such as this one involving complex protocols for measuring the outcome. If subjects are measured at different times, it could result in so much measurement error that the treatment effects are obscured. On the other hand, it is often not practical to measure all the subjects at the same time. Accordingly, ways need to be found of splitting the experiment up into smaller bits which can be handled

Table 24.1. Liver GST Levels Following Treatment with Diallyl Sulfide (arbitrary units)

Strain	Treatment	Replicate 1	Replicate 2
1	C	444.0	764.0
2	C	423.0	586.0
3	C	408.0	609.0
4	C	447.0	606.0
1	T	614.0	831.0
2	T	625.0	782.0
3	T	856.0	1002.0
4	T	719.0	766.0

Note: C = control, T = treated. The two replicates were separated by about two months, but the protocols were identical.

conveniently, as was done in this case using a randomized block design (see below). Similar spatial differences are often seen when the same subjects are measured in different laboratories, with some laboratories consistently scoring the subjects higher than others.

FIXED-EFFECT VARIATION

So called "fixed effects" also need to be taken into account at the design stage. These are factors which can affect the outcome, but unlike temporal or spatial effects, they can be controlled by the investigator. Apart from the treatments, factors such as sex, strain, prior treatments, and husbandry in the case of animals and culture medium, cell line, and culture conditions in the case of in vitro experiments can be controlled by the investigator and will often affect the outcome.

There are basically three ways of taking such factors into account in the experimental design:

1. The factor can be deliberately varied, as is done in the case of the treatment. Thus an experiment could contain both sexes of animals or some of the culture dishes could contain one medium and others a different medium. Assuming that the treatment is also being varied, the experiment can be done most effectively using a factorial experimental design (see below) with all combination of the treatment and the other factors.
2. If a decision is taken to use, for example, only a single sex, then the conclusions may be restricted to animals of that sex.
3. Another alternative is to use one sex, for example, but assume that the same effect would have been seen in the other one also. This is a matter of scientific judgment, but there is often an implicit assumption that this is the case. For example, an experiment might be done with two treatments and the species, strain, sex, caging, bedding, diet, and prior treatment of the animals being specified. It will usually be implicitly assumed that had the animals been fed, say, on a different diet or the cage design had been different, the differences between treatment means would have been about the same.

TYPES OF DATA

Scale of Measurement

Variables can be expressed on one of four scales. The nominal scale is a descriptive scale such as male/female, dead/alive, strains A, B, C, and so on. When there are only two classes, it is called a binary scale. An ordinal scale is similarly descriptive, but order is also important. Thus small/medium/large or 0, +, + + + are ordinal scales. It would be incorrect to convert scores of this last type to 0, 1, 2, 3, ... and then obtain means and standard deviations (SDs) because the difference between 0 and + may not be the same as between + and ++. Numerical data may be expressed on an interval or a ratio scale with equal spacing between units, but in the case of the interval scale without any implication about relative magnitude. For example, the Celsius temperature scale is an interval scale because there is equal spacing of temperature between 5, 6, and 7 degrees, but it is not possible to say that a temperature of 10°C is twice as hot as 5°C because these same temperatures are 41°F and 50°F, and the latter is not twice as high as the former. However, with the ratio scale relative magnitudes hold good, so that it is possible to say that a temperature of 200 K is twice as hot as a temperature of 100 K.

Numerical Data

Numerical data can be discrete whole numbers, such as counts, or continuous, such as most measurements. In practice, when counts are high, they can usually be treated as being continuous. For example, red blood cell counts per milliliter of blood are theoretically discrete but in practice are treated as being continuous measurements. As a broad generalization, it is usually better to measure things than to count them if the counts are low.

Frequency Distributions

Numerical data can have various frequency distributions, and it is important to know something about the distribution in order to carry out the correct statistical analysis. Sets of discrete data expressed as proportions or percentages will often have a *binomial* distribution in which individual values range between 0 and 1.0 (or 0% to 100%). If most of the proportions are less than about 0.2, then the distribution will be skewed to the left, while if most are over about 0.8, the distribution will be skewed to the right. Depending on circumstances, skewed distributions may not be directly suitable to statistical analysis, although it depends on individual circumstances. In some cases it is possible to transform the data so that the distribution is no longer skewed.

Counts of the presence of some attribute which is relatively rare may have a *Poisson* distribution. A good example is cell counts from a hemocytometer where there may be counts of 0, 1, 2, 3, ... cells with a particular characteristic per square but with higher counts being quite rare. This results in a frequency distribution which is skewed to the left. Where mean counts are high (e.g., over about 8–10), then the distribution tends toward the bell-shaped *normal*, or *Gaussian*, frequency distribution.

Measurement data often have a normal frequency distribution with the typical symmetrical bell-shaped curve. There are good theoretical reasons why this should be so, and many statistical methods assume that deviations of each individual from its group mean have such a distribution. However, a log-normal distribution is also common. This often occurs when measuring the concentration of some substance in biologic samples. A concentration cannot be less than zero, and commonly there are many samples with low concentrations but just a few with high concentrations, so that the distribution is skewed to the left (i.e, has a long tail to the right). When the data are expressed on a logarithmic scale, the distribution becomes normal.

POPULATIONS AND SAMPLES

A fundamental concept in statistics is the distinction between the population and the sample. The population may be real, such as the population of humans with a particular type of cancer, or it may be hypothetical, such as all the petri dishes growing a particular cell line. The population in this case may not actually exist, although potentially could do so. It is assumed that the experimental material represents one or more samples from the population, and the aim of the experiment is to make inferences about the population. For example, the mean of a sample may differ from the mean of the population due to chance sampling variation. Its accuracy will depend on how much the population varies and on the sample size. The aim of the statistical analysis is usually to estimate something about the population from the sample but also to give some indications of the possible error in the estimates. For example, a mean may be presented together with a 95% confidence interval (CI), which is the interval within which we can be reasonably sure (i.e., 95% of the time) that the true population mean lies. With experiments comparing two treatments, the aim will often be to test whether the two samples (say, treated and control) could (with a certain probability level) have come from the same population.

EXPERIMENTS AND OBSERVATIONAL STUDIES

An experiment is an investigation where the scientist has control over the variables which are being investigated, such as the effect of some treatment. This is in contrast with observation studies where the investigator has no such control. Thus it is not possible to do an experiment to determine whether smoking causes lung cancer in humans. Any such effect must be detected by epidemiologic studies, but these have many pitfalls because it is usually extremely difficult to eliminate any possible biases or confounding effects. People who smoke may drink more alcohol or have more environmental exposure to carcinogens, or cancer susceptibility and a desire to smoke may be linked through some unknown genetic or environmental mechanism. Experiments can only be conducted in situations where it is possible to assign experimental subjects to the experimental treatments. Ideally this is done by some form of randomization. However, the statistical methods used in the analysis of experiments and observational studies are usually very similar. But with observational studies it is much more difficult to be reasonably sure that the variable under study, such as smoking, *causes* the observed response. The reason why experiments are such a powerful research tool is that they can isolate causation, rather than just association.

Randomization and Blinding

Randomization is another key statistical concept and is particularly important when designing experiments. If the subjects are not assigned to treatment groups using appropriate randomization and maintained and measured in random order, it is never possible to be sure that the results are not biased in some way. For example, if the treated group is measured at a different time from the control group, then the differences between groups may be due to some unidentified environmental factor, as is seen in Table 24.1, where the protocols were identical in different weeks but the results differed in absolute terms.

Tables of random numbers and computer programs can be used to randomize subjects to treatment groups, or it can be done physically by putting pieces of paper labeled "treatment 1," "treatment 2," and so on, into a well-shaken receptacle and drawing these out in order to assign each individual. Once the individuals have been assigned, all subsequent procedures and measurements should be done in random order. Ideally the subjects should be coded and all manipulations should be done blind with respect to the treatment groups. If the treatment is being given to animals in the diet or water, then the diet or water containers should be coded so that the investigator and staff are blind with respect to the treatment group. With randomized block and crossover designs (see below) the randomization is done within each block or subject.

Pilot Studies

Pilot studies are useful for gaining experience of the logistics of a proposed experiment, for getting some idea of likely responses and variation among individuals, or simply to see if something works. This will be useful in sample size calculations (see below). Their use is strongly recommended before starting any experiment likely to use large numbers of animals or other scientific resources. They can range in size from a single subject to larger studies involving up to 20 or 30 subjects if the proposed definitive experiment is expected to be large and expensive or if a whole series of similar experiments are planned. In most cases the results of a pilot study will not be reported, although the fact that it was done should be.

DESIGNED EXPERIMENTS

There are a small number of formal experimental designs which cover virtually all experiments likely to be done in experimental cancer research. These designs, discussed in this section, vary in the way in which temporal, spatial, and fixed-effect variability is controlled and studied, and the choice of a design depends on the objectives of the study, the type of experimental material available, and any relevant temporal, spatial, or fixed-effect factors.

The names of these designs vary to some extent between disciplines. For example, a *completely randomized design* is usually called a *between-subjects design* in psychologic research. Also, many of the designs were developed for agricultural research and have names which are more appropriate to filed trials than to biomedical research.

Choice of Treatments

Treatments may consist of qualitative factors such as drug A, drug B, and so on, or there may be several dose levels of the same drug. In this case there may be statistical advantages in using equally spaced doses. One of the treatments may be a control group receiving no drug or receiving a standard treatment with which all the other groups are to be compared. It is still considered to be one of the "treatments" even if subjects are untreated.

Some factors such as the sex and strain of animals are known as *classification variables* because they cannot be assigned to subjects at random. Thus, if an experiment is done to compare different strains of mice, it is up to the investigator to ensure that the mice are comparable in age, origin, history, and so on. If they should differ in age, for example, then any observed differences may be due to that rather than being due to genetics.

Factorial experimental designs will involve two or more factors in all combinations. Table 24.1 shows a factorial design involving four mouse strains and two dose levels of antioxidant. Such designs are usually set up to study any potentiation of one drug by the other (known statistically as an interaction) or to find out whether the response to one factor is influenced by the other one. Four strains (Table 24.1) were used in order to determine whether there were large genetically determined differences in response to the antioxidant.

Completely Randomized (Between-Subjects) Design

This is the simplest possible design in which subjects are randomly allocated to the experimental treatments without regard to any characteristics of the experimental material. It is applicable when the experimental subjects are homogeneous and there are not thought to be any important temporal or spatial factors likely to influence the outcome. There can be any number of treatment groups and any number of subjects per treatment provided there is some within-group replication. Unequal group sizes present no statistical problems. However, where there are several treatment groups, thought needs to be given to how the resulting data are to be analyzed. With several dose levels it may be better to determine dose–response relationships rather than comparing each dose level with the control. Large numbers of post hoc comparisons (see below) should be avoided. The data in Table 24.2 come from a completely randomized design in which C3H strain mice were given various doses of chloramphenicol to study its effect on hematologic parameters (Festing et al., 2001). Only the data for white blood cell counts are given. These data can be analyzed using the analysis of variance (ANOVA, see below), possibly after a suitable transformation of the raw data (also discussed below).

The main disadvantage of the completely randomized design is that, if it is large, it may be difficult to obtain sufficiently homogeneous material. It is likely to be easier to find 20 than 100 mice with the same birth date. In such circumstances it may be better to break the experiment down into smaller, more homogeneous parts using a randomized block design.

Randomized Block Design

This design allows the experiment to be broken down into smaller parts which can be run separately, usually

Table 24.2. White Blood Cell Counts in C3H/He Mice given Chloramphenicol at Selected Dose Levels (mg/kg) by Gavage for 7 Days

Mouse	White Blood Cell Count ($\times 10^9$)			
Number	0.0 mg/kg	0.5 mg/kg	1.0 mg/kg	1.5 mg/kg
1	2.2	2.1	1.0	0.6
2	1.3	1.1	1.2	0.8
3	2.0	1.3	1.4	0.7
4	1.5	1.1	1.0	0.7

Source: Festing et al., 2001.

at a different time, in the same laboratory or in different laboratories. The aim might be to increase precision by making it possible to choose more homogenous material or exploit some natural structure in the experimental subjects or to ensure that the results are repeatable. This design is common with in vitro studies where the investigator will sometimes say "the experiment was repeated X times." In effect, what they are doing is using a randomized block design to ensure repeatability. However, some investigators seem to be unaware that they have used such a design and do not know how to analyze it. Another example of a randomized block would be if substances were applied topically to the skin of an animal. It might be possible, for example, to have a number of skin patches which receive different treatments on each animal. In this case the blocking factor would be the animal. Typically, the block size is equal to the number of treatments or some whole multiple of this number. Thus with five treatments the block size will consist of 5 or possibly 10 experimental units.

A simple example of a randomized block design is shown in Table 24.3. The experiment was to determine the effect of three treatments (two drugs and a vehicle-treated control) on apoptosis in rat thymocytes in vitro. Each rat only provided enough cells for three culture dishes, and these were assigned at random to the three treatments. After a suitable period the apoptosis score was determined. This was repeated for three weeks, so the blocking factor was time. Note that, although the relationship among the treatments was the same in each block, the apoptosis score varied between blocks. All the scores in block 3, for example, were substantially lower than those in block 2. This needs to be taken into account in the statistical analysis, which is discussed below. Note that Table 24.1 also shows a randomized block design but with a more complex factorial treatment structure.

Table 24.3. Randomized Block Design: Effect of Three Treatments[a] on Apoptosis in Rat Thymocytes

Week (Block)	Treatment	Apoptosis Score
1	1	365.0
1	2	398.0
1	3	421.0
2	1	423.0
2	2	432.0
2	3	459.0
3	1	308.0
3	2	320.0
3	3	329.0

[a]Treatment 1 was a control and the other two were different drugs.

Within-Subject or Crossover Designs

Sometimes it is possible to apply different treatments to the same subject sequentially in random order. For example, different anesthetics could be tried out one at a time on each subject. As an example, Table 24.4 shows preference of A2G strain mice for various test solutions when offered in a two-bottle-choice situation. In this case the cage containing a pair of mice was the experimental unit, and the mice were offered the test solutions in random order, as shown in Table 24.4. In this example, the cage is a random factor just like the block in a randomized block design, and the statistical analysis is very similar to that for a randomized block design. This analysis is discussed in a later section.

Latin Square Designs

In the taste preference experiment it was assumed that the mice do not alter their response to the test solutions over the period of time in which they were tested. Had the outcome been, say, body weight, then on average the mice might have weighed more at the end of the test period than at the beginning. This would lead to extra "noise" or intrinsic variability which would decrease the power of the experiment. It could be overcome by using a Latin square design. As there were six treatments, then the design

Table 24.4. Within-Subject Experiment

Cage	Saccharin (0.01%)	PTC[a] (0.01%)	Caffeine (0.005%)	NaCl (0.05M)	Ethanol (10%)	Distilled Water
1	77.5 (6)	45.0 (1)	58.4 (3)	63.4 (4)	27.1 (5)	46.6 (2)
2	66.8 (5)	46.9 (2)	63.3 (6)	60.9 (4)	7.2 (3)	63.3 (1)
3	69.4 (1)	45.3 (3)	43.6 (4)	34.3 (2)	20.0 (6)	51.2 (5)
4	78.3 (4)	26.8 (2)	40.0 (1)	31.9 (3)	6.9 (5)	49.4 (6)

Note: Percent fluid of six types consumed in a two-bottle-choice test in mice. Mice of A2G strain, housed in pairs, were presented with distilled water and the test substance in different bottles for one week. Following a rest period with only tap water they were then tested with another test solution. These were given in random order (shown in parentheses). The control treatment was two bottles of distilled water with one bottle arbitrarily being designated the "test" solution.
[a]PTC = Phenylthiocarbamide.

would have required six cages of mice, and it would result in $6 \times 6 = 36$ total observations. The randomization would then have been done in such a way as to ensure that at each time period exactly one cage would have received each treatment and over the course of the experiment each cage received exactly one of each treatment. A description of the methods of randomization and methods of analysis are beyond the scope of this chapter, but in appropriate circumstances this type of design can have considerable power with the use of quite limited resources. As the number of experimental subjects required with a Latin square design is the square of the number of treatment groups, the experiments are rather small for two to four treatments and become difficult to handle with over about eight treatments. The smaller squares can be replicated if necessary, and partial squares can be used when there are many treatments.

Other Formal Experimental Designs

There are a range of other formal experimental designs which are useful in some circumstance but should probably not be used without expert advice. *Incomplete block* designs are used in situations where the experimental material has some natural structure which dictates a natural block size but the experiment requires a greater number of treatments. For example, a study might involve a paired organ such as the eyes so that the natural block size would be one animal with two eyes but the experiment might want to compare four different treatments. In this case an incomplete block design might be used so that all treatments can be compared with the same precision. *Sequential designs* are used in situations where the results of an experiment can be obtained quickly and the aim is to minimize the total number of experimental subjects used. The up-and-down method for determining the median lethal dose (LD_{50}) is an example (Rispin et al., 2002). An animal is administered what is thought to be about the LD_{50} of the toxic compound. If it dies, then the next animal is given a lower dose. Had it lived, then the next animal would have been given a higher dose. There need to be rules for choosing doses and for stopping, but this method usually uses fewer animals than the normal fixed-sample-size method.

Factorial Designs

Factorial designs, discussed briefly above, are experiments where two or more factors (e.g., treatment, strain, sex, age) are varied simultaneously, usually in all combinations. They often provide a way of getting extra information out of an experiment without using extra experimental material, so when properly used, they can be extremely valuable. Some statisticians can almost not conceive of designing an experiment that is not a factorial design. Although factorial experiments are classified here as "designs," in fact they represent ways of combining treatments rather than being true designs. For this reason it is possible to have a completely randomized factorial design, a randomized block factorial design (as in Table 24.1), or even a Latin square factorial design.

As an example, in Table 24.1 the main aim of the experiment was to study the effect of diallyl sulfide on liver GST levels, and there were eight treated and eight control mice. These could all have been of the same strain or outbred stock. However, response may be genetically determined so that mouse strains differ in their response, so it was decided to use two mice of each of four strains. The comparison of treated and control animals is still based on eight animals in each group. A statistical analysis will show whether there are treatment and/or strain differences and whether the strains respond equally. The statistical analysis of these data is discussed later.

There are several reasons for using a factorial design. First, the aim may be to see whether a response to a treatment is seen over a wide range of conditions. So some factor such as strain, sex, or prior treatment will be deliberately varied even if that factor is not of primary interest. This was the aim with the experiment presented in Table 24.1. Alternatively, the aim may be specifically to see whether two factors interact. Table 24.5, for example, shows hemoglobin levels in two strains of mice treated with chloramphenicol. In the CD-1 mice chloramphenicol had virtually no effect while in C3H/He mice levels were significantly reduced. A similar approach could be used to see whether two drugs potentiate each other by having an untreated group, one receiving drug A, one receiving drug B, and one receiving both drugs A and B. When many similar experiments are planned, such as in screening potential anticancer drugs where a standard protocol is used and only the identity of the drug varies, factorial designs can be used to optimize the experimental conditions so that the experiments

Table 24.5. Hemoglobin Levels in Control Mice and Mice Treated with Chloramphenicol at Two Dose Levels

	Hemoglobin Level		
	0 mg/kg	2000 mg/kg	Difference
CD-1	14.7	14.5	0.2
C3H/He	13.9	12.1	1.8
Difference	0.8	2.4	

Note: CD-1 mice did not respond, but C3H/He mice did. This illustrates interaction in 2×2 factorial design. There were eight mice per group and the pooled SD was 0.51.

have maximum statistical power or are of a minimum size (Shaw et al., 2002).

SAMPLE SIZE ESTIMATION

There are two recognized methods of determining sample size. *Power analysis* should be used wherever possible, but in situations where it is not possible, the *resource equation* method can often be used. Power analysis depends on a mathematical relationship between a number of variables discussed by Dell et al. (2002). The equations are quite daunting and would be tedious to work out by hand, but fortunately a number of computer programs are available, and a number of web sites offer free calculators for the simpler situations. A power analysis is almost essential for clinical trials or studies involving a large number of animals in order to avoid ethical problems. However, for one-off small biologic experiments the resource equation method may be adequate.

Power Analysis

Power analysis will be discussed first in relation to a completely randomized design where the aim is to compare the means of two groups (measurement data) using an unpaired Student's t-test, one of the most commonly used statistical tests for such situations. In order to carry out such an power analysis the investigator must do the following:

1. Decide what significance level is to be used in analyzing the experiment. This is often set at 0.05, but there is nothing magic about this number, and in some cases other values can be used.
2. Decide on whether to use a one-sided or a two-sided test. In most studies the aim of the experiment is to study the null hypothesis that there is no difference in means between treatment groups. The alternative hypothesis may be that there is a difference between group means, in which case a two-sided test will be used because it is not known whether the treatment will increase or decrease the mean. However, in some situations there are good biologic reasons why the response can only be in one direction, in which case a one-sided test should be used. By default a two-sided test should be used.
3. Estimate the SD if the character of interest is a measurement outcome. As the experiment has not yet been done, this has to come from a previous study, the literature, or a pilot study. Unfortunately small differences in the postulated SD can make a large difference to the estimates of sample size. Where the aim of the experiment is to compare two proportions, an estimate of the SD is not necessary as it is calculated from the two proportions.
4. Determine the effect size of scientific interest. This is the minimum size of the difference between the means which would be of scientific interest. For example if a compound is being tested for its ability to shrink a tumor, it may be of little interest if it only reduces tumor diameter by, say, 10% but would be of great potential interest if it reduced it by 50%. Somewhere between these two values would be a cutoff point for deciding an effect size to be used in the proposed experiment. Similar reasoning is used for comparing two or more proportions.
5. Decide on the power of the experiment. This is the probability that the experiment will successfully detect the specified effect when the other variables are as specified. Thus a power of 80% would imply that the experiment will, on average, detect the effect 80% of the time.

Once these variables have been specified, they can be entered into the computer program and an appropriate sample size will be estimated. Deciding the effect size of scientific interest may be a problem. One practical way is to judge this in relation to the SD. With a two-sample t-test an effect of about the size of one SD or more will be easy to detect with a relatively small experiment. However, if the effect size is a fraction of a SD, then a larger experiment will be required. This is illustrated in Table 24.6, which shows the sample size required in an experiment to compare two groups as a function of the effect size D expressed in SDs, assuming a power of 90%, a significance level of 0.05, and a two-sided test.

Suppose, for example, that a series of experiments are to be done to determine the effects of various compounds

Table 24.6. Required Sample Size as Function of Effect Size in Two-Sample Unpaired t-Test

Effect Size[a]	Required Sample Size per Group[b]
0.5	86.0
0.8	39.0
1.0	23.0
1.3	15.0
1.5	11.0
1.8	8.0
2.0	7.0

[a] Given in standard deviations, i.e., an effect size of 1.0 means that the effect is the same size as the standard deviation.

[b] Assuming a power of 90%, a significance level of 5% and a two-sided t-test.

on liver GST levels, based on the factorial randomized block experimental design shown in Table 24.1. These experiments should be of an appropriate size so as not to waste scientific resources. A statistical analysis of these data (shown later) gives a SD of 54 units. The level of liver GST in control mice was 536 units. From Table 24.6 it can be seen that a change from the control level of 54 units (one SD) would require 23 mice per treatment group, and a change of two SDs or 108 units would require only 7 mice per group. In this experiment the mean GST level of the treated group was actually 238 units higher than in the control group, a change of more that four SDs. This is a massive change, but it was a carefully done experiment using isogenic, specific pathogen-free mice with a randomized block design so the SD would be expected to be quite small, giving a powerful experiment. An experiment using outbred mice and a completely randomized design may have needed substantially more mice. If, on reflection, it is decided to do future experiments with eight animals in each group using all four isogenic strains, then it can be expected that the experiment would be able to detect changes in liver GST levels of about 1.75 SDs, or about 80 units.

This example is of one of the simplest situations. In practice, experiments may involve several dependent variables (outcomes), in which case it will be necessary to decide which is the most critical one and design the experiment accordingly. There may also be several treatment groups to be analyzed by the ANOVA or the experiment may involve survival or regression analysis. For such situations expert advice, more specialized publications (Dell et al., 2002), and/or dedicated software such as nQuery Advisor (www.statsol.ie, www.statsolusa.com) are recommended.

Resource Equation Method

This method can be used when a power analysis is not possible, usually because there are no estimates of the SD. It is only appropriate when dealing with measurement data to be analyzed using the ANOVA and depends on the law of diminishing returns. Increasing the size of a small experiment by one experimental unit may give good returns, but increasing the size of a larger experiment may give little extra information. The method suggests that E, the error degrees of freedom (DF) in the ANOVA (see below under Statistical Analysis) should be somewhere between 10 and 20. In the case of a completely randomized design this is

$$\text{total number of experimental units} - \text{number of treatments}$$

So for the experiment in Table 24.2, there were 18 mice with six treatment groups, so $E = 18 - 6 = 12$. This experiment would be of an acceptable size. For a randomized block experimental design the formula would be the

$$\text{total number of subjects} - \text{number of treatments} - \text{number of blocks} + 1$$

For the experiment shown in Table 24.1, there were eight treatments, two blocks, and 16 mice altogether, so $E = 16 - 8 - 2 + 1 = 7$. This implies that this is a relatively small experiment. However, the method is not ideal for randomized block designs as it implies that such designs are worse (E is lower) than completely randomized designs when in practice such designs are usually more powerful because they give better control of the variation and therefore reduce the SD. It might be appropriate just to use the formula for a completely randomized design.

STATISTICAL ANALYSIS

A golden rule is that nobody should ever start an experiment unless they know how they are going to analyze the results. If in any doubt expert advice should be obtained at the design stage. The statistical analysis of designed experiments is a function of the design. Most experiments involve measurement data, and these can usually be analyzed efficiently by the ANOVA. The two main exceptions are where the outcome is a count or a proportion, which may require some variant of the χ^2 test, or data where the assumptions underlying the ANOVA are not met, which may require nonparametric methods. These are discussed briefly below.

Data Screening

The first step in any statistical analysis is to enter the data into a computer and then screen it for any obvious anomalies or errors. This is where graphical methods which show individual data points are particularly valuable. Figure 24.1 shows a dot plot of the chloramphenicol data from Table 24.2. In this case there are no obvious outliers or other anomalies, but such plots give a good idea of the distribution of the data. Where there are several characters being measured which are correlated, it is useful to plot one against the other, as this will often reveal transcription mistakes which otherwise might not be spotted. The plots of fits versus residuals and the normal probability plots presented in the next section will also highlight any seriously anomalous points. If outliers are found, then they need to be checked to ensure that they are not just transcription errors. If not, then the next step is to check them in the laboratory notebooks, which should

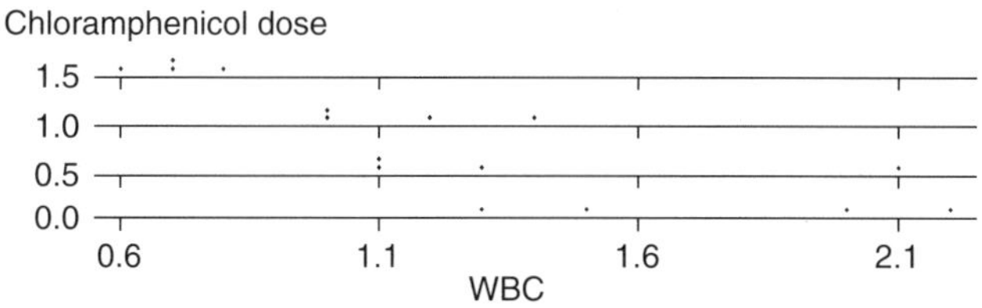

Figure 24.1. Dot plot of white blood cell counts (WBCs) for chloramphenicol data shown in Table 24.2. This type of plot and plots of one character against another (not shown here) are useful for screening the data for obvious errors and for getting a feel for the results.

have recorded any doubtful observations at the time that they were made. Outliers should never be discarded without explanation. In some cases they will disappear with a data transformation (discussed below). Alternatively, the analysis can be done both with and without the outlier to see if it makes any difference. If not, then the outlier can safely be retained. Note that in most cases outliers tend to reduce the significance of any differences among treatment means because they inflate the SD.

Nonparametric Versus Parametric Methods

Parametric statistical methods such as the *t*-test and the ANOVA depend on the assumption that the variation is the same in each treatment group and the deviations of each observation from its group mean (known as the residuals) have a normal distribution and the observations are independent, that is, they come from different experimental units. These assumptions are often met or can be met following a data transformation. Nonparametric methods do not depend on these assumptions. However, as nonparametric methods replace the actual observations by their rank order, they discard some of the data. As a result, they tend to lack statistical power so that they are unable to detect a true treatment response (known as a type II error) when it is present more often than the parametric methods. Nonparametric methods are only recommended when parametric methods are inappropriate

Need for Dedicated Statistical Software

It is inadvisable to attempt to analyze statistical data using a spreadsheet. These can sometimes give inaccurate results, they often produce nonstandard output, and they do not have the versatility of dedicated statistical programs which are readily available. Choice of software partly depends on availability. Some commercial companies specify the software to be used, and both commercial companies and academic organizations may have only a limited range of software available on their networks. Ideally training in statistical methods should be integrated with a software package as few packages are so easy to use that they can be used without formal training. Some packages such as SPSS are aimed at a particular discipline (in this case experimental psychology), others at the occasional user or the professional statistician. Widely used packages include SPSS (www.spss.com), MINITAB (www.minitab.com), SAS (www.sas.com), Statistica (www.statsoftinc.com), Graphpad (www.graphpad.com), and BMDP (www.statsol.ie). The "R" package can be downloaded free (www.r-project.org), but it is aimed mainly at professional statisticians. It has a number of associated packages aimed, for example, at analyzing data from microarray experiments. A list of software suppliers is given at www.stata.com/support/links/stat_software.html and reviews of some of the packages at www.stats.gla.ac.uk/cti/activities/reviews/alphabet.html. The analyses discussed below have been done using the MINITAB statistical package.

One-Way ANOVA for Completely Randomized Design

The ANOVA is the most widely used statistical technique for analyzing the results of designed experiments. Basically it determines whether the treatments (and any other specified factors) have increased the total variation, as would be the case if the treatment altered the observations in a treatment group, and it quantifies this variation. The text by Roberts and Russo (1999) is an excellent introduction, although it is aimed at psychology students. The results of an ANOVA are presented in a table with columns for source of variation (e.g, treatment, error, and total), DF, sums of squares (SS), mean square (MS), *F* (a test statistic), and *P* (a probability), as shown in Table 24.7 for the chloramphenicol data given in Table 24.2.

The DF indicate how many bits of information there are on the variation. Two observations are needed to obtain one estimate of variation (the difference between them); if there are three observations, then there will be two bits of information; and with N observations there will be $N - 1$ bits of information. So in Table 24.7, as there were four dose levels there are three DF for dose (treatment), and as there were 16 mice total there are 15 total DF. By subtraction that leaves 12 DF for error.

The third column headed SS is a quantification of the variation. In this case there were 5.117 units of total variation of which 2.752 units were associated with treatments and 2.365 units with error. The fourth column headed MS puts this variation on a "per-unit" basis by dividing column 3 by column 2. The column headed *F* is a test statistic obtained by dividing the MS for treatments by the MS for error. The last column is a

Table 24.7. One-Way ANOVA of Data in Table 24.2[a]: WBC Versus Dose

Analysis of variance for WBC 1					
Source	DF	SS	MS	*F*	*P*
Dose	3	2.340	0.780	6.99	0.006
Error	12	1.340	0.112		
Total	15	3.680			

Level	*N*	Mean	SD
0.0	4	1.7500	0.4203
0.5	4	1.4000	0.4761
1.0	4	1.1500	0.1915
1.5	4	0.7000	0.0816

Pooled SD = 0.3342

[a]Source indicates the source of variation (dose, error, total), DF stands for degrees of freedom, given by $N - 1$, where N is the number of objects Thus, with four dose levels there are three DF for dose. SS stands for sums of squares and gives a quantification of the various sources of variation. In this case there are 3.68 total units of variation, of which 2.34 are associated wit the dose levels. MS stands for mean square, which is the SS divided by the DF, so it can be considered as the variation per degree of freedom. F is a test statistics which is used for determining the relative magnitude of a source of variation, and P is the probability that differences among the means of the groups as great as or greater than those observed could have arisen by chance.

P value, or the probability that variation attributed to the treatments as great as or greater than that actually observed could have arisen by chance. As this probability is quite low (0.006), the treatments have probably resulted in more total variation. The critical value for P, below which the effects of the treatment are often assumed to be "significant," is usually 0.05 (5%), although this is entirely arbitrary. Now that computers give the actual P values, these should always be quoted as they give a good idea of how close the results are to this 0.05 level. An experiment which finds a difference to be significant at the 0.06 or even at the 0.10 level should not be dismissed as providing no evidence for a treatment effect. There may well be a real effect, but the experiment is too small or the material too variable to detect it reliably.

Assumptions Underlying ANOVA

There are three assumptions underlying the ANOVA:

1. The variation should be approximately the same in each treatment group.
2. The residuals (i.e., the deviations of each observation from its group mean) should have a normal distribution.
3. The observations should be independent. This depends on the correct identification of the experimental unit and appropriate randomization.

The first two of these assumptions should be routinely checked graphically. Most good statistical packages will provide *residual diagnostic plots* (another good reason for not using a spreadsheet).

A plot of fits (group means) versus residuals (deviations from group means) can be used to test the first assumption. It should show no pattern, with the points scattered in all regions. Figure 24.2 shows such a plot for the chloramphenicol data of Table 24.2. In this case there is no evidence of any clear pattern. If the variation increased with the mean, then the points would form a pattern like a V-shape on its side (e.g., <). The second assumption can be tested using a normal probability plot of the residuals. If the residuals have a normal distribution, then the points will fall on an approximately straight line, as shown in Figure 24.3. The ANOVA is said to be quite robust against departures from two of these assumptions, but it is not possible to give an exact indication of how seriously they have to be violated in order to make the tests

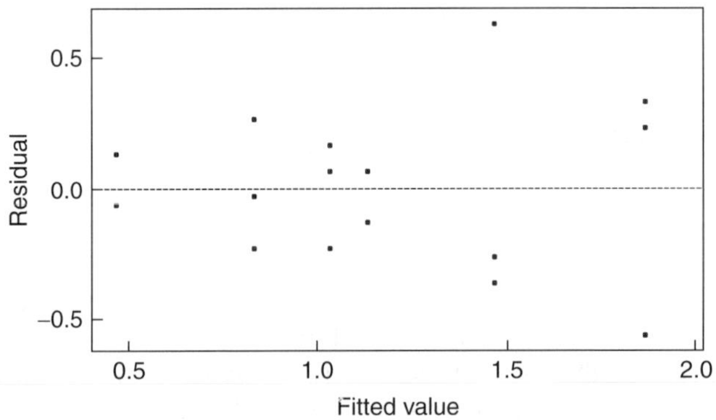

Figure 24.2. Plot of fitted versus residual values for chloramphenicol data. Absence of any pattern suggests that there is no evidence that the variation is greater for high values of fits. This type of plot and the normal probability plot in Figure 24.3 will also show any obvious outliers.

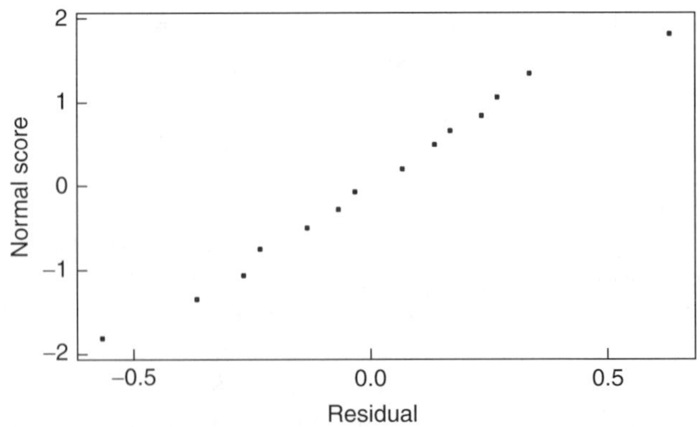

Figure 24.3. Normal probability plot of residuals from chloramphenicol data. The good approximation to a straight line shows that the residuals have an approximately normal distribution.

unreliable. Departures from the assumptions usually result in less powerful tests in which true differences are missed, so where these departures are serious either a transformation of the data or nonparametric tests should be used.

Post Hoc Comparisons

So far the ANOVA in Table 24.7 has only shown that the treatments, taken together, are probably altering group means. The next part of the analysis depends on the purpose of the experiment and should have been decided before the experiment was started. In this case the treatment is increasing doses of chloramphenicol, and the purpose of the experiment would probably be to determine a dose–response relationship. In other cases the experiment might have involved several different drugs to determine their effect on white blood cell (WBC) counts, so the aim would have been to compare each drug with every other drug. This can be done using so-called post hoc comparisons of the treatment means. There are a number of such tests for making all pairwise comparisons among means, such as Tukey's test, Scheffe's test, Fishers least significant different test, and Duncan's multiple-range test. Unfortunately, there is no general agreement on which is the best. A good discussion is given by Roberts and Russo (1999). In general, making lots of such comparisons is not a good idea. A few important comparisons should be decided on before starting the experiment, and one of the post hoc methods available on the computer package can then be used sparingly. Dunnett's test can be used when the aim is to compare several treatments with a control group.

DOSE–RESPONSE RELATIONSHIPS

It is more appropriate to analyze the data from Table 24.2 as a dose–response relationship by using regression to study the relationship between dose and the WBCs. In the ANOVA it is possible to break down the treatment SS based on three DF into three components representing a linear trend, a quadratic trend (i.e., a curve) and a cubic trend (i.e., a curve with two bends), each associated with one DF. This can be done by the method of orthogonal polynomials, details of which are beyond the scope of this chapter. However, linear regression analysis will separate out the linear trend. Most good statistical packages will produce a regression analysis which includes an ANOVA table like the one given in Table 24.8. In this case one of the treatment DF and the appropriate share of the SS associated with linear regression have been separated out and the other two DF (associated with any quadratic and cubic trend) have been incorporated into the error DF. Many good statistical packages will produce a fitted line

Table 24.8. ANOVA Associated with Regression of WBC Counts on Chloramphenicol Dose

Source	DF	SS	MS	*F*	*P*
Regression	1	2.31200	2.31200	23.6608	0.000
Error	14	1.36800	0.09771		
Total	15	3.68000			

Note: The total DF and SS are the same as for Table 24.7 and there is clearly a highly significant linear trend to the data, as shown in Fig. 24.1.

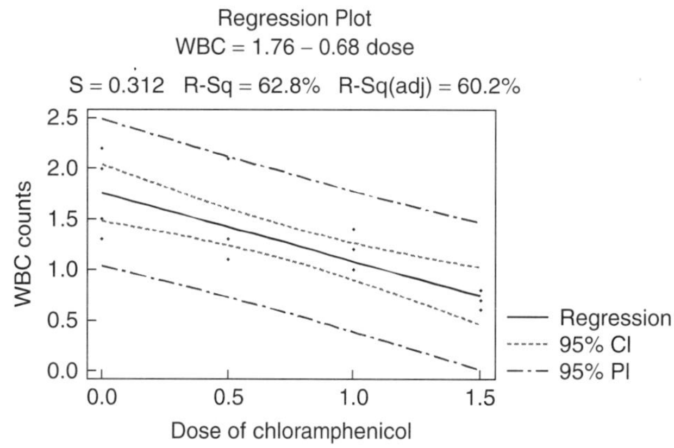

Figure 24.4. Dose–response relationship for WBC counts as a function of dose of chloramphenicol showing individual points, best-fitting straight line, 95% confidence interval for the mean of WBC counts for a given dose of chloramphenicol, and 95% prediction interval within which 95% of the points are expected to lie. The figure also shows the SD, the R^2 value, and the regression equation (see text).

plot like Figure 24.4, which shows the best fitting line, the 95% confidence interval for the mean WBC for any given dose level, and the 95% prediction interval within which 95% of the observations should fall. The graph also gives the regression equation of the form $Y = a + bX$, where a and b are constants estimated from the data and X is the independent variable. In this case the equation is WBC $= 1.76 - 0.68 \times$ dose, where the regression coefficient b is -0.68. The graph also gives the R^2 value, in this case 62.8, which is the percentage of the total variation which is accounted for by the regression relationship, and it gives the within-group SD of 0.312 units.

ANOVA for Randomized Block and Within-Subject Designs

The ANOVA for the data shown in Table 24.3 separates out variation due to both the treatment (a fixed effect) and the block (a random effect). In most statistical packages this involves having a column for the week, one for the treatment, and one for the score. The ANOVA for the data given in Table 24.3 is shown in Table 24.9. In this case

Table 24.9. ANOVA of Randomized Block Data from Table 24.3

Factor	Type	Levels	Values
Week	Random	3	1 2 3
Trt	Fixed	3	1 2 3

Analysis of variance for apop

Source	DF	SS	MS	*F*	*P*
Week	2	21764.2	10882.1	114.82	0.000
Trt	2	2129.6	1064.8	11.23	0.023
Error	4	379.1	94.8		
Total	8	24272.9			

Pooled SD = 9.74

Means for apop

Trt	Mean
1	365.3
2	383.3
3	403.0

Dunnett 95.0% simultaneous confidence intervals
Response variable apop
Comparisons with control level
Trt = 1 subtracted from:

Trt	Lower	Center	Upper
2	−8.315	18.00	44.31
3	11.352	37.67	63.98

Note: The first three lines of output show that there are two factors [week and treatment (Trt)], that week is a random effect while Trt is a fixed effect. Each has three levels, designated for this analysis as levels 1, 2, and 3. The next section is an ANOVA table for the apoptosis score (abbreviated apop) with four sources of variation (week, Trt, Error, and Total). The pooled standard deviation (SD) is given as 9.7 (this is obtained as the square root of the error mean square 94.8). The group means by treatment codes are given next, followed by Dunnett's test to compare the two treatment groups with the controls. In this case the control levels (Trt = 1) have been subtracted from the two treated group means and a 95% CI for the difference has been calculated.

Table 24.10. ANOVA for Taste Preference Data in Table 24.4

Analysis of variance for % fluid consumed

Source	DF	SS	MS	*F*	*P*
Cage	3	781.80	260.60	3.05	0.061
Trt	5	7069.66	1413.93	16.56	0.000
Error	15	1280.36	85.36		
Total	23	9131.83			

Means for obs

Trt	Mean	
1	73.00	Saccharin
2	41.00	PTC
3	51.33	Caffeine
4	47.62	Sodium chloride
5	15.30	10% Ethanol
6	52.62	Distilled water

Dunnett 95.0% simultaneous confidence intervals
Response variable obs
Comparisons with control level
Trt = 6 subtracted from:

Trt	Lower	Center	Upper
1	1.98	20.38	38.77
2	−30.02	−11.62	6.77
3	−19.70	−1.30	17.10
4	−23.40	−5.00	13.40
5	−55.72	−37.32	−18.93

Note: This ANOVA has virtually the same form as the one shown in Table 24.9. The first section showing that cage is a random effect and Trt is a fixed effect has been omitted; otherwise the explanation is the same as for Table 24.9. Note that in this case the control group was coded as group 6. A 95% CI which spans zero (e.g., with group 2) implies that the difference between the means of the control group and that group could well be zero, so any difference is not statistically significant at the 5% level. In this case the only treatments that differ from the control was saccharin, which they liked, and ethanol, which they avoided (strain C57BL/6 mice in the same experiment consumed large quantities of ethanol).

the ANOVA shows that overall there was a significant differences between the treatments ($F_{2,4} = 11.23$, $P = 0.023$). In this case, Dunnett's test has been used to compare the mean for each treatment group by subtracting the control mean from it and then calculating a 95% CI for the difference. For treatment 2 the difference was 18 units, with a 95% CI of −8.3 to 44.3. As the CI includes zero, we cannot have any confidence that the two means actually differ. However, the treatment 3 mean differs from the control by 37.7 units with a 95% CI of from 11.3 to 64.0, and as this does not span zero, we can be reasonably confident that the two means are different.

The ANOVA of the taste experiment (Table 24.4) is very similar (Table 24.10). In this case, the cage is a random effect and there were six treatments (including the control). Again Dunnett's test is appropriate for comparing each treatment mean with the control (distilled water).

Scale Transformations

When the assumptions for the ANOVA are not met, there are two alternatives: Either the scale can be transformed to produce a normal distribution of the residuals or a nonparametric method can be used (discussed later).

If the concentration of a substance is being measured, then the data will often have a log-normal distribution with a few high values but with most values being low or near to zero. In this case the data can often be normalized by taking the logarithms of the original observations. All the statistical analyses are then done on the log scale, although it may be necessary to transform the results back to the original scale for presentation. Percentages and proportions may have a skewed distribution if most of the values fall below 20% or over 80%. Such data can be transformed by a logit transformation (Altman, 1991):

$$\text{logit}(p) = \log_e \frac{p}{1-p}$$

A third commonly used transformation is when the data are counts with a relatively low mean of 2–3. In this case the data may have a Poisson distribution, which can be normalized by taking the square root of the counts. Further details of transformations are given in most statistical textbooks.

Statistical Analysis of Factorial Experimental Design

Table 24.1 shows a 2 × 4 factorial experiment with two treatments (presence or absence of diallyl sulfide, DS) and four mouse strains done as a randomized block experiment with the blocks being separated by approximately two months. The main aim of the experiment was to see whether DS altered liver GST levels (this is known as the "main effect" of the treatment). A secondary aim was to see whether there were important strain differences in response (this is an interaction effect). Blocking was used for convenience because the experiment also involved determining the activity of a number of liver and lung enzymes (not discussed here) and it was not easy to handle more than about eight subjects at a time. Blocking was also able to show whether or not the results were repeatable.

The ANOVA in Table 24.11 shows that there was a highly significant effect of the treatment with DS ($F_{1,7} = 76.9$, $P < 0.0005$), no significant strain differences ($F_{3,7} = 3.23$, $P = 0.09$), and a significant strain–treatment interaction ($F_{3,7} = 5.6$, $P = 0.028$). Inspection of the mean response of the individual strains given as differences between treated and control mice at the bottom of Table 24.11 shows that this was largely due to a greater response in strain 3 (129/Ola) than in the other three strains. Although statistically significant, it is debatable whether such a difference is of any great biologic interest. Statistical significance and biologic importance are not the same thing. However, it does mean that the magnitude of the response to DS is strain dependent and so can only be approximated by the observed difference, which in this case is 238 units (95% CI 174–303 units). When there are strong interaction effects, the size of any main effects will depend on the level of the other factors.

Table 24.11. ANOVA of Factorial Experiment Shown in Table 24.1

Analysis of Variance for GST

Source	DF	SS	MS	F	P
Block	1	124,256	124,256	42.02	0.000
Strain	3	28,613	9538	3.23	0.091
Trt	1	227,529	227,529	76.94	0.000
Strain*Trt	3	49,591	16,530	5.59	0.028
Error	7	20,701	2957		
Total	15	450,690			

Means

Strain	N	GST
1	4	663.25
2	4	604.00
3	4	718.75
4	4	634.50

Trt	N	GST
1	8	535.88
2	8	774.38

Strain	Trt	N	GST	Difference
1	1	2	604.0	
1	2	2	722.5	118.5
2	1	2	504.5	
2	2	2	703.5	199.0
3	1	2	508.5	
3	2	2	929.0	420.5
4	1	2	526.5	
4	2	2	742.5	216.0

Note: The character is liver GST levels (arbitrary units). There are six sources of variation. The block and Error terms are random effects, while Strain, treatment (Trt), and Strain*Trt are fixed effects. The means and N (number of observations) for the four strains (coded 1–4) and the two treatments (1 = control, 2 = treated with diallyl sulfide) and the Strain* Trt interaction are given below the ANOVA table. The pooled standard deviation obtained as the square root of the error MS (2957) is 54.38 units.

EXPERIMENTS WITH MULTIPLE CHARACTERS OR OUTCOMES

Some experiments involve measuring multiple characters. For example, routine hematology and blood biochemistry may result in 20 or more outcomes being measured. Often one outcome will be of main interest and the others can be treated as ancillary variables, in which case the methods discussed in this chapter would be appropriate. The ancillary variables can be used in several ways, including in the screening of the data and in developing new hypotheses to be tested in future experiments (Gaines Das, 2002). When just a few characters are being studied, it may be appropriate to analyze each one separately, but with many characters some sort of multivariate analysis may be needed. The chloramphenicol data in Table 24.1 come from a study using both approaches (Festing et al., 2001). Microarray experiments may involve several thousand outcomes, and their analysis involves specialized techniques (Nadon and Shoemaker, 2002). In such cases the main interest might be whether it is possible to distinguish between groups from the pattern of response and if so determine which characters differ most. Multivariate techniques include principal-component analysis, discriminant function analysis, factor analysis, and multivariate analysis of variance. A full discussion is beyond the scope of this chapter. Everitt and Dunn (2001) discuss all these and more techniques in detail.

NONPARAMETRIC METHODS

Parametric methods such as the t-test and the ANOVA are generally more powerful and versatile than nonparametric methods. However, there are occasions when nonparametric methods are necessary because the assumptions for parametric methods are not met and there is no suitable transformation available. These methods are not considered in any detail here, but, briefly, the Mann–Whitney test will give some indication of whether two groups are likely to be random samples from the same population. When there are two or more groups, the Kruskal–Wallace test can be used, followed by the Mann–Whitney as a post hoc comparison method. The Friedman test is the nonparametric equivalent of the randomized block ANOVA. Currently, there is no nonparametric alternative to the type of factorial ANOVA shown in Table 24.11. These nonparametric methods replace the actual observations by their ranks so the Kruskal–Wallace is equivalent to doing the usual ANOVA on the ranks of the observations. When there is concern about the normality assumptions or there are some outliers which may be influencing the results unduly and there is no nonparametric alternative to the ANOVA, it has been suggested that the ANOVA should be done on both the original data and the ranks. If the results agree, then the assumptions underlying the ANOVA are probably satisfactory. If they differ, then the ANOVA based on the ranks should probably be favored but the reasons for the differences should be explored. Other transformations might also be tried (Montgomery, 1997).

Categorical Data

This type of data is only considered briefly here as most designed experiments produce measurement data. Table 24.12 shows the results of an experiment in which mutations in transgenic mice were classified by type. The treatment with a toxic agent had already been shown to increase the number of mutations, but the investigators wanted to know whether they were qualitatively different. Approximately 50 mutants were sequenced and classified from each group. A table of this sort is known as a contingency table, and it is usually analyzed using a χ^2 test of the null hypothesis that there is no difference in the frequencies in the two treatment groups:

Table 24.12. Data Illustrating Analysis of 2 x 3 Contingency Table

Mutation Type	Control	Treated	Total
G:C to T:A	7	20	27
G:C to A:T	15	6	21
Other	31	26	57
Total	53	52	105

Note: The data are counts of mutation types in a reporter gene in control animals and those treated with a toxic agent.

$$\chi^2 = \sum \frac{(O - E)^2}{E}$$

where χ^2 is a test statistic, O is the observed and E the expected number in each cell of the table, and the χ^2 statistic has $(r - 1)$ $(c - 1)$ DF, where r and c are the number of rows and columns, respectively. The expected cell frequencies are the products of the rows and columns divided by the grand total. So, the expected number in the first cell is $(27 \times 53)/105 = 13.6$, and the contribution to the final χ^2 value is $(7 - 13.6)^2/13.6 = 3.20$. However, all the calculations are normally also easily done using a statistical package. In this case the total χ^2 value is 10.55, with two DF, giving a P value of 0.005. So the null hypothesis should be rejected, and it will be concluded that there do seem to be qualitative differences in the types of mutations in the two treatment groups.

PRESENTATION OF RESULTS

In presenting the results, the aim should be to describe each experiment briefly but clearly. Each experiment should be numbered and its design (completely randomized, randomized block, factorial, etc.) noted. The nature and numbers of the experimental subjects and any manipulations should be described. Methods of deciding sample size such as power analysis or the resource equation method and methods of assigning subjects to treatments (e.g., randomization) and whether blinding was used should be stated. When animals are used, this should be ethically justified and methods used to alleviate suffering such as anaesthesia, analgesia, and humane endpoints should be mentioned. The strain (using correct genetic nomenclature), sex, age, source, health status, diet, and brief details of their husbandry should be given. Any losses during the experiment should be noted and the numbers used should be clearly stated. Graphs showing individual observations are preferred to ones just showing means and standard errors (SEs). The latter should never be shown without also indicating the numbers in each group. Statistical significance should never be quoted without also giving an indication of the size of any response. Means and other sample parameters such as regression coefficients should always have some estimate of their variation such as a SD, SE, CI. Current best practice is to express means and SD, SE, or CI in the form 10.3 (SD 2.1) rather than using the plus or minus symbol, which does not indicate whether it is a SD, SE, or CI being quoted (Festing and Altman, 2002). Exact *P* values should be quoted unless they are vanishingly small. Do not place undue emphasis on the 5% level of significance. A well-designed experiment which gives negative or nonsignificant results should not be rejected by a journal editor. Such studies are often useful, and rejecting them can lead to publication bias.

REFERENCES

Altman DG (1982): Statistics in medical journals. *Stat Med* 1:59–71.

Altman DG (1991): *Practical Statistics for Medical Research*, Chapman and Hall. London.

Altman DG (2002): Poor quality medical research: What can journals do? *JAMA* 287:2765–2767.

Cox DR (1958): *Planning Experiments*, Wiley, New York.

Cox DR, Reid N (2000): *The Theory of the Design of Experiments*, Chapman and Hall/CRC, Boca Raton, FL.

Dell R, Holleran S, Ramakrishnan R (2002): Sample size determination. *ILAR J* 43:207–213.

Everitt BS, Dunn G (2001): *Applied Multivariate Data Analysis*, 2nd ed. Arnold, London.

Festing MFW (1994): Reduction of animal use: Experimental design and quality of experiments. *Lab Anim* 28:212–221.

Festing MFW, Altman DG (2002): Guidelines for the design and statistical analysis of experiments using laboratory animals. *ILAR J* 43:233–243.

Festing MFW, Diamanti P, Turton JA (2001): Strain differences in haematological response to chloramphenicol succinate in mice: Implications for toxicological research. *Food Chem Toxicol* 39:375–383.

Festing MFW, Overend P, Gaines Das R, Cortina Borja M, Berdoy M (2002): *The Design of Animal Experiments*, Laboratory Animals Ltd. London.

Gaines Das R (2002): The role of ancillary variables in the design, analysis and interpretation of animal experiments. *ILAR J* 43:214–222.

Howard BR (2002): The control of variability. *ILAR J* 43: 194–201.

Johnson PD, Besselsen DG (2002): Practical aspects to experimental design in animal research. *ILAR J* 43:202–206.

Markel P, Shu P, Ebeling C, Carlson GA, Nagle DL, Smutko JS, Moore KJ (1997): Theoretical and empirical issues for marker-assisted breeding of congenic mouse strains. *Nature Genet.* 17:280–284.

McCance I (1995): Assessment of statistical procedures used in papers in the *Australian Veterinary Journal*. *Aust Vet J* 72:322–328.

Mead R, Curnow RN, Hasted AM (1993): *Statistical Methods in Agriculture and Experimental Biology*, 3rd ed, Chapman and Hall. London.

Montgomery DC (1997): *Design and Analysis of Experiments*, 4th ed. Wiley, New York.

Nadon R, Shoemaker J (2002): Statistical issues with microarrays: Processing and analysis. *Trends Genet* 18:265–271.

Rispin A, Farrar D, Margosches E, Gupta K, Stizel K, Carr G, Greene M, Meyer W, McCall D (2002): Alternative methods for the LD50 test: The up and down procedure for acute toxicity. *ILAR J* 43:233–243.

Roberts I, Kwan I, Evans P, Haig S (2002): Does animal experimentation inform human healthcare? Observations from a systematic review of international animal experiments on fluid resuscitation. *BMJ* 324:474–476.

Roberts MJ, Russo R (1999): *A Student's Guide to the Analysis of Variance*. Routledge, London.

Shaw R, Festing MFW, Peers I, Furlong L (2002): The use of factorial designs to optimise animal experiments and reduce animal use. *ILAR J* 43:223–232.

25

CANCER DRUG DEVELOPMENT IN THE MODERN ERA

WILLIAM R. SELLERS

Department of Medical Oncology, Dana-Farber Cancer Institute, and Department of Medicine, Brigham and Women's Hospital, Harvard Medical School, Boston, Massachusetts

ALEX MATTER

Novartis Institute for Tropical Diseases, Novartis International, Basel, Switzerland

For several decades the development of anticancer therapeutics has focused largely on the identification of compounds capable of interfering with the process of cell division. The completed human genome sequence, the use of comprehensive genetic studies in lower organisms, and the robust definition of signaling, death, motility, metabolic, hypoxia, DNA repair, stress response, and other pathways have enabled and are reinvigorating both the selection of rational targets and the application of small-molecule drug design to the problem of cancer therapeutic development.

INTRODUCTION

For several decades the development of anticancer therapeutics has focused largely on the identification of compounds capable of interfering with the process of cell division. Often referred to as *cytotoxic agents*, these drugs, including DNA-crosslinking agents (e.g., cyclophosphamide), DNA-intercalating agents (e.g., adriamycin), topoisomerase inhibitors (e.g., etoposide), antimetabolites (e.g., 6-mercaptopurine), and microtubule perturbants (e.g., taxanes), were designed to exploit perceived differences in the rate of cellular division between cancer and normal cells. A number of such cytotoxic compounds have achieved a degree of clinical success, but *in toto* these agents have had little impact on the majority of adult epithelial tumors. Indeed, even among cancers in which such agents often elicit substantial tumor responses (regressions), including adult acute myelogenous leukemia (AML) and non-Hodgkin's lymphoma, the majority of patients are not cured (Ries et al., 2003; Stone, 2002). The lack of widespread success for this paradigm may not be surprising given that the difference in cellular proliferation is only one of many attributes attained by cancer cells, and in many cases the proliferative differences are minor if they even exist.

The lack of progress in the arena of cytotoxic drug development has led to substantial investment in developing novel nonmedicinal anticancer approaches. For example, efforts are ongoing to develop anticancer vaccines [e.g., granulocyte–macrophage colony-stimulating factor (GM-CSF) transduced autologous tumor vaccines (Dranoff, 2003)], antibody-based therapeutics (e.g., Herceptin and Rituximab) (Davis et al., 1999; Maloney et al., 1994; Slamon et al., 2001), biologic agents (e.g., interferon-gamma), and nucleic acid–based therapeutics (e.g., antisense oligonucleotides directed against bcl-2 (Webb et al., 1997). Here, there has clearly been encouraging progress, and in particular antibody-based therapeutics have had notable success. Nonetheless, one must take note of the fact that the successful treatment of most medical illnesses, including cardiovascular and infectious diseases, has been based largely upon the targeted and rational design of small-molecule inhibitors. Thus, in the pharmaceutical industry there is a robust platform for the elaboration of highly effective, safe, and orally available drugs.

Mouse Models of Human Cancer, edited by Eric C. Holland
ISBN 0-471-44460-X

The completed human genome sequence, the use of comprehensive genetic studies in lower organisms, and the robust definition of signaling, death, motility, metabolic, hypoxia, DNA repair, stress response, and other pathways have enabled and are reinvigorating both the selection of rational targets and the application of small-molecule drug design to the problem of cancer therapeutic development. The description of this process will be the central focus of this chapter.

This new-found transition away from empiric selection and testing of cytotoxic agents to rational target selection, lead discovery, and lead optimization will necessitate changes in clinical testing of such targeted therapeutics. Specifically, clinical drug development (clinical trials) will require a shift away from broad organ-specific clinical trials to those focused on specific tumor subsets thought to be highly target "dependent." Thus, the current and emerging paradigms for clinical testing of oncologic therapeutics will be discussed briefly.

PROPERTIES OF A CANCER CELL

Cancer as Hyperproliferative Disease

The simplistic notion that cancer results from the uncontrolled and excessive proliferation of cells has gradually given way to a more sophisticated view of the process by which clonal population of cells gain the characteristics associated with cancer (Hanahan and Weinberg, 2000). Indeed, the "Petri dish" view of cancer cells growing in the incubator belies the complexity of the tumors that arise in humans. Human cancers are a heterogeneous collection of diseases that range from rapidly proliferating, highly malignant tumors such as Burkitt's lymphoma to the slowly growing but nonetheless lethal low-grade sarcomas that remain locally invasive and never metastasize.

Cancer as Stem Cell Disorder

While a single model is unlikely to fully account for the above-mentioned cancer heterogeneity, there are several common unifying notions that likely encompass the diversity of human cancers. First is the notion that cancers are diseases that arise from precursor stem cells. The relationship between stem cells and cancer is best demonstrated for leukemia where the subtypes of leukemia (e.g., myeloid vs. lymphoid acute lymphoblastic leukemia, or B-ALL vs. T-ALL) are precisely aligned with the known stem cell hierarchy of the hematopoietic cell lineages. While in solid tumors the stem cell nature of cancer has yet to be fully evident, presumably some form of a stem cell hierarchy will emerge with the elaboration of improved stem cell markers.

In this regard it is clear that most tumors remain to some extent beholden to the tissue or tissue stem cell from whence they originate. For example, among the sarcomas there are specific tumors that originate from nearly every mesenchymal tissue, including rhabdomyosarcomas from muscle and liposarcomas from fat. Thus tumors retain many of the features of the original tissue. This distinction is manifest not only by the retained histologic features but also by the retention of specific markers that the cancer holds in common with the normal tissue of origin. Indeed, the retention of these tissue-specific markers can be used to classify tumors based on mRNA profiles alone (Ramaswamy et al., 2001; Su et al., 2001) and can be used clinically when such a marker becomes excessively abundant in the body, as is the case for the prostate-specific antigen (PSA) used in detecting prostate cancer.

While tumors retain properties of the starting tissue, the degree to which they preserve such properties, that is, the extent to which differentiation is preserved, varies widely. For example, among the common adenocarcinomas such as colon cancer, the state of differentiation ranges from undifferentiated to well differentiated. This link between cancer and the state of differentiation is more than of basic biologic interest. Across many cancer types patient prognosis is strongly linked to the histologic grade, or state of tumor differentiation, and thus this information is used to stratify the delivery of existing therapeutics. Moreover, there are at least two therapeutic approaches that could derive from this model.

The first is that tumors often retain some of the same molecular dependencies found in the tissue of origin. Androgen receptor (AR), a notable example, is required for survival and growth of both normal and cancerous prostate epithelial cells. Taking advantage of the continued dependency of prostate cancer on AR led Huggins to discover the role of androgen withdrawal therapy in the treatment of patients with prostate cancer (Huggins and Hodges, 1972). Similarly, the utility of antiestrogen therapies such as tamoxifen in breast cancer relies on preservation of estrogen receptor (ER) function in breast cancer epithelial cells.

The second therapeutic approach linked to a stem cell model of cancer is based on the finding that cancers acquire specific genetic alterations that block further differentiation. This is most notably the case in murine models of leukemia, where induction of inappropriate kinase activation within the hematopoietic compartment results in a proliferative but nonmalignant state unless accompanied by a specific block to differentiation (Gilliland, 1998). In adult leukemia, the elucidation of this paradigm in acute promyelocytic leukemia has led to the clinical development of all-*trans* retinoic acid (ATRA), which specifically targets the function of a chimeric protein containing the PML protein and the retinoic acid receptor (RAR) produced by a translocation between chromosomes

15 and 17. Targeting the so-called PML-RAR fusion protein results in differentiation of the clonogenic tumor cell and a marked therapeutic response (Huang et al., 1987; Warrell et al., 1991).

An improved understanding of both the stem cell of origin for each cancer and genetic alterations that alter or subvert the differentiation pathway of that stem cell will likely lead to additional target discoveries worthy of consideration in the drug development pipeline.

Cancer as Genetic Disorder

The second and more widely recognized commonality among cancer states is the acquisition of genetic alterations either as rare hereditable variations present in the constitutional DNA or, more often, as somatically acquired genetic changes, including point mutations, deletions, insertions, translocations, and amplifications. In general, such alterations result in unregulated gains in the function of the gene products of oncogenes or in the loss of the function of the gene products of tumor suppressor genes.

Genetic alterations conspire to endow cancer cells with properties such as growth factor independence, immunity to cell death signals, failure to differentiate (as described above), induction of angiogenesis, invasiveness, and metastasis. More importantly, it is likely that the acquisition of one set of genetic alterations requires the coacquisition of a second set. Such genetic codependency has been best illustrated in model systems where, as an example, loss of function of the retinoblastoma protein (pRB) induces apoptosis in the presence of wild-type p53; however, coinactivation of pRB and p53 results in tumorigenesis (reviewed in Sellers and Kaelin, 1997). The acquisition of multiple complementary and codependent mutational events may result in cancer cells that are unable to shift or respond to specific chemical stresses in the same way a normal cell might. It is likely that these phenomena will lead to a substantial therapeutic window if exploited with suitable pharmacologic agents.

That cancer cells do indeed become "wired" to or dependent upon specific genetic alterations has been remarkably demonstrated over the past few years by the targeting of oncogenic proteins whose function has been altered by either a translocation or an activating mutation. The first example, as stated above, was the elucidation of the therapeutic utility of ATRA in patients with acute promyelocytic leukemia.

More recently, the success of a kinase inhibitor (imatinib, Gleevec) has dramatically highlighted the dependence of chronic myelogenous leukemia (CML) on the fusion protein BCR-ABL that leads to inappropriate and constitutive activation of the Abelson tyrosine kinase (Ben-Neriah et al., 1986; Konopka et al., 1984). Imatinib treatment of patients with this genetically homogeneous disease has led to dramatic and sustained responses, particularly in the so-called stable phase of the disease (Druker et al., 2001). Following variable periods of treatment, a number of responding patients have subsequently progressed. In many cases, sequence analysis of the resistant tumor cells has revealed the gain of new mutations within the ABL kinase, now rendering such cells resistant to imatinib (Gorre et al., 2001). These data strongly suggest that continued activation of the ABL kinase is required for the tumorigenic phenotype of CML cells.

LESSONS LEARNED FROM CYTOTOXIC CHEMOTHERAPY

While the empiric discovery of effective chemotherapeutics has not been a tremendously successful enterprise, cytotoxic agents have nonetheless brought cures to a majority of patients afflicted by a number of diseases, including pediatric leukemia, testicular cancer, and subtypes of non-Hodgkin's lymphoma. From these examples there are certain principles that may continue to apply in the current era of drug discovery (Table 1).

With few exceptions the elaboration of curative therapy first requires the discovery of highly active individual drugs that in the same disease can each produce complete remissions, although typically do not result in high cure rates. These agents become the foundation or basis for the development of curative combination therapies. The treatment of testicular cancer, once a universally fatal scourge when metastatic, was revolutionized by the discovery of the DNA-crosslinking activity of cisplatin (Roberts and Pascoe, 1972). This alkylating agent has remarkable activity in this disease; nevertheless, it is the combination of cisplatin with bleomycin and etoposide that is effective in eliciting high-percentage cures (reviewed in Einhorn, 2002).

A major difference between the treatment of oncologic diseases and other medical illnesses is the extent to which significant toxicities can be tolerated if a chemotherapeutic agent has a significant therapeutic effect. Agents routinely administered to cancer patients are associated with hair loss, gastrointestinal side effects, and profound bone marrow suppression. While such side effects are clearly

Table 25.1. Principles Learned from Chemotherapy

Highly active single agents with a favorable pharmacologic profile are required.
Combinations are required for curative therapy.
Toxicity is tolerable and can be ameliorated.
Schedule matters.
Suprapharmacologic dosing does not help.

not desirable, therapeutic regimens that provide significant patient benefit are not necessarily excluded by such side effects from successful drug development. Major improvements in antiemetics [in particular H3-receptor antagonists (Marty, 1990)] and in ameliorating marrow suppression through the use of hematopoietic growth factor support, including erythropoietin and G-CSF, have significantly reduced the untoward consequences of the classical cytotoxic chemotherapeutic agents.

Finally, in oncology suprapharmacologic dosing of cytotoxic agents was long thought to be a way to overcome tumor resistance to such agents. Randomized trials comparing "dose-intensive" regimens to standard regimens in intermediate-grade lymphomas (Fisher et al., 1993) and standard- to high-dose regimens in breast cancer have dramatically failed to validate this model (Hortobagyi et al., 2000; Schrama et al., 2002; Stadtmauer et al., 2000). These findings are of significance as we move to testing a generation of targeted therapeutics. Here, it will be critical to determine whether the activity of the target has been successfully ablated at any given dose of a targeted agent and to use this as the endpoint for dose escalation.

DRUG DEVELOPMENT: THE PROCESS

In the past, whole animal-based or cell-based screens designed to survey libraries for compounds with antigrowth or prodeath attributes have been used with very little success in finding truly innovative compounds. This was due to the fact that the endpoints measured were relatively crude and the real drug target was hidden in a "black box."

These empiric efforts have been in large part replaced by drug development focused on specific target molecules. The vast majority of targets under attack today are cellular proteins. In addition, most drug development efforts are focused on identifying selective low-molecular-weight compounds (at most 1000 daltons, usually less than 500 daltons) that typically can be developed into molecules with highly desirable druglike properties. As a result, these relatively small-sized molecules typically act as catalytic site–directed inhibitors or as ligand-displacing agents at the level of receptors.

The focus on a specific protein target allows for rationally designed chemical alterations leading to improvements in the affinity of lead and secondary compounds for the target molecule while at the same time improving the "druglike" qualities of a molecule such as oral bioavailability, protein binding, and solubility. These optimizations are either often based on reiterative in vitro screening for inhibition of target activity or can be based on structural information. Indeed, cocrystallization of the target protein with small-molecule inhibitors often is developed in tandem with medicinal chemistry efforts.

The successful development of small-molecule inhibitors is likely linked to the "drugability" of the selected target. In practice, this translates into protein surfaces that form hydrophobic pockets, such as those found in catalytic sites of molecules with enzymatic activity. While this is the current trend, the future of medicinal chemistry may allow us to envision pushing drug development into the realm of molecules that act via allosteric sites or, even more difficult, via disruption of protein–protein interactions that are notoriously "hard targets." This may allow us to greatly expand the number of drugable targets.

The components of the drug development process (Fig. 25.1) will be discussed.

TARGET SELECTION: DISCOVERY AND VALIDATION

As a result of our incomplete understanding of the process that gives rise to cancer, target selection remains an imprecise science. Ultimately one must satisfy the condition that inhibition of the target leads to cytotoxic or cytostatic effects on tumor cells in excess of the potential "side effects" of inhibiting the target in normal cells.

Targets have been selected based upon their role in the cellular processes thought to have gone awry in cancer cells, including cell cycle regulation, DNA replication, DNA synthesis, cell death regulation, angiogenesis, cytokinesis, and others. Illustrative of the difficulty in predicting where therapeutic windows might be found is the recent finding that inhibition of the proteasome appears to have differential effects on certain tumor types (e.g., myeloma) as compared to normal tissues (Orlowski et al., 2002).

CANCER GENOMICS

More recently, there has been an explosion in the application of genomic technologies to satisfying the goal of systematic target discovery. Such technologies, including differential display, subtractive hybridization, and array-based RNA expression analysis, have initially focused on elucidating the mRNA differences between cancer and normal tissues, between different stages of cancer, and between cancers having widely disparate clinical outcomes (e.g., survival vs. cure) (Golub et al., 1999; Perou et al., 2000; Rosenwald et al., 2002; Singh et al., 2002; van de Vijver et al., 2002). Such expression-based efforts have had success in the elucidation of specific subclasses of cancers, yet it has been difficult to hone in on novel therapeutic targets emerging from these analyses. On the other hand, it is still early in the deployment of such technologies, and the recent identification from expression-profiling data of Flt-3 as a therapeutic target in mixed-lineage leukemia may augur a wave of such discoveries (Armstrong et al., 2003).

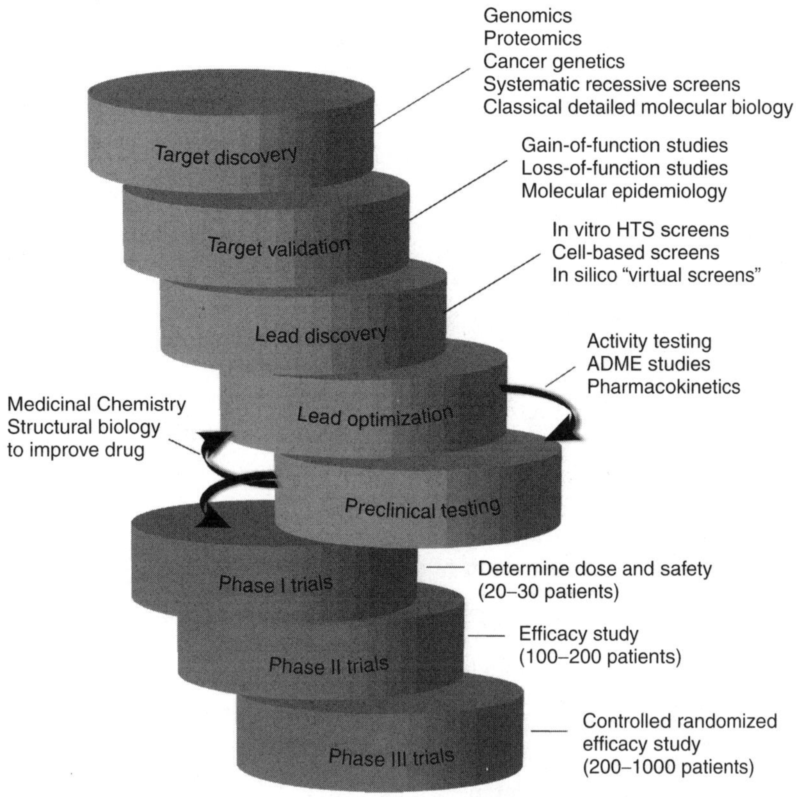

Figure 25.1. Components of drug development.

More recently, similar efforts are being made to define the global and specific protein differences that constitute the cancer cell. Here, it is thought that directly measuring the abundance or state of phosphorylation (as an example) of proteins is more likely to directly reflect an alteration that has direct functional consequence to the cell. It is safe to say that proteomic approaches are still undergoing technological improvements and have been insufficiently tested in the mode of target discovery.

CANCER GENETICS

While expression abundance and protein levels might ultimately lead one to understand the functional dependencies within cancer cells, the study of cancer genetics has led to the discovery of genes whose functional relationship to the process of tumorigenesis is not in doubt. Loss of function of tumor suppressor genes such as *TP53* and *RB1* and gain of function in the oncogenes such as *RAS* conspire to create the cancer state, and in many model systems these events are sufficient to drive cancer progression. For many years, however, few if any novel therapeutics had been developed based upon the elucidation of these or other specific cancer genetic pathways.

The successful targeting of the oncoprotein BCR-ABL and of the mutant and therefore oncogenic forms of c-Kit and PDGF-R by the tyrosine kinase inhibitor imatinib (Gleevec^{TM}) has restimulated interest in the use of cancer genetics to discover drug targets (Demetri et al., 2002; Druker et al., 2001; Heinrich et al., 2003; Tuveson et al., 2001). In particular, a number of new small-molecule inhibitors targeted against kinases that are activated by tumor-derived mutations are currently in development. Somatic mutations that activate the tyrosine kinase RET, FGFR3, and FLT3 have been discovered in cancers of the thyroid and bladder as well as in myeloma and AML (Cappellen et al., 1999; Chesi et al., 1997; Hofstra et al., 1994; Nakao et al., 1996). Moreover, in extending and broadening the BCR-ABL/imatinib paradigm, inhibitors of FLT3 have shown promise in early-stage clinical trials that are ongoing in AML patients whose cancer harbors a sequence-verified FLT3 mutant kinase (Smith et al., 2003; Stone et al., 2003).

These examples have spurred systematic efforts to undercover the complete spectrum of activating events both in human genes in general and in the kinase family of genes in specific (Bardelli et al., 2003; Davies et al., 2002; Strausberg et al., 2003). These studies have yielded the remarkable finding of high-frequency mutations in the serine–threonine kinase B-raf in melanoma and to a lesser extent in colon cancer (Davies et al., 2002). This discovery along with the availability of Raf kinase inhibitors, already in clinical development at the time of the initial findings, has led to the rapid testing of Raf inhibitors in the patients.

RECESSIVE GENETICS AND TARGET VALIDATION

Target selection has been fundamentally blocked on two fronts. The first has been the inability to rapidly test the effects of loss of function of specific proteins (the typical consequence of a drug) on a phenotype relevant to human cancers. The second has been the lack of model systems that accurately reflect the genetics, the tumor–host interactions, and the stem cell nature of human cancer. These limitations have been, at least in part, the driving force behind the development of genetically engineered mouse models of human cancer. These limitations have also motivated attempts to use chemical libraries to probe cellular phenotypes. The goal of this chemical genetic approach is to use relatively crude chemical inhibitors to generate loss-of-function phenotypes typically scored in high-throughput assays. Small molecules scoring in such assays might then be used to "fish" out the relevant target molecules, or alternatively, such assays could be used directly in further development of initial lead compounds. This approach was used in a phenotypic screen relying on the generation of astral spindles to identify monasterol as a novel inhibitor of the eg5 mitotic kinesin (Mayer et al., 1999).

The development of siRNA-mediated gene knockdown has added a potent new recessive genetic technology to the process of target discovery and validation. The recent success in applying this technology in mammalian cells grown in culture and in mammalian tissues treated in vivo represents a fundamental step forward in our ability to both functionally validate proposed targets and discover new targets (Brummelkamp et al., 2002; Elbashir et al., 2001; Paddison et al., 2002a, b). For the first time one can reliably target and selectively inactivate a specific gene or mRNA species of interest. This technology does not mimic the effect of a small-molecule inhibitor that can often simultaneously inactivate multiple proteins (e.g., family members of a gene class) and can disrupt protein function over a relatively short period of time—a period of time that can be limited by pharmacokinetics and in some instances may lead to gain-of-function effects (as is the case with FK506). Nonetheless, siRNA represents a breakthrough in target validation. The next one to two years will witness the generation of genomewide libraries of siRNA, the discovery of targets through high-throughput siRNA screens and through the use of siRNA libraries that can be selected for both in vitro and in vivo.

Despite these advances, a significant limitation remains in the area of target discovery. Namely, there are no model systems that accurately reflect human disease and that can successfully predict for the in vivo efficacy of therapeutics. Surmounting this obstacle will remain a formidable challenge for the foreseeable future.

LEAD DISCOVERY: HIGH-THROUGHPUT SCREENING

Lead discovery is designed to find the structural basis or initial chemical backbone upon which a therapeutic inhibitor will ultimately be built. There are two general types of high-throughput assays: biochemical in vitro and cell-based assays. While the former are the mainstay of lead discovery in most major pharmaceuticals, the latter are beginning to stage a revival of sorts.

In vitro High-Throughput Screening (HTS)

In vitro biochemical assays typically require purified proteins and a measurement of protein activity. Protein activities, including enzyme activity or protein binding, can be measured at a cost of one dollar or less per compound (full cost including materials and assay development) and with throughputs of 100,000 compounds per day for "classical" HTS, but up to 1,000,000 compounds per day in the most advanced ultra-HTS systems. The in vitro technologies applicable to such screens have shifted increasingly to fluorescent-based detection systems, including fluorescence polarization (FLP), which detects the differences in polarized fluorescent light output between unbound and hence rapidly rotating fluorescent molecules and bound and hence more slowly rotating molecules, and homogeneous time-resolved fluorescence (HTRF), which detects the transfer of fluorescent energy between two fluorophores when they come into close proximity (Fernandes, 1998).

The in vitro biochemical approach has the advantage of a single target, a quantitative read-out, low cost, and high throughput. There are however several disadvantages. An in vitro assay relies on the assumption that the protein, usually produced in bacteria or in insect cells through baculovirus transduction, is assayed in the same conformation as when the target protein is found in the cell. This can be difficult when or if posttranslation modification is required for proper activity or when folding is guided by specific chaperones. Finally, in vitro assays can require large-scale protein production and require substrates that can be measured in vitro. Large-scale protein production and establishment of a reliable assay system can require significant resources.

Cell-Based High-Throughput Screening

Cell-based screens were the primary initial mode of cancer drug discovery, where cytotoxic agents were identified by their ability to either kill or limit the proliferation of cancer cells in culture. A problem with such screens is that without knowing the specific drug target, it is difficult to improve upon novel active chemical entities that, while active in an initial screen, lack proper druglike properties. As a result, such assays have been largely abandoned in favor of in vitro target-directed screens.

The rebirth of cell-based assays has come about as generalized phenotypic endpoints (e.g., cell death) have given way to read-outs that interrogate specific cellular processes (e.g., mitosis) or those that are designed to interrogate a small number of targets. In some cases, cell-based assays can be generated that are "rigged" to sense the activity of a single protein or the activity of a specific signaling pathway.

Cell-based assays can take two general forms requiring either direct cellular imaging or assaying a protein activity in a cell lysate. The former can analyze complex cellular processes, as in the aforementioned example of astral spindles (Mayer et al., 1999), or cellular localization of shuttling proteins, as our group has done in detecting novel regulators of the phosphoinositol-3 kinase (PI3K) signaling pathway (Kau et al., 2003), although the requirement for image acquisition time limits the throughput.

The generation of high-throughput luciferase-based modalities of cell-based screening has dramatically increased the throughput of such assays so that in some instances the throughput rivals or outperforms in vitro assays. Cell-based assays have the advantage that molecules that score in such assays are already cell permeable, that the target protein is more likely in a native conformation, and that multiple mechanisms of target inhibition can be scored. This latter point can be underscored by considering an enzyme that is activated in cancer but that typically cycles between the inactive and active states. Moreover, let us assume that, once activated, this enzyme is degraded. In vitro assays would often score only the inhibition of the active enzyme, while an in vivo cell-based screen against the same target would score not only this class of compounds but also compounds that stabilize the enzyme in an inactive state (shifting the equilibrium), block conversion of the inactive enzyme to the active one, or accelerate degradation of the active kinase. These types of dynamic equilibria are difficult to establish in vitro, and thus opportunities in this area may be missed. Surprisingly, the structure of imatinib bound to BCR-ABL revealed that it is a prime example of this type of compound and it inhibits BCR-ABL blocking the inactive kinase (Schindler et al., 2000).

On the other hand, cell-based assays tend to require substantial amounts of cell culture, are generally more expensive, are more subjected to nonspecific drug effects, and will exclude compounds that might in fact be very good leads that are not cell permeable. In short, cell-based and in vitro methods are complimentary, and it is likely that they will find increasing concomitant use.

In silico LEAD DISCOVERY

Methodologies to allow in silico docking of three-dimensional small-molecule structures with known crystal-derived or model-derived structures have been developed to try and filter virtual libraries of small molecules leading to smaller enriched libraries that can then be used in biochemical screening. Here, hundreds of thousands of compounds can be reiteratively fitted to a three-dimensional model of the target and scored for the appropriateness of the fit. There are a growing number of docking programs now in use that employ a variety of scoring methods for ranking the "virtual hits" (reviewed in Lyne, 2002). While these methods are not sufficiently robust to directly select lead compounds, when compound libraries are first preselected based on an initial "virtual prescreen" and then screened in a standard assay format, in some cases a 10-fold improvement in the hit rate over random compound libraries has been seen. Ongoing efforts in this area are aimed at allowing the robust virtual screening of millions of structural entities at high speed and low cost.

LEAD OPTIMIZATION AND MEDICINAL CHEMISTRY

Lead optimization is the process by which crude inhibitors are gradually shaped and molded into molecules with druglike properties that are worthy of testing in humans. It is useful to conceive of the lead compound or compounds as simply providing a platform upon which one can build the final product. It is therefore beneficial to have several distinctly different platforms demonstrating activity in the primary screen. Without such diversity one runs the risk of having intractable problems relating to the starting platform itself and not "correctable" with additional chemistry. The lead optimization process in general serves two distinct purposes: improving the selective inhibition of the target and rendering the resulting molecules "druglike."

Improving the efficacy of target inhibition is presumably, although not always, related to (1) increasing the potency of the inhibitor against the target, (2) limiting or at the least understanding cross inhibition against highly related targets, (3) limiting nonspecific toxicity or off-target effects, and (4) improving the druglike properties of the molecule, including stability, ease of synthesis, solubility, formulability, and others. Steps 1 and 2 are addressed by reiterative cycles of chemical synthesis of modified derivatives of the initial leads and retesting such derivatives in one or more of three types of assay: (1) the assay used as the primary screen, (2) virtual docking of derivatives (prior to synthesis) against the target, and (3) assaying the inhibitors both in limited screening assays and in silico against related family members. Derivatives can be generated by chemical craftsmanship based on structural information or by combinatorial chemistry where entire libraries of the initial leads can be synthesized in high-throughput assay format and used for rescreening. Off-target activities can be quite difficult to quantify (since "off" targets can be anything) and thus by definition cannot be measured in vitro. A useful concept that will take hold as siRNA becomes more widely available is to generate or identify cell lines that are sensitive (i.e., demonstrate a measurable phenotype) upon selective inhibition of the target and cells that are immune to or simply do not "care" about the target when the target is knocked down. Such lines can then act as indicator cells for efficacy and for off-target activity. A robust example where this strategy was applied was in the development of Flt-3 inhibitors. Here, Tse and co-workers in the Small laboratory rendered IL-3-dependent Baf-3 cells dependent on activated Flt-3 by generating stable Flt-3-expressing lines selected in the absence of IL-3. These cells became dependent or wired to FLT-3 activity and could thus be used to "sense" Flt-3 inhibition, and conversely, the IL-3-dependent Baf-3 lines, while totally dependent on IL-3, were unperturbed by FLT-3 inhibition and thus could serve as indicators of off-target activity (Tse et al., 2001).

Endowing small molecules with druglike properties is again the work of medicinal chemists. Drugs must be rendered cell permeable, they must have an appropriate half-life in vivo; they must avoid excessive binding to serum proteins; the volume of distribution into fat compartments must be considered; metabolism and activation of cytochrome p450 activities must be avoided; there must be synthesis methods that allow reasonable cost and ease of large-scale production synthesis; of course they must have limited toxicity to major organs such as liver, kidney, brain and heart; and the drug clearance and half-life must be appropriate for human administration and timely target inhibition. The lack of a detailed understanding of the relationship between chemical structure and these properties has hampered the successful development of many interesting lead molecules and has spurred directed research into solving these problems more globally.

A number of the factors that determine the performance of a drug in a given individual may relate to host factors rather than to tumor factors. Thus, while tumor cell genetics may be critical for target identification and selection, understanding the genetics of host–drug interactions applies to almost all, if not all, drug development. Thus, there is a great interest in the field of pharmacogenomics, where efforts are ongoing to identify the host mediators of the pharmacokinetics and toxicology. There are abundant examples where individual variation has dire effects on drug toxicity. Among the best known links between genetic factors and drug response are mutations found in the skeletal muscle ryanodine receptor that are associated with the development of malignant hyperthermia upon exposure to anesthetic agents (MacLennan et al., 1990).

There are likely many more such genetic variants linked either to adverse drug events or to alterations in drug metabolism and pharmacokinetics. It is likely that such human variation will be accounted for, at least in part, by polymorphisms within the genome. The goal of pharmacogenomics is to relate such genetic variants to the drug toxicity or other pharmacologic events (such as response). Identification of such variations and the associated protein products should allow the stratification of individuals to avoid idiosyncratic responses but, more importantly, should help elucidate the molecular basis upon which these responses are based.

PRECLINICAL TESTING

Preclinical testing again serves two masters: determination or proof of in vivo efficacy and determination of toxicity. While the Federal Drug Administration (FDA) does not require efficacy demonstration in animals, and rather only requires toxicology studies, most pharmaceutical companies rely on preclinical efficacy testing to guide development decision making. Animal models, in particular mouse models, remain the mainstay of efficacy demonstration. Animal models include both xenograft and orthograft models, viral transduced models, and genetically engineered models (knockouts and transgenics). The genetically engineered models are increasingly elegant, although xenografts, for all their foibles, are heavily used in preclinical testing to determine efficacy. Recent advances in imaging, primarily luciferase-based systems imaged using sensitive cooled charge-coupled device cameras, have made these models and the orthograft versions more robust and more sensitive to inhibition when tumors are treated at a reasonable size (Contag et al., 1998). While the response of a xenograft to a specific therapeutic is not predictive of success in humans, lack of efficacy would be problematic.

A major hurdle in the field is to identify or develop models that allow for prediction of success in human trials. One can imagine a number of ways forward; however, in the short term, a focus on mechanistic-based models is likely to aid in the preclinical evaluation of therapeutics. A possibility here is to select cell lines that reflect the activity of the target or the genetics of the pathway under study. No comprehensive analysis of cell lines and their genetic relationship to human cancers has been done, and clearly expression-profiling data suggest that they can be quite different. Nonetheless, a genetic understanding of current cancer cells lines or cancers maintained exclusively as xenografts in combination with a more detailed understanding of cancer genetics may allow the selection of predictive models. Similarly, one could simply engineer cell lines to harbor the activating or inactivating lesions of interest.

In the absence of such pairings, models engineered to recapitulate even at a rudimentary level the target under study are worth considering. Returning to the example of the kinase oncogenes, elegant use has been made of retroviral transduction of primary hematopoietic cells to generate leukemia "look-a-likes" that are reflective of the transduced oncogene. For example, Flt-3 activation in such systems recapitulates a leukemic state and demonstrates marked response when treated with specific inhibitors (Kelly et al., 2002; Weisberg et al., 2002). Such mechanistic-based models suffer from a lack of genetic heterogeneity so common to human tumors and thus cannot be considered predictive of future clinical success in patients. However, failure of the inhibitor to work in this "perfect" or rigged model would give pause to further progression.

CLINICAL TESTING

Phase I clinical trials are designed to determine the relationship between dose and toxicity and to set the dose that will be used in phase II trials. In the areas of drug development outside of oncology, the optimal dose is often determined both by the effect on a therapeutic endpoint or a surrogate endpoint and by toxicity. In oncology the history of cytotoxic chemotherapy has left us with the concept of maximally tolerated dose (MTD). Since cytotoxic agents do not typically interact with a single protein, one cannot deliver a set dose based upon a plasma concentration sufficient to inhibit the target of interest. Rather, one incrementally increases the dose, typically testing each new dose in three patients, until dose-limiting toxicity (DLT) is reached, at which point the phase I trial ends and the maximally tolerated dose is defined. As we move into an era of targeted oncologic therapeutics, it is likely that there will be a refocusing of the aims of phase I oncology trials. Specifically, such trials will likely focus on using molecular endpoints to determine the biologically relevant dose that is associated with quantified target inhibition. A benefit of such an approach would be the ability to define the relationship between the dosing schedule and target inhibition and ultimately the relationship between duration of target inhibition and therapeutic efficacy versus toxicity.

Phase II trials are the first step in attempting to determine drug efficacy. Here, patient selection is driven by the tumor type thought to be amenable to treatment with the therapeutic under study. These data often come from the response data that might be seen in phase I testing or from the preclinical rationale that led to the development of the drug. More recently, there have been examples of patient trials in which selection was driven not only by the tumor type under study but more importantly by the presence of a genetic marker thought to be indicative of response. Specifically, inhibitors of the FLT3 tyrosine kinase inhibitors are being tested in subsets of patients with acute myelogenous leukemia whose cancer bears a mutated and hence activated FLT3 protein (Smith et al., 2003; Stone et al., 2003). Similarly, analogues of the mTOR inhibition rapamycin, thought to be particularly effective in downstream interruption of PI3K pathway signaling, may soon see selective testing in patients with PTEN-deficient tumors (Neshat et al., 2001; Podsypanina et al., 2001).

Phase III clinical trials are larger proof-of-efficacy experiments controlled by randomization. A positive result in such trials is usually, although not always, required for FDA approval of a new agent. In most cancers there is some form of therapy that is considered the standard even if such therapy is only marginally effective. Thus in the design of a phase III trial one must decide whether to add a new investigational agent to an established regimen and test the new combination against the old regimen or to simply test the new agent against the existing combination or existing single agent. The former approach requires that the proposed new combination be tested first in phase II trials to again test the notion that there might be efficacy. The decisions driving the design of such trials are often divorced from preclinical research. As a result, systematic evaluations of combinations and systematic investigations into the scientific basis for successful or unsuccessful combinations are rarely conducted prior to the implementation of such trials. Indeed, it is possible that while an existing regimen might be partially effective, it may antagonize the action of a more effective new agent, limit the dose of a new agent, or carry unanticipated drug–drug interactions leading to problems in appropriate dosing.

SUMMARY

The improvements in genomic methods for target discovery and the advent of siRNA-based methods for inactivation of putative targets portend an improvement in our ability to select rational targets for drug development. High-throughput in vitro and cell-based screens along with improvements in the in silico modeling of drugs with targets also portend well for the process of lead discovery. We remain limited in key areas: the ability to rapidly design molecules that will have desired drug-like properties specifically avoiding poor pharmacokinetic behavior and undesirable off-target toxicity and the ability to use preclinical models to predict for efficacy in humans. Solutions to these remaining problems would dramatically enhance our ability to make progress against the majority of human cancers.

ACKNOWLEDGMENTS

The authors would like to acknowledge grant support from the following: The Damon Runyon Cancer Research Foundation, CaP CURE, The Linda and Arthur Gelb Center for Translational Research, and the National Cancer Institute (RO1 CA85912, PO1 CA89021, P50 CA09038).

REFERENCES

Armstrong SA, et al. (2003): Inhibition of FLT3 in MLL. Validation of a therapeutic target identified by gene expression based classification. *Cancer Cell* 3: 173–183.

Bardelli A, et al. (2003): Mutational analysis of the tyrosine kinome in colorectal cancers. *Science* 300: 949.

Ben-Neriah Y, et al. (1986): The chronic myelogenous leukemia-specific P210 protein is the product of the bcr/abl hybrid gene. *Science* 233: 212–214.

Brummelkamp TR, et al. (2002): Stable suppression of tumorigenicity by virus-mediated RNA interference. *Cancer Cell* 2: 243–247.

Cappellen D, et al. (1999): Frequent activating mutations of FGFR3 in human bladder and cervix carcinomas. *Nat Genet* 23: 18–20.

Chesi M, et al. (1997): Frequent translocation t(4;14)(p16.3; q32.3) in multiple myeloma is associated with increased expression and activating mutations of fibroblast growth factor receptor 3. *Nat Genet* 16: 260–264.

Contag PR, et al. (1998): Bioluminescent indicators in living mammals. *Nat Med* 4: 245–247.

Davies H, et al. (2002): Mutations of the BRAF gene in human cancer. *Nature* 417: 949–954.

Davis TA, et al. (1999): Single-agent monoclonal antibody efficacy in bulky non-Hodgkin's lymphoma: Results of a phase II trial of rituximab. *J Clin Oncol* 17: 1851–1857.

Demetri GD, et al. (2002): Efficacy and safety of imatinib mesylate in advanced gastrointestinal stromal tumors. *N Engl J Med* 347: 472–480.

Dranoff G (2003): GM-CSF-secreting melanoma vaccines. *Oncogene* 22: 3188–3192.

Druker BJ, et al. (2001): Efficacy and safety of a specific inhibitor of the BCR-ABL tyrosine kinase in chronic myeloid leukemia. *N Engl J Med* 344: 1031–1037.

Einhorn LH (2002): Chemotherapeutic and surgical strategies for germ cell tumors. *Chest Surg Clin N Am* 12: 695–706.

Elbashir SM, et al. (2001): Duplexes of 21-nucleotide RNAs mediate RNA interference in cultured mammalian cells. *Nature* 411: 494–498.

Fernandes PB (1998): Technological advances in high-throughput screening. *Curr Opin Chem Biol* 2: 597–603.

Fisher RI, et al. (1993): Comparison of a standard regimen (CHOP) with three intensive chemotherapy regimens for advanced non-Hodgkin's lymphoma. *N Engl J Med* 328: 1002–1006.

Gilliland DG (1998): Molecular genetics of human leukemia. *Leukemia* 12 (Suppl 1): S7–12.

Golub TR, et al. (1999): Molecular classification of cancer: Class discovery and class prediction by gene expression monitoring. *Science* 286: 531–537.

Gorre ME, et al. (2001): Clinical resistance to STI-571 cancer therapy caused by BCR-ABL gene mutation or amplification. *Science* 293: 876–880.

Hanahan D, Weinberg RA (2000): The hallmarks of cancer. *Cell* 100: 57–70.

Heinrich MC, et al. (2003): PDGFRA activating mutations in gastrointestinal stromal tumors. *Science* 299: 708–710.

Hofstra RM, et al. (1994): A mutation in the RET proto-oncogene associated with multiple endocrine neoplasia type 2B and sporadic medullary thyroid carcinoma. *Nature* 367: 375–376.

Hortobagyi GN, et al. (2000): Randomized trial of high-dose chemotherapy and blood cell autografts for high-risk primary breast carcinoma. *J Natl Cancer Inst* 92: 225–233.

Huang ME, et al. (1987): All-trans retinoic acid with or without low dose cytosine arabinoside in acute promyelocytic leukemia. Report of 6 cases. *Chin Med J (Engl)* 100: 949–953.

Huggins C, Hodges CV (1972): Studies on prostatic cancer. I. The effect of castration, of estrogen and androgen injection on serum phosphatases in metastatic carcinoma of the prostate. *CA Cancer J Clin* 22: 232–240.

Kau TR, et al. (2003): A chemical genetic screen identifies inhibitors of regulation nuclear export of a forkhead transcription factor in PTEN-deficient tumor cells. *Cancer Cell* 4: 463–476.

Kelly LM, et al. (2002): CT53518, a novel selective FLT3 antagonist for the treatment of acute myelogenous leukemia (AML). *Cancer Cell* 1: 421–432.

Konopka JB, et al. (1984): An alteration of the human c-abl protein in K562 leukemia cells unmasks associated tyrosine kinase activity. *Cell* 37: 1035–1042.

Lyne PD (2002): Structure-based virtual screening: An overview. *Drug Discov Today* 7: 1047–1055.

MacLennan DH, et al. (1990): Ryanodine receptor gene is a candidate for predisposition to malignant hyperthermia. *Nature* 343: 559–561.

Maloney DG, et al. (1994): Phase I clinical trial using escalating single-dose infusion of chimeric anti-CD20 monoclonal antibody (IDEC-C2B8) in patients with recurrent B-cell lymphoma. *Blood* 84: 2457–2466.

Marty M (1990): A comparative study of the use of granisetron, a selective 5-HT3 antagonist, versus a standard anti-emetic regimen of chlorpromazine plus dexamethasone in the treatment of cytostatic-induced emesis. The Granisetron Study Group. *Eur J Cancer* 26 (Suppl 1): S28–32.

Mayer TU, et al. (1999): Small molecule inhibitor of mitotic spindle bipolarity identified in a phenotype-based screen. *Science* 286: 971–974.

Nakao M, et al. (1996): Internal tandem duplication of the flt3 gene found in acute myeloid leukemia. *Leukemia* 10: 1911–1918.

Neshat MS, et al. (2001): Enhanced sensitivity of PTEN-deficient tumors to inhibition of FRAP/mTOR. *Proc Natl Acad Sci USA* 98: 10314–10319.

Orlowski RZ, et al. (2002): Phase I trial of the proteasome inhibitor PS-341 in patients with refractory hematologic malignancies. *J Clin Oncol* 20: 4420–4427.

Paddison PJ, et al. (2002a): Short hairpin RNAs (shRNAs) induce sequence-specific silencing in mammalian cells. *Genes Dev* 16: 948–958.

Paddison PJ, et al. (2002b): Stable suppression of gene expression by RNAi in mammalian cells. *Proc Natl Acad Sci USA* 99: 1443–1448.

Perou CM, et al. (2000): Molecular portraits of human breast tumours. *Nature* 406: 747–752.

Podsypanina K, et al. (2001): An inhibitor of mTOR reduces neoplasia and normalizes p70/S6 kinase activity in Pten+/− mice. *Proc Natl Acad Sci USA* 98: 10320–10325.

Ramaswamy S, et al. (2001): Multiclass cancer diagnosis using tumor gene expression signatures. *Proc Natl Acad Sci USA* 98: 15149–15154.

Ries L, et al. (Eds) (2003): *SEER Cancer Statistics Review, 1975–2000*, http://seer.cancer.gov/csr/1975_2000. National Cancer Institute, Bethesda, MD.

Roberts JJ, Pascoe JM (1972): Cross-linking of complementary strands of DNA in mammalian cells by antitumour platinum compounds. *Nature* 235: 282–284.

Rosenwald A, et al. (2002): The use of molecular profiling to predict survival after chemotherapy for diffuse large-B-cell lymphoma. *N Engl J Med* 346: 1937–1947.

Schindler T, et al. (2000): Structural mechanism for STI-571 inhibition of abelson tyrosine kinase. *Science* 289: 1938–1942.

Schrama JG, et al. (2002): Randomized trial of high-dose chemotherapy and hematopoietic progenitor-cell support in operable breast cancer with extensive lymph node involvement: Final analysis with 7 years of follow-up. *Ann Oncol* 13: 689–698.

Sellers WR, Kaelin WG Jr (1997): Role of the retinoblastoma protein in the pathogenesis of human cancer. *J Clin Oncol* 15: 3301–3312.

Singh D, et al. (2002): Gene expression correlates of clinical prostate cancer behavior. *Cancer Cell* 1: 203–209.

Slamon DJ, et al. (2001): Use of chemotherapy plus a monoclonal antibody against HER2 for metastatic breast cancer that overexpresses HER2. *N Engl J Med* 344: 783–792.

Smith BD, et al. (2003): Single-agent CEP-701, a novel FLT3 inhibitor, shows biologic and clinical activity in patients with relapsed or refractory acute myeloid leukemia. *Blood* 13: 3669–3676.

Stadtmauer EA, et al. (2000): Conventional-dose chemotherapy compared with high-dose chemotherapy plus autologous hematopoietic stem-cell transplantation for metastatic breast cancer. Philadelphia Bone Marrow Transplant Group. *N Engl J Med* 342: 1069–1076.

Stone RM (2002): The difficult problem of acute myeloid leukemia in the older adult. *CA Cancer J Clin* 52: 363–371.

Stone RM, et al. (2003): PKC 412 FLT3 inhibitor therapy in AML: results of a phase II trial. *Ann Hematol.* 83 Suppl 1: S89–90.

Strausberg RL, et al. (2003): Sequence-based cancer genomics: Progress, lessons and opportunities. *Nat Rev Genet* 4: 409–418.

Su AI, et al. (2001): Molecular classification of human carcinomas by use of gene expression signatures. *Cancer Res* 61: 7388–7393.

Tse KF, et al. (2001): Inhibition of FLT3-mediated transformation by use of a tyrosine kinase inhibitor. *Leukemia* 15: 1001–1010.

Tuveson DA, et al. (2001): STI571 inactivation of the gastrointestinal stromal tumor c-KIT oncoprotein: Biological and clinical implications. *Oncogene* 20: 5054–5058.

van de Vijver MJ, et al. (2002): A gene-expression signature as a predictor of survival in breast cancer. *N Engl J Med* 347: 1999–2009.

Warrell RP Jr, et al. (1991): Differentiation therapy of acute promyelocytic leukemia with tretinoin (all-trans-retinoic acid). *N Engl J Med* 324: 1385–1393.

Webb A, et al. (1997): BCL-2 antisense therapy in patients with non-Hodgkin lymphoma. *Lancet* 349: 1137–1141.

Weisberg E, et al. (2002): Inhibition of mutant FLT3 receptors in leukemia cells by the small molecule tyrosine kinase inhibitor PKC412. *Cancer Cell* 1: 433–443.

26

PRECLINICAL TRIALS IN MOUSE CANCER MODELS

BRIAN WEISS AND KEVIN SHANNON

Department of Pediatrics University of California—San Francisco, San Francisco, California

The goals of this chapter are to address general principles involved in testing cancer therapeutics in the mouse with an emphasis on how murine cancer models might be deployed to overcome some of the difficulties involved in evaluating new agents in human beings. We compare xenograft systems with genetically engineered models and discuss the benefits and drawbacks of each. We suggest criteria that should be considered in designing and implementing preclinical trials in mouse models.

INTRODUCTION

Other sections in this volume describe how many of the mouse models that recapitulate biologic and clinical aspects of specific human cancers have been generated through systematic application of advanced technologies for introducing cancer-associated genetic lesions into the murine germline. Genetically engineered mice can serve multiple purposes, such as demonstrating if mutations that initiate tumorigenesis are required for maintenance (Chin et al., 1999; Felsher and Bishop, 1999), implicating the cell of origin in a complex tumor type (Holland et al., 2000), and identifying specific domains within dominantly acting oncoproteins that are essential for transformation (Kogan et al., 2000). Another exciting potential application involves harnessing mouse cancer models to evaluate cancer therapeutics. These systems provide investigators with the opportunity to rigorously test new compounds in controlled experiments that overcome some of the genetic heterogeneity and safety concerns inherent in studying new agents in human patients. While straightforward in principle, executing preclinical trials in the mouse is difficult in practice. Other chapters in this volume address issues related to biostatistics, pharmacokinetics/pharmacodynamics, and in vivo imaging of mouse tumors. In addition, Chapter 25 addresses the specific topic of developing therapeutics against signaling molecules such as Ras and Akt. The goals of this chapter are to address general principles involved in testing cancer therapeutics in the mouse with an emphasis on how murine cancer models might be deployed to overcome some of the difficulties involved in evaluating new agents in human beings. We compare xenograft systems with genetically engineered models and discuss the benefits and drawbacks of each. Based on our experience, we suggest criteria that should be considered in designing and implementing preclinical trials in mouse models. Finally, we review data from a number of studies that have been performed to date and attempt to draw some general conclusions.

EVALUATING NEW CANCER THERAPIES IN HUMAN PATIENTS

Testing potential cancer therapeutics in human patients is expensive and has certain inherent limitations (see Table 26.1). New agents are typically evaluated in three phases. Whereas the primary goal of phase 1 trials is to determine the maximally tolerated dose of a drug, these studies may also provide some preliminary indications of efficacy. Phase 1 trails are reserved for patients who have failed standard therapies. Phase 2 trials are designed to measure response rates in a group of patients with refractory or recurrent cancers that are treated with a

Mouse Models of Human Cancer, edited by Eric C. Holland
ISBN 0-471-44460-X

Table 26.1. Attributes of Mouse Systems and Human Cancers Relevant to Drug Testing

Feature	Xenograft	Engineered Model	Human Cancer
Initiating mutation	Usually unknown	Known	Usually unknown
Genetic characteristics	Uniform	Some variability	Heterogeneous
Cellular origin	Field	Field	Clonal
Metastases	Uncommon	Variable	Common
Tumor tissue	Available	Available	Often unavailable

dose that is based on the phase 1 data. Responses are traditionally reported as "complete" (objective regression of all detectable lesions), "partial" (some regression), or "mixed" (regression of some lesions with growth of others). The few compounds that show significant promise in phase 2 studies are advanced to randomized phase 3 trials, in which the new drug is compared, either alone of in combination with other agents, to the "standard" treatment for a specific cancer. In contrast to phase 1 and phase 2 trials, newly diagnosed patients are included in phase 3 studies. Phase 3 trials are frequently performed in the setting of cooperative multi-institutional networks and are logistically challenging, expensive, and time consuming. Some of these traditional strategies for evaluating new cancer therapeutics are being reconsidered as new agents are developed that are directed toward specific biochemical targets (Druker, 2002). For example, inhibition of the target, rather than overt clinical toxicity, may represent a better endpoint for phase 1 testing of drugs that have a well-defined biochemical target. Moreover, it may be appropriate to select patients who demonstrate a specific molecular abnormality in tumor cells for trials of targeted therapies. While monitoring the effects of a therapeutic agent on a molecular target in primary tumor cells is feasible in diseases such as leukemia, it is highly problematic in many other cancers. As discussed below, genetically engineered mouse cancer models may be particularly useful for evaluating targeted therapeutics.

MOUSE MODELS AS A PLATFORM FOR PRECLINICAL TRIALS

The mouse has been used extensively as an experimental system for testing the efficacy of cancer therapeutics on explanted human tumor xenografts. Recently, a few studies have been performed in genetically engineered models that spontaneously develop cancer. Some characteristics are shared between these systems. First, it is possible to investigate large numbers of animals on a uniform genetic background. This may be particularly important in testing potential therapeutics for relatively rare cancers. Second, although new therapeutics are almost invariably evaluated singly in the initial human phase 1 and 2 studies, potential synergism between agents can be explored during the early phases of drug development in the mouse. Third, murine tumors can be excised to assess the histologic and biochemical effects of specific agents. Finally, safety is much less of a concern in murine studies. Common shortcomings of xenograft and genetically engineered mice are that the absorption and metabolism of experimental agents as well as toxicities to normal tissues may be substantially different than in humans.

HUMAN XENOGRAFTS IN IMMUNODEFICIENT MICE

The severe combined immunodeficient (SCID) mouse arose from a point mutation in chromosome 16 in CB-17 inbred strain (Custer et al., 1985). This mutation causes a generalized DNA repair defect and results in interruption of lymphocyte maturation. SCID mice do not have functional T or B cells but possess an intact innate immune system with normal numbers of monocytes, macrophages, natural killer (NK) cells, and granulocytes. Therefore, depending on the human tissue implanted and the site of implantation, SCID mice retain some capacity to mount a complex, multistage immune response (Bankert et al., 2001). This residual innate immunity can limit long-term graft survival. In addition, SCID mice can spontaneously produce functional lymphocytes as they age. Some investigators have attempted to overcome these problems by crossing the SCID mutation onto a nonobese diabetic background (NOD/SCID); however, these mice also retain NK cells and, like the SCID strain, develop normal lymphocytes later in life (Prochazka et al., 1992). A newer model, the recombinase-activating gene 2 (RAG-2) mutant mouse developed by Shinkai et al. (1992) crossed with the common cytokine receptor gamma-chain mutant strain, produces a double mutant with no functional B or T lymphocytes and no NK cells, and these animals retain xenografts for extended periods (Cooper et al., 2001).

Since the first report that SCID mice could be engrafted with human tumor cells, this strain and other immunodeficient mice have been exploited for preclinical testing of cancer therapies. These xenograft models involve implanting or injecting human tumor cells or cell lines. Some investigators have implanted human tumor biopsy specimens with attendant stroma and other benign tissue into SCID mice in an effort to better recapitulate the tumor microenvironment in this model (Sakakibara et al., 1996). SCID mice provide a platform for efficiently screening potential therapeutics without the risk or expense of human trials. Paine-Murrieta et al. (1997) have shown that the responses of human tumor xenografts to conventional chemotherapeutic agents mirror the clinical situation in that drugs that are clinically active in people also tend to show activity in the xenograft model. Xenografts can also be used to test molecularly targeted agents. For example, human breast cancer xenografts were used to document synergism between a Bcl-2 antisense nucleotide (G3139) and doxorubicin and to rigorously study the pharmacokinetics of these two compounds (Lopes and Mayer, 2002). In another interesting study, the responses of human prostate cancer cell line xenografts to the mammalian target of rapamycin (m-TOR) inhibitor CCI-779 correlated with the presence of the *PTEN* tumor suppressor, which regulates signaling through the phosphoinositol-3 (PI3) kinase-Akt-mTOR cascade (Neshat et al., 2001).

However, while xenograft models are useful for screening compounds in a living organism, there are inherent deficits of models with impaired immunologic function, and xenografts incompletely model the complicated interplay between tumor and normal stroma. The ideal mouse model faithfully recapitulates the natural history, pathobiology, and biochemistry of the human disease (Hann and Balmain, 2001). This is not true of xenograft models, and explanted cancer cell lines may acquire mutations in culture that modulate responses to drugs. Xenograft models are also problematic for evaluating processes such as tumor angiogenesis that evolve through distinct stages as they develop in situ.

GENETICALLY ENGINEERED MOUSE MODELS

Strains of mice that develop cancer due to a mutation in a gene (or genes) that are altered in human malignancies are appealing in principle for performing preclinical therapeutic trials. The value of a given model is dependent on a number of factors, including penetrance, latency to tumor formation, the ability to detect disease with sufficient lead time to allow investigators to test specific interventions, and the genetic lesion used to initiate tumor formation. Advantages of many genetically engineered mouse cancer models include that (1) the initiating genetic lesion is known, (2) the mice are immunocompetent, and (3) tumors develop spontaneously in the appropriate tissue compartment. Complex processes such as tumor angiogenesis can be modeled and interrogated in these in vivo systems. Furthermore, the ability to readily investigate combinations of agents in naive tumors that have not relapsed after previous treatment addresses inherent limitations in human phase 1 and 2 trials.

However, these models also differ in important respects from de novo human cancers (Table 26.1). For example, mutant alleles that initiate tumorigenesis are typically expressed simultaneously in many cells throughout the animal or in a target tissue. Mouse models in which one allele of a tumor suppressor gene such as *Tp53, Nf1,* or *Brca1* is inactivated in the germline may accurately model human heritable cancer syndromes. However, tumors that arise in this context account for a relatively small percentage of human cancers, and most malignancies appear to arise de novo from a single cell or small population of mutant cells. It is not known if initiating cancer from a field of mutant cells rather than a single aberrant clone might change the requirement for cooperating mutations, and few experiments have addressed this issue. However, it is interesting that the ability of *Nf1*-deficient fetal liver cells to induce leukemia in irradiated recipients was suppressed by coinjecting normal competitor cells (Zhang et al., 2001). The use of strains in which mutations in more than one gene are induced together in a field of cells might be problematic, as such systems are less likely to model the genetic heterogeneity found in human cancers. In short, the responses of tumors that develop in genetically engineered strains of mice to therapeutics or their propensity to develop resistance may not mimic the corresponding human cancer if these lesions are genetically less complex. Another potential pitfall is that promoters that are used to express an oncogene or inactivate a tumor suppressor in mice might not target the cellular population that gives rise to the human cancer. This could, in turn, have dramatic effects on whether a tumor responds to a given drug. In addition, mice that are phenocopies of human cancers may not accurately model the underlying genetic lesions that contribute to tumor initiation and progression. For example, the use of viral oncogenes to drive tumor formation may not induce the same constellation of genetic and biochemical changes that are found in human cancers. At best, a given genetically engineered model perhaps reflects a subset of the human cancers with which it shares histologic and clinical features. This principle is perhaps best illustrated by retrovirally induced leukemias and lymphomas in which, like human cancers, a diverse mutational spectrum is observed (Li et al., 1999). Finally, the costs of performing preclinical therapeutic trials in genetically engineered mouse models are substantially greater than xenograft studies, particularly when there is prolonged latency and incomplete penetrance. Few mouse models approach the theoretical ideal of efficiently initiating tumorigenesis in one (or a few) target cells because of a mutation that is known to contribute to the same histologic type of cancer in human patients.

Despite these concerns, technical advances in engineering cancer-associated mutations into the mouse germline and innovative strategies for inducing gene expression in subpopulations of cells have fostered the development of a number of elegant cancer models. For example, homologous recombination was used to introduce a mutant allele of the *Kras* oncogene into the endogenous locus that was silenced by including a strong transcriptional stop sequence flanked by loxP sites (Jackson et al., 2001; Johnson et al., 2001; Meuwissen et al., 2001). In this system, expression of mutant K-*ras* can be achieved by using Cre recombinase to excise the stop element. Tracheal instillation of an adenovirus encoding Cre recombinase induced mutant K-*ras* in a patchy distribution and resulted in a robust model of lung cancer (Jackson et al., 2001). Genetically engineered mice are particularly attractive for testing therapeutics that interfere with biochemical pathways that are deregulated as a result of the cancer-associated mutations that were used to create the model. Thus, K-*ras*-driven models of lung cancer might be used to test an agent such as the MAP kinase/ERK kinase (MEK) inhibitor PD184352 (Sebolt-Leopold et al., 1999). This small molecule targets a component of the Raf-MEK-ERK (extracellular regulated kinase) effector cascade, which is downstream of activated Ras. PD184352 has shown impressive effects in xenografts; however, it is uncertain if tumor cell lines that are selected for in vitro growth on plastic will demonstrate similar dependence upon Ras-Raf-MEK-ERK signaling as cancers that arise de novo. A particular concern is that cell lines become progressively more "addicted to" the growth advantage provided by an oncogene with serial passage in culture, which renders them unusually susceptible to drugs that interfere with the biochemical aberrations that result from expressing the oncogene. Administering an agent such as PD184352 to mice with K-*ras*-driven lung cancers offers a way of addressing this question. Furthermore, tumors can be excised and investigated to determine if tumor regression (or lack thereof) correlates with pharmacodynamic inhibition of a relevant biochemical target in primary tumor cells. This in turn can provide biologic insights regarding whether a specific effector pathway is crucial for the proliferation and/or survival of the tumor in vivo. Finally, studies of this type can give some indication about the therapeutic index of a given compound (i.e., its relative effects on the growth of the tumor vs. any deleterious effects on normal tissues).

GENERAL CONSIDERATIONS FOR PERFORMING TRIALS IN GENETICALLY ENGINEERED MICE

A general feature of most mouse cancer models is that the inciting mutation is present throughout the life of the animal. Thus, if a tumor responds to a therapeutic agent and reappears, it may be difficult to distinguish relapse of the original clone due to acquired drug resistance from the development of an independent tumor. Other characteristics of specific models influence whether or not they are suitable platforms for preclinical trials. For example, strains that accurately express a cancer-associated mutation in the correct cell lineage but develop disease with low penetrance or prolonged latency might provide excellent biologic models of tumorigenesis but may be impractical for testing therapeutics. By contrast, highly penetrant cancer models that are driven by viral oncoproteins or by combining two mutations in the same animal may be extremely tractable but may differ at the molecular level from the corresponding human malignancy. Investigators who wish to exploit genetically engineered mice to test cancer therapies must consider additional issues related to detecting and monitoring tumors, therapeutic endpoints, pharmacokinetics/pharmacodynamics, and biostatistics (see Chapters 19–21 for a detailed discussion of some of these questions).

EXAMPLES OF PRECLINICAL TRIALS

Proof of Principle

A central problem related to assessing the predictive value of testing therapeutic agents in mouse cancer models is the lack of consistently effective treatments for most human malignancies. Thus, there are few precedents in which a specific mouse model can be interrogated to ascertain if it recapitulates the successful treatment of a human cancer. Furthermore, relatively few cancer-prone mouse strains have been engineered to express markers such as luciferase within tumor cells for the specific purpose of testing responses to conventional or targeted anticancer agents. The wealth of genetic data on the etiology of human leukemia and lymphoma, the tractable nature of the hematopoietic systems, and well-defined therapeutic targets in some of these cancers have facilitated the use of mouse models to test therapeutics. The initial experience in two such models suggests that treatment responses in genetically engineered mouse models mirror the clinical behavior of human cancers.

In acute promyelocytic leukemia (APL), blasts are blocked in the promyelocytic stage of differentiation secondary to chromosomal translocations of the retinoic acid receptor alpha gene to a variety of partner genes. All-*trans* retinoic acid (ATRA) induces complete remissions in ~80% of patients with APL who have a *PML-RARA* translocation (Tallman et al., 1997) by relieving the differentiation block but does not induce remission in those with *PLZF-RARA* fusions (Licht et al., 1995). Similarly, ATRA induces remissions in *PML-RARA* transgenic mice but is ineffective in a *PLZF-RARA* strain that also develops APL (He et al., 1998). These models have also been harnessed to test new therapeutic approaches such as arsenic trioxide (As_2O_3) and histone deacetylase inhibitors and to investigate the mechanisms of acquired ATRA resistance (He et al., 2001; Lallemand-Breitenbach et al., 1999; Rego et al., 2000). Imatinib, a small molecule that inhibits the Abl, Kit, and PDGFRα kinases, has proven highly successful in treating chronic myelogenous leukemia (CML), gastrointestinal stromal tumor (GIST), and idiopathic hypereosinophilic syndrome (Cools et al., 2003; Druker, 2002). Therapeutic responses correlate with inhibition of the relevant mutant kinases (BCR-ABL in CML, Kit in GIST, and PDGFRα FIP1L1 in hypereosinophilic syndrome). Imatinib also inhibits the growth of hematopoietic malignancies initiated by BCR-ABL in mice (Hoover 2002; le Coutre et al., 1999). Thus, effective treatments for patients with APL and CML show efficacy in genetically engineered mouse models of these diseases. These encouraging data suggest that preclinical trials in mouse cancer models might identify agents that are likely to also prove effective in corresponding human cancers.

FLT3 Inhibitors

The *FLT3* gene encodes a receptor tyrosine kinase that is mutated in ~30% of patients with acute myeloid leukemia (AML) (Gilliland and Griffin, 2002). Mutant FLT3 proteins are intriguing drug targets because the existing paradigm of AML pathogenesis is consistent with the notion that *FLT3* mutations do not initiate AML but are acquired during outgrowth of the abnormal clone (Kelly et al., 2002a). A number of small-molecule inhibitors of the mutant kinase are undergoing phase 1–2 testing in patients with relapsed AML. Recent experiments in which some of these compounds were administered to mice transplanted with hematopoietic cells expressing activated *FLT3* alleles suggest that these agents will prove useful in the treatment of AML (Kelly et al., 2002b; Levis et al., 2002; Sawyers, 2002; Weisberg et al., 2002). However, a recent experiment in which murine APL cells were transduced with a retrovirus encoding a mutant *FLT3* allele found that a FLT3 inhibitor enhanced the efficacy of ATRA but was ineffective as a single agent (Sohal et al., 2003). These data suggest that inhibiting the mutant kinase in human AML cells that contain multiple mutations will only be efficacious in the context of combination therapy. It will be most interesting to ascertain if studies in human patients give similar results or if FLT3 inhibitors will show substantial activity when administered as single agents. If combination therapy is required, mouse models may be an efficient way of identifying agents that are likely to be synergistic in human AML patients.

Angiogenesis Inhibitors

One of the hallmarks of cancer is that tumors promote the growth of new blood vessels, which are essential for continued neoplastic proliferation (Hanahan and Weinberg, 2000). Indeed, cancer cells begin to promote angiogenesis early in development. This "angiogenic switch" is characterized by proangiogenic oncoproteins such as vascular endothelial growth factor (VEGF), basic fibroblast growth factor (bFGF), and placentalike growth factor (PLGF) (Kerbel and Folkman, 2002). Tumor-specific neoangiogenesis is an attractive therapeutic target because agents that inhibit this process are likely to be selective for neoplastic growth; in addition, since the agents target relatively genetically stable endothelial cells instead of the tumor itself, resistance due to acquired mutations is less likely.

There are two classes of angiogenesis inhibitors: direct and indirect. Direct angiogenesis inhibitors (e.g., angiostatin and endostatin) prevent endothelial cells from migrating or proliferating in response to proangiogenic proteins. As predicted, murine tumor cell lines injected into C57BL/6 mice and treated with endostatin did not develop resistance (Boehm et al., 1997). On the other hand, indirect angiogenesis inhibitors such and Iressa and Herceptin generally prevent the expression or block the activity of a tumor protein that activates angiogenesis, many of which are oncoproteins. In a xenograft study, Udagawa et al. (2002) observed a balance of apoptosis and proliferation in implanted human osteosarcoma cells until these cells were transfected with either VEGF165 or an activated *Hras* oncogene. *Hras*-transfected cells had a shift in the angiogenic balance toward increased vascularity with reduced tumor cell apoptosis. These studies confirm that targeting oncoproteins such as Ras in conjunction with a direct angiogenesis inhibitor may produce a synergistic effect. Another xenograft model demonstrated that angiogenesis inhibitors, unlike conventional chemotherapy, are most effective when administered in a manner that maintains a constant concentration in the blood. The investigators showed that continuous administration of human endostatin resulted in sustained systemic concentrations of endostatin, leading to (a) increased efficacy manifested as increased tumor regression and (b) an 8- to 10-fold decrease in the dose required to achieve the same antitumor effect as the single daily bolus administration of endostatin (Kisker et al., 2001).

The initial results in human patients, however, have been disappointing. The studies of Hanahan and co-workers in the RIP1-Tag2 model of islet cell cancer have suggested an explanation for these findings. In these mice, simian virus 40 (SV40) T-antigen expression induces tumors that progress through well-defined pathologic stages that include a preangiogenic phase followed by an "angiogenic switch" (Hanahan, 1985). Importantly, preclinical studies in this model showed that individual antiangiogenic compounds inhibit discrete steps in this process and are ineffective at different stages of tumorigenesis (Bergers et al., 1999). The data infer that RIP1-Tag2 mice and other mouse tumor models that exhibit neoangiogenesis provide valuable platforms for identifying inhibitors that are likely to show activity against various types of human malignancies.

Farnesyltransferase Inhibitors (FTIs)

Ras proteins regulate cell fates by cycling between an active GTP-bound state and an inactive GDP-bound state in response to extracellular stimuli (Boguski and McCormick 1993; Bourne et al., 1990, 1991). *RAS* genes are mutated in ~30% of human cancers, with mutations in *KRAS* detected commonly and *HRAS* mutations infrequently [reviewed in Bos (1989) and Rodenhuis (1992)]. Mutant alleles invariably encode proteins that accumulate in the GTP-bound conformation. In addition to *RAS* point mutations, other cancer-associated genetic lesions such as inactivation of the *NF1* tumor suppressor and mutations of the *FLT3* and *CKIT* receptors are thought to contribute to tumorigenesis, at least in part, by deregulating Ras signaling (Kelly et al., 2002a; Le and Shannon, 2002). These data establish hyperactive Ras as a major therapeutic target; however, disrupting Ras function may have adverse effects on normal tissues.

Ras proteins are processed at a common C-terminal *CAAX* sequence that results in targeting to the plasma membrane [reviewed in Gibbs and Oliff (1997), Le and Shannon (2002), and Sebti and Hamilton (2000)]. Ras processing is initiated by cytosolic prenyltransferases, which attach either a farnesyl or geranylgeranyl isoprenoid lipid to the thiol group of the cysteine. Geranylgeranyl transferase 1 (GGTase 1) and farnesyltransferase (Ftase) catalyze the transfer of isoprenoid groups, which are donated by geranylgeranyl pyrophosphate and by farnesyl pyrophosphate, respectively. FTIs have shown promise as anticancer agents (Karp, Lancet et al. 2001; Kohl et al. 1995a,b; Le and Shannon 2002; Sebti and Hamilton 2000). However, K-Ras and N-Ras are substrates for GGTase 1 and are processed by this alternative pathway when Ftase is inhibited. Indeed, extensive data now support the view that non-Ras *CAAX* proteins are critical in vivo targets of the Ftase inhibitors [reviewed in Le and Shannon (2002), Omer et al. (2000), and Sebti and Hamilton (2000)].

An important consideration in evaluating data from preclinical trials of FTIs in mouse cancer models is that therapeutic effects may be due to inhibition of any one of many farnesylated proteins. Furthermore, because the relevant target(s) of this class of drugs are uncertain, it is difficult to distinguish specific cancers that might be sensitive to FTIs in vivo. Preclinical studies of FTIs performed by researchers at Merck in transgenic mouse models of breast cancer initiated by using the mouse mammary tumor virus (MMTV) promoter to drive the expression of oncogenic *Hras* or *Kras* emphasize that FTIs should not be viewed as Ras-specific therapeutics. Whereas FTI L-744,832 induced regression of established tumors in the MMTV-H-*ras* mice (Kohl et al., 1995b), it inhibited further tumor growth in the *Kras*-driven model (Omer et al., 2000). Importantly, the authors demonstrated that processing of the farnesylated protein HDJ2 was significantly inhibited in MMTV-K-*ras* tumors but observed no changes in K-Ras processing. Thus, the antitumor activity of L-744,832 was independent of its ability to directly affect K-Ras (Omer et al., 2000). Another study of this agent in the MMTV-H-RAS model found that breast tumor regression was associated with apoptotic cell death that was partially independent of TP53 function (Barrington et al., 1998).

FTIs have also been tested in murine leukemia models with variable results. The FTI SCH66336 was administered to transgenic mice engineered to express a p190 BCR-ABL fusion protein that consistently develops pre-B-cell acute lymphoblastic leukemia (ALL) (Reichert et al., 2001). BCR-ABL expression deregulates multiple intracellular kinase cascades, including proteins downstream of Ras-GTP, and a number of studies infer that hyperactive Ras plays an important role in transformation by BCR-ABL (Mandanas et al., 1992; Sawyers et al., 1995). In this system, SCH66336 delayed or prevented the development of ALL for the duration of treatment (Reichert et al., 2001). In another study, Ba/F3 cells transfected with *BCR-ABL* were used for in vitro and in vivo studies of SCH66336 (Peters et al., 2001). This FTI inhibited colony formation by Ba/F3 *BCR-ABL* cells, blocked cell cycle progression at G2/M, and protected mice injected with these cells from a fatal CML-like disorder. Interestingly, the authors also showed that in vitro exposure of Ba/F3 cells to SCH66336 resulted in reduced RasGTP levels (Peters et al., 2001). Most recently, Hoover and co-workers (2002) reported synergy between SCH66336 and imatinib in inhibiting both the growth of Ba/F3 cells and the formation of progenitor colonies from CML patient bone marrows. Importantly, this FTI demonstrated effects on both imatinib-sensitive and imatinib-resistant cells.

In contrast to the encouraging preclinical data in models of *BCR-ABL*-associated leukemias, our laboratory found no beneficial effects of FTI treatment in a model of juvenile myelomonocytic leukemia (JMML) induced by injecting *Nf1* mutant fetal liver cells into irradiated recipients (Mahgoub et al., 1999). The *NF1* tumor suppressor encodes a GTPase-activating protein that interacts directly with Ras-GTP and negatively regulates signaling by accelerating GTP hydrolysis (Donovan et al., 2002). Mice with disease were assigned to receive one of two doses of FTI L-744,832. Animals treated with FTI showed no clinical responses, which were assessed by comparing white blood counts and spleen sizes in the treated and control groups. Biochemical investigation of primary bone marrow cells from mice that received L744,832 revealed an increase in the amount of unprocessed H-Ras and no effects on N-Ras or K-Ras (Mahgoub et al., 1999).

In summary, although experimental data link the therapeutic effects of FTIs to their ability to selectively inhibit Ftase in vivo, the relevant substrates that account for the antitumor effects reported in some studies have not yet been identified (Le and Shannon, 2002). It remains possible that FTIs inhibit hyperactive Ras in some cellular contexts either through direct effects on Ras proteins or by interfering with the processing of other farnesylated substrates. It will therefore be interesting to ascertain if trials performed in mouse models provide some indication regarding which human tumors are likely to respond to treatment. For example, preclinical data infer that FTIs will show efficacy in hematopoietic malignancies associated with BCR-ABL expression but will fail in juvenile myelomonocytic leukemia (JMML). Clinical trials of FTIs are underway in these diseases.

Chemotherapy Responsiveness and Resistance

Although cytotoxic agents comprise the primary treatment modality against many systemic malignancies, relatively few studies of conventional cytotoxic agents have been performed in genetically engineered mouse models. One reason for this is that is it difficult to generate and maintain cohorts of mice that spontaneously develop tumors; in addition, it is difficult to assess the responses of tumors that can only be visualized with CT scanning or other modalities (also see Chapters 22–25). Despite these issues, demonstrating that the responses of specific tumors that develop in mice to chemotherapeutic agents parallel those of the corresponding human cancers would validate the general approach of exploiting mouse models to develop better therapies.

A related question involves elucidating why some human tumors are inherently insensitive to chemotherapeutic agents while others respond initially but become resistant in patients who relapse. Most conventional agents damage cellular components such as DNA, and it was thought that this was directly responsible for their antitumor effect (Johnstone et al., 2002). Drug resistance was therefore thought to arise from changes that prevented the drug from reaching the target macromolecule (e.g., through induction of drug efflux pumps or intracellular detoxifiers). It is now appreciated that drug-induced damage is not invariably lethal but instead can initiate a series of postdamage responses, including apoptosis, cell cycle checkpoints, mitotic catastrophe, and a prolonged cell cycle arrest that resembles cellular senescence (Johnstone et al., 2002). Accordingly, pioneering studies in an Eμ-Myc B-cell lymphoma model performed by Lowe and colleagues have elucidated that disruption of these damage responses can underlie resistance to chemotherapeutic agents (Schmitt and Lowe 1999, 2001, 2002; Schmitt et al., 2000, 2002). These experiments assessed the ability of cyclophosphamide or doxorubicin to cure lymphomas in secondary recipients. An important discovery was that in vitro assays of tumor cell growth correlated poorly with in vivo responses. Further studies in this system identified several biologic (apoptosis, senescence) and genetic *(Tp53, INK4a, Bcl2)* determinants of drug action. They also showed that postdamage cellular responses are almost exclusively responsible for the efficacy of commonly used chemotherapeutic agents. These studies provide evidence that genetically engineered mouse cancer models can be harnessed to address mechanisms of drug action and inherent variations in treatment sensitivity. The hematopoietic system is tractable for studies of this type, and lymphoma is highly curable in both humans and mice. It will be important to devise strategies for uncovering mechanisms of chemotherapy responses and resistance in relevant mouse models of common epithelial malignancies.

CONCLUSIONS AND FUTURE PERSPECTIVES

Limited progress has been made in treating malignancies such as pancreatic cancer, lung cancer, and glioblastoma over the past few decades, and there is an urgent need to develop better cancer therapies. Evaluating new agents in human patients has important theoretical limitations related to difficulties in studying drug combinations in the phase 1 or 2 setting, and the fact that patients entered into these trials have refractory or relapsed disease. While the use of xenograft models to screen potential agents for efficacy remains a mainstay of modern drug development programs, these systems are biased toward cytotoxic agents and do not correlate well with clinical efficacy. Furthermore, xenografts are problematic for testing agents such as angiogenesis inhibitors that target the tumor microenvironment. These limitations have stimulated considerable interest in the potential use of genetically engineered mice for cancer drug discovery. Indeed, many strains have been created that develop cancers that accurately recapitulate genetic, biochemical, and phenotypic features of specific human malignancies. While the general strategy of testing new therapies in these models has considerable theoretical appeal, transgenic and knockout strains are expensive and cumbersome compared to traditional xenograft assays, especially since most lines were not engineered optimally for use in preclinical trials. Furthermore, the responses of the murine tumors to conventional regimens that show some efficacy in the corresponding human cancers are largely unknown (i.e., most models are not yet validated as platforms for drug discovery). Indeed, there is no precedent to date in which data from studies in a transgenic or knockout mouse cancer model were shown to predict a therapeutic response in the corresponding human tumor.

With these caveats in mind, engineered mouse models hold great promise and may prove particularly valuable for testing molecularly targeted cancer therapeutics. In this setting, pharmacodynamic monitoring of the relevant protein or pathway in primary tumor cells may yield biologic as well as preclinical insights. Studies in engineered mouse cancer models may be especially useful for determining whether inhibiting a specific target is associated with a beneficial therapeutic index in vivo (i.e., does an experimental agent preferentially block the growth of tumor cells without damaging normal cells?). Impediments to the widespread use of these models for drug discovery include cost, issues related to intellectual property and commercial value, intrinsic differences in pharmacokinetics and in drug metabolism and toxicity profiles between mouse and human, and difficulties in imaging mouse tumors. Other chapters in this volume discuss recent attempts to address many of these questions and the progress that is being made. When viewed in the context of the pressing need for new therapies and the challenges of evaluating these agents using current paradigms, harnessing genetically engineered mouse models represents perhaps the most promising strategy for translating basic knowledge about cancer pathogenesis into improved treatments

ACKNOWLEDGMENTS

We apologize to investigators whose work we were not able to discuss in this focused review. Our research in this area is supported by U.S. Army Neurofibromatosis Research project DAMD 17-02-1-0638, American Cancer Society grant RSG-96-104-06-LIB, and National Institutes of Health grants CA89318 and CA84221. We are grateful to our colleagues in the Mouse Models of Human Cancer Consortium for many helpful discussions.

REFERENCES

Bankert RB,Egilmez NK, Hess SD (2001): Human-SCID mouse chimeric models for the evaluation of anti-cancer therapies. *Trends Immunol* 22(7):386–393.

Barrington RE, Subler MA, Rands E, Omer CA, Miller PJ, Hundley JE, Koester SK, Troyer DA, Bearss DJ, Conner MW, Gibbs JB, Hamilton K, Koblan KS, Mosser SD, O'Neill T.J., Schaber MD, Senderak ET, Windle JJ, Oliff A, Kohl NE (1998): A farnesyltransferase inhibitor induces tumor regression in transgenic mice harboring multiple oncogenic mutations by mediating alterations in both cell cycle control and apoptosis. *Mol Cell Biol* 18(1):85–92.

Bergers G, Javaherian K, Lo KM, Folkman J, Hanahan D (1999): Effects of angiogenesis inhibitors on multistage carcinogenesis in mice. *Science* 284(5415):808–812.

Boehm T, Folkman J, Browder T, O'Reilly MS (1997): Antiangiogenic therapy of experimental cancer does not induce acquired drug resistance. *Nature* 390(6658): 404–407.

Boguski M, McCormick F (1993): Proteins regulating Ras and its relatives. *Nature* 366:643–653.

Bos JL (1989): *ras* oncogenes in human cancer: A review. *Cancer Res* 49:4682–4689.

Bourne HR, Sanders DA, McCormick F (1990): The GTPase superfamily: A conserved switch for diverse cell functions. *Nature* 348:125–132.

Bourne HR, Sanders DA, McCormick F (1991): The GTPase superfamily: Conserved structure and molecular mechanism. *Nature* 349:117–127.

Chin L, Tam A, Pomerantz J, Wong M, Holash J, Bardeesy N, Shen Q, O'Hagan R, Pantginis J, Zhou H, Horner JW, Cordon-Cardo C, Yancopoulous GD, DePinho RA (1999): Essential role for oncogenic Ras in tumour maintenance. *Nature* 400(6743):468–472.

Cools J, DeAngelo DJ, Gotlib J, Stover EH, Legare RD, Cortes J, Kutok J, Clark J, Galinsky I, Griffin JD, Cross NC, Tefferi A, Malone J, Alam R, Schrier SL, Schmid J, Rose M, Vandenberghe P, Verhoef G, Boogaerts M, Wlodarska I, Kandarjian H, Marynen P, Coutre SE, Stone R, Gilliland DG (2003): A tyrosine kinase created by fusion of the PDGFRA and FIP1L1 genes as a therapeutic target of imatinib in idiopathic hypereosinophilic syndrome. *N Engl J Med* 348(13):1201–1214.

Cooper RN, Irintchev A, Di Santo JP, Zweyer M, Morgan JE, Partridge TA, Butler-Browne GS, Mouly V, Wernig A (2001): A new immunodeficient mouse model for human myoblast transplantation. *Hum Gene Ther* 12(7):823–831.

Custer RP, Bosma GC, Bosma MJ (1985): Severe combined immunodeficiency (SCID) in the mouse. Pathology, reconstitution, neoplasms. *Am J Pathol* 120(3):464–477.

Donovan S, Shannon KM, Bollag G (2002): GTPase activating proteins: Critical regulators of intracellular signaling. *BBA Rev Cancer* 1602:23–45.

Druker BJ (2002): Perspectives on the development of a molecularly targeted agent. *Cancer Cell* 1(1):31–36.

Felsher DW, Bishop JM (1999): Reversible tumorigenesis by MYC in hematopoietic lineages. *Mol Cell* 4(2):199–207.

Gibbs JB, Oliff A (1997): The potential of farnesyltransferase inhibitors as cancer chemotherapeutics. *Annu Rev Pharmacol Toxicol* 37:143–166.

Gilliland DG, Griffin JD (2002): Role of FLT3 in leukemia. *Curr Opin Hematol* 9(4):274–281.

Hanahan D (1985): Heritable formation of pancreatic beta-cell tumours in transgenic mice expressing recombinant insulin/simian virus 40 oncogenes. *Nature* 315(6015):115–122.

Hanahan D, Weinberg RA (2000): The hallmarks of cancer. *Cell* 100(1):57–70.

Hann B, Balmain A (2001): Building "validated" mouse models of human cancer. *Curr Opin Cell Biol* 13(6):778–784.

He LZ, Guidez F, Tribioli C, Peruzzi D, Ruthardt M, Zelent A, Pandolfi PP (1998): Distinct interactions of PML-RARalpha and PLZF-RARalpha with co-repressors determine differential responses to RA in APL. *Nat Genet* 18(2):126–35.

He LZ, Tolentino T, Grayson P, Zhong S, Warrell Jr. RP, Rifkind RA, Marks PA, Richon VM, Pandolfi PP (2001): Histone deacetylase inhibitors induce remission in transgenic models of therapy-resistant acute promyelocytic leukemia. *J Clin Invest* 108(9):1321–1330.

Holland EC, Celestino J, Dai C, Schaefer L, Sawaya RE, Fuller GN (2000): Combined activation of Ras and Akt in neural progenitors induces glioblastoma formation in mice. *Nat Genet* 25(1):55–57.

Hoover RR, Mohan F-X, Melo JV, Daley GQ (2002): Overcoming STI571 resistance with the farnesyl transferase inhibitor SCH66336. *Blood* 100(3):1068–1071.

Jackson EL, Willis N, Mercer K, Bronson RT, Crowley D, Montoya R, Jacks T, Tuveson DA (2001): Analysis of lung tumor initiation and progression using conditional expression of oncogenic K-ras. *Genes Dev* 15(24):3243–3248.

Johnson L, Mercer K, Greenbaum D, Bronson RT, Crowley D, Tuveson DA, Jacks T (2001): Somatic activation of the K-ras oncogene causes early onset lung cancer in mice. *Nature* 410(6832):1111–1116.

Johnstone RW, Ruefli AA, Lowe SW (2002): Apoptosis: A link between cancer genetics and chemotherapy. *Cell* 108:153–164.

Karp JE, Lancet JE, Kaufmann SH, End DW, Wright JJ, Bol K, Horak I, Tidwell ML, Liesveld J, Kottke TJ, Ange D, Buddharaju L, Gojo I, Highsmith WE, Belly RT, Hohl RJ, Rybak ME, Thibault A, Rosenblatt J (2001): Clinical and biologic activity of the farnesyltransferase inhibitor R115777 in adults with refractory and relapsed acute leukemias: A phase 1 clinical-laboratory correlative trial. *Blood* 97(11):3361–3369.

Kelly L, Clark J, Gilliland DG (2002a): Comprehensive genotypic analysis of leukemia: Clinical and therapeutic implications. *Curr Opin Oncol* 14(1):10–18.

Kelly LM, Kutok JL, Williams IR, Boulton CL, Amaral SM, Curley DP, Ley TJ, Gilliland DG (2002b): PML/RARalpha and FLT3-ITD induce an APL-like disease in a mouse model. *Proc Natl Acad Sci USA* 99(12):8283–8288.

Kerbel R, Folkman J (2002): Clinical translation of angiogenesis inhibitors. *Nat Rev Cancer* 2(10):727–739.

Kisker O, Becker CM, Prox D, Fannon M, D'Amato R, Flynn E, Fogler WE, Sim BK, Allred EN, Pirie-Shepherd SR, Folkman J (2001): Continuous administration of endostatin by intraperitoneally implanted osmotic pump improves the efficacy and potency of therapy in a mouse xenograft tumor model. *Cancer Res* 61(20):7669–7674.

Kogan SC, Hong SH, Shultz DB, Privalsky ML, Bishop JM (2000): Leukemia initiated by PMLRARalpha: The PML domain plays a critical role while retinoic acid-mediated transactivation is dispensable. *Blood* 95(5):1541–1550.

Kohl NE, Conner MW, Gibbs JB, Graham SL, Hartman GD, Oliff A (1995a): Development of inhibitors of protein farnesylation as potential chemotherapeutic agents. *J Cell Biochem* 22:145–150.

Kohl NE, Omer CA, Conner MW, Anthony NJ, Davide JP, DeSolms J, Giuliani EA, Gomez RP, Graham SL, Hamilton K, Handt LK, Hartman GD, Koblan KS, Kral AM, Miller PJ, Mosser SD, O'Neill TJ, Rands E, Schaber MD, Gibbs JB, Oliff A (1995b): Inhibition of farnesyltransferase induces regression of mammary and salivary carcinomas in *ras* transgenic mice. *Nat Med* 1(8):792–797.

Lallemand-Breitenbach V, Guillemin MC, Janin A, Daniel MT, Degos L, Kogan SC, Bishop JM, de The H (1999): Retinoic acid and arsenic synergize to eradicate leukemic cells in a mouse model of acute promyelocytic leukemia. *J Exp Med* 189(7):1043–1052.

le Coutre P, Mologni L, Cleris L, Marchesi E, Buchdunger E, Giardini R, Formelli F, Gambacorti-Passerini C (1999): In vivo eradication of human BCR/ABL-positive leukemia cells with an ABL kinase inhibitor. *J Natl Cancer Inst* 91(2):163–168.

Le DT, Shannon KM (2002): Ras processing as a therapeutic target in hematologic malignancies. *Curr Opin Hematol* 9(4):308–315.

Levis M, Allebach J, Tse KF, Zheng R, Baldwin BR, Smith BD, Jones-Bolin S, Ruggeri B, Dionne C, Small D (2002): A FLT3-targeted tyrosine kinase inhibitor is cytotoxic to leukemia cells in vitro and in vivo. *Blood* 99(11):3885–3891.

Li J, Shen H, Himmel KL, Dupuy AJ, Largaespada DA, Nakamura T, Shaughnessy Jr. JD, Jenkins NA, Copeland NG (1999): Leukaemia disease genes: Large-scale cloning and pathway predictions. *Nat Genet* 23(3):348–353.

Licht JD, Chomienne C, Goy A, Chen A, Scott AA, Head DR, Michaux JL, Wu Y, DeBlasio A, Miller Jr. WH, et al. (1995): Clinical and molecular characterization of a rare syndrome of acute promyelocytic leukemia associated with translocation (11;17). *Blood* 85(4):1083–1094.

Lopes D, Mayer LD (2002): Pharmacokinetics of Bcl-2 antisense oligonucleotide (G3139) combined with doxorubicin in SCID mice bearing human breast cancer solid tumor xenografts. *Cancer Chemother Pharmacol* 49(1):57–68.

Mahgoub N, Taylor BR, Gratiot M, Kohl NK, Gibbs JB, Jacks T, Shannon KM (1999): In vitro and In vivo effects of a farnesyltransferase inhibitor on Nf1- deficient hematopoietic cells. *Blood* 94(7):2469–2476.

Mandanas RA, Leibowitz DS, Gharehbaghi K, Tauchi T, Burgess GS, Miyazawa K, Jayaram HN, Boswell HS (1992): Role of p21 RAS on p210 bcr-abl transformation of murine myeloid cells. *Blood* 82:1838–1847.

Meuwissen R, Linn SC, van der Valk M, Mooi WJ, Berns A (2001): Mouse model for lung tumorigenesis through Cre/lox controlled sporadic activation of the K-Ras oncogene. *Oncogene* 20(45):6551–6558.

Neshat MS, Mellinghoff IK, Tran C, Stiles B, Thomas G, Petersen R, Frost P, Gibbons JJ, Wu H, Sawyers CL (2001): Enhanced sensitivity of PTEN-deficient tumors to inhibition of FRAP/mTOR. *Proc Natl Acad Sci USA* 98(18): 10314–10319.

Omer CA, Chen Z, Diehl RE, Conner MW, Chen HY, Trumbauer ME, Gopal-Truter S, Seeburger G, Bhimnathwala H, Abrams MT, Davide JP, Ellis MS, Gibbs JB, Greenberg I,

Koblan KS, Kral AM, Liu D, Lobell RB, Miller PJ, Mosser SD, O'Neill TJ, Rands E, Rands E, Schaber MD, Senderak ET, Oliff A, Kohl NE (2000): Mouse mammary tumor virus-Ki-rasB transgenic mice develop mammary carcinomas that can be growth-inhibited by a farnesyl:protein transferase inhibitor. *Cancer Res* 60(10):2680–2688.

Paine-Murrieta GD, Taylor CW, Curtis RA, Lopez MH, Dorr RT, Johnson CS, Funk CY, Thompson F, Hersh EM (1997): Human tumor models in the severe combined immune deficient (scid) mouse. *Cancer Chemother Pharmacol* 40(3): 209–214.

Peters DG, Hoover RR, Gerlach MJ, Koh KY, Zhang H, Choe K, Kirschmeier P, Bishop WR, Daley GQ (2001): Activity of the farnesyl protein transferase inhibitor SCH66336 against BCR/ABL-induced murine leukemia and primary cells from patients with chronic myeloid leukemia. *Blood* 97(5):1404–1412.

Prochazka M, Gaskins HR, Shultz LD, Leiter EH (1992): The nonobese diabetic scid mouse: Model for spontaneous thymomagenesis associated with immunodeficiency. *Proc Natl Acad Sci USA* 89(8):3290–3294.

Rego EM, He LZ, Warrell Jr. RP, Wang ZG, Pandolfi PP (2000): Retinoic acid (RA) and As2O3 treatment in transgenic models of acute promyelocytic leukemia (APL) unravel the distinct nature of the leukemogenic process induced by the PML-RARalpha and PLZF-RARalpha oncoproteins. *Proc Natl Acad Sci USA* 97(18):10173–10178.

Reichert A, Heisterkamp N, Daley GQ, Groffen J (2001): Treatment of Bcr/Abl-positive acute lymphoblastic leukemia in P190 transgenic mice with the farnesyl transferase inhibitor SCH66336. *Blood* 97(5):1399–1403.

Rodenhuis S (1992): *ras* and human tumors. *Semin Cancer Biol* 3(4):241–247.

Sakakibara T, Xu Y, Bumpers HL, Chen FA, Bankert RB, Arredondo MA, Edge SB, Repasky EA (1996): Growth and metastasis of surgical specimens of human breast carcinomas in SCID mice. *Cancer J Sci Am* 2(5):291.

Sawyers CL (2002): Finding the next Gleevec: FLT3 targeted kinase inhibitor therapy for acute myeloid leukemia. *Cancer Cell* 1(5):413–415.

Sawyers CL, McLaughlim J, Witte ON (1995): Genetic requirement for Ras in the transformation of fibroblasts and hematopoietic cells by the Bcr-Abl oncogene. *J Exp Med* 18:307–313.

Schmitt CA, Fridman JS, Yang M, Lee S, Baranov E, Hoffman RM, Lowe SW (2002): A senescence program controlled by p53 and p16INK4a contributes to the outcome of cancer therapy. *Cell* 109(3):335–346.

Schmitt CA, Lowe SW (1999): Apoptosis and therapy. *J Pathol* 187(1):127–137.

Schmitt CA, Lowe SW (2001): Apoptosis is critical for drug response in vivo. *Drug Resist Updat* 4(2):132–134.

Schmitt CA, Lowe SW (2002): Apoptosis and chemoresistance in transgenic cancer models. *J Mol Med* 80(3): 137–146.

Schmitt CA, Wallace-Brodeur RR, Rosenthal CT, McCurrach ME, Lowe SW (2000): DNA damage responses and chemosensitivity in the Eμ-myc mouse lymphoma model. *Cold Spring Harbor Symp Quant Biol* LXV:499–510.

Sebolt-Leopold JS, Dudley DT, Herrera R, Van Becelaere K, Wiland A, Gowan RC, Tecle H, Barrett SD, Bridges A, Przybranowski S, Leopold WR, Saltiel AR, (1999): Blockade of the MAP kinase pathway suppresses growth of colon tumors in vivo [see comments]. *Nat Med* 5(7):810–816.

Sebti SM, Hamilton AD (2000): Farnesyltransferase and geranylgeranyltransferase I inhibitors and cancer therapy: Lessons from mechanism and bench-to-bedside translational studies. *Oncogene* 19(56):6584–6593.

Shinkai Y, Rathbun G, Lam KP, Oltz EM, Stewart V, Mendelsohn M, Charron J, Datta M, Young F, Stall AM, et al. (1992): RAG-2-deficient mice lack mature lymphocytes owing to inability to initiate V(D)J rearrangement. *Cell* 68(5):855–867.

Sohal J, Phan VT, Chan PV, Davis EM, Patel B, Kelly LM, Abrams TJ, O'Farrell AM, Gilliland DG, Le Beau MM, Kogan SC, (2003): A model of APL with FLT3 mutation is responsive to retinoic acid and a receptor tyrosine kinase inhibitor, SU11657. *Blood* 101(8):3188–3197.

Tallman MS, Andersen JW, Schiffer CA, Appelbaum FR, Feusner JH, Ogden A, Shepherd L, Willman C, Bloomfield CD, Rowe JM, Wiernik PH, (1997): All-trans-retinoic acid in acute promyelocytic leukemia. *N Engl J Med* 337(15):1021–1028.

Udagawa T, Fernandez A, Achilles EG, Folkman J, D'Amato RJ, (2002): Persistence of microscopic human cancers in mice: Alterations in the angiogenic balance accompanies loss of tumor dormancy. *FASEB J* 16(11):1361–1370.

Weisberg E, Boulton C, Kelly LM, Manley P, Fabbro D, Meyer T, Gilliland DG, Griffin JD (2002): Inhibition of mutant FLT3 receptors in leukemia cells by the small molecule tyrosine kinase inhibitor PKC412. *Cancer Cell* 1(5):433–443.

Zhang Y, Taylor BR, Shannon K, Clapp DW, (2001): Quantitative effects of Nf1 inactivation on in vivo hematopoiesis. *J Clin Invest* 108(5):709–715.

INDEX

Mouse Models of Human Cancer, edited by Eric C. Holland
ISBN 0-471-44460-X